BOTANY

An Introduction to Plant Biology

Shelley Herlich
Feb. 1989

BOTANY
An Introduction to Plant Biology

Mathew Nadakavukaren

Illinois State University

Derek McCracken

Illinois State University

West Publishing Company

Saint Paul New York Los Angeles San Francisco

This book is dedicated to all our teachers.

Printed in the United States of America

Library of Congress Cataloging in Publication Data

Nadakavukaren, Mathew J.
Botany, an introduction to plant biology.

Includes index.
1. Botany. I. McCracken, Derek. II. Title.
QK47.N24 1985 580 85-640
ISBN 0-314-85279-4

Typesetting: The Clarinda Company
Copyediting: Dr. Robert Bergad
Text Design: Bruce Kortebein

Cover Photo

The cover photograph shows the design seen in one of the mosaics located at the Dupont Subway Station in Toronto. Photo by Michael McCracken. Used by permission of the Toronto Transit Commission.

Photos that are not accredited to a specific source belong to Mathew Nadakavukaren.

Photo Credits

Fig. 2-5. Reprinted with the permission of Worth Publishers, Inc. From Raven, Evert, and Curtis: *Biology of Plants,* 3rd edition (Worth Publishers, Inc., New York, N.Y.). © 1985 Worth Publishers, Inc. **Fig. 12-1.** Photos reprinted with the permission of Ward's Natural Science Establishment, Inc. **Fig. 18-12 and 18-16(b).** Reprinted with the permission of the Field Museum of Natural History, Division of Photography. Copyright © 1984 Field Museum of Natural History, Chicago, IL. **Fig. 19-1.** Photograph by Daniel Todd. Reprinted with the permission of the U.S. Department of Interior. **Figs. 19-16, 19-17, and 19-18.** Reprinted with the permission of the Field Museum of Natural History, Division of Photography. Copyright © 1984 Field Museum of Natural History, Chicago, IL. **Fig. 23-6(a).** Reprinted with the permission of the Peabody Museum of Natural History, Yale University. **Fig. 25-5.** Reprinted with the permission of the Field Museum of Natural History, Division of Photography. Copyright © 1984 Field Museum of Natural History, Chicago, IL. **Fig. B-5.** Courtesy of Taurus Photos. **Fig. B-6.** Courtesy of Reichert Scientific Instruments.

Contents

Chapter 6 Stems 91

Chapter 7 Roots and Mineral Nutrition 117

Chapter 8 Leaves 143

Chapter 9 Photosynthesis 167

Chapter 10 Respiration 187

Chapter 15 Prokaryotes: Bacteria and Cyanobacteria 283

Chapter 16 Fungi 301

Chapter 17 Algae 323

Chapter 20 Angiosperms: Flowers and Life Cycle 401

Chapter 21 Angiosperms: Fruits, Seeds, and Seedlings 423

Chapter 22 Plant Genetics 443

Preface

BOTANY: AN INTRODUCTION TO PLANT BIOLOGY is the culmination of almost 30 years of combined experience in teaching introductory botany to a diverse student body. Over the years we have felt an acute need for a text appropriate for a one-semester course such as ours that is composed of a mixed group of both majors and nonmajors. We do not mean to imply that good botany texts are unavailable, but we have found that, for the most part, those currently on the market are more appropriate for a two-semester majors' course. We find that our students prefer reading assignments that complement the material discussed in lecture and are dissatisfied with the comprehensive, lengthy texts which contain excessive amounts of detailed information more pertinent to advanced courses. Introductory level students, particularly freshmen who compose the majority in general botany classes, are frequently overwhelmed by the encyclopedic volume of coverage in such books and consequently lose interest in the subject.

We were very pleased to be asked to write a botany text appropriate for use in a one-semester introductory course similar to ours or for an exclusively nonmajors course. We approached this project bearing the following guidelines in mind:

1. to include only the amount of material that can be covered adequately in a one-semester course
2. to write each chapter as an independent unit so that the text can be used conveniently by instructors who wish to present the topics in a sequence differing from that in our book
3. to give equal emphasis to all aspects of plant life covered
4. to provide a fundamental understanding of botanical principles that will give students sufficient preparation for taking higher level courses in botany

We hope that these features will make this text attractive even to those instructors whose philosophy of teaching general botany differs from ours.

During the preparation of this manuscript we remained aware that many general botany students have little or no science background. Since this may be the only science course they have taken, we have not assumed any prior knowledge on the part of the student for understanding any of the topics covered in this text. For this reason we have used nontechnical introductions to each chapter in order to highlight the material and to capture the students' interest. Wherever applicable, we have provided an historical background for these introductions. Through experience we recognize that students seldom look up terms in a glossary; therefore, technical or unfamiliar terms are defined wherever they appear in the text or in the margin on the same page.

Organization

We have organized the text in the conventional manner, beginning with a chapter on the cell as the basic unit of all living things. This is followed by chapters on cell metabolism and cell division. A chapter on cell types introduces the student to the complexity of plants. The next three chapters describe the external and internal structural details, the functions and the various modifications of stems, roots, and leaves. These are followed by chapters on photosynthesis and respiration, a logical sequence that shows how food produced in the leaves through photosynthesis subsequently is utilized to produce energy during respiration. Additional physiological aspects are completed in the two chapters on transport and plant growth. At this point reproduction and meiosis are introduced in order that life cycles can be presented in the subsequent chapters dealing with the various plant groups. In the chapter on plant classification we introduce the concept of a multi-kingdom system, which serves as an introduction to the diversity of plants. The six following chapters cover all the taxonomic groups traditionally included in a general botany course without emphasizing the kingdoms to which each group belongs. This approach gives individual instructors enough leeway to follow their own philosophy regarding the multi-kingdom system. We felt it was appropriate to introduce genetics after the chapter on flowers and angiosperm life cycle and prior to that on plant origin and evolution. Chapters on ecology and plants of special interest complete the main portion of the book.

Throughout the text we have attempted to explain the evolutionary relevance of the development of specific plant features, new forms, and ecological adaptations. A major stumbling block in understanding botanical principles for most nonscience majors is the section on plant physiology, which requires a rudimentary background in chemistry. To help overcome this problem, we have used a stepwise approach in explaining complex reactions and have also included an appendix on elementary chemical principles. In addition, the appendix on green thumb botany will be of practical use to many students because of the widespread interest in houseplants. It will also give students a chance to apply some of the information gained from the course, thus making the course more meaningful. The microscope appendix gives the student a brief introduction to light and electron microscopes, describes the basic differences between the two, and explains why electron microscopes are able to show the incredible details of a cell. It also describes the manner in which specimens are prepared for examination with these microscopes, information which is seldom presented in introductory texts. We feel such coverage is important because students in a botany laboratory use microscopes regularly, but generally have little understanding of these instruments or of their potential.

Chapter Format

Each chapter in the text follows the same basic format: an introduction, the main body, a summary in paragraph form of all important points, review questions, and one or more highlights. The book is thoroughly illustrated with labelled diagrams and photographs. In addition to the index, an extensive glossary at the end of the text defines all important terms. Throughout the book we have included practical aspects of botany to enable students to relate what they learn in the course to their everyday experiences. The text is accompanied by an instructor's manual and a student study guide.

Acknowledgements

Those who reviewed this book throughout its developmental stages made many helpful suggestions and necessary corrections from which this book has benefitted considerably. For this assistance we are very grateful to the following individuals:

Lydia Arciszewski—Bergen Community College
Bruce Bennett—Community College of Rhode Island
Linda Berg—University of Maryland
Murray W. Coulter—Texas Tech University
David E. Dallas—Northeastern Oklahoma A. & M. College
Jerry D. Davis—University of Wisconsin at LaCrosse
R. Edward Dole—University of Illinois
Richard E. Koske—University of Rhode Island
Irving B. McNulty—The University of Utah
William H. Miller—William Rainey Harper College
David W. Newman—Miami University
P. C. Pendse—California Polytechnic State University
S. N. Postlethwait—Purdue University
James L. Seago—State University of New York at Oswego
Tom Scott—The University of North Carolina at Chapel Hill
Wallace R. Weber—Southwest Missouri State University

We also want to extend our thanks to our colleagues at Illinois State University—Drs. Joseph Armstrong, Jerome Cain, Roger Anderson, Herman Brockman, Anthony Liberta, Tsan-Iang Chuang, and David Weber—whose expertise was of immense help when we needed ready answers to questions pertaining to topics in their respective fields of specialization.

We have nothing but praise for Mary Schiller, our editor, whose constant encouragement, interest, and advice have been of tremendous help to us throughout this project. We are also grateful for the expert advice we have received from John Orr, our production editor. We also want to thank Dr. Robert Bergad of The University of Minnesota for an excellent job of copy editing the text. We want to extend our thanks to the other members at West Publishing Company who have been involved with the production and marketing of this book. It has been a real pleasure working with these people. We want to thank Jean Nicolaides for doing the final typing of several chapters on short notice.

We are indeed indebted to our families for their constant support and understanding during the three years that it took us to complete this project when everything else had to be "put on the back burner." We are especially thankful to our wives, Anne (Mrs. M. J. N.) and Carol (Mrs. D. A. M.), for the numerous hours of work they put into the preparation of the manuscript, including typing and editing. A special thanks to Anne for writing the original drafts for the chapters on Algae and Plants of Special Importance.

Mathew Nadakavukaren
Derek McCracken

About the Authors

Mathew J. Nadakavukaren is currently Professor of Botany and Electron Microscopy at Illinois State University. After receiving a Ph.D. degree from Oregon State University, he was a Post-doctoral Fellow at Purdue University for three years. During the second and third years, he was also in charge of the Electron Microscope Laboratory in the Department of Biological Sciences at Purdue University and developed a course in transmission electron microscopy. Since joining the Department of Biological Sciences at Illinois State University in 1964, he has taught courses in Introductory Botany, Electron Microscopy, and seminars in Cell Ultrastructure. During the past several years, he has been chairman of the multi-section introductory botany course. He has also written a laboratory manual for the course. His research is focused on structure-function relationships in biological systems. Over the years he has published 51 research papers which have appeared in scientific journals such as Journal of Experimental Botany, Planta, Physiologia Plantarum, Phycology, The New Phytologist, Phytopathology, Journal of Cell Science, Journal of Cell Biology, Journal of Subcellular Cytology, Cell and Tissue Research, Tissue and Cell, Science, and Comparative Biochemistry and Physiology. In addition to writing this text, Professors Nadakavukaren and McCracken collaborate on ongoing research projects related to the ultrastructure and physiology of chloroplasts, as well as to the ultrastructure and physiology of fungal starch biosynthesis.

Derek A. McCracken is an Associate Professor of Biological Sciences at Illinois State University. He received his M.A. and Ph.D. in plant physiology from the University of Toronto. At Illinois State University he has taught General Botany to biology majors and nonmajors, Green Thumb Botany (a course he developed for nonmajors), Ecology of Man, graduate and undergraduate courses in plant physiology, and a graduate course in Enzymology. He also has taught General Biology in a maximum security penitentiary which he credits with helping to simplify the presentation of certain concepts in this book. His research has centered on plastid structure and function but has ranged widely from studies of algae and fungi to the development of the photosynthetic apparatus in 2,4-D-treated leaves. Articles detailing this research have appeared in American Journal of Botany, Analytical Biochemistry, Biochemical and Biophysical Research Communications, Journal of Phycology, Mycology, Mycopathologica, New Phytologist, Physiologica Plantarum, Plant Physiology, Planta, Science and Stärke. He currently is continuing his interest in the control of starch synthesis and in the development of photosynthetic ability in screening plastids.

I

Introduction

Outline

Origins and Development of Plant Science
The Nature of Plants

Origins and Development of Plant Science

Evidence for a long association between plants and humans can be found—plants have been eaten, worn as decoration, worshipped, made into weapons, used as medicines and poisons, and even served as currency. In order to survive, early humans had to be able to identify which plants were edible and which were poisonous, and they needed to know when and where edible plants could be found. Approximately 10,000 years ago, our ancestors discovered that their food supply could be increased substantially by deliberately planting seeds of wild grains, which previously had been gathered wherever people had happened to find them. This domestication of crop plants marked the beginnings of the agricultural way of life and prepared the way for the birth of civilization. In the Middle East and eastern Mediterranean region, wheat, barley, dates, figs, olives, apricots, and grapes have been cultivated for over 4000 years. The ancient Chinese domesticated rice, oranges, tea, and peaches. Native American Indians raised such crops as maize, squashes, white potato, cocoa, chili peppers, avocadoes, tomatoes, and tobacco.

The beginnings of botany as a science—i.e., attempts to collect, organize, and interpret facts about plants—can be traced to the Golden Age of Greek civilization during the 4th century B.C. Theophrastus, considered to be the "Father of Botany," laid the foundation for scientific botanical inquiry in *Historia Plantarum.* Whereas virtually all botanical writings before Theophrastus' time had been purely practical descriptions of medicinal plants, *Historia Plantarum* included sections on plant structure and classification, plant propagation, ecology, economic botany, therapeutic value, and culinary use of plants. In a subsequent volume, *The Causes of Plants,* Theophrastus described approximately 500 species of plants and introduced several descriptive terms still in use. Theophrastus made original observations on seed germination and was the first to distinguish between monocotyledonous and dicotyledonous plants (the two types of flowering plants). He also discussed methods of vegetative propagation* and described cross-pollination* in date palms. Many of his observations of the basic plant processes were the most accurate recorded until 2000 years later.

Vegetative propagation
(also called asexual reproduction) Reproduction of organisms by nonsexual means.

Cross-pollination
Transfer of pollen from the flower of one plant to that of another of the same species.

Contributions to botany during the Roman period included extensive illustrations of medicinally valuable plants. *De Materia Medica,* a volume of medicinal plants by Dioscorides, contained descriptions of 600 plants and was regarded as an authoritative book for 1500 years. Pliny's (A.D. 23–79) book, *Historia Naturae,* described 1000 species of plants and succeeded in arousing popular interest in nature. For hundreds of years it ranked second only to the Bible in volume of sales. Unfortunately very little of this work was original, and, in addition, it contained many inaccuracies that were perpetuated for centuries.

During the Middle Ages, botanical inquiry was kept alive in the Arab world where interest continued to be focused on the medicinal properties of plants. The Islamic rulers established botanical gardens and encouraged the exchange of seeds and plants among these institutions. Extensive translations of Greek and Roman botanical works were also made into Arabic during this period.

By the 16th century a renewed interest in the natural world had resulted in the growing popularity of botanical gardens and the publication of illustrated botanical books called "herbals." Beginning with the European age of exploration, new and exotic plant species from the Americas, Africa, and Asia were brought to Europe. One of the major achievements of this period was a two-word system, the binomial system, of naming plants. The Swiss docter Gaspard Bauhin, is credited with devising this system of nomenclature. This concept was later adopted and refined by the great Swedish botanist and taxonomist, Carolus Linnaeus. The binomial system has since been accepted universally as the scientific system for naming all organisms.

During the 17th and 18th centuries the foundations of modern botanical science were laid. The descriptive disciplines of form and structure, classification, and relationships between plant groups were the first to develop, since such studies required only careful observation of plants. After the invention of the light microscope and an understanding of the basic principles of chemistry and physics, the pace of botanical investigation accelerated. Some of the major developments of this period were the discovery by Robert Hooke that plants are made up of cells; the investigation of the microscopic structure of plants by Nehemiah Grew and Marcello Malpighi; the modern system of classification developed by Carolus Linnaeus; the experiments in plant function by J. B. von Helmont, who demonstrated that water is a major requirement for plant growth; and the experiments of Stephen Hales on water transport in plants which demonstrated that plant processes can be explained by physical laws.

The 19th century witnessed an ever-increasing tempo of botanical research and discovery, including the formulation of the Cell Theory by Mathias Schleiden and Theodor Schwann, elucidation of the nitrogen cycle by Boussingault, recognition of the importance of mineral nutrients for plant growth by von Liebig, and the discovery of the laws of heredity by Gregor Mendel. The publication of *Origin of Species* by Charles Darwin provided a basis for explaining plant evolution and the formation of new species.

In the present century the application of highly sophisticated techniques such as electron microscopy, differential centrifugation, chromatography, electrophoresis, radioisotope labeling, spectrophotometry, and x-ray diffraction has provided at least partial answers to some of the most basic questions regarding the structure and function of plants. These include the nature of the hereditary material, DNA, and its duplication; the structure of chromosomes; differentiation and growth; action of hormones; structure and function of membranes and organelles; and the mechanisms of cell division, respiration, photosynthesis, and transport of water, food, and mineral elements. In spite of the fact that new knowledge about plant life is accumulating from laboratories and field stations all over the world at a pace faster than ever before, by no means is our understanding of the plant world complete.

The field of botany presently includes a number of different basic areas of study such as **morphology** (external form, structure and development), **cytology** (cell structure), **anatomy** (origin and arrangement of cell types and tissues), **physiology** (mechanisms of plant function), **genetics** (mechanism of inheritance), **ecology** (interaction with environment), **taxonomy** (diversity, naming, and classification), **evolution** (origin and development of new forms), **paleobotany** (fossil and extinct plants), and **phytogeography** (plant distribution). In addition to these, there are a number of applied areas of botany including **forestry** (planting, developing, and managing forests), **agronomy** (production of field crops), **horticulture** (cultivation of gardens or orchards), and **plant pathology** (plant diseases). Because of the great diversity of organisms considered by botanists, specialized areas of study related to specific groups such as **phycology** (study of algae), **mycology** (study of fungi), **bryology** (study of mosses and their relatives), and **pteridology** (study of ferns) have become established. All of these areas have contributed much to our understanding of botany and the application of this information to practical use.

Many of these botanical discoveries, however, could not have been possible without the contributions from related fields such as chemistry, physics, and geology. For this reason, a present-day botanist usually is trained in a number of related fields, in addition to one or more specialized disciplines of botany. Botanical knowledge in recent years has increased to such an extent that it is not possible for any one person to be an expert in all areas of this science. Whereas major discoveries of the past centuries could be credited to single individuals, many such discoveries in the future will most likely be the result of teamwork.

Botany is a science and, therefore, the study of plants requires scientific methods of inquiry. The vast amount of knowledge about plant life that has accumulated thus far has been obtained using established scientific methods such as critical observation, experimentation, and analysis of results. Observations related to any aspect of plant life must be unbiased to be of any value in making generalizations or in comparing observations made by different individuals. A hypothesis may result from repeated observations about a particular phenomenon related to plant life; however, experiments need to be set up to prove or disprove such a hypothesis. An important aspect of scientific methodology is the use of controlled experiments designed to eliminate personal bias in interpreting results. Usually such experiments require the use of two identical or similar groups of plants maintained under constant conditions except for the variable that is being tested. Another aspect of scientific inquiry is the reproducibility of results. The validity of results obtained by one group of scientists is enhanced when their experiments are repeated by other independent researchers who obtain the same results. Finally when the validity of a hypothesis is proved by repeated experimentation by different groups of scientists, the observed phenomenon is accepted as a theory or principle. Exceptions to a theory are not unusual in biological science, however.

The Nature of Plants

Plants, like animals, are living organisms. Regardless of their size plants are composed of basic units called cells, each of which is able to carry out all characteristic processes of a living organism. These processes include synthesis of organic molecules, energy transformation, growth, and reproduction. In addition, plants, like all other living things, are able to adapt to changes in their environment through the process of natural selection, which is the basis for evolution of new forms of life. Living things can be divided into two broad categories based on their method of obtaining food: food producers, or autotrophs, and food consumers, or heterotrophs. The autotrophs synthesize food from carbon dioxide and water through the process of photosynthesis, releasing oxygen as a by-product. The heterotrophs, in turn, are dependent on the autotrophs both for their supply of food and for oxygen. Both autotrophs and heterotrophs utilize oxygen in the process of respiration, where food is broken down to generate usable energy; at the same time the carbon dioxide produced returns to the atmosphere. Traditionally all autotrophic organisms (algae, mosses, liverworts, whisk ferns, club mosses, horsetails, true ferns, conifers and their relatives, flowering plants) have been considered as plants. In addition, fungi, which are heterotrophs, also have been regarded as plants, as have the bacteria, most of which are heterotrophs.

In the past, living organisms were placed into either the plant kingdom (Figure 1–1) or the animal kingdom, mainly on the basis of the presence of a cell wall for plants or its absence in the case of animals. With our enhanced understanding of the chemical nature and function of living things, it is increasingly difficult to divide organisms into only two kingdoms. It is becoming more and more evident that there are a number of organisms that have both plantlike and animallike characteristics or traits distinct from either plants or animals. Fungi and bacteria are examples of cell wall-containing organisms that were originally included in the plant kingdom but are now assigned to their own kingdoms in the newer classification systems. Despite these changes in classification, botanists still study fungi and other organisms that have been moved out of the plant kingdom. In fact the term *plant* is still often used by many botanists to refer to all organisms that earlier were placed in the original, monolithic plant kingdom.

The simplest plantlike organisms are microscopic algae, too small to be seen with the naked eye. A number of them are composed of a single cell and most of the others are simple filaments in which cells are attached end-to-end. Rarely do the algae attain the complex arrangement and differentiation of cells to form tissues. There is strong evidence to support the hypothesis that land plants evolved from one of the algal groups. Therefore, the first land plants must have been small and structurally simple and probably required water for sexual reproduction. According to the fossil record, nonvascular plants, such as mosses, and primitive vascular plants*, now extinct, both appeared as land plants at about the same time. In one line of evolution the transitional organisms between algae and land plants must have possessed specialized cells to transport water and food through the plant body. Organization of cells into specific tissue types must have come about as the land plants gradually became larger and more complex, requiring specialized tissues to perform the various functions. The tissues responsible for transport are called vascular tissues, and the absence of vascular tissues in the algae and some primitive land plants provides the basis for the division of plants into two anatomical groups—vascular and nonvascular. A

Vascular plant
A plant that has specialized tissues (xylem and phloem) for conducting water and food.

Figure 1–1. The classical plant kingdom. Following groups of organisms were originally included in the plant kingdom. Recent classification schemes group them under three or more kingdoms. *(a)* Bacteria. *(b)* Cyanobacteria. *(c)* Fungi. *(d)* Algae. *(e)* Bryophytes. *(f)* Whiskferns. *(g)* Club mosses. *(h)* Horsetails. *(i)* Ferns. *(j)* Gymnosperms. *(k)* Angiosperms.

(a) (c) (b) (d) (e)

distinct division of labor in terms of tissue function can be recognized among vascular plants. Along with these anatomical changes during the evolution of the complex vascular plants, changes in the structure of reproductive organs and the mode of reproduction also took place. Sexual reproduction increased the possibility for genetic recombination or mixing of gene pools to bring about variation in a population, which in turn is essential for adaptation to a changing environment. Flowers are the most highly specialized reproductive structures in the plant world and are responsible for the production of seeds within a fruit. The fruit provides an efficient mechanism for the dispersal of the seed, assuring widespread distribution of the flowering plants. The evolutionary changes that led to the development of the

(f) (h) (j) (g) (i) (k)

seed also eliminated the necessity for motile sperm that swam through water to reach the egg. With these developments, the gymnosperms (conifers, et al.) and later the flowering plants achieved the equivalent of internal fertilization in animals, and, like the higher animals, the seed plants became independent of water for completion of their life cycle. The increase in reproductive success that resulted was important in establishing the seed plants as the dominant vegetation in terrestrial environments.

The seed contains an embryonic plant in a dormant state and a small amount of stored food that is used by the embryo as it develops into a seedling. When given the proper conditions for germination*, the embryo develops into a seedling capable of an independent existence. The under-

Germination
The process of renewed growth by a reproductive structure such as a seed or spore.

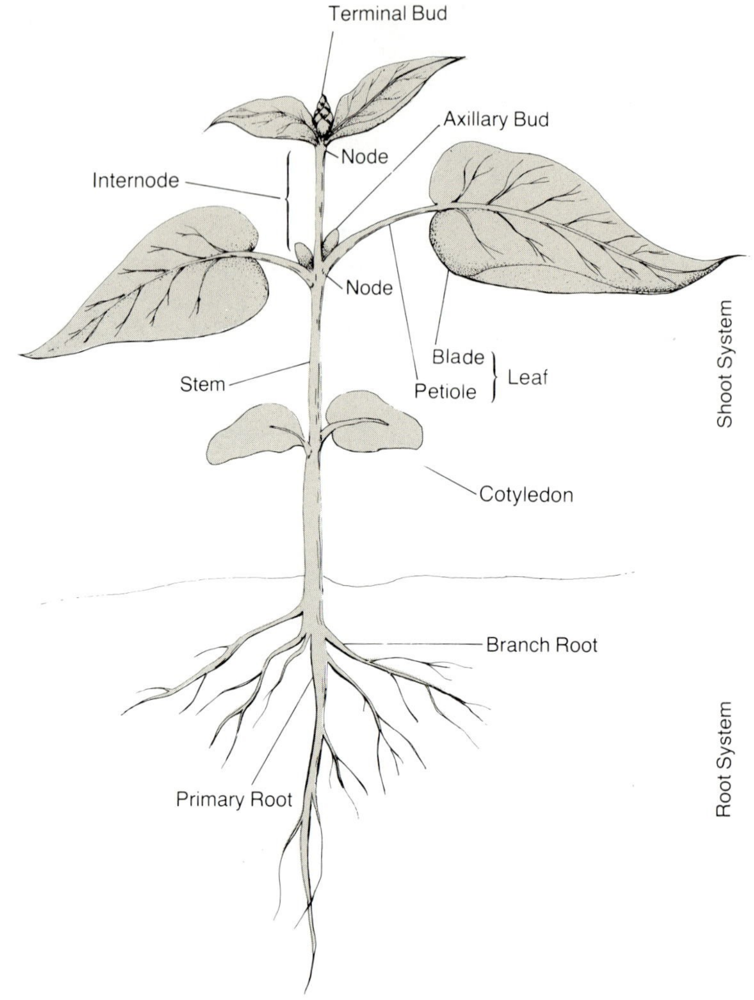

Figure 1–2. Drawing of a typical seedling. The aerial portion is the shoot system consisting of stems, leaves, and buds while the underground portion is the root system. A node is the point of attachment of a leaf to a stem. Future branches are produced by the axillary buds, which are present at the angle between the leaf and the stem.

ground portion of a seedling consists of an extensive and highly branched root system, whereas the above-ground portion makes up the shoot system, consisting of a stem from which are produced leaves and buds (Figure 1–2). The root system of a plant is responsible for water and mineral absorption and anchorage, while the shoot system is the site of food synthesis primarily in the leaves, which are highly specialized as photosynthetic organs. The vascular tissues of a seedling are continuous between the root and shoot systems, allowing for the transport of food and water throughout the plant.

The germination of the seed, the subsequent differentiation of the embryo into a seedling, and the continued development of the seedling into a mature plant are all influenced by growth-regulating substances called hormones synthesized by the plant. The general pattern and potential for growth of a plant, however, is predetermined by the genetic material, DNA, that is present in the nucleus of each cell. There are also a number of environmental factors such as light, temperature, precipitation, wind, soil, mineral elements, and interaction with other living organisms that can influence the growth of a plant.

2

The Plant Cell

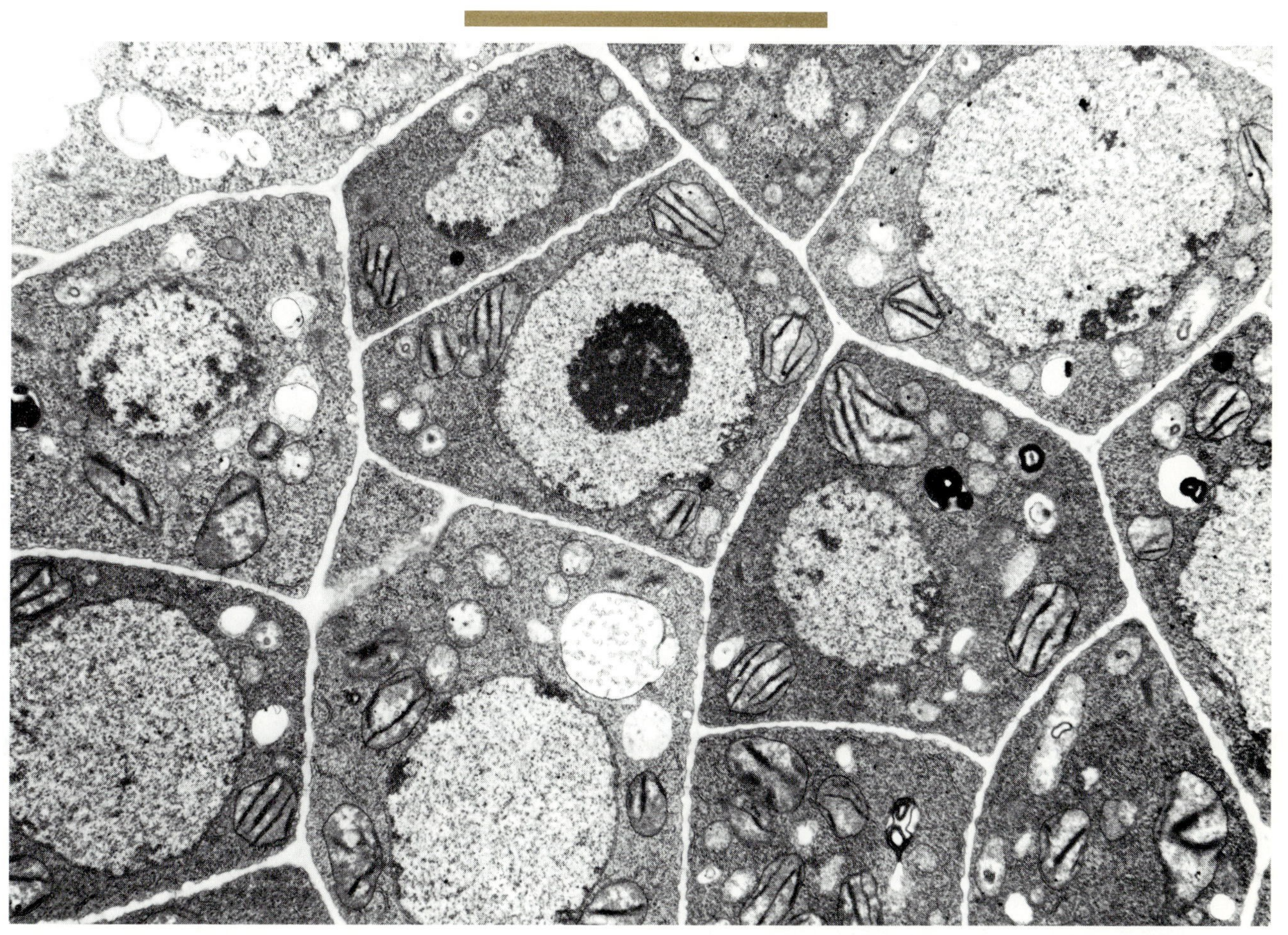

Outline

Introduction

A dog. A rose. A brick building. All of these are familiar as intact structures. It is readily apparent that the building is constructed of a large number of individual, small units called bricks, yet when most observers look at the edifice, all they see is the building as a whole, not its component parts. Only when the building is disassembled or viewed closely does it become apparent that it is constructed from many bricks. What about the dog and the rose—how are they constructed? Observation reveals that they too are made up of a number of distinct parts: legs, head, leaves, flowers, and so on; however, these are not the building blocks of such organisms. Closer inspection with the naked eye still does not reveal the building blocks. Only an examination with a microscope can reveal the building blocks of organisms.

In 1665 Robert Hooke used the word **cells** (derived from the Latin *cella,* meaning small room) to describe the "microscopical pores" that he had observed in several kinds of cork. It was not until almost 200 years later, however, that the true importance of cells was realized and even later before their structural and functional complexity became apparent. With improvements in light microscopes and techniques for making thin sections of tissues, more investigators came to study cellular detail in plants and animals until, in 1839, Mattias Schleiden, a botanist, and Theodor Schwann, a zoologist, independently proposed the **cell theory.** In its most basic form the cell theory states that all living organisms are composed of cells and that these cells have common structural and functional features. The cell theory was strengthened as the result of the observation by Rudolph Virchow, in 1858, that all cells come from preexisting cells. Subsequent investigations with even the most modern microscopes have failed to reveal any evidence that the cell theory is incorrect; hence, at present, the cellular nature of all organisms is universally accepted by scientists. Cells are the building blocks of plants and animals.

A cell is the smallest unit that can remain alive when removed from the living organism. This fact has been well established by isolating cells from plants or animals and growing them in cell and tissue culture.* If the cells are first disrupted and then the various subcellular structures, called **organelles,** separated from each other, these components quickly lose any metabolic* activity and cannot be maintained alive for more than a matter of hours. Isolated cells, on the other hand, have been kept alive in culture for many years. Plant tissue culture (Figure 2–1) has even become a commercially accepted way of propagating plants, with many houseplants getting their start in tissue culture factories (Chapter 12).

Cell/tissue culture
A technique for growing isolated cells or tissue in a sterile medium of known composition.

Metabolism
(*adj.,* metabolic) All of the chemical reactions that occur in a living cell.

With the exception of the simplest organisms, which are **Unicellular** (single-celled) or **colonial,** all organisms are **multicellular.** In certain multicellular organisms, such as some of the algae (Chapter 17), the cells are attached end-to-end, forming a filament* (Figure 2–2), whereas in maple trees billions of cells are arranged in a three-dimensional manner, producing the familiar shape of a tree. These various cells differ considerably in size,

Filament
A chain of cells linked end to end.

Figure 2–1. Tissue culture. A group of isolated cells *(a)* in a culture medium can differentiate to produce a new plant *(d)*.

shape, and structure as a result of differences in function (Chapter 5). Cell size ranges from cells that are barely visible under a light microscope, such as bacterial cells, to those that can be seen with the unaided eye, such as certain algal cells. One of the largest plant cells is the juice cell of a citrus fruit (Figure 2–3). Most cells, however, fall between these extremes, although they are still usually only visible under a microscope. In general, cells range in size from 0.2 micrometer (0.0002 millimeter) to 3 centimeters. Objects smaller than 0.1 millimeter in diameter are not usually visible without the aid of enlarging lenses. Although the shape of cells in a multicellular plant varies, most cells are multifaceted and of varying degrees of elongation. Filamentous organisms such as algae and fungi are usually made up of cylindrical cells (Figure 2–2).

Structural variations of plant cells are due to functional differences of the diverse cell types found in a multicellular plant body. The diversity of cell types results from the high degree of correlation between structure and function in plants. For example, the water-conducting tissue of a plant is composed of many elongated, dead cells that are stacked end-to-end to form pipelike structures (Figure 2–4a). The walls of these cells are distinctive because of the extensive thickening necessary to withstand considerable amounts of pressure or tension. By way of contrast, food synthesis occurs in living, thin-walled cells whose shape varies as a function of where in a plant they are located (Figure 2–4b). Much of the study of plant anatomy consists of learning to identify plant cell types by either their characteristic structure or shape or both. Although this diversity exists at the cellular level, there are many features that are common to most living cells.

The living plant cell is composed of the **protoplast** surrounded by a **cell wall.** The protoplast is the site of most of the metabolic activities in the cell, while the cell wall, a product of the protoplast, allows the cell to withstand pressures developed inside. **Protoplasm** is a term that describes the sum total of all protoplasts from the individual cells of a multicellular organism.

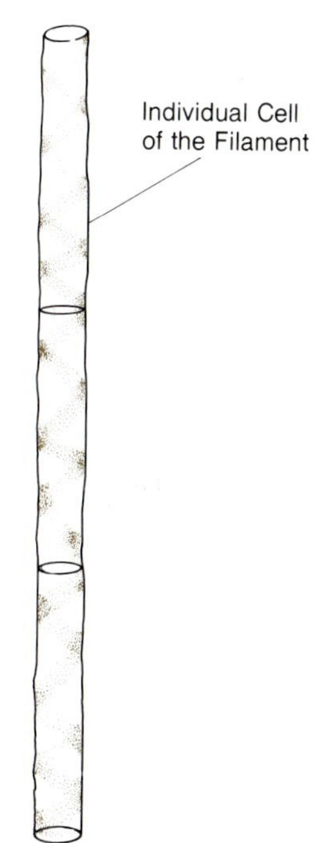

Figure 2–2. Filament. Most filamentous forms consist of many cells attached end to end.

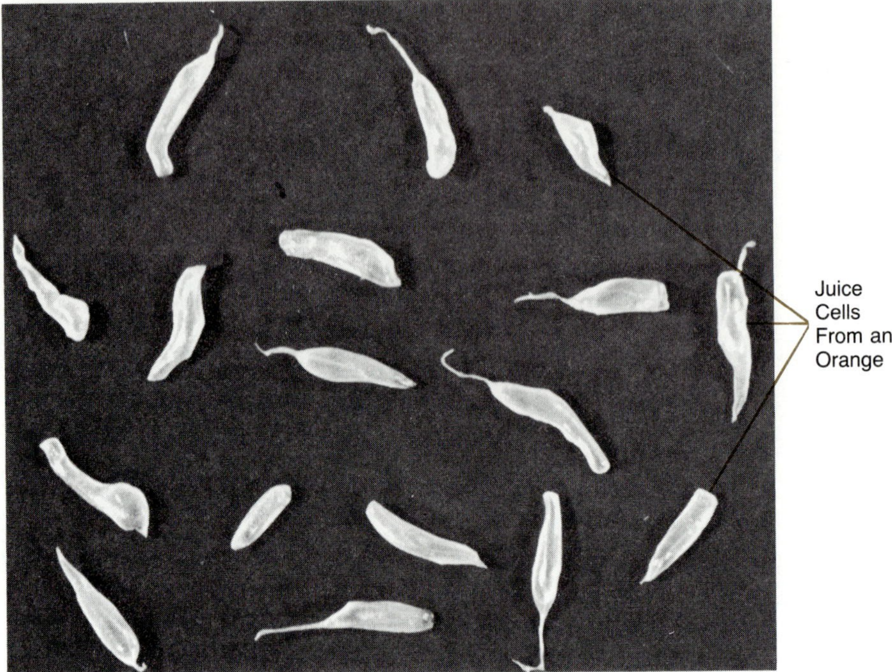

Figure 2–3. Juice cells of an orange. These are some of the largest plant cells and are easily visible without the aid of a microscope.

Cell Wall

All plant cells form a **primary wall** first, with some specialized types of cells subsequently adding to this a **secondary wall.** Primary walls are made up predominantly of **cellulose, hemicellulose,** and **pectin,** all of which are

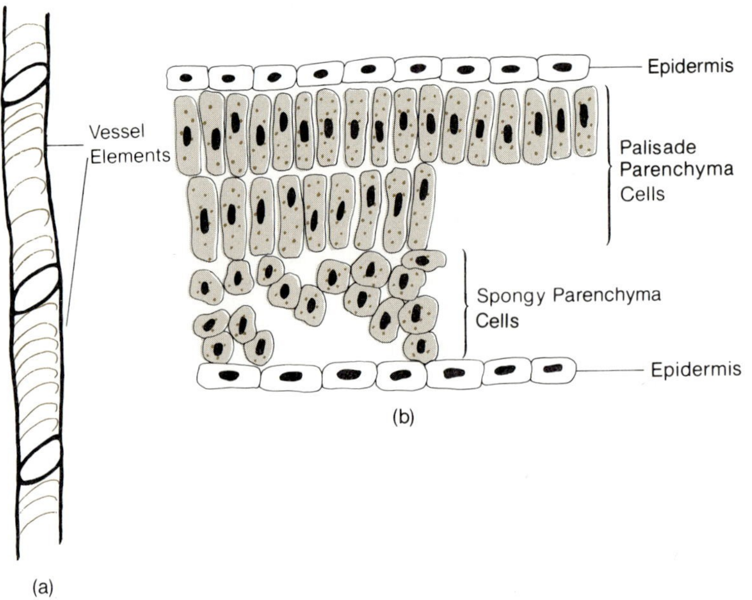

Figure 2–4. Cell types. *(a)* Vessel elements, attached end to end, form a pipe system in the water-conducting tissue of plants. *(b)* Cells in the food-synthesizing tissue of a leaf show differences in size and shape of the same cell type.

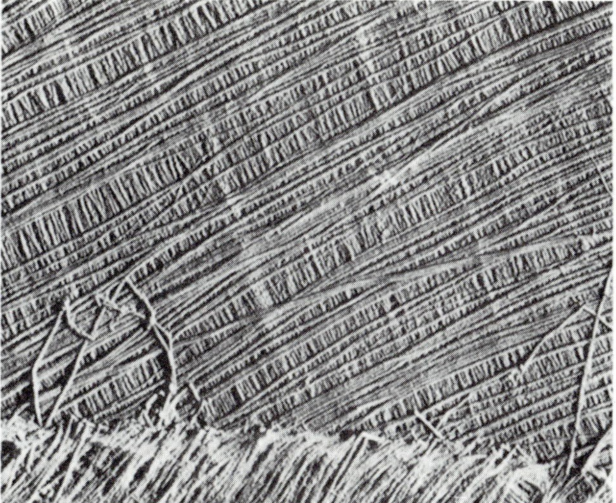

Figure 2–5. Cell wall structure. This electron micrograph shows cellulose microfibrils oriented at different angles in the cell wall.

large molecules (Chapter 3). The cellulose molecules in the wall are linked together forming minute rods called **microfibrils** that are embedded in a "glue" composed of hemicellulose, pectins, and proteins. The structure of the primary wall can be likened to that of a reinforced concrete wall with metal rods held in place by the surrounding cement except that, unlike concrete, the cell wall is a flexible structure. In actuality the microfibrils are arranged in layers or bands with each layer oriented at a different angle from the adjoining layers (Figure 2–5). In some cells, in addition to the primary wall, a secondary wall is formed by the deposition of new wall material on the inside face of the primary cell wall (Figure 2–6). This new wall

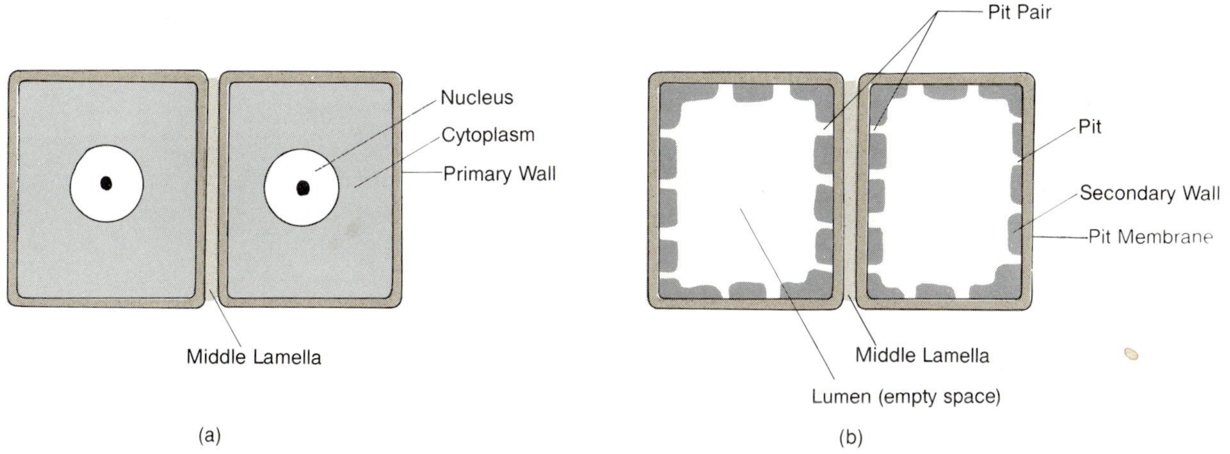

Figure 2–6. Cell wall formation. *(a)* Two adjacent living cells sharing a common middle lamella. *(b)* After the formation of the secondary wall, the protoplast disintegrates, usually leaving an empty space surrounded by the walls. Pits are small gaps in the secondary walls. In adjacent cells pits often face each other forming pit pairs.

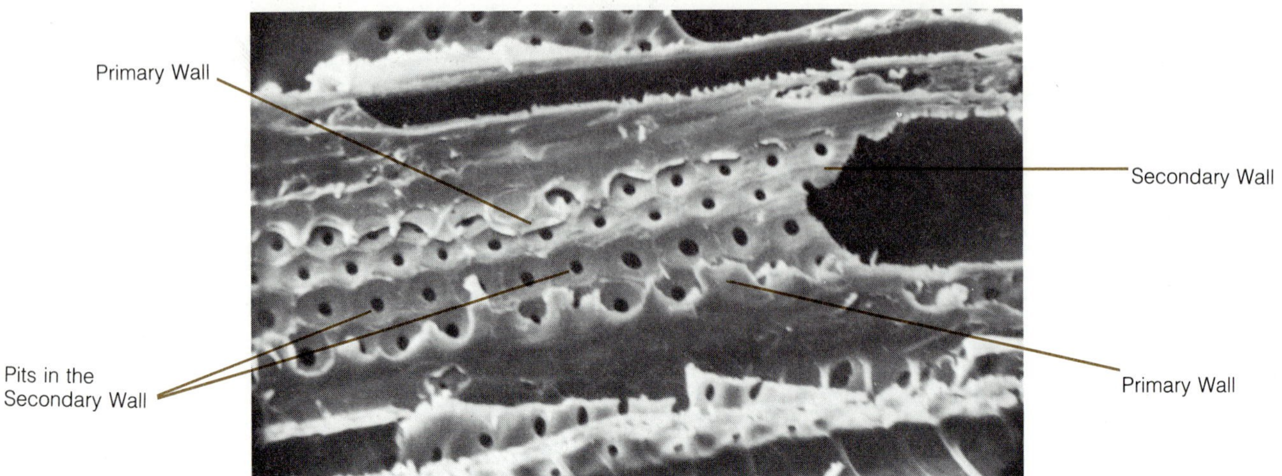

Figure 2–7. Pits Scanning electron micrograph of a tracheid showing pits that are small openings in the secondary wall. The pit membrane, which is the primary wall that covers the secondary wall is removed to show the pits. The jagged edges of the primary wall are still visible.

material consists initially of cellulose microfibrils embedded in a matrix of hemicellulose, cellulose, and protein, but the protoplast quickly begins to produce either **lignin** (a carbohydratelike material that strengthens and rigidifies the cell wall) or **suberin** (a waxy material) or both, which are added to both the primary and secondary walls to produce a lignified or suberized cell. In certain special cases a cell may lignify or suberize its primary wall without having produced a secondary cell wall (Chapter 5). The secondary wall is rarely uniform in thickness and often has many small cavities or openings called **pits** (Figure 2–7). Pits are often distributed randomly on the secondary wall. As a result of their presence, the primary wall is not covered by the secondary wall over its entire surface. Usually the pits of adjacent cells, called pit pairs, are oriented opposite each other so that water and dissolved materials can pass between cells (Figure 2–6). That part of the primary wall which covers the pit is known as the **pit membrane** (Figure 2–6). In certain kinds of water-conducting cells the secondary walls are deposited in a helical or ringlike pattern (Figure 2–8). Since movement of water into a cell having a secondary wall is obstructed, a cell usually dies after a secondary wall is laid down, leaving only a framework made up of primary and secondary cell wall material. In such dead cells there will be a central empty space, the lumen, that was formerly occupied by the protoplast (Figure 2–6). In contrast to animal cells, which break down after they die, most dead plant cells carry out functions such as water conduction, mechanical support, and protection for embryonic tissue.

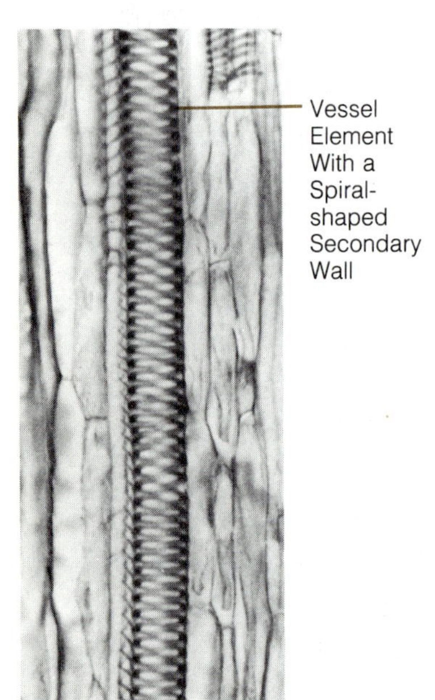

Figure 2–8. Secondary wall. The spiral structure inside this water-conducting cell is the secondary wall, which gives the cell mechanical support to withstand the negative pressure known to develop within these cells.

A multicellular plant body made up of individual cells is like a building made of bricks: there must be a mortar to hold the cells together or, like the building, the plant would disintegrate into individual cells at the first breeze. The mortar that holds the cells together is a gluelike layer called the **middle lamella** (Figure 2–9), whose major chemical component is pectin. The middle lamella is formed from the cell plate, which is a partition that separates the cytoplasm of a dividing cell into two compartments. A primary wall is then produced on either side of this partition by each of the two newly

Figure 2–9. Structure of young cell. *(a)* Diagram of an immature plant cell. *(b)* Electron micrograph of an immature plant cell.

Mitochondrion
Endoplasmic Reticulum
Primary Cell Wall
Middle Lamella
Plasmalemma
Plasmodesma
Chloroplast
Microtubule
Cytoplasm
Golgi Body
Nucleolus
Vacuole
Nucleus
Ribosomes
Microbody
Nuclear Pore
Golgi Vesicle

(a)

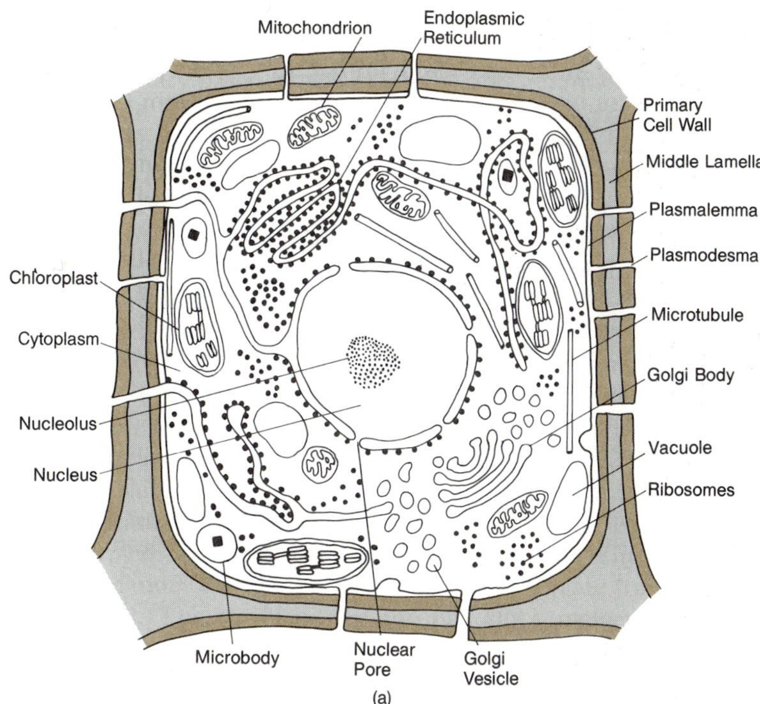

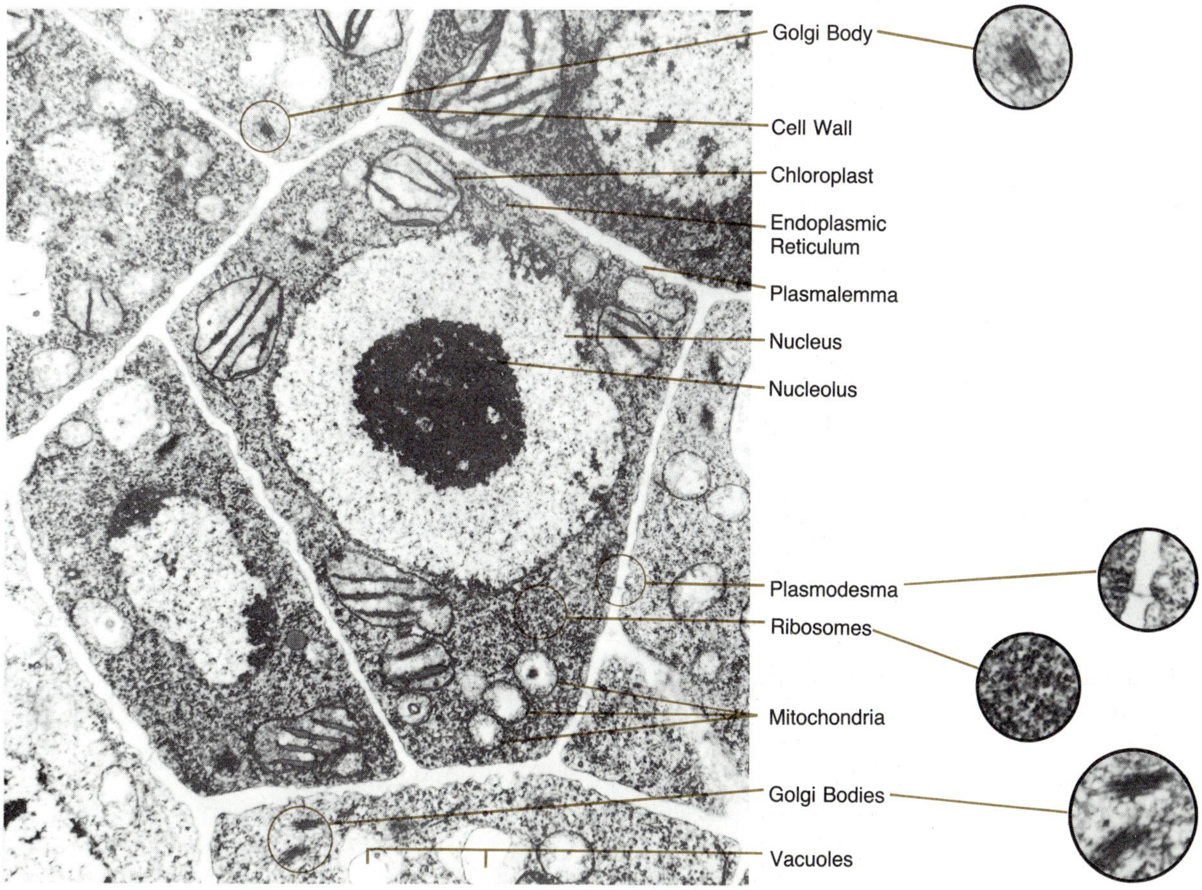

Golgi Body
Cell Wall
Chloroplast
Endoplasmic Reticulum
Plasmalemma
Nucleus
Nucleolus
Plasmodesma
Ribosomes
Mitochondria
Golgi Bodies
Vacuoles

(b)

formed cells. Thus, the middle lamella is always found between the primary walls of adjacent cells. Although neighboring cells are separated by cell walls, cytoplasmic connections called **plasmodesmata** pass through the walls facilitating chemical communication between cells (Figure 2–9).

Protoplast

Colloidal
Consisting of a suspension of very fine particles.

Organic molecules
Molecules containing carbon atoms.

Inorganic molecules
Molecules that lack carbon atoms.

Ion
An atom or molecule with a positive or negative charge.

Pigmented
Containing a pigment—a molecule that absorbs a particular color of light.

The living part of a cell, called the **protoplast,** is composed of a complex colloidal* fluid containing dissolved proteins, sugars, organic* acids, and inorganic* substances, as well as fats and other molecules in suspension. The chemical nature of the living cell is discussed in detail in the next chapter; therefore, the remaining discussion will focus on the structural organization of the protoplast. The outermost boundary of the protoplast is the cell membrane, more commonly known as the **plasmalemma** (Figure 2–9). This membrane is **differentially permeable.** This means that the membrane is selective in the types of ions* and molecules it will allow to pass in either direction, and it also regulates the rate of movement of these substances. Thus the plasmalemma regulates the movement of materials in and out of the protoplast. The cell wall, on the other hand, has no selective capacity and is completely permeable to substances dissolved in water.

Although the protoplast of a living cell is transparent when examined with a standard light microscope, there are a variety of structures called **organelles** (literally small organs) present in it (Figure 2–9). To be able to see even the largest of these organelles, except the pigmented* ones, with the light microscope, the cell has to be killed and stained. Although many of these organelles can be recognized in their living condition under a light microscope with phase contrast illumination, the structural details can be revealed only when cells are examined under an electron microscope. Present-day electron microscopes are able to magnify objects to 500,000 times or more of their actual size. Magnification alone, however, is not sufficient for the study of cell structure; the optical system must also allow the user to distinguish minute details (see Appendix on microscopes for further explanation). Electron microscopes can reveal details as small as 0.2 nanometer (1 nanometer = 1/1,000,000 millimeter or 1/25,000,000 inch). This resolving power of electron microscopes is at least a thousand times better than that of ordinary light microscopes. While the number and kind of organelles present in a cell will depend on the specific type of cell and its function, the organelles most commonly found in a plant cell are the **nucleus, plastids, mitochondria, ribosomes, endoplasmic reticulum, golgi bodies, microtubules,** and **microbodies.** All organelles except the ribosomes are membrane bounded; thus they are separated into discrete units within the protoplast. Functionally the protoplast is divided into two areas—**cytoplasm,** which is the area of the protoplast found outside the nucleus, and **nucleoplasm,** which is the area found within the nucleus. Thus all organelles except the nucleus are part of the cytoplasm proper. In a mature plant cell the nucleus and most of the cytoplasm are usually limited to the periphery of the cell with the remaining volume of the cell being taken up by a large, centrally located **vacuole** (Figure 2–10). In the past, the vacuole was treated as an

Figure 2–10. Structure of mature cell. *(a)* Diagram of a mature plant cell. *(b)* Electron micrograph of a mature plant cell.

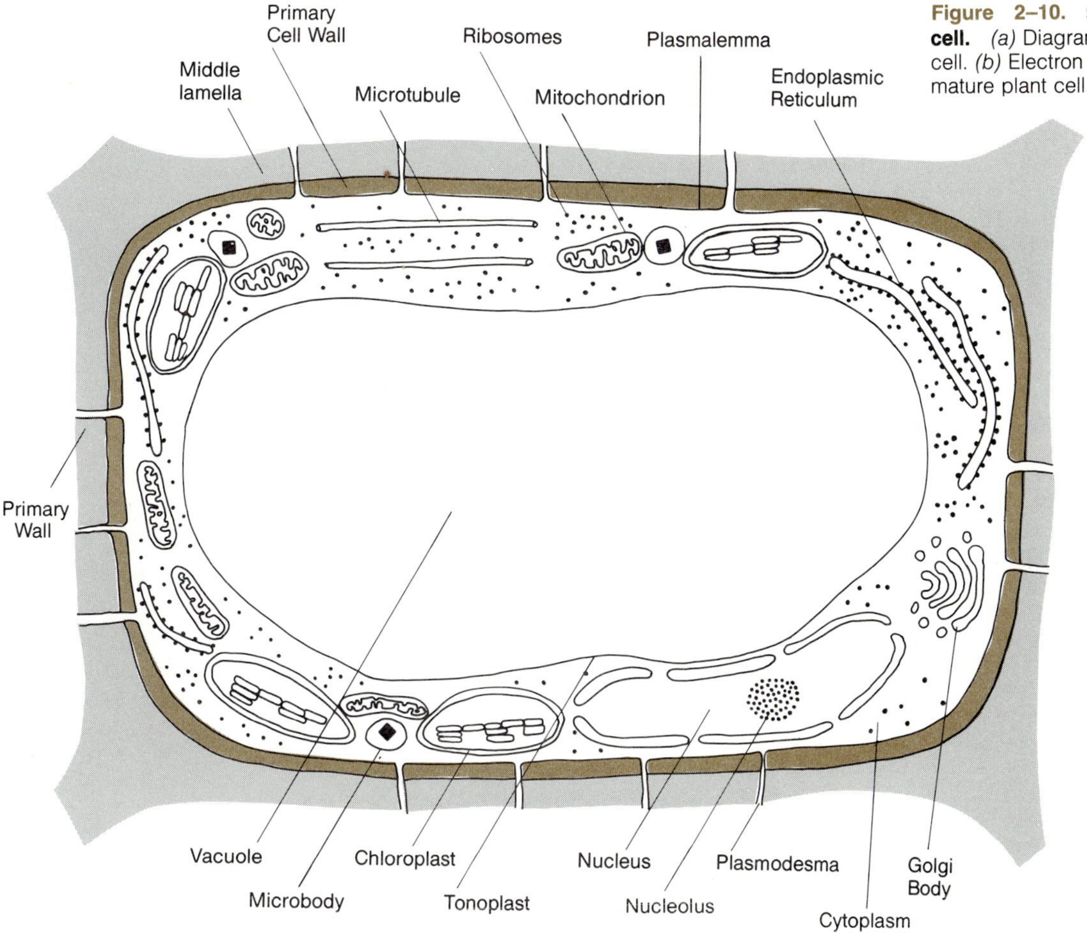

Primary Cell Wall
Ribosomes
Plasmalemma
Middle lamella
Microtubule
Mitochondrion
Endoplasmic Reticulum
Primary Wall
Vacuole
Chloroplast
Nucleus
Plasmodesma
Golgi Body
Microbody
Tonoplast
Nucleolus
Cytoplasm

Nucleus
Nucleolus
Middle Lamella
Cytoplasm
Plasmodesma
Mitochondrion
Chloroplast
Tonoplast
Plasmalemma
Cell Wall
Vacuole

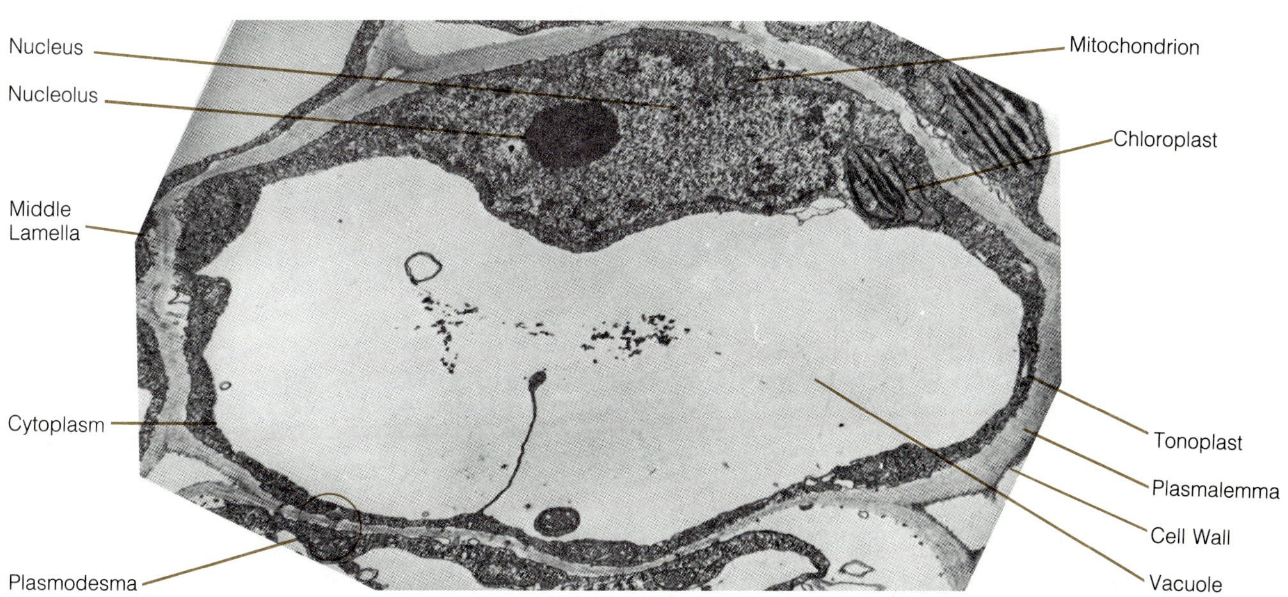

essentially inert region that played no part in cellular metabolism. Although presently there is controversy among botanists concerning whether a vacuole is an organelle, current research involving isolated vacuoles has shown definite metabolic roles for the vacuole.

In recent years investigators have discovered, with the help of high-voltage electron microscopes, that the cytoplasm itself has a complex internal structure, the **cytoskeleton,** consisting of microtubules and microfilaments held together in an intricate network by interconnecting strands. The cytoplasmic streaming that is exhibited by different types of plant cells is believed to be the result of a rearrangement of these microtubules and microfilaments.

Membrane Structure

The term *membrane* has been used frequently above in connection with cell structure. Most of the cell structures are basically made up of membranes. Although there are differences in the thickness of different kinds of membranes, they all seem to have the same fundamental structure. All membranes consist of proteins and lipids. Electron micrographs of a sectioned membrane show two dense lines separated by a less dense space (Figure 2–11). Based mainly on this evidence, a **unit membrane model** was originally proposed to describe membrane structure. According to this model, two layers of lipids are thought to be in the middle with a continuous layer of proteins on either side (Figure 2–12a). However, newer techniques for examining fractured surfaces of membranes as well as higher resolution of sectioned membranes strongly suggest that the outer protein layers are not continuous. The **fluid mosaic model,** which is generally accepted at present as a functional hypothesis for membrane structure, proposes a mosaic distribution of lipid and protein molecules in a membrane, as shown schematically in Figure 2–12b. A significant fact concerning the lipid-protein nature of membranes is their ability to allow both water-soluble and fat-soluble substances to pass through them. In addition to the function of regulating the passage of molecules, the membranes also provide a surface to which enzymes and other molecules can be attached in the proper sequence for chemical reactions to take place or electrons to be transported during energy transformations. The membrane systems of the various organelles also increase the surface area where these reactions can take place.

Nucleus

The **nucleus** is a distinct area of the protoplast surrounded by a double membrane called the **nuclear envelope** (Figure 2–13). In this envelope there

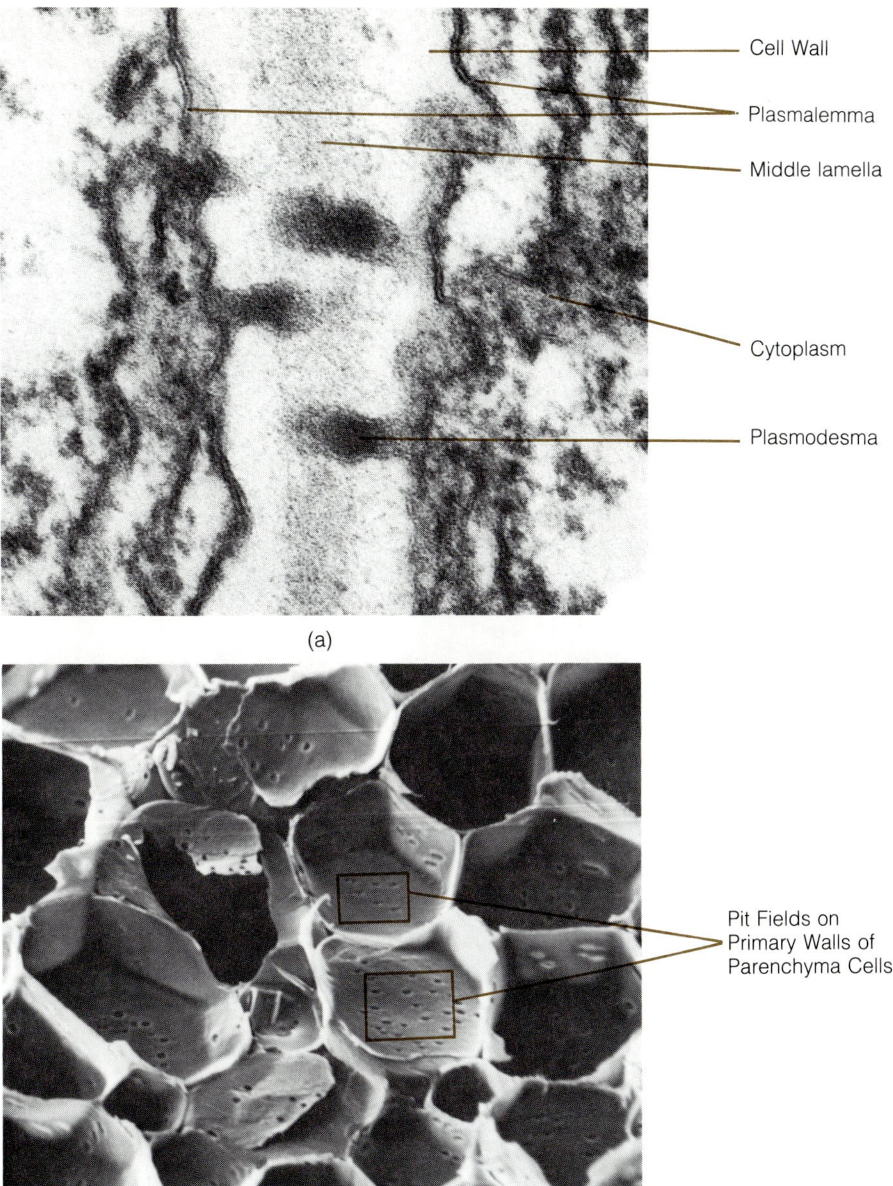

Figure 2–11. Plasmalemma. *(a)* This electron micrograph shows the membranous plasmalemma as the outermost boundary of the protoplast. *(b)* This scanning electron micrograph of the primary cell walls shows pit fields. The plasmalemma pass through pit fields of adjacent cells.

are a number of circular openings called **nuclear pores.** Although the nuclear pore appears blocked by a complex of fibrils, experiments have shown that molecules are able to pass readily through these pores. The nuclear envelope is continuous with the endoplasmic reticulum and also has ribosomes attached to the surface facing the cytoplasm. The part of the protoplast enclosed by the nuclear envelope is the nucleoplasm in which are found elongated strands of **chromatin** and one or more nucleoli. Chromatin is made up of the genetic material* DNA (deoxyribonucleic acid) and protein. In a dividing nucleus, each chromatin strand shortens and thickens by coiling forming a visible **chromosome** (Chapter 4). The **nucleolus** consists

Genetic material
The DNA molecules that contain coded instructions for the synthesis of all cellular proteins.

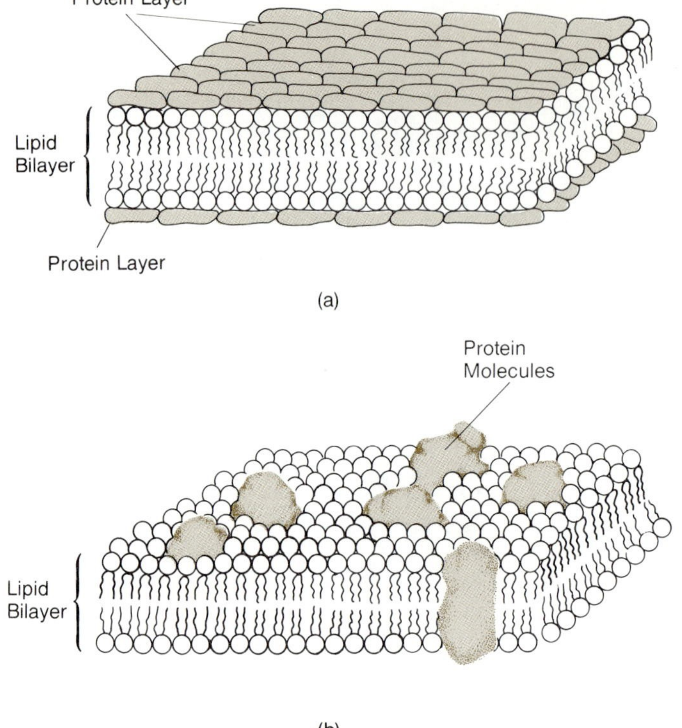

Figure 2–12. Membrane structure. *(a)* The unit membrane model shows two layers of phospholipids in the middle with a continuous layer of protein on either side. *(b)* The fluid mosaic model shows protein molecules of different sizes distributed in the phospholipid bilayer. Some of the protein molecules occur either on the inner surface or on the outer surface, while others occupy the full width of the membrane.

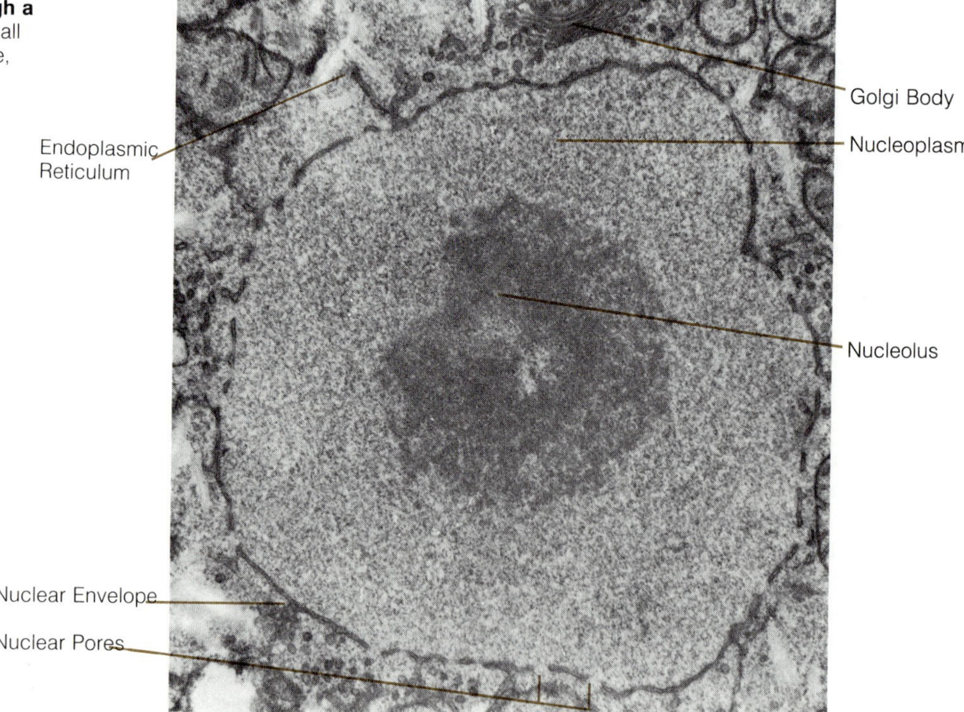

Figure 2–13. Electron micrograph of a section through a nucleus. Nuclear pores are small openings in the nuclear envelope, which usually has ribosomes attached to its outer surface.

of particles rich in RNA (ribonucleic acid) and is the site of synthesis of the precursors of ribosomes. Usually there is only one nucleus per cell.

Two major functions of a cell nucleus are 1) to regulate the metabolic activities of the cell through protein synthesis and 2) to transmit the stored genetic information to newly forming cells. The first function is carried out through a chemical communication feed-back system where information reaches the nucleus from the cytoplasm in the form of chemical molecules and the nucleus sends out messages in chemical form also. Messages that go out from the nucleus to the cytoplasm contain coded information for the synthesis of enzymes that regulate nearly all of the chemical reactions taking place in the cell. Thus the nucleus becomes the control center for the cell. Details of how the nucleus regulates cellular functions are presented in Chapter 3. The second function is carried out during nuclear division when the genetic information stored in the chromatin is duplicated precisely and passed on to the newly formed nuclei. The process of nuclear division is described in Chapter 4.

Plastids

Plastid is the name for a group of membrane-bounded organelles that are characterized by their ability to produce the carbohydrate starch under either natural or experimental conditions. There are a number of functionally different plastids that are correspondingly different in structure. The most important type of plastid is the **chloroplast,** which is the **chlorophyll**-containing plastid. Since chlorophyll is a green pigment, the chloroplast is green and is responsible for the green color of plants. Chlorophyll is found only in chloroplasts; there is no chlorophyll in the remainder of the cytoplasm. Because of their size and green color, chloroplasts can be seen even without staining. There may be as many as forty or more chloroplasts in a cell; however, in many algal cells, there is usually just one chloroplast per cell. In most plants the chloroplasts are biconvex lens-shaped structures surrounded by two membranes that make up the chloroplast envelope (Figure 2–14). Enclosed by the envelope is a transparent region called the **stroma** and a group of interconnected, highly organized photosynthetic membranes called **thylakoids.** Within the chloroplast the thylakoids are arranged in such a way that they form stacks called **grana** at various places throughout the stroma. The number of thylakoids per granum, as seen in a section through a chloroplast, varies from two to many. The chlorophylls and other photosynthetic pigments, such as the **carotenoids** (yellow to orange pigments), are localized in these thylakoids. Many particles are seen attached to the thylakoid membranes when they are fractured to expose their surfaces (Figure 2–15). These particles are the sites where light energy is converted to chemical energy, which is then used in sugar synthesis (Chapter 9). Most of the sugar precursors are exported from the chloroplast to the cytoplasm where they are either used in cellular metabolism (Chapter 3) or are converted to sucrose, common table sugar, for transport to the nonphotosynthetic regions

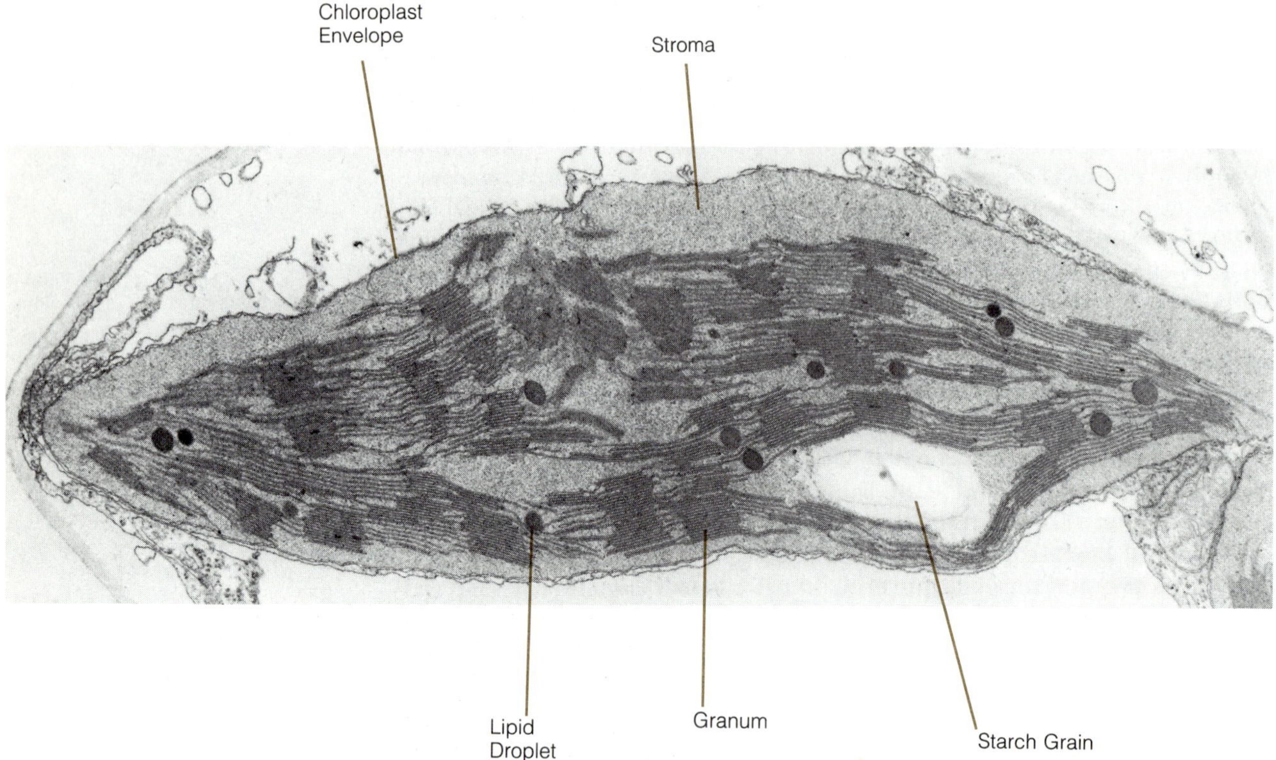

Figure 2–14. Electron micrograph of a chloroplast section. A granum is a stack of flattened membranous sacs, each called a thylakoid. The different grana of a chloroplast are interconnected. The fluid portion of the chloroplast surrounding the grana is the stroma, which contains many ribosomes.

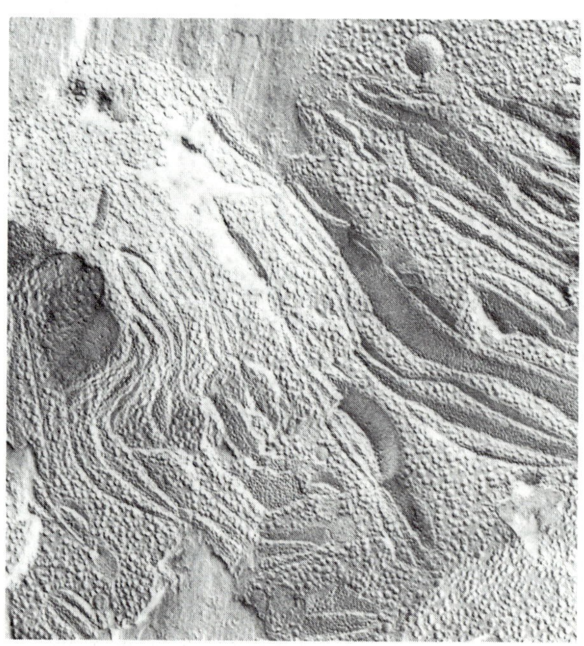

Figure 2–15. Electron micrograph of a freeze-fractured surface of a chloroplast. The particles that are part of the thylakoid membranes are associated with energy transfer, which takes place during photosynthesis.

of the plant. When there is an excess of sugar precursors, simple sugars are produced in the chloroplast and then linked together forming starch, which accumulates in the form of granules inside the chloroplasts (Figure 2–14).

Chromoplasts are plastids that have a disorganized internal membrane system and lack chlorophyll. Usually they are formed from chloroplasts by a reduction or complete cessation of chlorophyll synthesis and a breakdown of the grana. They contain yellow and orange (and sometimes also red) carotenoid pigments that are responsible for the color of carrots, ripe tomatoes, some flowers, and some leaves during autumn. **Leucoplasts** are colorless plastids, lacking any pigment. Although they are bounded by typical plastid outer membranes, these plastids usually lack an internal membrane system. They store starch and other food reserves such as lipids and proteins in many stems, roots, and other underground organs of a plant. The large amount of starch found in potato tubers, for example, is stored in leucoplasts.

Plastids are formed either by division of existing plastids or by development from undifferentiated plastids known as **proplastids.** It has been reported that plastids can change their function in response to chemical signals from the nucleus, so that the various mature plastid types can change their metabolic role as part of an overall change in the function of the cell. Since plastids arise from the division of preexisting plastids and the differentiation of proplastids, it is interesting to note that plastids contain their own genetic material (DNA) and ribosomes (Figure 2–16). Thus, plastids are capable of synthesizing some of their own proteins without relying on information encoded in nuclear DNA. The importance of this observation will be evident in later chapters, but for now suffice it to say that conventional wisdom up until the 1960s held that only the nucleus of a cell contained DNA and could direct protein synthesis. The logical extension of this statement is that any cellular structure containing DNA, other than the nucleus, must represent a separate cell. By this reasoning, a hypothesis was advanced that the chloroplast, and therefore all plastids, originated as free-living photosynthetic organisms that were trapped by the nonphotosynthetic ancestors of the present-day green plants (Chapter 23). This hypothesis of the **endosymbiotic** origin of plastids is widely, but not universally, accepted.

Mitochondria

Mitochondria are spherical or rod-shaped organelles that are also enclosed by an envelope made up of two membranes (Figure 2–17). A characteristic feature of most mitochondria are **cristae,** fingerlike or tubular invaginations of the inner membrane of the envelope. The number of cristae is dependent directly on the energy-requiring activity of the cell, as is the number of mitochondria in the cell of a multicellular plant: the greater the energy requirement of a cell, the greater the number of mitochondria present in it. This number could vary from a few to a thousand or more. Some recent evidence suggests that in unicellular plants there may be only a single large

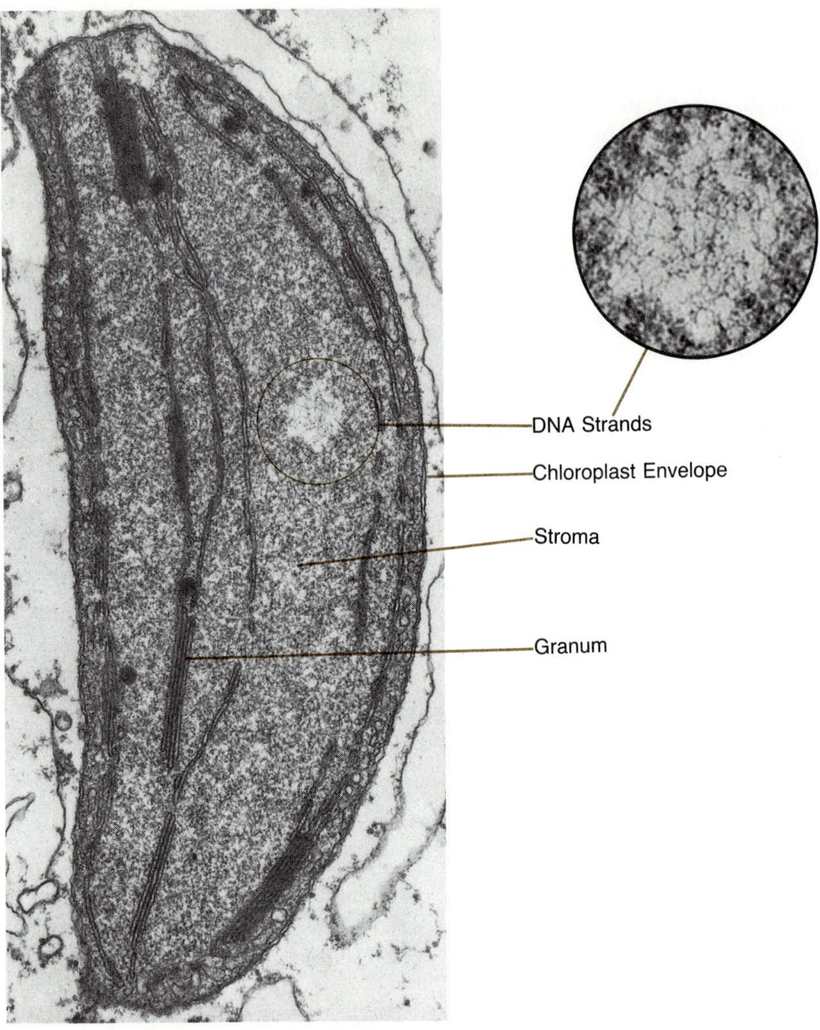

Figure 2–16. Electron micrograph of a chloroplast section. DNA can be seen as thin fibrils within the stroma. The small particles are chloroplast ribosomes.

Aerobic respiration
The breakdown of organic molecules in the presence of oxygen to release stored energy that can be used to make ATP.

mitochondrion located at the periphery of the cell. The physiological role of mitochondria is to perform the major steps of **aerobic respiration*** that result in the breakdown of food molecules and the production of ATP (adenosine triphosphate) molecules, which serve as a ready source of energy for cellular metabolism (Chapter 3). Many of the enzymes related to the synthesis of ATP are localized in the cristae, but others are found in the **matrix,** the inner compartment of the mitochondrion. New mitochondria appear to be produced by the division of mature mitochondria and, unlike the situation with plastids, there are no known mitochondrial precursors in immature cell types. The most probable reason for the absence of undifferentiated mitochondria is the need for energy (ATP) production even by immature cells. Whereas there are many living cells in a plant that cannot synthesize food for lack of chloroplasts, no living cell can survive without the synthesis of ATP molecules by the mitochondria. Mitochondria, like plastids, have their own DNA and ribosomes. There is speculation, therefore, that mito-

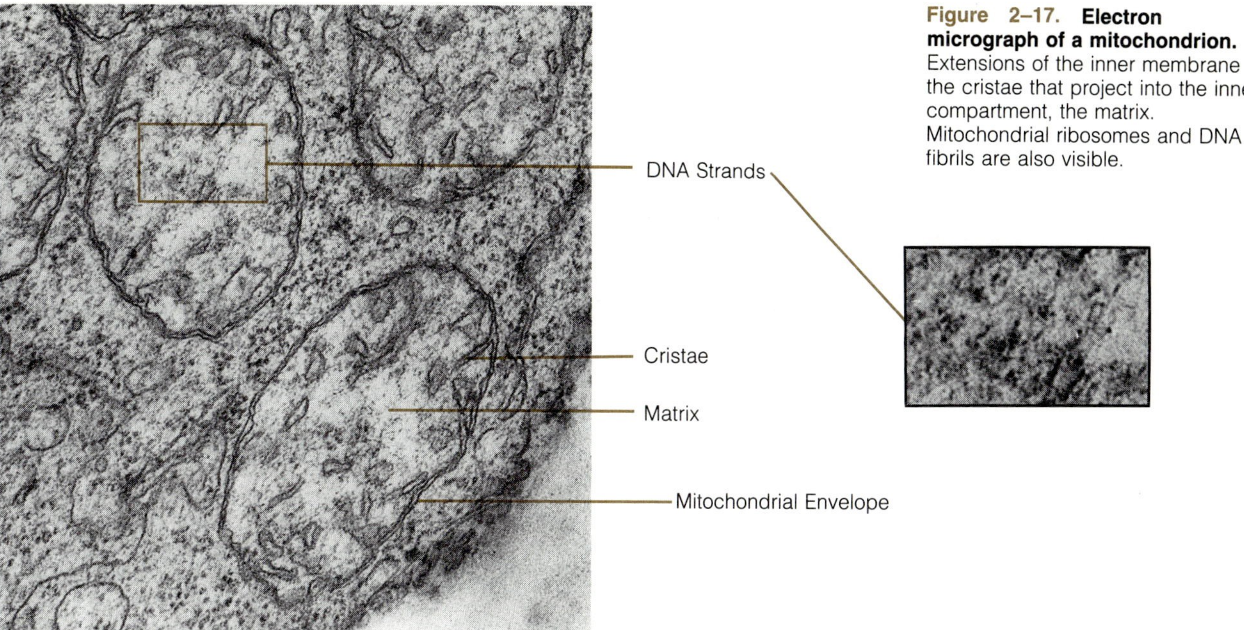

Figure 2–17. Electron micrograph of a mitochondrion. Extensions of the inner membrane are the cristae that project into the inner compartment, the matrix. Mitochondrial ribosomes and DNA fibrils are also visible.

chondria also may have evolved as free-living bacterialike organisms before they started a symbiotic* association with the ancestors of the present-day plants and animals.

Symbiotic
Association between two dissimilar organisms that is advantageous to one or both.

Ribosomes

Ribosomes are extremely small (17–23 nanometers), somewhat spherical structures made up of two subunits, one of which is larger than the other (Figure 2–18). Each ribosome is composed of approximately equal amounts of protein and ribonucleic acid (RNA) (Chapter 3). After the ribosome subunits are synthesized in the nucleolus, they are transported to the cytoplasm where they are assembled into functional ribosomes. In the cytoplasm ribosomes may be found free or attached to the **endoplasmic reticulum** (see below) or attached to the outer surface of the nuclear envelope. The function of the ribosome is protein synthesis, which occurs when nuclear instructions for the assembly of a single protein are "read" by the ribosome (see Chapter 3 for details). Ribosomes are also present in the plastid stroma and mitochondrial matrix where they take part in the synthesis of certain plastid and mitochondrial proteins. These ribosomes, however, are smaller than the ones present in the cytoplasm and are similar in size to the ones present in bacteria and cyanobacteria, a fact that adds further support for the endosymbiotic hypothesis of the origin of plastids and mitochondria.

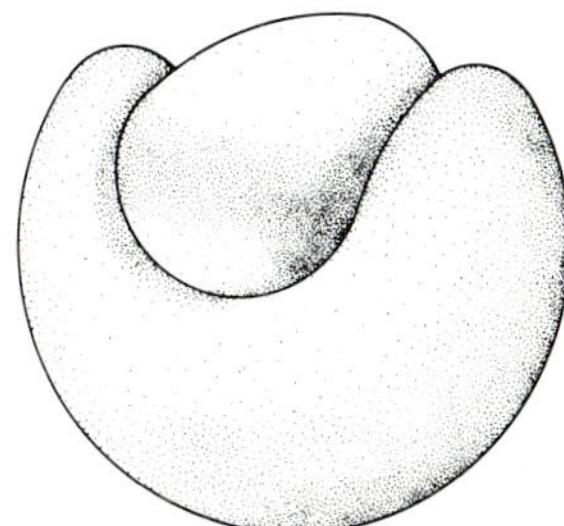

Figure 2–18. Ribosome. Schematic drawing representing the two subunits of unequal size and shape that make up a ribosome.

Endoplasmic Reticulum

The **endoplasmic reticulum,** commonly called **ER,** is a complex network of membranous channels in the cytoplasm (Figure 2–19). The ER membranes are arranged in the form of flattened sacs, known as **cisternae,** and in a section through a cell they appear as parallel lines with a space (lumen) between lines. The width of the lumen may vary considerably. The portions of the ER where ribosomes are attached to the outer surface of the cisternae are the **rough ER,** while the portions without attached ribosomes are the **smooth ER.** It is not uncommon to find that the ER is connected to the nuclear envelope. There is also some evidence to suggest that ER connections between adjacent cells run through plasmodesmata. The endoplasmic reticulum of a cell constitutes a complex system where synthesis and transport of proteins and other molecules take place. Therefore, cells that are very active in synthesis and transport often have an extensive network of ER cisternae.

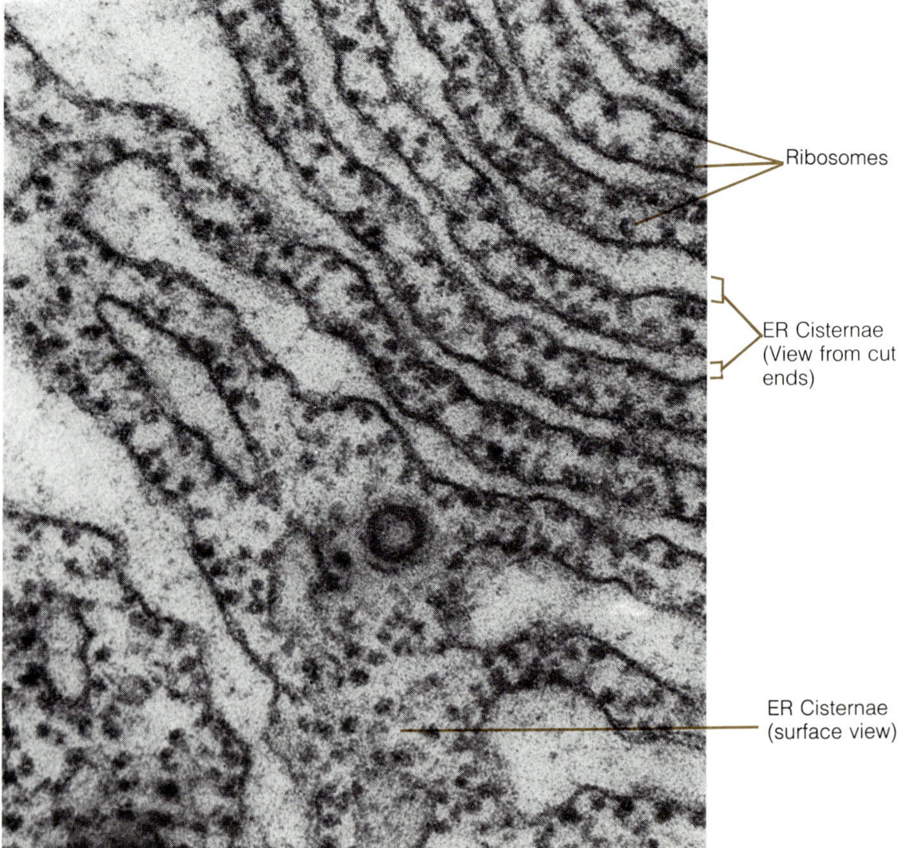

Figure 2–19. Endoplasmic reticulum. The tubular outlines seen in this electron micrograph are sections through the endoplasmic reticulum. These outlines are elongated, flattened saclike structures consisting of membranes. Often ribosomes are attached to the membrane surface on both sides.

Golgi Body

The **Golgi body** is made up of two to many stacks of flattened, circular, membrane-bounded sacs also called **cisternae** (Figure 2–20) and are believed to originate from the cisternae of the smooth ER. The margins of the Golgi cisternae are often swollen, and in surface view the outer margin has a highly lacelike, or fenestrated, appearance. It is believed that the Golgi vesicles are formed from these fenestrated margins by a pinching-off process (Figure 2–20). A major function of Golgi bodies in plant cells is that of secretion of cell wall materials. It has been suggested that materials synthesized in the ER are accumulated in the Golgi cisternae and transported inside vesicles to other areas in the cell or outside the cell. Export of Golgi contents such as cell wall materials occurs as the result of the fusion of the membranes of a Golgi vesicle with the plasmalemma, followed by the release of the contents to the outside (Figure 2–20).

Microtubules

Most commonly **microtubules** are seen as straight cylinders of indeterminate length (Figure 2–21a). In cross section each microtubule appears as a small circle composed of nine subunits that surround a hollow core (Figure 2–21b). The microtubular subunits, composed of a specific type of protein called **tubulin,** are arranged in a helical pattern in the microtubular wall. In plant cells microtubules appear to serve diverse functions. They are often found associated with formation of the cell plate and new cell walls, cytoplasmic streaming, and motility of certain gametes, algal cells, and fungal cells. The typical cilia* or flagella* of motile cells are made of microtubules arranged in an orderly manner (Figure 2–21c). The spindle fibers that are associated with chromosome movement during mitosis* and meiosis* in a dividing cell are also made of microtubules.

Cilia
(*sing.*, cilium) Short, hairlike cell structures that function in cell motility.

Flagella
(*sing.*, flagellum) Long, hairlike cell structures that function in cell motility.

Mitosis
The process of nuclear division that results in the production of two genetically identical daughter nuclei.

Meiosis
The process of nuclear division that results in the production of daughter nuclei, each containing one-half the original number of chromosomes.

Microbodies

Microbodies are single membrane-bounded organelles of varying size serving diverse functions. The contents of the microbodies usually appear as granules or crystals. The predominant types of microbodies in plant cells are the **peroxisomes** and **glyoxysomes.** Peroxisomes are present mainly in leaf tissue where they are seen in close association with chloroplasts and mitochondria (Figure 2–22). In these cells peroxisomes take part in the process of photorespiration, which is a series of metabolic reactions associated with

Figure 2–20. The Golgi Body. *(a)* Reconstruction of a Golgi body based on electron micrographs of Golgi bodies sectioned along different planes. Each Golgi body consists of several flattened sacs (cisternae) with extensive fenestrations. *(b* and *c)* Electron micrographs of sectioned Golgi bodies viewed from the side and surface, respectively.

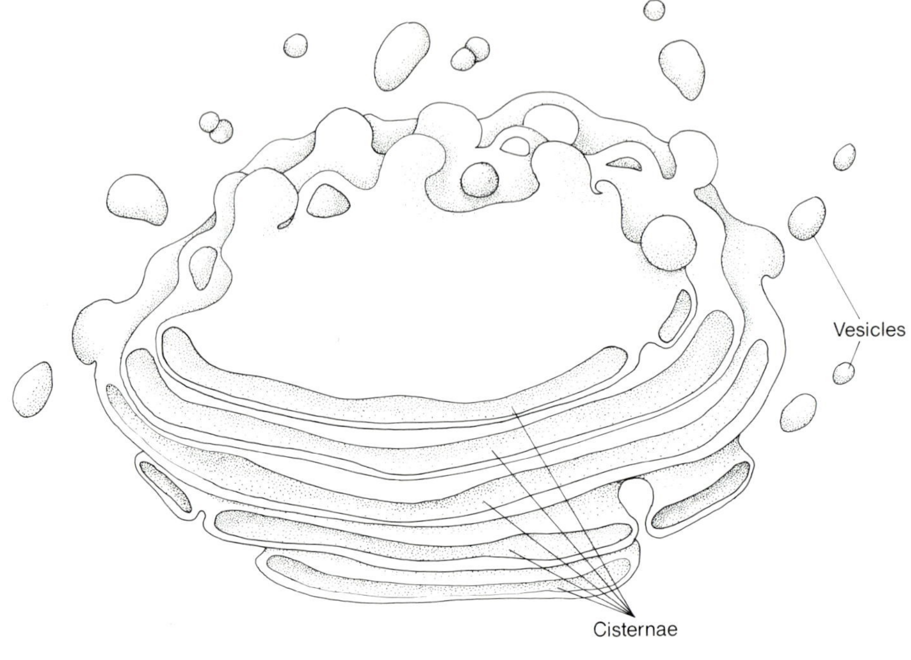

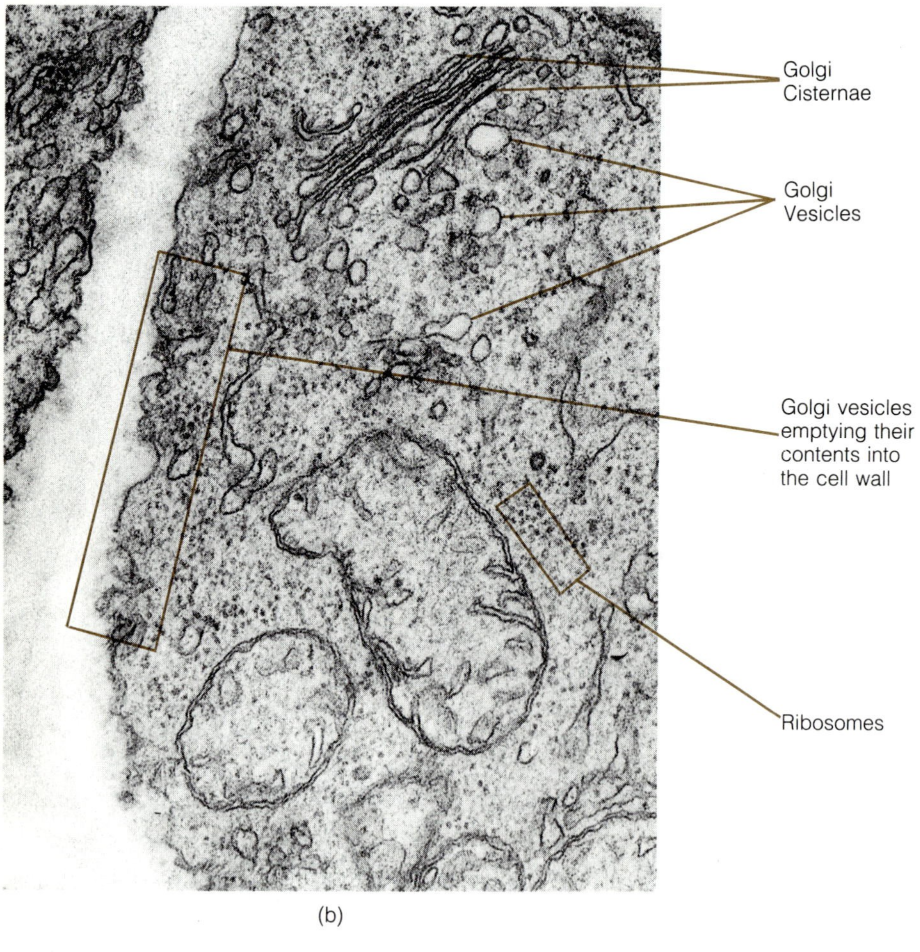

Vesicles
Cisternae
(a)

Golgi Cisternae
Golgi Vesicles
Golgi vesicles emptying their contents into the cell wall
Ribosomes
(b)

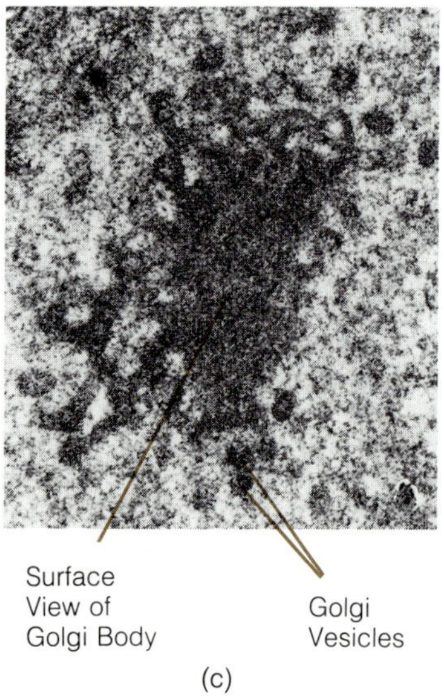

Surface View of Golgi Body
Golgi Vesicles
(c)

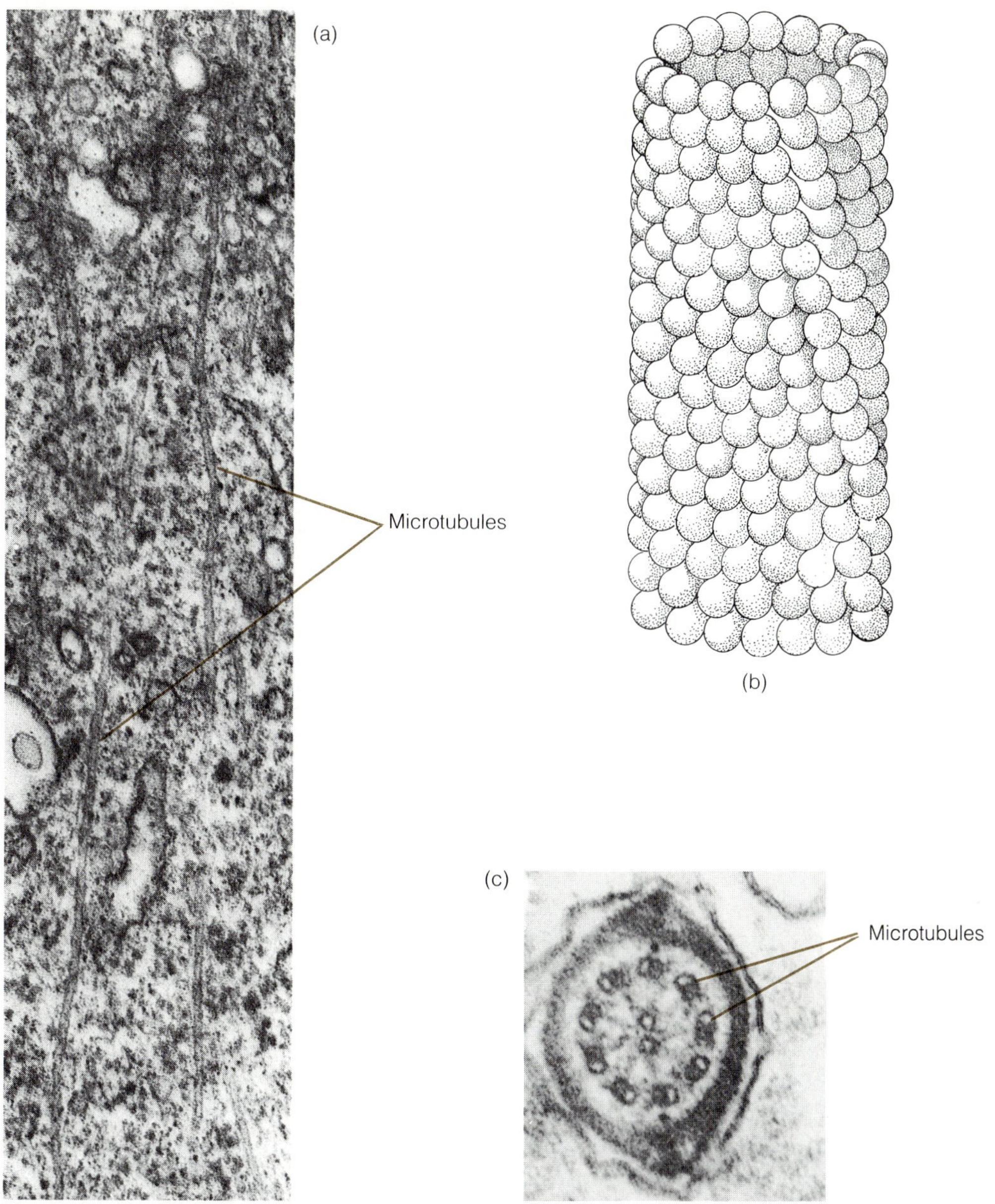

Figure 2–21. Microtubules. *(a)* Electron micrograph of a dividing plant cell showing many microtubules in the cytoplasm. *(b)* Each microtubule is a hollow cylinder of small diameter and of undetermined length, consisting of many spirally arranged spherical subunits as shown in this diagram. *(c)* The cross section of a flagellum shows 9 pairs of microtubules around the periphery and 2 in the center.

photosynthesis* in many types of plants (Chapter 9). The role of glyoxysomes is in lipid metabolism, and thus they are typically found in the storage tissue of plants that use lipids as stored food. A third type of microbody is the **lysosome.** These contain digestive enzymes that can break down nonfunctional and damaged cell organelles, the chemical components of which can then be recycled. Although there is evidence to suggest that lysosomes are produced by the Golgi body, the origin of the other types is speculative at present.

Photosynthesis
The metabolic process by which carbohydrates are produced from CO_2 and a hydrogen source such as water, using light as the source of energy.

Figure 2–22. Peroxisome. Electron micrograph of a leaf cell showing the close association between a peroxisome, mitochondrion, and a chloroplast. Whereas the envelopes of the mitochondrion and chloroplast consist of two membranes, the peroxisome envelope is a single membrane as is characteristic of all microbodies.

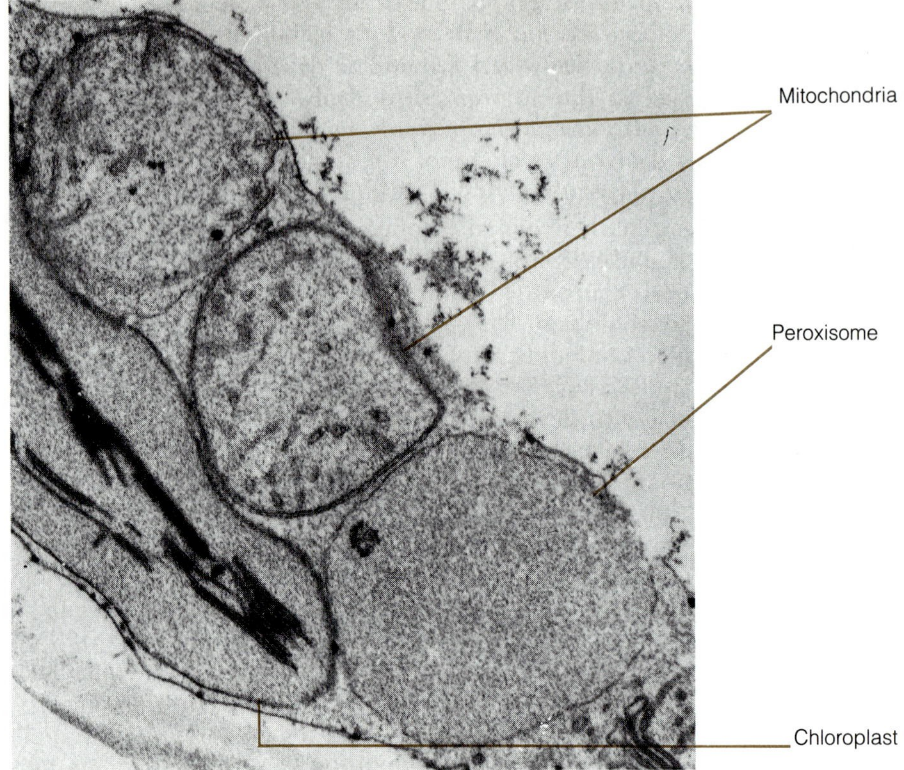

Vacuole

The central vacuole of a mature plant cell is a membrane-bounded region in the protoplast, occupying the greatest portion of the cell (Figure 2–10). Separating the vacuole from the cytoplasm is a differentially permeable membrane called the **tonoplast** whose function is to regulate the movement of materials in and out of the vacuole. Since plants do not ingest food, they do not produce large volumes of waste materials. Consequently, an excretory system did not evolve in plants. However, even plant metabolism produces some by-products that cannot be used or degraded and must be disposed of by the cell. The vacuole, which is a water-storing region within the cell, serves as a dumping site for these sometimes toxic metabolic wastes. In addition, the vacuole serves an important role in the water relations of cells because it contains by far the largest amount of available water in the cell. Usually in a newly formed plant cell there are several small vacuoles that are formed from the endoplasmic reticulum. As the cell enlarges during maturation, these small vacuoles fuse to form a large centrally located vacuole. It is the pressure of the enlarging vacuole on the cell wall that is responsible for cell enlargement (Chapter 12). Furthermore, it is now being recognized by botanists that the vacuole is more than an inert reservoir; the vacuole is metabolically active and contains characteristic enzymes. Finally, in some

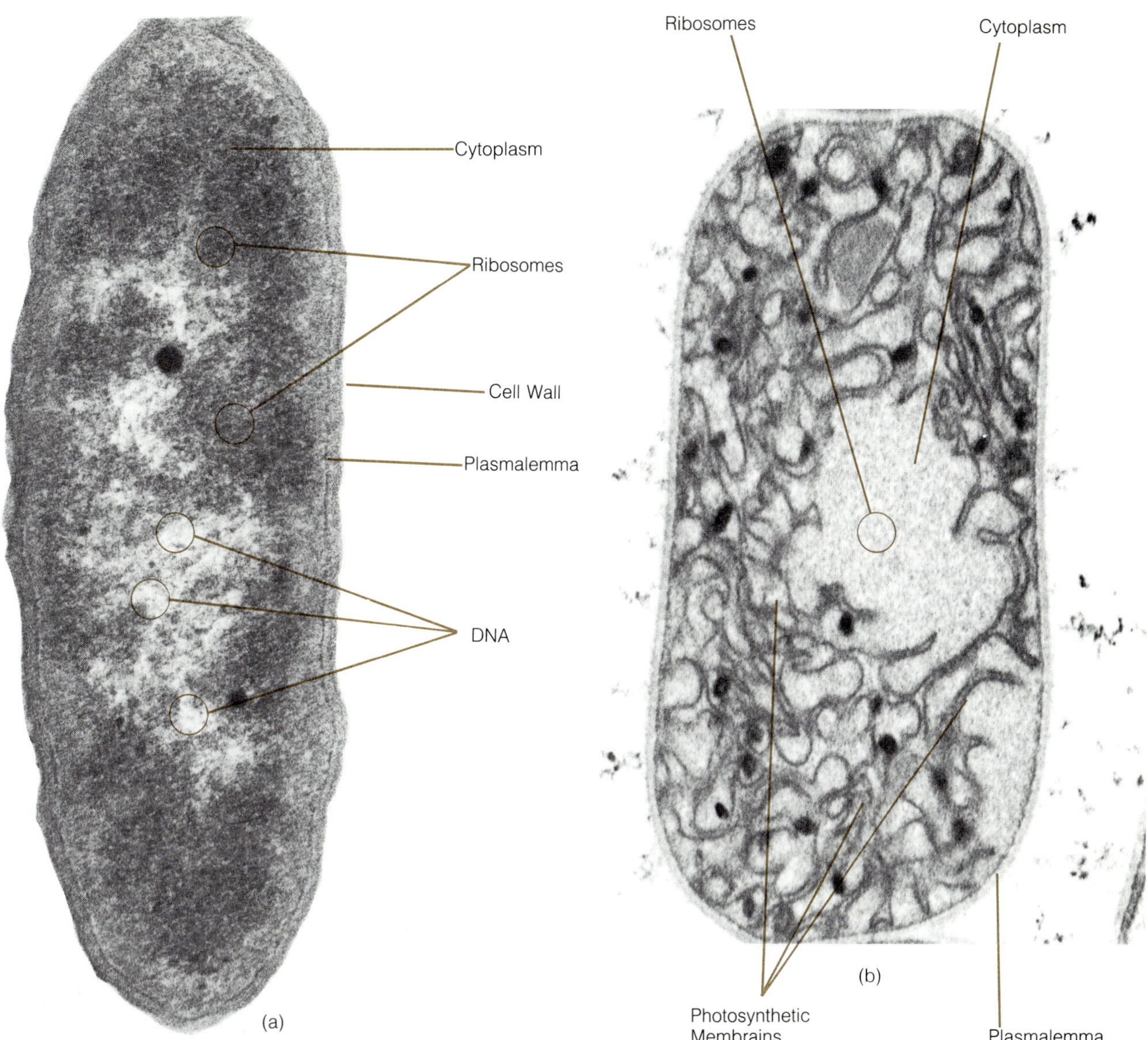

Figure 2–23. Prokaryotic cells. Electron micrographs of a bacterium *(a)* and a cyanobacterium *(b)* showing the absence of membrane-bounded organelles typical of prokaryotic cells.

cells vacuoles also contain water-soluble pigments such as **anthocyanins, flavones,** and **betacyanin.** The color of many flowers, fruits, and some leaves is due to the presence of one or more of these pigments. Beets contain large amounts of betacyanin, a reddish pigment, in their vacuoles so that eating many beets results in highly pigmented urine and feces.

Animal cells contain all of the organelles except the plastids described above. In addition, they contain certain organelles that are not seen in typical plant cells. The presence of a cell wall, plastids, and a large vacuole further distinguish a typical plant cell from a typical animal cell. The basic similarity of plant and animal cells, however, strongly suggests that both plants and animals must have evolved from a common ancestor (Chapter 23).

The type of cell described thus far in this chapter is characterized by the separation of the nuclear material, photosynthetic apparatus, and respiratory mechanism by membranous envelopes. Such cells are called **eukaryotic cells** since they possess true *(eu-)* nuclei. This type of cellular organization is typical of all living organisms except bacteria and cyanobacteria discussed in Chapter 15. These organisms do not have internal compartmentalization of metabolism, and so the only organelles that they possess are non-membrane-bounded, such as the ribosomes. Since they do not have nuclei, the bacteria and cyanobacteria are **prokaryotic,** meaning before *(pro-)* the nucleus (Figure 2–23). The prokaryotic organization of cells is considered to be a primitive characteristic which, during the course of evolution, developed into the eukaryotic condition (Chapter 23). As indicated earlier, there is considerable evidence that plastids and mitochondria originated from symbiotic cyanobacteria and bacteria, respectively. Photosynthetic membranes, when present in a prokaryotic organism, are not surrounded by an envelope.

Summary

The Cell Theory, which states that all living organisms are made up of cells and that these cells have common structural and functional features, was originally proposed jointly by a botanist and a zoologist in the mid-19th century and has become one of the major unifying principles in biology today. That a cell is the basic unit of a living organism and that new cells are formed from preexisting cells have now been universally accepted. A plant can be unicellular, colonial, or multicellular, with great variation in the size, shape, structure, and function of cells. Most cells are too small to be seen without the aid of a light microscope, and the ultrastructural details can be seen only through an electron microscope. In animals dead cells are usually sloughed off, whereas in plants both living and dead cells play significant functional roles.

A living plant cell contains a protoplast surrounded by a cell wall. A primary wall is present in all cells while a secondary wall, deposited on the inner side of the primary wall, is a characteristic feature of cells that provide support and flexibility or transport water. Cells usually die after the addition of the secondary wall, leaving a space in the place of the protoplast. Pits are commonly present in the secondary walls of many cells. A middle lamella that is initiated during cell division glues adjacent cells together.

The organelles such as the nucleus, plastids, mitochondria, ribosomes, endoplasmic reticulum, Golgi, microtubules, and microbodies are suspended in the protoplast of the cell. Functionally, the protoplast is divided into nucleoplasm contained within the nucleus, and cytoplasm, which is the remaining area excluding the nucleus. The plasmalemma, the outer boundary of the protoplast, regulates movement of materials in and out of the cell. The central region of a mature plant cell is a vacuole surrounded by the tonoplast, which prevents the vacuolar contents from mixing with the cytoplasm. The nucleus of a cell contains the genetic material DNA, and it controls all metabolic activities of the cell through a complex chemical communication system. Plastids such as chloroplasts synthesize carbohydrates using light energy and carbon dioxide, while leucoplasts store starch or lipids in different parts of the plant body. Aerobic respiration takes place in the mitochondria where energy contained in food is converted to usable energy in the form of ATP molecules. Organelles such as nuclei, plastids, and mitochondria are surrounded by envelopes consisting of two membranes. Ribosomes are the site of protein synthesis and are present in the cytoplasm as well as in the plastids and mitochondria. The endoplasmic reticulum is a complex network of membranous channels where synthesis and transport of different molecules take place. Rough endoplasmic reticulum consists of parts of the endoplasmic reticulum with attached ribosomes. The Golgi, a membranous organelle made up of flattened sacs, functions for the secretion of wall materials. Microtubules are hollow cylindrical organelles associated with protoplasmic streaming, chromosome movement, and motility of certain algal and fungal cells. Microbodies such as glyoxysomes, peroxisomes, and lysosomes have envelopes made up of one membrane and are associated with fat metabolism, photorespiration, and digestion of cellular components, respectively.

It should be evident from the above descriptions that the varied functions of a cell are closely associated with membranes. According to the most recent structural model of the membrane, all membranes consist of proteins and lipids which are distributed in a mosaic manner, allowing for a nonrigid, fluid nature of the membrane. Membranes regulate the passage of molecules through them as well as provide a surface on which chemical reactions can take place in proper sequence.

There is a considerable similarity between plant and animal cells that strongly suggests a common ancestry for these diverse groups of living organisms. The bacteria and cyanobacteria, however, have a distinctly different cellular organization from all other living things, reflected in the absence of membrane-bounded organelles. This prokaryotic cellular organization of bacteria and cyanobacteria is more primitive than the eukaryotic organization of the other forms of life.

Review Questions

1. Draw a typical plant cell and label all parts.
2. List all organelles present in a typical plant cell and write a function for each.
3. How does the nucleus regulate cellular metabolism?
4. What is the structure of a differentially permeable membrane as we understand it now?
5. What is the functional significance of the differentially permeable membrane?
6. How is a prokaryotic cell different from a eukaryotic cell?

Frost Damage

Have you wondered why certain plants die after the first frost while others remain green all winter? In most cases involving plants of temperate regions, temperatures around 0° C (32° F) have no effect on the sensitivity of enzymes or other cellular chemicals. Plants that are harmed by cold are so affected because a drop in temperature to freezing causes water in the intercellular spaces to form ice crystals. Because of this, additional water moves out of the cells and into the intercellular spaces owing to the higher concentration of water inside the cells, resulting in the death of these cells. It has been observed recently that certain bacteria can act as centers for crystal formation, known as nucleation centers. These bacteria are present on crop plants and fruit trees, and are responsible for the frost damage that occurs periodically in the southern states. When these bacteria are absent, water in plants will not freeze; instead it becomes supercooled as the temperature drops a few degrees below 0° C. Plants are known to withstand temperatures of −8° C for several hours if ice crystals are not formed. More recently, strains of bacteria have been created through genetic engineering that do not act as nucleation centers. Initial results indicate that frost damage of vegetable crops and fruit trees can be prevented by spraying the plants with these new bacterial strains.

Unprotected structures such as flowers, leaves, and herbaceous stems are very susceptible to winterkill, whereas the bark of woody plants provides protection against freezing temperatures. Covering garden plants at night when frost is expected prevents the formation of ice crystals in the leaves, thus postponing winterkill. Investigators have shown that plants can be conditioned to withstand freezing or below-freezing temperatures by gradually exposing them to increasingly cold temperatures. Plants such as Brussels sprouts, grass, and the various evergreens have "antifreeze" compounds that prevent nucleation and thereby inhibit the formation of ice crystals. In addition to these defenses against freezing, the cells also need protection against the effects of cold temperatures on the various molecules that make up cellular structures, particularly the enzymes that drive cellular metabolism and the lipids that are important constituents of cellular membranes (Chap. 3). The enzymes of plants that are adapted to extreme winter temperatures are able to function at lower temperatures than those of tropical plants. Similarly, the cell membranes of cold-hardy plants retain their essential characteristic of differential permeability.

You may already know that bananas do not keep well in a refrigerator. This is due to the fact that low temperatures can cause damage even before the temperature reaches the freezing point of water. Chilling damage in cold-intolerant plants can be caused by a change in the fluidity of the lipids of the cellular membranes. As the temperature drops, the normally fluid membrane lipids become solid, resulting in increased permeability of the membranes to water and other cellular chemicals and in the collapse of the cells because of their inability to maintain their proper internal composition.

Potatoes also should not be kept in the refrigerator, but not because they are damaged by cold temperatures. Potatoes stored at cool temperatures will last for a long time; however, they rapidly will become unpalatable, even though no visible changes have taken place. During the cold treatment, the potato cells convert stored starch to sugar, a physiological adaptation that protects the cells from possible freezing, since progressively lower temperatures are required to freeze water as the concentration of solutes increases. Unfortunately, this leaves the potato tuber with an uncharacteristic sweet taste that is unacceptable in french fries or baked potatoes.

3

Cellular Metabolism and its Regulation

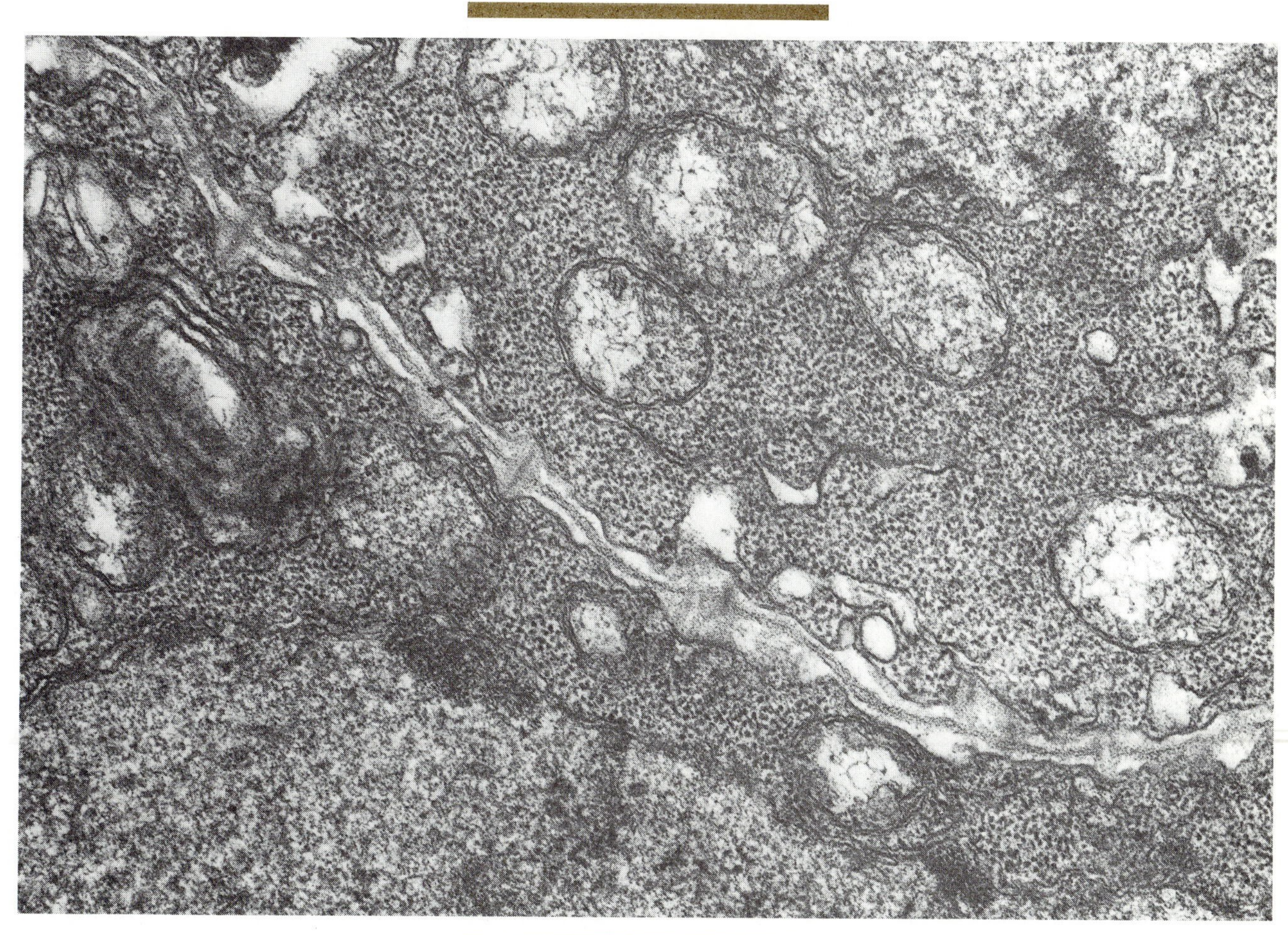

Outline

Plastid
A class of membrane-bounded organelles characterized by the ability to make starch.

Mitochondrion
(*pl.,* mitochondria) A membrane-bounded organelle that functions in respiration.

Nucleic acid
A complex organic molecule involved in the storage and expression of genetic information.

Golgi
A cytoplasmic organelle composed of stacks of flattened, circular, membrane-bounded sacs.

Metabolism
All of the chemical reactions that occur in a living cell.

Introduction

The cell is the basic building block of an organism. But what are the building blocks of the cell? What is the organizing force that assembles the cell from its building blocks? Is this different from the "vital force" that nineteenth-century philosophers supposed was the very essence of life itself?

According to the cell theory (Chapter 2), cells are produced from existing cells, not from a stockpile of cell components. This focuses attention on the cell as the place where the answers to all of these questions must lie. At one level it is clear that the plant cell is formed from a cell wall, cytoplasm, and nucleus, but this leaves open all of the same questions. A study of cell division (Chapter 4), however, indicates that during the formation of new cells, the nucleus, plastids*, and mitochondria* reproduce themselves, and a new cell wall is produced. The cell produces these, as well as its other components, from chemicals. The cell, therefore, is a unique combination of chemical molecules: the organizing force leading to the assembly of these molecules into new cell components resides in the genetic information contained in the nucleus; and the "vital force" behind all of this is the inherent property of nucleic acid* molecules to reproduce themselves (Chapter 23). Nucleic acids contain the genetic blueprint for assembling simpler molecules into more complex ones and associating them into the cell wall material, or the Golgi*, or the various membrane systems, or the cytoplasmic fluid, and so forth. In conjunction with similar, but much smaller, genetic blueprints in the plastids and mitochondria, the nucleus is able to direct all of cellular metabolism.* Each cell, then, is a chemical factory directed by its own nucleus as part of a concerted effort to facilitate the survival of the organism and the transmission of its nucleic acids to subsequent generations.

It is sometimes claimed that the human body is composed of about one dollar's worth of chemicals. A similar statement can be made about plants, though the dollar amount will vary with their size. The point of this statement is that organisms are composed of fairly common chemical elements, not rare or exotic ones. Because organisms are composed of complex molecules, not simple chemical elements, the real value of an organism would run to hundreds or thousands of dollars. This distinction between the complex molecules produced by living organisms and the simpler compounds found in the nonliving world led scientists in the 1700s and 1800s to divide the study of chemistry into two different areas: organic and inorganic. Organic chemistry dealt with the compounds produced by living organisms, whereas inorganic chemistry dealt with the remainder of chemical substances. There was also believed to be a fundamental difference between the two types of compounds such that organic compounds only could be produced in a living cell. Today we still retain the distinction between organic and inorganic chemistry, though we recognize that organic compounds can be synthesized in the laboratory in greater variety than ever existed in nature. The current basis for this distinction is that organic chemistry is the chemistry of a unique element, carbon. Carbon is an unusual element in a

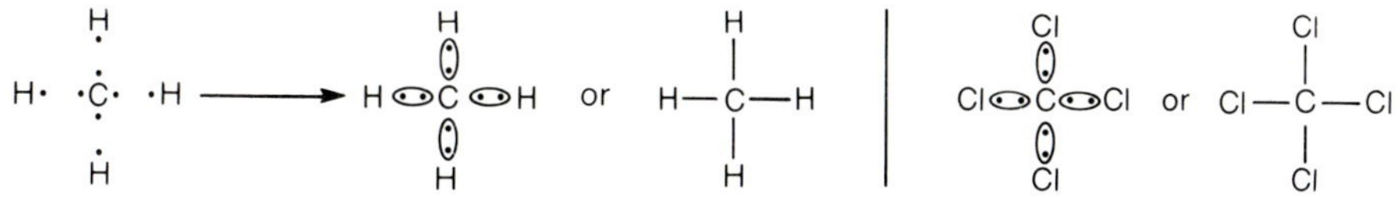

Figure 3–1. Covalent bonding by carbon. Each carbon atom (C) has 4 electrons (·) that can participate in covalent bonds. Each hydrogen atom (H) or chlorine atom (Cl) has 1 electron that can enter into covalent bonds.

number of ways. Each carbon atom forms four covalent bonds* (see Appendix); for example, a carbon atom can combine with four hydrogen atoms thus forming methane, CH_4 (Figure 3–1), or with four chlorine atoms thus forming carbon tetrachloride, CCl_4 (Figure 3–1). Of more importance for organic chemistry and the chemistry of life is the fact that carbon atoms form covalent bonds with other carbon atoms thus forming a great variety of stable molecules of different sizes and shapes all with a carbon backbone or skeleton (Figure 3–2).

Covalent bond
A chemical bond that results from two atoms sharing a pair of electrons.

Of the many types of complex organic compounds, each built up from simple organic molecules, found in living cells, the lipids, carbohydrates, proteins, and nucleic acids predominate. All of these have a carbon skeleton to which are attached other elements, such as hydrogen, oxygen, nitrogen, and phosphorus, to complete the covalent bond requirements of the carbon.

Lipids

Lipids are water-insoluble organic compounds, which can be divided into several major classes such as fats, phospholipids, and waxes. Most lipids are complex molecules composed of **fatty acids** combined with another molec-

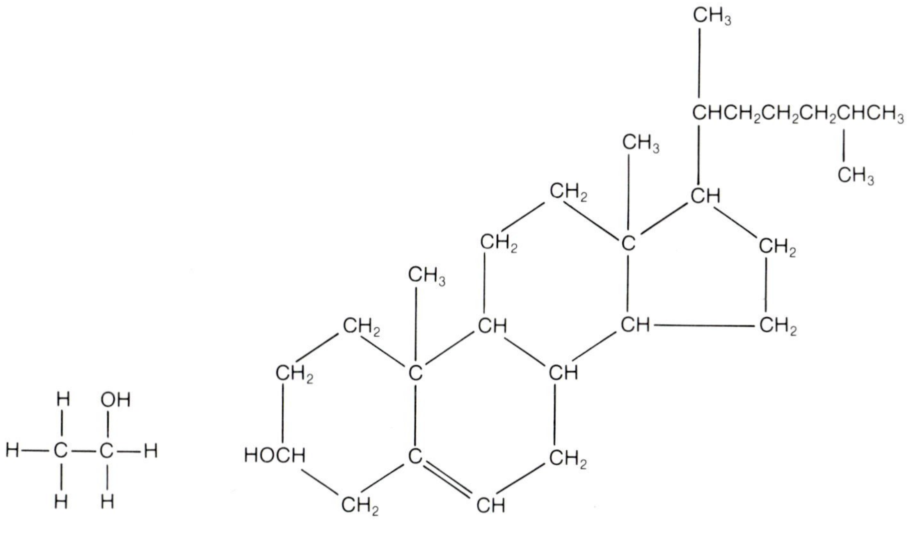

Figure 3–2. Carbon skeletons. Covalent bonding between carbon atoms allows the formation of complex organic molecules.

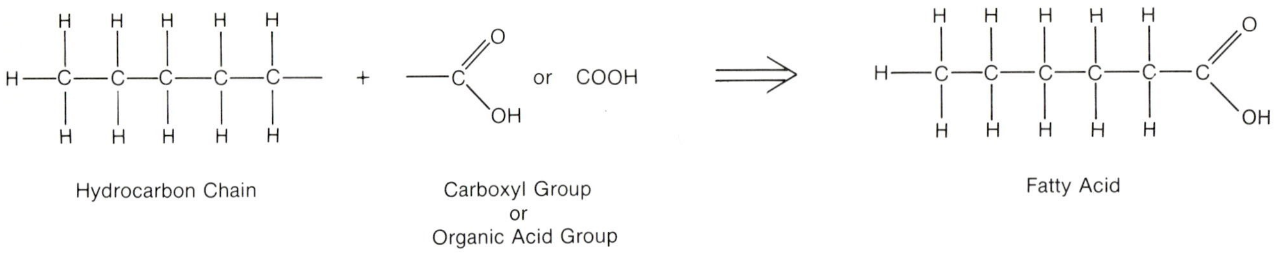

Hydrocarbon Chain

Carboxyl Group
or
Organic Acid Group

Fatty Acid

Figure 3–3. Fatty acid formation. A fatty acid is formed by the combination of a hydrocarbon chain and a carboxyl group.

ular component forming the final biologically active molecule. A fatty acid is a relatively simple molecule that only consists of carbon, hydrogen, and oxygen. The carbon atoms of a fatty acid are linked together forming a linear **hydrocarbon*** chain, at one end of which is an acidic or carboxyl group (Figure 3–3). Typically, fatty acids in plant lipids occur as chains between 14 and 22 carbon atoms long, though shorter and longer fatty acids are found in special situations.

Hydrocarbon
A molecule that consists exclusively of a carbon skeleton with attached hydrogen atoms.

Although all fatty acids contain a hydrocarbon chain, there are differences in the number of hydrogen atoms bound to the individual carbon atoms. In a **saturated** fatty acid the carbon atoms are said to be saturated with hydrogen, because each carbon atom in the chain is bound to the maximum number of hydrogen atoms (Figure 3–4), that is, each carbon atom has only a single bond to each neighboring carbon with the remainder of the four covalent bonds being made with hydrogen. An **unsaturated** fatty acid has at least two adjacent carbon atoms that are not saturated with hydrogen and so have formed two covalent bonds with each other instead of the usual one (Figure 3–4). Unsaturated fatty acids have a lower melting point than saturated ones. Therefore, molecules such as fats that contain unsaturated fatty acids tend to be liquid above 5° C (40° F) whereas saturated fats remain solid at physiological temperatures (up to 38° C or 100° F).

Fats are lipids that contain fatty acids attached to a three-carbon alcohol called **glycerol** (Figure 3–5). Plant fats are liquid because they have a higher proportion of unsaturated fatty acids than do animal fats. A liquid fat is usually referred to as an **oil** (not to be confused with petroleum oil), and is found either in plastids or in the cytoplasm of food storage cells where it

Figure 3–4. Saturated and unsaturated fatty acids. A saturated fatty acid has only single covalent bonds between adjacent carbon atoms whereas the unsaturated fatty acid has at least one pair of carbon atoms that have formed two covalent bonds between them (C=C).

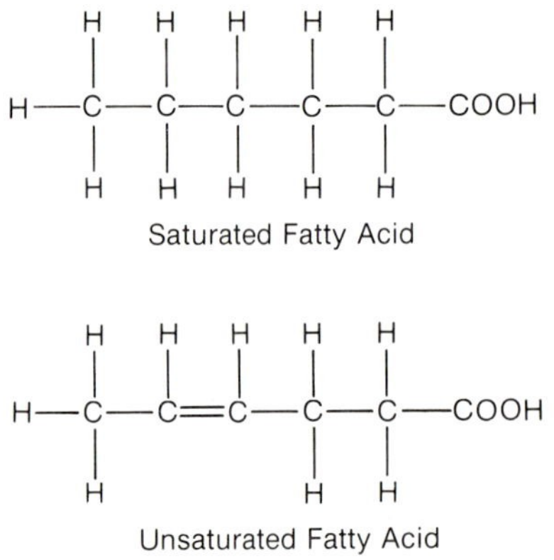

Saturated Fatty Acid

Unsaturated Fatty Acid

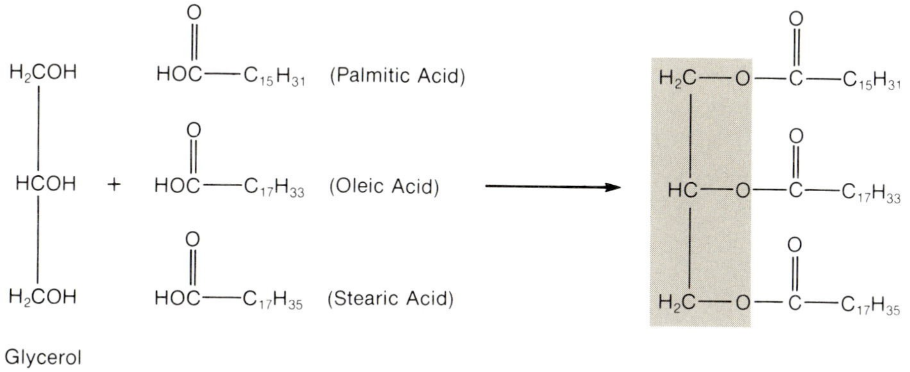

$$\begin{array}{l} H_2COH \\ | \\ HCOH \\ | \\ H_2COH \end{array} + \begin{array}{ll} HOC(=O){-}C_{15}H_{31} & \text{(Palmitic Acid)} \\ HOC(=O){-}C_{17}H_{33} & \text{(Oleic Acid)} \\ HOC(=O){-}C_{17}H_{35} & \text{(Stearic Acid)} \end{array} \longrightarrow \begin{array}{l} H_2C{-}O{-}C(=O){-}C_{15}H_{31} \\ | \\ HC{-}O{-}C(=O){-}C_{17}H_{33} \\ | \\ H_2C{-}O{-}C(=O){-}C_{17}H_{35} \end{array}$$

Glycerol

Figure 3–5. Fat. A molecule of fat is formed by the combination of 1 molecule of glycerol with 1 to 3 molecules of fatty acid.

serves as a reserve food in the form of droplets of various sizes. Most frequently oil is stored in seeds either in the endosperm* or the cotyledons.* Peanut, sunflower, corn, palm, and soybean are all grown for the oil contained in their seeds, which can be extracted for the manufacture of soaps, paints, and lubricants or for cooking and the production of margarine. When fatty acids are incorporated into fats, they become of interest as components of human diet. A major controversy exists at present regarding the merits of saturated and unsaturated fats. The proponents of unsaturated fats point to an apparent link between saturated fats in the diet and the occurrence of heart disease. Consequently, the use of vegetable oils that are low in saturated fats is being encouraged by many nutritionists.

Endosperm
A non-embryo food storage tissue in a seed.

Cotyledon
A leaflike structure in the seed that may contain stored food.

A class of lipids closely related to the fats is the **waxes.** Waxes are formed by the combination of long chain fatty acids with long chain alcohols rather than with the short chain glycerol molecule. As a result, waxes remain solid until high temperatures are reached. Unlike the fats, waxes are not used as a food reserve; instead they form waterproof layers in the epidermis (cutin in the cuticle), endodermis, and cork (suberin in the cell wall). The role of waxes in preventing the free movement of water and associated solutes should not be confused with the role of other lipids in the differential permeability* of cell membranes.

Differential permeability
The selective passage of molecules across a membrane.

Differential permeability of cell membranes is a function of their protein and lipid components specifically, **phospholipids.** Structurally they are similar to the fats except that one of the three fatty acid molecules attached to a glycerol molecule has been replaced by a phosphate group (Figure 3–6).

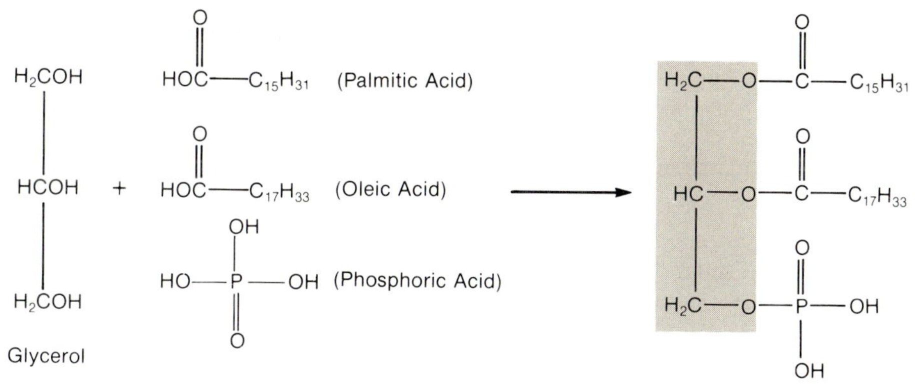

$$\begin{array}{l} H_2COH \\ | \\ HCOH \\ | \\ H_2COH \end{array} + \begin{array}{ll} HOC(=O){-}C_{15}H_{31} & \text{(Palmitic Acid)} \\ HOC(=O){-}C_{17}H_{33} & \text{(Oleic Acid)} \\ HO{-}P(OH)(=O){-}OH & \text{(Phosphoric Acid)} \end{array} \longrightarrow \begin{array}{l} H_2C{-}O{-}C(=O){-}C_{15}H_{31} \\ | \\ HC{-}O{-}C(=O){-}C_{17}H_{33} \\ | \\ H_2C{-}O{-}P(=O)(OH){-}OH \end{array}$$

Glycerol

Figure 3–6. Phospholipid. A phospholipid molecule is formed by the combination of 1 molecule of glycerol with 1 molecule of phosphoric acid and 1 or 2 fatty acid molecules.

Polar
Molecules that have a positive and/or negative charge.

Hydrophilic
Molecules that will associate with the aqueous phase; literally "water-loving."

Hydrophobic
Molecules that will associate with the lipid phase but not the aqueous phase; literally "water-fearing."

Lipid bilayer
A double layer of lipid molecules formed when the nonpolar tails associate.

Carbohydrate
An organic molecule consisting of carbon, hydrogen, and oxygen in an approximate ratio of 1:2:1; the simplest carbohydrate is known as a sugar.

The presence of the phosphate group converts such lipids into **polar*** molecules, whereas fats and waxes are **nonpolar.** Actually, the phospholipid molecule has a polar head and a nonpolar tail. Because the polar heads are **hydrophilic*,** they will associate with water while the **hydrophobic*** nonpolar tails will not. Therefore, polar lipids placed in water will form into layers with the hydrophilic heads toward the water and the hydrophobic tails away from the water. This property of phospholipids results in the spontaneous formation of a **lipid bilayer*** (Figure 3–7) similar to that in cell membranes (Chapter 2). Furthermore, even artificial membranes created just from phospholipids possess a rudimentary differential permeability.

Carbohydrates

Carbohydrates* are entirely different from lipids in their chemical and physical properties even though they both consist of carbon, hydrogen, and

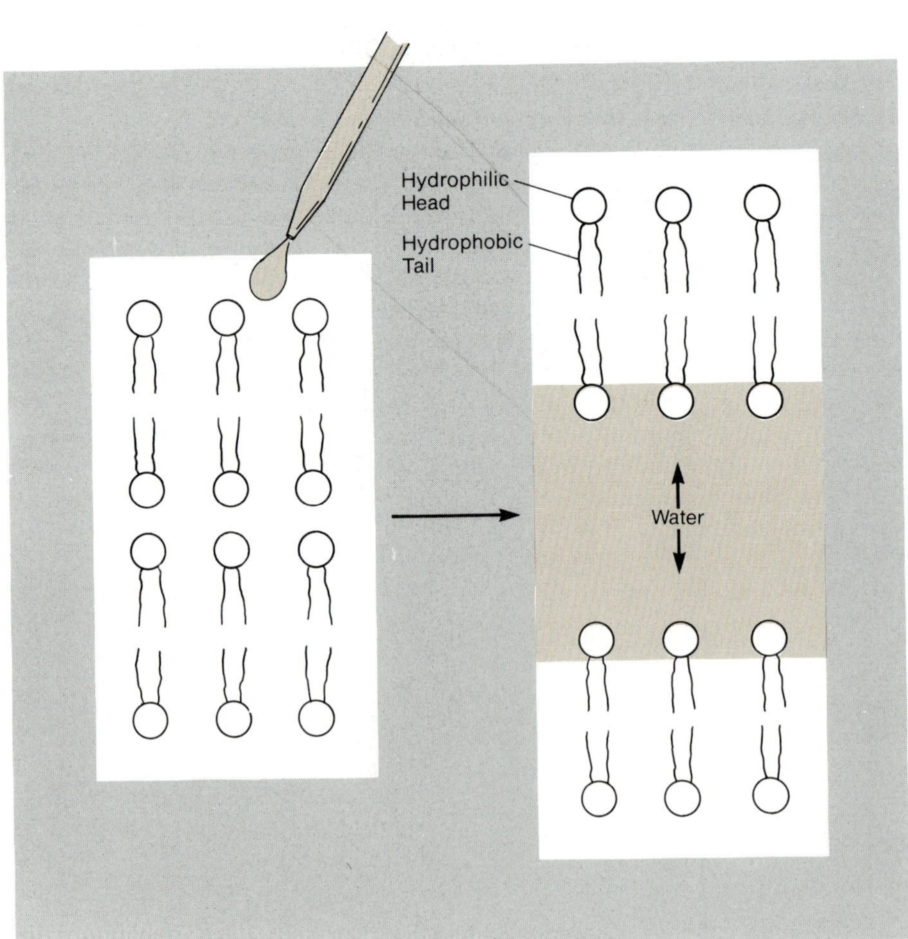

Figure 3–7. Formation of lipid bilayers. In the absence of water phospholipid molecules line up head to tail in adjacent layers. The addition of water causes the spontaneous separation of these layers into bilayers with the hydrophilic head in the aqueous phase.

oxygen. The difference relates to how these atoms are arranged to form the molecules and in the relative amounts of hydrogen and oxygen in the molecules. Although lipids contain only limited amounts of oxygen, carbohydrates (literally hydrated carbon) contain oxygen in the general ratio of $C:H_2:O$. This makes the lower molecular weight carbohydrates readily soluble in water. The basic carbohydrate molecule is a simple **sugar,** a small molecule with a 3– to 8–carbon atom skeleton. Simple sugars are used as building blocks or precursors for larger carbohydrates. In addition, the simple sugars are important compounds in cellular metabolism as is evidenced by their involvement in photosynthesis*, respiration*, and nucleic acid synthesis (see section in this chapter). Sugars are typically classified according to the number of carbons in their skeleton: 3 carbons = triose, 4 carbons = tetrose, and so on through pentose, hexose, heptose, and octose. The simple sugars with five or more carbons tend to form ring structures instead of remaining in linear chains (Figure 3–8). The formation of the more complex carbohydrates involves the combination of sugar molecules that are in the ring form. As a result, the individual sugars retain their identity even when combined to form **oligosaccharides*** and **polysaccharides*,** much as individual beads in a necklace retain their separate identity.

Glucose and **fructose** are simple hexose sugars that can combine to form **sucrose,** commonly known as table sugar (Figure 3–9). Sucrose, a carbohydrate central to plant metabolism, is the form in which most plants transport the excess food resulting from photosynthesis in the leaves to the nonphotosynthetic parts of the plant. Once sucrose is taken in by a cell, it is broken down to glucose and fructose again, and these simple sugars are used in respiration, metabolism, or the synthesis of stored food in the form of **starch.** Starch is a polysaccharide that is a polymer* of glucose (Figure 3–10). If the glucose molecules are linked together in a slightly different way, the resulting polysaccharide is **cellulose** (Figure 3–10). Because of this difference in linkage, these two polymers of glucose serve two very different functions in a plant. Most of the stored food in a plant is in the form of starch granules contained in plastids. Cellulose, by contrast, is not digested readily by plants, so it does not represent a form of stored food. The individual cellulose molecules become associated into units called **microfibrils*,** which are the basic structural components of the cell wall (Chapter 2). The

Photosynthesis
The metabolic process by which carbohydrates are produced from CO_2 and a hydrogen source such as water, using light as the source of energy.

Respiration
The metabolic process that results in the release of chemical energy from simple organic molecules.

Oligosaccharide
A carbohydrate consisting of a small number of linked sugar molecules.

Polysaccharide
A carbohydrate consisting of a large number of linked sugar molecules.

Polymer
A molecule made up of a large number of repeating subunits.

Microfibril
A structure made up of strands of cellulose molecules.

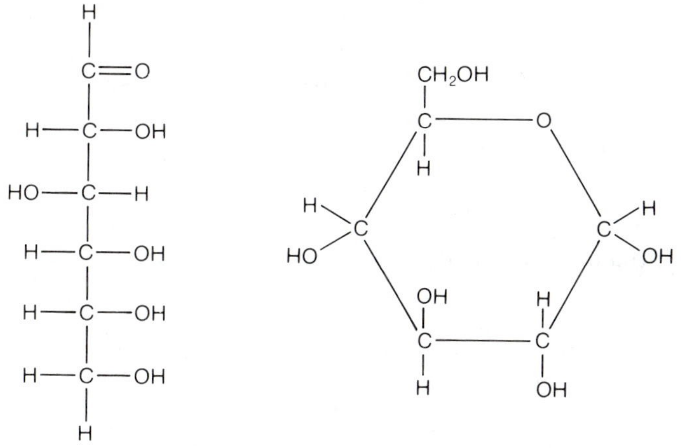

Figure 3–8. Simple sugar. Simple sugar molecules can exist in a linear form but usually exist in a ring form, the form that is metabolically most active. The ring form results when the —CHO group becomes linked to a carbon atom further down the chain.

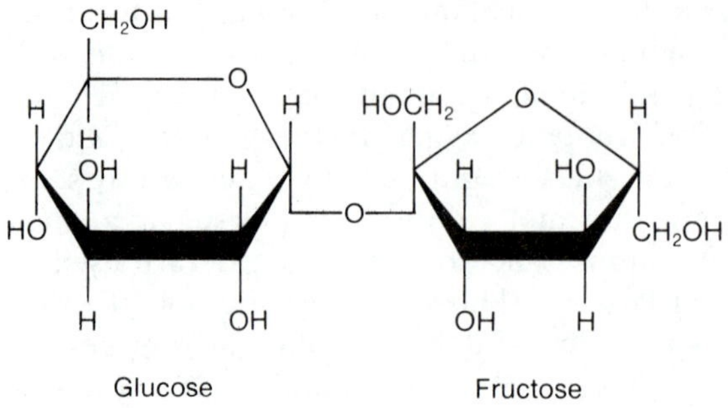

Figure 3–9. Sucrose. The sucrose molecule is formed by the linkage of a molecule of glucose to a molecule of fructose. Both the glucose and the fructose molecules retain their identity in the combined form.

matrix in which the cellulose microfibrils are embedded is also composed of carbohydrates (**hemicellulose** and **pectin**). Consequently, carbohydrates constitute more than half of the organic material in the living world because cell walls comprise the greatest amount of the dry weight of a plant.

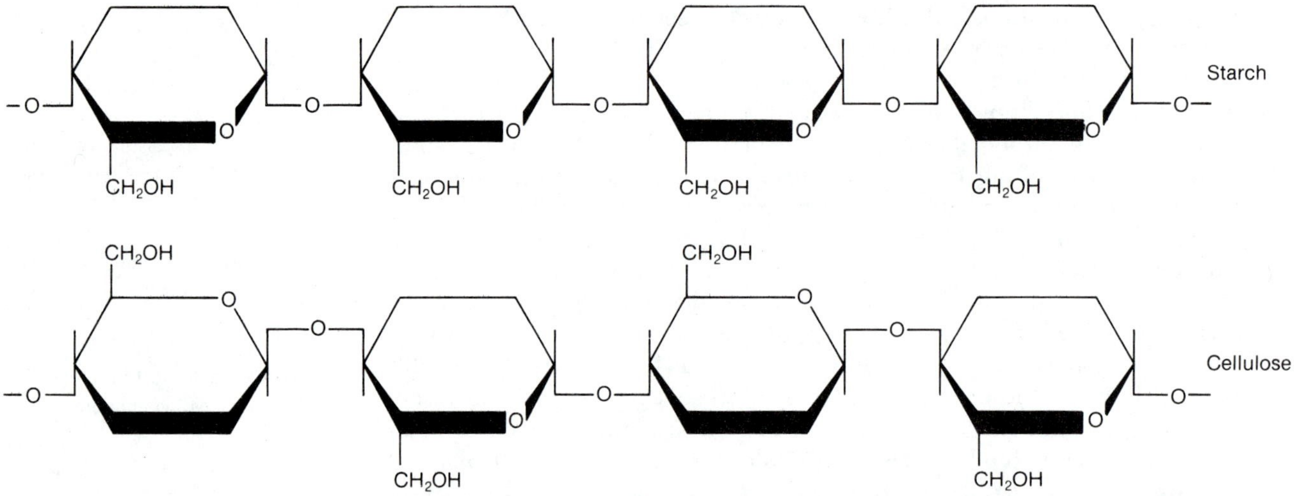

Figure 3–10. Polysaccharides. Starch and cellulose are molecules formed by the end-to-end linkage of large numbers of glucose molecules. In cellulose every other glucose molecule is flipped over, which makes the cellulose molecule rigid and suits it for a structural role.

Proteins

Proteins, like polysaccharides, are polymers, but not of sugar molecules. A protein consists of a linear chain of **amino acids.** The basic formula for an amino acid is

$$H_2N-\underset{\displaystyle H}{\overset{\displaystyle R}{\underset{|}{\overset{|}{C}}}}-COOH.$$

The H_2N- or $-NH_2$ represents the *amino* group, and the *acid* is represented by the –COOH (carboxyl) group.

There are 20 different amino acids commonly found in proteins, each differing from the others in the nature of the –R group. In theory, R is the symbol for any atom or group that can form a covalent bond with carbon; however, there are only 20 R groups that have been incorporated into the amino acid building blocks of proteins. For example, the amino acid glycine has a hydrogen atom (–H) as its R group; the amino acid alanine has a methyl group ($-CH_3$). Since the amino acids differ only in the R group, it is the R group that gives each amino acid its special chemical properties. The special biochemical and metabolic properties of proteins, in turn, are a function of the amino acids that are linked together to form each individual type of protein. Proteins can contain hundreds of the 20 amino acids in a specific sequence (Figure 3–11). If the sequence of amino acids is altered even slightly or if the wrong amino acid occurs anywhere in the chain, then the resulting protein is likely to have properties different from those it should have. Because the function of each type of protein depends on its having a particular set of chemical properties, any alteration of these properties will alter the ability of the protein to perform its metabolic function. Typically, an altered protein will be less efficient, or even nonfunctional; this will lower the capacity of the cell or plant with the defective protein to carry out that specific metabolic function. As will be described in a later section, proteins function as membrane components that regulate permeability, as **enzymes***, and as structural components of various subcellular structures. In addition, certain seeds, such as legumes, store protein as a food supply for germination. When proteins are digested, they are always broken down to their individual amino acids, and the amino acids usually are recycled into a new protein molecule that is being constructed rather than being degraded completely as in respiration (Chapter 10).

Enzyme
A protein that functions as a biological catalyst.

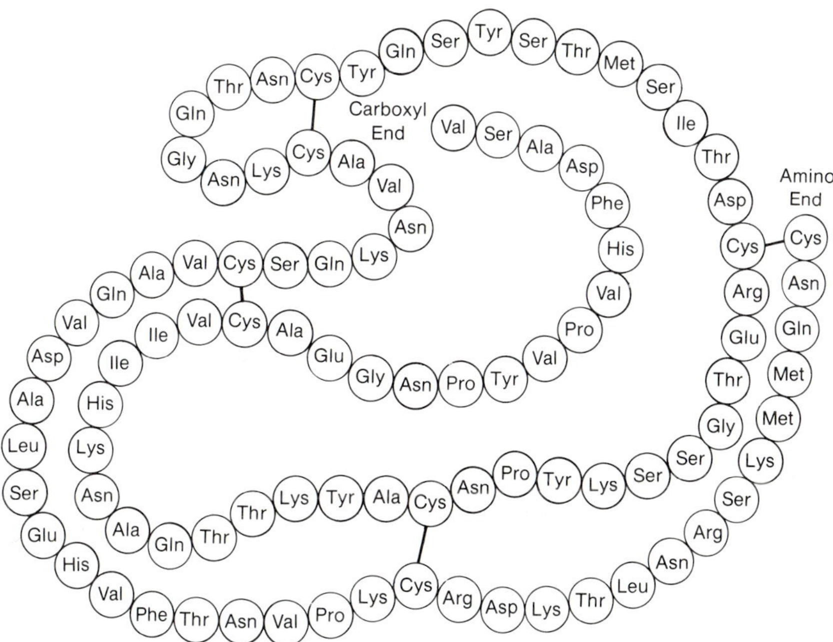

Figure 3–11. Amino acid sequence of a protein. The folding of a protein is determined by the amino acids that are in the protein. For example, absence of the amino acid *Cys* from the amino end of this protein would prevent formation of the sulfur—sulfur bridge (S—S) and leave a long, unfolded end.

Nucleic Acids

DNA

Each chromosome contains one large nucleic acid molecule known as **DNA** (deoxyribonucleic acid) that functions to store genetic information. The protein associated with the DNA in the chromosome is only a structural component and has nothing to do with the storage of genetic information. Subsequent chapters explain the central role of genetic information in growth and development (Chapter 12), transmission of inherited characteristics from generation to generation (Chapter 22), and evolution of species (Chapter 23). From this brief list, though, it should be obvious that DNA as a store of genetic information is of central importance in the study of plants, much more so than any of the other types of molecules found in plants. It is correct to say that living cells require the other molecules that have been discussed above, but without the genetic information "cells" would be of limited existence and unable to reproduce themselves (Chapter 23).

DNA plays a crucial role in the life of the cell and of the plant because it alone carries the genetic information—information containing instructions for making proteins and determining the types of protein an organism can make. Each DNA molecule contains the instructions, encoded in the structure of the DNA molecule, for the synthesis of hundreds of different proteins. Therefore an understanding of the DNA molecule itself is necessary before beginning a description of the process of protein synthesis.

The DNA molecule is a polymer whose repeating subunit is the **nucleotide.** Each nucleotide in turn consists of three parts: a **phosphate** group, a pentose **sugar** (deoxyribose), and a nitrogenous **base.** The phosphate and sugar are used to connect adjacent nucleotides and thus form the backbone for the DNA molecule; however, they are not involved in the storage of genetic information. Only four nitrogenous bases are found in DNA—**adenine (A), cytosine (C), guanine (G),** and **thymine (T).** Although these are relatively simple molecules (Figure 3–12), their sequence in the DNA molecule functions to encode the genetic information. Figure 3–13 shows a portion of a DNA molecule indicating how the nucleotides are linked together.

In 1953 Watson and Crick assembled all of the known information about the DNA molecule to produce a theoretical model. According to their model, the DNA molecule consists of two separate linear chains of nucleotides that are held together by weak attractions, called hydrogen bonds (see Appendix), between bases in the two strands (Figure 3–14). Furthermore, the double-stranded DNA was predicted to be twisted into a helix (Figure 3–14), the famous double helix of molecular biology. Subsequent experimental work has borne out their predictions, and the double-stranded helix model of DNA is accepted as being correct. In order for the helix to form correctly, according to Watson and Crick, the two strands must complement each other, that is, bases opposite each other must just fit the available space. This restricts which bases can be opposite each other because, as shown in Figure 3–12, adenine and guanine are larger than cytosine and thymine. Consequently, adenine could not be opposite guanine without a bulge appearing in the helix, and cytosine and thymine could not be opposite each other as they would not be close enough for hydrogen bonds to form. For this and other reasons Watson and Crick deduced that adenine and thymine were complementary bases, as were cytosine and guanine. Thus in the DNA double helix, the sequence of bases in one strand determines the base sequence in the complementary strand. This ensures that both strands contain the same genetic information. Prior to mitosis* and meiosis* (Chapters 4 and 13), it is necessary to duplicate the DNA so that each chromosome will consist of the normal two chromatids* instead of only one. In order to duplicate a DNA molecule, the cell must have a mechanism for making an exact copy of the base sequence in each strand, because the sequence represents the coded genetic information. It is not possible, for example, to make a brand new double-stranded DNA molecule from free nucleotides in the manner that a new starch molecule is made from glucose. The only way to produce an exact copy of the existing base sequence in each strand is to use the existing strands as templates or guides. This is where the required complementary base pairing plays its role. During DNA duplication, the existing double helix gradually becomes "unzipped", starting at one end of the molecule, as hydrogen bonds are broken between complementary bases. As the two strands separate, each strand is used as the template for the production of a new strand. The result is that the two DNA molecules produced by duplication each contain one old and one new strand (Figure 3–15). Each template strand directs the formation of its new complementary strand because nucleotides can only be added to the growing strand if they contain a

Mitosis
The process of nuclear division that results in the production of two genetically identical daughter nuclei.

Meiosis
The process of nuclear division that results in the production of daughter nuclei, each containing one-half the original number of chromosomes.

Chromatid
One of the two longitudinal halves of a chromosome.

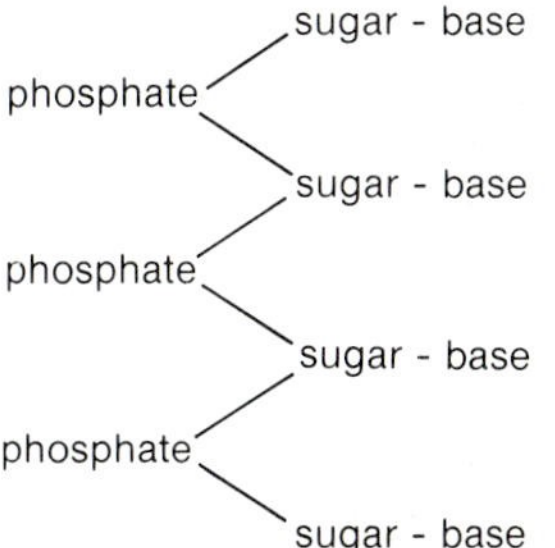

Figure 3–13. Structure of DNA. A DNA molecule is composed of nucleotides each of which consists of a phosphate group linked to a sugar that is linked to a base. The nucleotides are joined together by a bond between the sugar of one nucleotide and the phosphate group of the nucleotide below it. This arrangement of sugar and phosphate forms the backbone of the DNA molecule.

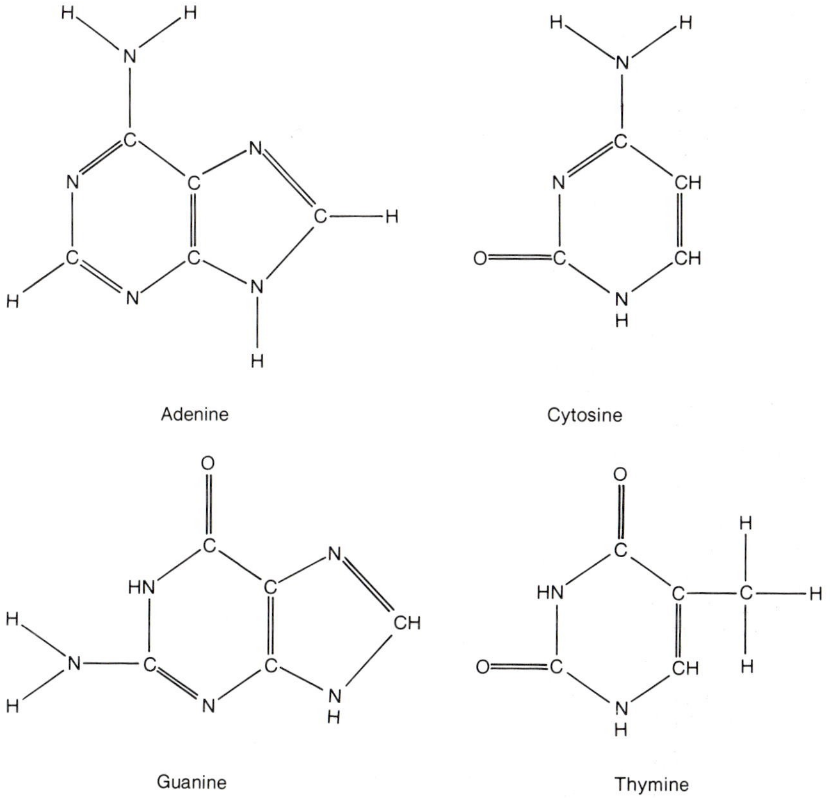

Adenine

Cytosine

Guanine

Thymine

Figure 3–12. Nitrogenous bases of DNA. Genetic information is encoded in the sequence of the bases adenine, cytosine, guanine, and thymine in DNA.

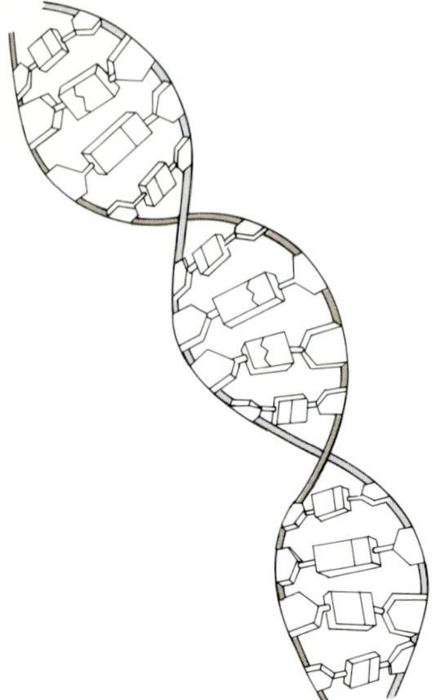

Figure 3–14. The double helix. A DNA molecule consists of two linear chains of nucleotides held together by hydrogen bonds. The bases in the two chains are paired so that adenine in one chain is always opposite thymine in the other; similarly, guanine is always opposite cytosine. The two chains are twisted around each other forming the double helix.

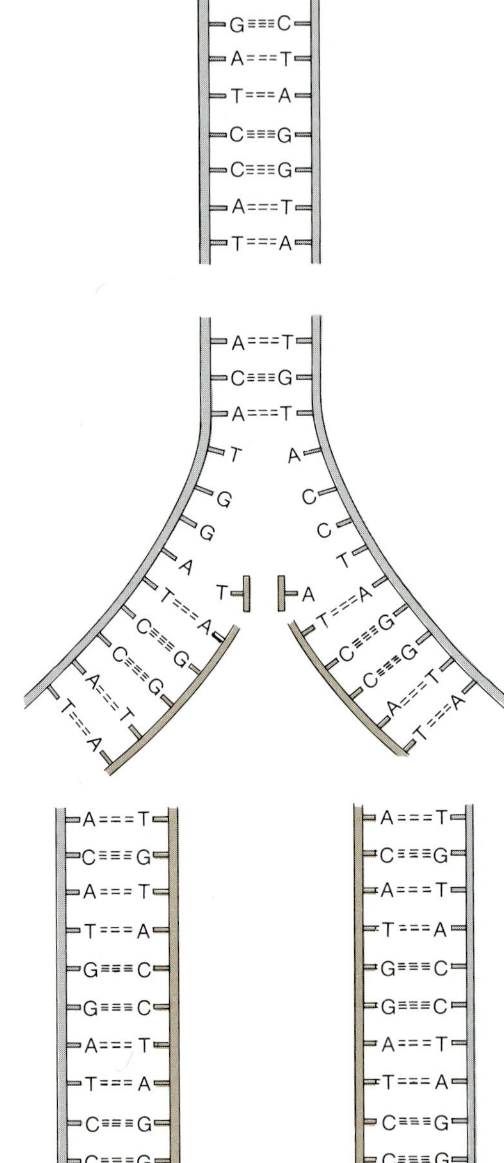

Figure 3–15. DNA duplication. During the duplication of the DNA molecule each of the existing strands is used as a model, or template, for the formation of a new strand. Since complementary base pairing must occur, the two new DNA mulecules, each composed of one old and one new strand, are indistinguishable from the original molecule.

base complementary to the base in the existing strand; that is, an A in the existing strand forces the new strand to incorporate a T at this position or else hydrogen bonding and double helix formation cannot take place. This type of self-guided molecular synthesis is essential for an organism to maintain a reliable information store.

RNA

Cells contain another type of nucleic acid that is more intimately associated with the actual synthesis of proteins: **RNA** (ribonucleic acid). There are three types of RNA—**messenger RNA (mRNA)**, **transfer RNA (tRNA)**, and **ribosomal RNA (rRNA)**—each of which has a special function in protein synthesis. All three types of RNA are produced by copying specific portions of a DNA molecule, although each type of RNA is transcribed from different DNA regions. RNA differs from DNA in having the following: (1) the pentose sugar ribose instead of deoxyribose, (2) the base uracil (U) in place of thymine, and (3) a single strand as opposed to a double strand. Synthesis of RNA takes place on a DNA template; however, in this case, only a small part of the DNA double helix becomes "unzipped" and only one of the two DNA strands is used as a template. As the DNA template becomes exposed, nucleotides bearing the ribose sugar are linked together in a sequence that is complementary to the DNA base sequence: C, G, and T in the DNA would result in G, C, and A, respectively, in the RNA; whereas A in the DNA would result in U in the RNA because RNA does not contain thymine (Figure 3–16). After this portion of the DNA has been copied into RNA, it is "zipped" back up with hydrogen bonds.

Metabolism

Protein Synthesis

As is indicated in other sections, proteins play a central role in cell function as enzymes and as membrane components with a major responsibility for differential permeability. Furthermore, protein structure is crucial to correct protein function. Protein synthesis, therefore, is a carefully regulated process whereby a long chain of amino acids must be assembled into the correct sequence. The instructions for assembling a particular protein are encoded in the DNA, each DNA molecule containing the instructions for many proteins. That portion of the DNA molecule containing the instructions for a single protein is referred to as a **gene.** Actually, since a protein can be composed of two or more subunits called polypeptides*, each gene only codes for a single polypeptide. If the protein is composed of only one polypeptide, then one gene does indeed code for one protein. In the case of a protein made of two different polypeptides, two genes would be required to code for the complete protein.

Polypeptide
A molecule composed of amino acids.

Nuclear DNA, which codes for the majority of the cellular proteins, remains in the nucleus, although protein synthesis also occurs in the cytoplasm. The instructions needed in the cytoplasm cannot leave the nucleus, so the cell must make a copy of the instructions that can be read in the cytoplasm. This situation is analogous to one that we may encounter in a library. For instance, a diagram, book chapter, or magazine article is needed for study at home, but the library will not let the original circulate. One solution to the problem is to copy the part of the original that is needed and then take the copy from the library. A second solution is to cut out the desired part of the original and take that from the library. It should be obvious what the second solution does to information storage and transfer and also what will happen to the piece of the original that is taken from the library. Once this segment is out of the protective storage of the library, it is much more likely to become destroyed or damaged, thus reducing its information content. For similar reasons DNA cannot be cut into individual genes that are transferred to the cytoplasm, used to make proteins, and then returned to the nucleus. This would lead to a garbling of the instructions with subsequent loss of the ability to translate this set of instructions into the correct protein. Consequently, the DNA instructions are copied, as in the first solution above, in coded form into an RNA molecule in the process of **transcription** (Figure 3–16). Since the RNA is only a copy, it can be used for a while in the cytoplasm and then destroyed. When the same instructions are needed again, another copy can be made from the protected original.

Messenger RNA (mRNA) is the copy of a gene produced during transcription. For the mRNA information to be used in the assembly of the amino acid sequence of a specific protein, the message needs to be translated. The **translation** of mRNA, which involves the assembly of amino acids into a protein molecule, occurs on cytoplasmic structures called ribosomes, composed of ribosomal RNA (rRNA) and proteins. A ribosome binds to the mRNA in such a way that the message is read from beginning to end in the correct sequence. On the mRNA a series of three bases, called a **codon,** represents the code for a specific amino acid. The amino acids themselves, however, cannot "recognize" codons, but other bases can. Therefore another nucleic acid is required—one that carries an **anticodon*** as well as an attachment site for a specific amino acid (Figure 3–17). Because this nucleic acid transfers amino acids, it is called transfer RNA (tRNA). The tRNA is folded in such a way that the amino acid attachment site and the anticodon are at opposite ends of the molecule. Because there are 20 types of amino acids in proteins, there are at least 20 different types of tRNA molecules, one or more for each different amino acid (Table 3–1). Enzymes in the cytoplasm recognize a particular tRNA and attach the correct amino acid to it from the pool of free amino acids.

Anticodon
A sequence of three bases complementary to the triplet codon.

On the basis of this information, we see that protein synthesis proceeds in two steps—transcription and translation. A gene is transcribed into an mRNA molecule that leaves the nucleus and becomes associated with a ribosome. The ribosome positions the mRNA molecule so that a codon is exposed and then guides a tRNA molecule with the complementary anticodon into place. After the tRNA is positioned, its amino acid is linked to the growing protein. Once the amino acid has been removed, the tRNA molecule is released to the cytoplasm where it can pick up another amino acid. At the same time, the ribosome moves along the mRNA to the next codon in the reading sequence and the whole process is repeated again and again until the complete protein has been assembled (Figure 3–18). Since there are many ribosomes, many different proteins can be synthesized at the same time.

Only a limited number of genes actually are being transcribed into mRNA at any one time. Most of the genes in a cell are turned off. Exactly which

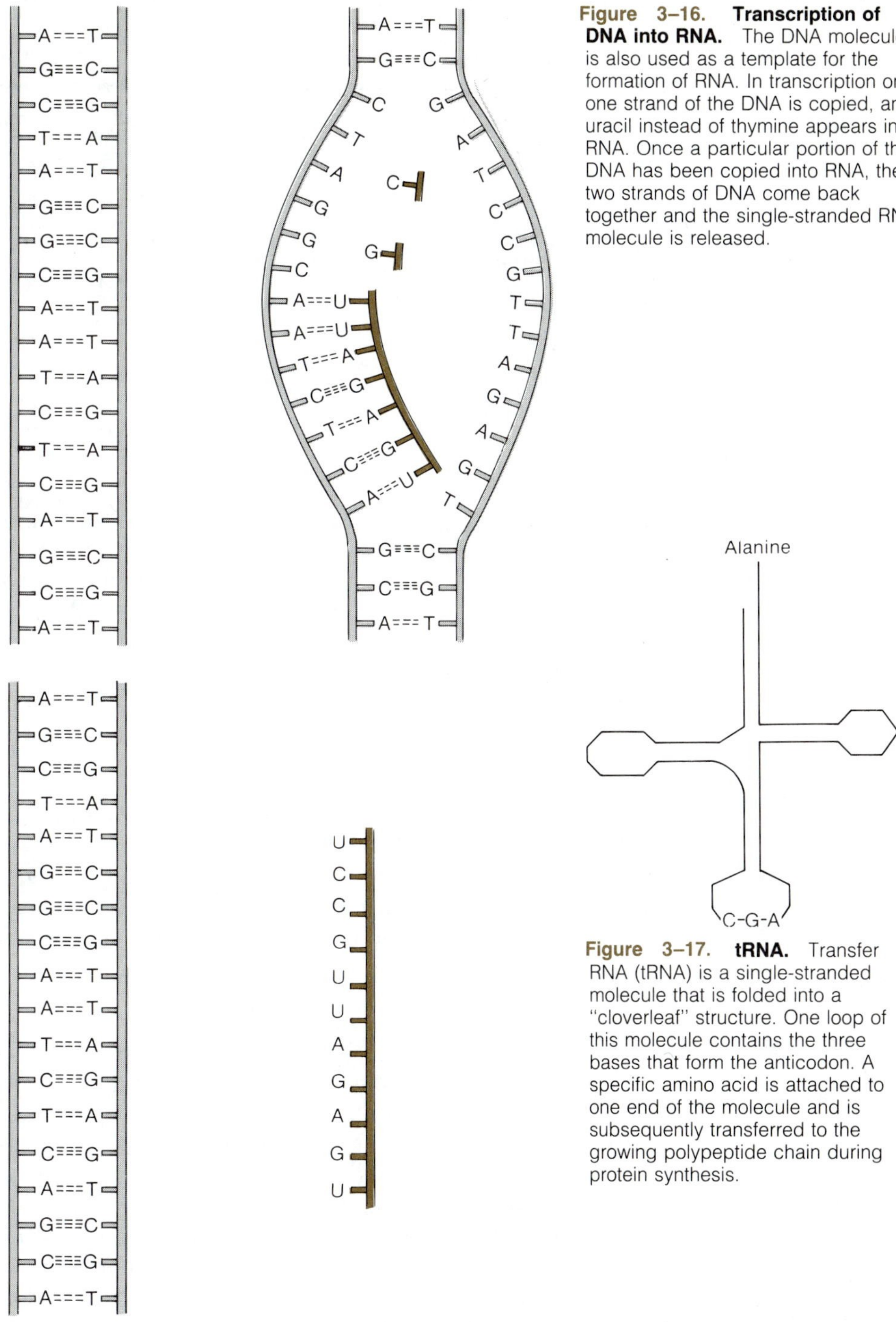

Figure 3–16. Transcription of DNA into RNA. The DNA molecule is also used as a template for the formation of RNA. In transcription only one strand of the DNA is copied, and uracil instead of thymine appears in RNA. Once a particular portion of the DNA has been copied into RNA, the two strands of DNA come back together and the single-stranded RNA molecule is released.

Figure 3–17. tRNA. Transfer RNA (tRNA) is a single-stranded molecule that is folded into a "cloverleaf" structure. One loop of this molecule contains the three bases that form the anticodon. A specific amino acid is attached to one end of the molecule and is subsequently transferred to the growing polypeptide chain during protein synthesis.

genes are active is a function of the physiological condition of the cell. Hormones (Chapter 12) exert a specific effect on the genome* with the result that new combinations of genes become active after hormone treatment. Despite the fact that for any specific cellular specialization (Chapter 5) there is a particular combination of genes that will be active, genes are not necessar-

Genome
The complete genetic program in a nucleus, plastid, or mitochondrion.

Table 3–1.
Amino Acids and Their mRNA codons

Amino Acid	mRNA Codon(s)
Alanine	GCU, GCC, GCA, GCG
Arginine	CGU, CGC, CGA, CGG, AGA, AGG
Asparagine	AAU, AAC
Aspartic acid	GAU, GAC
Cysteine	UGU, UGC
Glutamic acid	GAA, GAG
Glutamine	CAA, CAG
Glycine	GGU, GGC, GGA, GGG
Histidine	CAU, CAC
Isoleucine	AUU, AUC
Leucine	CUU, CUC, CUA, CUG, UUA, UUG
Lysine	AAA, AAG
Methionine	AUG
Phenylalanine	UUU, UUC
Proline	CCU, CCC, CCA, CCG
Serine	AGU, AGC, UCU, UCC, UCA, UCG
Threonine	ACU, ACC, ACA, ACG
Tryptophan	UGG
Tyrosine	UAU, UAC
Valine	GUU, GUC, GUA, GUG

ily clustered in functional groups. On the contrary, genes for a specific metabolic pathway could be dispersed over several chromosomes. On the chromosome, genes are in a linear sequence (Chapter 22) with very specific start and stop codes in the sequence of bases. Thus, transcription can be restricted to just one gene out of many along a specific portion of the DNA.

Enzymes and Their Reactions

All enzymes are proteins, though there are many proteins that do not function as enzymes. Enzymes are biological catalysts—they accelerate the rate of the biochemical reactions that constitute **metabolism.** Because most of the metabolic reactions that occur in a cell would proceed exceedingly slowly by themselves, a catalyst, such as an enzyme, is needed for the reactions to occur at a sufficiently fast rate. The role of the enzyme is to overcome the energy barrier that prevents the reaction from occurring at an appreciable rate in the absence of a catalyst. All enzyme-catalyzed metabolic reactions are chemical reactions, or, in other words, they involve the making and breaking of chemical bonds. Both the formation and the rupture of bonds require that the atoms and molecules involved be in the correct energy state. Chemists know that there are certain atoms and molecules that are highly reactive and readily form new chemical bonds. The molecules that compose the living cell, however, are not very reactive. In fact, they cannot be very

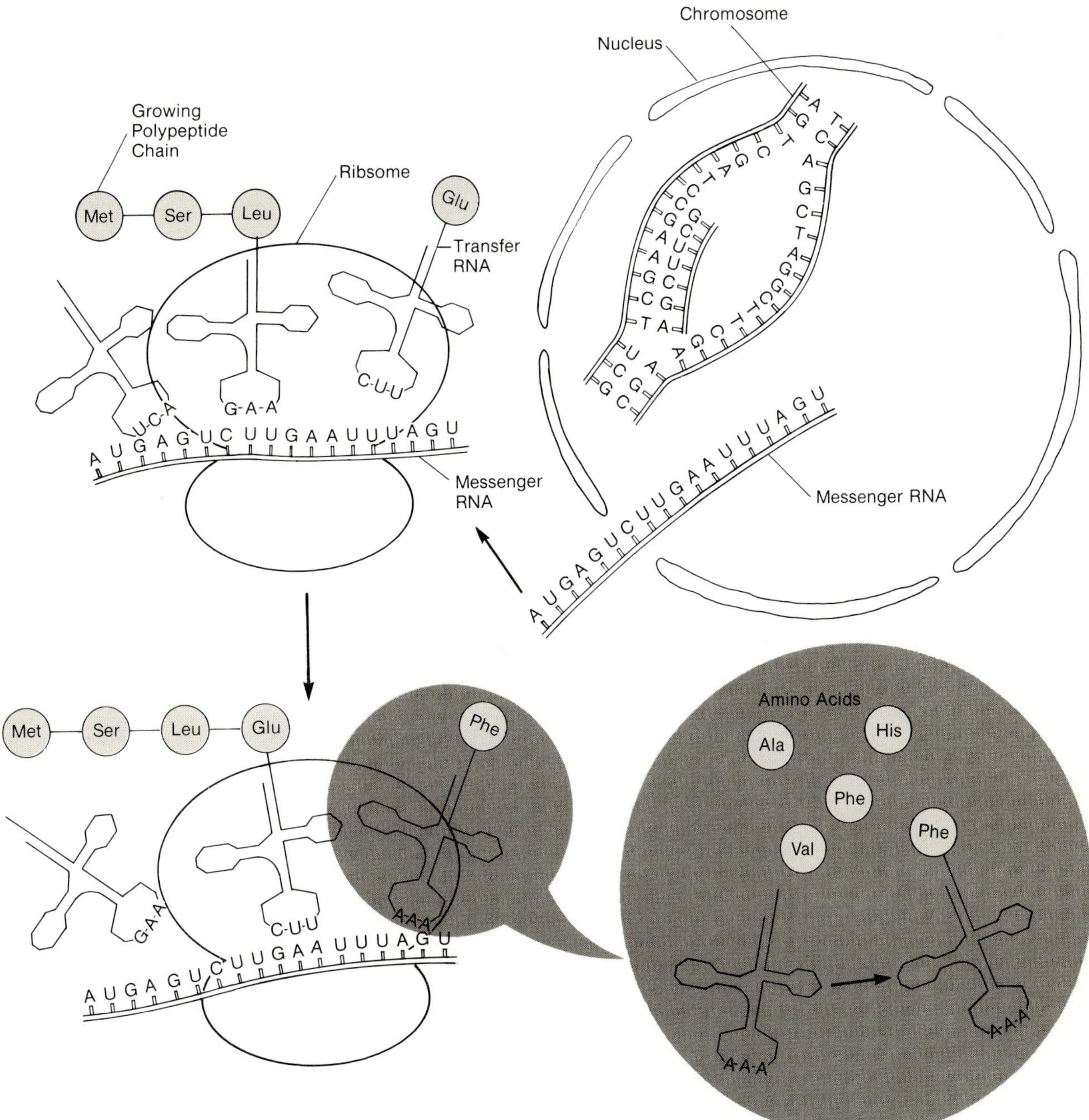

Figure 3–18. Protein synthesis. In a cell the genetic information contained in the DNA of the chromosomes is selectively transcribed into RNA forming ribosomal RNA (rRNA), transfer RNA (tRNA), and messenger RNA (mRNA). Each mRNA molecule is a copy of a segment of the DNA, called a gene, which contains the coded instructions for the assembly of a particular polypeptide (protein subunit). The mRNA leaves the nucleus through a pore in the nuclear membrane and associates with a ribosome. As a triplet codon of mRNA is exposed in the ribosome, a tRNA molecule bearing an amino acid and the complementary anticodon line up above it and the amino acid is added to the growing polypeptide chain. Now the ribosome reading frame moves one codon further along the mRNA and the appropriate tRNA is brought into place. At this point the previous tRNA molecule (minus its amino acid) is released. After diffusing away from the ribosome a tRNA molecule binds another amino acid and is ready to participate in protein synthesis again.

reactive or else the cell would not have control of metabolism, and the molecular components of the cell would change as the result of random chemical reactions. Enzymes are able to convert the relatively unreactive molecules of the cell to reactive ones by providing an environment that activates the molecules (Figure 3–19). This activation energy can result from changes in the configuration of the enzyme. In order for fatty acids to be attached to a glycerol in fat synthesis, or for glucose and fructose to be combined forming sucrose, or for amino acids to be linked during the formation of a protein, an enzyme has to bring the two substrate* molecules close enough together for the correct new chemical bonds to form. Once the substrate molecules have become attached to the enzyme surface, a change in the configuration of the enzyme would bring them closer together causing a bond to form between them (Figure 3–20). On the other hand, the breakdown of a molecule could occur as the result of the change in enzyme configuration. By creating a pull on the weakest bond, it would break it and release the two fragments of the original molecule (Figure 3–20). When sucrose is metabolized, the first reaction is the enzyme-catalyzed rupture of the bond between glucose and fructose, which fragments sucrose into glucose and fructose.

Substrate
A molecule that can be acted upon by an enzyme.

Enzymes are catalysts that are essential for metabolism; however, enzymes are much more specific than ordinary chemical catalysts. Each enzyme will catalyze only one type of chemical reaction involving one or a limited number of substrate molecules. For instance, the enzyme invertase is responsible for degrading sucrose (glucose-fructose), but invertase could not synthesize sucrose from glucose and fructose, nor could invertase degrade maltose (glucose-glucose) or any other sugar. Life is possible only because of this specificity of enzymes. In the absence of enzyme specificity, any enzyme would be able to catalyze any reaction and chaos would result. Life requires

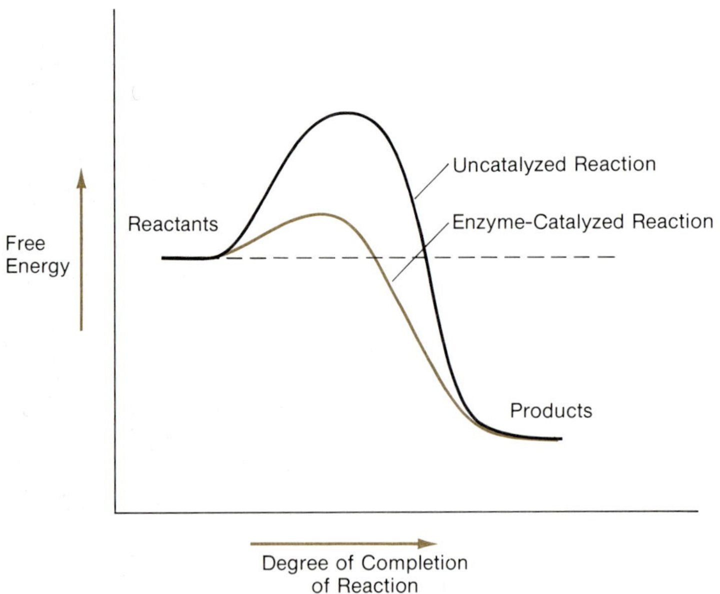

Figure 3–19. Activation energy. For a chemical reaction to occur the energy level of the products of the reaction has to be lower than that of the starting molecules unless there is an input of energy from the environment as in photosynthesis. Therefore chemical reactions are essentially downhill reactions. Unless the reaction proceeds rapidly and spontaneously, there is a significant energy barrier to be overcome. Enzymes reduce this energy barrier to a very low level, which allows enzyme-catalyzed reactions to occur hundreds or thousands of times faster than uncatalyzed reactions.

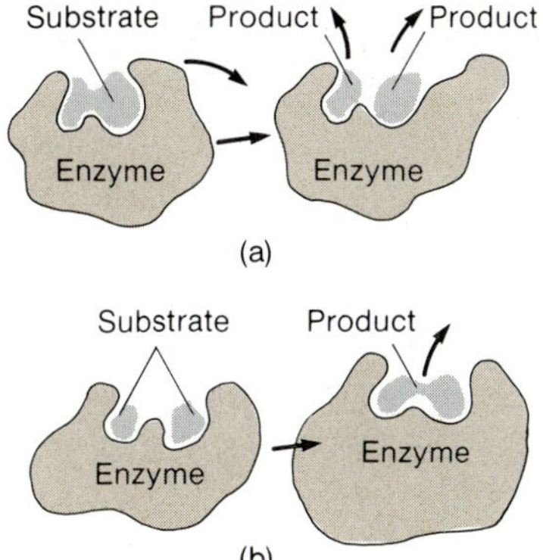

Figure 3–20. Mechanism of enzyme action. As the result of binding substrate, an enzyme changes its shape. *(a)* A conformational change in the enzyme that puts a strain on a bond within the substrate can result in the breaking of the bond and the separation of the substrate into two molecules of product. *(b)* This conformational change can result in two substrate molecules being brought closer together thus overcoming the energy barrier to bond formation between them.

the ordered set of chemical reactions we call metabolism. Enzyme specificity guarantees order in metabolism as a particular enzymatic reaction can proceed only in the presence of the required enzyme.

Regulation of Metabolism

The more that is known about cellular metabolism, the more it is obvious that it is a complicated, but well-regulated process. One aspect of the regulation of metabolism is the compartmentalization of different types of metabolism into the different organelles.* For example, photosynthesis and respiration are two major forms of metabolism that are segregated into separate organelles, chloroplasts and mitochondria, respectively. The overriding control that regulates metabolism within each cell, however, is the genetic information in the nucleus. The nucleus functions as the control center of the cell in much the same way as a computer does in an automated factory. The genetic information is the program, and, as chemical signals are received from the cytoplasm, the program responds by sending out messages to the cytoplasm in the form of mRNA. In the cytoplasm, the messages are deciphered and used to alter metabolism. Since enzymes are proteins, metabolism can be redirected as the result of the production of a new set of enzymes from a new set of mRNA molecules. As a means of preserving flexibility in metabolic control, both mRNA and protein molecules have fairly short lives. The mRNA molecules often exist only for minutes or hours before they are broken down. Proteins tend to last a little longer (hours to days) before they too are degraded. In the case where the same metabolism needs to continue, the cell remakes the same type of mRNA and thus the same protein. Cellular metabolism is also regulated by the nucleus through the control of the differential permeability of the various cellular membranes because the membrane proteins, like all proteins, are coded for in the nucleus and are regularly broken down and replaced. To change the internal composition of the cell, the nucleus merely needs to change the protein composition of the cell membrane.

Organelle
A body in the cytoplasm with a specific function, e.g., a plastid or mitochondrion.

Summary

Cellular metabolism is the set of chemical reactions, involving organic compounds, that occurs in a living cell. The four major types of organic compounds found in plant cells are lipids, carbohydrates, proteins, and nucleic acids.

Lipids are water-insoluble molecules typically consisting of fatty acids combined with another molecule. Unsaturated fat molecules contain unsaturated fatty acids. They occur as liquids over the range of physiological temperatures and are referred to as oils. Waxes are a group of lipids similar to the fats and have excellent waterproofing qualities (e.g., cutin in the cuticle and suberin in cork and endodermis). In the phospholipids, another group similar to the fats, a fatty acid has been replaced by a phosphate-containing group. When placed in water, these molecules readily form lipid bilayers similar to membranes.

Carbohydrates differ from lipids, in part, because carbohydrates contain a significant number of oxygen atoms and many are water soluble. The basic carbohydrate is a sugar molecule with a carbon skeleton containing 3 to 8 carbon atoms. More complex carbohydrates are formed by the combination of simple sugars. Excess carbohydrate is stored in the plant in the form of starch granules in plastids. Cellulose, although chemically similar to starch, is the main structural component of the plant cell wall and is unavailable to the plant as food.

Proteins are polymers of amino acids ($H_2N-\overset{\overset{\displaystyle R}{|}}{\underset{\underset{\displaystyle H}{|}}{C}}-COOH$). Their biochemical and metabolic properties are a function of the amino acids that are linked together to form a particular type of protein. Consequently, the amino acid composition and sequence are extremely important for correct functioning of a protein as an enzyme or structural component of a membrane. Changes in this composition or sequence would likely reduce the effectiveness of the resulting protein. Proteins also serve as stored food in seeds of some plants such as the legumes.

A cell contains two types of nucleic acid: deoxyribonucleic acid (DNA) and ribonucleic acid (RNA). DNA is an information-storing molecule that is a major component of the chromosome. Its genetic information is stored in the sequence of its bases. The DNA molecule forms a double-stranded helix with complementary base pairs—adenine and thymine, cytosine and guanine—holding the strands together. DNA duplication requires the "unzipping" of the double helix and the use of both existing strands as templates for the production of two new complementary strands. After this process is complete, both DNA molecules are in a double helix composed of one old strand and one new one.

The genetic information of the DNA is transcribed into RNA—messenger RNA (mRNA), transfer RNA (tRNA), and ribosomal RNA (rRNA). The RNA molecule contains the base uracil in place of thymine and is single stranded. Transcription of DNA into RNA is done selectively on small regions of DNA by complementary base pairing. At the end of RNA synthesis, the RNA molecule is released and the DNA molecule reassociates.

All metabolism is mediated by special proteins called enzymes. An enzyme is a biological catalyst that is highly specific for the chemical reaction it will catalyze. Enzymes mediate their particular reaction by accelerating either the formation of chemical bonds, which will hold two substrate molecules together as one molecule, or the rupture of a molecular bond so that two fragments result.

All enzymes are not present simultaneously in a cell or else chaos would result. Each gene in the DNA contains the information for the synthesis of one polypeptide of a protein, and this is transcribed into mRNA, which leaves the nucleus for the cytoplasm. Translation of the mRNA message into protein occurs on the ribosome. By controlling the production of all proteins, the nucleus regulates metabolism both through enzyme production and control of membrane differential permeability.

Review Questions

1. Distinguish between a carbohydrate and a lipid; include a description of the physiological role of each.
2. Fresh pineapple juice contains a protein-digesting enzyme. Explain what would happen to the enzymatic activity of amylase (a starch-digesting enzyme) if amylase and pineapple juice were mixed.
3. Explain what would happen if DNA duplication did not result in perfect copies; i.e. what would be the effect on the functioning of the cell with the altered DNA?
4. Distinguish between transcription and translation and state where within the cell each occurs.
5. Explain how a phospholipid is formed and how its structure relates to its cellular function.

Water

Water is the major chemical component of cells, most metabolism occurs in the aqueous phase, and life would be impossible without it. The special properties of water are usually overlooked, however. One particularly relevant property of water is that it is an extremely good solvent. Many different compounds including enzymes dissolve in water, which then serves as a common medium for chemical reactions. In addition to all components being dissolved in water, many of the enzyme reactions of metabolism involve water either as a substrate or as a product. Another important property of the water molecule is its ability to form hydrogen bonds (see Appendix). Although the hydrogen bond is much weaker than a covalent bond, it is the presence of hydrogen bonds between water molecules and individual solute molecules, such as sugar or protein, that is responsible for these molecules going into solution. The sugar in coffee is in solution because each sugar molecule is isolated from all of the other sugar molecules by a shell of water molecules, which are hydrogen bonded to it (Figure 3–21). Hydrogen bonds between water molecules are also responsible for liquids moving up a drinking straw in response to suction or up from the roots to the leaves of a plant. In this case the bonds cause the water molecules to stick together and move as a unit.

Figure 3–21. Sucrose in solution. For sucrose, or any other molecule, to go into solution in water, it must be surrounded by a shell of water molecules. This effectively separates the individual solute molecules and prevents the formation of large clusters of solute molecules that will settle out in response to gravity.

Coffee, Tea, or War!

For many people a morning cup of tea or coffee is essential to their functioning. The coffee or tea gives them a lift, makes them more alert, and generally makes them more able to face the new day and its problems. All this from a hot water extract of plant material. The common ingredient in tea and coffee responsible for these and other effects on the human body is caffeine. Caffeine is a mild central nervous system stimulant chemically similar to the nitrogenous base adenine that is found in DNA and RNA. Why coffee, tea, and other plants produce caffeine is unknown, although it is accepted widely that caffeine performs no essential role in basic metabolism, as it is not present in the majority of plant species. Many of these secondary plant metabolites* appear to have a defensive role and thus serve to protect the plant from disease, insects, or animals. Caffeine, however, has not been shown to have this protective function. Regardless of the reason that plants contain caffeine, many human cultures have discovered the value of beverages produced from them. Actually, caffeine-producing plants are not native to Europe or North America and were unknown there until about the middle of the seventeenth century. Until that time, the only stimulating beverage generally available in the Western world was alcohol, produced in various forms as beer, cider, wine, and distilled liquor.

Portuguese traders were the first to send tea from China to Europe as the result of their having developed a taste for this beverage during business dealings with Chinese officials. Tea did not really catch on in Portugal except with the aristocracy for whom it became a fashionable drink. Shortly afterwards, the Dutch forced the Portuguese out of their trading enclaves in China and took over the tea trade. Because of differences in their respective merchant marines and trading attitudes, the Dutch, unlike the Portuguese, began the large scale import of tea into Europe, in the early seventeenth century, with the Dutch people becoming tea drinkers. To promote the consumption of tea, the Dutch East India Company paid doctors to make extravagant claims about the health benefits of drinking tea. A similar approach was adopted by the British East India Company after the British government gave them a monopoly on the British tea trade as the result of mutual hostility between Great Britain and the Netherlands.

Before tea reached Great Britain, however, coffee was introduced, and by 1660 was consumed widely. In fact, the consumption of coffee became a social occasion that triggered the formation of coffee houses, which served the same function as the modern pub or tavern. Lloyd's of London, an international insurance company specializing in marine insurance, which is also willing to insure the unusual, such as a movie star's legs or a singer's voice, originated in a coffee house where a group of businessmen gathered regularly. Coffee houses were men-only institutions that served for the conduction of business and political discussion. Similarly, widespread coffee drinking in France led to the establishment of coffee houses as centers of political and business life. In fact, the coffee house was seen as a center for political dissidence and attempts were made to close them. In the Arab regions where coffee drinking originated in the middle of the fifteenth century, one of the major complaints voiced about coffee also was related to the coffee houses that sprang up to serve the demand. In all cases, the coffee house survived and coffee drinking spread.

Both coffee and tea were advertised widely as medically beneficial beverages capable of curing all manner of ills and promising many benefits. Today we would not credit either of these beverages with such properties, yet at the time that tea and coffee were being popularized in Europe, they most likely did deliver on many of these promises. No, coffee and tea have not changed over the centuries; however, society has. In the 1600s the germ theory of disease was unknown. Although people of all ages consumed beer, cider, or wine in preference to water, this was not because a connection had been drawn between sickness and the consumption of water. As tea and coffee became popular, they reduced disease rates due to water-borne infections, because the preparation of these beverages used boiled water, which killed many disease-causing organisms. Furthermore, replacing beer, cider, and wine with tea or coffee reduced the incidence of complaints such as kidney stones and gout, which are related to the consumption of alcoholic beverages and a high protein diet. Replacement of alcohol with the water in tea or coffee would prevent these painful afflictions and make the new beverages appear miraculous, particularly at a time when there was no anaesthesia for surgery to remove kidney stones. In addition, tea became so popular in England that the lower classes switched from gin to tea as their mood-modifying substance.

Given all the real and imagined benefits of tea and coffee as well as their social value, there is little wonder that governments saw economic advantages in controlling their trade. Great Britain and the Netherlands competed for the tea trade from China, until Great Britain found that tea was native to a part of India. Because this part of India was not part of the British Empire, troops were sent in, the land was annexed, and a plentiful supply of tea for the home market was assured by war. The Boston Tea Party, which was an expression of displeasure over the tax placed on tea, was one of the more colorful incidents of the American Revolution. A consequence secondary to this and related events is that the revolutionaries and the newly independent nation were unable to obtain tea for a number of years, so they switched to coffee with such vehemence that coffee is now considered by the rest of the world to be an American drink. The British, Dutch, and French colonial expansion and the associated wars of the eighteenth and nineteenth centuries resulted, in part, from their efforts to extend tea and coffee cultivation into new areas. All of this was taking place to obtain a source of caffeine and associated secondary plant metabolites for the mild, mood-altering potion on which the masses had come to depend. What self-respecting Briton would be without his "cuppa," that magic brew that puts all calamities in perspective, including World Wars? Or, equally hard to imagine, what American would go without the morning coffee that makes the day survivable?

4

Cell Division

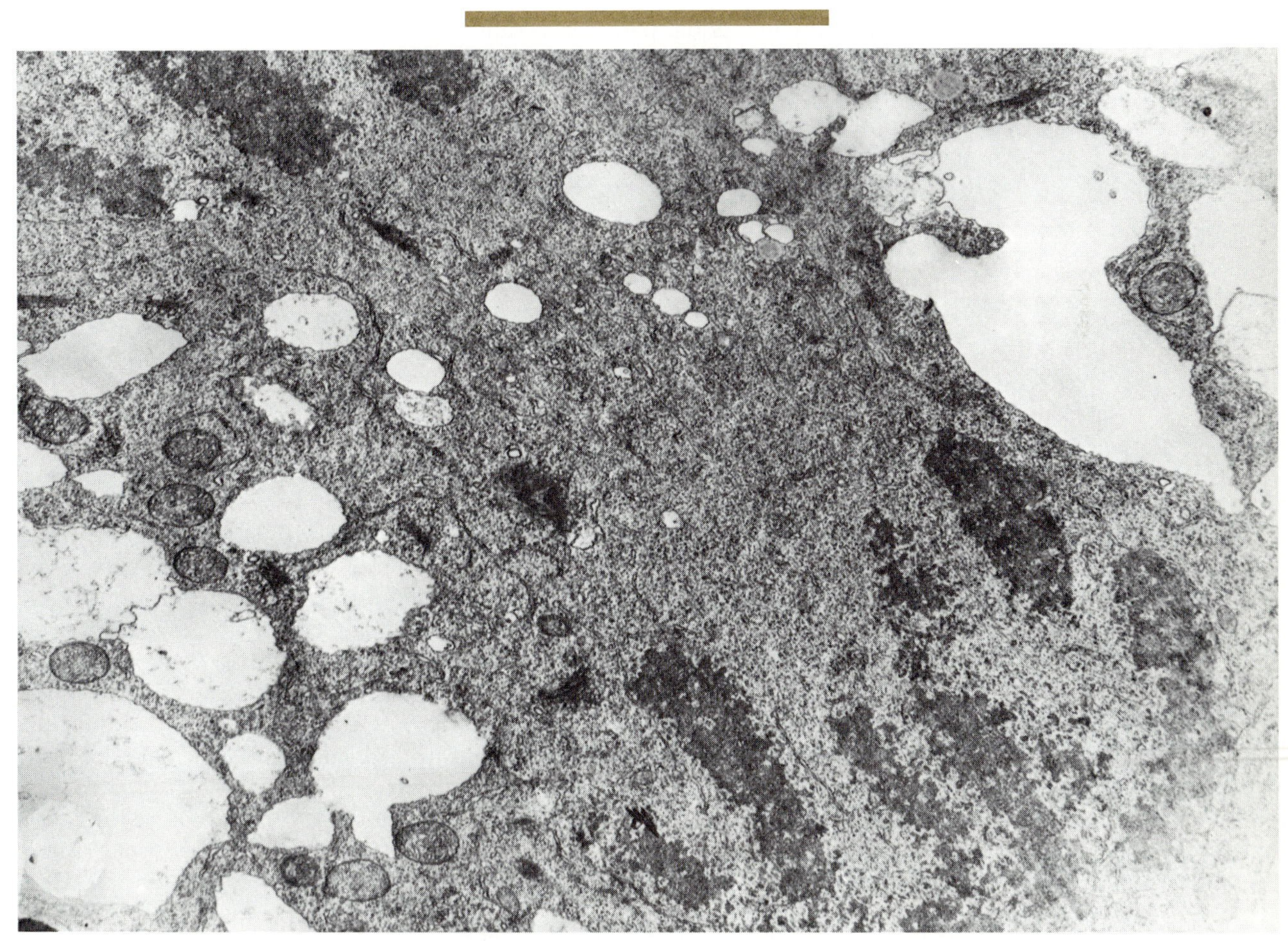

Outline

Introduction

One of the great principles of biology described in Chapter 2 is the cellular nature of all living organisms. Rudolf Virchow in 1858 was the first to suggest that all cells come from preexisting cells, although the mechanism whereby this occurs was not known to him. Subsequent improvements in the microscope and new methods of staining cells made possible the exploration of the nucleus and the discovery of the actual method of cellular reproduction. Robert Brown in 1831 had already recognized the **nucleus** as a regular feature of all plant cells, and the division of a nucleus was noticed by Eduard van Bouchen in 1875. The significance of the nucleus within the cell was completely unknown, however, as was the internal organization of the nucleus. **Chromosomes** were not mentioned until 1876, and the term *chromosome* was not even used until 1888 when it was introduced to describe the colored (chromo-) bodies (-somes) that could be observed after fixing* and staining certain cells. Usually chromosomes are seen only in dividing nuclei. They are flexible, rodlike structures of varying lengths that are divided longitudinally into equal parts, each called a **chromatid.** The two chromatids of a chromosome are attached to each other only at one point, the **centromere** (Figure 4–1). The location of the centromere along a chromosome varies from one chromosome to another. The length of chromosomes and the location of the centromeres are two characteristics commonly used to distinguish different chromosomes in a dividing cell (Figure 4–2). As will be shown in Chapter 22, each chromosome contains the genetic information, in the form of **genes,** for a particular group of characteristics. Genes, the basic units of heredity, are arranged linearly on a chromosome so that the position of a particular gene on a specific chromosome will always be the same. In addition, the two chromatids of a chromosome have identical sets of genes. A simple separation of the chromatids, therefore, will result in the production of two chromosomes that carry identical sets of genes. A chromosome consists of a long double helix of DNA intricately coiled with proteins, and a gene is a segment of this DNA molecule. The number of chromosomes in a nucleus varies with the plant species (Table 4–1).

Fixing
The act of killing and preserving cells.

New cells are produced from existing cells by **cell division,** without which the development of a multicellular plant cannot take place. A plant begins its life cycle from a unicellular **zygote,** which is formed by the fusion of an egg and a sperm during fertilization (Chapter 13). Cell division in the zygote and in the subsequent embryo results in a highly complex, multicellular plant body. Continued cell division in such a plant is restricted to groups of perpetually young cells known as **meristems,** which are present at the tips of roots and stems, in lateral buds, and in cambial tissue. New cells are also produced in specialized tissues prior to sexual reproduction. There are significant differences, however, between the latter and cell division in the meristems. Cell division relative to sexual reproduction is described in Chapter 13.

The process of cell division constitutes three separate events: **DNA repli-**

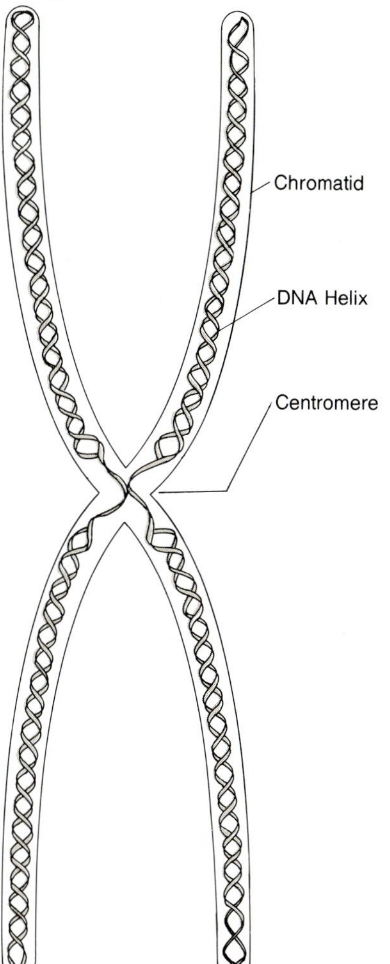

Figure 4–1. Chromosome structure. A single chromosome consists of two chromatids that are attached at a single point, the centromere. The position of the centromere may be anywhere along a chromosome. Each chromatid contains a DNA double helix. Genes are represented by segments of the DNA molecule.

cation, mitosis (division of the nucleus), and **cytokinesis** (division of the cytoplasm). Mitosis is a significant step of cell division inasmuch as the genetic information contained in the nucleus has to be proportioned equally between the two new nuclei. On the other hand, cytokinesis is a much less precise division of the cytoplasm into two units. Since the genetic information is stored in the DNA molecules, it is imperative that they be replicated prior to mitosis. The mechanism of DNA replication is already described in Chapter 3.

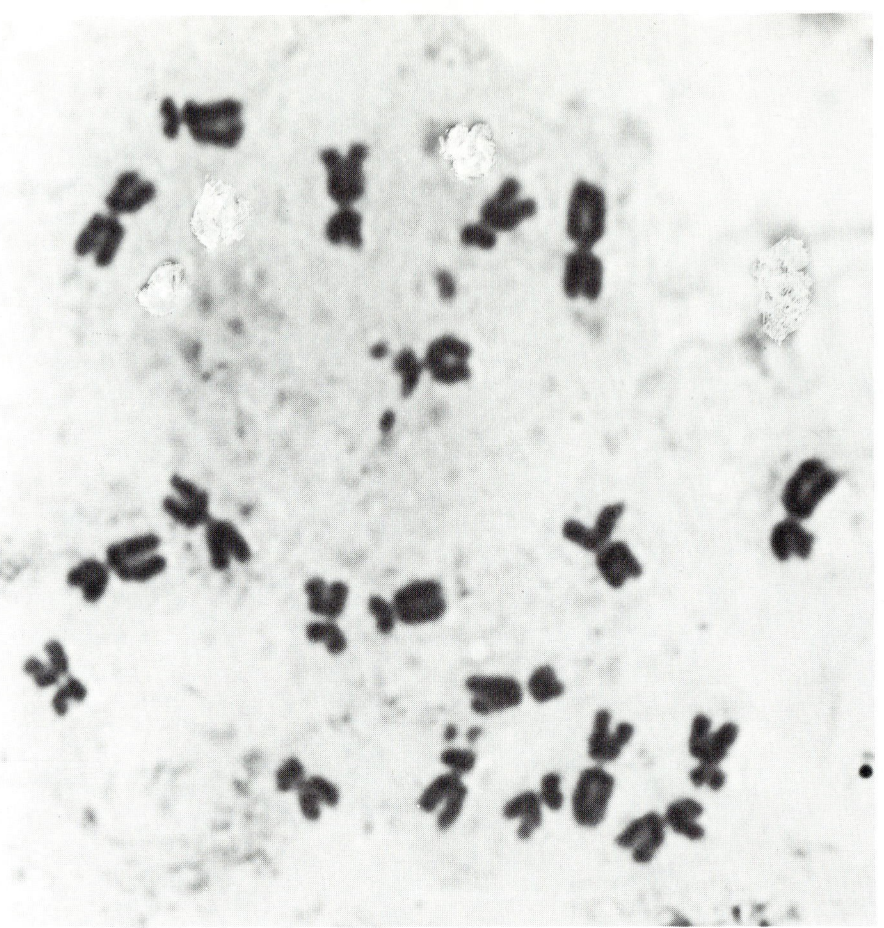

Figure 4–2. Photomicrograph of 20 chromosomes (10 pairs) from a root cell of corn. The differences in the lengths of the chromosome arms are due to the position of the centromere on the various chromosomes. (Light micrograph by Dr. David Weber)

Table 4–1
Chromosome Number of Some Common Plants

burroweed	4	cotton	52
tomato	48	tobacco	48
carrot	18	alfalfa	32
radish	18	rose	14
lettuce	18	lily	24
cucumber	14	snapdragon	16
pumpkin	40	peach	16
onion	16	apple	34
garden pea	14	American elm	56
kidney bean	22	white oak	24
broad bean	12	black walnut	32
corn	20	white ash	46
sugar cane	80	redwood	22
bread wheat	42	pine	24
rice	24	grape fern	90

Cell Cycle

The chromosomal behavior and the structural changes during mitosis and cytokinesis are easily recognized under a light microscope. On the other hand, the biochemical and physiological changes taking place between divisions are much less apparent. Nevertheless, they are just as significant as are the changes that occur during the process of division and are, indeed, a prerequisite for mitosis. Cells in meristems are known to go through repeated divisions. The sequence of events these dividing cells pass through is called the cell cycle, a process consisting of four distinct phases—gap one phase (G_1), synthesis phase (S), gap two phase (G_2), and mitotic phase (M) (Figure 4–3). The G_1, S, and G_2 phases together constitute **interphase,** which at one time was believed to be a resting stage between divisions. It presently is known, however, to be a most active stage when the newly formed cells complete their growth. More specifically, during the G_1 phase, which starts immediately after cell division, the cells synthesize the different kinds of ribonucleic acids needed for protein synthesis (Chapter 3). Whether a cell is to become a specialized cell or to remain a meristematic cell is also programmed during this phase, although the determining factors are not fully known at present. If the cell is to undergo further division, it enters the S phase, during which the DNA replicates itself from precursors synthe-

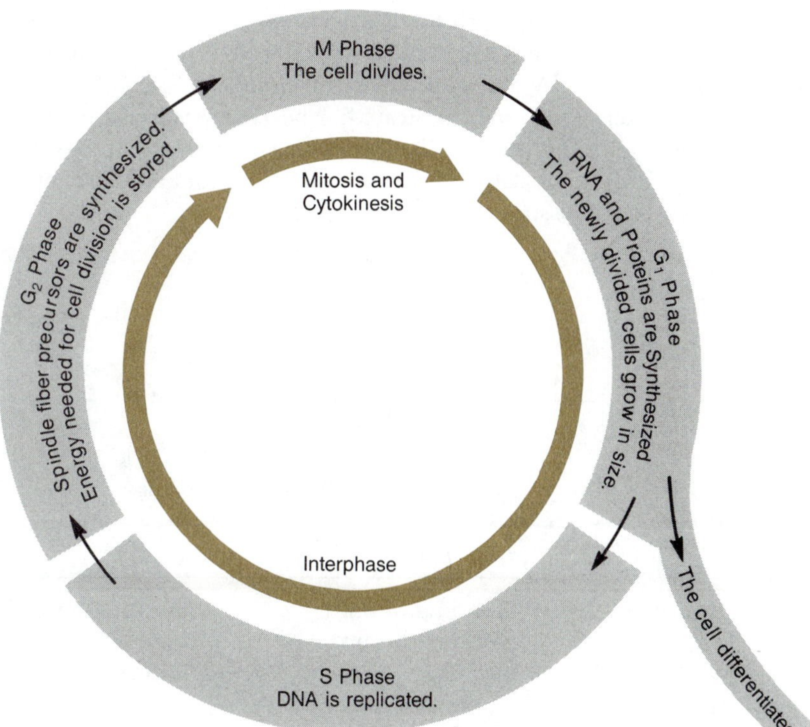

Figure 4–3. The cell cycle. This diagram represents the different phases through which dividing cells pass. The cells in the cycle remain meristematic. The cells that differentiate and become specific cell types leave the cycle after the G_1 phase and prior to the S phase.

sized during the G_1 phase. The G_2 phase that follows is characterized by the synthesis of components, such as microtubular proteins necessary for the assembly of spindle fibers. Energy needed for the process of division is also produced by respiration* (Chapter 10) during this phase. Mitosis and cytokinesis then complete the cycle. The time necessary for the completion of a cell cycle is influenced by a variety of factors, one of which is temperature (Table 4–2).

Respiration
The metabolic process that results in the release of chemical energy from simple organic molecules.

Interphase

A nucleus of a cell that is not dividing is said to be in **interphase,** and is characterized by the presence of an intact **nuclear envelope,** one or more **nucleoli,** and an extensive network of fine threadlike **chromatin,** the material of which chromosomes are composed (Figure 4–4). There are, however, two distinctly different groups of interphase cells—a larger group arrested in the G_1 phase of the cell cycle and differentiated into the various cell types, and a second group consisting of meristematic cells that are between divisions. During interphase in meristematic cells, the DNA strands forming the chromatin are duplicated in preparation for the division of the nucleus (Chapter 3). Apparently, duplication of the DNA strands in the chromosome is initiated simultaneously at several points along the molecule and proceeds in both directions (Figure 4–5). Duplication stops when the DNA-duplicating enzymes reach the centromere, which is left unduplicated. The centromere holds the four strands of DNA together. At this stage, the two new strands of DNA are associated in a double helix with the original strands from which each was duplicated. The centromere also serves as a point of attachment for the spindle fibers that will separate the daughter chromatids during mitosis. At any given time approximately 90% of the nuclei in a meristematic tissue will be at interphase.

Table 4–2
Difference in Time Intervals for the Completion of a Cell Cycle in Pea Root Meristematic Cells at Two Different Temperatures

Phase	15° C (59° F)	25° C (77° F)
Interphase	22 hours 36 min.	14 hours 30 min.
Prophase	2 hours 6 min.	54 min.
Metaphase	24 min.	14 min.
Anaphase	5 min.	3 min.
Telophase	22 min.	11 min.

Mitosis

For the convenience of explaining nuclear division, mitosis is divided into four stages. It should be understood that these stages are artificial steps and that in nature mitosis takes place as a continuous process. It should also be realized that it may be difficult sometimes to recognize the end of one stage and the beginning of the next. There are, however, certain configurations of chromosomes, commonly called mitotic figures, that can help to determine a particular stage of mitosis. These stages in sequence are **prophase, metaphase, anaphase,** and **telophase.** Although it is necessary to kill and stain the tissues to be able to recognize these stages with an ordinary light microscope, mitosis has been observed as a continuous process in living cells under phase contrast microscopes.

Prophase

Prophase is the first stage of mitosis, and a number of structural changes take place within the nucleus during this stage. The first and most apparent of these changes is the appearance of chromosomes formed by the shortening and thickening of the chromatin strands of which they are composed, a process known as **condensation.** This is a necessary prerequisite for mitosis because interphase chromatin strands are too long to separate within the small space of a cell. The relative length of a chromatin strand compared to that of a chromosome is one hundredfold or more.

While the chromosomes are being formed, the nucleolus gradually disappears and threadlike **spindle fibers** are formed by an aggregation of **microtubules** from the opposite poles of the nucleus. As prophase proceeds, the chromosomes with their conspicuous chromatids move toward the central region of the nucleus, the final step in prophase being the breakdown and disappearance of the nuclear envelope.

Metaphase

Metaphase is characterized by the orientation of the chromosomes at the equatorial region of the nucleus. The centromere of each chromosome is positioned along the equatorial plane with the chromatid arms lying randomly on either side of the equatorial plane. As metaphase continues, some of the spindle fibers attach themselves to the centromere of each chromosome in such a way that each centromere will have spindle fibers attached to it from opposite poles. The remaining spindle fibers run from pole to pole unattached to any chromosome.

Anaphase

Anaphase starts with the splitting of the centromere separating the two chromatids of each chromosome and a migration of all the chromatids on one side of the equatorial plane to one pole and the other set of chromatids to the opposite pole. There is evidence to suggest that the microtubules of the spindle fibers are responsible for this process. These separated chromatids are considered to be new chromosomes, often referred to as daughter chromosomes. Anaphase ends when the newly formed chromosomes reach the respective poles.

Telophase

Telophase is the last stage of mitosis, and is somewhat of a reversal of prophase. During telophase the chromosomes at each pole gradually lose their identity as they uncoil and revert to the chromatin network. At the same time, nucleoli are reformed from **nucleolar organizing regions** that are present on specific chromosomes. While the spindle fibers are disorganized, portions of endoplasmic reticulum* surround these areas producing nuclear envelopes, which results in the formation of two new nuclei at opposite ends of the cell.

Endoplasmic reticulum
A network of membranes within the cytoplasm.

One major aspect of mitosis that presently is not fully understood is the mechanism of chromosome movement. Over the past thirty years, numerous techniques, such as cinematography and electron microscopy, along with a variety of mitotic toxins*, have been used to elucidate the basic mechanism for chromosome movement. Accumulated data so far strongly suggest that microtubules are essential for the movement and that any physical or chemical factor that disrupts microtubular assembly can prevent the completion of mitosis.

Mitotic toxins
Chemicals that block or disrupt mitosis.

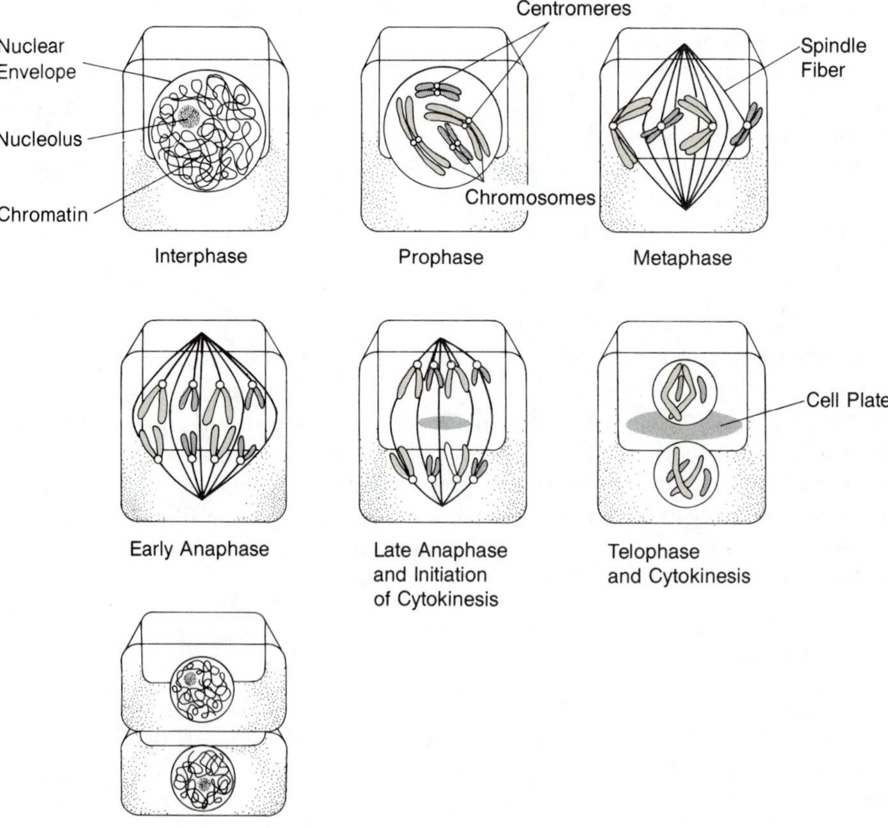

Figure 4–4a. **Cell division.** During prophase the chromatin material condenses forming chromosomes that, at first, are distributed randomly within the nucleus. As prophase proceeds, the nucleolus disappears, the nuclear envelope breaks down, and the chromosomes migrate toward the center. At metaphase the chromosomes are aligned along the equator, with spindle fibers from opposite poles attached to each chromatid. The chromatids separate during anaphase and move in opposite directions. Beginning with telophase, the two sets of chromosomes revert to the chromatin condition, while nucleoli reappear and two nuclear envelopes are reformed. Also during telophase, cytoplasmic division is initiated by a disklike cell plate forming at the center and expanding outward.

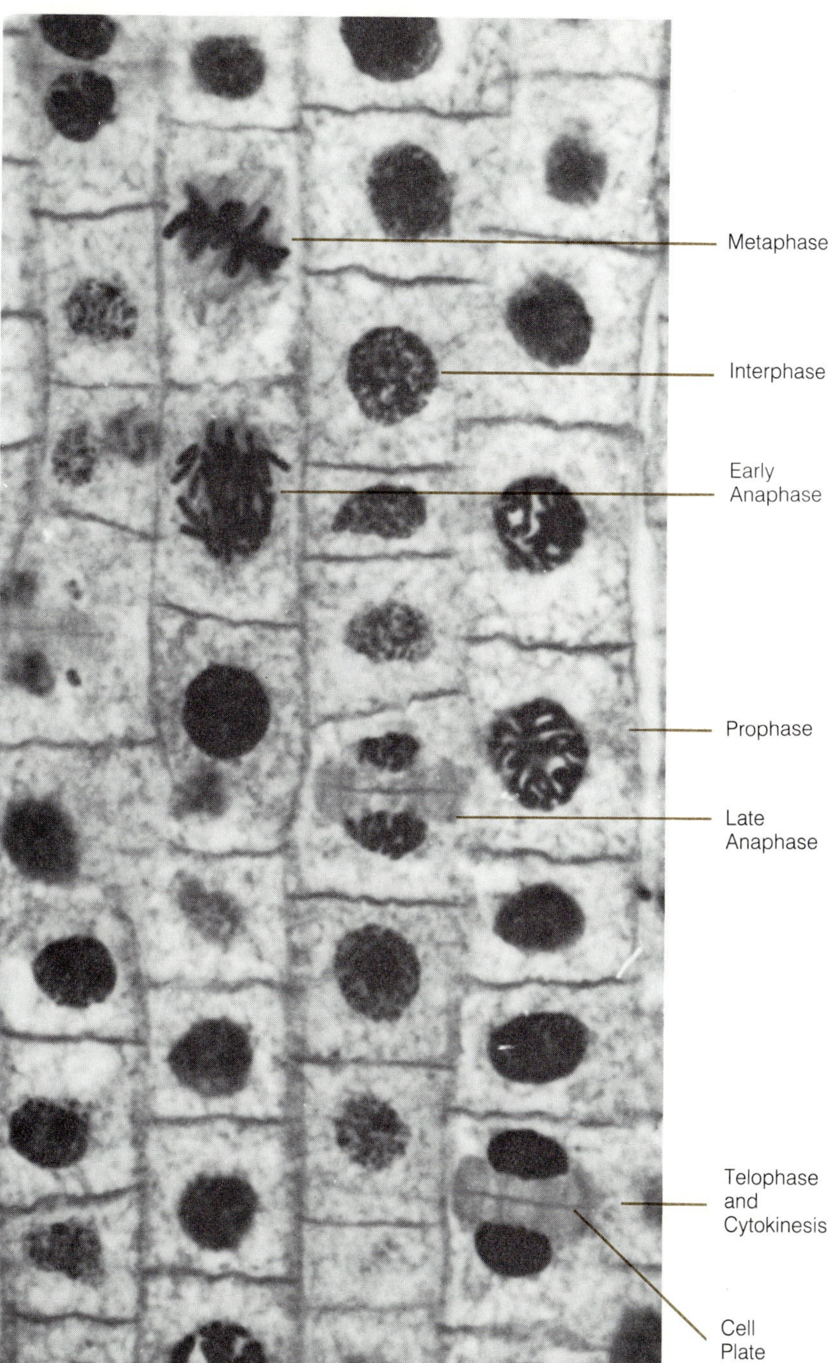

Figure 4–4b. Cell division-continued. Photomicrograph of mitotic stages in onion root cells.

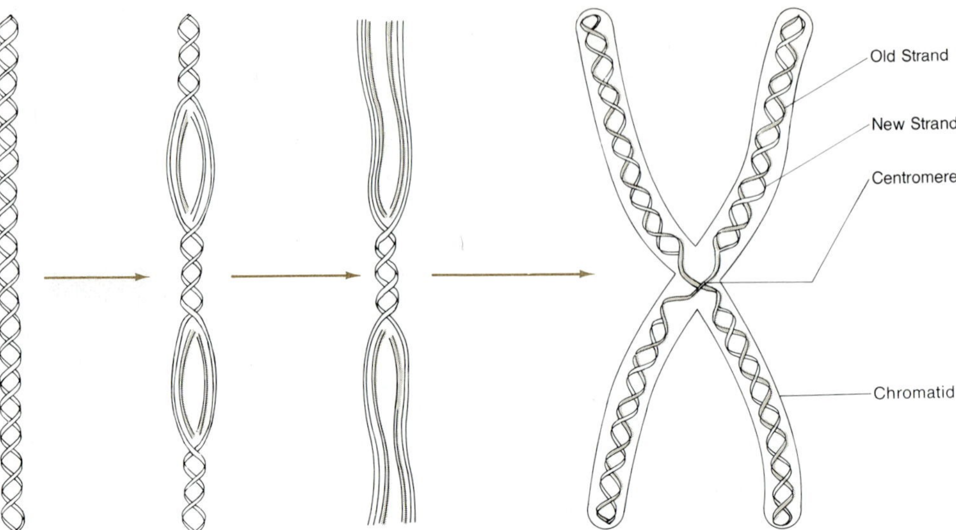

Figure 4–5. DNA replication. This sequence represents the replication of DNA prior to mitosis and the subsequent formation of a new chromosome. DNA replication is initiated simultaneously at several points along the double helix, resulting in the formation of two new strands joined at the centromere. Replication of the DNA segment at the centromere takes place just prior to anaphase.

Cytokinesis

In most plant cells the division of the cytoplasm is usually initiated before the completion of mitosis. The early stages of cytokinesis become recognizable with the appearance of the **cell plate** along the equatorial plane during telophase (Figure 4–6). The cell plate is formed by the fusion of vesicles pinched off from Golgi cisternae.* The cell plate starts as a small disc in the center and increases in diameter until it is in contact with the cell walls, thus dividing the cytoplasm into two parts, each containing a nucleus. New primary walls are deposited by the respective cytoplasms onto both sides of the cell plate, which by now has become the **middle lamella** that glues the walls together, thus completing the formation of two new cells.

Although the above described process of cell division is typical of most plants, there are a number of variations from this pattern in other organisms. One of the most common of these variations is found in many fungi and certain algae where the nuclear envelope remains intact during the process of mitosis. Another variation is seen in some cells where mitosis is known to take place without cytokinesis, resulting in the formation of cells with two nuclei.

It is most important to realize that the mechanism of cell division described in the preceding paragraphs is a means by which new cells will receive genetic information and a chromosome number identical to that of the other cells in the plant. This is made possible by the linear arrangement of genes on chromosomes, the longitudinal duplication of each chromosome into genetically similar chromatids, and the final separation of the chromatids to form new chromosomes. An essentially identical process also takes place in animal cells, suggesting a close relationship between plant and ani-

Golgi cisternae
Flattened, circular, membrane-bounded sacs that make up the Golgi body.

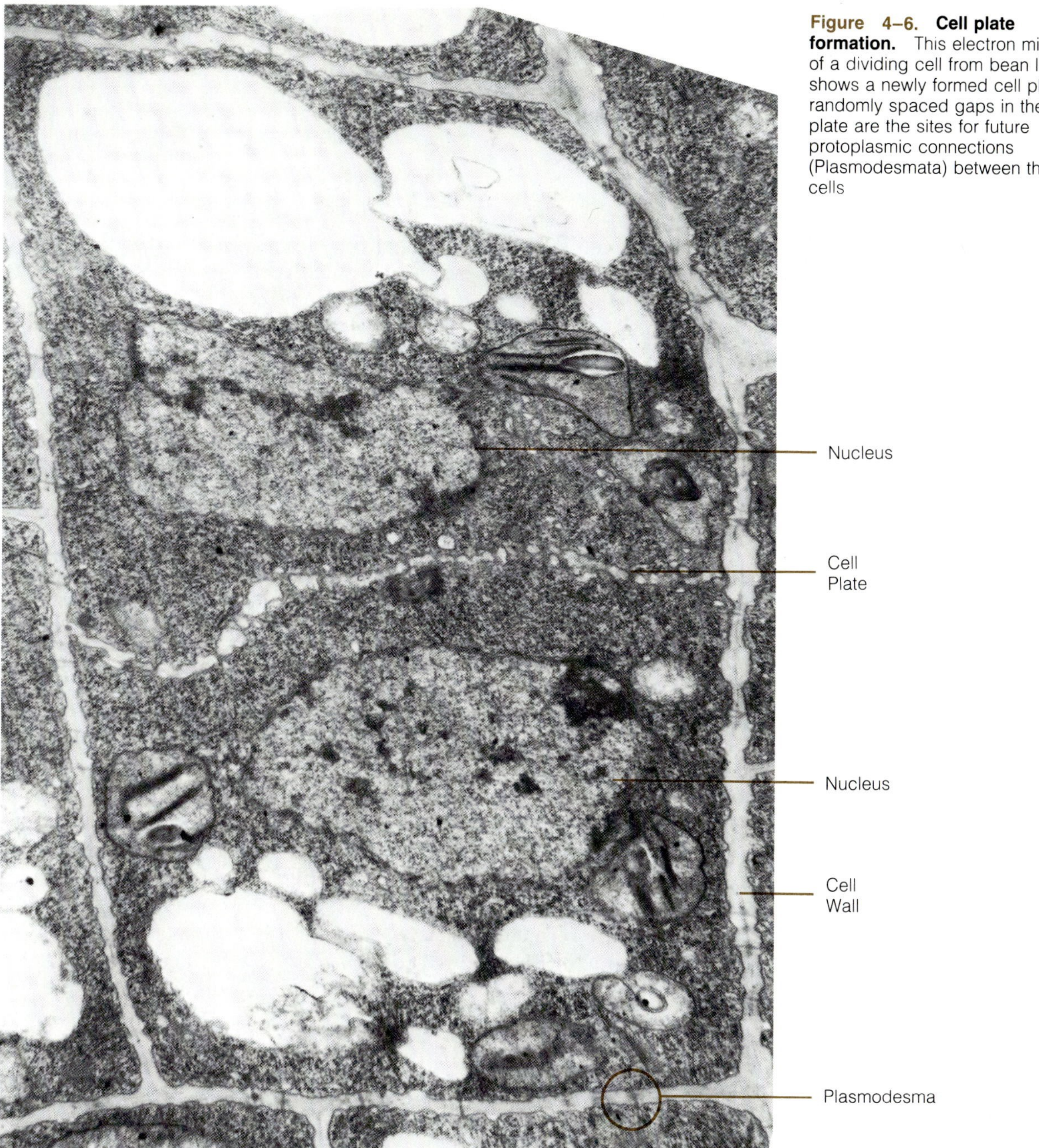

Figure 4–6. Cell plate formation. This electron micrograph of a dividing cell from bean leaf shows a newly formed cell plate. The randomly spaced gaps in the cell plate are the sites for future protoplasmic connections (Plasmodesmata) between these cells

mal cells. The presence of short, cylindrical bodies called **centrioles** as organizers of the spindle fibers during mitosis and an absence of a cell plate during cytokinesis in animal cells are the major differences in cell division between a typical plant cell and an animal cell. Centrioles are present, however, during mitosis of the motile cells that are common among certain groups of algae.

Summary

The process of cell division is essential for the growth and development of a multicellular plant from a zygote. Normally, cell division is restricted to meristems. In addition to the production of new cells, the basic mechanism of cell division ensures that the new cells get exact copies of the genetic information. Chromosomes that are visible in a dividing nucleus consist of two identical parts, the chromatids, attached to each other only at one point, the centromere. Genes, the basic unit of genetic information, are arranged linearly on the chromatids. During nuclear division the chromatids of each chromosome separate, resulting in two new sets of daughter chromosomes, one set for each new nucleus.

Cell division takes place in three steps: DNA replication, mitosis (the division of the nucleus), and cytokinesis (the division of the cytoplasm). DNA is replicated during interphase prior to mitosis. Although mitosis is a continuous process, successive stages such as prophase, metaphase, anaphase, and telophase can be recognized by the behavior of the chromosomes. Chromosomes become visible during early prophase, followed by a disorganization of the nucleolus, formation of spindle fibers, and the final breakdown of the nuclear envelope. At the beginning of metaphase, chromosomes are arranged at the equatorial region of the nucleus. Anaphase is initiated with a splitting of the centromeres and continues with the migration of the new daughter chromosomes toward the opposite poles. In telophase individual chromosomes lose their distinct nature, there is a reorganization of the nucleolus, and finally the nuclear envelope re-forms, completing the process of mitosis. Cytokinesis usually begins during telophase with the formation of a circular cell plate, which expands to separate the cytoplasm into two units, each with a newly formed nucleus. Cell division is complete when two new primary walls, one on each side of the cell plate, are formed. A similar process with certain changes takes place during cell division in animals.

Review Questions

1. Of what significance are the chromatids that make up a chromosome?
2. What is the role of a centromere on a chromosome?
3. Describe with appropriate diagrams the process of mitosis and cytokinesis.
4. Discuss interphase in relation to cell cycle.
5. What is the significance of mitosis?

Chromosome Number

There is considerable variation in the number of chromosomes present in different species of plants. Individuals of the same species have identical numbers of chromosomes, except in the case of certain cultivated and horticultural varieties in which there may be more than one number for the same species. Chromosome number is expressed as the "diploid number," alternately called the 2N condition. The diploid number is always twice that of the haploid number, or 1N condition. Sex cells of a plant or animal have only one set of chromosomes, and when two sex cells fuse, the resulting zygote will then have two sets of chromosomes—one set from the male parent and the other set from the female parent. The multicellular organism produced from the zygote, therefore, will have two sets of chromosomes in all, except for its sex cells. Later, when you study reproductive cycles of different plant groups, you will realize that there are certain exceptions to this general rule. However, if you count the number of chromosomes in the meristematic cells of the root or stem, you will find the diploid number, while the sperm and eggs of the same plant will have the haploid number. How the diploid number is reduced to the haploid number prior to sperm and egg production is described in Chapter 13. A list of the diploid number of chromosomes in a variety of common plants is presented in Table 4–1.

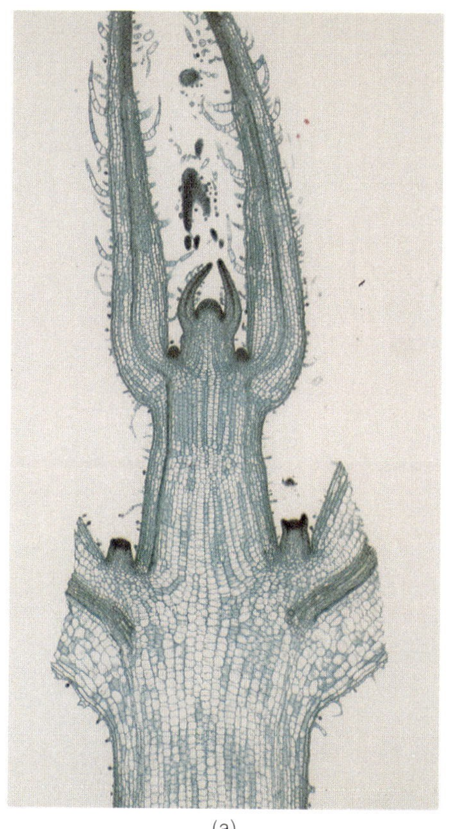

(a)

(b)

Photos: **(a)** Longitudinal section of shoot tip.
(b) Longitudinal section of dictotyledonous herbaceous stem.
(c) Cross section of monocotyledonous stem.
(d) Cross section of dicotyledonous stem.

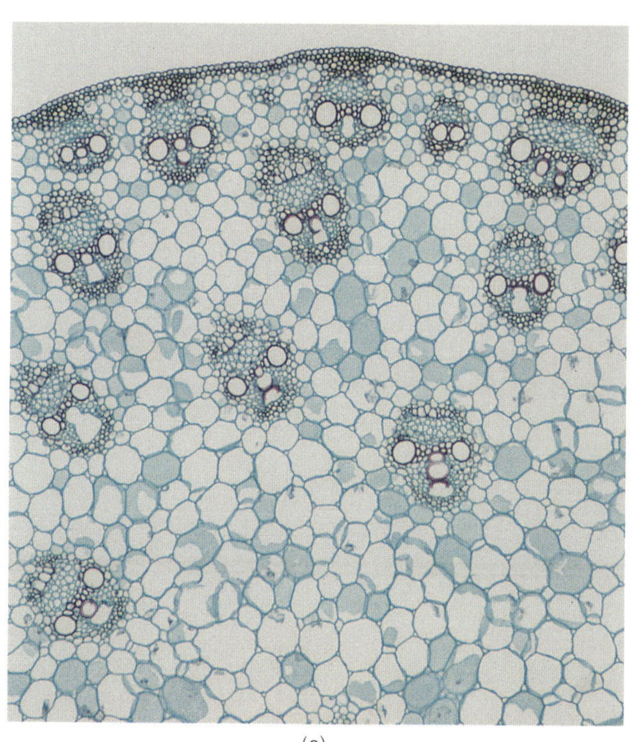

(c)

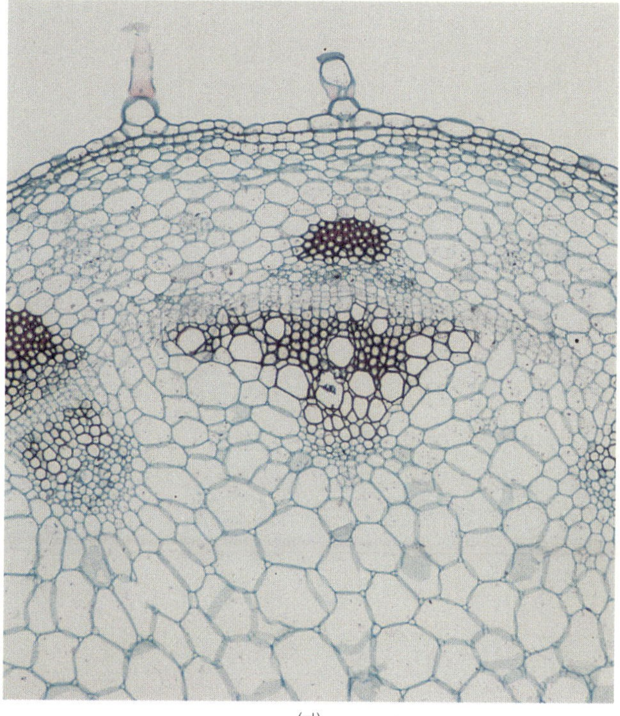

(d)

Color Plate 2

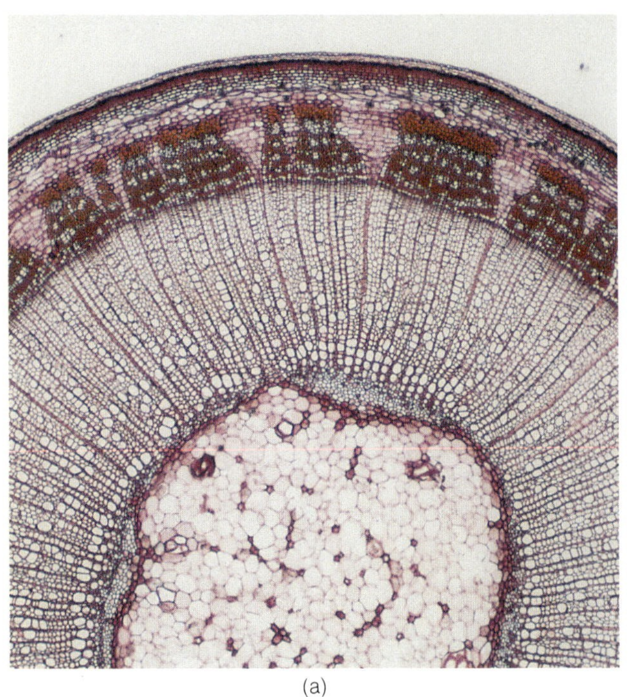

(a)

Photos: (a-c) Cross sections of angiosperm woody stem. (a) General view. (b) Secondary xylem and secondary phloem. (c) Cortex and cork. (d) Cross section of gymnosperm stem.

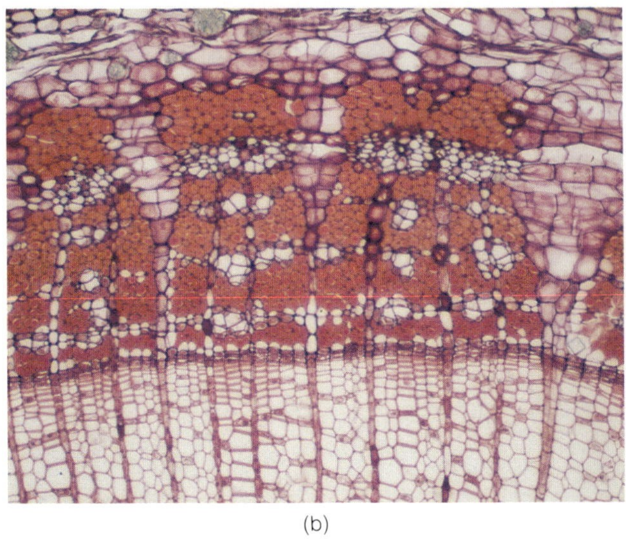

(b)

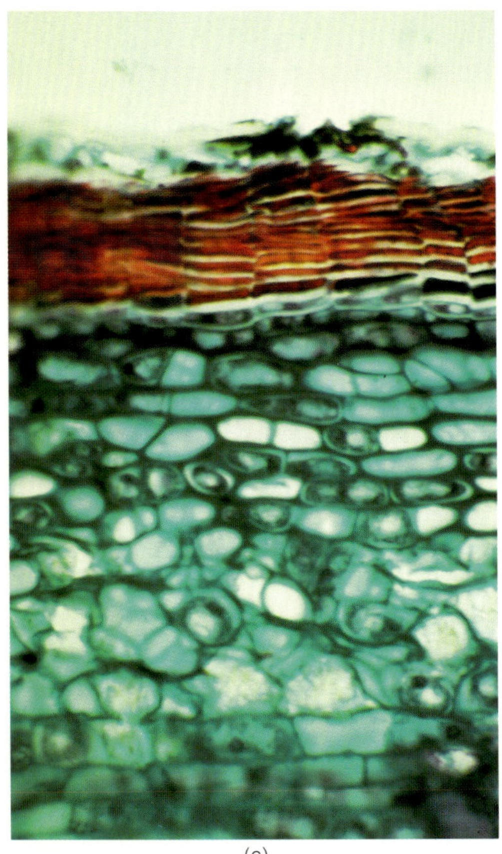

(c)

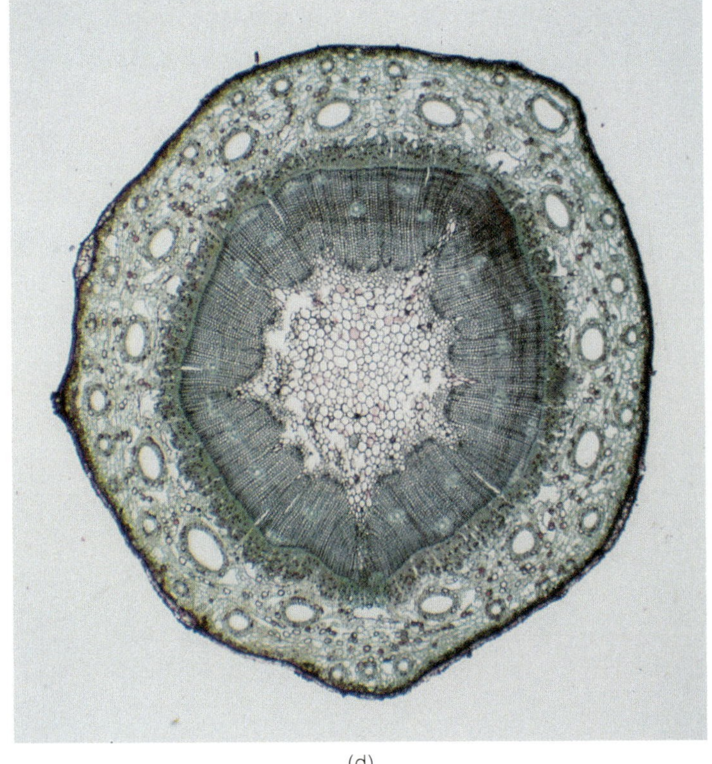

(d)

Photos: (a) Longitudinal section of root tip. (b-c) Cross sections of dicotyledonous root. (b) General view. (c) Stele.

(a)

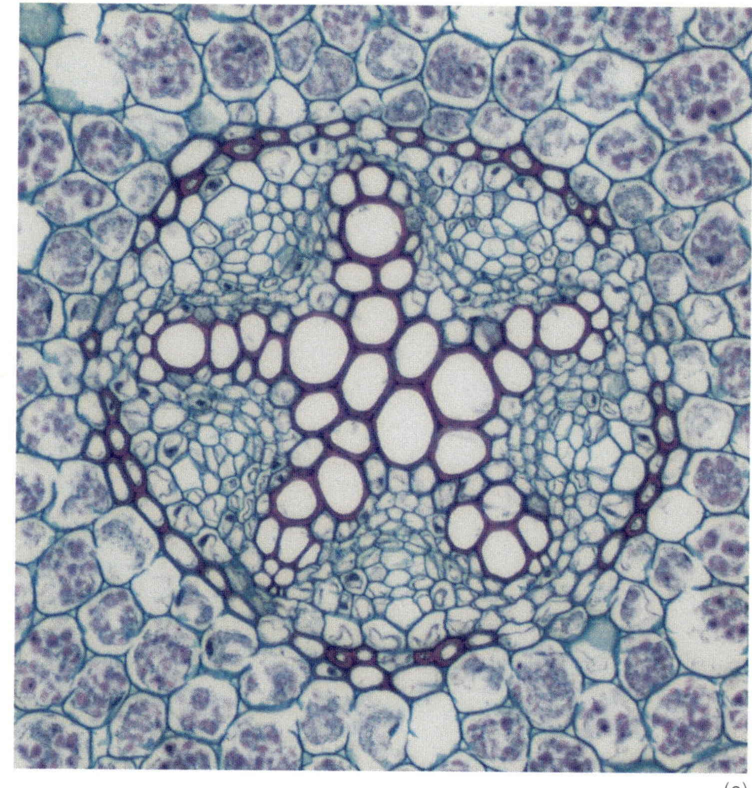

(c)

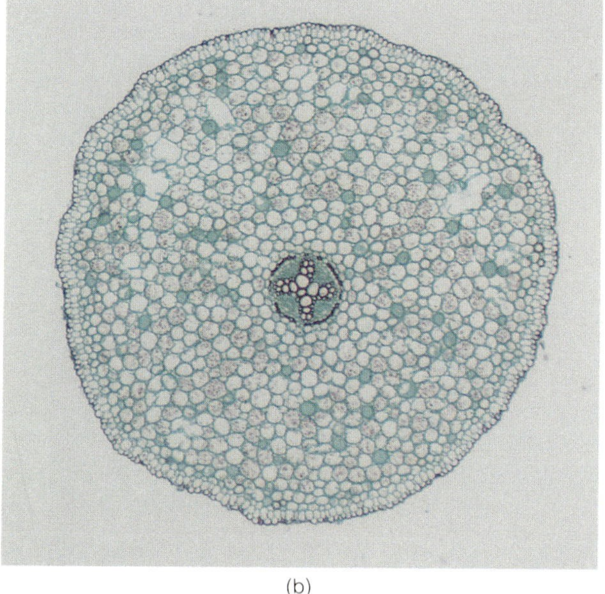

(b)

Color Plate 4

Photos: **(a)** Cross section of monocotyledonous root. **(b)** Branch root formation. **(c)** Haustorium attached to host. **(d)** Root system of a plant growing on a wall.

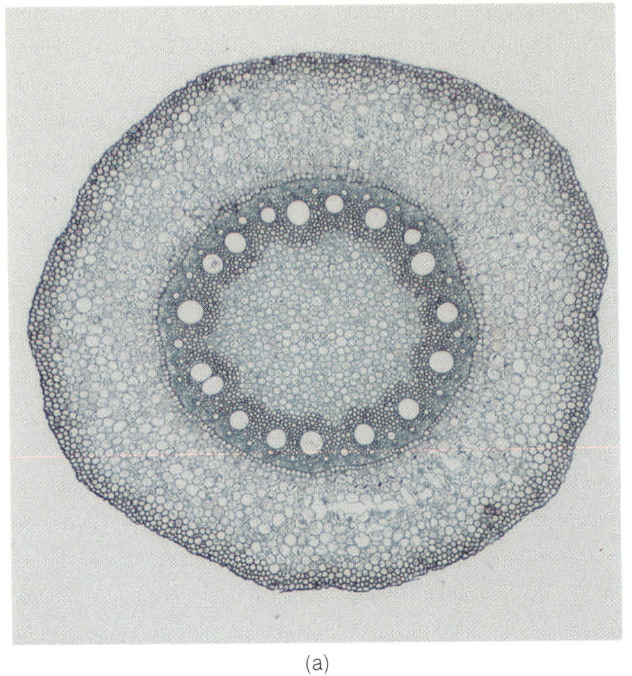

(a)

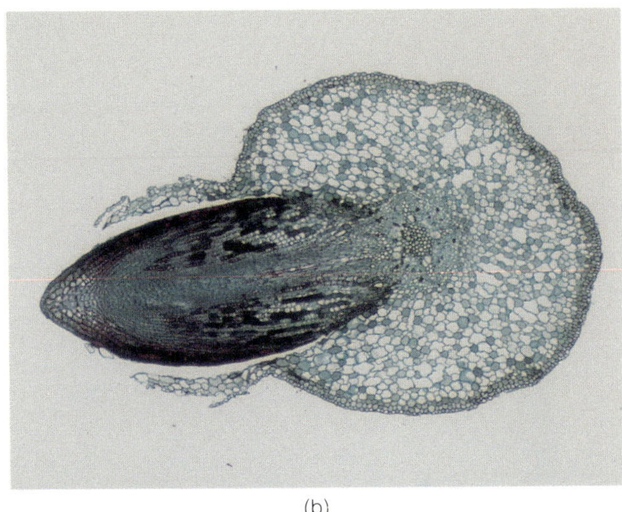

(b)

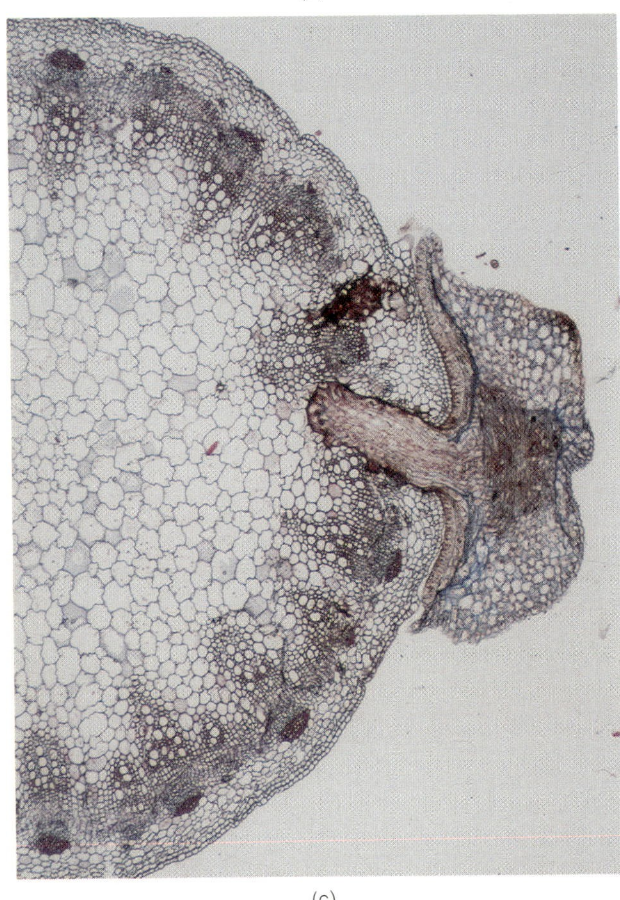

(c)

(d)

5

Tissues and Cell Types

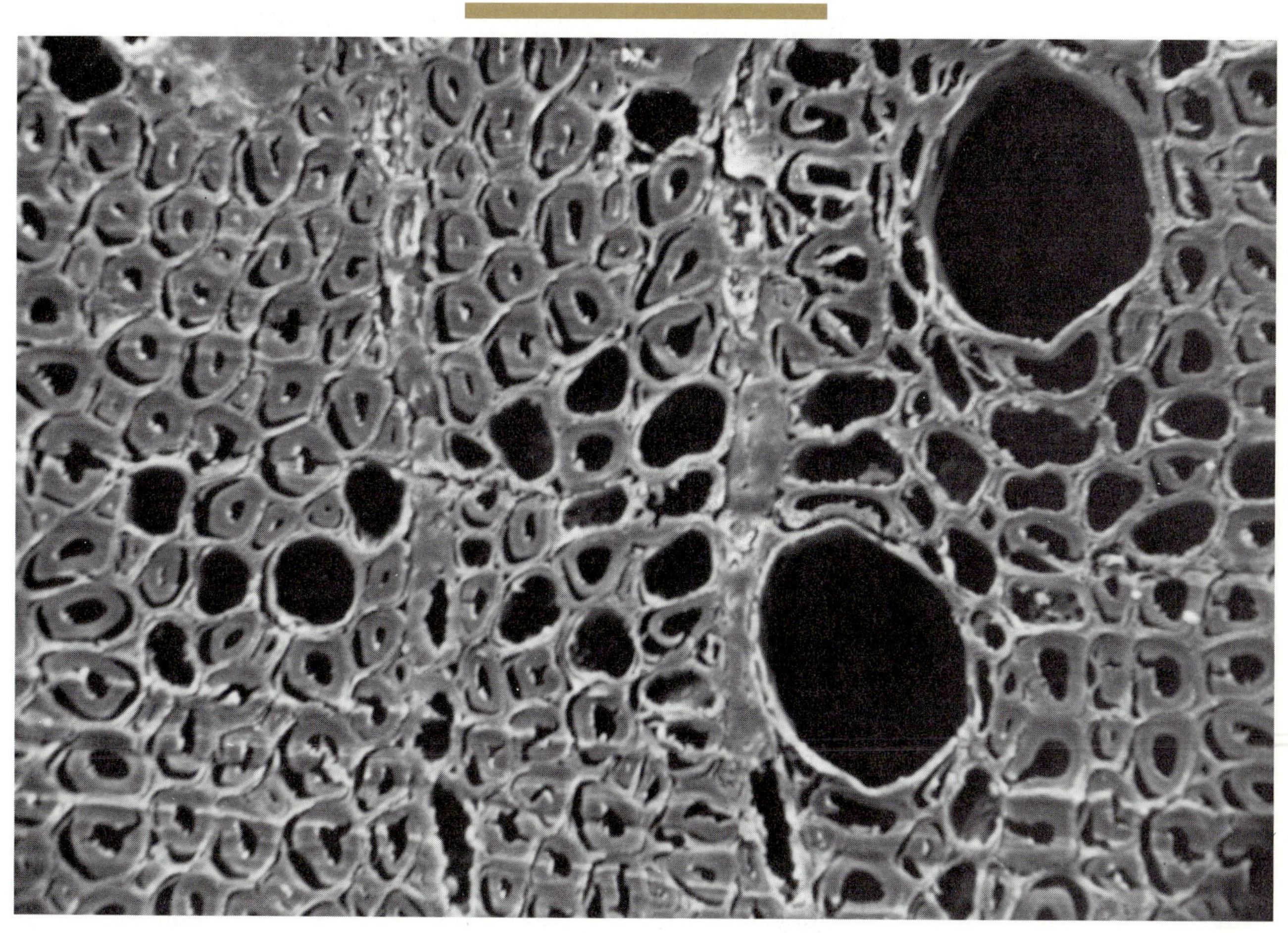

Outline

Introduction

There is evidence which suggests that the first living organisms were unicellular. Because each cell was an independent organism, it had to have the ability to carry out all of the various functions required for the completion of a life cycle: either food production or consumption or both, water and mineral absorption, reproduction, and so on. Subsequent evolution led first to the appearance of colonial and then of multicellular organisms. In both of these cases, the organism consists of several to many cells, not just one. With more cells came the opportunity for specialization and the increased efficiency that division of labor brings. Consequently, multicellular plant bodies are not composed of hundreds or thousands of cells that individually perform all the functions of an organism. Instead, the multicellular plant body is made up of a number of different types of specialized cells, each of which performs only one or two of the functions of an organism. This division of labor makes it possible for leaf cells in a tree to produce food by photosynthesis*, using water that was absorbed by root cells deep in the ground and then transported upward through water-conducting cells. Meanwhile, the root cells survive in the dark without any need to produce food because excess food is transported downward from the leaves through food-conducting cells. If one or more of these specialized cell types were to become nonfunctional, the organism could die. This chapter describes the cell types and the tissues into which they are organized as part of the background to an understanding of plant anatomy.

Photosynthesis
The metabolic process by which carbohydrates are produced from CO_2 and a hydrogen source such as water using light as the source of energy.

Casual observation of a plant is deceptive because it gives the impression that a plant is a relatively simple entity, made up of but a few different parts or organs. However, if the different organs, such as roots, stems, and leaves, are examined under a microscope, it becomes apparent that these are composed of cells and that groups of cells are arranged in regular patterns forming **tissues.** A tissue is a collection of cells that carries out a specific function, although certain tissues in a plant body are multifunctional. Thus a tissue is a functional unit and is the basis for the division of labor that must occur within a multicellular plant. In general, a specific type of tissue carries out a particular function whether it is part of a root, a stem, or a leaf. In addition, there is continuity of tissues between the different organs of the plant, providing for a high degree of integration of the diverse functions of the organs.

As stated in Chapter 2, cells were first recognized by Robert Hooke when a slice of cork tissue was examined under a microscope. The cork tissue, found on the outermost part of a tree, consists of only one kind of cell, the cork cells, all of which are dead when mature. A tissue such as cork, made up of only one kind of cell, is a **simple tissue,** whereas a **complex tissue** consists of two or more cell types. Certain kinds of cells die and lose their protoplasts when they reach maturity; in fact, the role of the protoplast during the development of these cell types is to produce a cell wall of the necessary size and architecture for the function of the tissue. Functional dead cells invariably have secondary walls, while most living cells have only primary walls. Unlike the situation in animals, plant tissues composed partially or exclusively of dead cells carry out certain important functions. The complex tissues consist of a mixture of both living and dead cells.

Meristems

The plant body contains many more cells than does the embryo from which it developed as the result of cell division (Chapter 4). Usually the production of new cells is restricted to certain regions of the plant called **meristems.*** Each stem and root has one growing tip containing an apical meristem to which all tissues in that organ can be traced. The apical meristems, in turn, can be traced to the embryo that gives rise to the plant. As the embryo develops into a plant and continues to grow, most of the cells lose the embryonic characteristic of cell division. A few cells at the tip of the embryonic stem and a few at the tip of the embryonic root, however, retain this characteristic and become the apical meristems. Division of cells of the apical meristem gives rise to three groups of meristematic cells referred to as **primary meristems** because they come directly from the apical meristem. Cell divisions in the primary meristems, the **protoderm,** the **procambium,** and the **ground meristem,** are responsible for the production of **primary tissues** found in a plant body, resulting in the continuous elongation of the shoot and root. Other meristems, lateral in position within the stem and root, are the **secondary meristems,** since they are not produced directly by the apical meristem. These are the **vascular cambium** and the **cork cambium.** Tissues produced by secondary meristems are known as **secondary tissues** and are responsible for the increase in the diameter of a plant stem or root. A third type of meristematic tissue, usually present only in the root, is the **pericycle,** which is responsible for the production of branch roots. Although these different meristems are the primary source of new cells and the continued growth of a plant, all living tissues are potential meristems and are able to undergo cell division in response to the correct stimulus. It is very common to see new buds produced on the living stump of a tree after the trunk is cut off. Normally, these buds would not have formed on the trunk if the tree had not been cut down. Similarly, roots form on plant cuttings placed in water as the result of renewed cell division in the stem tissues.

Meristem
A tissue whose cells are capable of repeated division.

Primary Tissues

Primary tissues are those tissues produced by the different primary meristems. The primary tissues that are found in a typical plant are the **epidermis, parenchyma, collenchyma, sclerenchyma, endodermis, pericycle, primary xylem,** and **primary phloem.**

Epidermis

Herbaceous plant
A plant composed mainly of primary tissue.

Woody plant
A plant that contains large amounts of secondary tissues, particularly xylem (wood).

Anthocyanin
A water-soluble pigment of variable color present in cell vacuoles.

Parasite
(*adj.*, parasitic) An organism that obtains its nutrition from a living host.

The epidermis is the outermost tissue on a herbaceous* plant or on the young stems and roots of a woody* plant (Figure 5–1). Usually it consists of a single layer of epidermal cells and is derived from the protoderm. Each epidermal cell is a tabloid or slightly flattened cell with the wall facing the outside thicker than the other walls. The epidermis of the aerial portions of a plant produces a waxy covering known as the **cuticle.** The function of the cuticle is to prevent loss of water by evaporation from the cell surface. Epidermal cells are alive when mature and usually lack chloroplasts; hence they are transparent. The green color of a leaf is due to the numerous chloroplasts in the underlying tissue that can be seen when looking through the transparent epidermis. However, many kinds of leaves, such as coleus, have anthocyanin* in the vacuoles of their epidermal cells, thus making them colored. On the aerial parts of a plant, particularly the leaves, there are many small openings, each one known as a **stoma** (plural, **stomata**). The size of the opening is regulated by two modified epidermal cells, the **guard cells,** which contain chloroplasts. Thus loss of water vapor and exchange of gases through the stomatal openings can be controlled by the guard cells (Chapter 8). Another modification of epidermal cells that is often seen on leaf and stem surfaces is the epidermal hair. These hairs are often multicellular, branched or unbranched structures that give the surface a velvety or fuzzy appearance. Some of these hairs are glandular and contain sticky or toxic substances that discourage animals from eating the plants. The only epidermal modification present on the unerground parts of a plant is the **root hair.** The epidermal cells on specific regions of very young roots produce elongated, tubular extensions called root hairs that increase the surface area for absorption of water from the soil. The primary functions of the epidermis are to prevent water loss from the internal tissues of shoots, to absorb water and minerals from the soil, and to minimize entry of parasitic* organisms such as bacteria and fungi.

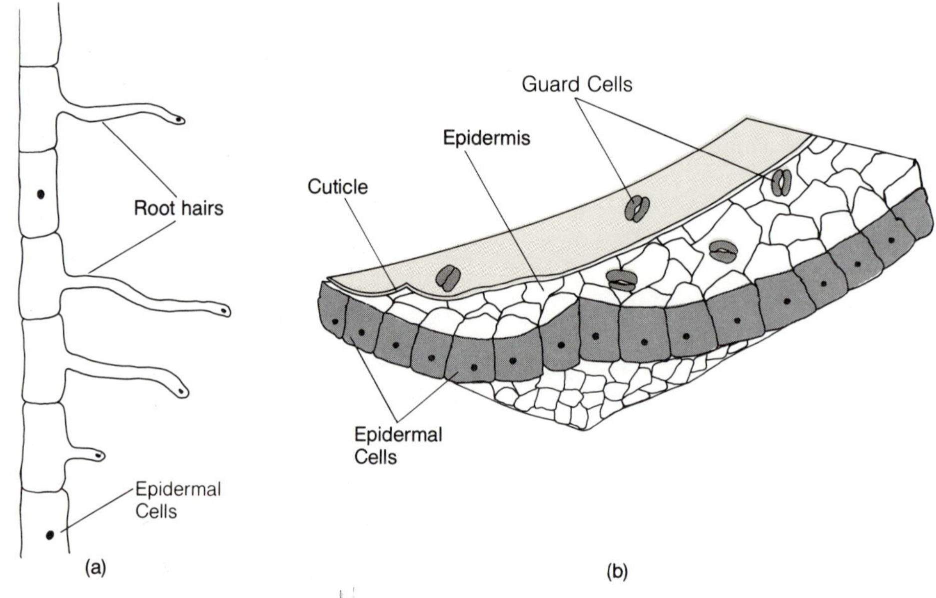

Figure 5–1. Epidermis. *(a)* Root hairs are tubular extensions of epidemal cells. *(b)* A sheet of epidermis in a cutaway view. The outside wall of each epidermal cell is thicker than all other walls. Cuticle is the waxy layer on top of the epidermis. Stomata are small openings surrounded by specialized epidermal cells called guard cells.

Parenchyma

Parenchyma tissue consists of **parenchyma cells** and is typically derived from the ground meristem. Parenchyma cells are found in large numbers in herbaceous plants and leaves of plants. In addition, parenchyma cells often differentiate further producing more specialized cell types such as collenchyma, fibers, sclereids, pericycle, and others. Parenchyma cells are alive with relatively thin primary walls; they are usually loosely arranged with considerable intercellular space, a feature that is used to recognize this tissue (Figure 5–2). Except for certain columnar cells that are present in many leaves, the shape of the majority of parenchyma cells is somewhat polyhedral or isodiametric. Most parenchyma cells of the leaves and some of them in the stems contain numerous chloroplasts and function in the synthesis of food by photosynthesis. All of the other parenchyma cells in the stems and those in roots have colorless plastids that, in many cases, function in the storage of starch or other reserve foods. In addition, parenchyma cells in succulent* plants, such as cacti, accumulate large amounts of water, making it possible for these plants to survive in a dry environment.

Succulent
A plant that stores water in its leaves or stem.

Collenchyma

Collenchyma tissue is made up of living **collenchyma** cells derived from the ground meristem and is found in relatively small amounts next to the epidermis in stems and leaves of certain plant species. Each collenchyma cell is elongated, with unevenly thickened primary walls (Figure 5–3). The uneven thickness is more apparent in a cross section of the tissue, showing that the walls of the collenchyma cells are thicker at the corners where two or more cells come together than they are at other parts. As a result of this irregular thickening of the walls, there are no intercellular spaces within this tissue, a feature that distinguishes it from parenchyma. The primary function of collenchyma is to provide mechanical support. Some collenchyma cells are known to contain chloroplasts, in which case they may also have a secondary function of food synthesis. Collenchyma cells are often present in stems and leaves in which other supporting tissues are absent or sparse as in the case of celery.

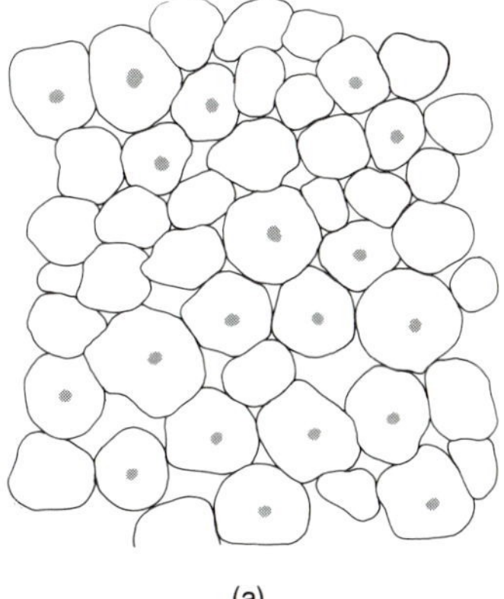

(a)

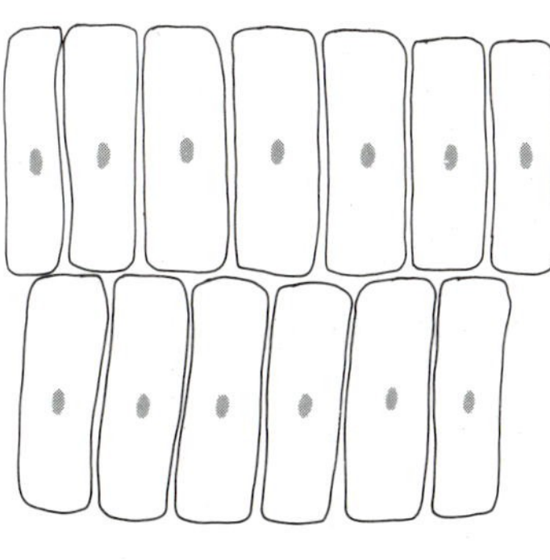

(b)

Figure 5–2. **Parenchyma tissue.** *(a)* Parenchyma cells are irregularly shaped. A large amount of intercellular space is characteristic of this tissue. *(b)* Vertically elongated parenchyma cells (palisade parenchyma) are typical of the leaves of dicotyledonous plants.

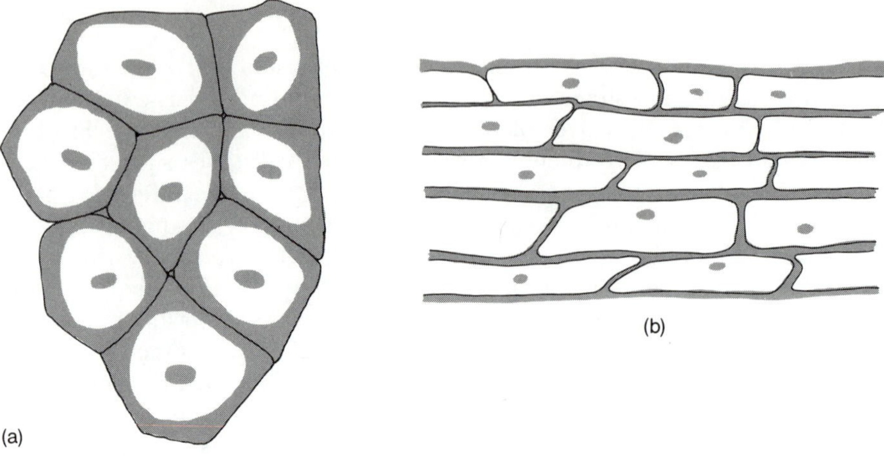

Figure 5–3. Collenchyma tissue. *(a)* The characteristically thickened corners of collenchyma cells are visible in a transverse section. *(b)* A longitudinal section shows that collenchyma cells are elongated vertically.

Sclerenchyma

The term **sclerenchyma** is applied to a simple tissue made entirely of either **fiber cells** or of **sclereid cells.** Notice that so far the tissue name and the cell type name have been the same, but here is a tissue name that is different from the cell type that makes up the tissue. Such differences usually will be encountered only in complex tissues. Sclerenchyma is still a simple tissue because wherever it is found, it consists of only one of the two cell types, fibers or sclereids.

Sclerenchyma is also derived from the ground meristem, but is a tissue whose cells die when they reach maturity. The most common sclerenchyma cell type is the fiber. Fiber cells are long and slender with tapering ends and a narrow **lumen** (Figure 5–4). The secondary wall of a fiber cell is very thick compared with its primary wall and their relative sizes can be recognized in a cross section. Since fibers can be several inches long, the presence of these cells give flexibility in addition to support to the plant body. Although similar in structure, the fibers associated with the vascular tissues have a different origin than the sclerenchyma fibers and will be explained later in this chapter.

Unlike fibers, sclereids are of varied shapes and sizes and are primarily responsible for the hardness and brittleness of certain plant structures such as the fruit walls of walnuts, coconuts,, almonds, and so forth. The gritty texture of pears is also due to numerous sclereids, generally known as stone cells, in the fruit tissue (Figure 5–5). In addition, sclereids are widely distributed in the stems and leaves of many plants. The sclereid cells have thick secondary walls and easily recognizable PITS.*

Pit
A small opening in the secondary wall of a cell.

Suberin
A wax produced as a cell wall component of cork and endodermal cells.

Transverse
A plane at right angles to the long axis.

Radial
A plane parallel to the long axis that passes through the center.

Endodermis

The endodermis usually occurs as a single layer of endodermal cells that is always present in the water-absorbing regions of roots; however, it is seldom seen in stems. All endodermal cells have a band of suberin*, the **casparian strip,** incorporated into the primary wall along the transverse* and radial* axes (Figure 5–6). Where the endodermal cells are arranged in the form of a hollow cylinder, as in a root, the Casparian strips of adjacent cells fuse. As

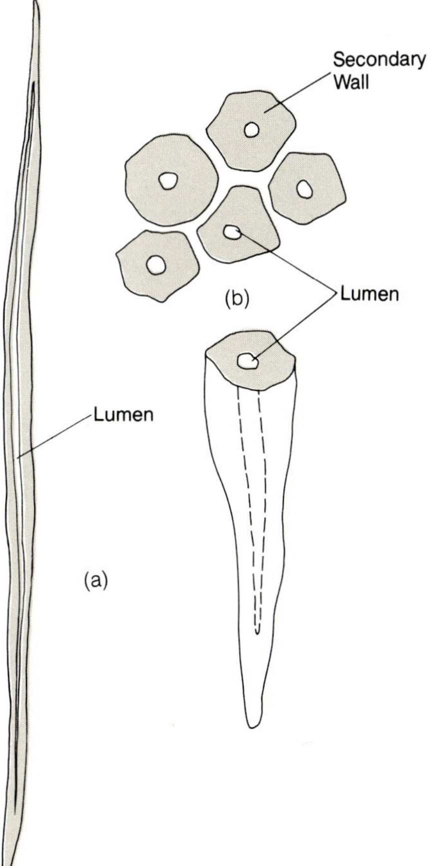

Figure 5–4. Fiber cells. *(a)* Longitudinal view. *(b)* Transverse view showing small lumen and thick secondary wall.

a result, any passage of materials from the outside to the inside of the endodermal cylinder has to be through the protoplasts of the endodermal cells and not through the space between cells (Chapter 11). Endodermal cells, having a three-dimensionally rectangular shape, are derived from the ground meristem. In a cross section of a root from a region where the primary tissues are differentiated, the endodermis often consists of a ring of thick-walled cells with a few thin-walled cells positioned opposite the water conducting tissue. The thickened cells with their suberized walls prevent any movement of water through them, while passage of water takes place through the thin-walled **passage cells.** Thus the function of the endodermis is to direct the flow of water and dissolved minerals into the water conducting tissue. See Chapter 7 for further discussion of the endodermis.

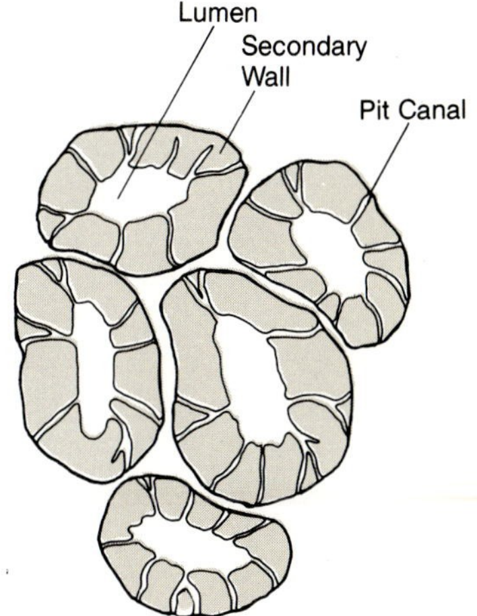

Figure 5–5. Sclereid cells. Each cell has many pit canals in its thickened secondary wall.

Pericycle

Pericycle is readily recognized in root sections and usually consists of a single layer of parenchymalike cells derived from the procambium. In the root it is found between the endodermis and the conducting tissues and serves to produce branch roots. Pericycle is not usually present in the stems of seed plants.

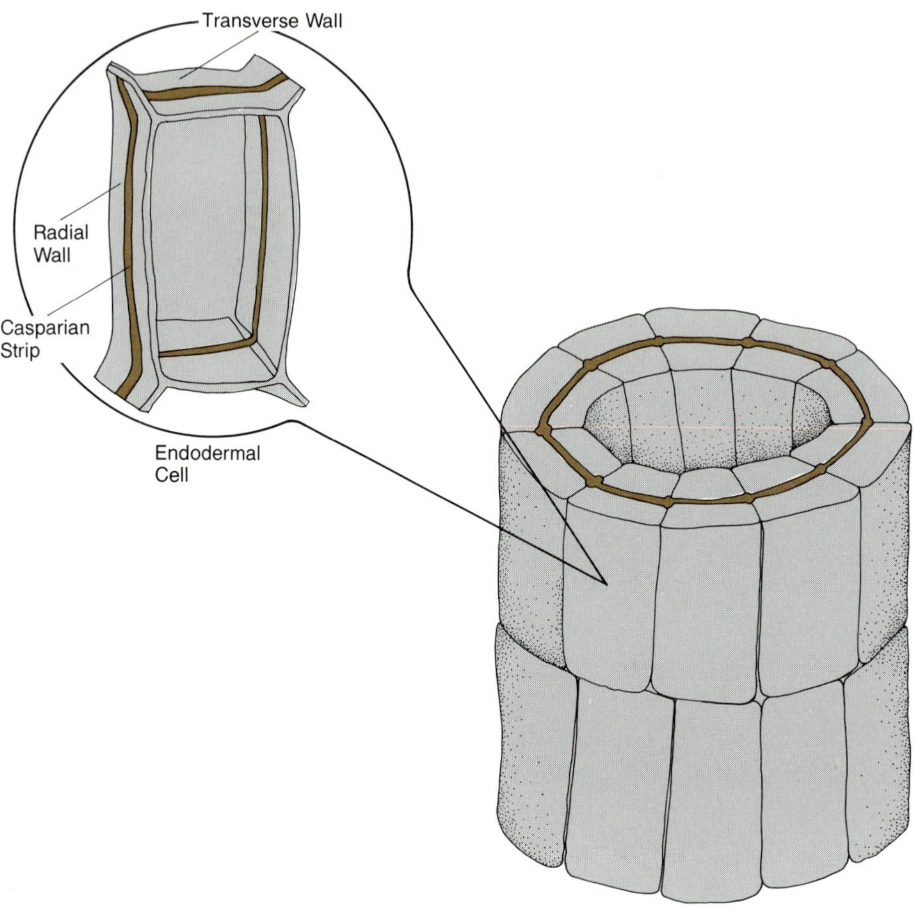

Figure 5–6. Casparian strip. The Casparian strips along the radial walls of adjacent endodermal cells fuse, forming a watertight barrier. Within the root, the endodermal cylinder has cells arranged on top of each other where the Casparian strips are also fused. This forces water to pass through the protoplasts of the endodermal cells and prevents it from passing between the cells.

Primary Xylem

The tissues discussed thus far are simple tissues because each tissue is made up of a group of cells that are of the same cell type. **Primary xylem,** on the other hand, is a complex tissue derived from the procambium and functions in water transport within the plant. Four different types of cells (not tissues) usually are seen in this tissue: **vessel elements, tracheids, fibers,** and **parenchyma cells** (Figure 5–7). Of these, the first three are dead at maturity, with the parenchyma being the only living cells in the xylem. Therefore, there are many more dead cells than living cells in mature xylem tissue.

Vessel elements are short, hollow, cylindrical cells with unevenly thickened secondary walls. One form of secondary thickening seen in vessel elements is helical. In the xylem tissue, vessel elements are arranged end-to-end to form a pipe system. Thus a functional unit is the **vessel,** made up of several vessel elements. Tracheids are narrower and longer than vessel elements, with more obliquely formed end walls. Often both vessel elements and tracheid have prominent pits on their side walls. The shape and structure of xylem fibers and parenchyma are similar to the ones described earlier. The different cell types are arranged in the tissue in such a way that the vertical transport of water takes place through the vessel elements and tracheids,

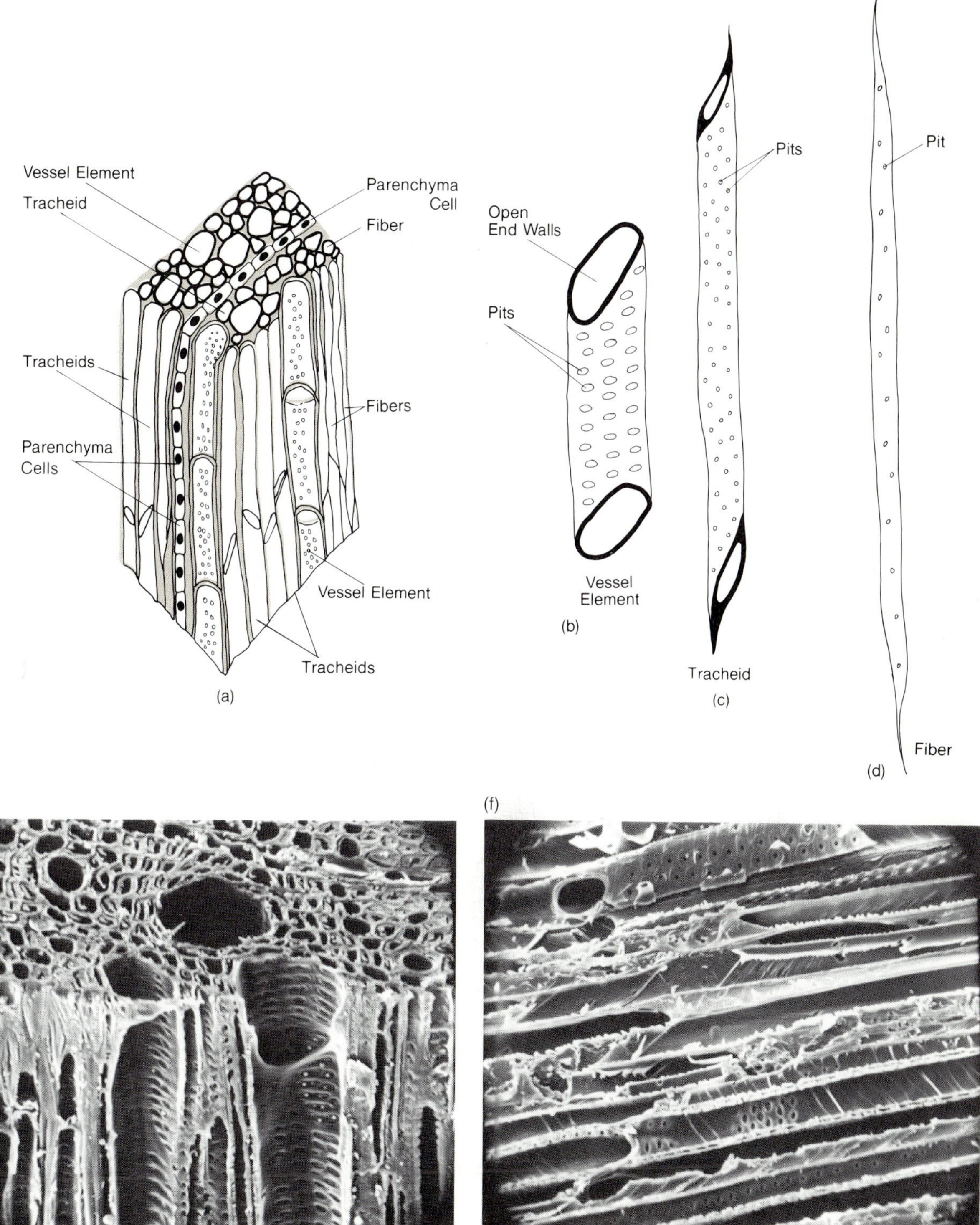

Figure 5–7. Xylem tissue. *(a)* Xylem tissue viewed in three dimension. *(b)* Vessel element. *(c)* Tracheid. *(d)* Fiber. *(e)* Scanning electron micrograph of xylem tissue. *(f)* Scanning electron micrograph of tracheids.

while the lateral movement of water takes place through the parenchyma cells.

Primary Phloem

The **primary phloem** is another complex tissue derived from the procambium. It functions in food transport and consists of four different cell types: **sieve tube members, companion cells, parenchyma cells,** and **fiber cells** (Figure 5–8). Phloem is primarily a living tissue with fibers being its only nonliving components.

Sieve tube members are elongated, tubular cells with **sieve plates** in the obliquely arranged crosswalls. Each sieve plate has many openings that permit extensive cytoplasmic bridges between cells thus facilitating movement of materials between sieve tube members. A sieve tube member is an unusual cell in the sense that when mature, the protoplast has no nucleus or other organelles except plastids and mitochondria. The metabolism of the sieve tube members is believed to be regulated by the companion cells closely associated with them. There are extensive protoplasmic connections between

Figure 5–8. Phloem tissue. *(a)* Phloem tissue viewed in three dimension. *(b)* A sieve tube member with associated companion cells.

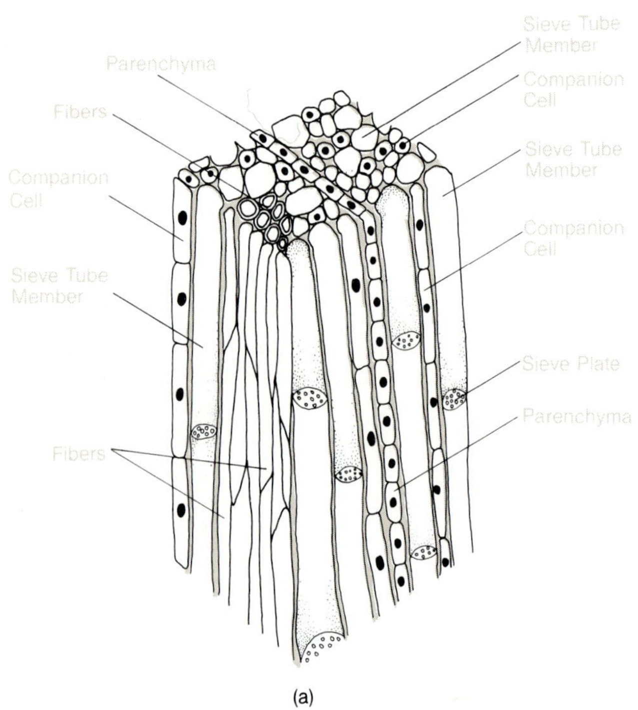

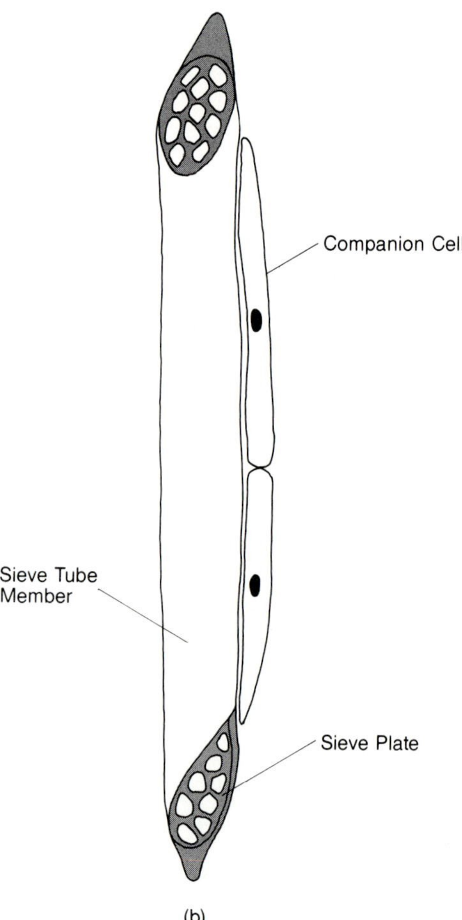

a sieve tube member and its companion cells. Although companion cells are elongated, they are much shorter than the sieve tube members. Several sieve tube members attached end-to-end make up a **sieve tube.**

The structure of fibers and parenchyma in the phloem is similar to the ones described earlier. The fiber cells in primary phloem are usually grouped together along the outside of this tissue, while in the secondary phloem they are present in clusters within the tissue. The parenchyma cells are scattered throughout the tissue and function in storage and lateral transport.

Secondary Meristems and Tissues

The above described tissues are all derived from the primary meristems and are therefore primary tissues that are produced in the growing tips of stems and roots. In woody plants, however, secondary tissues are continually produced along the full length of the plant by the two secondary meristems, the **vascular cambium** and the **cork cambium.**

Vascular Cambium

Vascular cambium consists of a single layer of brick-shaped, thin-walled cells between the xylem and the phloem (Figure 5–9). Especially in woody plants the cells of the vascular cambium undergo cell division regularly. Since un-

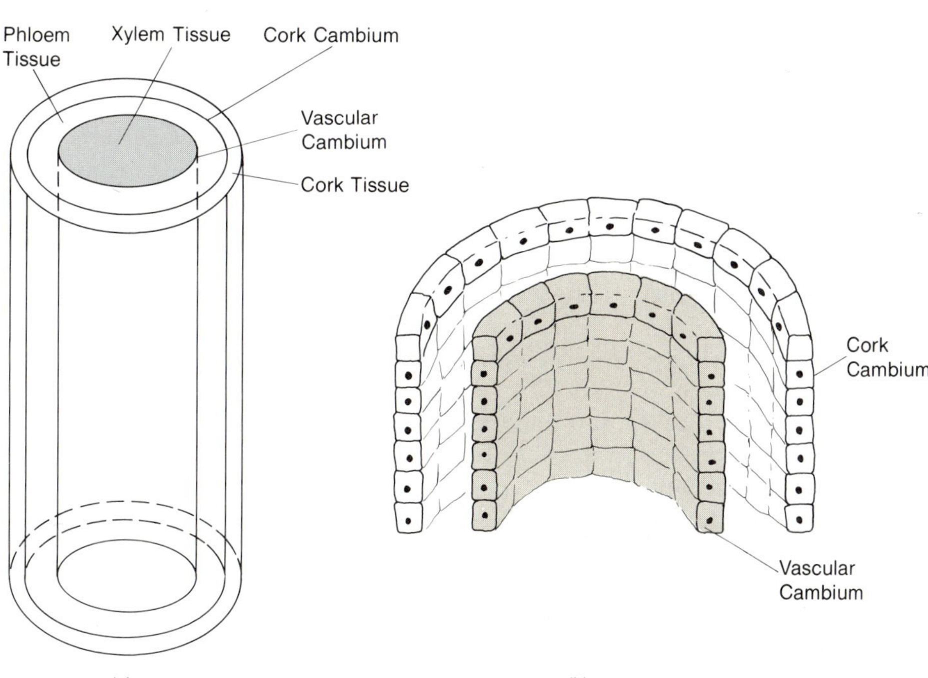

Figure 5–9. Vascular and cork cambia. Both the vascular cambium and the cork cambium may be thought of as two concentric cylinders, each one cell layer thick. The brick-shaped vascular cambial cells that produce xylem tissue toward the inside and phloem tissue toward the outside are arranged with their long axes parallel to the stem. The cork cambium produces compact rows of cells that form the cork tissue to the outside. To the inside, one or more rows of parenchyma cells are produced, forming the phelloderm.

differentiated newly formed cells are found on either side of the cambium, it is often difficult to recognize the true vascular cambial layer. Therefore, the area between the xylem and phloem is commonly called the vascular cambial zone. By definition, the vascular cambium is a primary tissue since it is produced by a primary meristem, the procambium. However, by virtue of the fact that vascular cambium is a meristem, but is not derived directly from the apical meristem, it is considered a secondary meristem. If it were not for the meristematic activity of the vascular cambium in trees, the lumber and other wood products that humans have come to depend on so much would not exist. The cambial cells that give rise to new cells are called **initials** and the cells derived from the initials are the **derivatives.** The derivatives produced toward the outside differentiate to become cells of the secondary phloem and the derivatives toward the inside differentiate to become cells of the secondary xylem. A cambial initial can only produce either a phloem derivative or a xylem derivative at one time—one of the new cells always remains as an initial to maintain the integrity of the vascular cambial layer.

The secondary xylem and secondary phloem are complex tissues with the same composition of cell types present in the primary xylem and primary phloem, respectively. The function of xylem, whether primary or secondary, is transport of water, and, similarly, the phloem functions in the transport of food.

Cork Cambium

Cork cambium is a secondary meristem composed of a single layer of brick-shaped, thin-walled cells arranged in a circular form toward the periphery of woody stems and roots (Figure 5–9). In the stem the cork cambium is derived from the parenchyma, epidermis, or even phloem, and in the root it is usually derived from the pericycle. As some stems and roots age, the original cork cambium is replaced by a newly formed cork cambium internal to the original one, resulting in the death of all tissues outside of the newly formed cork cambial cylinder. The cork cambial initials are capable of producing derivatives toward the outside and inside. The derivatives on the outside differentiate into brick-shaped **cork cells** whereas the ones toward the inside differentiate into parenchymalike **phelloderm cells**. Only a small number of phelloderm cells are usually produced, however, resulting in a phelloderm tissue seldom thicker than two layers. Although phelloderm cells are alive when mature, the function of the phelloderm is not understood at present.

Cork tissue, when present, is always the outermost tissue, consisting of many layers of brick-shaped cork cells arranged on top of each other with no space between cells. Mature cork cells are nonliving, with highly suberized walls that prevent water loss from the internal tissues. Cork may, in addition, protect the internal tissues from extremes of temperature.

Now that most of the tissues present in a plant body have been described, these diverse tissues can be grouped into four broad categories based on their function: the **meristems, ground tissues, vascular tissues,** and **dermal tissues.** The meristems, including both primary and secondary, are responsible for the production of other tissues. The ground tissues, made up of the parenchyma, collenchyma, and sclerenchyma, are responsible for synthesis, storage, and mechanical support. The vascular tissues, the xylem and phloem, have the primary function of transport and are organized in specific patterns or randomly scattered within the ground tissue. The dermal tissues, consisting of epidermis and cork, are found toward the outside of a plant organ and have several functions of which one is protection from water loss.

The specific arrangement of these tissues varies with different plant groups and different plant organs and is discussed in the next three chapters.

Summary

A multicellular plant body is made up of many different tissues arranged in specific patterns and with continuity between different plant organs such as roots, stems, and leaves. A tissue is a functional unit composed of a group of similar cells or a mixture of different cell types, the former being a simple tissue and the latter a complex tissue. All the cells in a simple tissue and one or more types of cells in a complex tissue may be alive or dead at maturity. Plant tissues composed of dead cells are still functional, unlike those in animal tissues.

The origin of plant tissues can be traced back to the apical meristem found in the shoot and root apices. The apical meristem gives rise to three primary meristems—protoderm, procambium, and ground meristem—which produce the primary tissues (Table 5–1). Epidermis, the outermost tissue on young parts, is derived from the protoderm. Parenchyma, collenchyma, sclerenchyma, and endodermis are produced by the ground meristem. The vascular tissues, primary xylem for the transport of water and primary phloem for the transport of food, are derived from the procambium. In addition, procambium also produces vascular cambium and pericycle.

Secondary tissues (Table 5–2) are produced by two secondary meristems, the vascular cambium and cork cambium. Secondary xylem and secondary phloem are derived from the vascular cambium, while cork and phelloderm are produced by the cork cambium. Mature cork cells have suberized walls and are dead at maturity. Xylem and phloem, whether of primary or secondary origin, are complex tissues composed of four cell types each: vessel elements, tracheids, fibers, and parenchyma in xylem; sieve tube members, companion cells, fibers, and parenchyma in phloem.

While primary tissues are responsible for growth in length, secondary tissues are produced throughout the mature regions of all woody plants and some herbaceous plants and result in growth in diameter.

Based on function, the four major categories of tissues are **meristems** (both primary and secondary) for the production of new tissues, **ground tissue** (parenchyma, collenchyma, sclerenchyma) for food synthesis, storage,

Table 5–1
Primary Tissues

Primary Tissue	Origin	Nature	Cell Types	Major Function
epidermis	protoderm	simple, living, usually one layer	epidermal cells	protection
parenchyma	ground meristem	simple, living, multilayered	parenchymal cells	storage, synthesis
collenchyma	ground meristem	simple, living, multilayered	collenchymal cells	support
sclerenchyma	ground meristem	simple, nonliving multilayered	fiber cells or sclereid cells	support, flexibility, hardness
endodermis	ground meristem	simple, composed of living and nonliving cells, one layer	endodermal cells	directs passage of water and minerals into water conducting tissue in roots
pericycle	procambium	simple, living, one layer	parenchymalike cells	production of branch roots
primary xylem	procambium	complex, mostly nonliving, multilayered	vessel elements tracheids parenchyma cells fiber cells	water transport
primary phloem	procambium	complex, mostly living, multilayered	sieve tube members companion cells parenchyma cells fiber cells	food transport
vascular cambium	procambium	simple, living one layer	cambial initials	production of secondary xylem and secondary phloem

Table 5–2
Secondary Tissues

Secondary Tissue	Origin	Nature	Cell Type	Major Function
secondary xylem	vascular cambium	complex, mostly nonliving, multilayered	vessel elements tracheids parenchyma cells fiber cells	water transport
secondary phloem	vascular cambium	complex, mostly living, multilayered	sieve tube members companion cells parenchyma cells fiber cells	food transport
cork	cork cambium	simple, nonliving, multilayered	cork cells	protection
phelloderm	cork cambium	simple, living, one to two layers	parenchymalike cells	unknown

and mechanical support, **vascular tissues** (both primary and secondary xylem and phloem) for transport of water and food, and **dermal tissue** (epidermis and cork) for protection.

Review Questions

1. Name the different primary meristems and the tissues produced by each.
2. What is the difference between a primary tissue and a secondary tissue?
3. Name the different secondary tissues present in a woody plant.
4. List the primary function of the following tissues: cork, phloem, collenchyma, epidermis, xylem, parenchyma
5. Mature plant cell must be living and capable of carrying out metabolism in order to perform their functions. Is this statement true? Explain.
6. Define the term tissue.

Tissues of Commercial Value

Plant tissues such as cork and xylem have had considerable economic importance throughout history. Cork used as stoppers for bottles is obtained from a species of oak *(Quercus suber)* that is native to the Mediterranean region, especially Spain and Portugal. These trees produce numerous cork cells that form a thick, spongy, deeply grooved layer that can be stripped once every 10 years without harm to the trees. The commercial life of a cork oak is 150 years or more.

Another tissue of high commercial value is secondary xylem, known popularly as wood. The lumber and paper industries depend on xylem, the former using the intact tissues to make lumber while the latter group utilizes macerated xylem or wood pulp to make paper.

Fiber cells are yet another commercially important cell type. A rope must have been the oldest device used to lift heavy objects or to hoist the sails on a ship or even to tie the ship to a pier. Different types of ropes are still very much in use. There was a time when all ropes were made of fiber cells and even today many types of ropes are made from plant fibers that give them high flexibility and relatively great tensile strength. Fiber cells from leaves and stems of hemp, sisal, and flax and from the fruit of coconut are commonly used in making ropes. Some of these fibers associated with the vascular tissue, called bundle fibers, reach lengths of many inches and are ideal for rope making. Some ropes and mats are fashioned out of fibers, referred to as coir, from the fruit wall of coconut. The word *coir* is an anglicized version of “kayar,” meaning “rope” in Malayalam, the language of one of the South Indian states where rope making has been practiced since ancient times. The fibers that are used in rope making are true fibers that have highly lignified secondary walls. In contrast, the fibers used in making cotton fabrics are dried-up hairs present on the seed coat of cotton plants and consist of cellulose primary walls only. The correct term to describe these so-called fibers is *lint,* the longest of which seldom attain lengths of more than 2–5 inches.

6

Stems

Outline

Introduction

The word *plant* conveys to most people a mental image of the stems and leaves that comprise the aerial **shoot system** of a plant body. Because of the growth pattern of most perennial plants inhabiting temperate zones, the stem is the most long-lived above-ground organ. Consequently, people have long known that the age of a particular tree can be determined by counting the annual growth rings near the base of the trunk. Only within the past several decades, however, has that knowledge been put to a more sophisticated use—that of dating the remnants of ancient civilizations and gaining new insights on past climatic patterns of a given region.

The relatively new science of dendrochronology, based on large-scale statistical tree-ring analysis, was pioneered by the late Edmund Shulman and has been continued at the Laboratory for Tree-Ring Research of the University of Arizona at Tucson. Basing their investigations on the observation that certain tree species exhibit marked variations in the width of their annual growth rings depending on climatic conditions during any given year, researchers developed methods for compiling complete climatological records extending more than 4000 years into the past. A sudden increase in the width of growth rings indicates unusually wet growing seasons or reduced competition, while a sudden decrease in width suggests periods of drought. Working primarily with bristlecone pines, the oldest of all tree species, Shulman extracted cores from living trees and found certain easily recognizable combinations of narrower and wider rings occurring in many different individual samples, an indication of specific climatic cycles occurring at various times in the past. Core samples are taken with a long drill equipped with a hollow, tubelike bit. The resulting rodlike cores contain a continuous sample of the bark and wood extending to the center of the stem and including every growth ring of the tree. By counting the number of annual rings from the outermost portion of the trunk back to the beginning of an unusual pattern of rings, the precise date of a given climatic episode could be established. In this manner it was possible to work out a complete climatological chronology of the southwestern United States extending four millennia back into time.

Other investigators have expanded on these dating techniques by comparing the distinctive ring patterns of ancient living trees with identical patterns found in the supporting timbers of long-abandoned, pre-Columbian Indian ruins. By this method scientists were able to extend the tree-ring calendar back more than 6,600 years and could determine precisely the dates when the ancient dwellings had been constructed. Such work also provided a plausible answer to the riddle concerning why the Pueblos were abandoned by their inhabitants: tree-ring analysis has revealed that in A.D. 1290 a severe drought, which lasted 24 years, devastated the American Southwest, probably forcing the ancient cliff dwellers to seek new homes elsewhere or perish.

Tree-ring analysis is done on the older parts of woody stems and is a measure of the growth in diameter of these stems. The initial growth and all elongation of the stem, however, is due to the activity of the **apical bud** found at the tip of the stem (Figure 6–1). The apical bud of the stem gives rise to all of the components of the shoot system: stem, leaves, reproductive organs, and axillary buds found in the axils* of leaves. While the apical bud is responsible for the continued elongation of a stem, the **axillary buds,** otherwise known as **lateral buds,** produce the **branches.** Each axillary bud, under the proper conditions, is capable of producing a branch that, in turn, becomes a shoot system with its own stem, leaves, buds, and reproductive organs.

Axil
The angle formed between the upper side of the petiole and the stem.

The shoot system of some plants can be extensive, covering a large area, whereas in others it is limited. The role of the stem is to maximize the exposure of leaves to sunlight, to conduct water and minerals from the roots to the leaves, and to transport food from the leaves to the roots. As can be seen from this description of its role, the stem is not a plant organ of fundamental importance; rather, the stem seems to function mainly as a place to hang leaves and as a connecting link between leaves and roots. As discussed in Chapter 18, the stem probably evolved as an adaptation that allowed plants to grow vertically instead of horizontally, thus allowing a larger plant body to occupy a smaller amount of surface space.

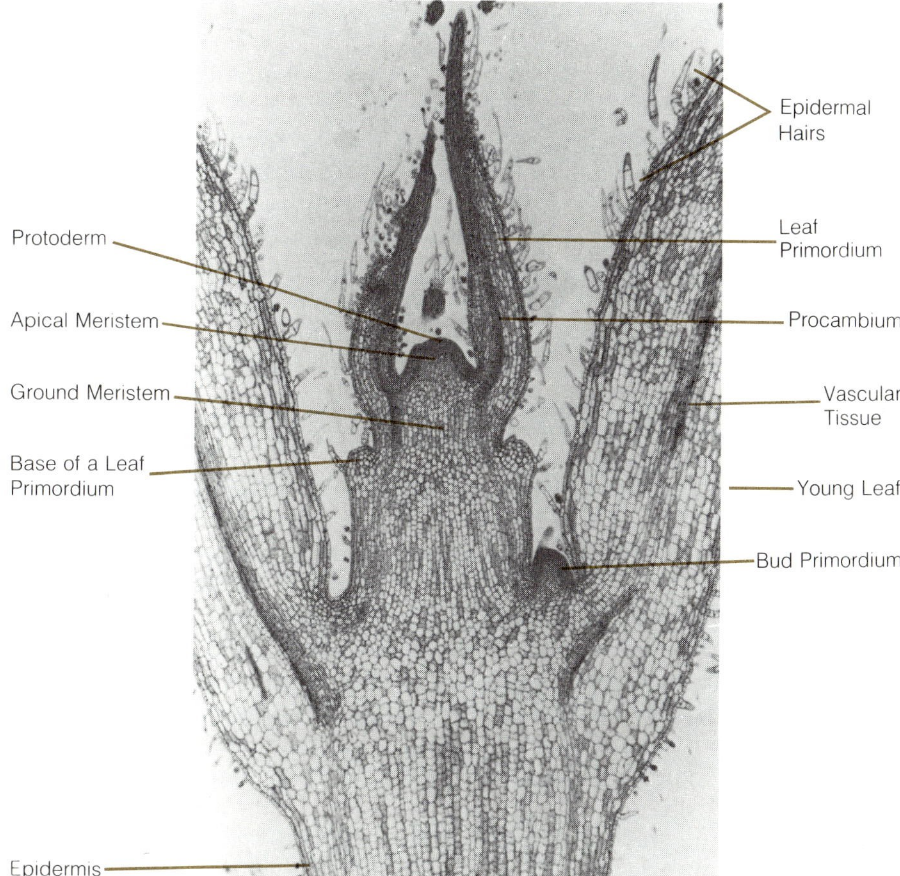

Figure 6–1. Dicotyledonous shoot tip. Longitudinal section through the shoot tip (bud) of *Coleus*. The arrangement of the apical meristem and the primary meristems as seen here is similar in all shoot buds of dicotyledonous plants.

Annual
A plant that completes its life cycle in one growing season.

Biennial
A plant that requires two growing seasons to complete its life cycle.

Perennial
A plant that lives for many years.

Xylem
Water-conducting tissue.

Vascular cambium
A secondary meristem that produces vascular tissues (xylem and phloem).

Secondary
Tissues produced by secondary meristems such as vascular cambium and cork cambium.

Phloem
Food-conducting tissue.

Cork cambium
A secondary meristem that produces cork and phelloderm.

Cork
A secondary tissue containing suberin that forms the outer layer of woody stems and roots.

Monocotyledonous plant
A flower-producing plant whose seeds have only one cotyledon.

Meristem
A tissue whose cells are cabable of repeated division.

Dicotyledonous plant
A flower-producing plant whose seeds have two cotyledons.

Longitudinal
A plane of section that passes through the long axis.

Primary meristem
A meristem responsible for the production of primary tissues.

In most plants the stems are strong enough to remain in an upright position without any additional support. It is not uncommon, however, to see weak-stemmed plants such as Swedish ivy, which grows along the soil surface, or morning glory, which grows upright only by becoming attached to an upright support. Although more often than not the entire shoot system is found above ground, there are some plants in which either a portion of the stem or the entire stem with highly reduced leaves is found below ground.

On the basis of their growth pattern seed-producing plants can be divided into **herbaceous plants** and **woody plants.** Many herbaceous plants of the temperate regions are annuals* or biennials*, although some are perennials*, and contain relatively small amounts of xylem* tissue, whereas the woody plants are all perennials and accumulate large amounts of xylem. Naturally most herbaceous plants are relatively small when compared to **shrubs** or **trees.** Shrubs are woody plants that have several stems of approximately equal size originating at soil level. Trees, by contrast, have one main stem, often called the **trunk,** which usually branches only after it rises several feet above ground level. The stems of woody plants increase in diameter as they grow older. This increased diameter is related to the activity of vascular cambium*, which produces secondary* xylem and phloem*, and to that of cork cambium*, which produces cork.* In herbaceous plants the vascular cambium is relatively inactive, and since most herbaceous plants do not have a cork cambium, they contain only a small amount of secondary tissue. In addition, the monocotyledonous* herbaceous plants, such as the grasses, lack a vascular cambium and hence do not produce any secondary tissues. Meristematic activity in plants is influenced by environmental factors, especially temperature. Plants growing in tropical and subtropical areas continuously produce new tissues, albeit at reduced rates during dry seasons, while plants of the temperate areas stop growing during the winter months; however, the meristems* in these plants resume activity in the spring.

Anatomy of the Shoot Apex

The shoot apex is the growing point of the shoot system, and, because it produces stem, leaves, and buds, its structure is more complex than that of the root tip (Chapter 7). Although a distinction is made between herbaceous and woody dicotyledonous* plants on the basis of the amount of secondary growth, the structure of the shoot apex is the same in both cases. Figure 6–1 is a longitudinal* section through the shoot apex of a typical dicot (also see Plate 1A). The **apical meristem** is the small group of cuboidal cells arranged in the shape of a dome at the very tip. This group of cells divides and gives rise to the primary meristems* as well as to **leaf primordia*** and **bud primorida.*** The layer of cells that covers the apical meristem, leaf primordia, and bud primordia is the **protoderm,** a primary meristem that differentiates into the epidermis of the mature regions. The **procambium** is a primary meristem formed as lateral outgrowths of the apical meristem that continues into the leaf primordia as well as forming a continuous cylinder extending downward from the shoot apex near the periphery of the stem. Subsequently, the procambium differentiates into the primary* vascular tissues of the stem: the xylem and phloem. The remaining central core of cells immediately below the apical meristem and the cells present between the procambium and the protoderm comprise the primary meristem known as the **ground meristem,** which produces the primary tissues of the cortex* and pith.* A critical examination of the shoot apex will show a progressive enlargement and differentiation of cells from the apical meristem downward. There is also a series of leaf primordia along the shoot apex, the youngest being closest to the apical meristem. Leaves always originate in the shoot apex as leaf primordia and the position of these primordia determines the position of the **nodes*** on a stem. The elongation of the shoot apex resulting in the growth of the stem is primarily due to the elongation and enlargement of the cells in the internodal* regions of the shoot apex.

In herbaceous plants the elongation of the shoot apex may continue throughout the growing season, while in some woody plants stem elongation may be completed within the first few weeks of the growing season. This newest segment of stem can be recognized easily because of the presence of leaves on it, whereas leaves are absent from the segments from previous years.

The shoot apex of monocotyledonous plants is similar to that of dicots with certain minor differences. One of these is the scattered distribution of the procambium within the ground meristem, which reflects the distribution of the xylem and phloem within the monocot stem.

Leaf primordium
(*pl.*, primordia) A group of meristematic cells that give rise to a leaf.

Bud primordium
A group of meristematic cells that give rise to a bud.

Primary
Tissues derived from a primary meristem.

Cortex
A region of the stem or root consisting mainly of parenchyma cells, located between the epidermis or cork and the vascular cylinder.

Pith
A region composed of parenchyma cells at the center of the vascular cylinder.

Node
The region of a stem to which a leaf is, or was, attached.

Internode
(*adj.*, internodal) Section of stem between adjacent nodes.

Anatomy of Mature Herbaceous Stems

The internal organization of tissues in the mature parts of stems varies considerably depending on whether a stem is from a monocotyledonous plant or from a herbaceous or woody dicotyledonous plant. A comparison of cross sections* made from mature regions of herbaceous monocotyledonous stems, herbaceous dicotyledonous stems, and woody stems provides the best way to understand the differences in tissue arrangements.

Cross section
A plane of section at right angles to the long axis.

Herbaceous Monocotyledonous Stems

Corn *(Zea mays)* is commonly used as an example of a herbaceous monocotyledonous stem (Figure 6–2, 6–3 and Plate 1C). The outermost tissue on a corn stem is the **epidermis** with its cuticular* covering. Next to the epidermis are two or more layers of thick-walled, dead **sclerenchyma fibers,** which provide mechanical support for the stem. The remainder of the stem consists of living, thin-walled **parenchyma** cells and a large number of **vascular bundles*** scattered within the parenchyma tissue. Because of the abundance of parenchyma in the corn stem, it is called the **ground tissue** of the stem. Each vascular bundle consists of **primary xylem** and **primary phloem** sur-

Cuticle
(*adj.*, cuticular) The outer, waxy layer produced by the epidermis.

Vascular bundle
An aggregation of vascular tissues (xylem and phloem).

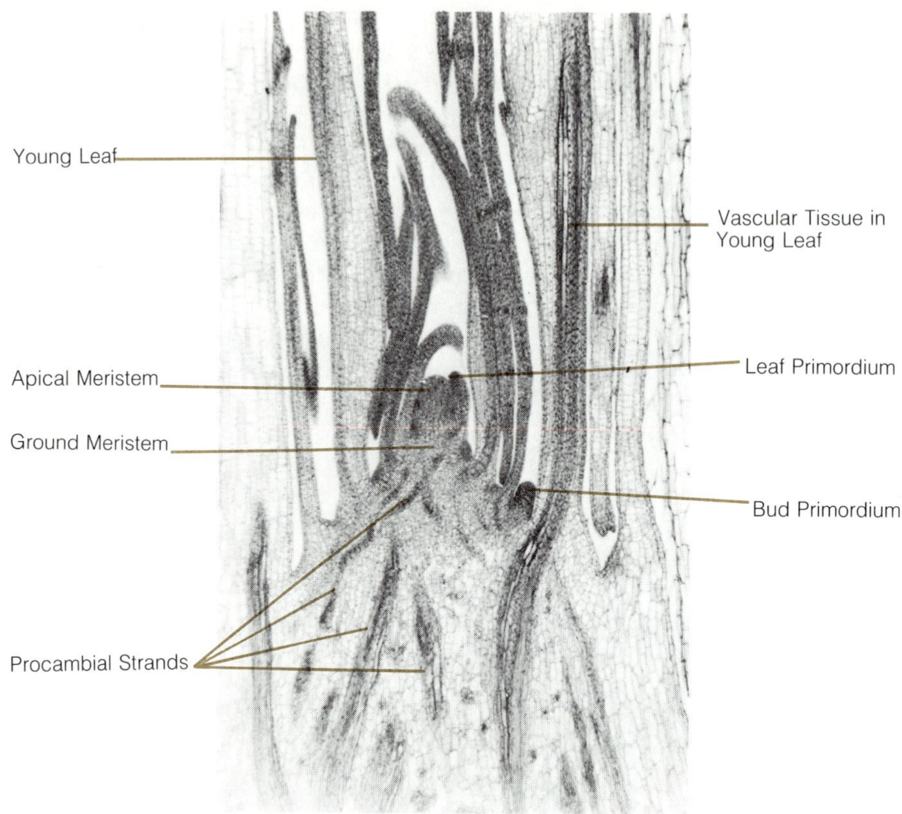

Figure 6–2. Monocotyledonous shoot tip. Longitudinal section through the shoot tip of corn, a plant belonging to the grass family. The difference between the dicot (Figure 6–1) and this section is due to the scattered distribution of the procambium seen here.

Peripheral
Toward the outside.

rounded by a **bundle sheath** consisting of fiber cells. The bundle sheath of the peripheral* vascular bundles is often continuous with the outer cylinder of sclerenchyma fibers, while the remaining vascular bundles are not connected to this cylinder. Within each vascular bundle (Figure 6–4) the phloem is arranged toward the outside and the xylem toward the inside of the stem. **Sieve tube members** and **companion cells** can be clearly recognized in the phloem, whereas only the large **vessel elements** of the xylem can be easily seen, even though **tracheids** and **parenchyma** cells also may be present. The three largest vessel elements are arranged in a triangular pattern with a conspicuous air space, the **lacuna,** below them. The lacuna results from the breakdown of a vessel element. With a bit of imagination, you can visualize the outline of a face in each of the bundles—the eyes and nose are represented by the large vessel elements, the mouth by the lacuna, forehead by phloem, and hair by the bundle sheath! The vascular system of the stem is continuous with the vascular systems of the leaves and roots, thus forming an interconnected transport system for water and food within the plant.

Since vascular cambium and cork cambium are absent from most monocotyledonous plants, such plants do not exhibit secondary growth. Nevertheless, there are monocotyledonous plants, such as the palms, in which the stems increase in diameter owing to division of a special cambium that produces thick-walled parenchyma cells rather than secondary xylem and phloem. In other monocotyledonous plants, such as yucca and aloe, second-

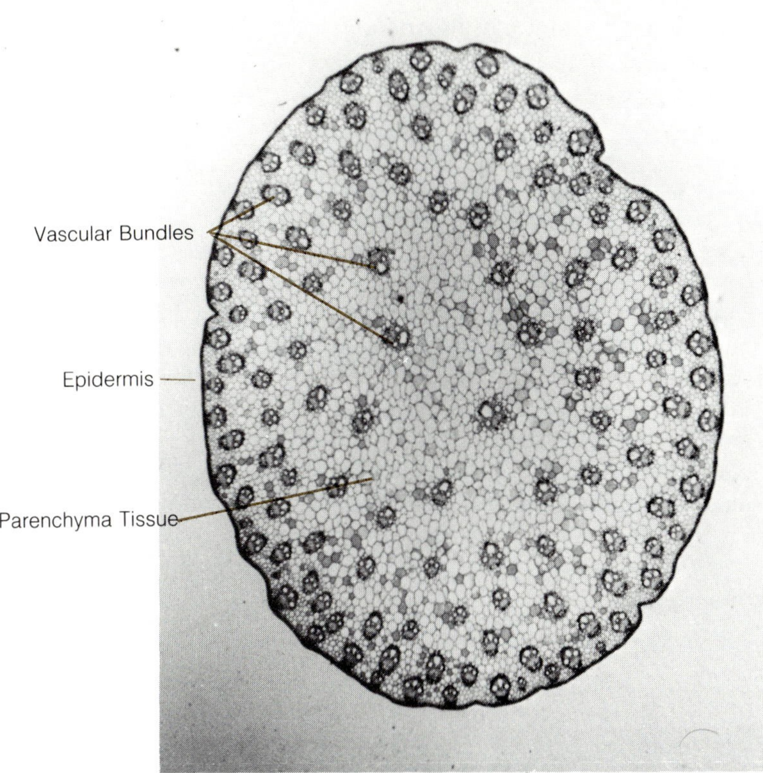

Figure 6–3. Transverse section of a herbaceous monocotyledonous stem. The vascular bundles are randomly distributed in this section of corn. Most of this stem consists of parenchymatous ground tissue.

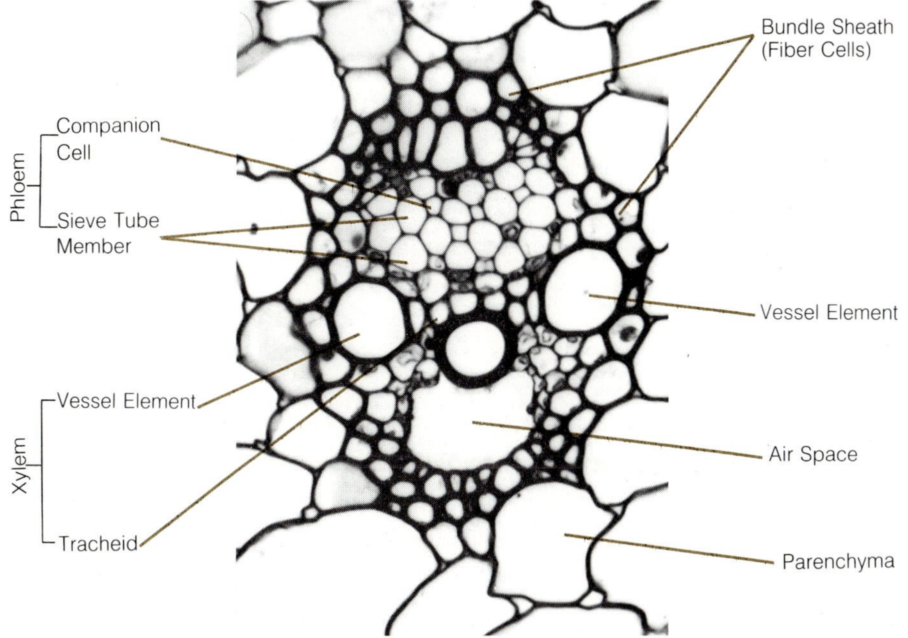

Figure 6–4. Transverse section of a herbaceous monocotyledonous stem vascular bundle. The phloem is on top and the xylem is below. Together they are surrounded by a bundle sheath of fiber cells.

ary growth through the activity of a nonspecific cambium produces secondary vascular bundles and parenchyma. Although most monocotyledonous shoot systems are annuals, the monocotyledonous shoots that exhibit secondary growth are perennials.

Herbaceous Dicotyledonous Stems

Sunflower *(Helianthus annuus)* stem is a classic example of a herbaceous dicotyledonous plant that completes its life cycle in one growing season (Figure 6–5 and Plate 1D). The outermost tissue of a sunflower stem is the **epidermis** with its cuticle. Next to the epidermis are one to several layers of **collenchyma** cells with their thickened corners, which provide support for the stem. The remaining area of the stem consists of **parenchyma** and a group of **vascular bundles** arranged in the form of a cylinder. Thus the cylinder of vascular bundles divides the stem into two distinct regions—**cortex,** a region between the vascular bundles and the epidermis, and **pith,** a central region inside the ring of vascular bundles. In the sunflower stem the cortex is made of collenchyma and parenchyma, while the pith consists of only parenchyma. The parenchyma that is continuous between the pith parenchyma and cortical parenchyma is the **pith ray,** which runs between two adjacent vascular bundles.

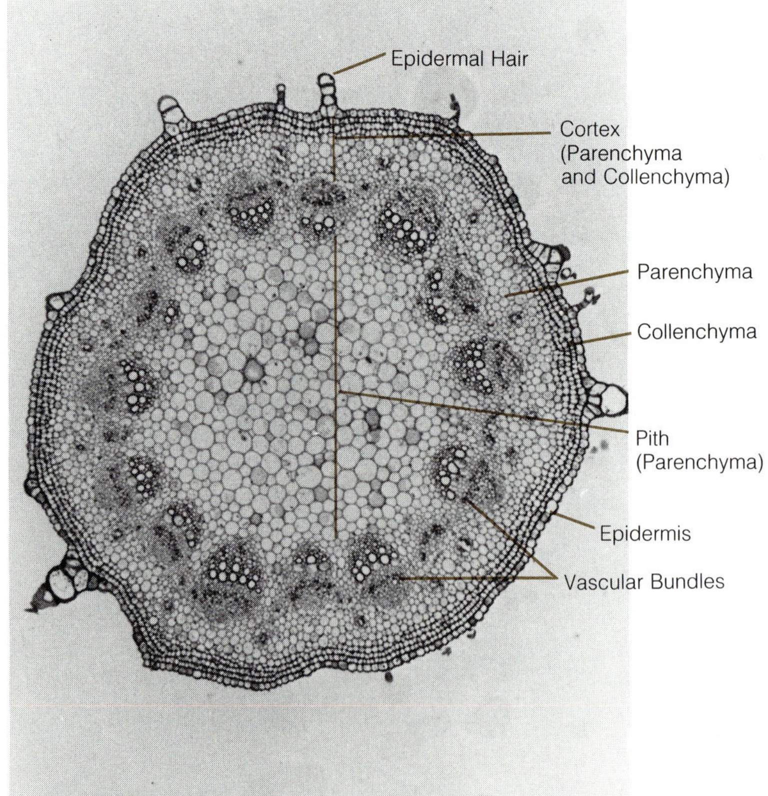

Figure 6–5. Transverse section of a herbaceous dicotyledonous stem. The vascular bundles are arranged in the form of a cylinder separating the stem into distinct regions, the outer cortex and a central pith.

The outermost part of the vascular bundle (Figure 6–6) is the **bundle cap** made of **phloem fibers.** Inner to the bundle cap is **phloem,** largely composed of **sieve tube members** and **companion cells.** The innermost part of the vascular bundle is the **xylem** with its **vessel elements, tracheids,** and **fiber cells.** Of these, the vessel elements are the easiest to recognize owing to their large size in a cross section. A row of brick-shaped cells found between the phloem and xylem is the **vascular cambium.** Since a few rows of cells on either side of the vascular cambium may appear similar in shape and size, it is very difficult to recognize which row specifically is the vascular cambium. To overcome this problem, several rows of brick-shaped cells between the xylem and phloem are called collectively the vascular cambial zone. In the more mature parts of a sunflower stem the vascular cambium is present in the form of a continuous cylinder. That portion of the vascular cambium present between the vascular bundles is called the **interfascicular** cambium while the **fascicular** cambium is the portion inside the bundles.

Many herbaceous dicotyledonous plants do not accumulate large amounts of secondary tissues. Buttercup is an example of a dicotyledonous herba-

Figure 6–6. Transverse section of a herbaceous dicot stem vascular bundle. The rectanglar cells of the vascular cambium are present between the phioem and the xylem. The bundle cap is above the phloem.

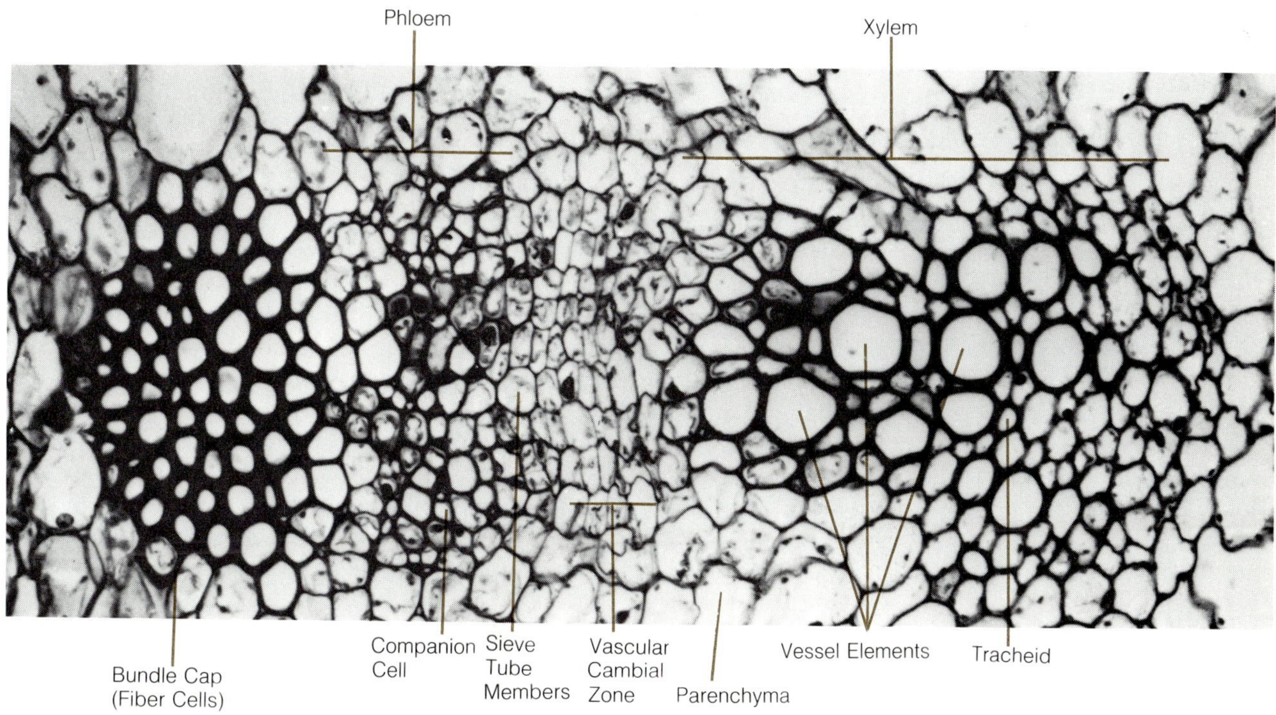

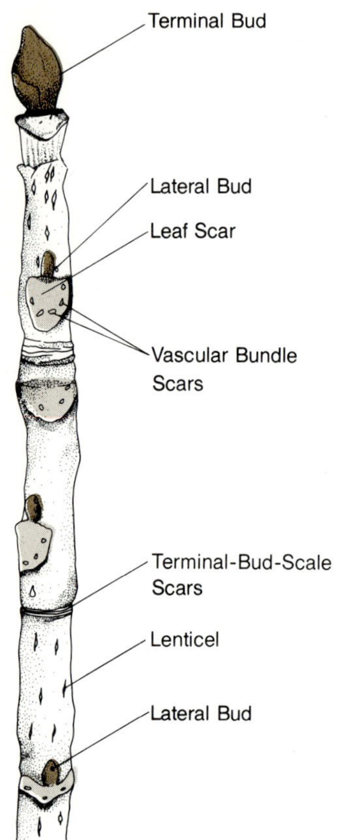

Figure 6–7. External characteristics of a stem collected during winter. The terminal-bud-scale scars represent the position of the terminal bud during previous years. Leaf scars and vascular bundle scars represent the positions of the leaves and the connection of the vascular tissue entering each leaf. Axillary buds give rise to future branches. Lenticels are small openings in the bark. The age of a branch can be determined by counting the number of growth segments between the terminal-bud-scale scars. This drawing represents three years of growth.

Angiosperm
One of a group of plants characterized by the production of flowers.

Gymnosperm
A plant that produces naked seeds, typically in cones.

ceous plant that produces no secondary tissues and thus is similar to a herbaceous monocotyledonous plant in this regard. On the other end of the scale are herbaceous dicotyledonous plants like certain species of hibiscus, the stems of which produce large amounts of secondary tissues and resemble those of woody dicotyledonous plants of similar age. Even many dicotyledonous annuals like the sunflower will become woody at the base of the stem by the end of the growing season. It is, therefore, difficult to draw a sharp distinction between dicotyledonous herbaceous plants and dicotyledonous woody shrubs. This is especially true in the case of many tropical and subtropical perennial herbaceous plants that accumulate secondary tissues and easily could be confused with shrubs. The term *herbaceous* is used by some botanical authorities in a broader sense than by others who would restrict it to a plant in which the cambial activity is lacking or limited to one season with greatly reduced accumulation of secondary vascular tissues.

Woody Stems

Almost all present-day woody plants are either dicotyledonous **angiosperms*** or **gymnosperms.*** In addition, some of the fossil representatives of more primitive groups were also woody (Chapter 18). The difference in the anatomy between the woody stem of angiosperms and that of gymnosperms is more in the presence or absence of certain cell types and structures than in the arrangement of tissues.

External Features of Woody Stems

If you examine a young branch of a temperate, woody plant such as walnut or buckeye, you will find a number of external features characteristic of a woody stem (Figure 6–7). At the very tip of the branch will be the **apical** or **terminal bud,** which is responsible for the continued elongation of the branch. During winter the bud is protected by several overlapping scales that fall off in the spring when the bud begins to produce new shoot growth. The scars left by the scales completely encircle the stem and are known as the **terminal-bud-scale scars.** The position of the terminal bud at the end of each previous year is demarcated by the terminal-bud-scale scars. Thus the age of branches on many species of woody plants can be determined by counting the number of segments separated by terminal-bud-scale scars (Figure 6–7). The length of segments produced during different years varies depending on the growth conditions for a particular year. This method may not work on older branches because, as bark develops, the outer tissues, including the scars, slough off. Also some species, such as pin oak, go dormant during the early summer and then a second set of scars results when growth resumes.

Leaf scars represent positions where leaves were attached to the stem during previous growing seasons (Figure 6–7). **Vascular bundle scars,** which can be seen within the leaf scars, are remnants of the conductive tissues, xylem and phloem, that connected the stem and leaf. **Axillary buds** also can be seen immediately above the leaf scars. These are the buds that may become branches and are always seen in the leaf axils. The slightly raised marks scattered on the stem are the **lenticels,** each composed of a group of loosely arranged cells. The role of lenticels is to allow gases such as oxygen and carbon dioxide to pass through the cork layer and thus facilitate respiration in the stem.

Secondary Growth

Primary growth in woody dicotyledonous plants is basically similar to that of herbaceous dicotyledonous plants. In both groups the primary vascular tissues are in discrete vascular bundles. In all woody plants, however, secondary growth is initiated soon after the completion of the primary growth and is continued for as long as the plant lives. In woody plants the two secondary meristems, **vascular cambium** and **cork cambium,** are responsible for the production of secondary tissues. In only a few herbaceous dicotyledonous plants does secondary growth take place to any appreciable extent. Once the primary tissues have differentiated, interfascicular segments are derived from the parenchymal cells present between the vascular bundles, resulting in a continuous cylinder of vascular cambium. Subsequently, secondary xylem is produced toward the inside of this cylinder and secondary phloem toward the outisde (Figure 6–8). Thus before the end of the first growing season there will be a continuous cylinder of vascular tissues formed in the woody stem (Plate 2A). Also during the first year of growth a cylinder of cork cambium is produced, usually by the cortical parenchyma. This cork cambium produces **cork** to the outside and **phelloderm** on its inner side (Plate 2C). Secondary growth only contributes to the thickness of a stem segment derived from the apical bud, not to its length. With continued secondary growth each stem segment increases in diameter year after year without any change in its length. Since growth in length takes place only at the tip of the main stem and branches, the base of a stem or branch will have a larger diameter than its tip. As a result, woody stems have a conical rather than a cylindrical shape.

Cambial activity in woody plants growing in the temperate regions of the world is intermittent: inactive during winter months, with resumed activity in the spring and tapering off toward the end of the growing season. During a given year, there is more secondary xylem produced than secondary phloem. In addition, only the xylem (wood) is permanent in a tree. Therefore, the amount of secondary xylem in a tree increases year after year while the other secondary tissues become crushed or slough off. Owing to the intermittent activity of the vascular cambium, the secondary xylem produced during a particular year can be recognized as a single ring, called a **growth ring** or **annual ring,** and can be distinguished from xylem of other years. The width of growth rings varies depending on the growing conditions during a given year. Within a single growth ring, the cells that are formed during the spring of the year when growth is vigorous are larger than the last several rows of cells formed during the summer months. The latter cells remain small and relatively undifferentiated, appearing similar in shape to the cells of the vascular cambial zone (Figure 6–9). The terms *spring wood*

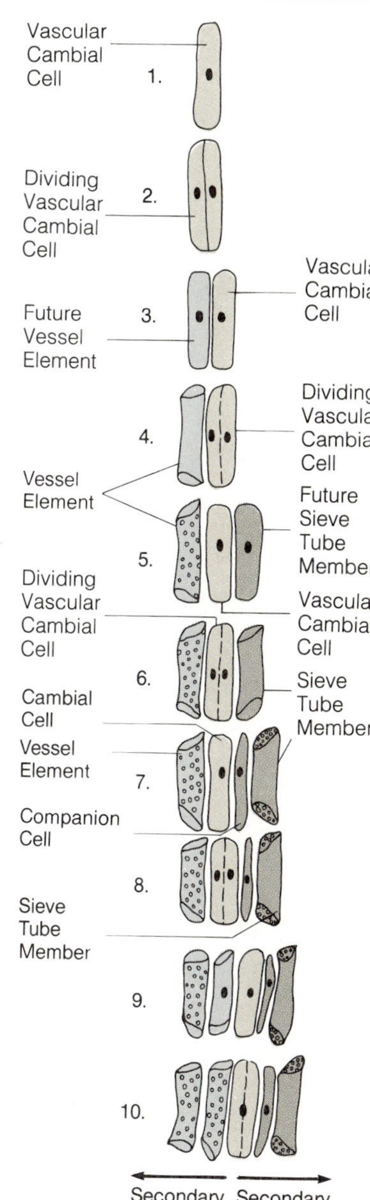

Figure 6–8. Vascular cambial activity. The vascular cambium is a cylinder extending the entire length of the stem. The fact that all cells of the cylinder are undergoing division results in the formation of xylem tissue toward the center of the stem and phloem tissue toward the outside of the stem. Thus the cambial cylinder remains between the xylem and phloem. At any given time, a vascular cambial initial can produce only either a xylem cell type or a phloem cell type. During any growing season more xylem than phloem is produced.

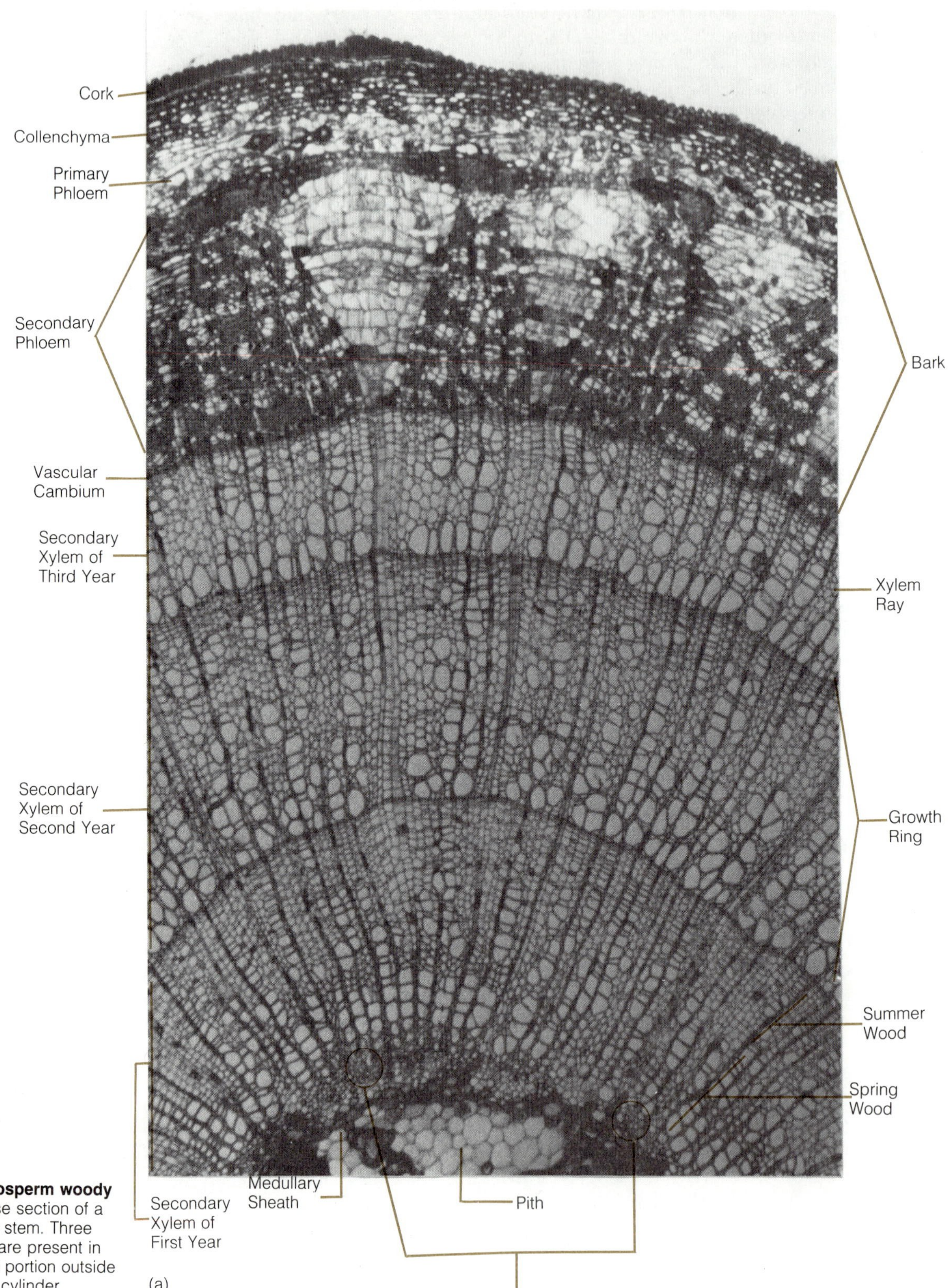

Figure 6–9. Angiosperm woody stem. *(a)* Transverse section of a three-year-old woody stem. Three distinct growth rings are present in the wood. Bark is the portion outside the vascular cambial cylinder.

Figure 6–9. ***(Continued)*** *(b)* The vascular cambrium is present between the newest secondary xylem and phloem. *(c)* An enlarged view of the bark shows details of the phloem tissue.

Companion Cell
Fiber
Sieve Tube Member
Secondary Phloem
Vascular Cambium
Secondary Xylem
Vessel Element

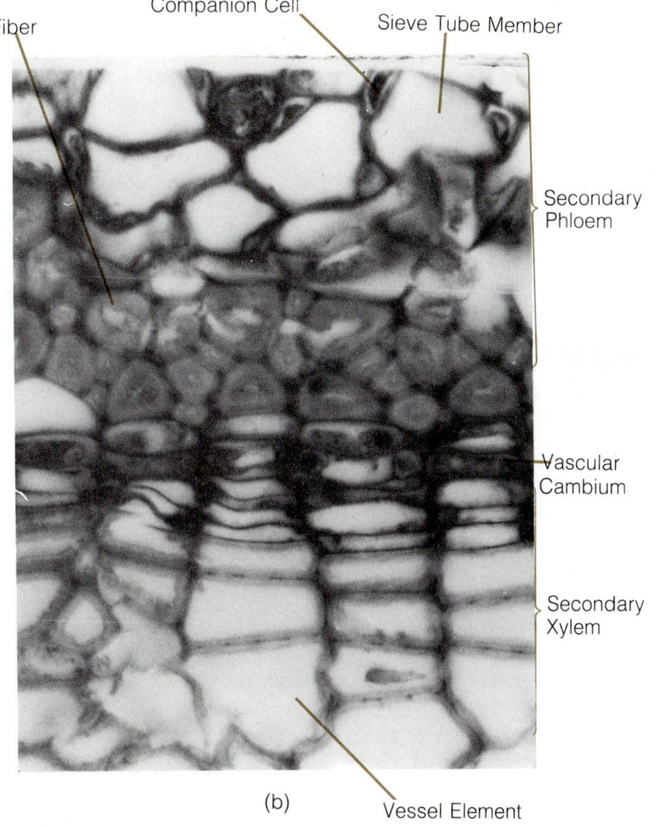

(b)

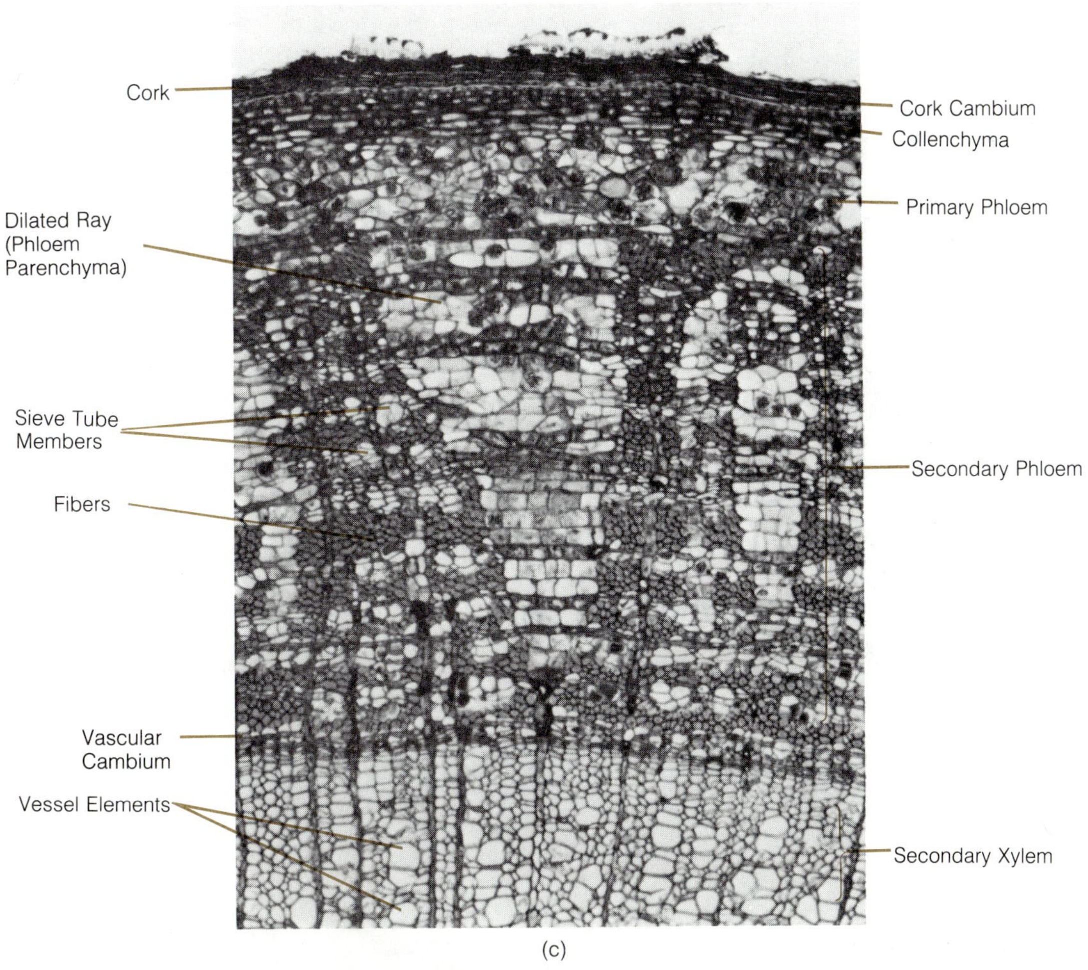

Cork
Cork Cambium
Collenchyma
Primary Phloem
Dilated Ray (Phloem Parenchyma)
Sieve Tube Members
Secondary Phloem
Fibers
Vascular Cambium
Vessel Elements
Secondary Xylem

(c)

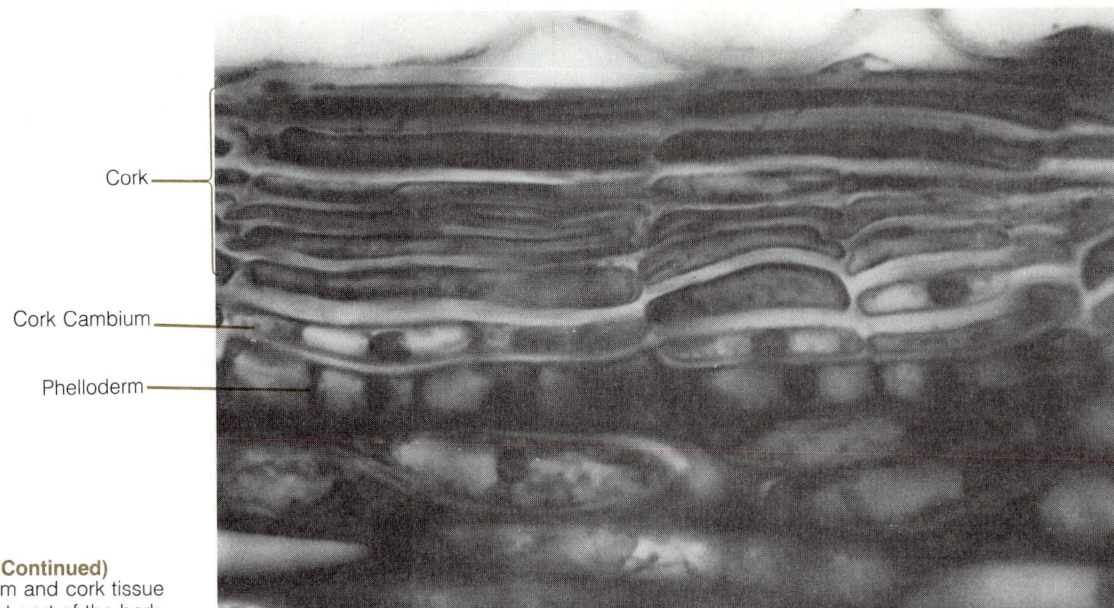

Figure 6–9. (Continued) *(d)* Cork cambium and cork tissue are the outermost part of the bark.

and *summer wood* are used to describe the secondary xylem produced during spring and summer of the year, respectively. As the vascular cambium begins renewed growth in the spring, new, large cells are produced adjacent to the small ones produced at the end of the previous season; it is the visual contrast produced by small cells next to large cells that creates the appearance of a ring.

Growth rings can be counted easily on a cross section of a stem, allowing determination of the age of that stem (Figure 6–15b). Care must be taken, however, to ensure that false growth rings are detected. All apparent growth rings are not annual rings, since all a growth ring represents is a visible boundary between small cells and large cells. Therefore a severe drought followed by plentiful rain will create a false growth ring, as would defoliation by insects early in the growing season. An experienced observer usually can distinguish the false from the true growth rings but the uninitiated should beware. Since woody trees are conical structures, accurate age determination requires counting the number of growth rings at the base of the main trunk. This can be done by taking a core of wood from the base of the tree. The resulting hole can be filled without any harm to the tree.

Anatomy of Dicotyledonous Woody Stems

Basswood *(Tilia americana)* is a good example for studying the arrangement of tissues in an angiosperm woody stem. The anatomical features seen in basswood are common to many other woody stems. A three-year-old stem will have all of the tissues found in a woody stem (Figure 6–9). The innermost area of such a stem is a column of **pith,** made up of parenchyma cells. Around the periphery of the pith the parenchyma cells are smaller than in the middle and form a layer in the shape of a cylinder called the **medullary sheath.** The next layer of tissue to the outside is the **primary xylem,** followed in sequence by three cylinders of **secondary xylem** and the **vascular cambium.** The innermost cylinder of secondary xylem is the **growth ring** produced during the first year of growth, with the newest growth ring being closest to the vascular cambium. The rows of xylem parenchyma radiating from the pith toward the vascular cambium in the form of spokes are **xylem rays.** The remainder of the xylem tissue consists of **vessel elements, tracheids,** and **fibers.** The cylinder of tissue outside the vascular cambium is the **secondary phloem,** which can be recognized easily in basswood because of the prominent **dilated rays** composed of **phloem parenchyma** (Plate 2B). The phloem rays are continuous with the xylem rays and together form the **vascular rays. Phloem fibers** are present in groups with the remainder of the secondary phloem made up of **sieve tube members** and associated **companion cells.** Next to the secondary phloem toward the outside is the **primary phloem,** consisting of only a few crushed layers of cells. A **cortex** composed of **parenchyma** and **collenchyma** is still present in a three-year-old stem, although some of the parenchyma cells will be broken or distorted. Outside of the cortex is the **periderm** with its three distinct tissues—**phelloderm, cork cambium,** and **cork.** The cork cambium gives rise to one or two rows of phelloderm cells to the inside and many rows of cork cells to the outside. In basswood and other trees certain segments of the cork cambium produce a mass of nonsuberized* loosely arranged cells that make up a **lenticel.** These appear as small breaks randomly scattered on the periderm surface. An **epidermis** is often still present as the outermost tissue in a three-year-old basswood stem. The mature, dead cork cells with their suberized walls eventually will prevent water from reaching the living epidermal cells that in turn will die and peel off. As a tree ages a new cork cambium is produced by parenchyma cells of the outer phloem; this results in the death of all cells further out. This process may be repeated as often as every one or two years or only after twenty years or more.

Nonsuberized
Lacking the wax suberin.

The term *wood* is applied to all tissues enclosed by the vascular cambial cylinder, while all tissues to the outside of the vascular cambium are collectively called **bark.** Wood is almost entirely made of secondary xylem and functions in the transport of water and minerals while providing major mechanical support for the tree. In an older tree, however, the xylem that actually transports water is limited to a band close to the vascular cambium, whereas in the remainder of the xylem the vessel elements and tracheids become plugged. Usually the conducting xylem, called **sapwood,** is lighter in color and density than the nonconducting xylem, **heartwood,** which is denser and darker in color (Figure 6–10). The increased density and dark color of heartwood result from an accumulation of resins, tannins, oils, and gums. Often **tyloses,** an invasion of parenchyma cells through the pits on the walls of vessel elements, also block the passage of water through them.

The inner part of the bark is mainly secondary phloem and the outer part is predominantly cork. Therefore, the function of bark is transportation of food and protection. Early settlers in North America used to remove a complete ring of bark from around the lower part of the trunks of large trees growing on land they wished to cultivate. The purpose of this deliberate girdling of the trees was to prevent the translocation of food from the leaves to the roots, thus causing the gradual death of the roots by starvation. Death of the shoot system soon followed, making removal of the trees easier, since

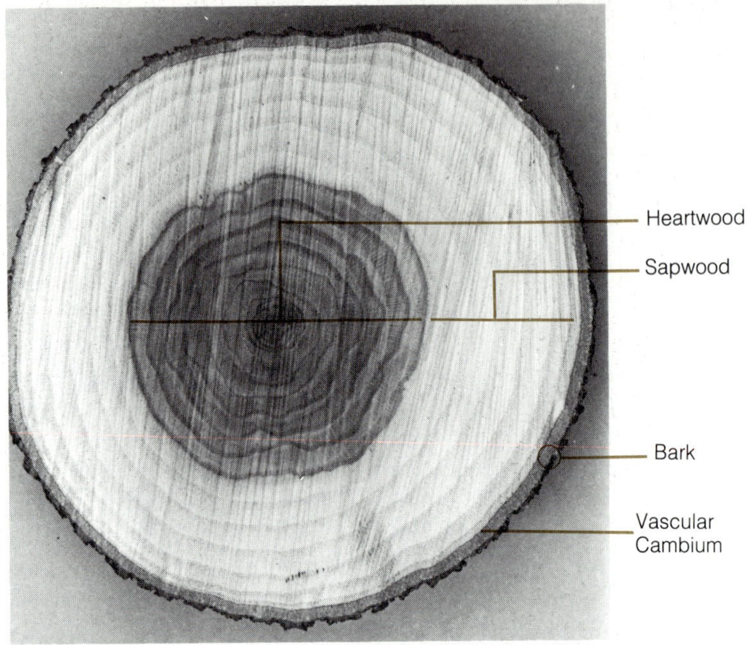

Figure 6–10. Heartwood and sapwood. Transverse section of a pin oak showing the dark-colored, nonconducting heartwood and the light-colored, functional sapwood. As a tree grows older, the ratio of heartwood to sapwood increases.

dead trees are easier to cut. Because it is easier to dig out a dead root system than a living one, girdling the trees meant less labor was needed to prepare the land for crops.

Anatomy of Gymnosperm Woody Stems

Because of its wide distribution pine (*Pinus* sp.) commonly is used to study gymnosperm woody stem structure (Figure 6–11 and Plate 2D). Since there is very little difference between a pine stem and a basswood stem in the arrangement of tissues, the anatomy of pine stems will not be presented in detail. Basswood stem anatomy and secondary growth already have been described in the two previous sections of this chapter, so only the major differences will be pointed out.

A three-year-old pine stem cross section will have the following tissues arranged in sequence from the innermost to the outermost regions: parenchymatous pith, primary xylem, three cylinders of secondary xylem, vascular cambium, secondary phloem, primary phloem, parenchymatous cortex, phelloderm, cork cambium, cork, and epidermis. The xylem of most gymnosperms differs from that of angiosperms in that it consists only of tracheids, fibers, and parenchyma; there are no vessel elements. In the phloem of pine, sieve cells* are present instead of sieve tube members whereas companion cells are missing and rays are not dilated. The most apparent difference between pine and basswood is the presence of **resin ducts** in the wood and bark of pine for the transport of resin. Each resin duct is a tubular structure surrounded by cells that secrete resin. Resins are complex metabolic by-

Sieve cells
The food-conducting cells of the phloem.

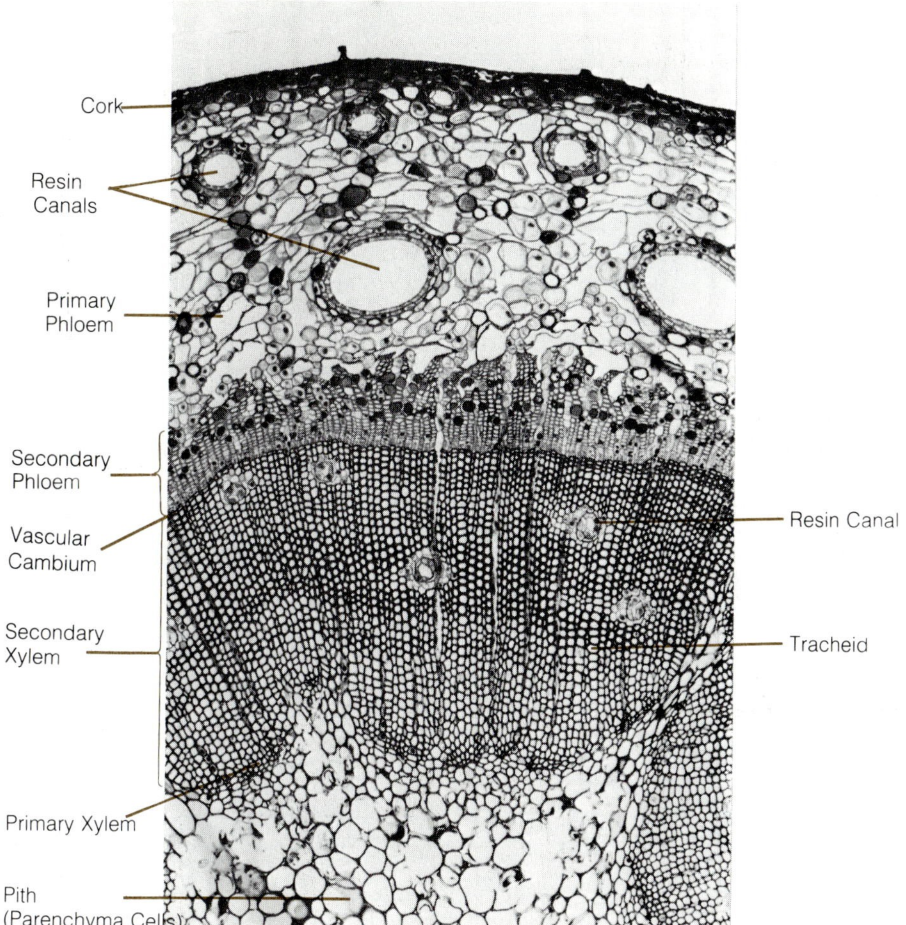

Figure 6–11. Gymnosperm woody stem. Transverse section of a one-year-old woody gymnosperm stem. The arrangement of tissues in both angiosperm and gymnosperm woody stems is similar; however, the presence of resin ducts in both wood and bark and the absence of vessel elements in the xylem of this pine section are characteristic differences.

products that prevent invasion by insects and pathogenic* organisms such as the fungi that cause wood rot. A crude extract of resin is known as pitch; when distilled it yields oil of turpentine and resin. The former is used as a thinner in paints and varnishes and the latter is used in the manufacture of a variety of products such as waxes, lubricants, and soaps. Amber, often used to make jewelry, is fossilized resin.

Pathogen
(*adj.*, pathogenic) A disease-causing organism.

Stem Modifications

In a number of plants, especially herbaceous plants, stems have become modified through evolution* so that their function is different from the normal functions of transporting water and food and providing an upright axis to support leaves for maximum exposure to the sun. These modified stems are often morphologically so different that they cannot be easily recognized

Evolution
The origin of new types of organisms from ancestral forms as the result of natural selection.

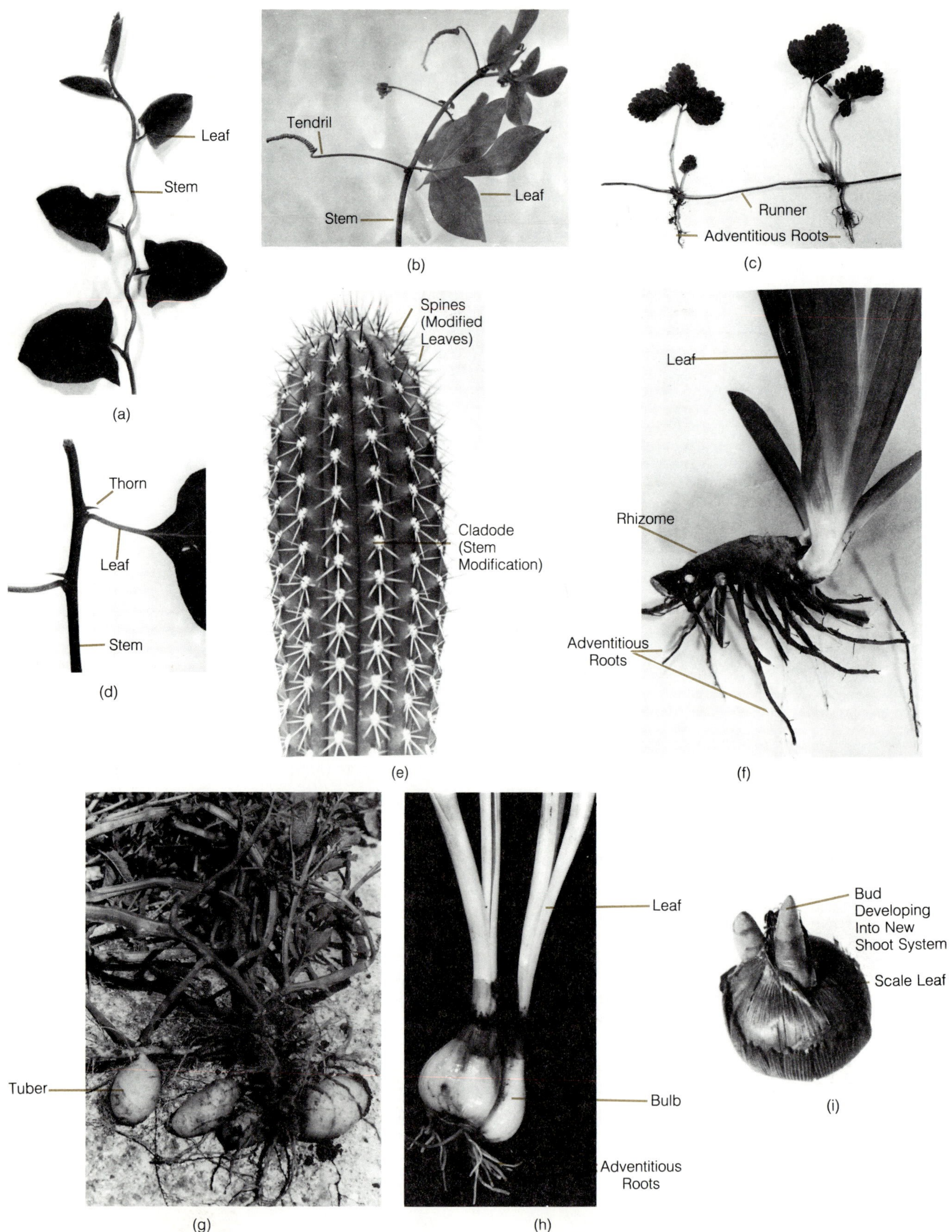
Leaf
Stem
(a)
Tendril
Leaf
Stem
(b)
Runner
Adventitious Roots
(c)
Thorn
Leaf
Stem
(d)
Spines
(Modified
Leaves)
Cladode
(Stem
Modification)
(e)
Leaf
Rhizome
Adventitious
Roots
(f)
Tuber
(g)
Leaf
Bulb
Adventitious
Roots
(h)
Bud
Developing
Into New
Shoot System
Scale Leaf
(i)

Figure 6–12. Stem modifications. *(a)* A twiner is a weak-stemmed plant that winds around a support. *(b)* In many weak-stemmed plants axillary buds give rise to tendrils that are used for attachment of the plant to a support. *(c)* A runner is a weak-stemmed plant that grows along the soil surface, producing new plantlets that are interconnected. *(d)* A thorn is produced from an axillary bud which normally would have given rise to a branch. *(e)* A cladode is a stem that serves the function of leaves; spines on a cladode are modified leaves. *(f)* A rhizome is a horizontally growing, underground stem that stores food; when the aerial stem dies during winter, the rhizome survives and produces a new shoot system in the spring. *(g)* A tuber is the terminal portion of a rhizome, swollen with stored food. The "eyes" of the tuber are buds that can produce new shoots. *(h)* A bulb has a disc-shaped, vertically compressed stem from which are produced adventitious roots and many leaves that store food and water. Bulbs are found underground and can produce new shoots. *(i)* A corm is a bulblike, vertically growing, underground stem that stores food. The leaves on a corm are scalelike.

as stems. However, the characteristic internal organization of stem tissues and the presence of either buds or leaves or both are criteria that can be used to determine the true nature of a modified stem.

Stem modifications can be grouped into two categories: **aboveground modifications** (twiner, tendril, stolon, thorn, cladode) and **underground modifications** (rhizome, tuber, bulb, corm).

Stem modifications seen among weak-stemmed plants often function to provide some means of support. One such modification is the **twiner,** where the stem coils around a suitable support (Figure 6–12a). Examples of twiners are morning glory, clematis, or sweet potato. In the case of grape or passionflower, the auxillary buds, instead of producing branches, develop into springlike climbing organs called **tendrils** (Figure 6–12b), which serve to attach the plants to a variety of supports. Weak horizontal stems of strawberry, called **stolons** or **runners,** grow along the soil surface (Figure 6–12c). They produce roots from the nodes and establish a series of interconnected plants, a common mode of propagation* in strawberry.

Propagation
The production of new plants by sexual or asexual means.

In plants like honey locust and osage orange some of the axillary buds develop into sharp pointed **thorns** that discourage grazing animals from eating them (Figure 6–12d). In most cacti the stems are modified into flat, broad, green structures called **cladodes** (Figure 6–12e) that are photosynthetic and store water. In these xerophytic* plants the leaves are highly reduced or absent to minimize surface area and water loss. To compensate, the stems carry out the function of leaves.

Xerophyte
(*adj.*, xerophytic) A plant capable of growing under dry conditions.

The remaining modifications are found underground and may be mistaken for roots. Closer examination will reveal the presence of buds and scale leaves, which are characteristic appendages of stems and not roots. Almost all of the modified stems found underground function in food storage and propagation. A **rhizome** is a horizontal stem that grows either completely below ground or slightly exposed (Figure 6–12f), with roots developing from its lower surface. In the case of iris, canna, and ginger, rhizomes store large amounts of food. Since the aerial portions of these plants die during winter, buds on the rhizome develop into new shoot systems during the spring. Propagation of these plants is almost entirely by planting segments of rhizomes containing buds. The rhizomes of potato plants do not store food along their full length; instead food storage is limited to the tips of rhizomes, resulting in swollen structures called **tubers** (Figure 6–12g). Since the "eyes" of a potato are axillary buds, a piece of a potato containing an "eye," when planted, will produce a new plant.

Bulbs of onion, lily, and narcissus are underground buds that store food in their fleshy leaves (Figure 6–12h). The stem portion of a bulb is a disclike structure at the base of the bulb from which are produced the roots and leaves. On the other hand, gladiolus and crocus store food in modified underground stems called **corms** (Figure 6–12i). Although corms may resemble onions superficially, they are entirely made up of stem materials, with only paper-thin leaves. Both bulbs and corms produce new shoot systems during spring.

Summary

Seed-producing plants are divided into two general categories, herbaceous and woody. Herbaceous plants contain little or no secondary tissues; woody plants produce large amounts of secondary tissues, especially secondary xylem.

A stem is the part of the shoot system that bears buds, leaves, and reproductive organs. The apical bud consists of the primary meristems—protoderm, procambium, ground meristem—and a series of leaf primordia with bud primordia in their axils. Elongation of a stem is a result of cell division in the primary meristems and the subsequent enlargement of the newly formed cells. Buds in the axils of leaves become branches or lateral shoot systems.

The arrangement of tissues in a cross section in the mature parts of stems can be used to distinguish between monocot stems and dicot stems as well as between herbaceous stems and woody stems.

External features usually present on a woody branch are leaf scars, vascular bundle scars, axillary buds, lenticels, and terminal-bud-scale scars. The age of a branch may be determined by counting the segments of yearly growth demarcated by the terminal-bud-scale scars.

Growth in diameter of a woody stem is the result of secondary xylem and secondary phloem produced by the vascular cambium, and cork and phelloderm produced by cork cambium during successive years. Of the secondary tissues produced during a given year, xylem is produced in the largest amounts and accumulates to form growth rings. The total number of growth rings at the base of a tree trunk is a measure of its age.

The arrangement of tissues from the center of a woody stem toward its outside is as follows: parenchyma, primary xylem, secondary xylem of the first year, secondary xylem of subsequent years, vascular cambium, secondary phloem, primary phloem, parenchyma and collenchyma, phelloderm, cork cambium, and cork. Both xylem and phloem are arranged in solid cylinders instead of in vascular bundles. The tissues enclosed by the vascular cambial ring of a tree cross section are collectively called wood, while those to the outside are termed bark. Periderm is composed of the outermost tissues of the bark—phelloderm, cork cambium, cork—that eventually replaces the epidermis as the protective layer.

The arrangement of tissues in a gymnosperm woody stem is similar to that of angiosperms; however, in the former vessel elements are lacking in the xylem and resin ducts are present in both wood and bark.

Stem modifications that carry out functions other than support and conduction commonly are found among many species, especially herbaceous plants. Stolons, twiners, tendrils, thorns, and cladodes are modifications of the aerial parts of plants, whereas rhizomes, tubers, bulbs, and corms are found underground. Underground stems can be distinguished from roots because of the presence of buds and scale leaves on the stems.

Review Questions

1. Describe how a stem grows in length.
2. Explain why leaf scars can not be seen on the basal segment of a ten year old branch.
3. List the tissues that are absent from a typical monocotyledonous herbaceous stem, but present in a typical dicotyledonous herbaceous stem.
4. Explain why leaves are normally produced only on the newest segment of a stem.
5. Explain the mechanism of secondary growth in a woody stem.
6. List in sequence, starting with the pith parenchyma, all tissues present in a three year old woody stem.
7. What are the anatomical differences between a woody angiospem and a woody gymnosperm?
8. Explain patterns found on a piece of lumber.

Patterns In Wood

Even a casual observation of a wooden tabletop, a wall panel, or a door will show varying patterns. The two major factors that determine the nature of these patterns in wood are the plane in which boards are cut from a log and the relative number, shape, size, and arrangement of cells in the wood. In a cross section, the growth rings appear as concentric circles with the vascular rays radiating from the center like spokes of a wheel. Most trees are not large enough in diameter to be cut into usable lumber in cross section. Besides, boards cut in cross sections may split along growth rings as well as along vascular rays. Usually trees are sliced into boards lengthwise, either parallel to the vascular rays, known as quartersawed (radial cut), or at right angles to the vascular rays, known as planesawed (tangential cut) (Figure 6–13). In quartersawed pieces of lumber, vertically cut growth rings occur as parallel lines with vascular rays running at right angles to them. The number of perfectly quartersawed pieces from a log is limited. Large and irregular patterns are found on boards from planesawed pieces (Figure 6–14). The irregular dark and light patterns found on these boards are due to

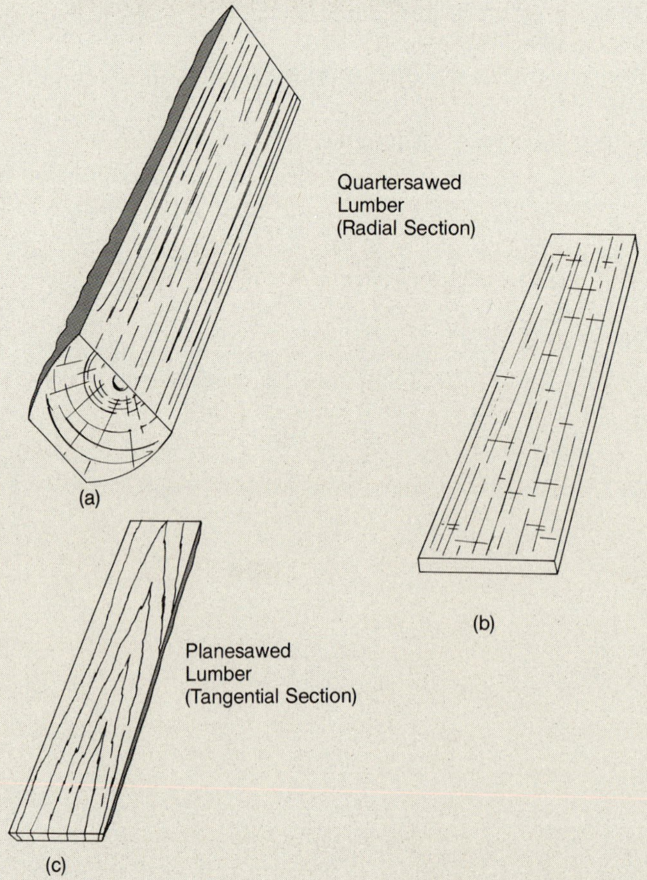

Figure 6–13. Planes of sectioning. The patterns seen on lumber depend on the plane of section. The patterns themselves are produced by the growth rings and the arrangement of vascular rays. *(a)* In a transverse section growth rings appear as concentric rings. *(b)* Radial sections show cut ends of growth rings arranged parallel to each other along the full length of the section. *(c)* The growth rings appear V-shaped in a tangential section.

(a)

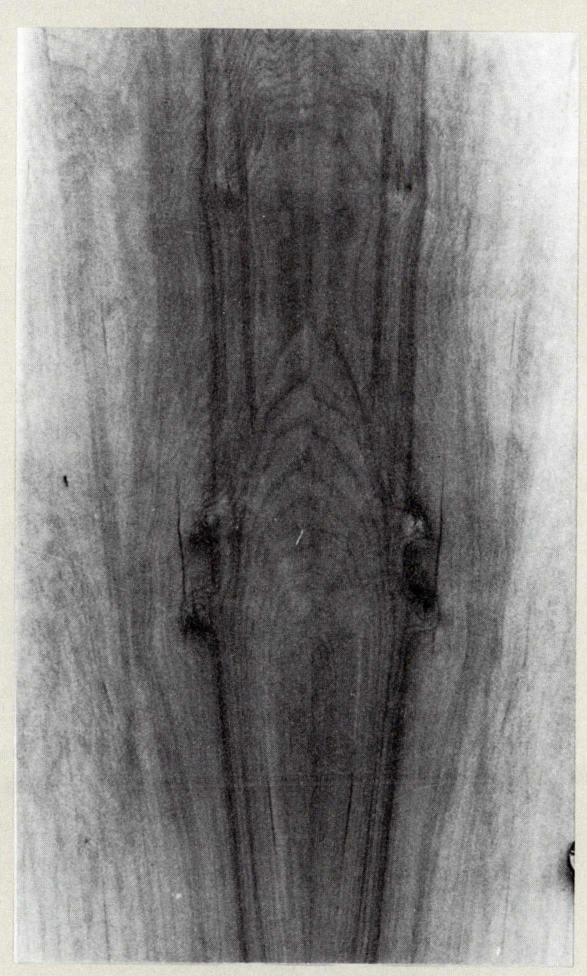

(b)

Figure 6–14. Patterns in wood. *(a)* Typical pattern on a planesawed piece of lumber. *(b)* Photograph of a door showing an unusual pattern of a planesawed piece of lumber. Can you recognize a ghost in this pattern?

Figure 6–15. Growth pyramid. *(a)* Critical examination of a tree shows that its shape is conical rather than cylindrical. This diagram represents the arrangement of annual growth rings in the form of cones within the trunk of a five-year-old tree. Each cone represents one year's growth of xylem, the outermost cone representing the newest growth. The length of each cone, starting with the innermost corresponds to the height of the tree at successive years. Normally there would be five growth rings at the base of a five-year-old tree and one at the tip. The vertical distance between the tips of cones and the diameter of each cone are influenced by growth conditions. *(b)* Growth rings on a 30-year-old woody stem. The differences in width of the growth rings are due to variations in environmental factors such as rainfall and wind.

alternating springwood and summerwood within each growth ring. The large cells of the springwood reflect less light, thus appearing darker than the summerwood.

It would be easier to understand patterns in wood if growth rings were considered as many long cones of varying wall thickness and diameter, placed one inside the other, making up the log (Figure 6–15). If this structure were sliced through the center (quartersawed), parallel lines would be seen along the cut surface, with varying distances between lines. If, on the other hand, the cut were made on a plane that did not pass through the center (planesawed), variations of v-shaped patterns running along the cut surface would be seen. The basis for the variations in the size and shape of patterns in lumber is the variation in thickness within each growth ring as well as between growth rings.

Knots in wood can add a characteristic pattern to lumber, especially in certain gymnosperms such as knotty pine; however, most lumber with knots has relatively less commercial value. Since the cambium of the trunk is continuous with the cambium of the branches, both the trunk and the branches will increase in diameter synchronously as long as both cambia are alive. Often branches die or they fall off, leaving only their basal portions attached. The trunk continues to increase in diameter owing to secondary growth and will eventually embed these dead branches or stubs. Living branches also may get buried. When trees are cut into lumber, these buried branches and stubs appear as knots in the wood. Embedded living branches form tight knots that remain in the lumber, whereas the dead branches produce loose knots that may fall out and leave holes in the boards, reducing their commercial value. Therefore, it is important when removing both living and dead branches to saw them off as close to the trunk as possible without damaging the trunk. This will increase the number of boards without knots that can be obtained from a tree.

7

Roots and Mineral Nutrition

Outline

Introduction

The roots of a typical seed-producing plant form an extensive underground system of major roots and their branches. Many of the branch roots are extremely thin; thus the true extent of a root system is not apparent because of its below-ground position and the fact that most small roots will break off and remain in the soil even when a plant is carefully pulled up. The lateral spread of a root system is often more extensive than the shoot system of the same plant. Many apple trees are known to have roots that spread 20–30 feet laterally, and the roots of many herbaceous and some woody plants also penetrate deep into the ground to distances equal to or greater than the height of the plants. In general the largest number of roots of most plants are found in the upper 4 feet of soil where the greatest amount of mineral nutrients are available. Soil factors, such as moisture, mineral composition, pH*, oxygen content, and temperature, influence the spread of a root system. Particularly impressive data from a rye plant, a member of the grass family, show that the total length of all roots from a single clump, if placed end-to-end, can stretch 387 miles! The roots of the rye plant in this example had a surface area of more than 2500 square feet, compared to the surface area of 52 square feet for the leaves. The large surface area provided by the root system of a plant is significant in its role of **water and mineral absorption.** The large surface area also increases the ability of the roots to **anchor** the plant and in turn to hold soil in place. A ground cover of grasses provides an ideal way to prevent soil erosion* of uncultivated land. Additional functions for roots are **storage** of food and **synthesis** of organic molecules, including certain hormones that are transported to the shoot system.

pH
A quantitative measure of the acidity of a solution.

Erosion
Loss of soil due to the action of wind or water.

Plants such as willow trees with deeply penetrating roots are able to survive even when the top layers of the soil become dry, provided that the roots have reached the water table or some other source of moisture. Unfortunately, in urban environments this often includes sewers and drainage tiles. Deep roots are of survival value to certain desert plants such as acacia and mesquite, while other desert plants, such as the different kinds of cacti with shallow, diffuse root systems, have other adaptations that overcome the dryness of the soil in which they exist. Among these are root cells specialized to store water. The cytoplasm of these cells also contains colloids* to which water molecules adhere strongly, thus reducing water loss.

Colloid
Very fine particles that will remain in suspension in a liquid.

Origin of Roots

In a seedling the primary root is formed by the elongation of the radicle, the lowermost portion of the embryonic axis (Figure 7–1) and the first struc-

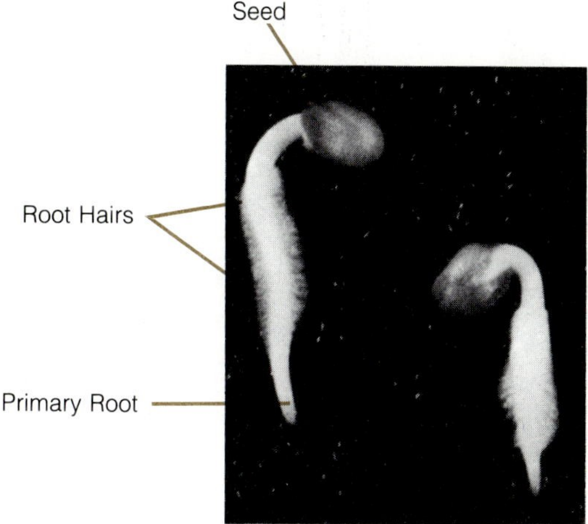

Figure 7–1. Primary root. The first structure to emerge from a germinating seed is the primary root.

ture to emerge from a germinating seed. Under normal conditions of growth, roots in general grow downward in response to gravity, a phenomenon known as geotropism. Stems, on the other hand, demonstrate a negative response to gravity and grow upward (Chapter 12).

In most dicotyledonous plants*, the primary root becomes the largest root of the root system and gives rise to branch roots. These branch roots in turn produce second order branch roots, the latter to third order branches and so on until a highly branched, extensive system with varying diameters of roots is formed. Such a root system is called the **tap root system,** with the primary root being the tap root (Figure 7–2a). In many dicotyledonous plants, such as carrot, beet, turnip, and radish, the tap root stores large amounts of food. In monocotyledonous plants* the primary root derived from the radicle is short-lived and is replaced by several new **adventitious roots*** of approximately equal size, derived from the base of the stem. Branch roots of different orders are then formed to establish a **diffuse root system,** otherwise known as a **fibrous root system** (Figure 7–2b). This type of root also can have a food storage function as in the case of the edible fibrous roots of sweet potato, a dicotyledonous plant. The difference between an adventitious root and a typical root is in their origin and not in their structure.

As indicated above roots are often produced from plant structures other than roots. Such adventitious roots form at the nodes* of many plants that grow horizontally along the ground instead of upright (Figure 7–3). Roots of most primitive vascular plants*, such as ferns, club mosses, and horsetails, are generally of the adventitious type because these roots are produced from rhizomes.* This ability of plants to produce roots from organs other than roots is the basis for vegetative propagation* both in nature and in horticultural practice. Many horticulturally important plants, as well as some crop plants, can be propagated easily by stem or leaf cuttings (Figure 7–4) because, under the right conditions, adventitious roots will form on the lower end of the stem or from the leaf in response to elevated levels of the plant hormone called auxin (Chapter 12).

Dicotyledonous plant
A flower-producing plant whose seeds have two cotyledons.

Monocotyledonous plant
A flower-producing plant whose seeds have only one cotyledon.

Adventitious root
A root that forms on an organ other than a root.

Node
The region of a stem to which a leaf is, or was, attached.

Vascular plant
A plant that has specialized tissues (xylem and phloem) for conducting water and food.

Rhizome
A modified underground stem that may store food.

Vegetative propagation
Reproduction of organisms by nonsexual means.

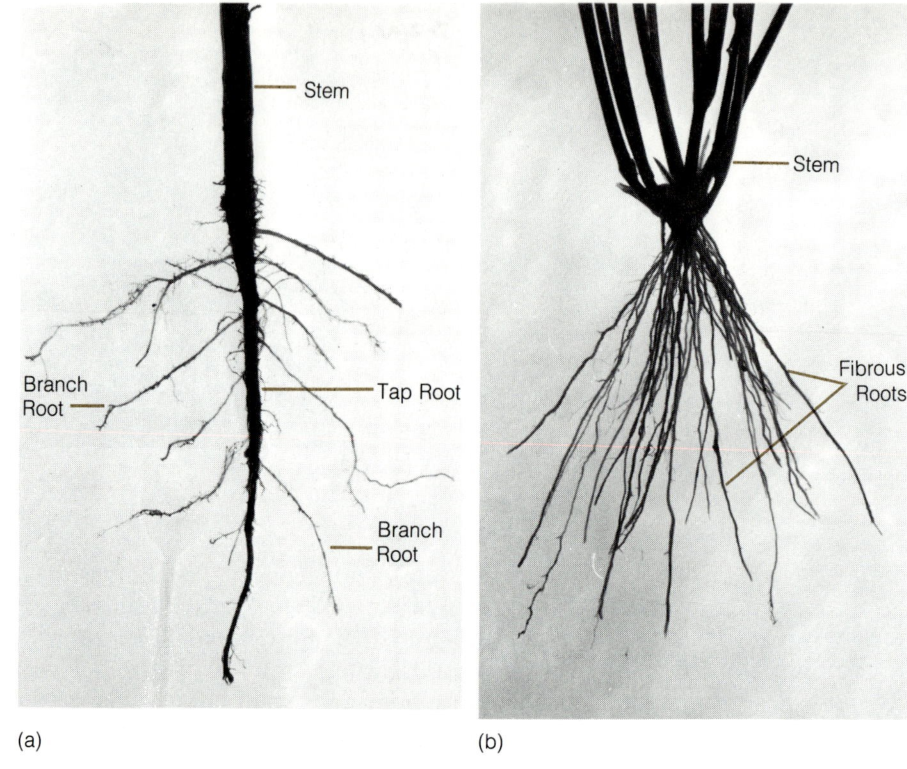

Figure 7–2. Root systems. *(a)* Tap root system characteristic of dicotyledonous plants. *(b)* Fibrous root system characteristic of monocotyledonous plants.

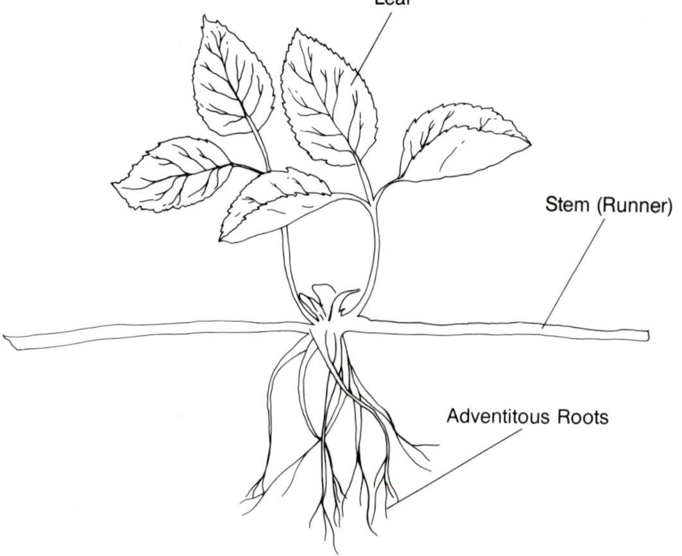

Figure 7–3. Adventitious roots. Roots produced from structures other than roots, such as the stem of a runner, are known as adventitious roots.

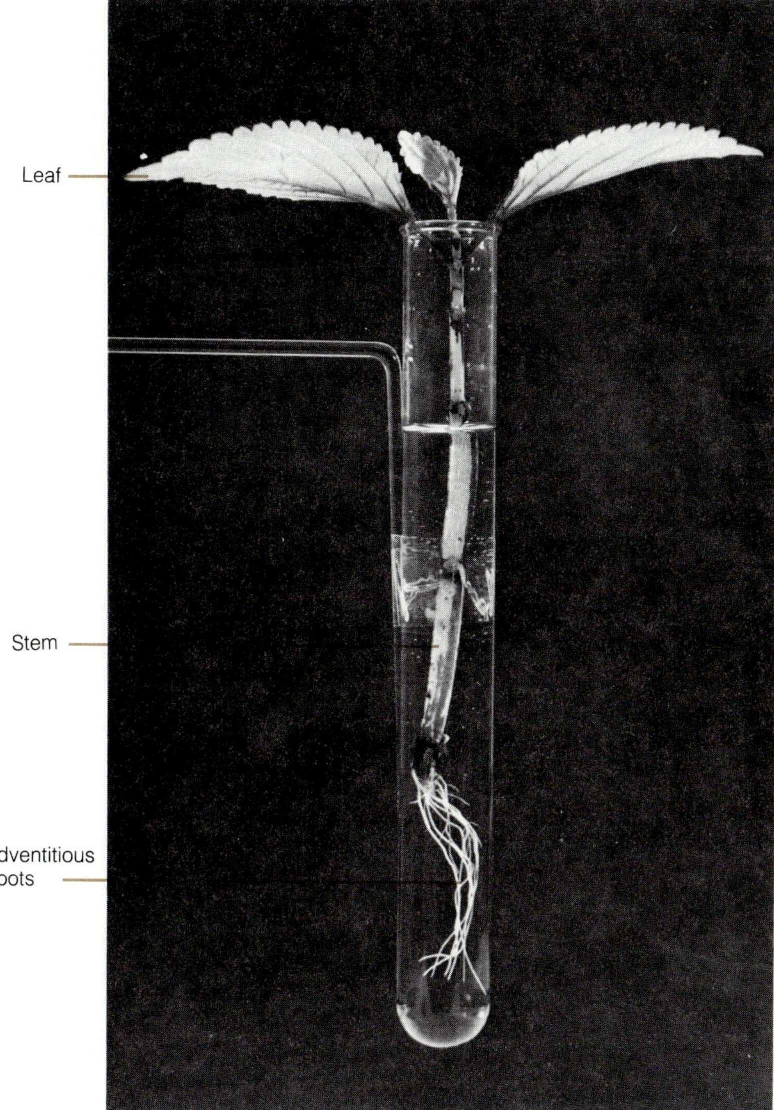

Figure 7–4. Stem cutting. Adventitious roots are produced from the base of a stem cutting.

Root Morphology

The terminal portions of all roots are perpetually young and constantly elongating, allowing the plant to come in contact with new sources of water and minerals. In contrast to stems, roots do not have buds or leaves, and, therefore, the growing tip of the root is simpler in appearance than that of the stem (Figure 7–5). Like stems, however, roots grow by division of cells in the apical and primary meristems* at the growing tip and the subsequent enlargement of cells in the adjacent region of elongation. The external features and internal organization of cells are similar in all root tips no matter

Meristem
A tissue whose cells are capable of repeated division.

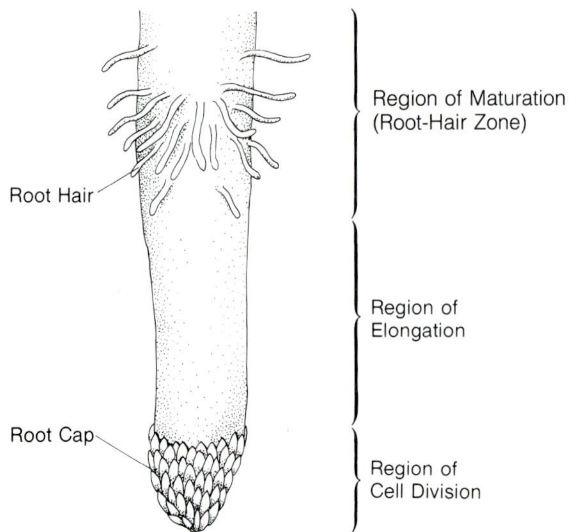

Figure 7–5. Root tip morphology.

where in a plant they are produced. In most roots, the growing tip is covered and protected by the **root cap,** a thimblelike arrangement of parenchymatous cells (Figure 7–5). The root surface is bare for several millimeters behind the root cap, but next to this bare region is the **root hair zone** in which most of the epidermal cells produce a single tubular projection called a **root hair.** These root hairs increase the surface area of a root many thousand-fold and are the primary absorptive surfaces of a root. Root hairs are short-lived, hence old ones farthest from the tip are constantly being replaced by newly formed ones closer to the tip. Because of this constant process of replacement, the length of the root hair zone and the distance between the root tip and the root hair zone always remain the same. Farther back from the root tip the **branch roots** begin to appear. Like the root hairs, the branch roots increase the absorptive surface of the root system but, unlike the root hair, the branch root is a multicellular structure formed deep within the root and has the same structure as the main root. Older portions of roots of perennial plants become woody* owing to accumulation of secondary tissues*, are large in diameter, and play no role in absorption.

Woody
Containing large amounts of secondary xylem (wood).

Secondary tissue
A tissue derived from a secondary meristem.

Internal Structure of Roots

Anatomy of the Root Tip

Longitudinal section
A section made along the long axis of an object.

Root tip anatomy is best seen in a longitudinal section* (Figure 7–6 and Plate 3A). The very tip of the root is composed of parenchyma cells arranged in the shape of a cone making up the **root cap,** which extends along the outer part of the root for a centimeter or more. The cells of its outer surface are sloughed off continuously as the root tip moves through the soil and new cells are added to its inner side by the **apical meristem** situated at

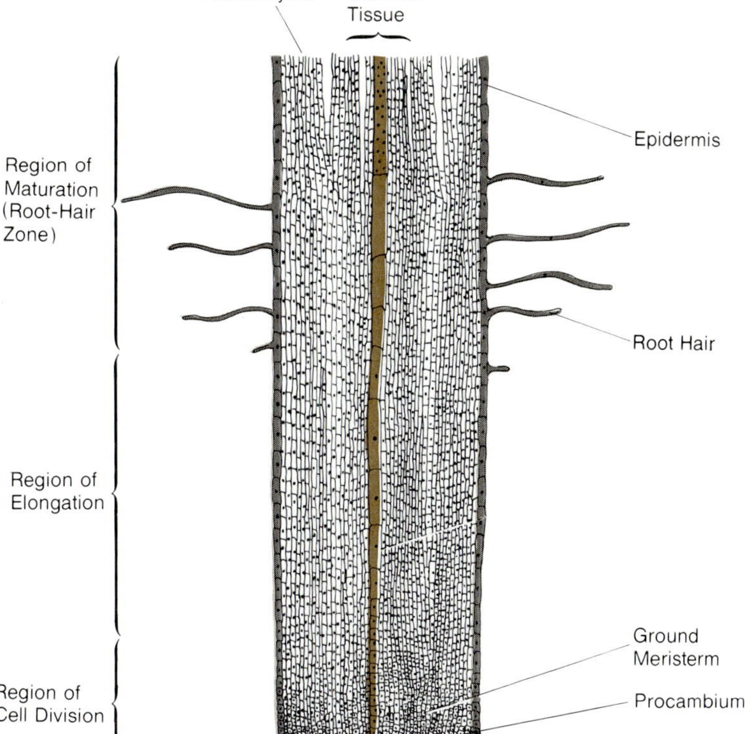

Figure 7–6. Root tip anatomy. Longitudinal section through root tip.

the tip of the growing region. The root cap has the dual function of protecting the meristematic tissues as well as of producing mucilage. Secretion of this slimy material by the root cap cells allows the growing region of the root to be pushed through the soil as the result of cell division and cell enlargement with minimum resistance and negligible injury to the internal tissues.

The primary meristems are derived from the apical meristem and are localized adjacent to it as indicated in Figure 7–6. It is from these meristems that the mature, differentiated* tissues eventually will form after cell division ceases. The central core of the root tip above the apical meristem is the **procambium,** which ultimately differentiates into the vascular tissue. Surrounding the procambium is the **ground meristem,** which produces all the tissues between the vascular cylinder and the epidermis. The epidermis is produced by the **protoderm,** the outermost meristematic layer. This portion of the root tip containing the above meristems is the **region of cell division,** where the cells are smaller and more cuboidal than mature cells. It is also in this region where the maximum number of cell divisions occur. Elongation and enlargement of the newly formed cells occur in an adjacent several-

Differentiated
Possessing a specialized function.

millimeter-long portion of the root called the **region of elongation.** Next to this is the **region of maturation,** where the cells complete their differentiation to become primary tissues.* The three regions described here are not distinctly separate regions in the root because the processes of differentiation and maturation proceed at a different pace in each of the three primary meristems. Thus, for example, cells of the procambium cease division and begin elongation while cells of the ground meristem are still dividing, resulting in considerable overlap of the regions (Figure 7–6). The external features of a root tip indicate the approximate position of each of these regions. The base of the root cap extends over the region of cell division; the root hair zone corresponds to the region of maturation, and the bare portion of the root between the root cap and the root hair zone represents the region of elongation. As a root grows older, the root tip moves farther from its origin, resulting in increased length of the root. This elongation is brought about solely by the addition of new cells by cell division in the meristems and by the subsequent elongation and enlargement of the newly formed cells. The only portion of the root that actually increases in length is the segment between the growing tip and the region of maturation; the length of all other regions remains unchanged. Therefore, a root accumulates more and more mature primary tissues as it grows. The root tip anatomy described above is common to roots of all seed-producing plants.

Primary tissue
A tissue derived from a primary meristem.

Anatomy of Mature Roots

The internal organization of mature primary root tissues is best seen in cross sections made slightly beyond the region of maturation. Two such sections, one from buttercup *(Ranunculus),* a herbaceous* dicotyledonous plant, and the other from corn *(Zea mays),* a herbaceous monocotyledonous plant, are presented in Figure 7–7 and 7–9 (also see Plate 3B, 3C and 4A). The outermost tissue is the single-layered **epidermis** derived from the protoderm. Unlike the epidermis of the stem, the root epidermis has only a very thin **cuticle.** The most significant role of the root epidermis is in the production of abundant root hairs, which increase the surface area for absorption; the amount of water absorbed through the epidermis lacking root hairs is negligible. Even though photosynthesis* normally does not occur in root cells, gas exchange must occur so that oxygen used in respiration* is available to all living root cells and carbon dioxide produced during respiration can escape from the tissue. Because the cuticle is not thick enough to function as a barrier to the diffusion of gases between the soil atmosphere and the root tissues, special openings in the epidermis are not present.

Herbaceous
Containing mainly primary tissue.

Photosynthesis
The metabolic process by which carbohydrates are produced from CO_2 and a hydrogen source such as water, using light as the source of energy.

Respiration
The metabolic process that results in the release of chemical energy from simple organic molecules.

Next to the epidermis is the **cortex,** a relatively wide area mostly composed of **parenchyma** tissue with numerous intercellular spaces. When roots function as storage organs, food, typically in the form of starch granules, is stored in the parenchyma cells of the cortex. The cross section through a buttercup root readily demonstrates the extensive nature of the root cortex. The inner boundary of the cortex is the **endodermis,** made of a single layer of endodermal cells. Both the cortical parenchyma and the endodermis are derived from the ground meristem. Each endodermal cell has a bandlike, suberized* thickening, the **Casparian strip,** that goes around the cell in a fashion similar to that of a rubber band placed around a rectangular box

Suberized
Containing the wax suberin.

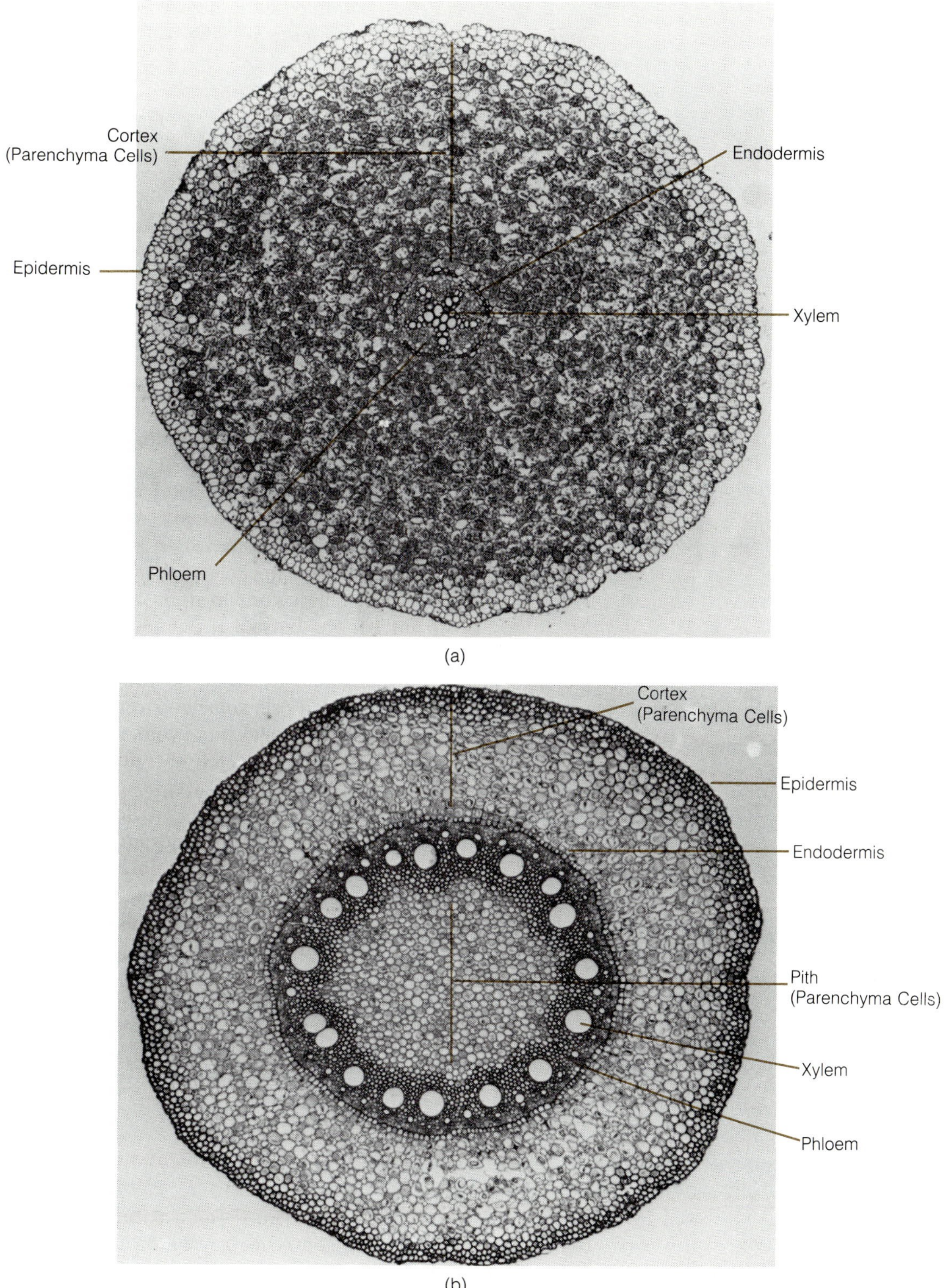

Figure 7–7. Herbaceous roots. *(a)* Transverse section of a herbaceous dicotyledonous root. Here the xylem forms a central plate with radiating points. *(b)* Transverse section of a herbaceous monocotyledonous root. The presence of a pith and the arrangement of xylem in rows are characteristics of monocot roots.

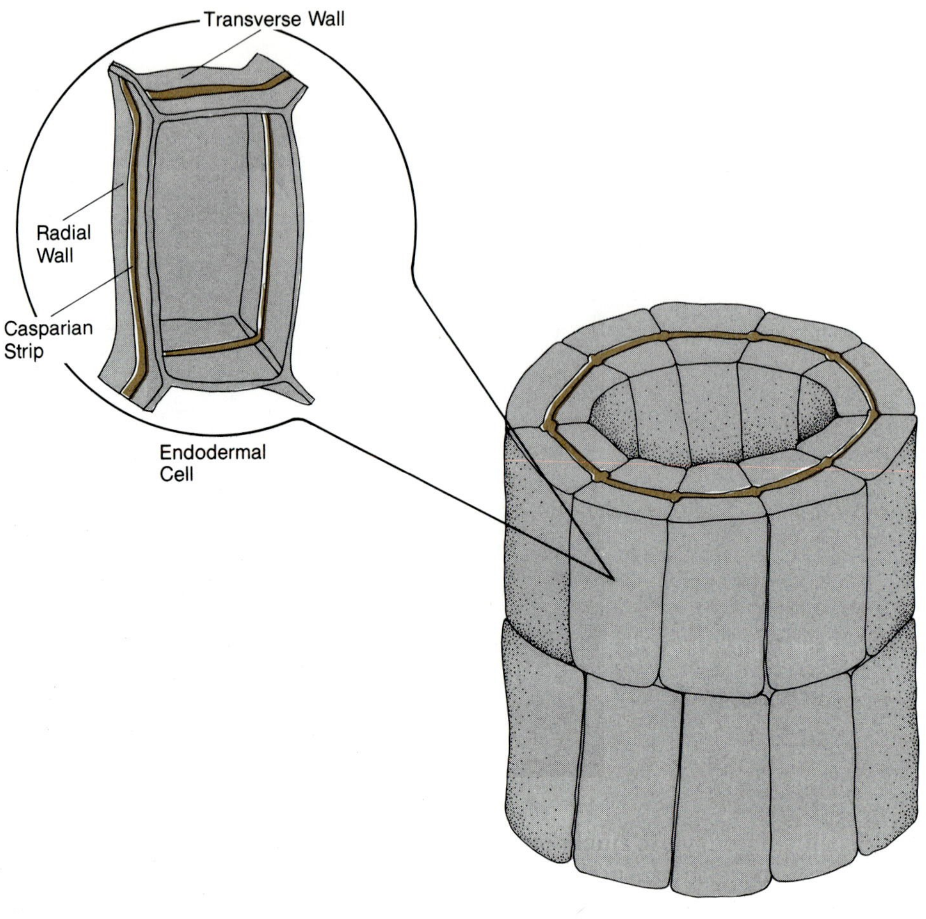

Figure 7–8. Casparian strip. The Casparian strips along the radial walls of adjacent endodermal cells fuse, forming a watertight barrier. Within the root, the endodermal cylinder has cells arranged on top of each other where the Casparian strips are also fused. This forces water to pass through the protoplasts of the endodermal cells and prevents it from passing between the cells.

along the long axis (see diagram, Figure 7–8). The Casparian strips of adjacent cells are tightly pressed together, thus creating a continuous suberized strip surrounding the vascular tissue. Only in the younger regions is the suberin restricted to a strip in the wall of endodermal cells. The endodermis in older regions is composed of cells with completely suberized cell walls, with the exception of **passage cells,** which have only a Casparian strip of suberin and are positioned opposite the xylem (Figure 7–9b).

Water and dissolved minerals move readily through the free space—the intercellular spaces and the normal cellulose cell wall—but not at all through the waxy, suberized regions (see Chapter 2 for details of cell wall construction). The endodermis thus serves a significant role in directing the flow of water into the xylem as well as in preventing the movement of dissolved substances from the soil to the xylem through free space. For water and minerals to enter the central conducting region of the root from the cortex they must move through the protoplasts of the endodermal cells. Differential selection takes place at the plasmalemma* of the passage cells or in any cell of the cortex and epidermis when dissolved substances are allowed to

Plasmalemma
The outermost portion of the cytoplasm that functions as a differentially permeable membrane.

Parenchyma

Endodermis

Pericycle

Vessel Elements

Phloem (Sieve Tube Members and Companion Cells)

(a)

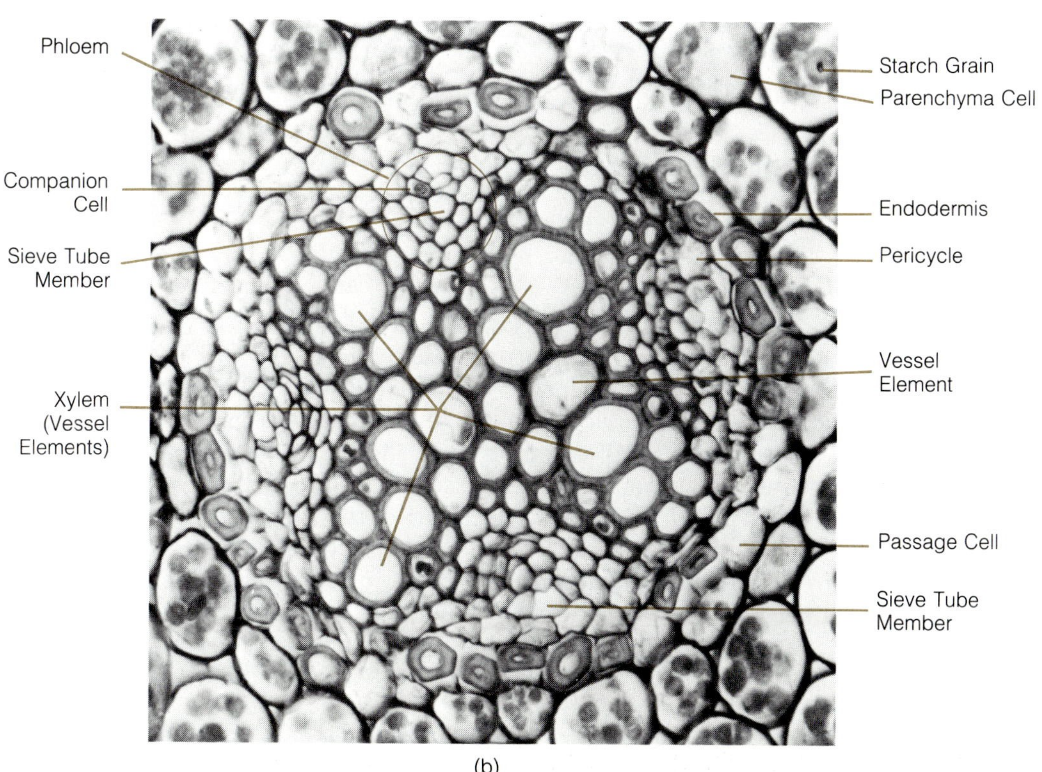

(b)

Figure 7–9. Root steles. The stele in a root is the central core of tissues enclosed by the endodermis. *(a)* Monocotyledonous root stele. *(b)* Dicotyledonous root stele.

pass through the plasmalemma. With the endodermal barrier in place, the only route leading to the xylem is from protoplast to protoplast, allowing the differential permeability* of the plasmalemma to be the major factor in determining the uptake of materials from the soil by the plant. If a cross section is made from a part of the root that does not absorb water, the entire ring of endodermal cells may be suberized, leaving no passage cells.

Permeability
The ability of molecules to penetrate a membrane or cell wall.

The central region of the root surrounded by the endodermis is the **stele.** All of the primary tissues* of the stele are derived from the procambium. The outermost tissue of the stele found next to the endodermis is the **pericycle,** consisting of a single layer of parenchymalike cells. The pericycle is a potential meristem in that some of its cells produce branch roots (Figure 7–10 and Plate 4B) while others give rise to the cork cambium* or segments of the vascular cambium* in roots that become woody.

Primary tissue
A tissue derived from a primary meristem.

Cork cambium
A secondary meristem that produces cork and phelloderm.

Vascular cambium
A secondary meristem that produces vascular tissues (xylem and phloem).

The most apparent difference between a dicotyledonous root and a monocotyledonous root is in the arrangement of the vascular tissues within the stele. In the dicotyledonous root (Figure 7–9b) the **xylem** is in the very center with 2 to 6 arms radiating like spokes of a wheel. The **vessel elements** can be easily recognized in the xylem. **Phloem** tissue is present in isolated groups between the xylem arms; the number of phloem strands is equal to the number of xylem arms. The phloem in the root is primarily made up of **sieve tube members** and **companion cells.** Parenchyma cells are often present between the xylem and phloem. What appears in cross section as a star-shaped central xylem and isolated groups of phloem are actually a central column of fluted xylem and strands of phloem originating in the procambium.

In a typical monocotyledonous root the center of the stele is a **pith** of parenchyma cells (Figure 7–9a). The pith parenchyma of some roots may become thick-walled and appear sclerenchymatous.* The xylem of these roots is arranged in isolated rows that radiate from the periphery of the pith. The largest vessel elements are closer to the pith and the smaller ones are toward the pericycle. Phloem is found in clumps between the rows of xylem.

Sclerenchymatous
Containing sclerenchyma fiber cells or sclereids.

Branch Root Formation

Since there normally are no buds on a root, the origin of branch roots, also known as lateral roots, is distinctly different from the origin of branches from buds on a stem. All branch roots originate from the pericycle in monocotyledonous and dicotyledonous plants as well as in gymnosperms.* Branch roots are initiated as root tips by cell division in groups of pericycle cells opposite the xylem arms within a segment of the root between the root hair zone and the older nonabsorbing part of the root (Figure 7–10). These tips contain the apical meristem and all primary meristems. As the young branch root elongates, it pushes its way through the cortex and epidermis and emerges to the soil. During the initial stages of its development and passage through the cortex, a portion of the endodermis acts as a protective covering until a root cap is formed. Also before the emergence of the branch root, the vascular cylinders of the original root and the branch root are connected by means of newly formed xylem and phloem derived from cells near the pericycle. All branch roots have an arrangement of tissues identical to

Gymnosperm
A plant that produces naked seeds, typically in cones.

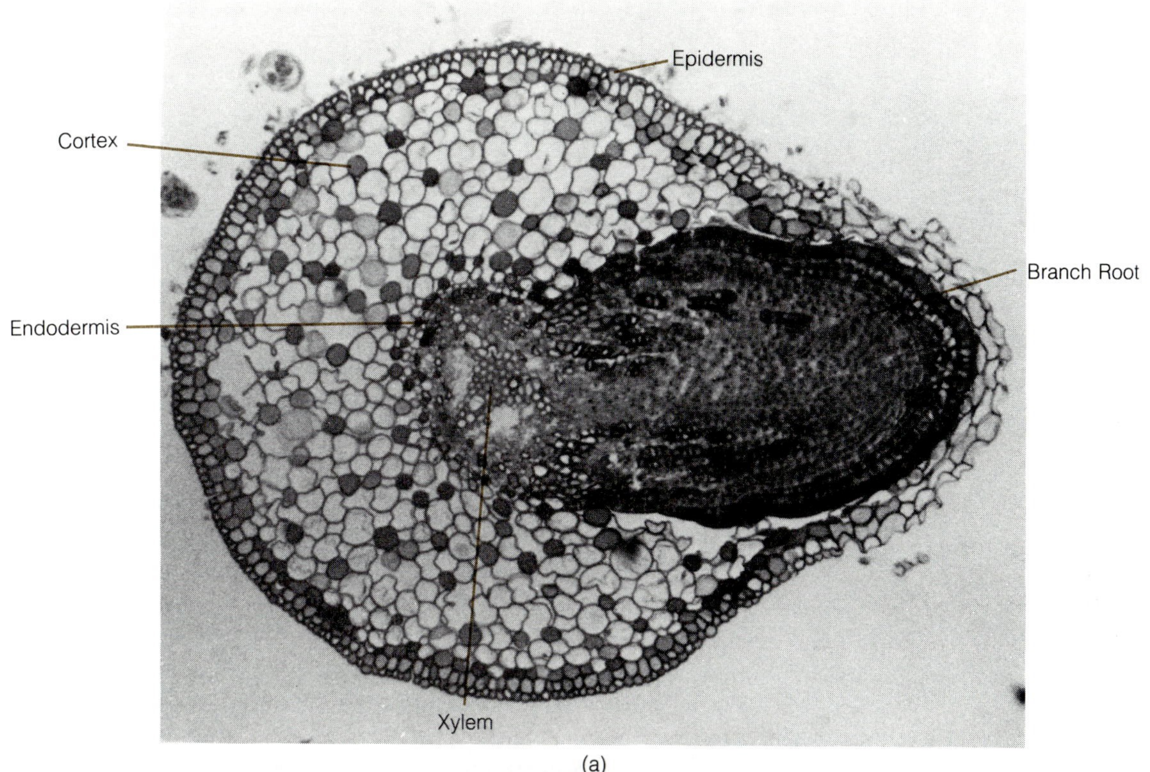

(a)

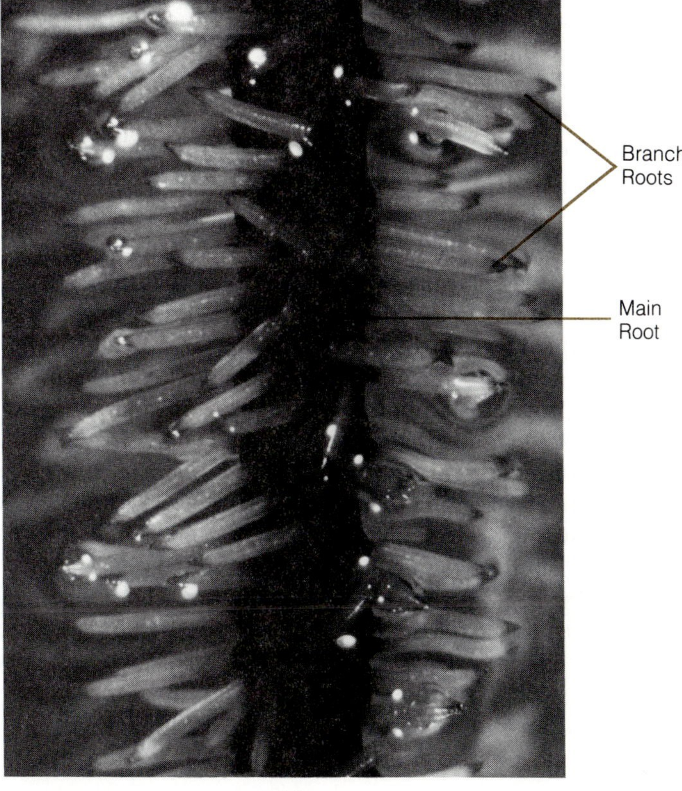

(b)

Figure 7–10. Branch root. Branch roots are formed from the pericycle allowing continuity of the vascular system of the two roots. *(a)* Cross section of a willow root showing the origin of a branch root. *(b)* Many branch roots produced by a main root of water hyacinth.

that found in any root tip. A single root is capable of producing many branch roots.

Secondary Growth in Roots

Because most monocotyledonous plants do not have a vascular cambium or a cork cambium, their roots do not produce secondary tissues. On the other hand, all trees and shrubs and some herbaceous dicotyledonous plants have woody roots, many of which may attain a significant diameter owing to accumulation of secondary tissues.

Secondary growth in roots is similar to that in stems, as described in detail in Chapter 6. Secondary tissues are produced by the **vascular cambium** and **cork cambium,** the former giving rise to secondary xylem and secondary phloem and the latter giving rise to cork* and phelloderm.* In roots that become woody, discontinuous strands of vascular cambium are initially formed between the primary xylem and primary phloem from procambial cells that had remained undifferentiated (Figure 7–11). Subsequently, additional strands of vascular cambium are derived from the pericycle to make a continuous layer of vascular cambium going around the xylem. As in a woody stem, the vascular cambium of a root produces secondary xylem toward the center and secondary phloem toward the outside. The pericycle continues to increase in diameter and remains as a ring outside the secondary phloem. While this is going on, a ring of cork cambium is derived from the pericycle and begins to produce cork toward the cortex and a limited amount of phelloderm cells in the direction of the pericycle. As in a woody stem, considerably more secondary xylem and cork than secondary phloem and phelloderm are produced during a given growing season. In the course of time, because of the accumulation of secondary vascular tissues and cork, the cortex and epidermis are pushed outward and are finally completely replaced by a bark.* Distinct growth rings*, similar to those present in woody stems, although not quite as obvious, can be recognized in the woody roots of plants growing in the temperate regions of the world.

Cork
A suberin-containing secondary tissue that forms the outer layer of woody stems and roots.

Phelloderm
Parenchyma cells produced by the cork cambium.

Bark
All tissues located outside the vascular cambium.

Growth ring
Secondary xylem produced during one growing season.

Root Modifications

Roots are typically present underground and function in absorption, storage, and anchorage. Nevertheless, it is not unusual for roots to be produced from the aerial* portions of plants or to have different functions.

Prop roots are adventitious roots that form on stems and function as an additional means of support for the plant. The prop roots of corn *(Zea mays)* originate from the stem above the soil level and anchor the plant, giving it additional support against the wind (Figure 7–12a). Prop roots of the banyan tree *(Ficus benghalensis)* are produced from horizontally growing branches and appear like pillars supporting the branches.

In plants such as poison ivy *(Toxicodendron radicans)* and English ivy *(Hedera helix)* adventitious roots produced along the stem attach these weak-

Aerial
Refers to the aboveground parts.

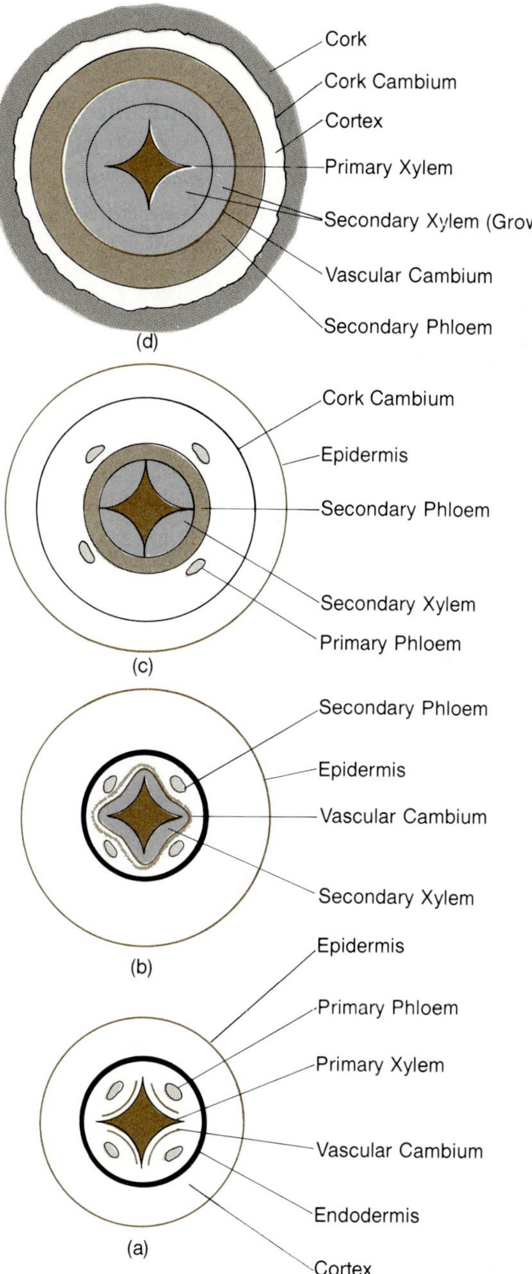

Figure 7–11. Secondary growth in roots. These diagrams represent cross sections from different levels of the same root of a woody plant. Section *a* is from a region closer to the tip and has only primary tissues; section *b* is from an area in which secondary growth has begun, while *c* and *d* represent areas at the end of one year and two years of secondary growth, respectively.

stemmed plants to a wall or to other support and are known as **climbing roots** (Figure 7–12b). Adventitious roots called **aerial roots** are common among various species of *Philodendron* as well as many orchids and other epiphytes* (Figure 7–12c). Such roots absorb moisture from the air. Aerial roots may also contain chloroplasts, as in the vanilla orchid and philodendron, in which case they will have the additional function of food synthesis.

Bald cypress trees *(Taxodium distichum)* have specialized roots called **breathing roots,** or "knees," which grow upwards until they reach above the

Epiphyte
A plant that grows on another plant without obtaining any water or minerals from it.

Figure 7–12. Root modifications. *(a)* Prop roots of corn. *(b)* Climbing roots of English ivy. *(c)* Aerial roots of orchid. *(d)* "Knees" of bald cypress.

ground (Figure 7–12d). Although their function is not fully understood, they are considered to provide aeration, especially oxygen, to the root system that is waterlogged owing to the swampy nature of the soil in which these trees grow.

Roots also can perform reproductive functions. Sweet potato *(Ipomoea batatas)* and dahlia *(Dahlia* sp.) usually are propagated naturally and in cultivation from pieces of roots that produce new shoot systems from adventitious buds.* Apple *(Pyrus malus)*, cherry *(Prunus* sp.), and teak *(Tectona grandis)* are examples of species that produce **suckers,** or new plants, from roots that can be separated and transplanted.

Adventitious buds
Buds produced from structures other than apical meristems.

One of the most highly specialized root types is the **haustorium** of parasitic plants* such as mistletoe *(Phoradendron* sp.) and dodder *(Cuscuta* sp.). These plants obtain either water or food or both from the host plants through their haustoria that penetrate the stems and establish vascular con-

Parasitic plant
A plant that derives part or all of its nutrients and water from another plant.

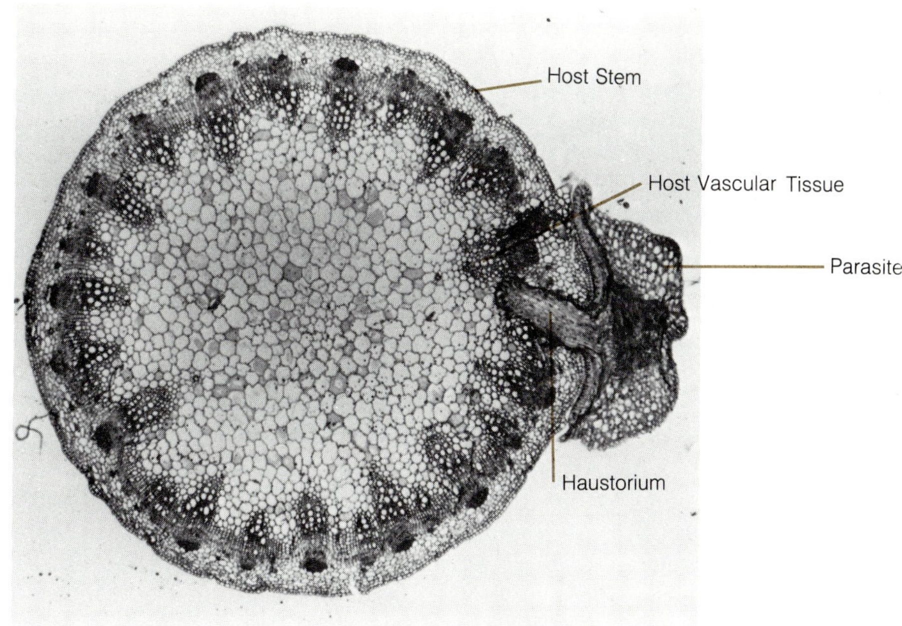

Figure 7–13. Attachment of haustorium. Transverse section of a host stem with an attached haustorium, the modified root of the parasitic plant dodder. The haustorium makes connection with the host vascular tissue, allowing the parasite to absorb water and food.

nection with the host (Figure 7–13 and Plate 4C). Some of these parasitic plants lack the ability to synthesize their food and are completely dependent on their hosts for survival.

Absorption of Minerals

A major function of the root is the absorption from the soil of minerals that are required for plant growth and metabolism. As indicated above, absorption occurs when dissolved minerals diffuse into the free space* of the root and pass through the plasmalemma of a root cell. While the actual movement of minerals into the protoplast of a root cell represents the first step in the translocation of minerals to the various regions of the plant where they are used, an often neglected aspect of mineral absorption by roots is the mechanism by which the dissolved minerals enter the root's free space. It was pointed out that root hairs vastly increase the surface area of the root system and the amount of soil with which the root comes into contact. For many plants, however, the surface area of the root hairs is not adequate to permit the absorption of a sufficient supply of minerals for optimal growth. In fact, it has been suggested that when plants invaded the land approximately 400 million years ago (Chapter 18) a major obstacle to life in a terrestrial environment, compared to the aquatic habitat from which they came, was a deficiency in mineral nutrients. Even when a soil contains more mineral nutrients than a body of water, the aquatic plant has greater access to

Free space
The cell walls and intercellular spaces of a tissue.

minerals because all of its surfaces are in contact with water and can absorb the minerals. In addition the large volume of water flowing past the plants continually bathes them with fresh supplies of dissolved minerals. A plant growing in soil has only its root system to absorb minerals from a limited volume of soil. A solution to this problem apparently evolved in the form of a symbiotic* association between plant roots and certain fungi, forming what is called a **mycorrhiza,** meaning "fungus-root." This mycorrhizal association is of nearly universal occurrence and has been reported in most groups of vascular plants, including the majority of economically important plants. It is becoming more and more evident that many forest trees are unable to grow well in the absence of the fungal partner. Although the role of the fungus is not fully understood, experimental data indicate that it plays a direct role in the absorption of certain essential elements such as phosphorus, potassium, calcium, nitrogen, zinc, copper, and molybdenum. It has also been suggested that the fungus even may provide some growth substances to the plant. The fungus, in return, may obtain carbohydrates and possibly vitamins from the plant. Mycorrhizae additionally are recognized as being significant in the recycling of soil nutrients owing to the role of the fungi in decomposing the organic remains, or humus, contained in the soil. Once this organic material has been broken down, the minerals of which it was composed again become available to the plants.

Symbiosis
(*adj.*, symbiotic) A close association between two different organisms.

There are two general types of mycorrhizal associations known—**endomycorrhizae** and **ectomycorrhizae**—the former being more common in nature. In the endomycorrhizal association fungal filaments extend from the soil into the cells of the root cortex and produce fungal structures called vesicles* and arbuscles* within the cells. In the ectomycorrhizal association the fungus either forms a mantle around the roots or invades the space between the cortical parenchyma cells or does both. Typically root hairs are absent from the ectomycorrhizae, their function being carried out by the fungal filaments. Because mycorrhizae are of particular importance in soils that are nutrient-poor, their role in the absorption of mineral elements normally unavailable to plant roots is being studied currently. Investigations are also being conducted to determine the efficiency of nutrient uptake in the presence and absence of mycorrhizal fungi with the intention of increasing crop production and reforestation by the incorporation of mycorrhizal fungi into nutrient-deficient soils. If successful, such methods could reduce the use of chemical fertilizers and increase the potential for crop production in many of the Third World nations. One of the major problems yet to be solved is how to culture mycorrhizal fungi on a large scale to be used as inoculum.* Methods being tested to introduce mycorrhizal fungi into soil include pelleting seed with fungal spores and adding pellets of infested soil along with seeds.

Vesicle
Saclike, food-storing fungal cell produced within a root cell or in the space between cells.

Arbuscle
Highly branched fungal structure within the root cell that functions for the exchange of nutrients.

Inoculum
A small amount of a culture of microorganisms that is used to start a new culture.

Common in different parts of the United States is the strip-mining of subsurface deposits of coal, a practice that is cheaper than underground mining. One of the problems related to reclamation of strip-mined lands is the poor growth of grasses and other crops due to the changes in the properties of the soil during storage, the mixing of topsoil and subsoil, and the reduced numbers or even the total absence of viable mycorrhizal fungi and other soil microorganisms in areas where the topsoil has been removed during mining operations. Presently extensive investigations are being conducted to determine the role of mycorrhizal fungi in land reclamation in strip-mined areas. Preliminary results indicate that incorporation of mycorrhizal fungi into these biologically inactive soils can enhance the growth of forage grasses and other crops through increased absorption of minerals.

The mineral nutrient required by plants in the largest amount is nitrogen—a building block for many important macromolecules such as proteins and nucleic acids (Chapter 3). Although nitrogen is plentiful, constituting approximately 78% by volume of air, it is frequently in short supply for plant growth since most living organisms are unable to utilize atmospheric nitrogen. In nature, plants usually get their nitrogen from nitrates, ammonia, or other nitrogen-containing compounds present in the soil. Certain species of soil bacteria, by contrast, are able to utilize atmospheric nitrogen and to convert it to nitrogen-rich compounds through a process known as nitrogen fixation. One such bacterium, *Rhizobium,* lives in a symbiotic association in the roots of leguminous plants such as soybean, clover, alfalfa, peas, and beans. These bacteria enter the root tissue through root hairs or other epidermal cells and invade the cortical parenchyma, the cells of which, in turn, proliferate producing **root nodules,** spherical outgrowths containing both bacteria and root tissue (Figure 7–14). This abnormal growth response by the cortical tissue is induced by the bacteria through the liberation of cytokinins, a class of plant hormone responsible for cell division (see Chapter 12 for a discussion of cytokinins). Through this symbiotic association the bacteria obtain carbohydrates from the plant and make available nitrogen-containing compounds to their host. Gaseous nitrogen is first trans-

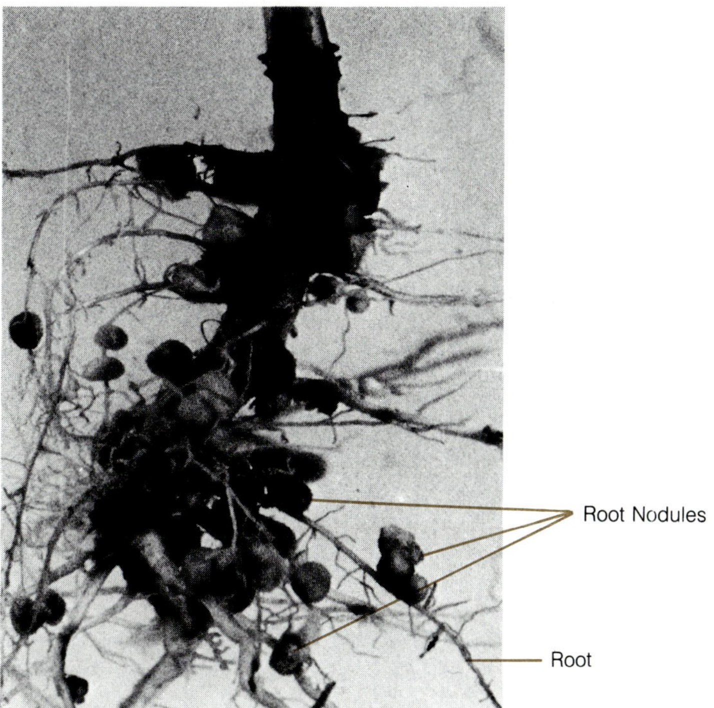

Figure 7–14. **Root nodules.** Root nodules of the soybean plant consist of proliferated root tissue and nitrogen-fixing bacteria, which live symbiotically within the roots of legumes and other plants.

formed by the bacteria into ammonia and later converted to amino acids or other nitrogen-containing compounds within the root nodules. Subsequently some of the fixed nitrogen is transported to other parts of the plant.

As a result of nitrogen fixation by these bacteria, large amounts of organic nitrogen are stored in the nodules and are subsequently incorporated into the soil when these roots are plowed under after harvest. It has been estimated that alfalfa roots plowed back into the soil can add approximately 300 pounds of nitrogen per acre. It is well known that crop rotation using leguminous plants is an efficient method for enriching the soil. It is interesting that the amount of nitrogen fixation in the root nodules is reduced if nitrogenous fertilizer is added to the soil and that the decrease in nitrogen fixation is proportional to the added amount of fertilizer. Since plants with the nitrogen-fixing bacteria in their roots are able to do well in nitrogen-deficient soils, attempts are being made presently to induce symbiosis of nitrogen-fixing bacteria with crop plants that normally lack them. To enhance the nitrogen-fixing capabilities of leguminous plants grown in soils deficient in the essential bacteria, some garden shops and seed companies are now marketing packages of soil inoculants, advising gardeners that by sprinkling the dry granules in the furrows where they sow their pea or bean seeds plant growth and crop yields can be increased.

Mineral Nutrition

Earlier in this chapter it was stated that the absorption of minerals is a major function of roots, but little has been said of the use of these minerals. Gardening books and plant stores refer to the minerals absorbed by plants as "plant food." In reality, food for plants is the same as food for animals—sugars, proteins, fats, and so on. Plants produce their food as the result of metabolic reactions beginning with the process of photosynthesis. Minerals absorbed by a plant are either present in the soil or added to it in the form of commercial fertilizer or manures. Fertilizer is merely a source of the minerals that are necessary for plant growth; the reason that fertilizer is referred to as plant food is the well-known observation that the addition of fertilizer to soil usually causes a dramatic increase in the growth rate and an improvement in the general appearance of the plant. The reason for such beneficial effects on plant growth is that the soil was lacking in one or more minerals required for plant metabolism and the added fertilizer compensates for this deficiency.

The major minerals required by a plant include nitrogen, phosphorous, potassium, calcium, magnesium, and sulfur. These are the **macronutrients** that are required in fairly large amounts for good plant growth. There are other minerals that are required for plant growth but in such small amounts that these are referred to as **micronutrients** or **trace elements.** Both macro- and micronutrients are necessary for healthy plant growth and collectively are referred to as essential minerals. All of these substances, with the sole exception of potassium, are necessary because they are components of one or more type of molecule in the plant cell. A deficiency in any of the essential minerals is usually manifested by recognizable symptoms in the plant's growth or appearance.

Nitrogen is the most abundant chemical element found in plant cells after carbon, hydrogen, and oxygen. Almost all of the molecules that make up protoplasm contain nitrogen (N)—proteins*, nucleic acids*, certain types of lipids*, chlorophyll*, and so forth (Chapter 3). Therefore a deficiency of nitrogen seriously affects plant growth in several ways. The nitrogen-deficient plant is stunted because reduced frequency of cell division results from the lack of nitrogen needed for the synthesis of new protoplasm. Reduced chlorophyll synthesis contributes to a drop in food production, enhancing the stunting effect and also leading to a yellowing of the older leaves, which is a characteristic sign of nitrogen deficiency in plants.

Phosphorus, although not a component of as many cellular molecules as nitrogen, is an integral part of certain molecules that are fundamental to cell function. DNA, the genetic material, has a molecular backbone held together by a phosphorus-containing group (Chapter 3). Therefore a phosphorus deficiency also stunts growth by preventing cell division. In addition, the phosphorus atom is part of the energy transferring molecule ATP (Chapter 10) that is generated during the process of respiration. As indicated in Chapter 10, respiration provides the energy for cellular metabolism and therefore for growth. Fertilizer sold for lawns and gardens has numbers listed on the bag representing the relative percentages of the "big three" mineral elements it contains. A listing of 10:10:10 indicates ten percent each of N (nitrogen), P (phosphorus), and K (potassium). Thus potassium, like nitrogen and phosphorus, is a major essential mineral element for plant growth, yet, unlike the other two, it is not a major component of cellular molecules. Despite this, potassium is as essential for the survival of the plant cell as it is for animal cells. The role and essential cellular function of potassium are not completely known; however, potassium is an activator of many enzymes and is involved in the maintenance of the correct electrical potential of the plasmalemma that, in turn, affects the permeability of this membrane. From the discussion in Chapters 2 and 3, it can be seen that the permeability of the plasmalemma significantly controls cellular metabolism and therefore growth of the whole plant. Another major function for potassium is in the opening and closing of the stomata, or pores, found in the epidermis of a leaf (Chapter 8). A significant factor causing the opening of stomata is the uptake of potassium from other epidermal cells by the guard cells surrounding each stoma. Subsequent loss of potassium by the guard cells results in the closing of the stomata.

The remaining three macroelements and the trace elements may be present in the same bag of fertilizer, but not mentioned on the label. These all are present in lesser amounts as deliberate additions to a premium fertilizer and often as desirable contaminants in cheaper grades. Calcium has a number of cellular roles, two of which are of major importance for plant growth. Calcium is an essential component of the pectin portion of the "glue" that holds the cell wall together, as well as forming the cell plate that initially separates the two halves of a dividing cell (Chapter 4). An even more crucial role of calcium is that of maintaining the stability of the cytoplasmic fluid. This fluid must possess the correct balance of liquid to gel if its properties

Protein
A large molecule consisting of a long chain or chains of amino acids.

Nucleic acid
A complex organic molecule involved in the storage and expression of genetic information.

Lipid
An organic molecule made up of one glycerol molecule and three fatty acid molecules.

Chlorophyll
A green pigment that functions in photosynthesis.

of metabolism, cytoplasmic streaming, and other processes are to continue. Calcium is the linking agent, so that a deficiency in calcium will cause the cytoplasmic fluid to be too "runny" and the cell will die. A deficiency in magnesium, like a deficiency in nitrogen, will affect growth because of a reduction in chlorophyll synthesis, as each chlorophyll molecule must contain one magnesium atom. In the absence of magnesium, chlorophyll production would completely cease. In addition to being a component of chlorophyll, magnesium is found in a number of enzymes crucial to cellular metabolism. Without magnesium these enzymes are unable to function, since magnesium plays a role in the catalysis of the metabolic reaction. Thus a cell with reduced amounts of available magnesium would have a smaller number of these enzymes and its metabolic rates would be correspondingly reduced. Although required by plants in fairly small amounts, sulfur is an essential component of certain amino acids.* A deficiency of sulfur will inhibit protein synthesis with consequent effects on growth and metabolism.

Amino acid
A simple organic molecule that is the building block of proteins.

Since the trace minerals have equally fundamental roles in cellular metabolism (Table 7–1), the effectiveness of mineral absorption by roots (including mycorrhizae) has a controlling influence on plant growth. The transport of minerals from the roots to the shoot will be described in Chapter 11; the observed effects of mineral deficiencies are detailed in the appendix on green thumb botany.

As the result of increased knowledge of the mineral requirements of plants, it is now possible to grow plants without soil. If the root system is suspended in a well-aerated solution containing all of the required minerals, a plant can be grown, with no ill effects, from seedling to maturity in water. This method of plant growth called **hydroponics** has important ramifications. In areas with poor soil or cities where no soil is available, crops can be grown that otherwise would be impossible to grow. Furthermore, plants grown hydroponically are less prone to disease organisms such as would be

Table 7–1.
Function of Micronutrients in Plant Metabolism

Iron	Electron transport reactions of photosynthesis and respiration; chlorophyll synthesis; nitrogen metabolism.
Chlorine	Enhances electron transport reactions of photosynthesis; unknown role in root development.
Manganese	Involved in the breakdown of H_2O in photosynthesis; component of many enzymes.
Boron	Possibly involved in sugar translocation; may be involved in DNA/RNA synthesis
Zinc	Component of various enzymes; probably involved in auxin synthesis.
Copper	A component of one of the electron carriers in photosynthesis; component of some enzymes.
Molybdenum	Involved in nitrogen fixation; may be involved in phosphorus metabolism.

found in soil. Finally, by providing the correct balance of mineral nutrients at each stage of plant growth, optimal growth and yield can be achieved more economically. The feasibility of hydroponics is effectively demonstrated in one of the exhibits at the EPCOT Center in Florida.

Summary

A root system consists of one or more main roots with its branch roots and is typically located underground. Distribution of a root system is influenced by the moisture and oxygen content, the temperature, and the physical nature of the soil. The primary functions of a root system are absorption of water and minerals, storage of food, and anchorage of the plant.

The primary root is derived from the radicle in plants grown from seed, while in plants grown from cuttings roots are derived from stems or leaves. Roots produced from organs other than roots are called adventitious roots. The direction of root growth is influenced by gravitational force, a phenomenon known as geotropism.

Many dicotyledonous plants have a tap root system in which there is only one major root, whereas the monocotyledonous plants have a diffuse root system consisting of several major roots. The tip of a root is usually covered by a root cap and is devoid of any other structures for several millimeters. The epidermal cells for a distance beyond this bare region produce root hairs that increase the surface area for absorption. Branch roots are produced only beyond the root hair zone. Internal regions corresponding to the external features are the region of cell division, consisting of the apical meristem and the primary meristems that are protected by the root cap; the region of elongation, corresponding to the external bare region; and the region of maturation, corresponding to the root hair zone.

The internal organization of mature tissues in a dicotyledonous root is as follows: a single-layered epidermis, a wide cortex consisting of many layers of parenchyma cells and a single layer of endodermis, a single layer of pericycle, and a centrally located fluted xylem with phloem between the arms of the xylem. The major difference in a monocotyledonous root is the presence of a central pith with the xylem arranged in the form of spokes radiating from the periphery of the pith. All branch roots originate from the pericycle as root apices and force their way out through the cortex.

Roots can increase in diameter either by the proliferation of cortical parenchyma, as found in certain storage roots, or by secondary growth similar to that in stems. Secondary xylem and secondary phloem are produced by a vascular cambium formed between the original xylem and phloem. Cork and phelloderm are produced by a cork cambium derived from the pericycle.

Root modifications are found among different plants. Many of these modified roots are adventitious because of their origin from the stem. Prop roots provide support for stems and branches; climbing roots help attach weak-stemmed plants to a support; aerial roots absorb moisture from the air and are often photosynthetic; breathing roots provide oxygen to submerged roots; suckers are new plants produced from roots; haustoria are roots of parasitic plants which absorb either water or food or both from the host.

The effectiveness of mineral absorption by roots is increased by the formation of mycorrhizal associations with fungi. Some plants are able to increase the amount of nitrogen available to them by forming symbiotic relationships with nitrogen-fixing bacteria. The various minerals that are absorbed by the roots are necessary for the growth of the plant. In some cases, such as nitrogen, the mineral is a major component of cellular chemicals and is needed in relatively large amounts, whereas other minerals are only needed in trace amounts.

Review Questions

1. What are the two major types of root systems? How do they differ from each other?
2. What is the origin of the primary root?
3. What are the different meristems present in a root? Where are they located?
4. How are mycorrhizae beneficial to plants?
5. What role do root nodules play in the nitrogen cycle?
6. Explain the role of root endodermis.
7. What is the significance of branch root formation from the pericycle rather than from one of the outer tissues?
8. How is a root morphologically and anatomically adapted for its role as:
   a) an absorbtive organ,
   b) a storage organ
9. List the macronutrients and micronutrients and write one function for each on your list.

Roots as Food

Anthropologists devote considerable effort to defining the components of the diet of the remaining preagricultural peoples of the world, as well as of our earliest human ancestors. Similarly, zoologists have been able to elucidate the diets of most types of extinct and extant animal species. From these studies comes a general picture of the common use of roots as food by animals and early human beings. In fact, a common description of the diet of our early ancestors and of modern-day hunter-gatherers is one that includes roots in addition to fruits, nuts, game animals, fish, various grubs, and insects. Although people have selected and improved the agriculturally important root-crop plants, the fact remains that in nature plants do produce roots whose function is food storage instead of just anchorage and absorption. In certain species with large, fleshy taproots, starches, sugars, some protein, vitamins, and minerals can be found in the root cortex or in the xylem and phloem parenchyma. Plants possessing this stored food are mostly biennials, accumulating nutrients during their first year of growth and then using these reserves for the energy needed during the second year when they produce flowers and fruit.

Among the better-known edible roots are the carrot *(Daucus carota)*, cultivated for at least 2000 years, and the beet *(Beta vulgaris)*, grown both as a vegetable and as a source of sugar. Radishes, parsnips, rutabagas, and turnips are also edible taproots used extensively both for human consumption and as food for livestock.

Two of the most important root crops, in terms of numbers of people who consider them dietary staples, are the sweet potato *(Ipomoea batatas)* and manioc *(Manihot esculenta)*. The former, a native of the Central and South American tropical lowlands, was taken by Spanish explorers to Europe, from whence it spread to China, India, Japan, Indonesia, and the Pacific islands. The sweet potato, which grows as a trailing vine, does not produce a taproot; instead, some of its roots become greatly swollen with starch deposits in the xylem and phloem parenchyma. Sweet potatoes, containing up to 5% protein, as well as large amounts of iron, calcium, and minerals, are by far the most nutritious of all the edible roots. Manioc (also known by such names as tapioca, cassava, or yucca), by contrast, has considerably less food value since it contains almost no protein. Nevertheless, it is a major source of carbohydrate in the tropics where the easily cultivated roots are boiled, baked, or dried and ground into a meal. Although lacking the prestige value of the more expensive grains, root crops have long provided a cheap, filling source of food for millions of people.

8

Leaves

Outline

Introduction

About 400–500 million years ago the ancestors of the land plants began the evolutionary transition from an aquatic to a terrestrial environment (Chapter 18). Among the two major groups of land plants, the bryophytes and vascular plants, leaves or leaflike structures evolved in at least some species. True leaves are characterized by the presence of vascular tissue* and thus are found only among club mosses, horsetails, ferns, gymnosperms*, and flowering plants. The leaflike structures found on bryophytes, such as mosses and liverworts, function as leaves but lack vascular tissue; hence they are not classified as true leaves. Although it is not possible to trace the evolution of the leaflike structures of bryophytes, fossil records and other evidence clearly show the evolutionary development of leaves within the vascular plant group. As described in Chapter 18, it appears that the first vascular plants had bodies that consisted of little more than stems similar to the whisk ferns. Since vascular plants, like the vast majority of other plants, produce their own food through the process of photosynthesis*, these stems were photosynthetic. A rod-shaped stem is not well suited for photosynthesis, however, because it does not provide a sufficiently large surface area for the absorption of light. Fossil evidence indicates that there was an evolutionary trend among the early vascular plants toward the production of small expansions on a stemlike axis. From this beginning, leaves have evolved to the efficient photosynthetic organs of the higher plants whose stems typically have only a minor role in photosynthesis.

Vascular tissue
Xylem and phloem tissues in which water and food, respectively, are transported.

Gymnosperm
A seed plant that produces naked seeds, typically in cones.

Photosynthesis
The metabolic process by which carbohydrates are produced from CO_2 and a hydrogen source such as water, using light as the source of energy.

Many leaves, particularly those found among the flowering plants, have extensive surface-to-volume ratios that maximize light absorption, and in some species leaves rotate on their stalks while tracking the sun across the sky. Leaf cells contain numerous chloroplasts*, which give leaves their characteristic green color. Leaf surfaces also feature numerous minute surface openings that facilitate diffusion of atmospheric carbon dioxide into the leaf. Rapid diffusion of gases throughout the tissue is also aided by the fact that most leaves are relatively thin, consisting usually of only a few layers of cells with large intercellular spaces.

Chloroplast
The type of plastid in which photosynthesis occurs.

In addition to their primary role in photosynthesis, leaves contribute to the recycling of water from the soil to the atmosphere as the result of the evaporative loss of water in transpiration.* The extensive vascular system of the leaf carries the water needed for food production and subsequently transports newly synthesized food to different parts of the plant for utilization or storage.

Transpiration
The evaporation of water from leaf and stem surfaces through stomatal openings.

Leaves are the most conspicuous structures on a stem and are present on all vascular plants, except on whisk ferns and certain xerophytic species such as cacti, where the leaves are modified into spines or are totally missing. There is extreme variation in size, shape, form, and coloration of leaves. Some of the smallest leaves are the scale leaves of primitive vascular plants such as the club mosses, horsetails and certain gymnosperms. These leaves measure only 1 or 2 millimeters wide and less than 5 millimeters long. At the other end of the scale are the leaves of many palms that are 6 meters (20 feet) long and almost half as wide (Plate 5B). One of the most interesting leaves is that of a native tropical South American water lily *(Victoria regia)* that has circular floating leaves with upturned edges. These leaves reach a diameter of 2 meters or more and are strong enough to support a small child. The diverse shapes of leaves include needlelike, spinelike, cylindrical, spindle-shaped, ovoid, heart-shaped, fan-shaped, pitcherlike, feathery, and circular types (Figure 8–1). It is not unusual for the same plant to have leaves of different shapes, as is the case in mulberry and sassafras (Figure 8–2). Leaf surfaces can be smooth, hairy, glossy, or sticky.

Although the typical color of leaves is green due to an abundance of chlorophylls, leaves can have a wide range of colors. During most of the growing season, the great majority of leaves are green; however, many of these same leaves display a beautiful array of colors during autumn. Trees such as elms, birches, and poplars usually turn yellow, while shades of red are predominant in oaks and maples.

Yellow colors are produced by the non-water-soluble carotenoids in the chloroplasts. Although the carotenoids are also present when leaves are green, their color is masked by the predominance of the chlorophylls. With the onset of cooler weather and shorter days, the synthesis of chlorophylls

Figure 8–1. **Leaf shapes.** These drawings represent some of the more common shapes of leaves.

Figure 8–2. Leaf dimorphism. It is not uncommon to have different shapes of leaves on the same plant as seen on this mulberry branch.

diminishes and the existing chlorophylls are degraded. As the chlorophyll content decreases, the leaves turn yellow.

The purple and red colors are due to water-soluble pigments such as anthocyanins and flavones present in the vacuole. Most trees do not synthesize anthocyanins and flavones until autumn. At that time, the drop in temperature reduces the transport of sugars out of the leaves. In some maples, oaks, and other trees the excess sugar accumulating in the leaf cells is used to synthesize anthocyanins, thus producing hues of red. The different shades of yellows, reds, and oranges are due to the varying proportions of these pigments present in leaves (Plate 5A). The ideal conditions for the development of brilliant foliar colors are bright, sunny days and a sudden, dry, cold spell during early autumn.

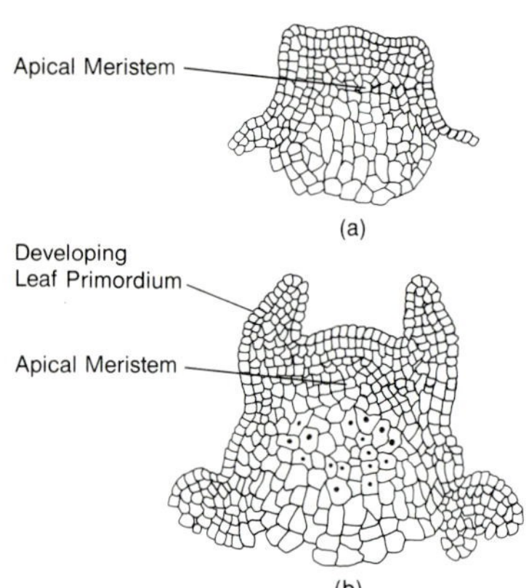

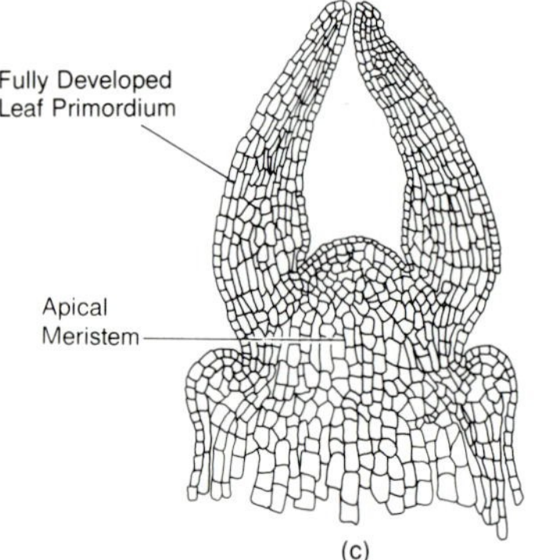

Figure 8–3. Leaf origin. Leaves originate as lateral outgrowths of the apical meristem as shown in this sequence of drawings.

Origin of Leaves

Leaves of flowering plants originate as lateral projections of the apical meristem in terminal and lateral buds (Figure 8–3). These outgrowths of the apical meristem, known as **leaf primordia,** subsequently give rise to mature leaves by cell division and cell enlargement. The extent and direction of growth during the initiation of leaves by the apical meristems varies considerably from one species to another, a factor that accounts for the wide variation in size and shape found among mature leaves. The position of these leaf primordia in a bud determines the relative position of leaves on a stem. The first pair of leaves, called primary leaves, on a plant are produced during the development of the seedling from a seed. Unlike stems, mature leaves typically do not possess meristematic tissues, hence show a determinate* type of growth. Exceptions are members of the grass family, such as lawn grasses, whose leaves continue to grow despite mowing because of the presence of a meristem at the base of the leaf. Only in rare cases can a leaf increase in size once it has reached maturity. Since leaves do not have the capacity to produce new tissues, they have a relatively short life span and the old ones are regularly shed. There are some notable exceptions to this general rule, however. Among evergreen conifers* leaves typically remain on the plant for a period of 3–5 years. Cycads* and many tropical flowering plants widely used as houseplants also display this tendency to retain their leaves for several years. *Shefflera, Dieffenbachia* (dumb cane), and *Monstera* (split-leaf philodendron) are familiar examples of tropical species that keep their leaves for a relatively long period of time. Among temperate flowering plants there are also examples of evergreen species—holly *(Ilex), Magnolia,* and some species of oak *(Quercus)* among others. Evergreen species of flowering plants are more common in areas with less severe winters. Most flowering plants, however, are deciduous with an individual leaf persisting only for one growing season before being shed. In perennial plants new leaves are produced on new stem segments replacing old leaves that have been shed. Each growing season the total number of leaves increases as the plant grows larger. Because leaves only originate at a shoot apex, however, only the younger regions of a perennial shoot will have leaves. An exception to this can be found on many temperate-zone trees and shrubs where leaves are observed on regions of stems that are several years old. In this case the leaves are produced by new branches that have just begun to develop and thus the leaves did indeed originate at a shoot apex, but not the apex of the main stem.

Determinate
Of limited extent; a type of growth pattern characteristic of flowers and leaves.

Conifers
The major group of gymnosperms; examples include pines, firs, spruces.

Cycad
A palmlike gymnosperm native to the tropics.

Leaf Morphology

A typical leaf has two parts, a stalk called the **petiole** and a flat, expanded portion, the **blade,** which is attached to the tip of the petiole (Figure 8–4). The petiole, although it can vary in size and shape, is commonly cylindrical with its base attached to the stem at the **node.** The **axillary bud,** also known as a **lateral bud,** is a characteristic structure usually present on the stem immediately above the point of attachment of a leaf. Since leaves are attached to the stem at the node, the segment of the stem between two adjacent nodes is called the **internode.** Paired leaflike structures known as **stipules** are present at the base of petioles in many flowering plants such as pea, rose, hibiscus, and sycamore (Figure 8–5). Despite their leaflike appearance, stipules are of minor importance in photosynthesis in most plants and simply may represent a structure that has lost its major function but which has not yet been lost through evolutionary change. The base of a petiole may encircle the stem to varying degrees. Leaves without petioles are referred to as **sessile** leaves, their blades being directly attached to the stem. Among the grasses the base of the leaf blade completely encircles the stem,

Figure 8–4. **Typical leaf.**

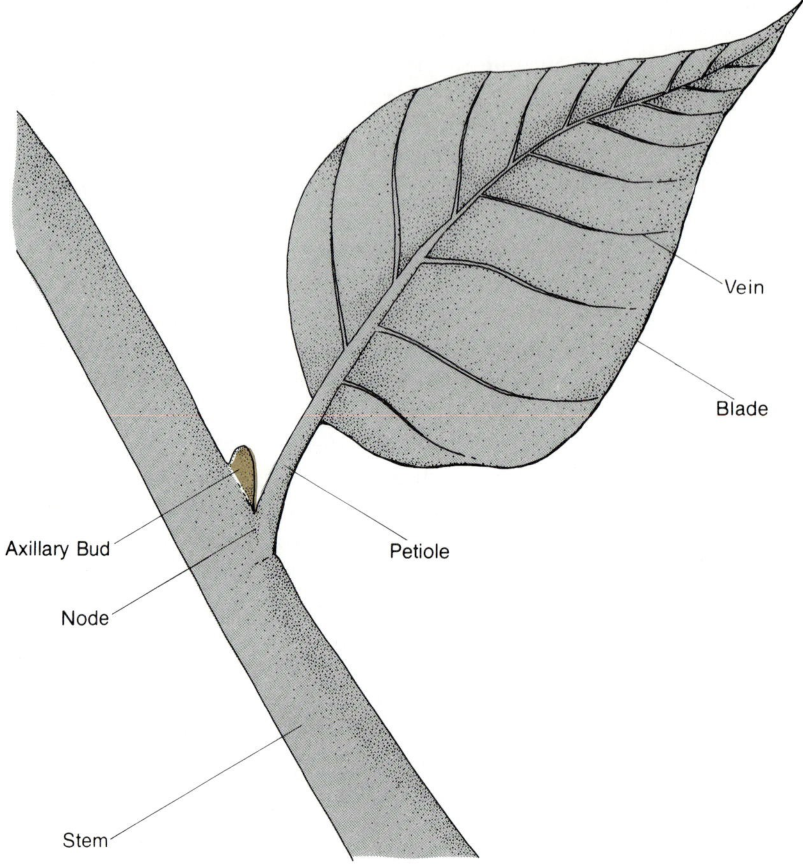

Figure 8–5. **Stipules.** Stipules are small appendages present at the base of the petiole as on the rose leaf seen here. On plants such as peas and sycamore, stipules are produced as small leaflike expansions at the node.

forming a **leaf sheath** that extends from that node almost to the next higher node, thus enclosing a major portion of the internode (Figure 8–6). **Veins,** which provide the structural framework of a blade, consist of vascular tissues (Figures 8–7). They are often prominent especially on the lower surface where a complex pattern of veins of different sizes can be recognized. This pattern of **venation,** or the arrangement of veins, is a common characteristic used in plant identification. In particular, the classification of a flowering plant as either monocotyledonous* or dicotyledonous* usually can be made on the basis of venation—dicots have **netted venation** while monocots have **parallel venation.**

Monocotyledonous plant
A flower-producing plant whose seeds have only one cotyledon.

Dicotyledonous plant
A flower-producing plant whose seeds have two cotyledons.

Leaves with **pinnately netted** venation have a blade that is bisected along its long axis by a large vein called the **midrib** that is continuous with the petiole. Smaller branch veins are produced from the midrib, each of which gives rise to a succession of progressively smaller branches that ultimately form a network. **Palmately netted** leaves have several veins of equal size diverging from the tip of the petiole instead of just a single midrib. Secondary veins then complete the network. Leaves with **parallel venation,** common among the grasses, have many major veins running parallel to each other along the length of the blade. The vascular system of the blade is continuous with the vascular system of the stem through one or more con-

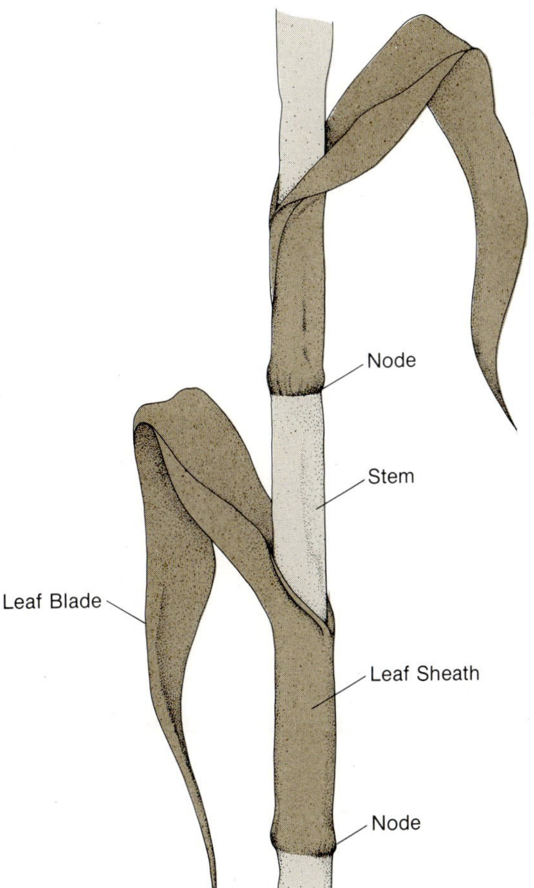

Figure 8–6. Leaf sheath. The leaf stalk in grasses forms a sheath that encloses a portion of the internodal segment of the stem.

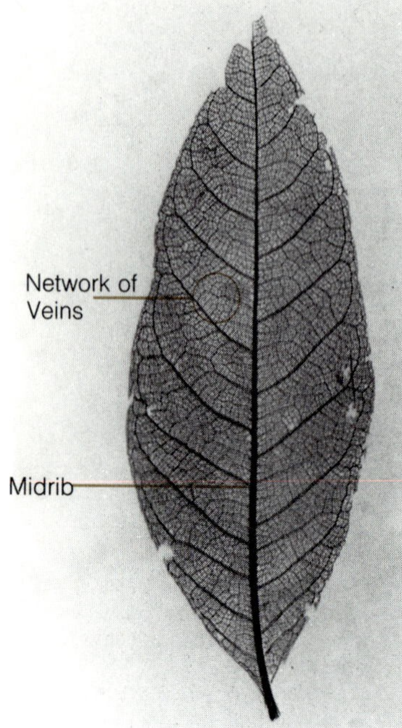

Figure 8–7. Leaf vascularization. This photograph of a cleared leaf shows that the vascular tissue (veins) make up the structural framework of a leaf.

nections known as **leaf traces** that branch off from the vascular tissue of the stem, which is in the form of a continuous cylinder. Leaf traces enter the blade through the petiole. As a leaf trace branches off, it leaves a space called a **leaf gap** in the vascular cylinder of the stem (Figure 8–8).

Simple and Compound Leaves

Leaves of plants such as redbud, elm, and maple are considered to be **simple** because there is only a single blade per leaf on these plants (Figure 8–9a). The blade margins of these plants can be **smooth, toothed,** or **lobed,** respectively (Figure 8–10). In contrast, when the blade is dissected into several **leaflets,** as in walnut, ash, hickory, and many other plants, the leaf is **compound** (Figure 8–9b, 9c, 9d), with each leaflet attached to a primary or secondary petiole, depending on the type of compound leaf. Since axillary buds are present only in the axils of primary petioles and absent from the axils of leaflets, the position of the axillary bud is used as a characteristic to determine whether a leaf is simple or compound. Another diagnostic characteristic that can be used to distinguish simple and compound leaves is the orientation of the leaves or leaflets. All the leaflets of a compound leaf are oriented in the same plane whereas if each leaflet were to be a simple leaf instead, they would be oriented in different planes.

If the leaflets of a compound leaf are attached along the length of a primary petiole, the leaf is said to be **pinnately compound,** as in ashes, walnuts, and roses (Figure 8–9b). In some plants the leaf is doubly pinnately compound or **bipinnately compound,** with leaflets attached to secondary petioles, as in Kentucky coffee tree, honey locust, and some ferns (Figure 8–9d). **Palmately compound** leaves have all of the leaflets attached to the tip of the primary petiole, as in horse chestnut, buckeye, and Schefflera (Figure 8–9c).

Figure 8–8. Anatomical features of a node. The vascular tissue of a leaf originates as a branch (leaf trace) from the vascular cylinder of the stem, resulting in a gap in the cylinder (leaf gap).

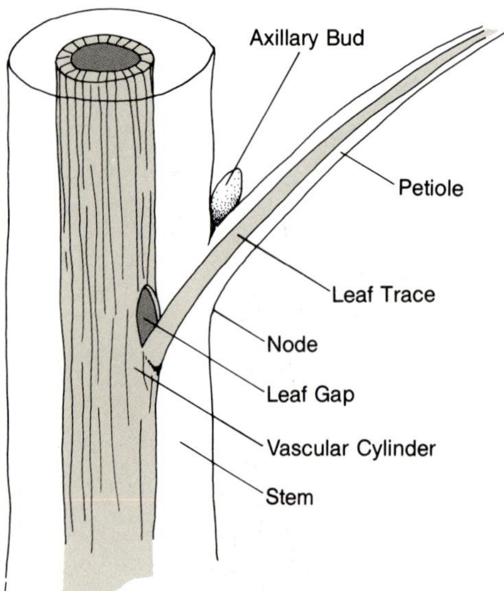

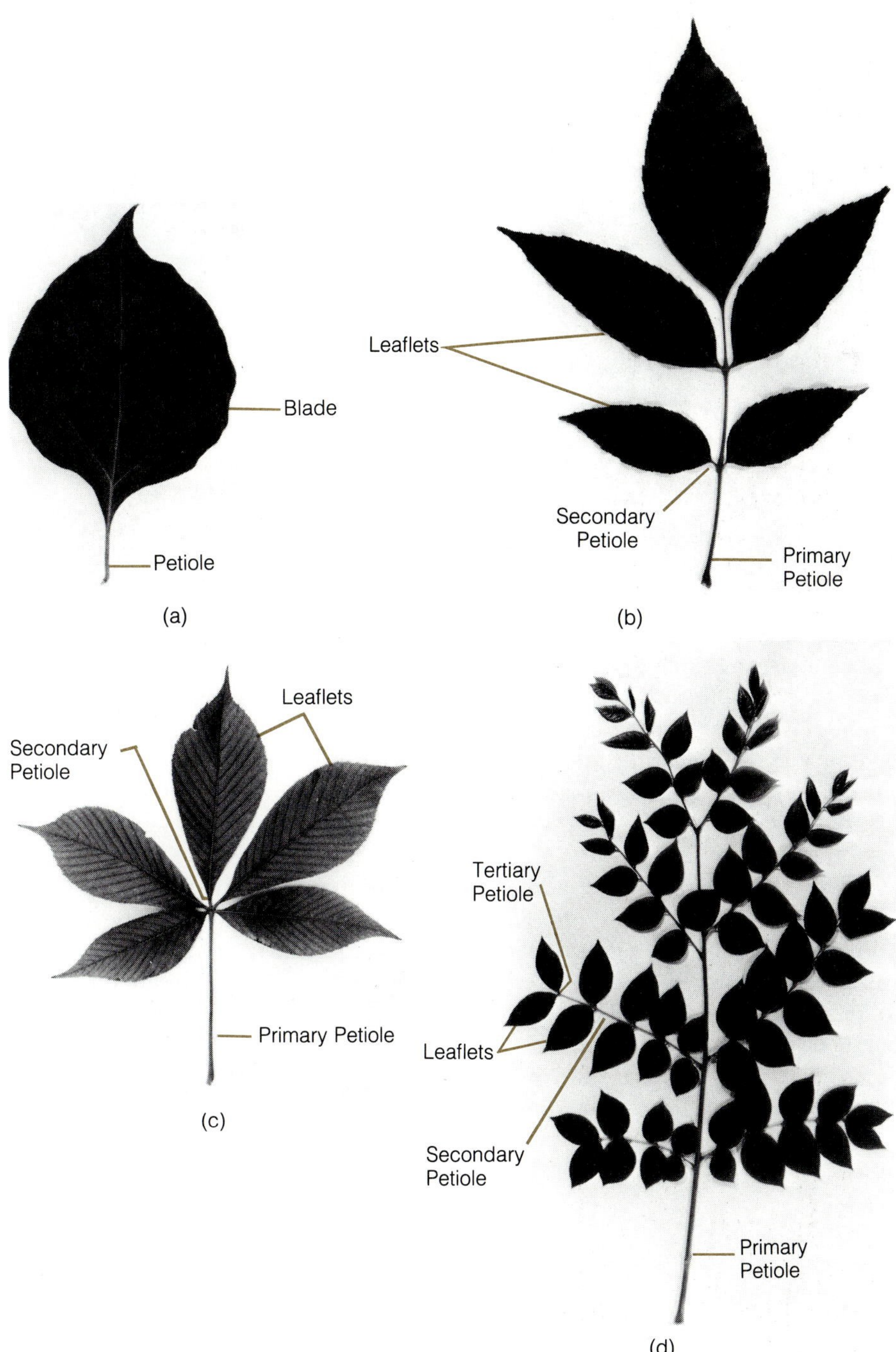

Figure 8–9. Simple and compound leaves. A simple leaf has one blade whereas a compound leaf has many leaflets. Differences in compound leaves are based on the position of the leaflets or the branching of the petiole. (*a*) Simple leaf. The blade is attached to the primary petiole. (*b*) Pinnately compound leaf. A number of leaflets are attached along the primary petiole. (*c*) Palmately compound leaf. A number of leaflets are attached to the tip of the primary petiole. (*d*) Bipinnately compound leaf. The leaflets are attached to the secondary petiole.

Figure 8–10. Leaf margins and venation. (*a*) A leaf with smooth, or entire, margin and pinnately netted venation. (*b*) A leaf with toothed margin and pinnately netted venation. (*c*) A leaf with lobed margin and palmately netted venation.

Arrangement of Leaves

Among many of the primitive vascular plants leaves may be in a spiral or a whorled arrangement on the stem. Pine leaves typically are clustered in groups with the number in each group being constant for a species. Among the angiosperms* three distinct patterns of leaf arrangement can be found. Irrespective of the differences in arrangement, it is apparent that leaves are distributed on a plant in such a way that the maximum number of leaves can be exposed to the sun at any given time, and a balanced weight distribution is achieved.

Angiosperm
One of a group of plants characterized by the production of flowers

An **alternate** leaf arrangement occurs when there is only one leaf at each nodal region along the stem (Figure 8–11a). If a line were to be drawn touching the base of each alternately arranged leaf, it would represent a spiral. When two leaves are present at a nodal region, the arrangement is called **opposite** (Figure 8–11b), with adjacent pairs of leaves on such plants being positioned at right angles to each other. Plants with a **whorled** leaf arrangement have more than two leaves at a nodal region (Figure 8–11c). The whorled configuration, however, is much less common than alternate and opposite arrangements in angiosperms.

Leaf Anatomy

Because of the highly specialized nature of leaves, the internal organization of their tissue is different from that of stems and roots. In this chapter only the anatomical details of leaves from the flowering plants will be described. As in stems and roots, there are observable differences in the anatomical features of leaves of monocotyledonous plants and dicotyledonous plants, although the basic organization is similar in both. Unlike stems and roots, leaves have two distinct epidermal layers—an **upper epidermis** and a **lower**

(a)

(b)

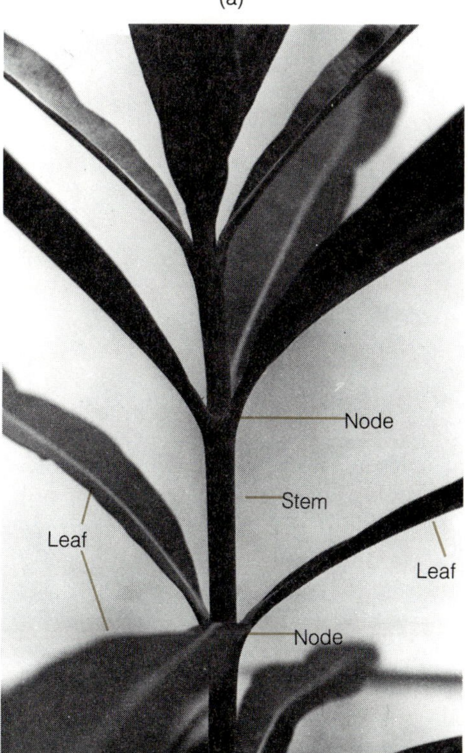

(c)

Figure 8–11. Leaf arrangements. (*a*) Alternate—one leaf at a node. (*b*) Opposite—two leaves at a nodal region. (*c*) Whorled—three or more leaves at a nodal region.

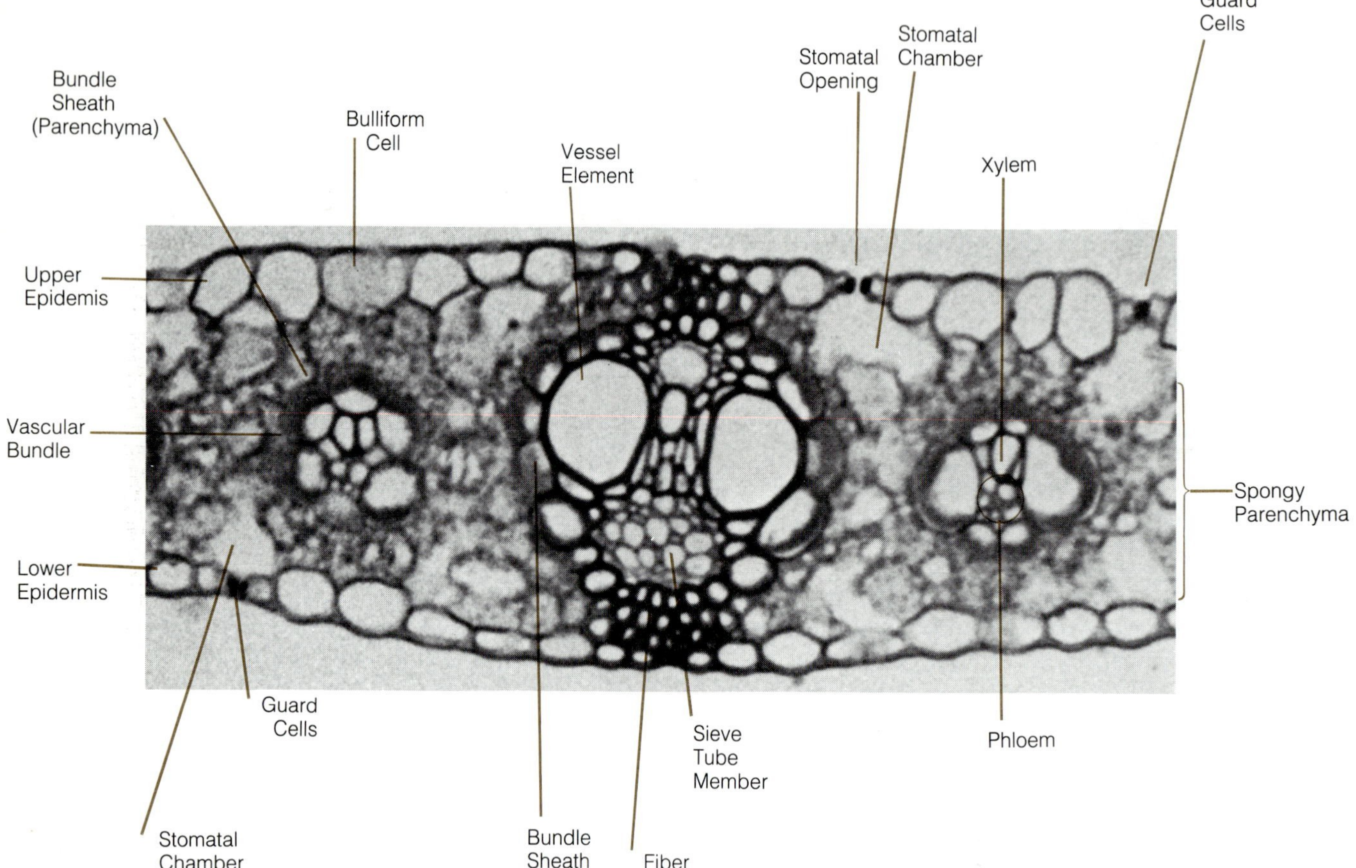

Figure 8–12. Transverse section of a monocotyledonous leaf. The mesophyll of a monocot leaf consists of spongy parenchyma tissue. Bulliform cells, which are specialized epidermal cells, allow the leaf to roll on itself during dry conditions.

epidermis (Figures 8–12, 8–13 and Plate 6). Epidermal cells usually are not pigmented and so do not interfere with the penetration of light to the adjacent photosynthetic tissues. The green color of a leaf is due to the chlorophyll-containing photosynthetic tissues visible through the transparent epidermis; however, leaves that appear red or purple, such as horticultural varieties of coleus and crabapple, contain anthocyanin, a water-soluble pigment, in the vacuoles of the epidermal cells (Chapter 2). The presence of anthocyanin reduces the light intensity reaching the photosynthetic tissues, a fact that probably explains why so few wild plants have red or purple leaves. There is evidence, however, that in a few species the anthocyanin is produced as a sunscreen that enables these species to survive in an environment where the sunlight, particularly its ultraviolet component, would otherwise be too strong. In other species the anthocyanin appears to render the leaf less attractive to herbivores. It would seem that for some insects and

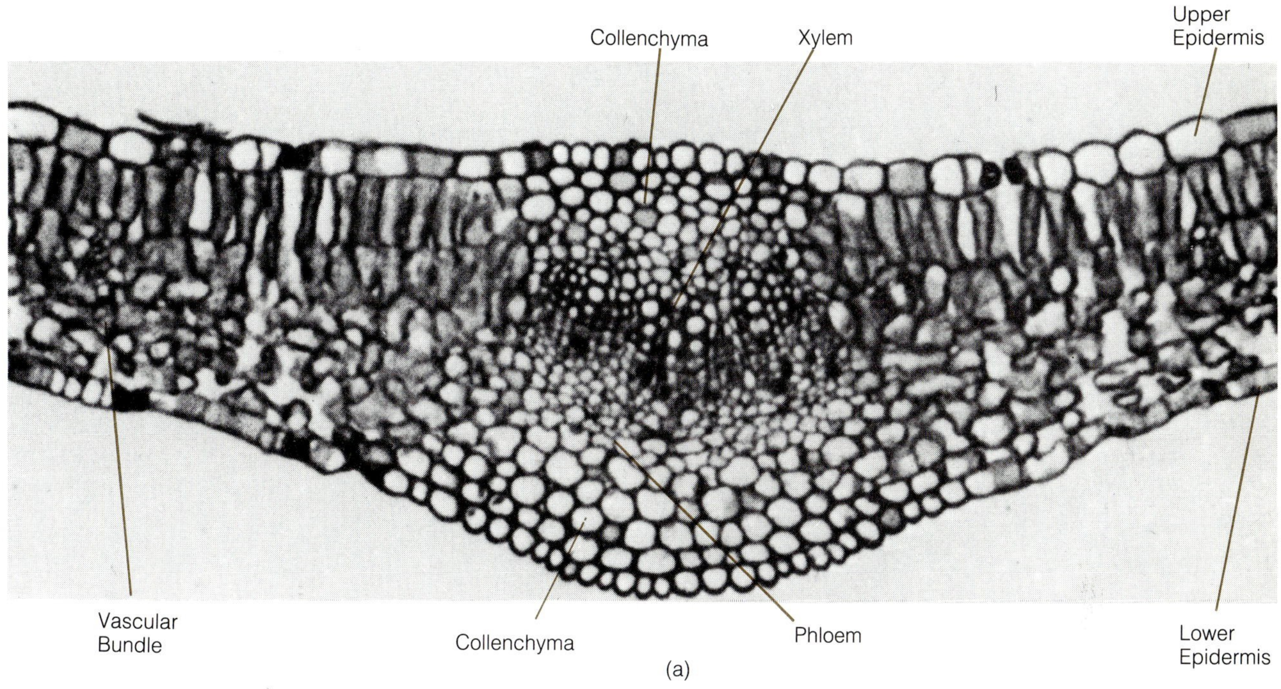

(a)

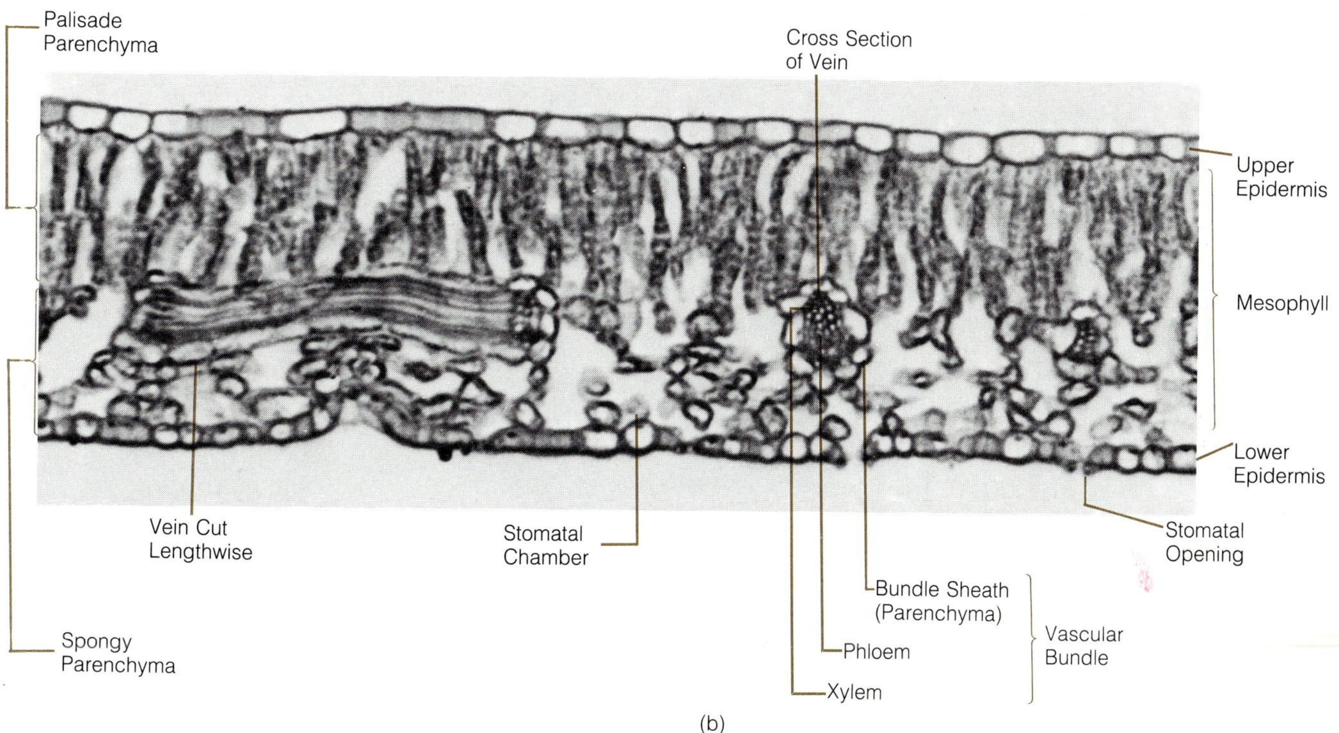

(b)

Figure 8–13. Dicotyledonous leaf. (*a*) In a transverse section of a dicotyledonous leaf, the mesophyll consists of both palisade and spongy parenchyma tissue. (*b*) Because there is a network of veins in a leaf blade, both transversely and longitudinally cut veins are seen in a transverse section of a leaf.

higher animals that a leaf must be green or it is avoided as unpalatable. A similar situation exists among humans who have definite color preferences in food such that blue eggs, yellow bacon, and pink coffee could only be consumed in the dark!

There is always a buildup of waxy materials on the outer surface of the epidermis forming the **cuticle.** In addition, modifications of the epidermal cells in the form of hairs and glandular cells commonly are present on many leaves (Figure 8–14). These structures play significant roles in minimizing water loss, in preventing herbivores from eating the leaves, or in the trapping and digestion of insects by insectivorous plants as described later. There are also many pairs of chloroplast-containing, modified epidermal cells called **guard cells** that are scattered randomly on both surfaces of leaves of many monocotyledonous plants such as corn. In most dicotyledonous plants, however, these are much more numerous on the lower epidermis than the upper. When the guard cells are turgid*, they become kidney shaped, leaving a small opening, called a **stoma,** in the center. This opening facilitates the diffusion of gases into and out of the leaf and also allows the escape of water vapor in the process of **transpiration.** The number of stomata on a leaf surface ranges between 50,000 and 200,000 per square inch. In corn and other grasses there are certain unusually large cells called **bulliform cells** on the upper epidermis. These are also modified epidermal cells that collapse when they lose water, resulting in a rolling of the leaf that minimizes evaporation of water from the leaf surface. Although the usual number of cell layers in an epidermis is one, there are plants, such as *Peperomia* and the rubber tree, that have a multilayered epidermis used for storage of water.

Turgid
The condition in which the cell vacuoles are filled with water.

Sandwiched between the two epidermal layers is the parenchymatous photosynthetic tissue, collectively called the **mesophyll,** which makes up the greatest volume of a leaf. In many monocotyledonous leaves the mesophyll consists of **spongy parenchyma,** characterized by the irregular shape of each cell and a loose arrangement with many intercellular air spaces that facilitate diffusion of carbon dioxide to every cell of the mesophyll tissue (Figure 8–12). The mesophyll of many dicotyledonous leaves consists of two distinct types of parenchyma cells—the **palisade parenchyma,** which con-

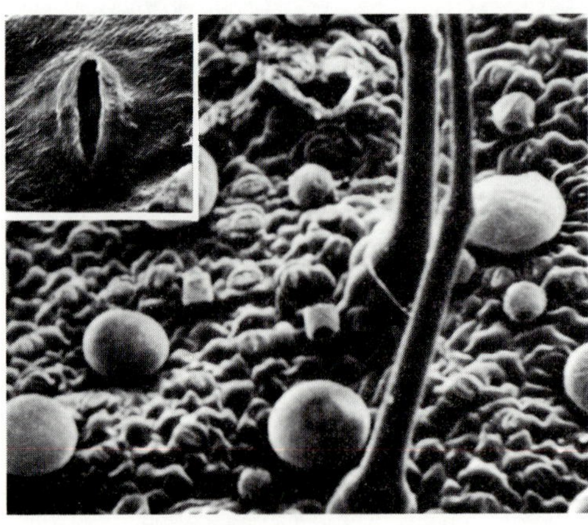

(a)

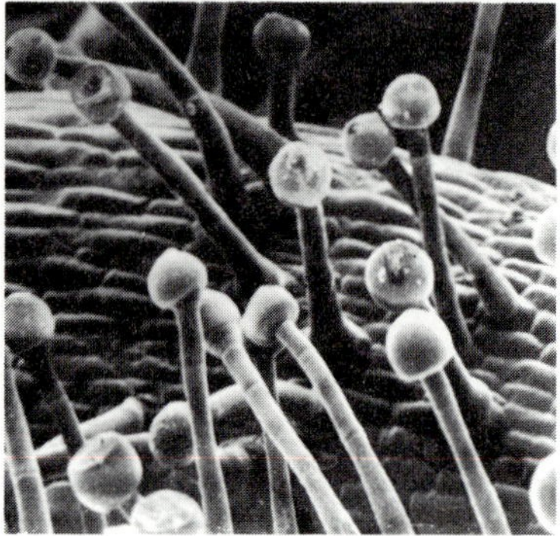

(b)

Figure 8–14. Scanning electron micrographs of leaf surfaces. (*a*) Surface showing glands without stalks and hairs, which are modified epidermal cells. Inset shows a stomatal apparatus consisting of two guard cells and the stomatal opening. The guard cells regulate the size of the opening. (*b*) Surface showing many stalked glands.

sists of vertically elongated columnar cells arranged immediately below the upper epidermis, and the **spongy parenchyma,** similar in shape to those in the monocotyledons, which is loosely arranged above the lower epidermis (Figure 8–13). The number of layers of palisade parenchyma varies from one to four, depending on the species. The concentration of chloroplasts is always higher in the palisade cells than it is in the spongy cells. In both monocotyledonous and dicotyledonous plants the arrangement of spongy parenchyma cells in the vicinity of stomatal openings creates large air spaces called **stomatal chambers.**

The veins are the **vascular bundles** in a leaf. Since they form a network in the leaf blade, both cross-sectional and longitudinal-sectional views of vascular bundles are present in a cross section of a leaf. Nevertheless, cross-sectional views of the major veins are always seen in any leaf cross section. In both monocotyledonous and dicotyledonous plants the vascular bundles are present midway between the upper and lower epidermal layers. Except for the vascular bundles represented by the major veins, all others consist of primary **xylem** and primary **phloem** surrounded by a **bundle sheath** of parenchyma cells. In certain plants the parenchyma cells of the bundle sheath are unusually large and contain chloroplasts, features that are related to basic differences in the utilization of carbon dioxide during photosynthesis between these plants and others with normal-size cells in the bundle sheath (see Chapter 9 for details of differences in photosynthesis).

In a cross section of the vascular bundle, the xylem, primarily composed of **vessel elements,** is located toward the upper epidermis while the adjacent phloem, consisting of **sieve tube members** and **companion cells,** is toward the lower epidermis. In addition to the larger amounts of xylem and phloem, the major veins also have either sclerenchymatous **fibers** or **collenchyma cells** above and below each vein. Vascular cambium, however, is absent from the vascular bundles of leaves, accounting for the lack of secondary growth in leaves.

Because of the extensive network of veins and their close spatial relationship to the mesophyll tissue, all photosynthetic cells of a leaf have access to water that is needed for food synthesis and to the phloem for transporting newly synthesized sugars out of the leaf. In addition, the stomatal openings allow diffusion of carbon dioxide into the leaf as well as regulating the process of transpiration and, in turn, the water economy of the plant. The anatomical features of a leaf are closely related to its functional aspects, making the leaf an efficient organ for photosynthesis.

Leaf Modifications

Although the primary function of leaves is to synthesize food, there are plants in which leaves carry out additional functions and may look completely different from normal leaves. Many of these modifications are adaptive features for the survival of these plants under harsh environmental conditions. A specific example of such an adaptation is often present in plants

Figure 8–15. Leaf modification in cactus. The leaves on this cactus are modified into spines to minimize surface area, while the cladode (stem) carries out the function of the leaf.

Xerophyte
(*adj.*, xerophytic) A plant capable of growing under arid conditions.

that grow in arid regions. The leaves of these plants are generally thick, with fewer stomatal openings or stomata located in depressions on the lower epidermis; in addition, such plants may have dense coverings of epidermal hairs on their leaves. The leaves of these plants also have parenchymatous water-storage cells. Such leaves are referred to as being **succulent.** In extreme cases of xerophytic* adaptation, plants such as cacti have leaves modified into **spines** while the function of the leaves is taken over by a leaflike stem called a **cladode** (Figure 8–15).

Another unusual adaptation is found among the insectivorous plants where the leaves are used as traps for insects (Plate 7). These plants grow in boggy or swampy soils deficient in nitrogen and seem to have evolved this adaptation to compensate for this lack of nitrogen by absorbing nitrogen-containing molecules from trapped insects. In Venus's flytrap the leaves are hinged along the midrib and triggering hairs are located on the upper surface. When two or more of these hairs are touched by an insect, the leaf blade closes quickly, trapping the insect, which is gradually digested and absorbed. The leaves of pitcher plants are modified into elongated pitcher-like traps containing a digestive liquid at the bottom with a slippery rim at the upper end and bristles pointing downward along the neck. An insect landing on the rim slides into the bottom and is unable to crawl back because of the bristles. It is subsequently digested and absorbed. The leaves of sundew have sticky glands to capture insects which are then digested and absorbed. Bladderwort is an example of an underwater plant which has leaves modified into traps that effectively capture small aquatic insects. In all of these examples the leaves also function in the manufacture of food through the process of photosynthesis.

Other leaf modifications are common in nature. Vegetative propagation* is carried out by leaves in several kinds of plants. Certain species of *Bryophyllum* (air plant) and *Kalanchoe* (maternity plant), for example, produce small but complete plants along the notched leaf margins (Figure 8–16). When these plantlets drop to the soil they establish new plants. When the tip of the leaf of walking fern comes in contact with soil, it produces a new plant. Leaf modifications that are found among weak-stemmed plants such as peas, where the terminal leaflets of the compound leaf are modified into coiled structures called **tendrils,** can attach the plant to a support (Figure 8–17). In **bulbs** such as onion, lily, tulip, and daffodil the underground leaves serve as storage organs and are called **fleshy scale leaves** (Figure 8–18). Paper-thin scale leaves are also often present on underground stems. One of the most common types of scale leaf is the **bud scale** of woody plants, which functions to protect the bud, particularly during the winter months.

Vegetative propagation Reproduction of organisms by nonsexual means.

Leaf Senescence and Abscission

Senescence or aging is a process that goes on in all organisms, both at the cellular level and at the organismal level. All of the factors that are respon-

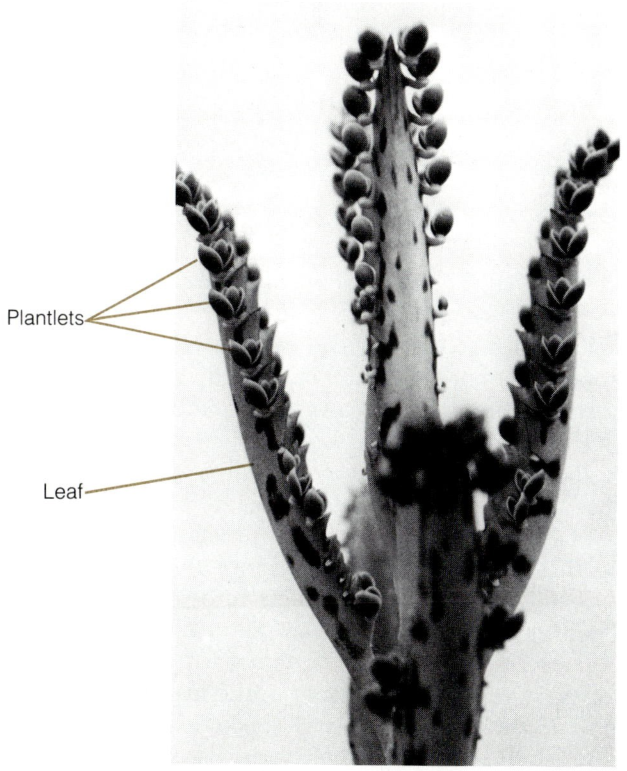

Figure 8–16. Leaf modification in *Kalenchoe.* The leaves of *Kalenchoe* produce plantlets along their margins as a means of vegetative propagation.

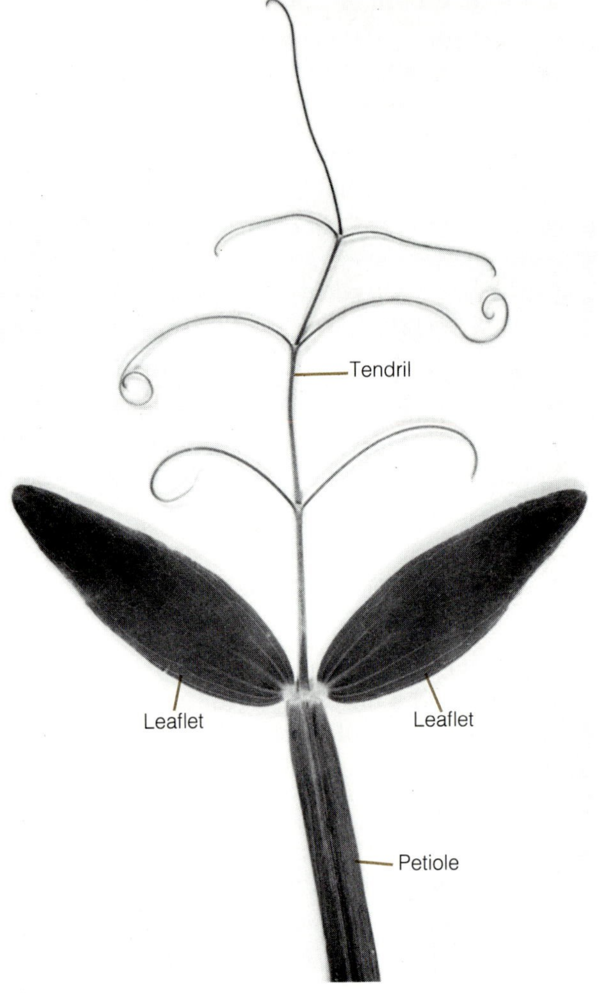

Figure 8–17. Leaf modification in peas. The terminal leaflet of the compound leaf of a pea is modified into a tendril, which helps the plant attach to a support.

sible for senescence or the exact mechanism involved are not fully understood. It is known, however, that diverse factors, including genetic factors, hormonal interactions, and environmental factors, are collectively responsible for aging in plants. Leaves of many trees have a shorter life compared to the potential for the tree to live for many years. The formation of an **abscission zone,** composed of one or more layers of parenchyma cells, at the base of the petiole near its junction with the stem is responsible for leaf death and subsequent fall (Figure 8–19). During this period considerably reduced transport of nutrients into the leaf and a degradation of cellular components in the leaf take place. Many of the degraded substances and existing nutrients usually are recycled to other parts of the plant. Abscission zones commonly are present in the leaflets of a compound leaf as well. During the final stages of senescence, a layer of cork cells derived from the parenchyma cells is formed at the base of the abscission zone where it serves

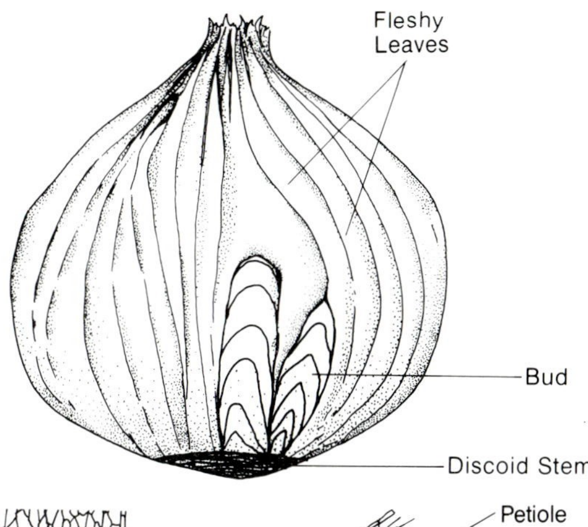

Figure 8–18. Leaf modification in onion. The fleshy leaves of an onion bulb are used for storage of food and water.

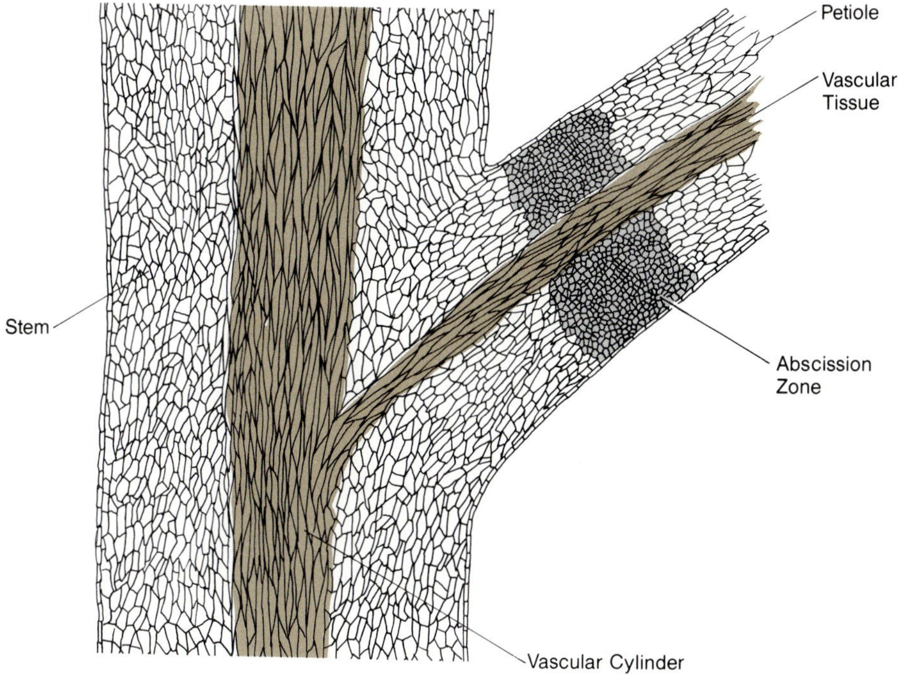

Figure 8–19. Abscission zone. Formation of a layer of parenchyma and cork cells at the base of the petiole results in death and subsequent fall of the leaf.

to protect the plant from invasion of pathogens* and from excess water loss through the wound. At this time the middle lamellae* of the cells in the abscission layers are digested, causing the leaf to fall. In plants such as the different kinds of oaks, however, the leaves remain attached even after they are dead until they finally drop off by the action of mechanical forces of wind and rain during the winter and early spring. When a leaf falls off, it leaves a mark on the stem called the **leaf scar** in which the scars of the leaf traces also can be found. Among deciduous plants in temperate latitudes, formation of the abscission zone and subsequent leaf drop occur at approximately the same time for most leaves on a plant, resulting in the common name for the autumn season—"Fall." In the tropics, by contrast, this process is not synchronized but is continuously taking place year-round. As a result, plants in the tropics lose a few leaves on a daily basis and also produce new ones regularly.

Pathogen
A disease-causing organism.

Middle lamella
(*pl.*, lamellae) The outermost region of a cell wall that joins adjacent cells.

Summary

True leaves, characterized by the presence of vascular tissue, are present on most vascular plants. All leaves originate as leaf primordia from the apical meristems in terminal and lateral buds. Primary leaves of a seedling originate as embryonic leaves in the seed.

A typical leaf consists of a cylindrical petiole and a flat, broad blade. The vascular cylinder of the stem continues through the petiole and forms the characteristic venation pattern in the blade: pinnately netted, palmately netted, or parallel. Parallel venation is typical of many monocotyledonous plants. Among many monocotyledonous plants the petiole is modified into a leaf sheath that encircles the stem.

A single blade, which may be toothed or lobed to varying degrees, is characteristic of simple leaves, whereas compound leaves have blades that are divided into leaflets. In a pinnately compound leaf, the leaflets are attached along the length of the primary petiole, while in a palmately compound leaf all leaflets are attached to the tip of the primary petiole. Compound leaves can be distinguished from simple leaves by the absence of axillary buds in the axils of their leaflets or by the orientation of all the leaflets in one plane.

Leaves may be arranged in clusters or in definite patterns. There is only one leaf at a nodal region on the stem in alternate arrangement, two in opposite arrangement, and more than two in whorled arrangement.

The basic organization of tissues is similar for all angiosperm leaves—unilayered upper and lower epidermises with cuticle and stomatal openings, parenchymatous photosynthetic mesophyll, and vascular bundles. Stomatal openings regulated by guard cells are present on both surfaces in many monocotyledonous plants but occur mainly on the lower surface in most dicotyledonous plants. the mesophyll in monocotyledonous plants consists of loosely arranged spongy parenchyma cells, while in most dicotyledonous plants spongy parenchyma is limited to the lower half of the leaf with columnar palisade parenchyma cells toward the upper half. The vascular bundles are located midway between the two surfaces and are closely associated with the mesophyll tissue. Parenchymatous bundle sheaths surround xylem and phloem of the vascular bundles. Within each bundle the xylem, composed mostly of vessel elements, is toward the upper surface and the phloem, consisting of sieve tube members and companion cells, is toward the lower surface. In addition, fiber cells or collenchyma cells are often present on the upper and lower sides of the major veins.

Leaves are often modified to carry out functions other than photosynthesis. Plants that live in dry conditions frequently have succulent leaves or leaves reduced to spines, the photosynthetic and storage functions being taken over by leaflike stems. Leaves are variously modified as traps on insectivorous plants such as Venus's flytrap, sundew, pitcher plant, and bladderwort. Leaves play a role in vegetative propagation by producing plantlets from the leaf margins in species of *Bryophyllum* or give rise to new plants when the leaf tip comes in contact with the soil, as in walking fern. Leaflets are modified to tendrils for support in pea. Bud scales are modified leaves serving a protective function, while fleshy scale leaves of bulbs store both food and water.

Since meristems are absent from most leaves, they have a determinate type of growth, unlike stems and roots, and are periodically replaced. Before a leaf falls off, reduced transport of water and nutrients and increased degradation of cellular components take place, along with the formation, at the base of the petiole, of an abscission zone composed of parenchyma cells. Subsequently the middle lamellae of the cells in the abscission zone are digested by enzymatic action causing the leaf to fall.

Review Questions

1. How will you decide if a leaf is simple or compound?
2. Explain the significance of the arrangement of leaves on a plant.
3. Why is it that there are leaves on the newest segment of growth only?
4. Explain the leaf gap.
5. Explain the different kinds of compound leaves and give an example of each that you have known.
6. What are the anatomical differences between a monocotyledonous leaf and a dicotyledonous leaf?
7. What role does the absission layer have in the life of a leaf?
8. How is a leaf specially adapted for its role in food synthesis.

Drugs From Leaves

The economic value of leaves as a source of vitamin-rich food is recognized by almost everyone. Spinach, cabbage, kale, lettuce, Swiss chard, and many other species too numerous to mention have nourished human populations since prehistoric times. The leaves of such herbaceous plants as parsley, dill, tarragon, oregano, thyme, mint, sage, and others are used extensively as seasonings to improve the taste of other foods. Essential oils, aromatic distillation products from plants such as wintergreen, geranium, peppermint, and lemon grass, are valuable for use in such diverse products as chewing gum, toothpaste, hand lotions, throat lozenges, mosquito repellents, and perfume.

Less well recognized by many people today is the importance of certain leaf extracts for medicinal or mind-altering purposes. Prior to the introduction of modern medicines, most illnesses were treated with home remedies concocted from plant extracts, many of which were derived from leaves.

For hundreds of years, leaves of the foxglove plant *(Digitalis purpurea),* pressed against the skin, were used to cure a variety of ailments. Juice squeezed from the leaves was used in ointments applied to cuts, bruises, and leg ulcers, and was credited with curing epilepsy. Although foxglove had been used in the treatment of dropsy during the eighteenth century, its role as a heart stimulant was recognized only in the last century, and today the drug is used widely to assist in regulating heart contractions, pulse, and blood pressure. Recent studies have shown that digitalis is also useful for treatment of glaucoma, asthma, and neuralgia.

Chemical compounds known as alkaloids are responsible for the poisonous or medicinal properties of many plants. A relative of the common tomato plant, belladonna or deadly nightshade, contains in its leaves several different alkaloids utilized for centuries. One of these, atropine, has long been used to stimulate circulation, as a local anesthetic, and to dilate the pupils of the eye—a property for which it won favor among 16th-century Italian ladies who put drops of diluted belladonna (literally, beautiful lady) in their eyes to evoke a dark, mysterious gaze. Belladonna's properties as a hallucinogen were familiar to ancient Greeks and Romans who used small amounts of the drug during bacchanalian orgies to spike their wine.

Coca leaf *(Erythroxylon),* native to the Andean region of South America, is known today chiefly as the source of cocaine. Natives of the region, however, have chewed the leaves from time immemorial to give them increased physical endurance in a harsh environment, increasing their heart and respiratory rates and dilating their arteries. That they do not become addicted to the drug is due to the fact that by simply chewing on the leaves, the amount consumed daily is only extremely small.

Tobacco, perhaps the leaf with the greatest economic importance in monetary terms, has been used since ancient times by native Americans for both medicinal and ceremonial purposes. Indians held wads of folded tobacco leaves inside their cheeks to dull hunger pangs while on hunting expeditions, as well as to lessen their thirst. Indian women consumed powdered tobacco to ease the pains of childbirth. When tobacco was first shipped to Europe by Spanish explorers, it was praised as a cure for abscesses, sores, convulsions, and snakebite—and as a stimulus to lovemaking! Today the adverse health impacts of tobacco are known to overshadow greatly its supposed medical benefits, but the sensation of "pleasant stupefaction" described by early European users, referring to the leaf's slightly hallucinogenic properties, results in tobacco's continuing popularity and its retention on the list of "basic crops" of the U.S. Department of Agriculture.

9

Photosynthesis

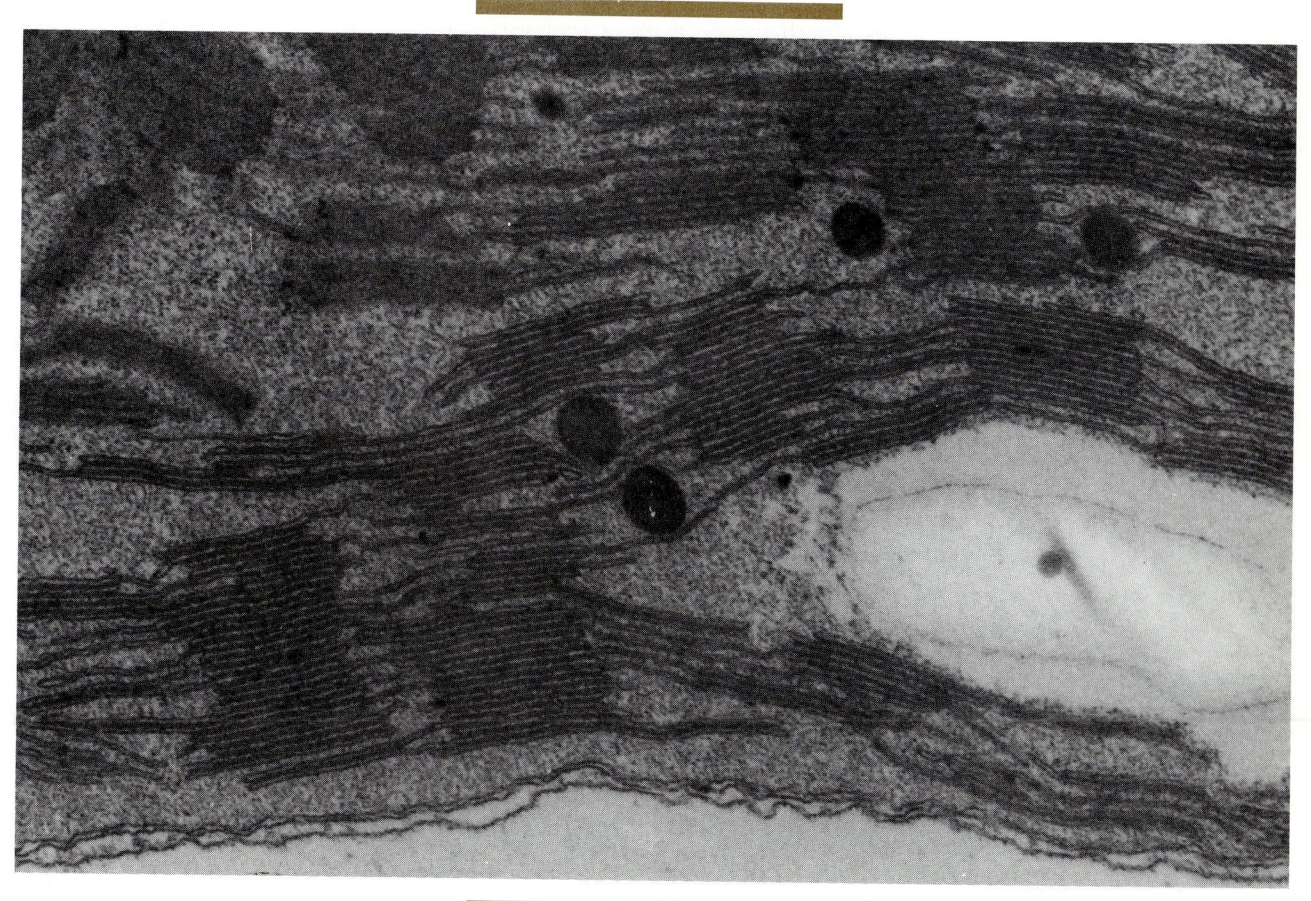

Outline

Heterotroph
An organism that cannot produce its own food from inorganic molecules.

Autotroph
An organism that can produce its own food from inorganic molecules.

Photosynthesis
The metabolic process by which carbohydrates are produced from CO_2 and a hydrogen source such as water, using light as the source of energy.

Transpiration
The evaporation of water from leaf and stem surfaces through stomatal openings.

Introduction

To be alive requires a constant expenditure of energy. **Heterotrophs*** obtain this energy from the food that they consume. **Autotrophs*** such as green plants are the only organisms that are capable of producing their own food. Most autotrophs produce their food through the process of photosynthesis.* Prior to the 1600s it was believed by European scholars that plants obtained their food from the soil in which they grew. In particular it was thought that plants obtained their food from the decaying plant and animal material in the soil. This idea was supported by the observation that the application of manure to fields dramatically increased crop yield. In fact, this belief is the basis for our present incorrect usage of the term *plant food* to refer to fertilizer preparations for house plants.

During the Renaissance, scholars challenged many of the old teachings concerning the nature of plant and animal life and sought to determine the truth from direct observation and experiment. In the early 1600s Jan van Helmont performed a simple experiment designed to test the theory that plants obtained their food from the soil. He planted a small willow seedling in a tub containing a known weight of soil. The willow was grown for five years, and the only thing that was added to the soil was rainwater. At the end of the experiment the willow weighed 75 kilograms but the soil had lost only a few grams. Van Helmont concluded that the plant did not increase in weight because of anything it took out of the soil; obviously the plant must have obtained from the water all that it needed to remain alive and grow.

In 1699 it was shown that water was not the source of the food that plants use to make new protoplasm since most of the water passed through the plant in **transpiration.*** Consequently, the question of the source of plant food was left open. By 1772 the gaseous nature of the atmosphere was becoming established by Joseph Priestley and others. One of Priestley's contributions was to show that a plant could purify the air in a jar in which an animal had suffocated. He also showed that if a plant and an animal were placed in the same jar, the animal would survive for a longer period than an animal kept in a jar with no plant. Shortly afterwards Jan Ingenhousz, a Dutch physician, reported that the purification of air only occurred in light and that only the green parts of plants were able to purify the air. He also recognized that plants were able to respire, which led to the old wives tale that plants poison the air at night and should be removed from sickrooms at night. Julius Mayer, who established the law of conservation of energy in 1842, stated that the ultimate source of energy used by plants to purify the air was the sun and that during the purification process they were converting light energy into chemical energy—food.

By the 1840s the fundamental significance and outline of the process of photosynthesis were known. The contribution of scientists since then has been to work out the details of the reactions. In photosynthesis green plants convert light energy, water, and carbon dioxide into carbohydrates and oxygen. As indicated in Chapter 3, a plant cell contains a variety of organic molecules, the major groups being proteins, lipids, nucleic acids, vitamins, and carbohydrates. Photosynthesis, however, produces only certain types of carbohydrates. All of the other organic molecules in a plant cell are produced by metabolic* rearrangement of the carbohydrates produced in photosynthesis. A certain amount of the carbohydrate produced in photosynthesis is used in respiration* to produce ATP molecules (energy), which are needed to maintain metabolic activities. The excess carbohydrate beyond that needed in respiration and other types of metabolism is stored typically in the form of starch in plastids throughout the plants.

Metabolism
(*adj.*, metabolic) All of the chemical reactions that occur in a living cell.

Respiration
The metabolic process that results in the release of chemical energy from simple organic molecules.

Preview of Photosynthetic Reactions

The general chemical reaction for photosynthesis can be written as:

$$\underset{\text{carbon dioxide}}{6CO_2} + \underset{\text{water}}{12H_2O} \xrightarrow[\text{chlorophyll}]{\text{light}} \underset{\text{sugar}}{C_6H_{12}O_6} + \underset{\text{water}}{6H_2O} + \underset{\text{oxygen}}{6O_2}$$

This equation is simply a statement of the relative quantities of raw materials and end products; it reveals nothing about the individual reactions of photosynthesis. The value of such an equation is to provide an overview of the whole reaction and to facilitate making comparisons between photosynthesis and respiration. This equation is, in fact, a gross oversimplification: the process of photosynthesis consists of two distinct sets of reactions—one set requires light (the **light reactions**) and the other does not (the **dark reactions**). Both of these reactions take place simultaneously when photosynthesis occurs in plants. The dark reactions get their name from the fact that, in the laboratory, they will occur in the dark in isolated chloroplasts supplied with the appropriate chemicals.

The initial reaction of photosynthesis is the absorption of light energy by a molecule of chlorophyll in a photosynthetic membrane of the chloroplast (Figure 9–1). This energy is used in reactions that break apart a water molecule (H_2O) and release hydrogen atoms (H) from it. The oxygen that was part of the water molecule is not used in photosynthesis and so is released as a by-product. The hydrogen atoms that are released by the breakup of the water molecule do not float around free in the chloroplast. Instead they are carried by a special molecule (a hydrogen carrier) whose only function is to carry hydrogen atoms from where they are produced to where they are used during the dark reaction. This is why the dark reaction only occurs during the daytime; at night there is no source of energy that can be used to break apart a water molecule. During the dark reactions enzymes combine carbon dioxide with existing organic molecules and then the hydrogen atoms from the water molecules are incorporated, sugar being the ultimate product.

The Light Reactions of Photosynthesis

As indicated above, the initial reaction of photosynthesis is the absorption of light by a molecule of chlorophyll. Because light is a form of energy, when

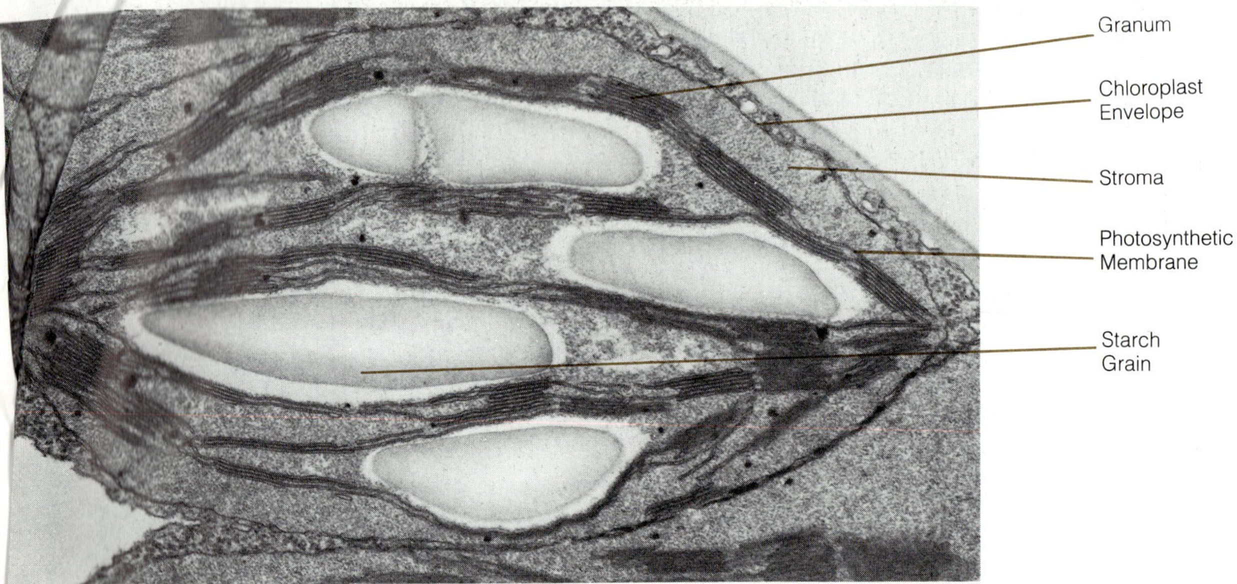

ıre 9–1. **Typical chloroplast.** The light reactions of photosynthesis occur in the photosynthetic membranes that contain the tosynthetic pigments. The dark reactions occur in the stroma.

otosystem
roup of chlorophyll molecules t function to trap light energy l funnel it to a strategically ated chlorophyll molecule that is olved in electron transfer ctions.

a molecule of chlorophyll absorbs light, it absorbs energy. The chlorophyll molecules are organized into **photosystems*** within the chloroplast membranes, and within each photosystem the energy from light is transferred to one specific chlorophyll molecule. The energy that is transferred to this chlorophyll molecule results in one of its outermost electrons being boosted out of its normal orbit. An electron orbiting around a molecule is like a satellite orbiting around a planet—it remains in orbit because it does not have the energy to escape. With the additional energy from light an electron can escape from the chlorophyll molecule, leaving the chlorophyll molecule positively charged:

Actually, the electron does not fly off into the cell; rather it is transferred to another molecule that functions as an electron carrier in photosynthesis. The electron subsequently is passed from this carrier to another until finally it is donated to $NADP^+$ (nicotinamide adenine dinucleotide phosphate)—a complex molecule that, in effect, functions as a carrier of two hydrogen atoms (a hydrogen atom is made up of a proton, H^+, and an electron, e^-; see Chapter 10 for a discussion of hydrogen carriers). Although each chlorophyll molecule can donate only one electron in this reaction, two electrons are carried by each NADPH molecule. The extra electron comes from another chlorophyll through the same reaction, or from the same chlorophyll molecule after it has been recycled in the reactions to be described later in this chapter.

Photos: (a) Foliar colors during autumn. (b) Coconut palms. (c) Photosynthetic pigments separated on a chromatogram.

Dr. John Nadakavukaren

(a)

(b)

(c)

Photos: **(a)** Cross section of monocotyledonous leaf. **(b)** Cross section of dicotyledonous leaf. **(c)** The visible spectrum is a part of the continuous electromagnetic spectrum shown here.

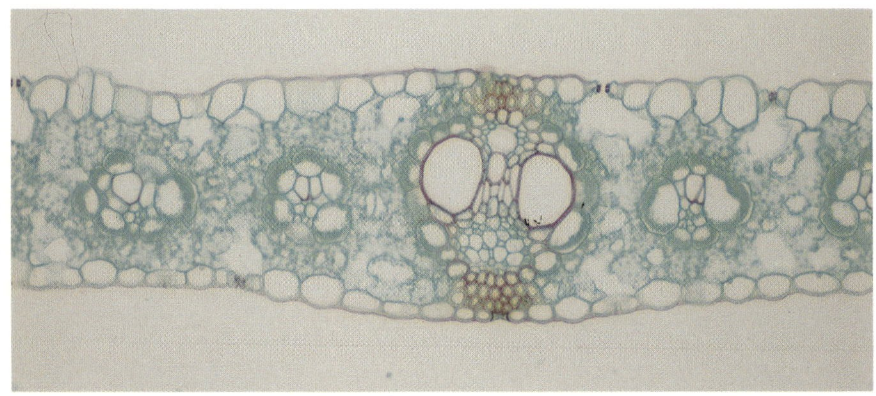

(a)

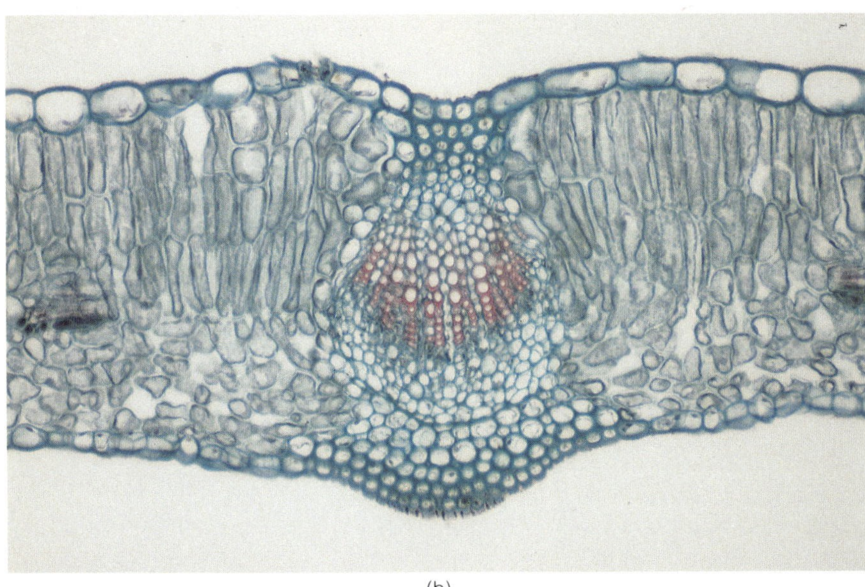

(b)

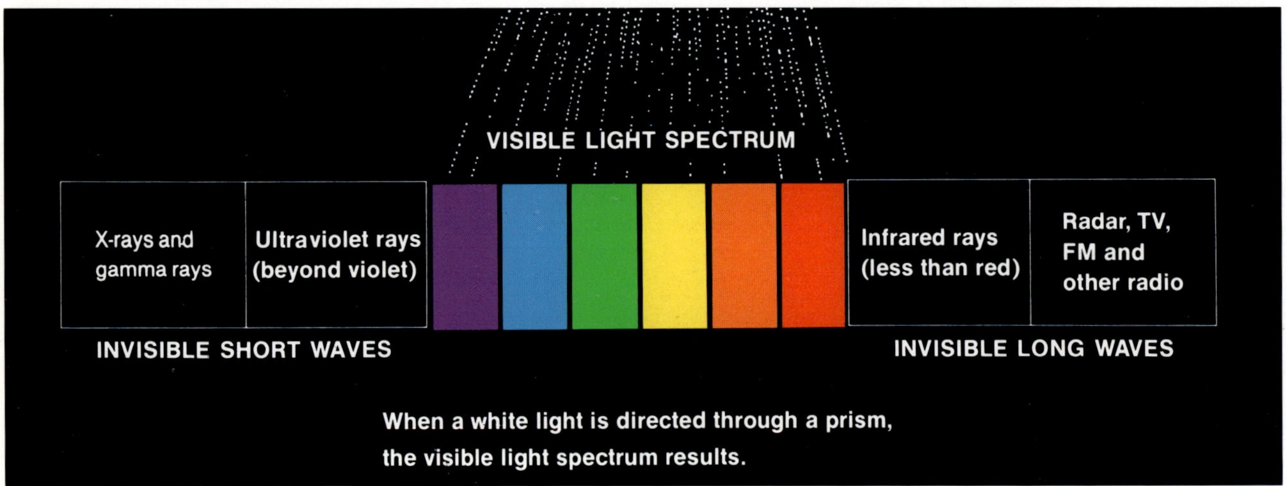

(c)

Photos: **(a)** Foliar colors during autumn. **(b)** Coconut palms. **(c)** Photosynthetic pigments separated on a chromatogram.

Dr. John Nadakavukaren

(a)

(b)

(c)

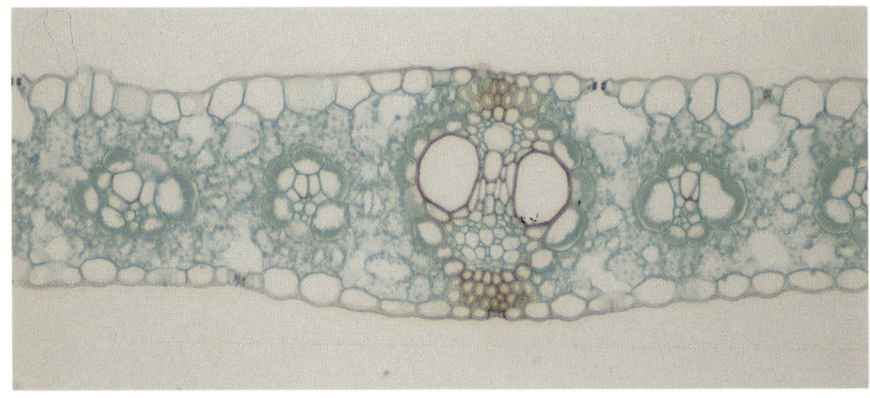

(a)

Photos: **(a)** Cross section of monocotyledonous leaf. **(b)** Cross section of dicotyledonous leaf. **(c)** The visible spectrum is a part of the continuous electromagnetic spectrum shown here.

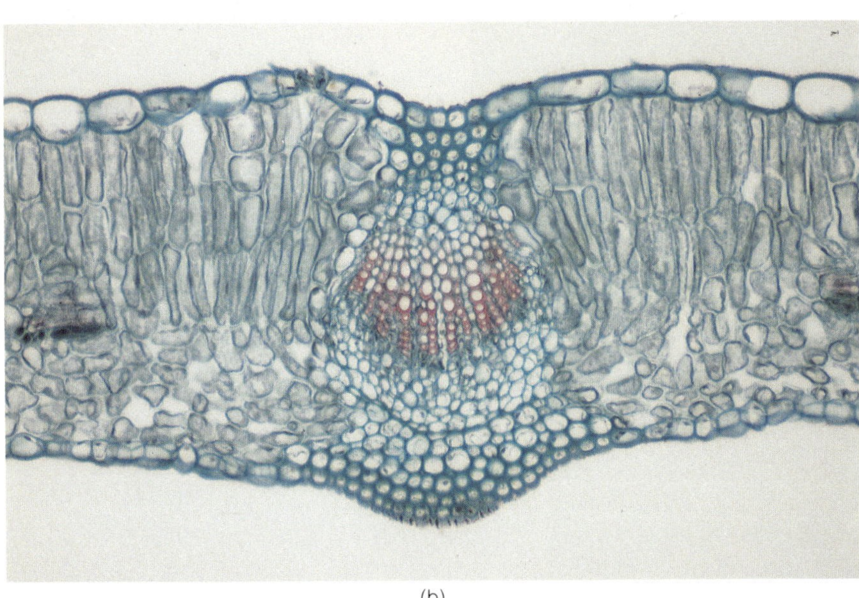

(b)

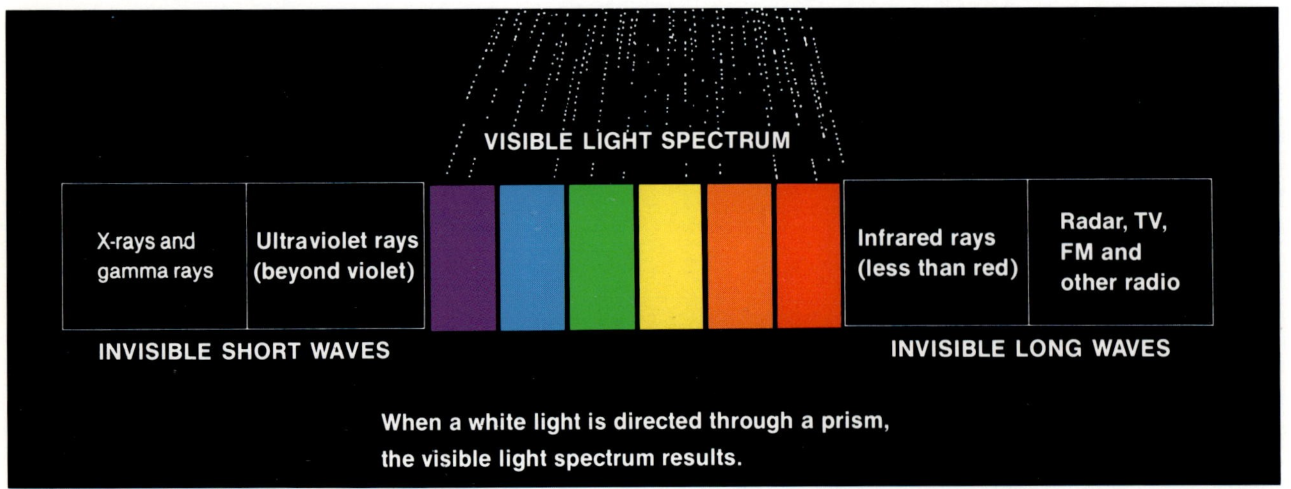

(c)

Photos: **(a)** Venus fly trap (*Dionaea* sp.). **(b)** Sundew (*Drosera* sp.) **(c)** Pitcher plant (*Sarracenia* sp.).

(a)

(b)

(c)

Photos: **(a)** Rockweed (*Fucus* sp.). **(b)** Morel (*Morchella* sp.). **(c)** Bracket fungus (class Basidiomycetes). **(d)** Lichen. **(e)** Whisk fern (*Psilotum* sp.). **(f)** Club moss (*Lycopodium* sp.).

Dr. Dale Brikenhole

(b)

(a)

Dr. John Nadakavukaren

Dr. John Nadakavukaren

(c)

Dr. John Nadakavukaren

(d)

(e)

(f)

The reaction so far is:

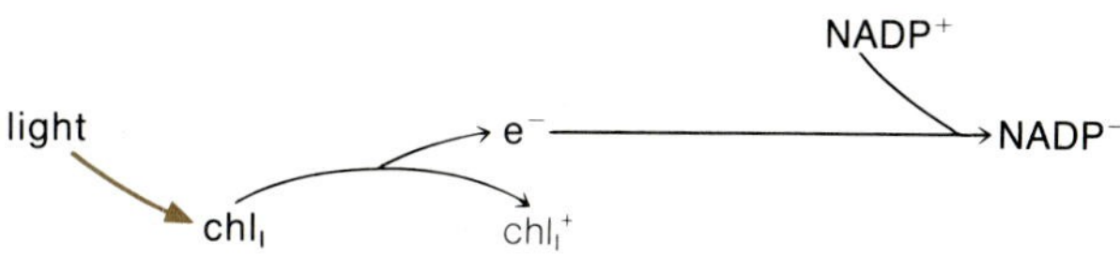

The chlorophyll is designated chl_I because there are two different photosystems involved in photosynthesis; the reaction described above is carried out by the chlorophyll molecule of **photosystem I.**

$NADP^+$ cannot carry only electrons; there must also be protons (H^+) available. These are obtained from the water in the chloroplast:

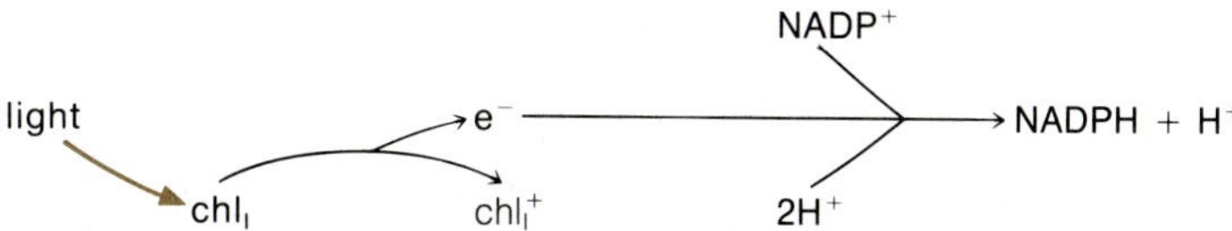

The NADPH + H^+ are needed for reactions that occur later in photosynthesis, the dark reactions. The breaking apart of water molecules that will produce protons (H^+) requires the additional light energy absorbed by the chlorophyll molecules of **photosystem II.** Once again, the energy transferred to one specific chlorophyll molecule causes one electron to leave this chlorophyll molecule. This results in a chlorophyll molecule with a positive charge that needs to be neutralized by an electron. The electron that is donated to this chlorophyll comes from a protein that breaks apart two water molecules and rearranges their atoms into two hydrogen ions (H^+) and two electrons (e^-), an oxygen atom, and a re-formed water molecule:

$$2H_2O \rightarrow 2e^- + 2H^+ + 2OH \rightarrow 2e^- + 2H^+ + H_2O + \tfrac{1}{2}O_2.$$

The electrons are donated singly from this protein to the $chl^+{}_{II}$ molecule returning it to the neutral chl_{II} condition. Since only one electron can be donated to one chl_{II}^+ molecule, each of these protein molecules will actually recycle two chl_{II}^+ molecules.

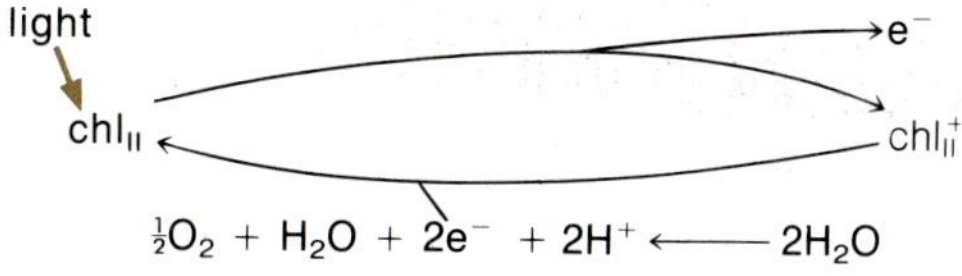

The electron that was lost by the photosystem II chlorophyll molecule is transferred via a series of electron carriers to the positively charged $chlorophyll_I$ molecule. This returns the $chlorophyll_I$ molecule to its original state, which allows photosystem I to absorb more light energy leading to the transfer of another electron to NADP. In the process of transferring electrons and protons from water to NADP, the chloroplast is able to use some of the energy from light to make ATP in a process called **photophosphorylation** (for a discussion of the metabolic significance of ATP see Chapter 10).

The photosynthetic reaction so far is outlined below:

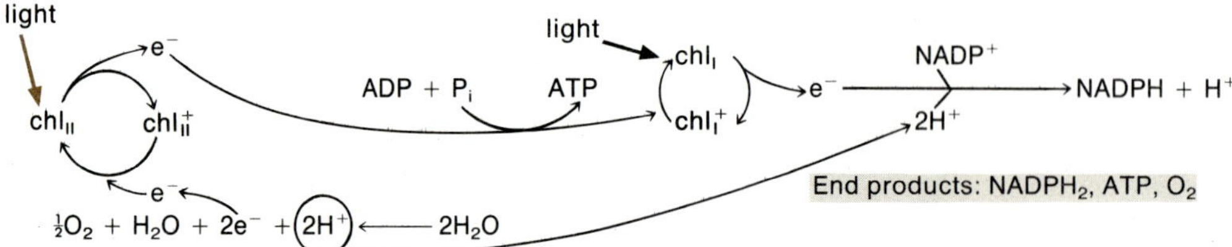

In this reaction the electrons follow a noncyclic path from water to $NADP^+$; ATP produced as the result of noncyclic electron flow is by **noncyclic photophosphorylation.** Under conditions where $NADP^+$ is unavailable to accept electrons, an alternate pathway of electron flow will occur, which also results in the formation of ATP. Electrons in this pathway move from chl_I in a cyclic direction back to chl_I:

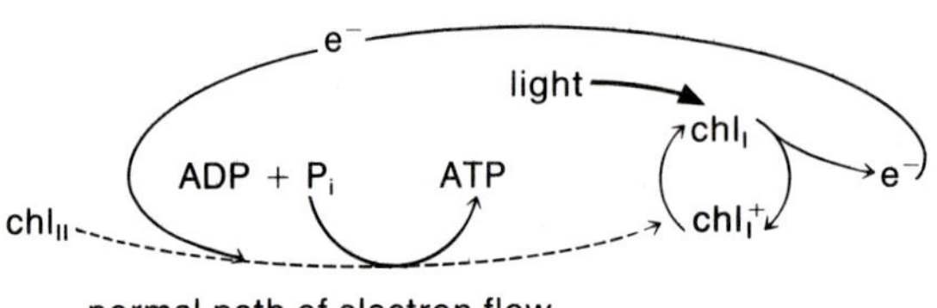

-------normal path of electron flow
End product: ATP

Since the electron flow is cyclic, the formation of ATP is called **cyclic photophosphorylation.**

The function of the light reactions of photosynthesis is to take hydrogen atoms from water and transfer them to $NADP^+$. To do this the plant must break apart the water molecule, a process requiring considerable energy. This energy is obtained from light. The chlorophyll molecules are essentially electron pumps that are fueled by solar energy. The light reaction of photosynthesis is usually drawn in such a way as to indicate the changing energy levels of the electrons as they move from water to $NADP^+$ (Figure 9–2). Chlorophyll of photosystem II pumps electrons from water to chlorophyll of photosystem I, which in turn pumps them to $NADP^+$. $NADP^+$ is a "reluctant" electron acceptor that can only accept high-energy electrons because its low-energy electron orbits are already occupied. $NADP^+$ could not accept electrons directly from water as they would not be energetic enough. The energy from light absorbed by photosystem II partially overcomes this energy barrier with photosystem I providing the final boost. This system is analogous to one with pumps in a deep well where one pump is not enough to pump the water all the way to the top, so two pumps are needed. The electrons travel this route to $NADP^+$ and the protons arrive at $NADP^+$ by diffusion in the water of the chloroplast.

The products of the light reactions of photosynthesis are NADPH + H^+, ATP, and O_2. The NADPH + H^+ and ATP are needed for the completion of photosynthesis. Oxygen, on the other hand, is only produced because water (H_2O) is the source of the hydrogen atoms and after the hydrogen is stripped from the water molecule, all that remains is an oxygen atom. The oxygen molecules produced diffuse out of the chloroplasts and, if not used in cellular metabolism, eventually reach the atmosphere. Photosynthetic bac-

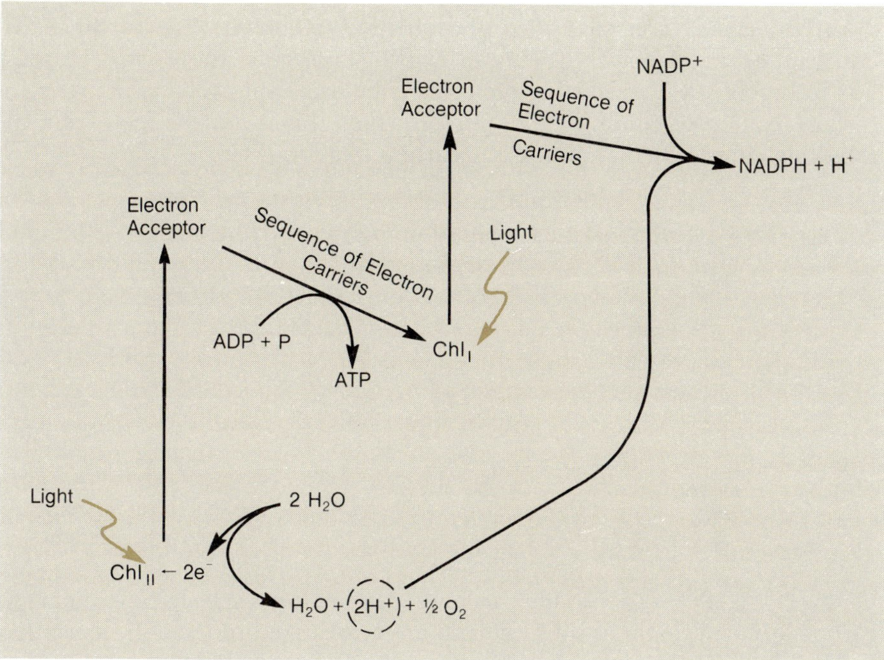

Figure 9–2. Light reactions of photosynthesis. During these reactions electrons and protons are transferred from water to the hydrogen carrier molecule $NADP^+$. Light energy absorbed by the photosynthetic pigments of photosystem II raises the energy level of a chlorophyll molecule high enough that the electron can be transferred to an electron acceptor. The lost electron is replaced by transfer from an electron donor molecule that ultimately obtains electrons from the breakdown of water. Since there is no further input of energy, the electron is transferred downhill (energetically) from one electron carrier to the next until it is donated to a chlorophyll molecule (Chl_I) of photosystem I. This electron will be accepted by $chlorophyll_I$ only after it has lost an electron to an electron acceptor after light absorption by photosystem I. After this initial rise in energy in photosystem I, electrons lose energy as they are transferred to the final electron acceptor, $NADP^+$. Since $NADP^+$ is a hydrogen carrier, not an electron carrier, protons (H^+) from the breakdown of water also associate with it. During electron transfer between photosystem II and photosystem I, ATP is formed from ADP and inorganic phosphate (P_i).

teria do not use water as the source of hydrogen atoms and therefore do not produce oxygen. Instead of oxygen they produce sulphur as a by-product of photosynthesis because they are able to use hydrogen sulfide (H_2S) as the source of hydrogen atoms for the light reaction.

The Dark Reactions of Photosynthesis (Carbon Fixation)

The light reactions of photosynthesis produced NADPH + H^+ and ATP, but these are not the carbon-containing raw materials that are required for the synthesis of new protoplasm. The production of such carbon-containing building blocks takes place during the dark reactions.

Simply stated, during the dark reactions carbon dioxide (CO_2) is enzymatically combined with an existing organic molecule. This process is often called carbon fixation. In the first step carbon dioxide is added to a pentose (5-carbon sugar) molecule called ribulose bisphosphate, resulting in the formation of a six-carbon compound. This six-carbon compound is not the end product of photosynthesis. In fact this compound has never been isolated since the same enzyme that added CO_2 to ribulose bisphosphate immediately rearranges this six-carbon compound into two identical three-carbon molecules. At this point the NADPH + H^+ produced in the light reaction is used as a hydrogen atom donor to convert this three-carbon molecule to a triose

(3-carbon sugar) molecule called phosphoglyceraldehyde (PGAL). Since carbon fixation combines CO_2 with an existing molecule, there must be some way to regenerate this starting molecule (ribulose bisphosphate) in order for carbon fixation to continue. The way in which this is done is seen from the following simplified diagram of the carbon fixation cycle:

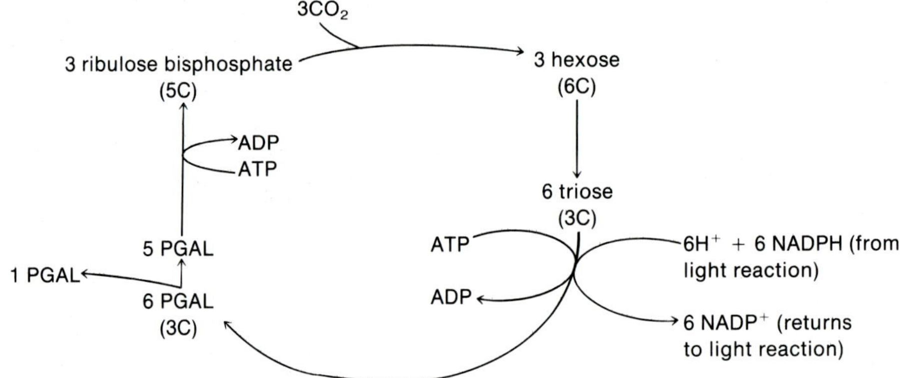

Of every six PGAL molecules accumulated in this cycle five of them (15 carbon atoms) remain in the cycle and are rearranged to form three ribulose bisphosphate molecules (15 carbon atoms). The sixth PGAL molecule represents the three carbon atoms which entered the cycle as three carbon dioxide molecules. This PGAL is used up in the synthesis of sugars such as glucose that, in the general equation stated above is considered to be the principal end product of photosynthesis. The PGAL molecules produced in the dark reaction are the real end products of photosynthesis and PGAL is the form in which carbon atoms enter the metabolic pathways of a plant. All metabolic pathways ultimately begin with PGAL produced in photosynthesis. Starch production by leaves is often used as an indicator of photosynthesis but starch, a polymer of glucose, is produced only when sufficient PGAL molecules have been converted into glucose that can then be converted into starch.

Light Absorption

At the beginning of this description of photosynthesis it was stated that light causes the chlorophyll molecules to lose electrons. Actually only certain wavelengths (colors) of light are effective in photosynthesis. Photosynthesis only utilizes visible radiation; all other wavelengths including ultraviolet and infrared are ineffective. The visible light spectrum (Plate 6) goes from violet to red (390–700 nm), the wavelength of light usually being expressed in nanometers (nm). Chlorophyll absorbs light mainly in the blue and red regions, that is, light with wavelengths of 390–460 nm and 630–680 nm (Figure 9–3). Since chlorophyll absorbs the blue and red from white light, the remaining green and yellow light passes right through a solution of chlorophyll molecules. All objects take on the color of light they do not absorb

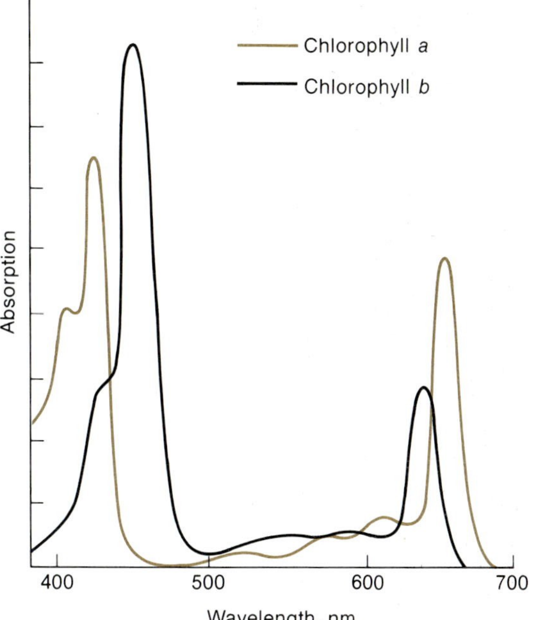

Figure 9–3. Chlorophyll absorption spectrum. All plants and certain algae contain two different types of chlorophyll molecules (*a* and *b*) that differ slightly in the wavelengths of light that they absorb.

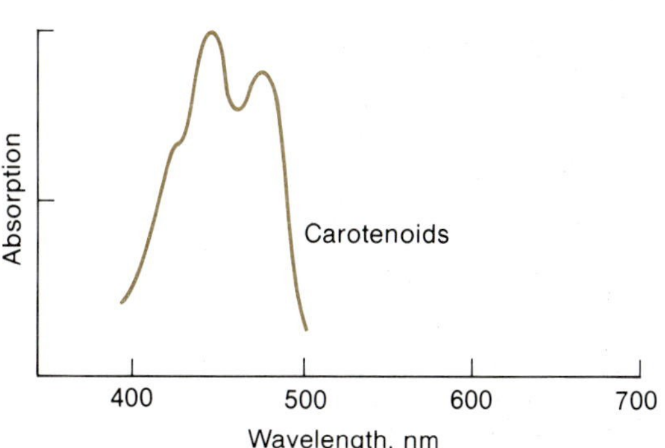

Figure 9–4. Carotenoid absorption spectrum. There are several different types of carotenoid pigments that might be found in a chloroplast. The absorption spectrum of the carotenoid pigment shown in this figure is typical of the range of wavelengths absorbed.

because our eyes use the light that comes from an object to see that object. Therefore chlorophyll molecules appear green. If chlorophyll absorbed all colors of light, it would appear black. Since chlorophyll is inefficient in absorbing light in the middle region of the visible spectrum, this would make the process of photosynthesis inefficient. Yet the process of evolution leads to increased efficiency and elimination of organisms that are less efficient at performing a particular function than the species with which they are competing directly (Chapter 23). Photosynthetic organisms have evolved a number of accessory pigments, such as carotene and xanthophyll, which absorb light at wavelengths different from chlorophyll (Figure 9–4). These pigments appear yellow-orange and so absorb some green and blue light. The amounts of these pigments in typical leaves is not sufficient to absorb all of the green light available and so normal leaves still appear green and not yellow. The energy of the light absorbed by the accessory pigments is passed to the chlorophyll molecules, which can use this energy for photosynthesis.

The accessory pigments alone cannot initiate photosynthesis; hence chlorophyll is absolutely essential for photosynthesis.

Photosynthesis and the Chloroplast

The reactions of photosynthesis occur in the chloroplast. The chlorophyll and accessory pigments along with the electron acceptors and enzymes involved in the transfer of electrons and protons from water to $NADP^+$ are located in the internal membrane system (**grana** and their interconnecting membranes) of the chloroplast (Figure 9–1). Chloroplast membrane organization is an active area of research today. This research informs us that the photosynthetic membrane is highly organized with photosystems and electron acceptors in precise spatial arrangements (Figure 9–5). Such precision allows efficient electron transfer reactions to occur. It should be pointed out that there are many photosystem I and photosystem II centers in a chloroplast.

The reactions of carbon fixation do not require a highly structured system; instead these reactions can take place in solution in the **stroma** (membrane-free regions of the chloroplast). Since the chloroplast is the food factory for the whole plant, the excess carbohydrates are exported from the chloroplast to the cytoplasm and ultimately are transported out of the leaf cell to the phloem (Figure 9–6). At times of active photosynthesis more carbohydrate is produced by a chloroplast than can be exported. This sugar is converted to starch that appears in the chloroplast stroma as starch granules.

Alternate Forms of Carbon Fixation

The process of photosynthesis, in essence, is the way in which an individual plant converts light energy into chemical energy in the form of ATP and new carbon-containing molecules. Even though the light reaction is common to all photosynthetic plants at least three different versions of the dark reactions have evolved. The version of carbon fixation that is utilized by the majority of plants is the one described above. This has been called the **C_3 pathway** of carbon fixation because the initial product of carbon fixation is a three-carbon compound. In response to different environmental pressures variants of the C_3 pathway have evolved. One variant, **Crassulacean Acid Metabolism (CAM),** evolved in certain plants adapting to arid environments. Basically CAM plants differ from C_3 plants only in the source of the CO_2 for photosynthetic carbon fixation. Whereas CO_2 for other plants comes from the atmosphere during the day, the stomata of CAM plants are closed during the day, which reduces water loss from transpiration. While this adaptation enables the plant to survive in a dry environment, it creates

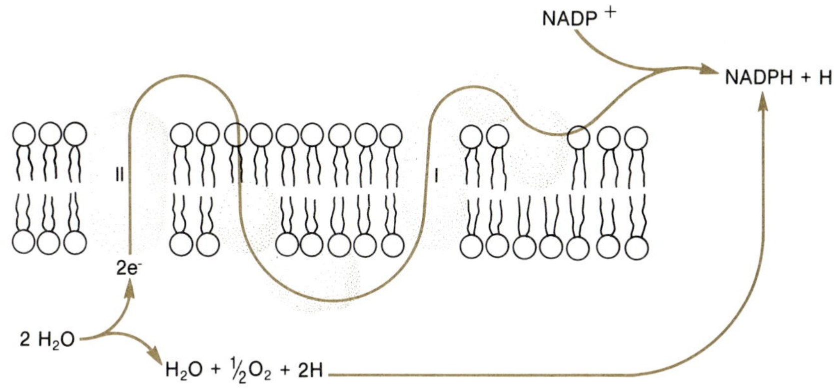

Figure 9–5. Photosynthetic membrane structure. The photosynthetic membrane is formed from a phospholipid bilayer in which proteins are embedded. Light absorption in the protein-pigment complex of photosystem II (II) results in electron transfer to an electron acceptor that is dissolved in the lipid phase of the membrane. Subsequent electron transfer reactions also occur in the membrane, sometimes involving electron carriers associated with membrane proteins. Light absorption by photosystem I (I) causes electron transfer to protein-associated carriers and on to $NADP^+$. The protons (H^+) generated by the breakdown of water associated with photosystem II cross the membrane through pores not shown in the diagram.

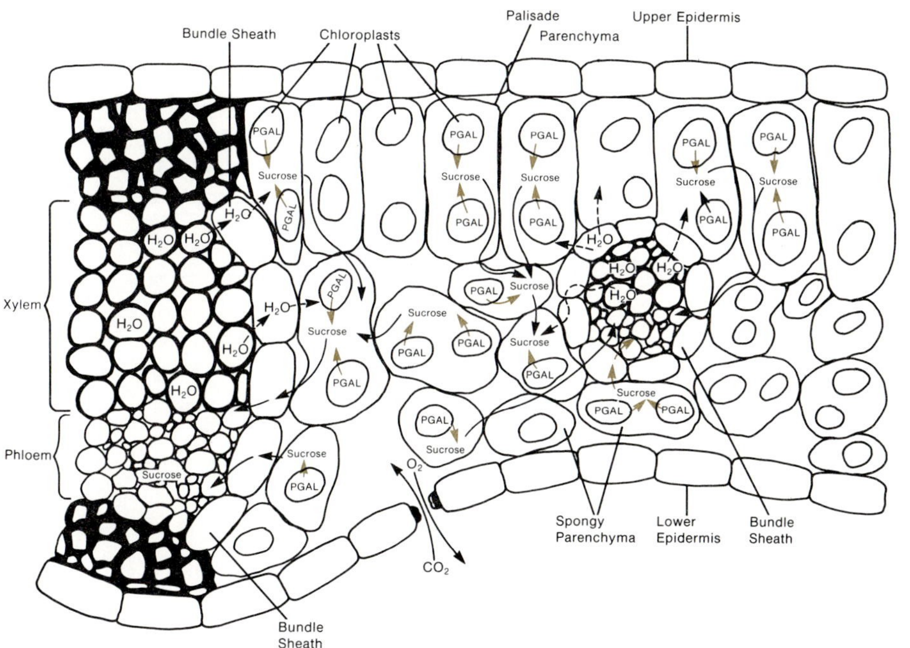

Figure 9–6. The leaf and photosynthesis. Photosynthesis in the chloroplast leads to the export of phosphoglyceraldehyde (PGAL) into the cytoplasm where it is metabolized into sucrose. Excess sucrose is transported from cell to cell until it reaches the bundle sheath cells. From here the sucrose is transported into the conducting cells of the phloem and exported from the leaf. Water enters the leaf in the xylem and moves by osmosis into the mesophyll cells. The consumption of CO_2 in photosynthesis leads to the diffusion of CO_2 into the leaf from the atmosphere through the stomata. Oxygen moves in the opposite direction because it is being produced by the mesophyll cells.

difficulties for the completion of photosynthesis. In order for photosynthesis to occur in a chloroplast the light reactions and the dark reactions must go on simultaneously; the chloroplast cannot store up the products of the light reactions until CO_2 is available for the dark reactions. This problem has been resolved in CAM plants by obtaining CO_2 for photosynthesis during the day from the breakdown of organic acids stored in the vacuole of each mesophyll cell. At night, when the stomata are open, the plant rebuilds its supply of organic acids from atmospheric CO_2. This temporary storage of CO_2 in the form of organic acids occurs in a limited group of arid land plants, mainly members of the family Crassulaceae. The only plant of agricultural importance to utilize CAM is the pineapple.

A more widespread variant of carbon fixation is the **C_4 pathway** where the initial product of carbon fixation is a four-carbon organic acid. The C_4 pathway should not be confused with CAM even though they both involve organic acids. C_4 photosynthesis requires a different organization of the reactions of photosynthesis. In both C_3 and CAM pathways a single mesophyll cell carries out the complete reaction of photosynthesis. In C_4 photosynthesis the carbon fixation pathway is divided between two cells: mesophyll cells and bundle sheath cells. The mesophyll cell incorporates atmospheric CO_2 into an organic acid. This organic acid is transported to a bundle sheath cell where the CO_2 is released and then incorporated in the standard C_3 carbon fixation pathway (Figure 9–7). C_4 plants are typically different from C_3 plants in bundle sheath anatomy and chloroplast morphology. On the surface the C_4 pathway appears to be less efficient than the C_3 pathway because it requires two types of cells to produce phosphoglyceraldehyde, with these two cell types being involved in shuttling molecules back and forth. In fact C_4 photosynthesis is much more efficient than C_3 photosynthesis for a variety of reasons. A major source of inefficiency for C_3 plants is the process of **photorespiration***, which does not occur in C_4 plants. In photorespiration, mesophyll cells break down molecules that have just been produced in photosynthesis, but the energy stored in these molecules is not used to make ATP as it would be in true respiration (Chapter 10). Instead the energy is wasted, resulting in inefficient photosynthesis. Photorespiration occurs in all

Photorespiration
Light-induced consumption of O_2 and release of CO_2 that does not result in ATP production.

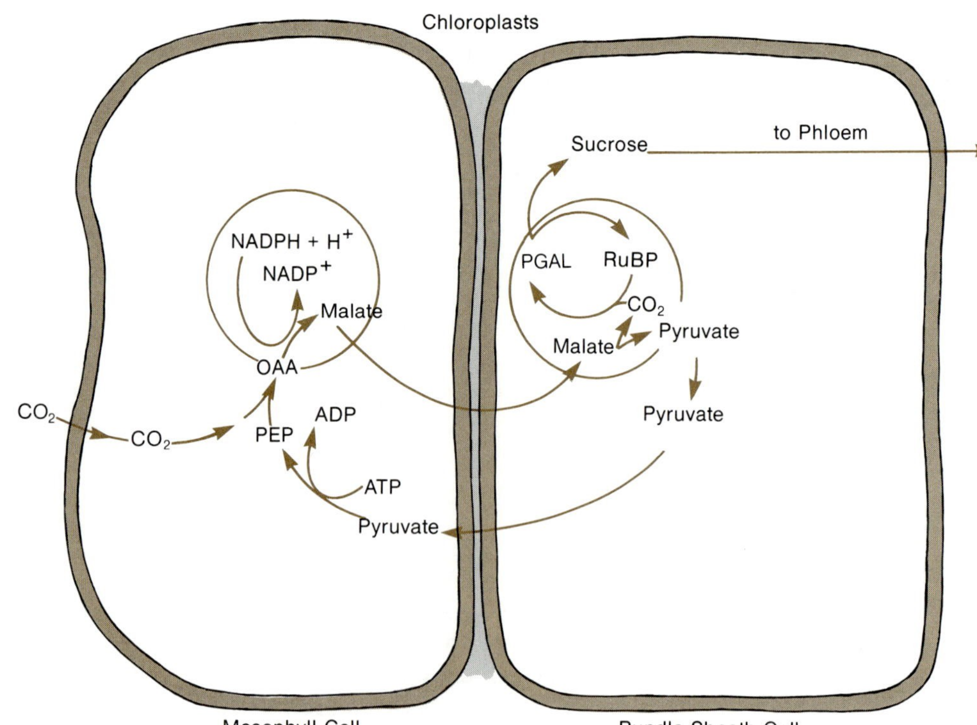

Figure 9–7. C_4 photosynthesis. This photosynthetic pathway involves two different cell types in the process of carbon fixation. The chloroplasts of both cell types carry out a complete and normal set of light reactions. Carbon fixation in C_4 plants begins in the cytoplasm of mesophyll cells where CO_2 is combined with phosphoenolpyruvic acid (PEP) to form oxaloacetic acid (OAA). This enters the chloroplast where hydrogen atoms are added, converting it into malate. The malate leaves the chloroplast and is exported from the mesophyll cell to a bundle sheath cell. Once in the bundle sheath cell malate enters the chloroplast where it is broken down, releasing CO_2 that is combined with ribulose bisphosphate (RuBP) in a standard C_3 carbon fixation cycle. The pyruvate that is formed in the breakup of malate is returned to the mesophyll cell where it is phosphorylated by ATP regenerating PEP, the starting compound in this carbon fixation pathway.

C_3 plants with a resultant loss in efficiency of carbon fixation of up to 25%. The value of photorespiration to C_3 plants is unknown. Another major difference is that the enzyme that makes the C_4 intermediate has a much higher affinity for CO_2 than does the first enzyme in the C_3 pathway. The net effect of this increased efficiency of C_4 plants is that the C_4 plant can incorporate more CO_2 than a C_3 plant in the same amount of time. Therefore a C_3 plant would have to keep its stomata open for much longer in order to incorporate the same amount of CO_2 as a C_4 plant. In dry environments, where water is in short supply, there is a survival advantage to being able to incorporate CO_2 while paying a minimum penalty in water loss. Plants from a wide variety of families have evolved the C_4 strategy in response to arid or saline environments. Two prime agricultural examples are corn and sugar cane, both tropical grasses that evolved in hot, arid areas.

Factors Affecting the Rate of Photosynthesis

In all leaves the rate of photosynthesis is regulated by many factors outside of the chloroplast so that even though the typical chloroplast has all of its chemical machinery intact, there is no guarantee that the photosynthetic factory will be operating at full capacity. Some of the major external factors regulating the rate of photosynthesis are light intensity, CO_2 concentration, temperature, and availability of water.

Light Intensity

Photosynthesis cannot occur in the dark. As day dawns the light intensity gradually increases, reaching a maximum during the middle of the day. Plants respond to the increasing light intensity by increasing their rate of photosynthesis (Figure 9–8). There are basically two groups of plants with regard to light intensity: shade plants and sun plants. (These will be discussed further in Chapter 24.) Concerning photosynthesis it can be pointed out that the shade plants become saturated with light long before the maximum light intensity is reached. This is part of their adaptation to a forest-floor environment. Sun plants, on the other hand, are adapted to full sunlight so that they do not become saturated with light until much higher light intensities are reached. C_4 plants are the most efficient sun plants since they do not become light saturated until just before the light intensity reaches its highest value on a summer day. This ability to use more of the available sunlight is part of the survival advantage that C_4 plants have over C_3 plants in their native environment. It is also one of the reasons that corn and sugar cane are such good crop plants.

For all plants, sun and shade, there is an increase in the rate of photosynthesis with increased light intensity until maximum photosynthesis is reached (light saturation). In order to increase their efficiency and extend the range

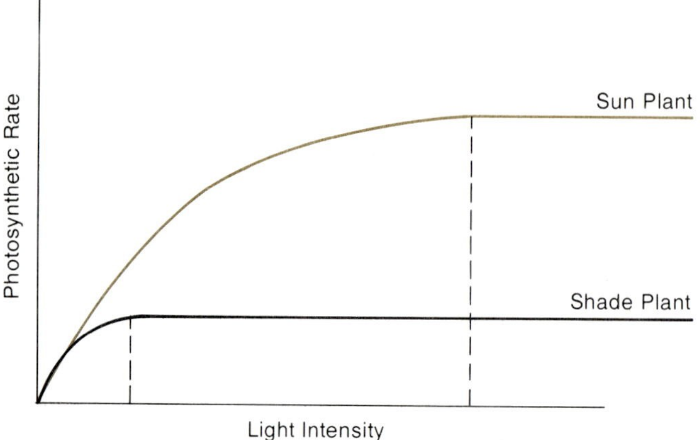

Figure 9–8. Light saturation. The rate of photosynthesis increases as light intensity increases until light saturation is reached. From this point light is no longer the rate-limiting factor in photosynthesis. Plants that are adapted to different environments have different saturation points even though the rates of photosynthesis at lower light intensities may have been the same.

of useful light intensities, plants take advantage of the shape of their chloroplasts, which are lens shaped. At low light intensities the chloroplasts are oriented so that their large, flat surface is toward the light to increase light absorption. As light intensity increases the chloroplasts change their orientation so that at light saturation they present their narrow edge toward the light. This cuts down their light-absorbing surface and reduces the destruction of chlorophyll molecules that occurs when excessive amounts of light energy are absorbed.

Concentration of CO_2

Since the end product of photosynthesis is a carbon-containing compound, it is to be expected that the rate of photosynthesis is dependent on the amount of available CO_2 (Figure 9–9). Experimentally it has been found that the rate of photosynthesis increases with increasing CO_2 until a maximum rate is achieved (saturation). The amount of CO_2 required to achieve saturation varies from plant to plant but it is clear that air containing the normal 0.03% CO_2 does not saturate photosynthesis. This supports other lines of evidence that indicate that photosynthesis evolved at a time when the atmosphere contained significantly more CO_2. Agronomists have used this observation to increase crop yield by growing crops in greenhouses with elevated levels of CO_2. In a number of trials it has been shown that the yield of potatoes, carrots, and beets, for example, can be increased up to 300% if the air is enriched with CO_2.

Temperature

Temperature affects the rate of photosynthesis because temperature affects the rate of all metabolic reactions. In general the optimum temperature for photosynthesis is 30°–35° C (85°–95° F). Photosynthesis can occur at extremely low temperatures in plants such as conifers during the winter because their leaf cells are protected from freezing. On the other hand, pho-

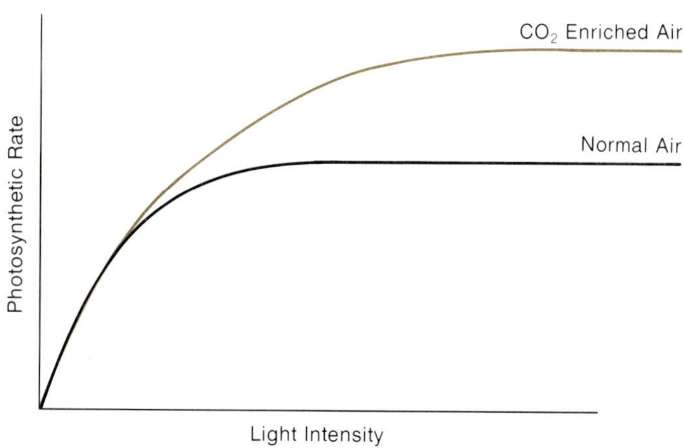

Figure 9–9. Effect of CO_2 concentration on photosynthetic rate. When light intensity is no longer the rate-limiting factor in photosynthesis, the concentration of CO_2 often becomes rate limiting. Increases in CO_2 concentration after (but not before) light saturation has been achieved will increase the photosynthetic rate until light saturation is reached again.

tosynthesis in cyanobacteria, which live in hot springs, can occur at temperatures above 50° C (120° F). Most plants, however, are not adapted to these extremes. Consequently most photosynthesis occurs between 5° C and 38° C (40°–100° F). Plants that grow in early spring or in cool environments have special adaptations that allow them to survive and metabolize under these conditions, but even these plants will photosynthesize more rapidly as temperatures rise. Similarly desert plants are able to metabolize at temperatures higher than other plants, but once again photosynthetic rates will most likely be higher if temperatures are lower (closer to the 30°–35° C optimum)

Availability of Water

Since water is the source of the electrons and protons used in photosynthesis, the chloroplast must have an adequate supply of water. The rate of photosynthesis, however, is not typically regulated by the water supply in the chloroplast. The role of water in regulating photosynthesis is more indirect. A shortage of water in the plant will cause the stomata to be closed that, in turn, will stop photosynthesis because of a lack of CO_2, long before the chloroplast water supply becomes rate limiting.

Concerning these and the other factors that regulate the rate of photosynthesis, it must not be assumed that they act independently. For instance, in the experiments done to determine the effect of light intensity on photosynthesis all other factors were kept constant. The same experiments conducted at a lower temperature would yield the same curve but the observed rates would be lower. If CO_2 concentration had been increased after light saturation had been achieved, then the rate of photosynthesis would have gone higher still. Thus the rate of photosynthesis, like that of any other metabolic reaction, is controlled by the limiting factor—the factor not present in optimal amounts. Under natural conditions the rate of photosynthesis will be controlled by a limiting factor such as one of the four listed above.

Summary

By the 1840s it was known that plants converted light energy into chemical energy in the process of photosynthesis. Since that time photosynthesis has been found to consist of light reactions and dark reactions. During the light reactions light energy absorbed by chlorophyll molecules of photosystems I and II is used to transfer electrons and protons (hydrogen atoms) from water to the hydrogen-carrier molecule $NADP^+$. These hydrogen atoms subsequently are used in the dark reactions in the process of carbon fixation that produces the triose, phosphoglyceraldehyde, as the end product of photosynthesis. ATP produced in the light reaction is also used in the process of carbon fixation. In order to absorb the light necessary for photosynthesis the chloroplast contains chlorophyll plus carotenoid pigments in the grana where the light reactions occur. Carbon fixation occurs in the stroma or membrane-free region of the chloroplast.

Although the majority of plants fix their carbon using the C_3 pathway, which produces three-carbon compounds, there are significant numbers of plants from many different angiosperm families that use an alternate pathway, the C_4 pathway, which forms four-carbon compounds as the initial products of carbon fixation. The C_4 pathway also involves changes in leaf anatomy and chloroplast morphology as well as increased efficiency in the arid, native environment of these plants. Another, smaller, group of plants has evolved a different photosynthetic adaptation to solve the problem of an arid environment. This solution is called Crassulacean Acid Metabolism (CAM) and involves the uptake and storage of CO_2 at night that is later used during the day in photosynthesis. Both CAM and C_4 photosynthesis reduce the water demand of the plant by allowing the plant to reduce the amount of time the stomata are open during the heat of the day.

The rate of photosynthesis in all plants is regulated by four major factors: light intensity, CO_2 concentration, temperature, and water availability. Each of these in addition to other conditions can be the rate-limiting factor for photosynthesis. The observed rate of photosynthesis is always a function of the prevailing conditions with the theoretical maximum rate seldom, if ever, being achieved by any plant in nature.

Review Questions

1. The process of photosynthesis consists of light reactions and dark reactions. Describe what happens during each of these sets of reactions.
2. Explain how a plant and an animal can be kept alive in a sealed glass chamber when the chamber is in the light but not when it is in the dark.

3. Compare and contrast C_3, C_4, and CAM forms of carbon fixation including statements on their ecological roles.
4. Describe the structure of the leaf and show how this structure is related to the photosynthetic function of the leaf.
5. Explain how light intensity can limit the rate of photosynthesis when light is not needed for carbon fixation.

Photosynthesis, the Cornerstone of Civilization

Human social development owes a great debt to the processes of photosynthesis, beyond the production of food and oxygen. The industrial society is based on fossil fuels—oil, coal, and natural gas, the remains of both microscopic and macroscopic organisms from past geologic eras. When these organisms died, for various reasons they did not decay, but over the subsequent millions of years, heat and pressure within the earth converted their accumulated remains into our present-day supplies of fossil fuels. With depletion of our oil reserves, plants through photosynthesis may be called upon to produce crude oil today, not for future generations. Melvin Calvin, a Nobel prize winner for photosynthesis studies, and other people are carrying out research with plants of the spurge family (relatives of poinsettia and crown-of-thorns) that produce a white latex, an emulsion of hydrocarbons (oil) in water. The hydrocarbons produced by some of these plants are similar chemically to those in gasoline. If this research is successful, we may find a dramatic change in the production of gasoline—instead of being pumped out of the ground, it may be "grown" as a commercial crop.

Present gasoline supplies in some areas are being extended by the addition of alcohol produced from starch, forming the product known as gasohol. In the United States corn is fermented to alcohol and the alcohol concentration is increased by distillation much as moonshiners and legal distillers have been doing for centuries. One part of this alcohol (ethanol) is added to nine parts of gasoline forming gasohol. Automobile engines are able to run on gasohol without any modifications; if the economics of alcohol production are favorable this results in a decrease in dependency on foreign supplies of fossil fuels. In Brazil the balance of payments deficit induced by paying for the import of crude oil has been so devastating that the government is supporting a program to produce large quantities of ethanol from sugar cane and other starchy plant residues. Furthermore, this program will involve modification of automobiles to run on 100% ethanol to further reduce oil consumption. A similar program was adopted in Europe during the Second World War with automobiles, buses, and trucks being modified to run on methanol derived from wood because crude oil imports were blocked by the Allies.

In order to be socially acceptable, a program of producing fuel from plants must not result in a reduction in food supplies. For countries such as Canada and the United States with a grain surplus, diversion of crops to fuel would be feasible. For the net importers of food, fuel would have to be produced from agricultural wastes such as bagasse from sugar cane. The other alternative is to use land that cannot be used for crops, which is why the work of Calvin is so interesting. The spurges he is employing grow naturally in poor soil under arid conditions and would not have to compete with agricultural crops.

Consequently photosynthesis is central to our survival as animals and human beings. This fact makes more immediate the concern over the effects of urban sprawl and air pollution on plants.

10

Respiration

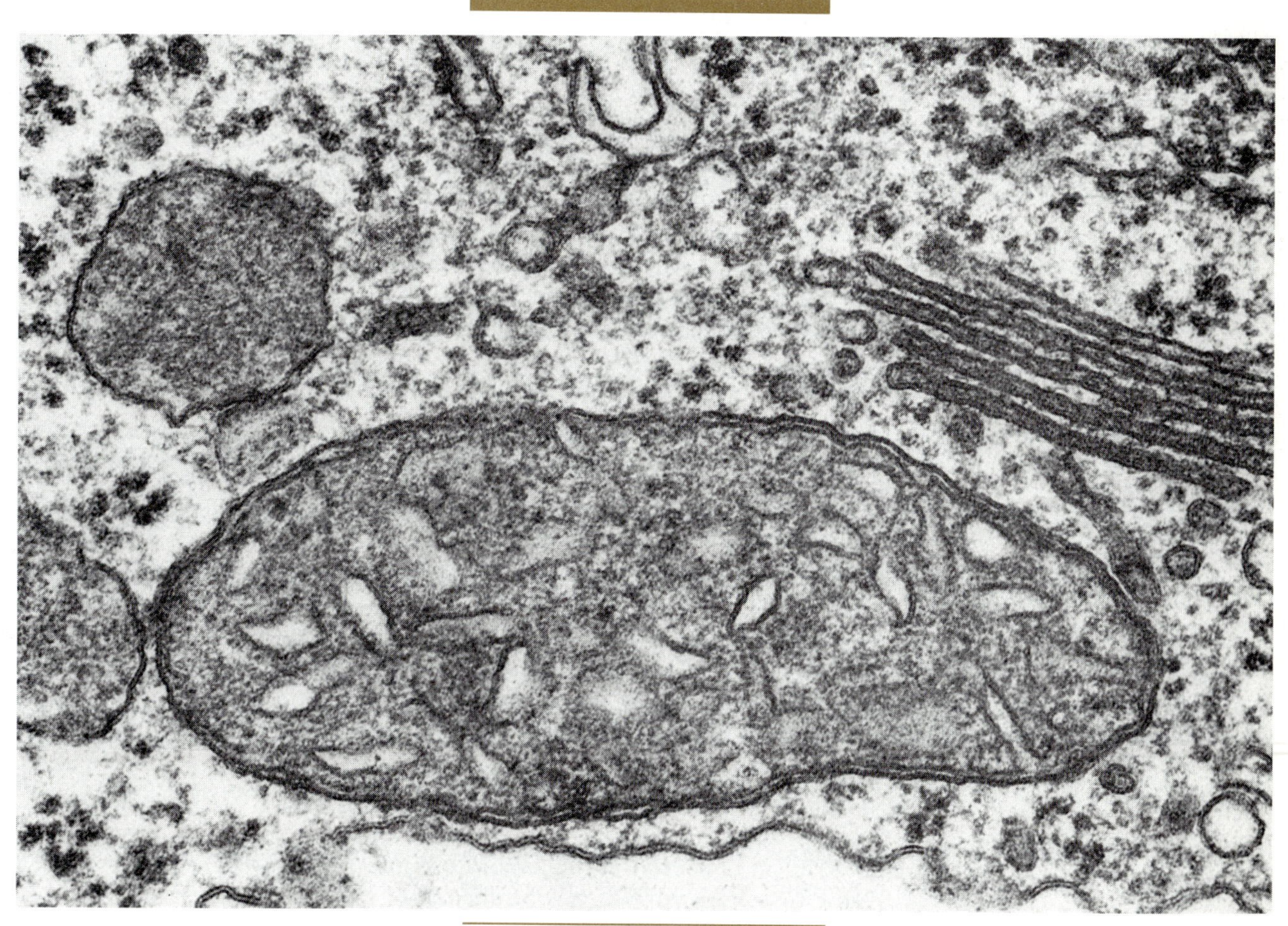

Outline

Introduction

Animals are often categorized as either warm-blooded or cold-blooded, the distinction being based on whether the organism has the ability to regulate its internal temperature. Cold-blooded animals lack this ability so their body temperature is determined by the temperature of their environment. Using these categories plants would be considered "cold-blooded" since the temperature of plant leaves and stems is a function of air temperature and the amount of solar radiation received. Both plants and animals, however, produce heat as a by-product of **respiration*** but plants and cold-blooded animals lack heat retention mechanisms and a temperature regulating center. Respiration of glucose results in the conversion of only a fraction of the stored chemical energy into ATP.* The remainder of the energy of the glucose molecule is released as heat. In most situations the process of radiation prevents heat death of plants because it dissipates both metabolic heat and solar heat from plant organs. There are certain circumstances when the heat produced in respiration is retained and causes significant changes in temperature. A common example would be the heat generated in a pile of grass clippings—the act of separating part of a leaf blade from the plant does not kill the cells in the clipping and so they continue to respire. In the pile there is little opportunity for dissipation of heat generated by respiration, which causes the temperature of the pile to rise significantly. A similar process occurs in a compost pile, only in this case the heat is generated by the respiratory activities of the bacteria and fungi, which are decomposing the organic material, and this heat further stimulates the growth of these organisms. The heat generated in a compost pile is also of value to the gardener because the elevated temperature kills weed seeds. Certain flowering plants have adapted respiratory heat production to their advantage. In particular, the skunk cabbage inflorescence* respires so rapidly as it comes up in early spring that it can actually melt any snow that is around it; internal temperatures of 40°–45° C (104°–113° F) have been recorded. The high temperature of the inflorescence has been postulated to be an advantage by helping the flowers withstand the low air temperatures that can occur at this time and also by attracting insects to the flowers to pollinate them.

The metabolic pathway known as respiration did not evolve as a means of producing heat for organisms, however. In Chapter 9 it was stated that to be alive requires a continuous expenditure of energy—energy that could be obtained by an organism from the food molecules that it contains. **Heterotrophs*** such as animals obtain this energy from food that they consume, while **autotrophs*** such as green plants obtain this energy from food that they produce by photosynthesis. Regardless of how an organism obtains its food molecules the basic processes involved in converting them to energy are the same: respiration. Therefore, the statement above implies that respiration must occur in all organisms if food is to be converted into energy. It is obvious that animals require energy in order to move their bodies but most plants do not move from place to place in their environment. Why,

Respiration
The metabolic process that results in the release of chemical energy from simple organic molecules.

ATP
A high-energy molecule that is involved in the transfer of chemical energy in cells.

Inflorescence
A group of flowers on a common stalk.

Heterotroph
An organism that cannot produce its own food from inorganic molecules.

Autotroph
An organism that can produce its own food from inorganic molecules.

then, must plants respire? The answer for plants is the same as it is for animals: a living cell must expend energy merely to remain alive. This is a consequence of the second law of thermodynamics which states that systems tend to move toward a condition of maximum disorder unless there is a continued input of energy to maintain the organized systems. The high level of organization in living cells is only maintained through the input of energy derived from respiration. In order to maintain the cell a variety of biochemical reactions (metabolism) are needed (Chapter 3), and most of these metabolic reactions require energy to drive them. Consequently for a cell to remain alive or grow, food must be converted into energy through the process of respiration. Furthermore, respiration must occur in each cell because the energy produced in respiration cannot be transported from cell to cell. It is for this reason that the plant possesses phloem, the food conducting tissue, which functions to supply each nonphotosynthetic, living cell with sugar that serves as the source of stored energy that can be released by respiration.

Cellular Energy

The importance of respiration is that it releases useful energy from food molecules. Energy in the form of heat or light would be of no value to the cell because the enzymes that are responsible for metabolism have no way of capturing these forms of energy and putting them to use to drive chemical reactions. For the energy to be of use to the cell for its metabolism it must be in the form of chemical energy—energy stored in the chemical bonds of a molecule. All molecules possess chemical energy by virtue of the fact that they have chemical bonds, yet only a few molecules are constructed in such a way that they can function as carriers of chemical energy. One such molecule that is produced during respiration is **ATP (adenosine triphosphate),** sometimes referred to as the energy currency of the cell. The production of sugar in photosynthesis required the input of energy in the form of light. The absorption of light by a photosynthetic pigment resulted in the conversion of light energy into chemical energy some of which was ultimately used to form the bonds that hold the various atoms together as a sugar molecule (Chapter 9). However, the chemical energy stored in the sugar molecule or any of the molecules derived from it is not readily available to do metabolic work. For this stored chemical energy to be used in metabolism it must be transferred to ATP. Respiration represents a withdrawal from the chemical energy bank to convert some of the stored energy into more readily available energy.

ATP is often referred to as a high-energy compound because it carries a large amount of chemical energy in one chemical bond. The ATP molecule is a triphosphate which, in this case, means that there are three phosphate groups linked together linearly (Figure 10–1). The high-energy bond is the one linking the third phosphate group to the molecule. Without this terminal phosphate group, the molecule is a diphosphate **(ADP)** and is unable to serve as a source of chemical energy for metabolism. The addition of the

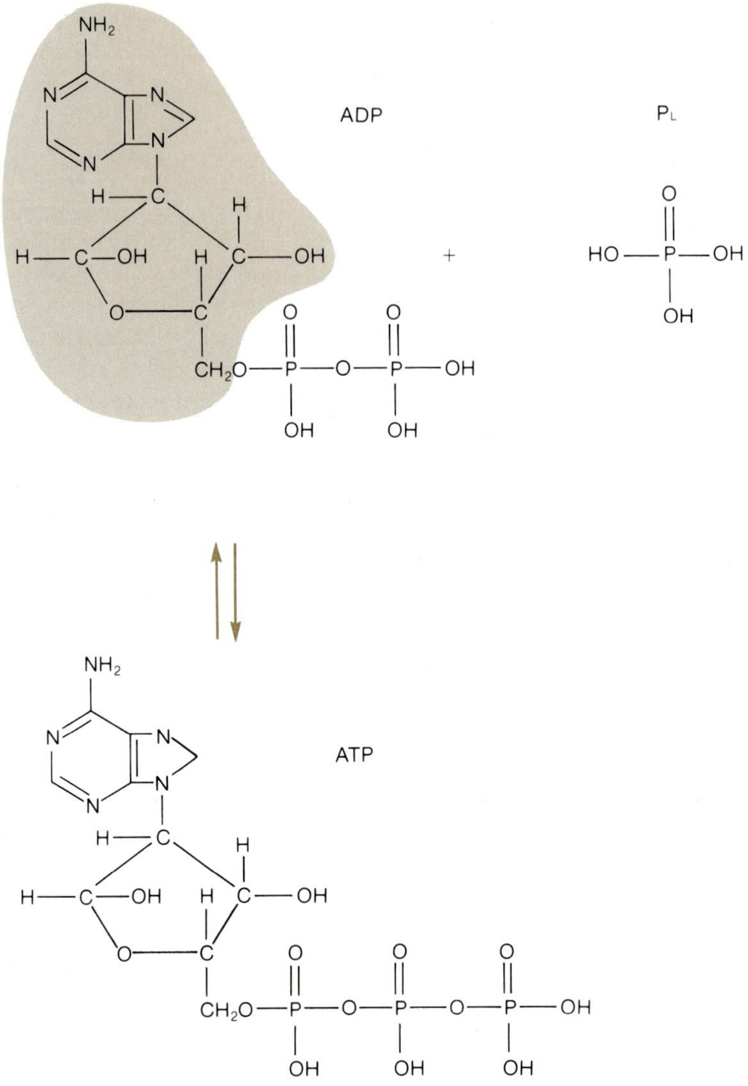

Figure 10–1. ATP. Adenosine triphosphate (ATP) is formed when a third phosphate group is added to adenosine diphosphate (ADP). The shaded portion of this molecule is adenosine derived from the base adenine.

extra phosphate group to convert ADP into ATP (ADP + P → ATP) requires a considerable input of chemical energy; liberation of this final phosphate (ATP → ADP + P) releases this energy for metabolic work. The energy is not really released; instead it is transferred to another molecule. The following two examples show how ATP may be used in a reaction. ATP can provide the energy required to induce an enzyme to change its configuration slightly and thus either push two molecules close enough together that they join or pull on a molecule and break a bond splitting the molecule into two fragments (Chapter 3); or ATP can provide the energy required to cause another molecule to accept a phosphate group (glucose + ATP → glucose-phosphate + ADP).

In the process of dismantling the sugar molecule during respiration the carbon, oxygen, and hydrogen atoms that formed the sugar are released. The carbon and oxygen are released as free CO_2 but the hydrogen atoms are transferred to a specialized molecule called a **hydrogen carrier** that functions to carry hydrogen atoms from one metabolic reaction to another. In the case of respiration, the primary hydrogen carrier is **NAD^+ (nicotinamide adenine dinucleotide).** Although each molecule of NAD^+ is involved in the transfer of two hydrogen atoms, there is actually only one complete hydrogen atom added to the NAD^+ molecule. The other hydrogen atom (a proton plus an electron, see Appendix) is split, with the electron being added to NAD^+ and the proton being released to the cellular water where it becomes part of the pool of available H^+ ions ($NAD^+ + 2H \longrightarrow NADH + H^+$). When $NADH + H^+$ donates hydrogen atoms in metabolism the electron is combined with any available H^+ to form a new hydrogen atom. In photosynthesis the hydrogen carrier was the related molecule $NADP^+$, which carries hydrogen atoms in exactly the same way as NAD^+.

Digestion

In both plants and animals digestion* must precede respiration if large food molecules are involved—for instance, starch must be broken down to glucose by the action of an enzyme called amylase; proteins must be broken down to amino acids by enzymes called proteases; and so on. During digestion complex molecules are broken down to simple ones, but no energy is released in the process. Digestion in plants is an intracellular event as each cell contains its own supply of stored food, mainly in the form of starch. Consequently, plant respiration most commonly involves the breakdown of glucose. Even though starch is the major source of material for respiration, other complex molecules, such as proteins and fats, are continually being broken down (Chapter 3) and would be available to the respiratory pathway. The amino acids and fatty acids, however, are usually recycled into new protein and fat molecules, respectively. When a cell contains an excess of amino acids or fatty acids, or when fats and proteins are stored as foods in certain seeds, then alternate respiratory pathways are utilized that will liberate the energy from these molecules. Because the main respiratory pathway in plants involves glucose, this will be described in detail before the other pathways.

Digestion
The enzyme-mediated breakdown of complex organic molecules into their simple components.

Overview of Respiration

Two different metabolic pathways that will liberate energy from glucose have evolved in response to different environmental conditions (Chapter 23). The first to evolve was **anaerobic respiration,** which is the respiratory pathway that occurs in the absence of oxygen. It was only after the earth's atmosphere came to contain oxygen that **aerobic respiration** evolved. Both anaerobic and aerobic respiration pathways have a common starting point in the process of **glycolysis*** after which they diverge.

Glycolysis
The first stage in the respiration of glucose resulting in the production of two molecules of pyruvic acid.

Glycolysis

Mitochondrion
(*pl.*, mitochondria) A membrane-bounded organelle that functions in respiration.

Even though the process of respiration is always listed as the function of **mitochondria***, glycolysis, the first portion of respiration, occurs not in the mitochondria but in the cytoplasm. In glycolysis (Figure 10–2) a molecule of glucose (a sugar with six carbon atoms) is broken down in a series of reactions mediated by enzymes to produce two molecules of pyruvic acid (each containing three carbon atoms) and two molecules of ATP. During this process four hydrogen atoms are stripped off the glucose and transferred to the **hydrogen carrier NAD$^+$,** converting NAD$^+$ into NADH + H$^+$. There is no release of carbon dioxide or consumption of oxygen during glycolysis. Since the glucose molecule has not been completely broken down to carbon dioxide and water, considerable chemical energy is still stored in the resulting pyruvic acid molecules.

As indicated above, the process of glycolysis is common to both types of respiration; however, after glycolysis the two pathways separate. Anaerobic and aerobic pathways provide different solutions to the problem, which exists at the end of glycolysis, of how to remove the hydrogen atoms from NADH + H$^+$ so that this carrier can be used again in glycolysis. Because the cell contains a limited supply of NAD$^+$ that must function in different metabolic reactions requiring hydrogen transfer, it is essential that a mechanism exist to recycle the NADH + H$^+$ into NAD$^+$ so that glycolysis does not stop for the lack of a hydrogen carrier.

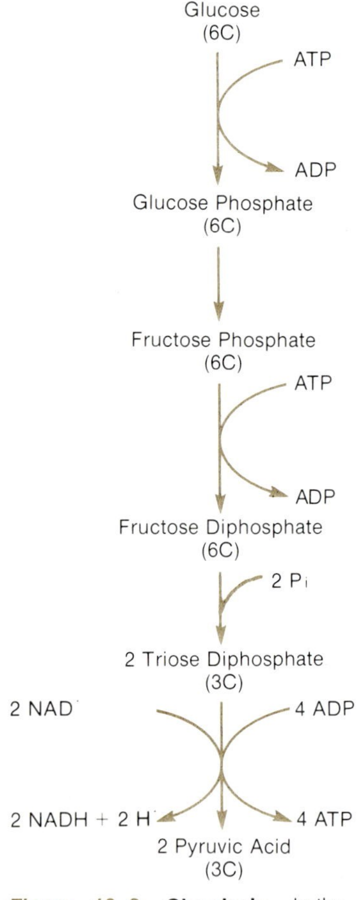

Figure 10–2. Glycolysis. In the initial stages of glycolysis two molecules of ATP are required for each molecule of glucose that enters the pathway. After molecular rearrangements and addition of further inorganic phosphate, two molecules of NADH + H$^+$, two molecules of pyruvic acid, and four molecules of ATP are produced from each molecule of glucose. Glycolysis, thus, yields a net gain of two molecules of ATP for each glucose molecule metabolized.

Anaerobic Respiration

Cells employing anaerobic respiration have evolved various metabolic solutions to the hydrogen-carrier problem. Of the two most common solutions, the first is **lactic acid fermentation.** In this process the cell uses an enzyme that transfers the hydrogen atoms from NADH + H$^+$ to pyruvic acid, which is the end product of glycolysis. This reaction allows NAD$^+$ to return to glycolysis while the addition of hydrogen to pyruvic acid converts it to lactic acid (Figure 10–3), a waste product of this fermentation process. A familiar example of lactic acid fermentation involves the lactic acid bacteria, *Lactobacillus.* Using milk sugar as their energy source, these bacteria grow in milk and produce lactic acid as a waste product. Because lactic acid is a cellular poison, it is released into the environment (the milk). As the bacterial population increases so does the lactic acid content of milk. Since acids have a sour taste, the milk begins to taste sour once the lactic acid concentration reaches sufficient levels. Curdling of milk occurs at the same time because the higher acidity causes the milk proteins to coagulate.

The other common solution to the hydrogen-carrier problem of anaerobic respiration is **alcoholic fermentation,** the process that produces alcohol. This kind of fermentation is popularly associated with yeast, which are uni-

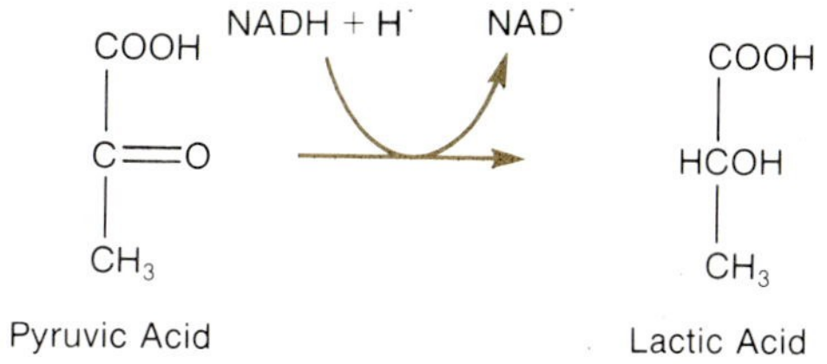

Figure 10–3. Lactic acid production. Organisms that carry out lactic acid fermentation contain an enzyme that takes hydrogen atoms from the NADH + H^+ produced in glycolysis and transfers them to pyruvic acid, also produced in glycolysis. This regenerates NAD^+ and permits glycolysis to continue since NAD^+ is required as a hydrogen acceptor in the reactions of glycolysis.

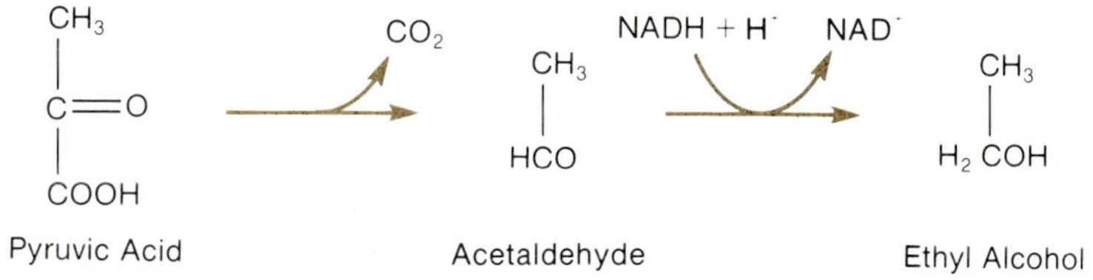

Figure 10–4. Ethyl alcohol production. Organisms that carry out alcoholic fermentation contain enzymes that convert pyruvic acid to ethyl alcohol. In the first step pyruvic acid from glycolysis loses one carbon in the form of CO_2. The resulting two-carbon acetaldehyde molecule functions as an acceptor of hydrogen atoms from NADH + H^+ produced in glycolysis. This converts acetaldehyde into ethyl alcohol.

cellular fungi (Chapter 16). In alcoholic fermentation the yeast cells regenerate NAD^+ from NADH + H^+ in a way different from that employed by the lactic acid bacteria. They break down pyruvic acid (a three-carbon compound) to a two-carbon compound plus CO_2. Subsequently, the hydrogens are transferred from NADH + H^+ to this two-carbon compound, which regenerates NAD^+ and produces alcohol as a waste product (Figure 10–4). Like lactic acid, alcohol is a cellular poison that must be excreted by the cell if the organism is to survive. As the alcohol content of the environment increases, growth and metabolism of yeast is slowed until, at 12% alcohol concentration, all growth ceases.

Note that in both forms of anaerobic respiration discussed above the hydrogen carrier (NADH) loses its hydrogens and can be reused in glycolysis, but the glucose molecule is not completely broken down to carbon dioxide and water. At the end of fermentation two ATP molecules have been produced per glucose molecule and a two- or three-carbon compound is released as a waste product that cannot be degraded any further by that organism. This waste product still contains stored chemical energy; however, the anaerobic organism is unable to extract this energy because it has no alternate way of disposing of the hydrogen atoms that are removed from the molecule being respired. Obviously this is an inefficient way to make a living; it is almost like having a one-hundred-dollar check and cashing it for only two dollars.

Aerobic Respiration

The majority of organisms respire aerobically. In this process glycolysis is followed by a different solution to the problem of how to regenerate NAD^+

from NADH + H^+. Organisms that carry out aerobic respiration have an enzyme-controlled pathway in their mitochondria (Figure 10–5) that transfers hydrogen atoms through a series of molecules to oxygen, producing water as the end product of the pathway. This is the **electron transport chain,** so named because, while each hydrogen atom consists of one proton and one electron, the energy of a chemical bond is associated with the electron, not the proton. Although the net result is the transfer of hydrogen atoms from NADH + H^+ to oxygen, in actual practice some of the carriers only accept electrons and so the protons (H^+) are released into the water. However, in a transfer of electrons from an electron-only carrier to a hydrogen carrier, a proton is regained from the surrounding solution for each electron, and the hydrogen atom is reconstituted. During electron transport, the hydrogen atoms are passed from one carrier to the next in the following fashion:

$$\mathrm{NADH} + \mathrm{H}^+ \rightarrow \mathrm{NAD}^+ \qquad \mathrm{A} \rightarrow \mathrm{AH_2}$$

(where A is used just as the symbol for the actual carrier)

After A receives the two electrons from NADH + H^+, it immediately transfers them to B which transfers them to C and so on to oxygen as shown in the following schematic diagram:

$$\mathrm{NADH} + \mathrm{H}^+ \rightarrow \mathrm{NAD}^+;\quad \mathrm{A} \rightarrow \mathrm{AH_2};\quad \mathrm{AH_2} \rightarrow \mathrm{A},\ \mathrm{B} - 2e^- + 2\mathrm{H}^+;\quad \mathrm{B} - 2e^- \rightarrow \mathrm{B},\ 2\mathrm{H}^+,\ \mathrm{C} \rightarrow \mathrm{C} - 2e^-;\quad \mathrm{C} - 2e^- \rightarrow \mathrm{C},\ 2\mathrm{H}^+ + \tfrac{1}{2}\mathrm{O_2} \rightarrow \mathrm{H_2O}$$

As the hydrogen atoms are transferred between carriers in the electron transport chain, the electrons lose some of the extra energy that they possessed when they were removed in glycolysis. This energy is used to convert ADP + P into ATP. It should be obvious, then, that an aerobic organism

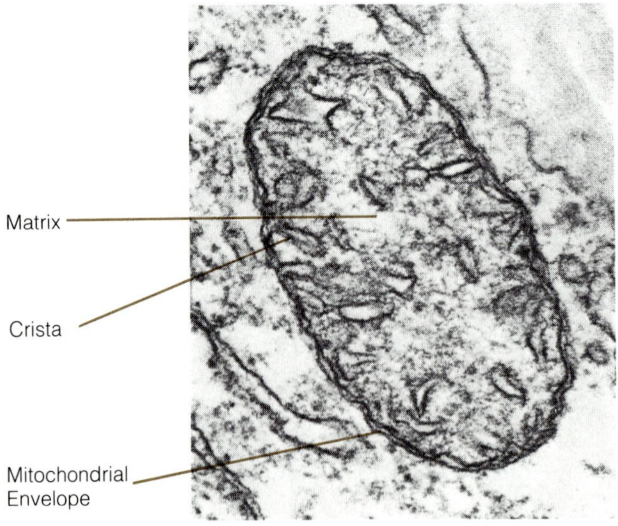

Figure 10–5. Typical mitochondrion. The electron transport chain is found in the cristae and the reactions of the Krebs cycle occur in the matrix.

obtains more ATP from the reactions of glycolysis because energy is obtained from the hydrogen atoms as well as from glycolysis itself. In addition, the aerobic organism is not forced into the position of using the end product of glycolysis, pyruvic acid, as a metabolic waste product. Pyruvic acid contains stored energy, as indicated above, which is now available to the aerobic organism because enzymes are present to metabolize pyruvic acid. The remainder of aerobic respiration is associated with the complete breakdown of pyruvic acid to carbon dioxide and water.

Initially pyruvic acid (from glycolysis) loses one carbon and associated oxygen as CO_2 (Figure 10–6) forming acetic acid (a two-carbon compound). It might seem that the cell should be able to break down the acetic acid in the same way, i.e., by using an enzyme to split the carbon skeleton to release CO_2 and transfer the hydrogen atoms to NAD^+. The cell does not have this type of enzyme, however, and is unable to dismantle the acetic acid molecule in such a straightforward manner. Instead, a more elaborate pathway must be used—the **Krebs cycle**—that occurs in the mitochondria. The role of the Krebs cycle in aerobic respiration is to break down acetic acid to CO_2 while transferring hydrogen atoms to a hydrogen carrier. As the acetic acid enters the Krebs cycle it first combines with a four-carbon organic acid (oxaloacetic acid) to form a six-carbon organic acid called citric acid (Figure 10–7). This step occurs because the mitochondria contain the enzymes that will dismantle six-carbon acids to four-carbon acids, thus disposing of the two-carbon acetic acid. The four-carbon oxaloacetic acid, in essence, functions as a carrier of the two-carbon acetic acid during its disassembly. In a subsequent step citric acid loses one carbon as CO_2 while hydrogen atoms are trans-

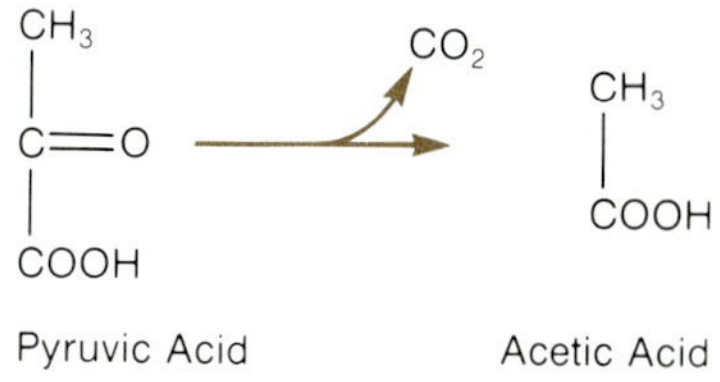

CH_3 | C=O | COOH → (CO_2) → CH_3 | COOH

Pyruvic Acid   Acetic Acid

Figure 10–6. Acetic acid production. Organisms that carry out aerobic respiration contain an enzyme that converts pyruvic acid to acetic acid by the removal of one carbon in the form of CO_2.

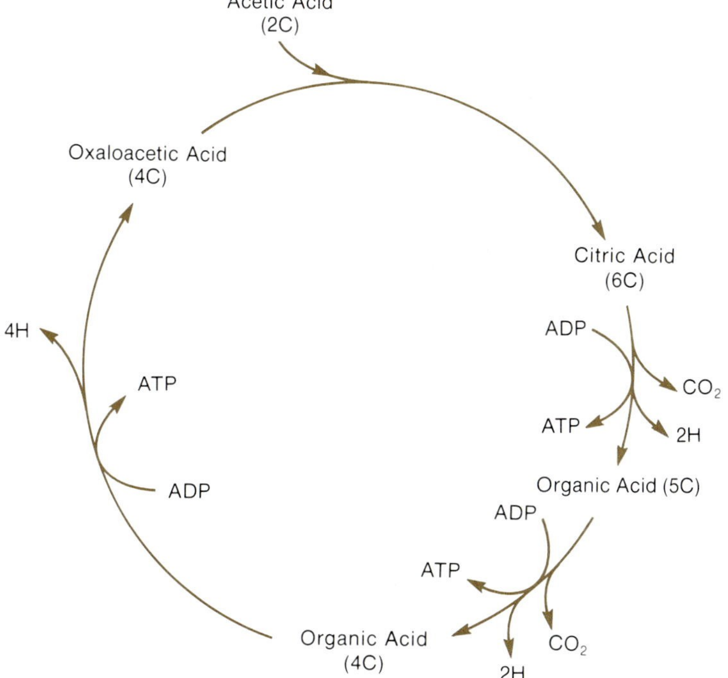

Figure 10–7. Krebs cycle. Acetic acid is enzymatically combined with oxaloacetic acid in the mitochondrion forming citric acid. In a series of molecular rearrangements the two carbon atoms and associated hydrogen and oxygen atoms of acetic acid are liberated in the form of CO_2 and hydrogen attached to a carrier molecule. In the process, ATP is formed and oxaloacetic acid is regenerated.

ferred to NAD^+, thus converting citric acid to a five-carbon organic acid. This process is repeated again releasing the second carbon of the acetic acid and regenerating a four-carbon organic acid. As the result of this molecular rearrangement, energy becomes available that is used in the production of ATP. Since this organic acid contains more hydrogen atoms than the oxaloacetic acid that was used in the first step, these hydrogen atoms must be transferred to a hydrogen carrier if oxaloacetic acid is to be regenerated. Once again this results in a molecular rearrangement that yields energy that is used to produce more ATP. At this point the Krebs cycle has come full circle with the starting molecule (oxaloacetic acid) having been regenerated and the acetic acid molecule having been completely dismantled. Since NAD^+ is needed continuously in the Krebs cycle, the hydrogen atoms are donated from $NADH + H^+$ to the electron transport chain generating more ATP.

Aerobic respiration (glycolysis + Krebs cycle + electron transport) results in the complete breakdown of glucose to CO_2 and H_2O, and the transfer of energy to ATP. Aerobic respiration, therefore, is more efficient than anaerobic respiration—in fact, aerobic respiration produces 36–38 molecules of ATP from each glucose molecule compared to the 2 molecules of ATP obtained from anaerobic respiration. Thus an anaerobic organism needs at least 18 times as much glucose as an aerobic organism to produce the same amount of ATP. Without the evolution of aerobic respiration the evolution of complex, multicellular plants would have been impossible because the energy demands of nonphotosynthetic, anaerobic structures would necessitate 18 times more photosynthetic area than is required by aerobic plants. Multicellular anaerobic plants would also have the problem of disposing of large quantities of what would be a metabolic poison for them (lactic acid or alcohol), which would most likely restrict these organisms to aquatic environments. Given all of these apparent advantages of aerobic respiration it is interesting to note that the evidence indicates that aerobic respiration evolved not as a means of increasing efficiency but as a means of detoxifying oxygen, which is another cellular poison (Chapter 23).

Respiration of Fatty Acids and Amino Acids

Digestion of fats and proteins releases fatty acids and amino acids, respectively, thereby making them available for metabolism. Typically these compounds will be recycled into new fats and proteins but, if there is an excess of these in the cell or if there are fats and proteins as stored food, both of these can be respired generating ATP. Fatty acids are subjected to further digestion, which consists of chopping the hydrocarbon chain (Chapter 3) into two-carbon fragments (acetic acid). Acetic acid from fatty acid degradation becomes part of the cellular pool of acetic acid that is disposed of through the Krebs cycle thus generating ATP from the respiration of fatty acids.

If amino acids are not recycled into new proteins they also can be disposed of via the Krebs cycle. In this case the amino acid must lose the amino group (Chapter 3) to convert it into an organic acid that can be rearranged and then enter the Krebs cycle.

Facultative Anaerobes

A **facultative anaerobe** is an organism that can switch from aerobic respiration to anaerobic respiration if oxygen becomes absent. In the discussion of alcoholic fermentation above it was pointed out that while yeast would repire anaerobically in the absence of oxygen, it would switch to aerobic respiration if oxygen were available. Anaerobic respiration in this instance has survival value for the organism during unfavorable conditions. Similarly, many plants can also become facultative anaerobes under certain conditions where oxygen is not readily available. This does not mean that whole trees respire anaerobically but that particular tissues or organs with inadequate supplies of oxygen temporarily could respire anaerobically thus preventing death from lack of the energy to maintain themselves. Alcoholic fermentation typically occurs in roots in waterlogged soil or in flooded regions either until oxygen becomes available or until the tissue dies from a lack of ATP or the accumulation of toxic amounts of alcohol. Germinating seeds of many species also respire anaerobically under conditions of oxygen deficiency during early germination while the seed coat is still only slightly permeable to oxygen. As indicated earlier, however, most plants cannot survive under totally anaerobic conditions for long, though the length of time that individual tissues or organs could survive would be at least partially related to the rate of metabolic activity of the tissue and to its sensitivity to alcohol.

Factors Affecting the Rate of Respiration

As with other forms of metabolism, the overriding influence on the rate of respiration is the genetic program in the nucleus of each cell. Within the constraints imposed by the DNA a variety of internal and external factors affect the observed rate of respiration.

Internal Factors

The physiological condition of a tissue is a major influence on the rate of respiration. Dormant structures such as seeds and buds have a low respiratory rate because the metabolic rate, and therefore the demand for ATP, is low. Actively growing regions have a high respiratory rate in order to supply adequate amounts of ATP. The metabolic rate controls respiratory rate directly by controlling the availability of ADP which is needed in all of the reactions of respiration that produce ATP.

External Factors

Oxygen is not usually a rate limiting factor for respiration of aerial plant parts such as leaves and stems because the atmosphere contains sufficient oxygen. For root respiration oxygen can become a rate-limiting factor since the soil atmosphere will contain less oxygen than the air as a result of respiration by soil microorganisms and roots themselves. Furthermore, air spaces in soil can become filled with water, reducing the availability of oxygen even further. Water availability also affects the respiration rate by its influence on the degree of hydration of cells. As the water content of a cell decreases below a certain point all metabolism including respiration is slowed down. This is one of the ways in which seed dormancy is induced and explains why germination requires water.

The major external factor that regulates respiration rate is temperature, however. Respiration is quite sensitive to changes in temperature with daily fluctuations in temperature being sufficient to cause significant changes in respiration rate. The exact temperature-response curve varies from organism to organism and, as might be expected, the normal environment of the plant does have an influence on this response. Tropical species will slow or stop respiration at higher temperatures than temperate plants that are adapted to cool temperatures during the growing season. Evergreens of temperate and northern latitudes are able to respire even at subzero temperatures.

Because of this sensitivity of respiration to temperature, most people today are able to store food for much longer periods than their ancestors could. Fresh food in a refrigerator respires slowly and as a result maintains its nutritional value longer. Bacterial and fungal growth are similarly slowed as a consequence of their slowed rates of respiration, which retards food spoilage. A simple demonstration of the effectiveness of temperature in controlling respiration rates is to store equal amounts of milk at room temperature and in a refrigerator. Lactic acid fermentation (souring) will occur more rapidly at room temperature. This same demonstration could be done with cut flowers, whose freshness is a function of the rate of respiration. Obviously, then, respiration is of fundamental importance at the cellular level for living organisms and also in human dealings with the botanical world.

Summary

Respiration releases the chemical energy stored in the bonds of organic molecules. This energy is used to generate the high-energy compound called ATP, which functions as the energy currency of the cell. Prior to respiration complex foods must be broken down by digestion to their simpler components. Plant respiration mainly involves the breakdown of glucose through either the anaerobic or aerobic pathway. A common component of both types of respiration is the process of glycolysis in the cytoplasm, during which glucose is broken down in a series of steps to two molecules of pyruvic acid. In glycolysis two molecules of ATP are produced and some of the hydrogen atoms from the glucose molecule are transferred to a hydrogen carrier, NAD^+, but no CO_2 is released or O_2 consumed. After glycolysis the cell must regenerate the NAD^+ by removing the hydrogen atoms. This is accomplished in different ways in anaerobic and aerobic respiration. In anaerobic respiration the hydrogen atoms are transferred either to pyruvic acid, converting it to lactic acid (lactic acid fermentation), or to a two-carbon compound derived from pyruvic acid, converting it to alcohol (alcoholic fermentation). In aerobic respiration the electrons of the hydrogen atoms are transferred through a series of hydrogen carriers (electron transport chain) to oxygen, forming water and liberating energy to generate ATP. Since pyruvic acid does not have to be used as a hydrogen acceptor in aerobic respiration, it is available for further degradation to liberate more of its locked-up energy. Removal of CO_2 from pyruvic acid converts it to acetic acid that is combined with a four-carbon acid to form citric acid in the first step of the Krebs cycle. Subsequent steps in the cycle result in the complete breakdown of acetic acid to CO_2 with the hydrogen atoms transferred to hydrogen carriers. The net gain of ATP in the aerobic respiration (glycolysis, Krebs cycle, electron transport) of a molecule of glucose is 36–38 ATP molecules as opposed to only 2 ATP molecules in anaerobic respiration. Fatty acids are broken down to acetic acid, which enters the Krebs cycle. Amino acids from protein digestion, after losing the amino group, can be rearranged and also enter the Krebs cycle.

The rate of respiration is controlled by the metabolic condition of the cells, with a high metabolic rate requiring a high rate of respiration. Oxygen, temperature, and water availability influence the rate of respiration.

Review Questions

1. Describe the general sequence of events occurring in aerobic respiration of glucose. What are the end products of aerobic respiration?
2. Explain what is happening in respiration, (disregard the exact chemical

events) and explain why an organism must perform this type of metabolism.
3. Distinguish between aerobic and anaerobic respiration and explain why aerobic respiration is the prevalent form.
4. Explain how the rate of electron transport in the electron transport chain will limit the rates of both glycolysis and the Krebs cycle.
5. Where does the ADP come from that is used to make ATP in respiration?

Alcoholic Beverages

In Chapter 3 tea and coffee were described as modern replacements for beer and wine as mealtime beverages. Beer, wine, and distilled liquor continue to be popular, however, which is not surprising considering that beer and wine have been with us at least since earliest recorded history, and distilled liquor is also ancient, having been mentioned by Aristotle. All types of alcoholic beverages are produced by the anaerobic respiration of various yeasts. Wine typically is produced by fermentation of the sugars contained in the juice obtained by crushing grapes, though wines can be made from many other plant materials including rhubarb, dandelion flowers, bananas, peaches, cherries, elderberries, blueberries, and more. In each case the plant material is crushed before wine yeasts are added. At this time the winemaker will also add extra sugar if the juice does not contain enough because, even though the flavor of the wine is mainly determined by the acids and other soluble organic compounds contained in the juice, it is the sugar content that influences the final alcoholic content. Grapes seem to be ideally suited for the production of wines for the cultivated taste. The quality and flavor of grape wine is controlled by many variables—cultivar, weather, soil, climate, cultivation techniques, time of harvesting, pressing, fermentation, as well as the steps involved in maturing and bottling. Consequently, because there are so many different kinds of wine, only a few can be mentioned in this limited space. White wine is derived from grape juice that has had the skins removed immediately after pressing, whereas for red wine the skins remain with the juice during fermentation thus allowing the wine to pick up anthocyanins that were present in the skin cells (epidermis); a rosé is obtained if the skins are removed after the initial stages of fermentation. Dry wines are those whose sugar content is low as the result of fermentation being allowed to continue until essentially all of the sugar has been converted to alcohol. A perennial favorite is champagne, a sparkling wine produced in the Champagne district of France. Champagne and other sparkling wines are produced by adding sugar solution to the bottled wine and then corking it securely or by adding sugar solution to a batch of wine in a closed vat under pressure. The yeast that remains in the wine ferments this additional sugar; because it is in a closed container the CO_2 produced as a by-product accumulates. When the pressure is released upon the removal of the cork, the dissolved CO_2 forms bubbles and the wine foams out of the bottle just as other carbonated beverages such as pop and beer.

Beer is even more widely distributed and consumed than wine. Differences in flavor and quality are also a function of many variables. Any source of carbohydrate can be used to make beer though most European and North American beer is made from barley. Japanese beer (sake) is made from rice, Mexican beers are made mainly from corn, some German beer is made from wheat, and there are novelty beers made from a variety of other materials to produce root beer, ginger beer, spruce beer, and so on. The production of beer begins with barley seeds, the source of the sugar for fermentation.

Before fermentation can begin the stored barley starch must be digested by amylase to release the glucose. Digestion of the starch is most easily accomplished by allowing the seeds to germinate during which they produce amylase to mobilize the sugar for their own growth. After germination has begun the seeds are dried and ground into a powder called malt. Malt contains abundant amylase activity so that when it is mixed with water to make the mash, all of the starch is digested and made available for fermentation. After yeast growth to produce alcohol the final product either can undergo a second fermentation in closed vats to produce the carbonation or, in order to save time, CO_2 can be pumped into the beer after the first fermentation.

Among the side benefits of the Crusades can be counted the art of distilling beer and wine to produce liquor. Whiskey, rum, gin, brandy, and vodka all contain a considerably higher concentration of alcohol than wine or beer; however, this is not due to the use of special yeasts that can tolerate more alcohol. After the initial fermentation the mash or wine is heated in a still to vaporize the alcohol which is subsequently liquified in the cool condenser yielding a liquid that is enriched in alcohol. A second distillation could follow to further increase the alcohol content without additional fermentation. Beer and wine labels usually indicate the percentage of alcohol that the beverage contains but liquor labels do not. A liquor-bottle label indicates the proof of the liquor. Proof can be converted to percentage of alcohol by diving the number by 2—thus 80 proof vodka is 40% alcohol. It is interesting to note the derivation of the term *proof* and why it has become the standard of measurement for the alcoholic content of liquors. In the time of few government regulations regarding quality of goods, the consumer had to beware of the unscrupulous merchant. Watering liquor has been for some time a favorite way of increasing profits for a tavern owner. In the American colonies and later on the American frontier, particularly in the gold fields and other boom towns where prices were so high that fortunes could be made by "stretching" the liquor supply, there had to be some way of proving the offense of watering liquor. Since the patron did not want to wait for samples to be sent for chemical analysis, a simple, rapid test came into use. A liquor-gunpowder mixture burns if there is about 50% alcohol in the liquor; less alcohol and there is no combustion whereas a higher alcohol concentration causes the mixture to flash instead of burn. Therefore any drink that burned with gunpowder was 100 proof, and heaven help the tavern owner whose drinks were less than 100 proof.

11

Transport of Water, Minerals, and Food

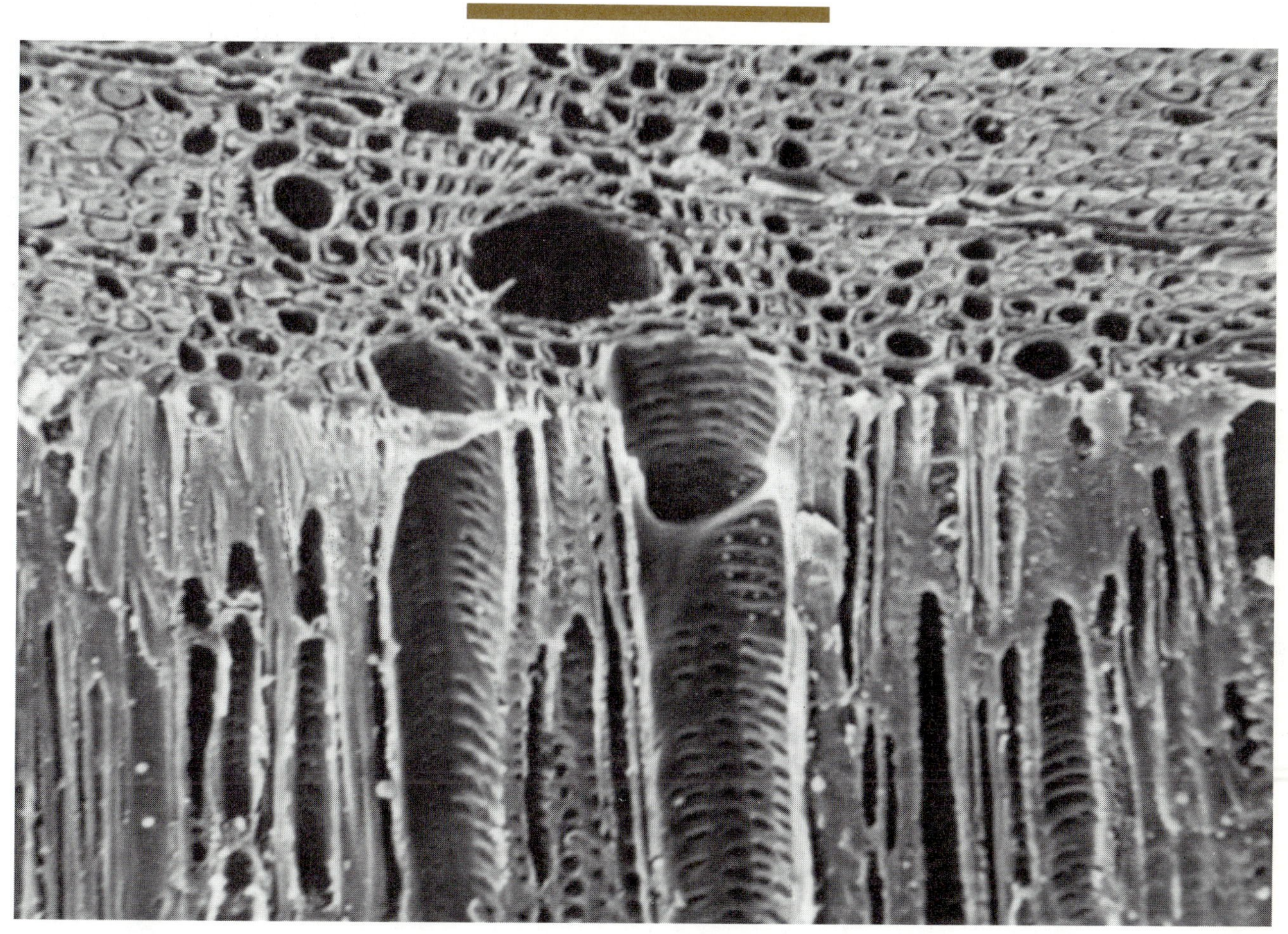

Outline

Introduction

About 400–500 million years ago a major step in the evolution of plants occurred: the transition from water to land. As plants made this transition in habitat they faced a number of challenges. In particular the land environment is dry, yet water is the major component of plant cells (Chapter 3). The plant species that successfully made the transition to land were those that had adaptations to prevent dehydration of their cells. An adaptation that is found in vascular plants*, the most successful group of land plants, is an outer, waterproof layer. Once the outer surface of a plant is waterproofed, water loss is prevented; however, water and dissolved-mineral absorption are also prevented. Consequently some form of water-absorbing structure would have had survival value, which resulted ultimately in the evolution of a root system. At the same time stems and leaves also evolved in plants. The effect of the evolution of these organs was to separate the site of water and mineral absorption (the root) from the aerial organs (the stems and leaves). However, a root that specializes in water and mineral absorption from soil is not in a position to receive light and therefore cannot produce food by means of photosynthesis. In order for this new plant body to be an evolutionary success, one final adaptation was necessary—food and water conducting tissue. The presence of the conducting, or vascular tissues, is in large part responsible for the dominance of vascular plants over the bryophytes* in colonizing the land environment.

Vascular plant
A plant that has specialized tissues (xylem and phloem) for conducting water and food.

Bryophyte
A member of a group of primitive terrestrial plants, typified by mosses and liverworts.

Transport of Water

In Chapter 3 it was indicated that water is essential to all living cells, with plant cells typically containing 80%–90% water; a reduction in water content is accompanied by a reduction in metabolism.* Since some water is consumed in metabolism and some is lost from the cell by other processes, it is necessary that water from the environment be able to enter the cell and thus keep each cell fully hydrated. Root cells or the cells of an aquatic plant obtain water directly from the soil or water. Aerial portions of a plant, however, are not in direct contact with a source of water. To replenish water lost from aerial cells in vascular plants, water is transported from the roots to the shoots through a "pipeline" system known as **xylem** (Chapter 6). This transport requires that there be a mechanism for loading and unloading the pipeline, as well as for moving water in the pipeline. Loading and unloading of the pipeline can be understood by analysis of how cells gain and lose water, i.e., water relations of cells.

Metabolism
All of the chemical reactions that occur in a living cell.

Water Relations of Cells

Water enters and leaves cells as the result of **diffusion*** across the differentially permeable **cell membrane,** or, in other words, through the process of **osmosis.*** Since osmosis is a specific type of diffusion, the general mechanism of diffusion will be explained first. Diffusion depends on the fact that all molecules are constantly in random motion, with the degree of motion being dependent on whether the molecule is contained in a solid (least motion), or a liquid, or a gas (most motion). Plant cells contain aqueous solutions; consequently individual molecules have considerable freedom to move. If the path of either an individual water molecule or a **solute*** molecule could be followed, it would resemble the flight pattern of a drunken fly (Figure 11–1). Each molecule in the solution would be moving in a similar fashion with absolutely no coordination among molecules. As a result of this random motion the various types of molecules in a solution will be distributed uniformly. This seemingly orderly result of a disorderly process can be demonstrated in a diffusion experiment where a dye crystal carefully is placed in water and then left undisturbed. After a short time the water near the crystal will have changed color, indicating that the dye molecules are dissolving in the water and diffusing away from the crystal. Ultimately the water will become uniformly colored. At the molecular level the process is not as straightforward. Dye molecules at the boundary of the crystal can move in any one of four directions. (A two-dimensional diagram is used for simplicity.)

Diffusion
The net movement of molecules along a concentration gradient.

Osmosis
The diffusion of water through a differentially permeable membrane.

Solute
A substance that can go into solution.

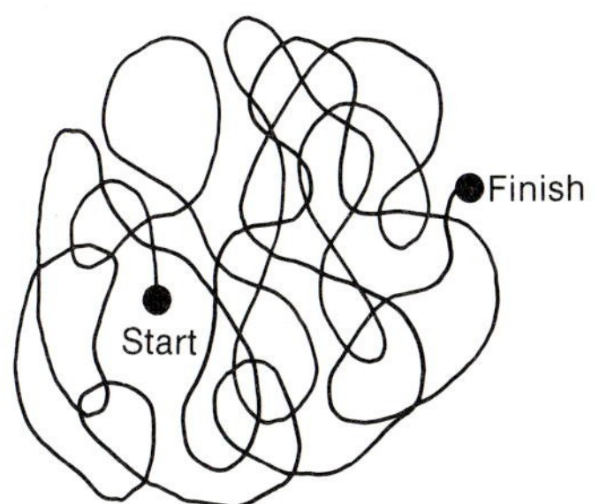

Figure 11–1. Random molecular motion. All molecules are in motion with their path being determined by random collisions and interactions with other molecules.

$$\begin{array}{ccc} & \uparrow & \\ \text{crystal} & \leftarrow \bullet \rightarrow & \text{water} \\ & \downarrow & \end{array}$$

As each of these four directions are equally likely, only 25% of the molecules will actually move into the water as a result of this random molecular motion. At the next instant, a dye molecule in solution can again move in any of the four equally likely directions.

$$\begin{array}{ccc} & \uparrow & \\ \text{crystal} & \leftarrow \bullet \rightarrow & \text{water} \\ & \downarrow & \end{array}$$

Again only 25% of the dye molecules will actually move further away from the dye crystal. In other words, only 25% of the molecules actually will be diffusing, moving from an area where they are in high concentration to one where they are in low concentration. In fact, at the same instant 25% will be moving from low to high concentration whereas 50% will be staying at the same concentration. However, as there are always some molecules moving from high concentration to low concentration, the net effect of the overall random molecular motion is diffusion and the uniform distribution of dye molecules throughout the solution. Once this uniform distribution has occurred, that is, when a state of equilibrium is reached, diffusion stops because there is no longer a **concentration gradient***, but molecular motion still continues as this is independent of gradients.

Concentration gradient
An unequal distribution of molecules in a solution.

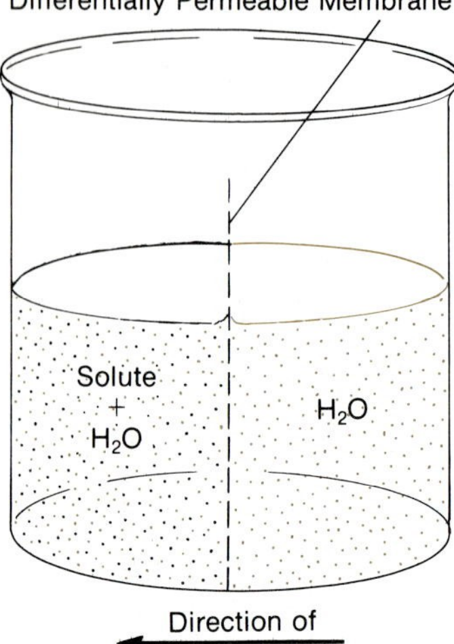

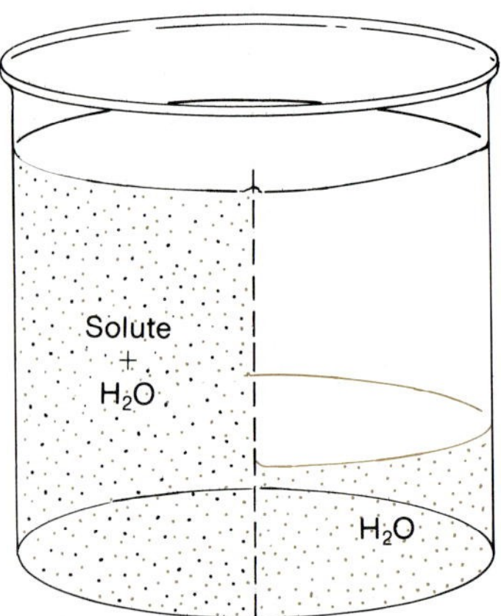

Figure 11–2. Diffusion in the presence of a differentially permeable membrane. A beaker containing a solution (solute plus water) on the left side and pure water on the right will have two concentration gradients. The concentration of free water on the right side will be higher than that on the left and the concentration of solute will be higher on the left than on the right. In the presence of a differentially permeable membrane that prevents the passage of solute only the water molecules will be free to diffuse along their concentration gradient. This will lead to an increase in the volume of solution on the left side and a decrease on the right. Diffusion stops because the free-water gradient has disappeared.

Diffusion as described above occurs in bodies of water and inside cellular compartments, but not routinely across cellular membranes. All cellular membranes are **differentially permeable** and regulate the molecules that can move through them by diffusion. Most molecules other than water cannot cross cellular membranes freely and therefore cannot diffuse in or out of a cell. Thus a special mechanism is necessary to transport these molecules across the plasmalemma (discussed later in this chapter), and this also makes it possible for plant cells to contain 80% to 90% water even though the environment usually contains much less. Since the cellular membranes are completely permeable to water, water can diffuse across a membrane in response to a water-concentration gradient; in other words, osmosis can occur.

Cells do not contain pure water; they contain solutions. This is an important fact for the understanding of osmosis because the presence of solute molecules affects the concentration of water that is free to diffuse (free water). When a solution is made, the solute molecules associate with some of the water molecules making these water molecules unable to diffuse. Since the water-concentration gradient is determined by the concentration of free water on both sides of a membrane, the addition of solute to water on one side of the membrane will alter the gradient. The following is a hypothetical example of this effect. A beaker with a differentially permeable membrane dividing it in half contains 50 water molecules, 25 on each side of the membrane. Because there are equal concentrations of free water molecules on both sides, osmosis will not occur. The addition of 5 solute molecules to the left side causes 20 of the water molecules in this compartment to asociate with the solute molecules, in effect reducing the free-water concentration in this compartment to 5. Now a free-water concentration gradient exists with a higher concentration on the right side than on the left and water will diffuse from the right compartment into the left until there are equal numbers of free water molecules on both sides (Figure 11–2). At the end of this process of osmosis there will be more *total* water molecules on the left side but there will be no water-concentration gradient because both sides will contain equal concentrations of free water molecules. In the absence of the membrane both the sugar and the free water molecules would diffuse independently until each concentration gradient had been abolished (Figure 11–3).

The process of osmosis described above is between pure water and a solution. However, osmosis occurs in any situation where a differentially permeable membrane separates two regions that differ in their free-water concentrations. Therefore osmosis will occur betwen two solutions of different sugar concentration under the same rules as between pure water and sugar solution (Figure 11–4). Similarly, osmosis can occur between sugar and salt solutions, cells and water, cells and sugar solutions, vacuole and cytoplasm, two cells, or cells and air.

For the purpose of discussing cell water relations, the plant cell can be considered to be just a compartment, containing many dissolved substances, that is separated from the environment by a differentially permeable membrane (the plasmalemma). If a cell is placed in water, osmosis will take place because of the free-water concentration gradient and water will enter the cell. However, a plant cell is a closed system; that is, it is enclosed by a rigid cell wall and it cannot expand indefinitely. As water enters, the vacuole enlarges until further increase in volume is prevented by the cell wall. At this point the pressure from the walls on the protoplast is squeezing water out of the cell as fast as the concentration gradient is causing it to enter; thus osmosis stops. The situation is analogous to that of a balloon being inflated inside a bottle: once the balloon is pushing against the bottle walls, there can be no further net input of air into the balloon. The pressure of the protoplast on the cell wall is called **turgor pressure.** Normally all cells contain sufficient water that they exert some pressure on the walls; cells that are exerting turgor pressure are said to be **turgid,** those that are not are **flaccid.** When turgor pressure is causing as much water to leave the cell as is entering the cell because of the free-water concentration gradient, the cell is at osmotic equilibrium. In this situation osmosis stops, not because the gradient has disappeared, but because the number of water molecules moving into the cell owing to diffusion is equal to the number moving out of the cell owing to turgor pressure. Osmosis is merely the net movement of water molecules through a differentially permeable membrane; therefore, when two processes result in the movement of water molecules in equal numbers in opposite directions, there can be no net movement of water molecules and no osmosis.

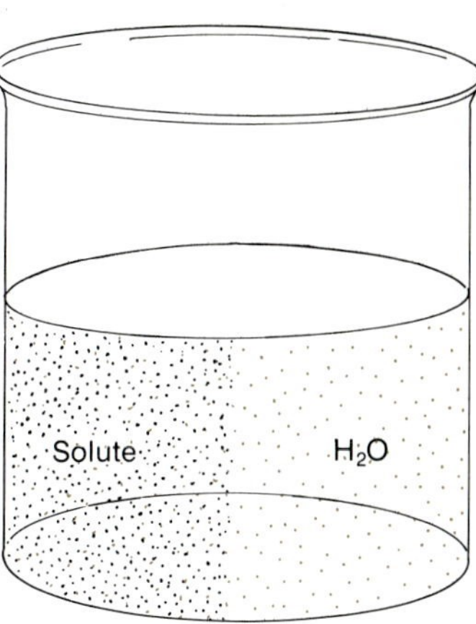

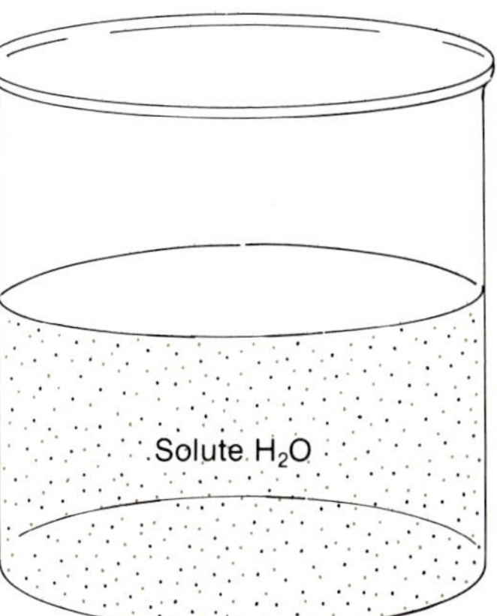

Figure 11–3. Diffusion in the absence of a differentially permeable membrane. If the right side of a beaker contained a solution and the left side contained pure water with no barrier between them, concentration gradients for solute and water would exist. In the absence of a differentially permeable membrane both would diffuse until the gradients were abolished.

If a turgid cell is placed in a strong sugar solution, water will move out of the cell by osmosis because the external solution has a lower free-water concentration than the cell sap. As the cell loses water from the vacuole, it loses turgor and the protoplast shrinks away from the cell wall (Figure 11–5). Water will continue to be lost from the cell until the internal free-water concentration equals that of the surrounding solution. A cell which has lost turgor is said to be **plasmolyzed.** This process of **plasmolysis** usually can be reversed if the cell is transferred back to water—now the free-water concentration gradient causes the osmotic uptake of water until full turgor pressure develops.

These types of situations demonstrate the principle of cell water relations but do not represent very realistic situations for most organisms. Aquatic organisms such as algae are the only ones that are routinely faced with the osmotic consequences of growth in water (fresh water) or salt solution (marine). For most plant cells osmotic interactions are between two adjacent cells, with only root epidermal cells and leaf mesophyll cells interacting directly with the environment.

Osmosis and the Functioning of Guard Cells

Osmosis is of central importance to mechanisms of water uptake and transport by plants. The role of osmosis in the functioning of the guard cells provides a simplified model for the role of osmosis in the water relations of plants and cells. The guard cells of the leaf epidermis function to open and close the stomata that, in turn, regulates gas exchange in the leaves. Opening of the stomata occurs when the guard cells are turgid, whereas closing of the stomata occurs as the guard cells lose turgor. In the morning, potassium ions (and perhaps other solutes) move into the guard cells from surrounding epidermal cells through the process of active transport (described below), which increases the solute content of the guard cells and therefore decreases their free-water concentration. This establishes an osmotic gradient that

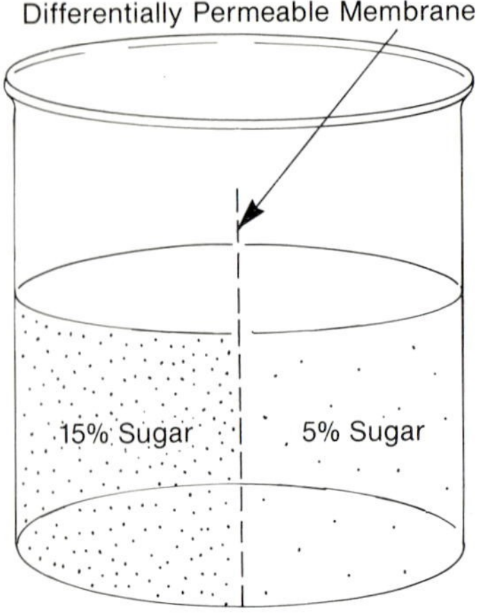

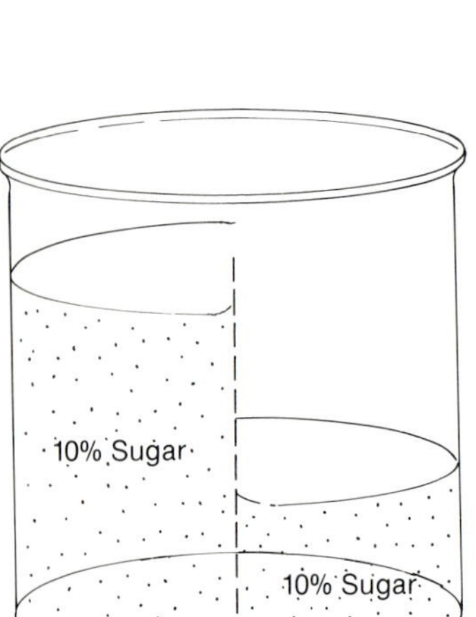

Figure 11–4. Diffusion between solutions. In this example there is a high concentration of sugar on the left side of the membrane and a low concentration on the right. The free-water gradient is the reverse being high on the right and low on the left. Because of the presence of the differentially permeable membrane only the water can diffuse. At equilibrium there will have been sufficient water diffuse to the left that both the free water and the sugar concentrations will have been abolished.

Angiosperm
One of a group of plants characterized by the production of flowers.

causes, by osmosis, a net movement of water into the guard cells thus increasing their turgor pressure. Since the cell wall of a guard cell is not of uniform thickness, an increase in presure on the wall can cause a bulge in a weaker region and alter the cell's shape. Guard cells are constructed so that the thickest part of the wall is on the side adjacent to the other guard cell. Therefore, the change in shape of the guard cell occurs in such a way that they separate (Figure 11–6) and open the stoma. At night the extra solutes diffuse from the guard cells to the surrounding epidermal cells, thus increasing the free-water concentration of the guard cells, which results in the osmosis of water from the guard cells. This process reduces the turgor pressure of the guard cells somewhat, allowing their shape to return to normal. Wilting of plants, which is the result of a general reduction in the turgor pressure of internal cells, will be accompanied by loss of turgor pressure in the guard cells, resulting in the closing of the stomata and preventing further water loss. The source of energy (ATP) for the active uptake of potassium by guard cells has been suggested to be photosynthesis in guard cell chloroplasts.

Water Uptake by Plants

Water uptake by plants occurs by osmosis. In the following discussion, water absorption by vascular plants such as angiosperms* will be emphasized, but the same general principles apply to other types of plants. For most angiosperms the root is the organ of water absorption. If a root were immersed in water, there would be an osmotic gradient between the root and water, with the root having the lower free-water concentration (see discussion on water relations in cells). Consequently, water would diffuse readily into the root cells. Plants in nature do not grow in pure water, however; instead they are rooted in soil. But soil contains solutes and particles that form associations with water molecules, resulting in soil having a lower free-water concentration than pure water. Therefore, the roots must have a lower free-water concentration than the soil if they are to absorb water from the soil by osmosis. Under most normal growing conditions, the root is, indeed, able to absorb water from the soil, indicating that the osmotic gradient favors the root cells. Immediately after a rain the free-water concentration of the soil is highest and water uptake by plant roots is maximal. Consequently, there is a spurt in plant growth after a rainfall. As the soil dries out, the osmotic gradient becomes less steep and water uptake by the root cells decreases. Finally, if there is a sufficiently long, dry interval, it is possible for the soil free-water concentration to drop below that of the root cells. Under these conditions there would be an osmotic gradient favoring the movement of water out of the root cells and into the soil by osmosis. Any time that a plant does not absorb enough water from the soil or loses water to the soil, it will suffer water stress and will show signs of wilting. Prolonged wilting leads to plant death. An unfavorable root-soil osmotic gradient that results in plant death and the failure of seed germination is often an unexpected side effect of ice-removal efforts. City work crews spread salt on roads and individuals do the same on their sidewalks in order to melt ice and improve traction. Unfortunately, salt that ends up in soil will dissolve in the soil water and reduce further the free-water content of soil. If there is sufficient salt in the soil, plants are unable to absorb water by osmosis and they die, leaving a brown zone of dead vegetation adjacent to a road or sidewalk.

Under conditions where water is absorbed by the root because of an osmotic gradient, the water continues to move within the root as a result of the osmotic gradient (Figure 11–7). Rapid upward transport of water in the xylem results in a low free-water concentration in the vascular cylinder so that water diffuses towards the xylem from the epidermal and cortical cells. Some of this water never enters the protoplasts of the epidermal and cortical cells but instead diffuses through the cell walls, which are completely permeable to water. By this pathway, diffusion (not osmosis) will cause a net movement of water from the soil toward the xylem. However, once the water reaches the endodermis, osmosis is necessary for its further movement into the stele. As is explained below, the endodermis prevents the diffusion of aqueous solutions from the cell walls of cortical parenchyma into the vascular cylinder. The net result of these diffusional processes is the movement of water from the soil into the xylem of the roots. Once in the xylem, water is moved up the root and into the stems and leaves by a completely different mechanism.

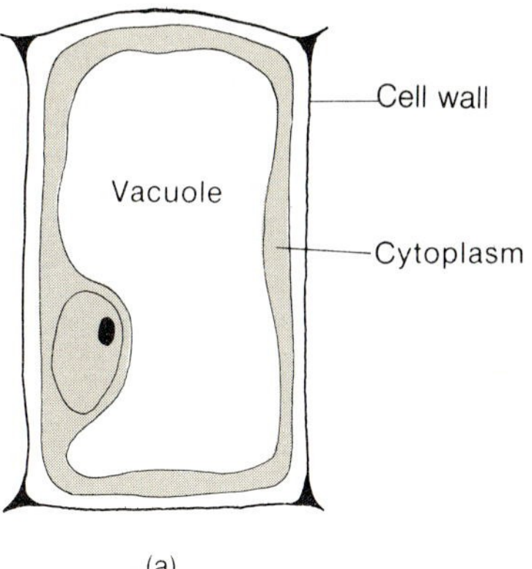

(a)

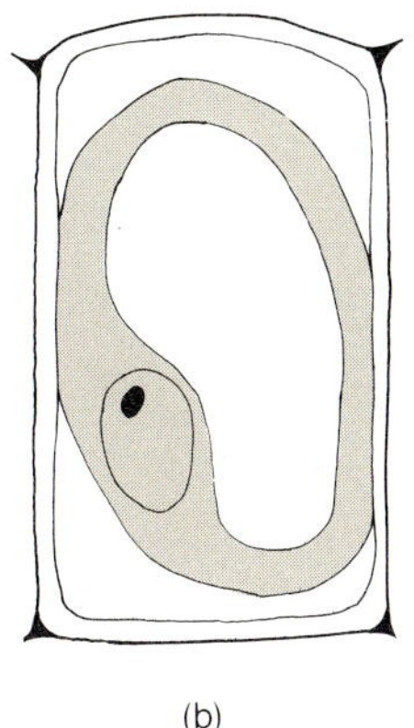

(b)

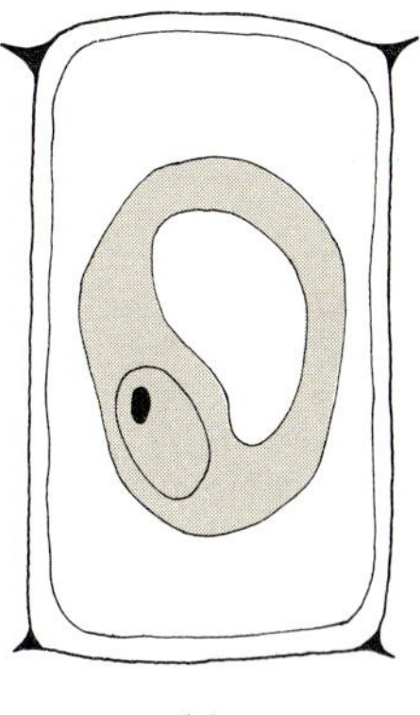

(c)

Figure 11–5. Plasmolysis. A cell placed in a solution that has a lower free-water concentration than the cell will lose water by osmosis. This will cause the protoplast to shrink and pull away from the cell wall. The space between the cell wall and the protoplast in this case would be occupied by the bathing solution.

Water Movement in the Xylem

The movement of water into the xylem from the soil occurs in response to an osmotic gradient; however, water does not flow up the xylem because of an osmotic gradient. The rate of diffusion of water in the xylem is much slower than actually observed rates of water movement. The only possible effective means of moving water upward in the xylem is either to push it or to pull it. The push, or **root presure,** hypothesis states that the root is acting as a pump to force water up the xylem. According to this hypothesis, water enters the base of the xylem column as a result of the osmotic gradient and, since water is continually diffusing into the root, the water that is already in the xylem is pushed up by the new water.

Experimentally it can be shown that water will, indeed, move up the xylem in response to root pressure. If the shoot is removed from a plant, water will exude from the cut end attached to the roots (Figure 11–8). Root pressure is inadequate, however, to move water to the tops of tall trees, such as the giant redwoods, which can reach heights of 300 feet. Furthermore, root pressure is effective only under certain conditions such as high soil-water content and high humidity. Few plants experience these conditions throughout their growing season. But root pressure does have a significant role to play in water movement in deciduous trees at one time of the year. The rising of the sap in trees is a sign of spring and is the result of root pressure forcing water up the xylem. Sugar maples are probably the best known example of this phenomenon.

Since root pressure is inadequate to push water far up the xylem all the time, water must be pulled up. The only significant pull on the water in the xylem results from the process of **transpiration** that mainly occurs in the leaves. Consequently, the **transpirational pull-cohesion** hypothesis of water movement has been developed. This hypothesis is based on the fact that water molecules are highly cohesive and, as a result, they are held together in a continuous chain that will be pulled up the xylem thereby replacing the

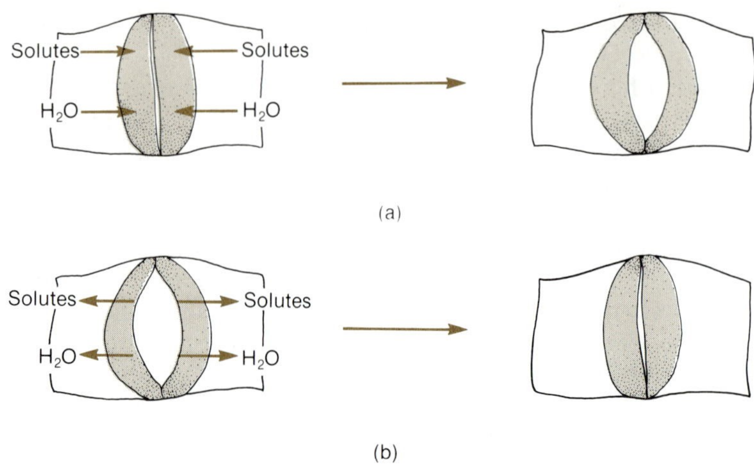

Figure 11–6. Stomatal opening and closing. (*a*) The active uptake of solutes by guard cells from surrounding cells causes water influx by osmosis. The resulting increase in volume of the guard cells forces them apart leaving a gap called the stoma. (*b*) Loss of solutes to the surrounding cells by the guard cells is followed by water efflux from the guard cells by osmosis. The reduction in guard cell volume closes the stoma.

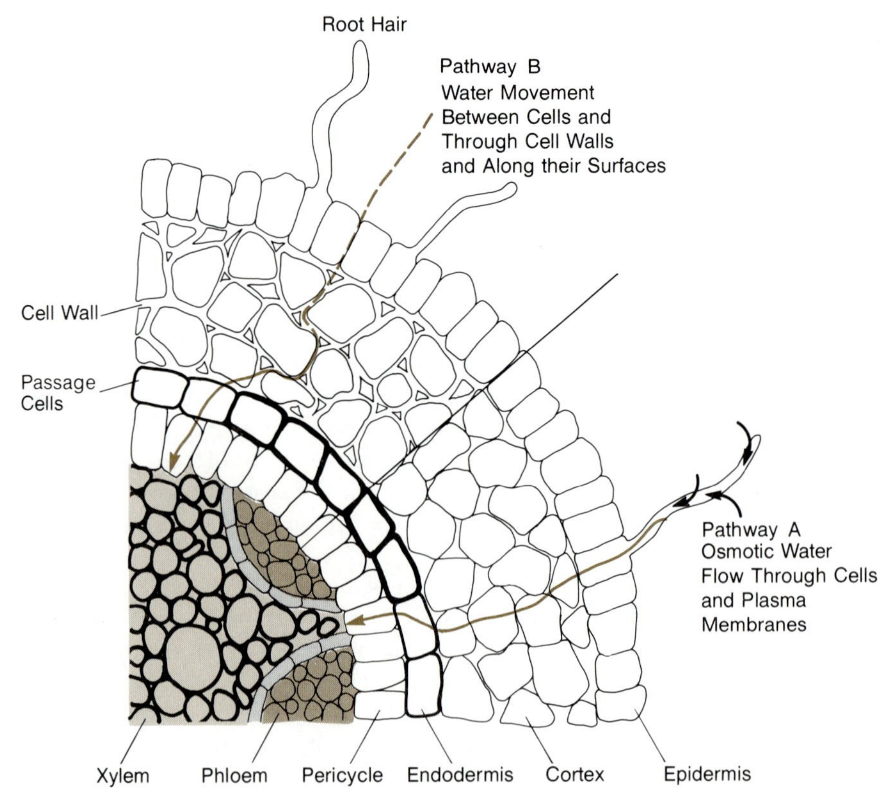

Figure 11–7. Pathway of water movement in the root. Water enters the root from the soil in response to a diffusion gradient. It may diffuse within the cell walls (free space) or it may enter an epidermal cell by osmosis. Water diffusing through free space may enter a cell by osmosis or continue along the gradient toward the xylem. Actual entry of water into the vascular cylinder from the free space is blocked by the Casparian strip in the endodermis. All water that enters the vascular cylinder does so by movement through the protoplast of an endodermal cell after having been taken up by osmosis by a cell of the epidermis or cortex and then moving between protoplasts by osmosis.

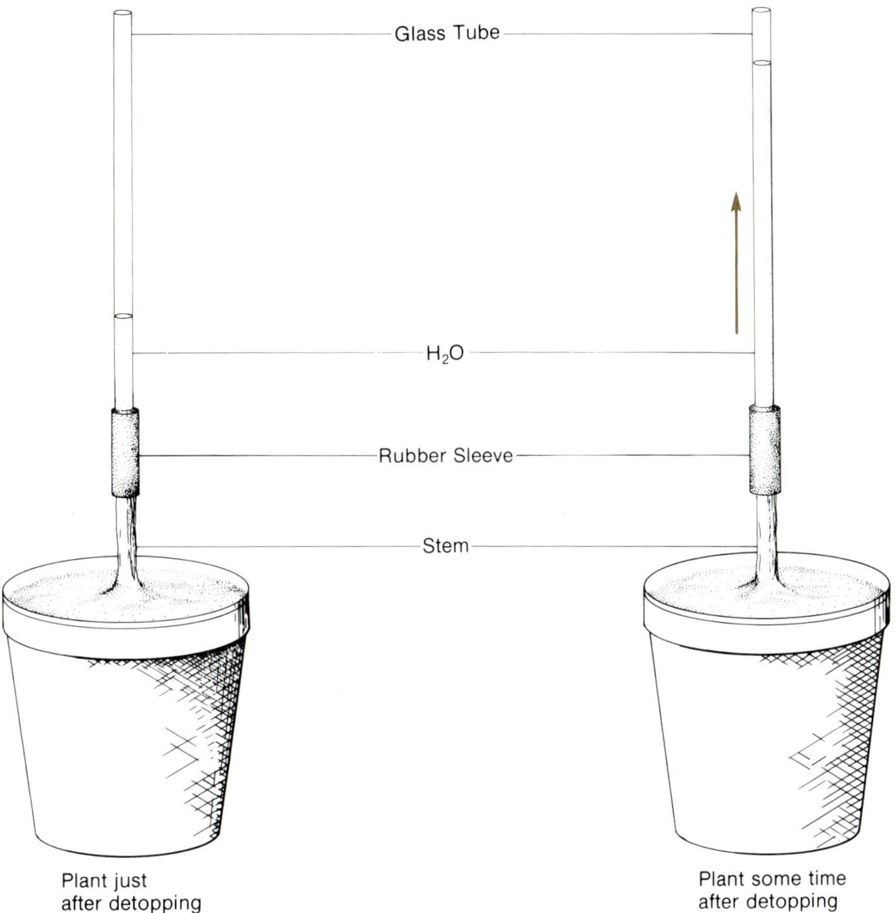

Figure 11–8. Root pressure. Root pressure can be demonstrated by removing the shoot system close to ground level. Fluid will then rise in a glass tube attached to the cut stump.

water molecules lost in transpiration. It is important to stress the cohesion aspect of the hypothesis because if water molecules did not stick together, a strong pull would break the water columns in the xylem-conducting elements. Investigators who have measured the cohesiveness of water molecules have found that water could be pulled up thin tubes to heights greater than the height of the tallest tree before the water column breaks. Clearly, water molecules in the tracheids* and vessel elements* are cohesive enough to be pulled from the roots to the leaves. One final aspect of water transport that must be emphasized is the adhesion of water molecules to the walls of the vessels and tracheids. The nature of the cell walls and the narrowness of the conducting cells allows capillary action* to be a major component of water movement in the xylem. Transport of water in the xylem is thus dependent on capillary action and transpiration that create sufficient tension on a cohesive water column to raise it from roots to leaves. Because of the concentration gradient of free water in the soil-plant-atmosphere system, the movement of water essentially will be unidirectional from soil to atmosphere through the plant.

Tracheid
One type of water-conducting cell in xylem, characterized by its relatively narrow diameter and the presence of end walls.

Vessel element
One type of water-conducting cell in xylem, characterized by its relatively large diameter and lack of end walls.

Capillary action
The movement of water up narrow tubes as a result of the surface tension of water.

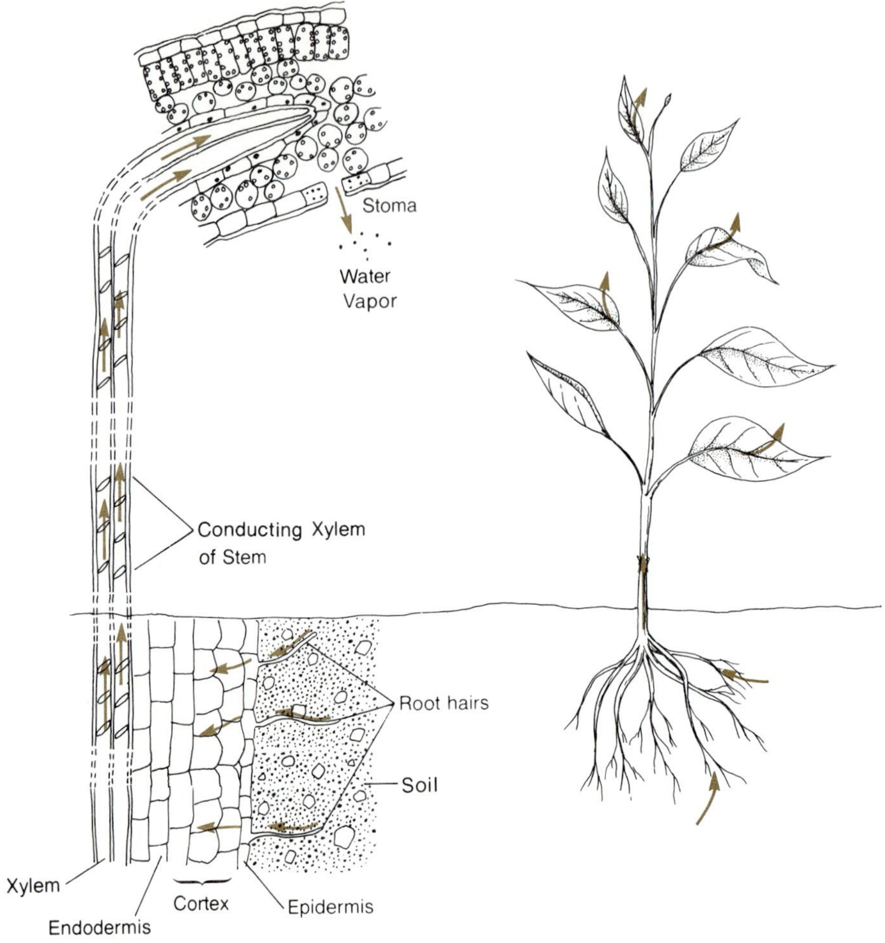

Figure 11–9. Pathway of water movement in plants. Water enters the roots by diffusion and moves to the vascular cylinder by osmosis and diffusion. Once within the xylem, water is pulled to the leaves by transpirational pull-cohesion forces. Water leaves the xylem in the leaf by osmosis into the bundle sheath cells. Evaporation of water from cell walls into the internal air spaces of the leaf creates the osmotic gradient that continually pulls water from the bundle sheath into the mesophyll and epidermal cells. Diffusion of water vapor from the leaf air spaces through the stomata creates the gradient that drives evaporation.

Transpirational pull results from the process of transpiration. A leaf contains considerable intercellular air space (Figure 11–9) and, as a result, all of the mesophyll cells have access to CO_2, which is used in photosynthesis. A leaf's stomata allow CO_2 to diffuse into the intercellular air space and O_2 to diffuse out. Even though the presence of stomata allows gas exchange associated with photosynthesis, it also allows the diffusion of water molecules along a concentration gradient from the internal atmosphere of the leaf to the external atmosphere. As water molecules, in the form of water vapor leave the intercellular air space, other water molecules take their place as a result of evaporation from the walls of mesophyll cells. This results in water movement out of the protoplasts of the cells by osmosis. All mesophyll cells are close to xylem conducting cells that contain water (Figure 11–9). From the principles explained earlier one can predict that the loss of water from one mesophyll cell will cause an osmotic gradient to occur between it and a neighboring cell that has not lost water (Figure 11–9). In a chain reaction, water will diffuse into this neighboring cell by osmosis from an adjacent cell, which has a higher concentration of free water. This chain reaction will continue until water from the xylem moves into the last cell in the chain. The net result of this process is that a continuous stream of water is pulled from the xylem into mesophyll cells, from which it evaporates to the atmosphere through the stomata. The continuous pull of water from the xylem causes an equal number of water molecules to be pulled up the xylem from the root, in turn resulting in the uptake of water from the soil (Figure 11–9).

Possible Significance of Transpiration

Botanists have come to realize that evolutionary processes appear to place a premium on efficiency (Chapter 23). During competition between organisms, the best adapted will have the advantage, particularly if the competition involves a scarce resource such as water. It may come as a shock to realize that 99% of the water that enters a plant from the soil is never used in growth and metabolism—rather it is lost to the atmosphere through transpiration. Some plant physiologists suggest that transpiration is a necessary evil, a deleterious side effect unavoidably tied to a vital process. Still other plant physiologists suggest that transpiration does serve a useful function for the plant. Of two suggestions, the first is that transpiration is involved in cooling the leaves. While leaves are absorbing light from the sun, they are also absorbing heat. If the leaves are not to be killed by this solar heat, there must be a mechanism to dissipate the heat. Evaporation of water uses heat energy, and since transpiration involves evaporation, transpiration, like sweating in animals, will cool plant leaves. Other forms of cooling, such as convection and radiation, also occur: the relative contribution of each depends on environmental conditions. Other evidence suggests that cooling by transpiration is unnecessary as the other forms alone are sufficient.

A second suggested useful function of transpiration relates to the evolution of the plant body. Since transpirational pull is the only force that can move water up the xylem more than a short distance above ground level, transpiration was essential to the survival of taller plants, particularly trees. In this case the process of transpiration would only seem to be inefficient because it was not being viewed in the proper context, that is, as a pump to pull water up plants. This would be analogous to considering the automobile engine an inefficient way to heat water until it was realized that the real function was to move people. A final suggestion for the role of transpiration is that it is necessary for the absorption of water from soil. Soil is not saturated with water, so a plant has to compete with soil for water. Since water enters the roots only by osmosis and since the steepness of the osmotic gradient is determined by the difference between the free-water concentration of the soil and of the xylem, then transpiration by removing water from the xylem makes root cells "drier" and, therefore, water is more likely to diffuse in from the soil.

Transport of Minerals

Various mineral elements are required for plant growth (Chapter 7) and these, like water, enter the plant through the root system. Unlike the uptake of water, most mineral uptake is not by diffusion even though a favorable concentration gradient may exist between the root and the soil. Diffusion is considered to be a passive process because no energy has to be expended by the cell, whereas other uptake mechanisms that use energy are classified as **active transport.** The uptake of minerals by active transport can be either along the concentration gradient or against it. However, a common feature of active transport is that there is a **carrier molecule** in the membrane that binds specifically to a particular mineral ion, with each different mineral being transported across the membrane by a different carrier molecule (Figure 11–10).

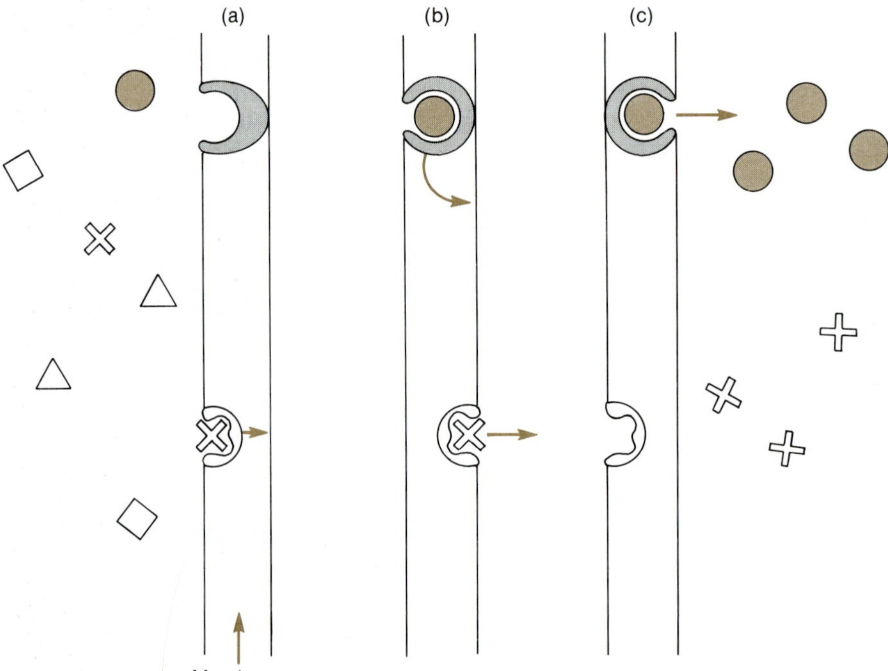

Figure 11–10. Active transport. Membranes contain carrier molecules that are specific for the different molecules to be transported. A large carrier such as a protein may extend across the membrane and once it binds its specific molecule, the protein changes its conformation so that this molecule is transported to the other side of the membrane and is released. Other carriers may be small proteins or low-molecular-weight compounds that migrate from one side of the membrane to the other after they have bound the molecule for which they are specific. In the absence of a carrier molecule, other molecules cannot pass through the membrane.

The differential permeability of cellular membranes severely restricts diffusion of molecules into and out of the cell. However, there are many molecules that a cell must be able to import or export if the organism is to survive. Consequently, active transport has evolved. Through control of the carriers in the cell membrane, the nucleus can control the permeability of the membrane. For roots this is important because they are in contact with the soil water that contains different types of minerals, beneficial and deleterious, in varying concentrations. A plant requires only certain minerals for growth and these must be in specific concentrations. Differential uptake of minerals by the root cells would ensure that these requirements are met and could prevent the uptake of deleterious minerals. This selectivity of the root system can only exist, however, if the minerals are taken into a living cell before they reach the xylem. In fact, dissolved minerals can diffuse into the intercellular spaces and cell walls (free space) in the outer portion of the root without having to enter a living cell. Since the mineral ions are dissolved in water, they can enter the cell walls, which are completely permeable and contain large amounts of water. A mineral ion traveling through the aqueous portion of cell walls could move from the soil through the walls and into the nonliving xylem without passing through a differentially permeable membrane if there were no diffusion barrier in the root. The **endodermis** separates the vascular cylinder from the outer portion of the root and prevents the nonselective uptake of minerals via this route. As a layer of cells whose walls are impregnated with **suberin*** (Chapter 7), the endodermis is completely impermeable to water and molecules dissolved in water. The only way a molecule can get past the diffusion barrier in the endodermal cell walls is to leave the cell wall system and enter the protoplast of a cell of the cortex (including the endodermis) or epidermis. Once a molecule is in the protoplast system it can pass from cell to cell until it reaches the xylem. Movement of minerals within the xylem is the result of the movement of water. As water is pulled up the xylem from the roots to the leaves the dissolved minerals are carried with it. Therefore, the transport of minerals through the xylem is also due to transpirational pull.

Suberin
A wax produced as a cell wall component of cork and endodermal cells.

Transport of Food

Leaves incorporate CO_2 into organic compounds in the process of photosynthesis. In order for the nonphotosynthetic portions of the plant to survive they must receive food from the leaves. The **translocation** of food occurs in the **phloem,** with most plants translocating the majority of their food in the form of sucrose. Within the phloem only the **sieve tubes** are directly involved in translocation. **Companion cells** function somehow to assist the sieve tube members in this process. Food is loaded into the phloem by active transport from mesophyll cells into bundle-sheath cells and ultimately into the sieve tubes (Figure 11–11). Once in the sieve tubes, the dissolved food does not move by diffusion, even though there is a concentration gradient between the leaves and roots. There is some dispute about the exact mech-

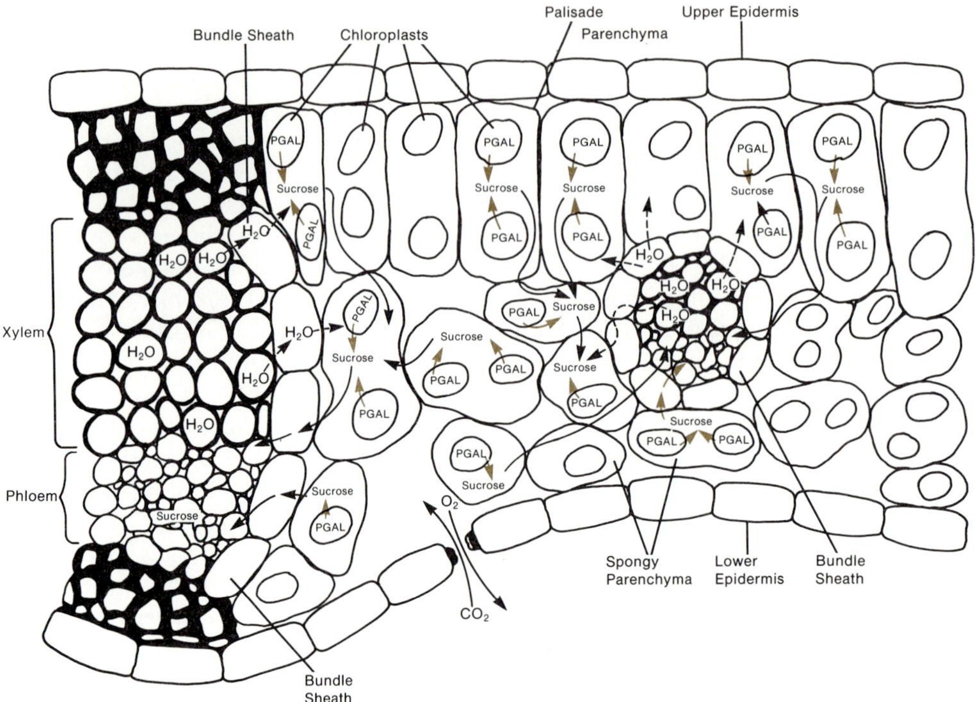

Figure 11–11. Phloem loading. Sucrose produced by photosynthesis is actively transported from mesophyll cells into the bundle sheath cells. Loading of the phloem sieve tube member is also by active transport. The selective nature of the carrier molecules ensures that sucrose is the main organic compound that enters a sieve tube member and therefore is the main food translocated around the plant.

anism, but all plant physiologists agree that some sort of active process is involved. One hypothesis, which has received considerable experimental support, is the **mass flow** or **pressure flow hypothesis.** According to this concept, the active accumulation of sugar in the sieve tube members in the leaf creates an osmotic gradient favoring water uptake by the sieve tube member. The increase in water content results in an increase in turgor pressure within the sieve tube member. This forces the sugar solution out of this sieve tube member into the next one and so on down through the phloem along a turgor pressure gradient within the phloem. This presure gradient is caused by sucrose being actively transported laterally out of the phloem and into surrounding nonphotosynthetic tissues. As the sucrose content of the sieve tube decreases, water leaves by osmosis and turgor pressure decreases. Areas of active metabolism, called **sinks** (e.g., growing shoot apex, ripening fruit), remove the most sucrose from the phloem, resulting in a strong pressure gradient in the sieve tubes leading to the sinks. Therefore, continued translocation from the **source** (leaves) to the sinks is promoted (Figure 11–12). When a region becomes less active metabolically (e.g. the apical meristem becomes dormant or the fruit has matured), then its function as a sink is reduced and translocation to this region diminishes. Sucrose in excess of that needed by metabolism in the various sinks is translocated

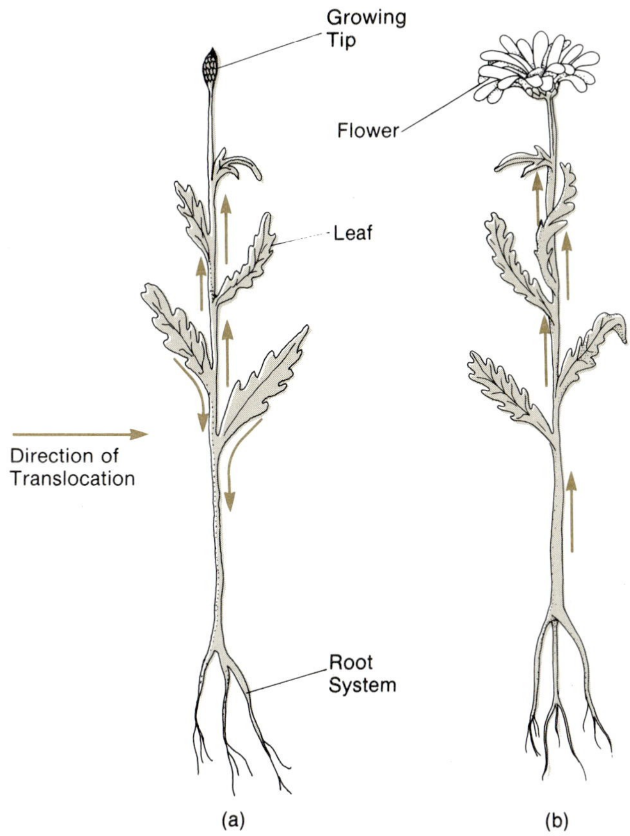

Figure 11–12. Translocation patterns. (*a*) During vegetative growth, mature photosynthetic leaves export sucrose in the phloem. This sucrose is translocated within the plant from a source (leaf) to a sink (area of active metabolism). A leaf at the base of the stem is closest to the root system and lower stem and so most of its sucrose goes to those sinks. A leaf midway up a stem would be midway between two sinks and therefore some sucrose from this leaf would be translocated up toward the upper stem and growing tip; the remainder would move down the stem toward the roots. A mature leaf near the top of the stem would be too far away from the roots for that sink to exert a strong pull. Therefore all of the sucrose would be translocated up to the growing tip and upper stem. Any immature leaves would also be sinks and import food via translocation. (*b*) Reproductive growth (a flower, for instance) produces an extremely strong sink and in many plants the presence of flowers shuts down all other sinks and food is translocated to the flowers almost exclusively.

to the parenchyma cells of the stem and root where it is converted to starch, a storage carbohydrate (Chapter 3).

Summary

Water enters and leaves individual cells as the result of diffusion across the plasmalemma, that is, osmosis. Normally cells contain sufficient water that they exert turgor pressure on the cell walls; cells that lose too much water become plasmolyzed and have no turgor pressure. The osmotic regulation of the water content of the guard cells is responsible for opening and closing the stomata as a result of changes in turgor pressure.

The absorption of water from the soil by the roots occurs by diffusion and osmosis. Once in the xylem, water moves as the result of a transpirational pull on the cohesive water column in the narrow tubes, tracheids and vessels, created by the xylem. Movement of water out of the xylem and into mesophyll cells is by osmosis as the result of transpiration. Mineral ions typically are absorbed by the root by active transport but once in the xylem minerals move passively with the water column.

Food synthesized in the leaves is translocated, in the form of sucrose, to the roots through the phloem as the result of pressure differentials between source and sink (mass flow or pressure flow hypothesis). Once the sucrose reaches a sink, it is either used in metabolism or is converted to starch for storage. Changes in the metabolic activity of different sinks will result in changes in the translocation pattern.

Review Questions

1. Explain the transpirational pull-cohesion theory for the movement of water in a plant.
2. Can a dead cell be plasmolyzed? Explain your answer.
3. Explain why the use of salt to remove ice from sidewalks causes problems for the plants growing beside the sidewalks.
4. Explain what is meant by differential permeability and state its roles in maintaining a living cell.
5. What is free space? Explain why an endodermis is necessary in a root as a barrier to movement in free space but not in a leaf where there also is movement of materials in free space.

Root Pressure and Maple Syrup

Maple syrup is produced from sap collected from sugar maple trees in early spring. Collection of the sap begins by boring a hole in the young wood near the base of the tree. Nowadays the sap is pulled from the trees by pumps instead of being allowed to flow from wooden spouts into buckets. The sap is converted to maple syrup by boiling, which concentrates it and enhances its flavor. Approximately 40 gallons of sap are required to produce 1 gallon of syrup.

Sap begins to rise early in the spring as the ground thaws and the sun warms the tree trunk. Under these conditions parenchyma cells of the stem and root become metabolically active, converting stored starch into sugar that is exported to the xylem. The resulting increase in sugar content of the xylem in the root causes water to move from the soil into the xylem along the osmotic gradient. As water enters the xylem pipeline, it displaces the sap already present, forcing the sap to rise. Continued application of the osmotically generated root pressure leads to the transport of sugar solution from the roots and lower trunk regions to the dormant buds. Because the xylem sap contains an energy source (sugar) as well as hormones produced by the root, bud dormancy is broken and shoot growth begins anew. Gradually, as the buds break dormancy, the physiology of the plant changes and less and less sugar becomes available for transport to the xylem. This reduces root pressure and the whole process winds down. By the time root pressure diminishes, the leaves are sufficiently mature for transpirational pull to take over the transport of water. Subsequently, phloem translocation of sugars produced by photosynthesis in the leaves results in the trunk and root parenchyma cells replenishing their depleted supplies of stored starch.

12

Plant Growth and Development

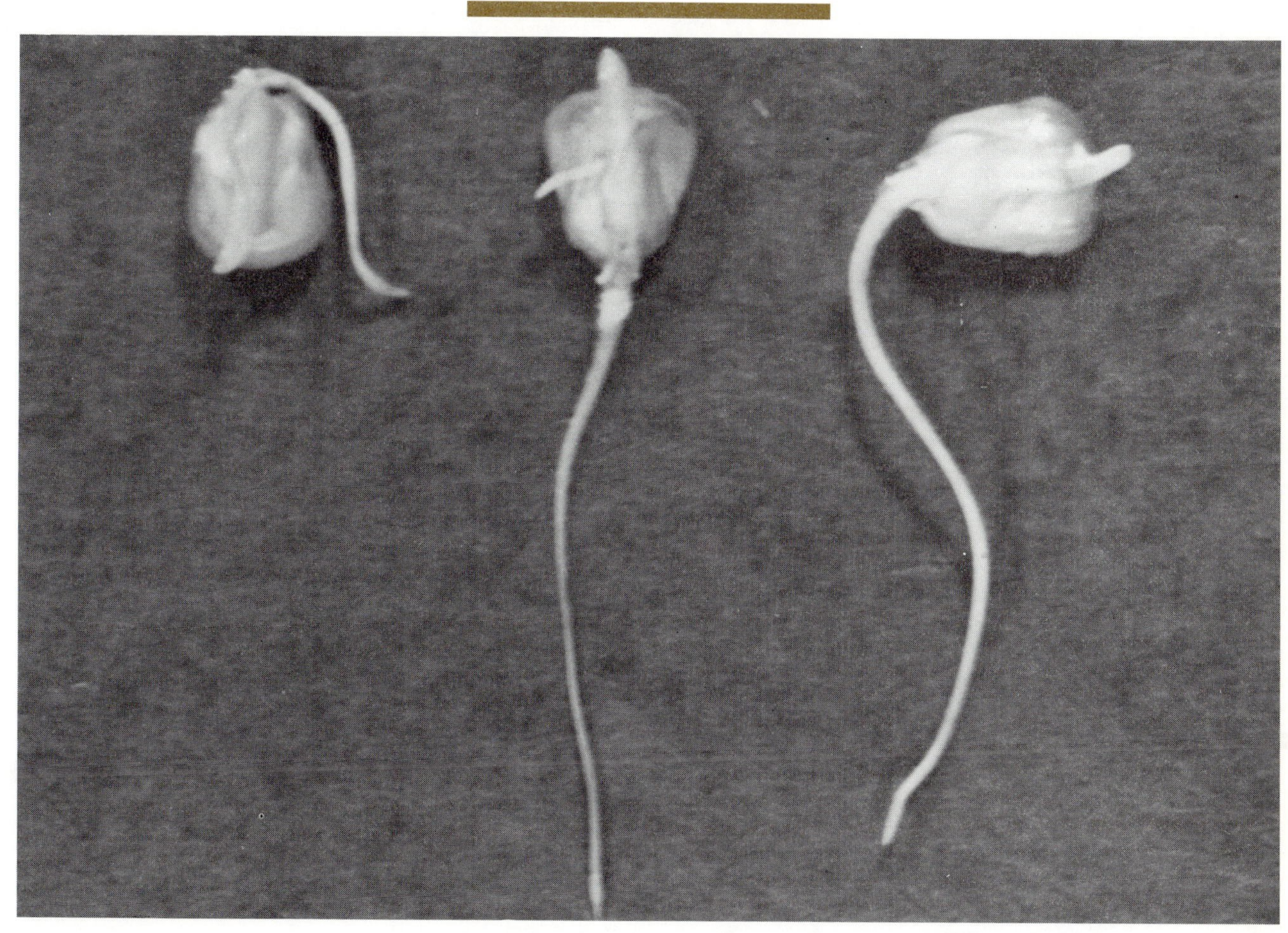

Outline

Introduction

Science fact or science fiction: Carrots and human beings can be grown in the test tube from single cells? At present, the only scientists capable of growing humans in the test tube are found in novels such as George Orwell's *1984*. However, the growth of plants from cultured cells is routine. Currently many of the commercially available house plants such as *Dieffenbachia* (dumbcane), Boston fern, and *Dracena,* as well as orchids and strawberries, are produced from **tissue cultures*** in large factorylike laboratories reminiscent of those described in *1984*. In the absence of plant propagation by tissue culture these house plants would cost two or three times their current price.

Plants, but not higher animals, can be grown from tissue culture. A major reason for this must be that the plant body is simpler than an animal's: for instance, the plant contains only four different organs (roots, stems, leaves, flowers) as compared to the large number in an animal; furthermore, plants have fewer different cell types than animals. The procedure for producing carrot tissue cultures is shown in Figure 12–1. Using tissue culture to produce plants reproduces, in the laboratory, the type of events that must occur

Tissue culture
A technique for growing isolated cells or tissue in a sterile medium of known composition.

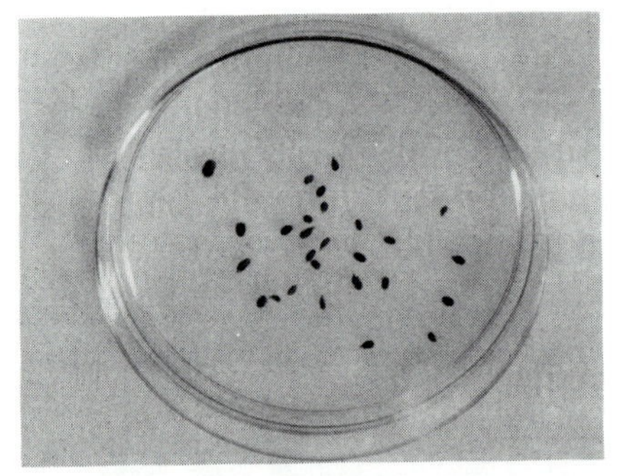
(a)

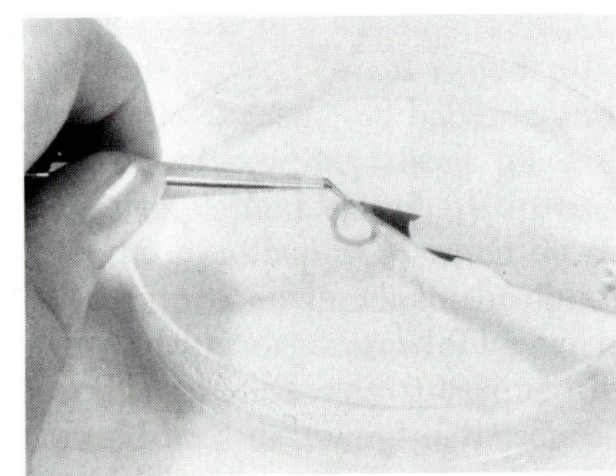
(b)

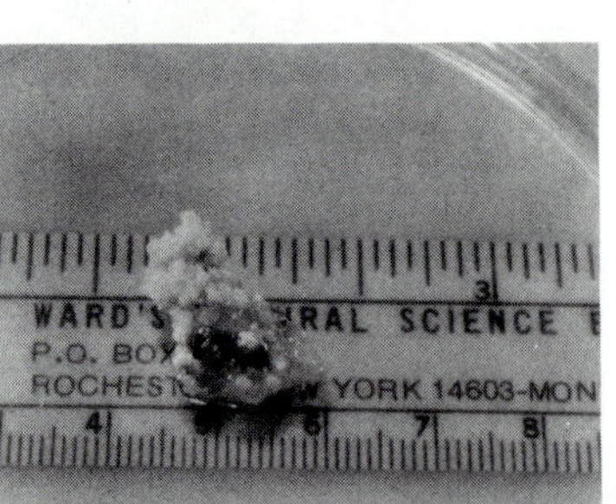

(c)

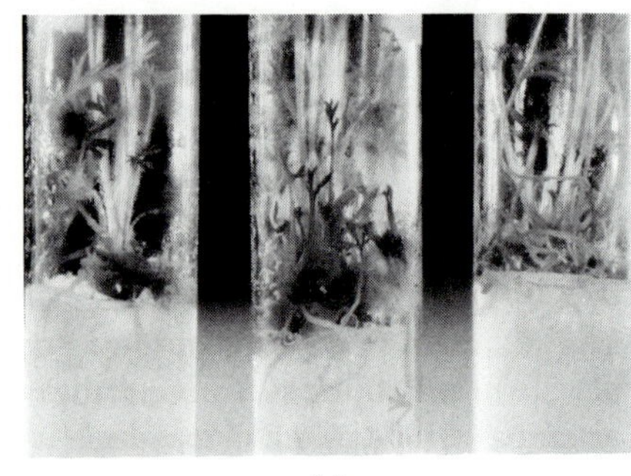
(d)

Figure 12–1. Carrot tissue culture. (*a*) Surface-sterilized carrot seeds are germinated in a sterile environment. (*b*) A small portion of the resulting seedling is cut out and placed on an agar growth medium containing auxin (a plant hormone). (*c*) After a mass of tissue (a callus) has developed, small portions are transferred to a different agar growth medium without auxin and grown in dim light. (*d*) Plantlets that have formed from the callus tissue. The complete process from the start of callus growth to plantlets requires approximately 12 weeks for this material.

as the zygote develops into an embryo and then into a seedling. First, cell division must produce more cells—all of them relatively unspecialized—but, if organized growth is to occur, these new cells must become specialized, i.e. differentiation occurs. The differentiation of a mass of cells in tissue culture or in an embryo results from the exposure of the cells to plant hormones. As indicated below, one set of conditions favors root development and another set produces shoots. Careful adjustment of the hormonal balance in a tissue culture will either result in a formless mass of cells or a group of plants each composed of roots, stem, and leaves (Figure 12–2).

Growth, as implied above, results in the increase in mass of an organism due to two related factors: (1) an increase in the number of cells and (2) an increase in the size of the cells. Increase in mass alone, however, is not sufficient to produce a functional organism; growth must be accompanied by developmental processes that regulate and organize growth so that the final result is the proper structure. One of the major developmental processes is **differentiation***—the production of structures that are specialized for specific functions. At the cellular level differentiation is the process by which the different cell types are formed, while at the organismal level differentiation is responsible for the formation of tissues and organs. Development also involves the organization of plant growth so that organs form in the correct places and at the correct times thus ensuring that leaves form on stems but not on roots.

Figure 12–2. Tissue culture. This *Kalenchoe* plant developed from a callus growing on an agar medium.

Increase in Mass

As indicated above, a plant increases in mass (grows) as the result of an increase in the number of cells **(cell division)** and an increase in the size of cells **(cell enlargement).** Cell divisions in plants occur almost exclusively in the meristematic* regions, each of which contributes to plant growth in a different way. Apical meristems* add cells to make the stems or roots elongate whereas the cambia* increase the diameter of the plant body.

After cell division has ocurred, the cells enlarge to reach their mature size and shape (Figure 12–3). In general, all mature cells are wider and longer than the meristematic cells that produced them though the extent of the elongation and enlargement depends on the cell type under discussion. For example, xylem* cells are more elongated than the parenchyma* cells of the cortex. During maturation, the cells increase in size because they develop a central vacuole. One of the most striking differences between mature cells and cells of the apical meristem is the lack of a large, central vacuole in the meristematic cells. As indicated in Chapter 2, the vacuole develops from the endoplasmic reticulum by the accumulation of water and solutes. The increase in volume of the developing vacuole causes the protoplast to exert pressure on the cell wall and to stretch it. Cell growth, then, is the result of pressure from an expanding vacuole on a stretchable cell wall. If the protoplast stops exerting pressure on the wall or the wall becomes rigid, cell en-

Differentiation
The development of specialized structure and function by cells, tissues, or organs.

Meristem
(*adj.*, meristematic) A tissue whose cells are capable of repeated division.

Apical meristem
Undifferentiated, dividing cells found at the tip of a stem or a root.

Cambium
(*pl.*, cambia) A secondary meristem.

Xylem
Water-conducting tissue.

Parenchyma
Relatively undifferentiated cells that frequently function in food storage.

largement stops. Both of these factors must be present. Under normal conditions the protoplast exerts pressure on the cell wall in all cells, but only in the region of cell enlargement are the cell walls stretchable instead of being rigid. Therefore, cell growth can only occur in limited regions of a plant—those regions immediately adjacent to a meristematic zone.

Differentiation

At the time that enlargement of the developing cells is taking place, differentiation or specialization is also occurring. Differentiation, the development of a special structure and metabolism, of individual cells is a necessity in a multicellular organism. Without the division of labor, which is a consequence of differentiation, each cell would need to perform all of the functions necessary to complete the life cycle. In other words, each cell of the plant independently would be involved with photosynthesis, food storage, water and mineral absorption, sexual reproduction, and so forth. Such a plant would be a colony of cells, not a multicellular organism, and would be restricted in size so that all cells would have equal access to the environment. The upper

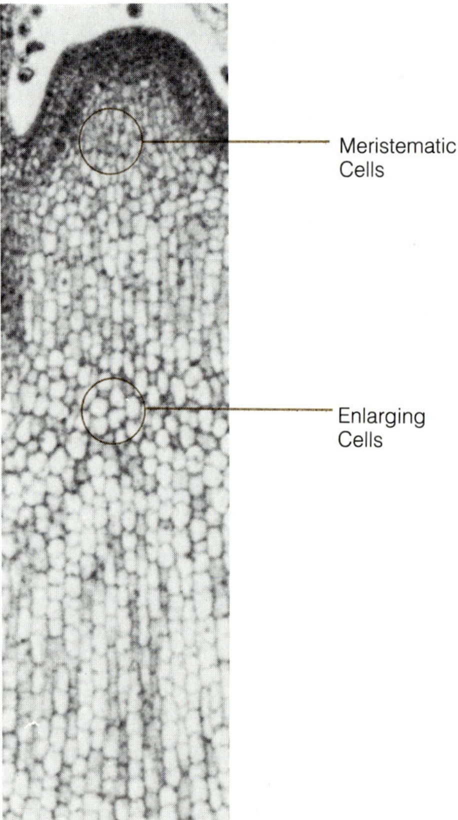

Figure 12–3. Cell enlargement. As part of the differentiation process, cells enlarge. The extent of this enlargement is a function of the cell type involved.

size limit for this type of organism is determined by the fact that diffusion of all materials is very slow except over extremely short distances. Therefore, the nutrients and gases needed for cellular metabolism would not be able to diffuse rapidly enough to keep interior cells alive in a large colonial organism. Certain primitive algae typify the colonial organism, but even among larger colonial algae such as *Volvox* some differentiation takes place with only certain cells being involved in reproduction, for example.

During development of a plant from a zygote, differentiation results in the production of mature cell types, tissues, and organs. The zygote itself, then, is **totipotent.*** This ability to differentiate into any type of mature cell is retained to the greatest extent by the cells of the apical meristems of roots and stems. Even here, however, future differentiation is controlled by a major factor in plant development—the position of a meristematic cell within the plant. Cells of the root apical meristem will not give rise to bud* and leaf primordia* just as the stem apical meristem will not produce a root cap.* As the apical meristem divides, it produces new meristematic cells, with the cells on the perimeter becoming part of one of the three primary meristems (protoderm, procambium, and ground meristem) (Chapter 6). These cells now have their future differentiation even further determined by their position in the stem or root. Cells of the primary meristems actively divide but they are also starting to differentiate. Protoderm cells are physically different from those of the ground meristem with the differences becoming magnified as the cells complete their differentiation. The extent of differentiation is least for cells derived from the ground meristem, mainly parenchyma cells in the cortex and pith.* Consequently, it is relatively easy for parenchyma cells to revert to the meristematic condition, as they do when cortical cells form the cork cambium.* An undifferentiated cell potentially is capable of carrying out all of an organism's metabolic reactions, but once it is differentiated it actually only carries out certain of these. Thus, a root epidermal cell would function in water and mineral absorption but it would not produce lignin* or suberin.* Similarly, a leaf parenchyma cell would be specialized for photosynthesis*, not water absorption from the soil.

Since all organisms start their life cycle as a zygote, or fertilized egg, it potentially must be able to develop into any cell type. In other words, the zygote potentially must be capable of carrying out all of the organism's metabolic sequences. Because there is no further input of genetic information after fertilization, the DNA of the zygote nucleus must contain the complete genetic program for all of the proteins that the particular plant will need. Proteins in living cells have a variety of functions (Chapter 3) but their role as enzymes* is particularly relevant to a discussion of differentiation. The zygote DNA contains the program for the formation of all of the enzymes the organism needs to catalyze the whole range of metabolic reactions. A differentiated cell only produces the limited set of enzymes that will allow it to perform its specialized functions. Somewhere between the totipotent zygote and the differentiated, mature cell the capacity to produce many enzymes is lost; however, this should not be interpreted as representing a selective loss of the unneeded portions of the genetic program. All evidence indicates that living, differentiated cells contain the complete genetic program—no chromosomes* or parts of chromosomes have been lost. Rather, differentiation leads to the selective turning on of only the portions of the

Totipotent
Possessing the potential to develop into any mature cell type.

Bud primordium
(*pl.,* primordia) A group of meristematic cells that give rise to a bud.

Leaf primordium
A group of meristematic cells that gives rise to a leaf.

Root cap
A group of cells that forms a protective layer around the growing tip of a root.

Pith
A region composed of parenchyma cells at the center of the vascular cylinder.

Cork cambium
A secondary meristem that produces cork and phelloderm.

Lignin
A complex organic molecule that adds mechanical strength to cell walls.

Suberin
A wax produced in cork and endodermis.

Photosynthesis
The metabolic process by which carbohydrates are produced from CO_2 and a hydrogen source such as water, using light as the source of energy.

Enzyme
A protein that functions as a biological catalyst.

Chromosome
A structure composed of DNA that contains all or part of the genetic information of a cell.

genetic program appropriate to the specific cell type that is being produced. The study of development is the study of how cells get the message as to which portions of the genetic program are to be turned on and of how this process of differentiation is integrated so that tissues and organs are formed correctly and only in the right place. For plants this is a more difficult area to investigate than it is for animals because, as will be discussed later, plants do not have an obvious system of endocrine glands, as do animals, to control development.

Growth-Regulating Factors

Growth and differentiation in plants are highly organized processes. If they were not, organisms would not develop into such complex forms as mosses, ferns, or trees; instead, they would be deformed or even just mounds of cells. Growth of organisms is regulated by internal and external factors. External growth-regulating factors are those features of the environment that influence the growth pattern, whereas internal growth-regulating factors are produced by the organisms themselves to regulate their own growth. This distinction is not as simple in natural systems, however, because there is a continual interaction between external and internal factors.

The major factor that regulates growth is the genetic information contained in the nucleus of each cell. The genetic program establishes the limits within which an organism can develop and, therefore, determines the general appearance of the mature organism. The plant that actually develops is the result of the influence of environmental factors on the expression of the genetic program. For example, tumbling mustard, a weedy plant, can reach maturity at 1 or 2 inches in height under unfavorable conditions or can mature at 3 to 4 feet under more favorable ones. Both plants would be recognizable as tumbling mustard; the difference between them would be in the size of the plants and their organs, not in their inherited characteristics. Because the general pattern for an organism is encoded in the genetic information, it is impossible for environmental conditions to induce an acorn to develop into a dandelion.

Hormone
A molecule produced by one group of cells within an organism that regulates the growth and development of other cells.

The various internal growth-regulating factors are chemicals produced in specific cells and regions of the plant. In practice, the internal growth-regulating factors act as chemical messengers, or **hormones.*** It is important to realize that there is no one central location within a plant where the genetic program is read and instructions set out to all other regions so that the plant body is built correctly. Every living cell has its own copy of the complete genetic program in its nucleus. The role of the internal growth-regulating factors (hormones) is to direct the "attention" of the cell to specific portions of the program that need to be turned on or off at that particular time in that particular cell in order for it to develop correctly. Furthermore, there is no region, tissue, or organ in a plant that is responsible for the coordination of growth through its ability to produce all of the plant hormones. Instead, different regions, tissues, and organs produce the various plant hormones with no central control. Hormone production, and therefore growth and development, is controlled by the physiological condition of the plant with the amount of hormone production being determined by the amounts of other hormones, sugar, water, and minerals reaching each hormone-producing cell. The response of a target cell* to a particular hormone is also determined by its physiological conditions, most particularly by which part of the genetic program is already turned on because of past hormonal signals.

Target cell
The cell whose development is controlled by the hormone under consideration.

There are a limited number of known plant hormones with each causing a characteristic response or set of responses: **auxins, gibberellins, cytokinins, abscisic acid,** and **ethylene.** Auxins, gibberellins, and cytokinins are also the names of classes of plant growth regulators each of which contains both naturally occurring compounds and man-made, or synthetic, compounds. Even though the designation *hormone* is properly reserved only for naturally occurring compounds, plant physiologists* tend to refer to both the natural and synthetic members of these classes of plant growth regulators as hormones because in many experimental situations the synthetic compound is either more readily available, cheaper, or more effective than the natural one.

Plant physiologist
A botanist who studies plant structure, function, and biochemistry.

For simplicity the mechanism of action at the cellular level will be discussed first followed by a discussion of whole-plant growth responses in which these hormones are involved. Since the auxins have been studied more intensively, more is known about how they interact with the cell to modify its physiology and its development. This discussion begins with the auxins using them as an example of how hormones can influence a cell, with the implication that the less well-understood hormones will probably work in similar ways.

Auxins

The characteristic response of a plant cell to an auxin is cell enlargement. Of course, in the plant auxins control much more than cell enlargement, but the assay* for auxins is a measure of their ability to cause cell enlargement in a test system, the oat coleoptile.* The most commonly occurring natural auxin is indoleacetic acid (IAA), the first plant hormone to be discovered (in 1928). When nonlignified* cells are treated with auxin, they enlarge. In an intact plant this is evident as a growth curvature if the auxin is applied to only one side of a young stem (Figure 12–4). At the cellular level IAA and other auxins have a variety of effects that culminate in cell enlargement (Figure 12–5). For a hormone to affect a cell it must interact with some component of that cell. Auxin binds to the plasmalemma* resulting in several events that aid cell enlargement. One early effect is the alteration of the permeability of the plasmalemma allowing H^+ ions to leak into the cell wall. Since an increase in H^+ ions is an increase in acidity, the acidity of the cell wall is increased. Within the cell wall are cell-wall digesting enzymes called glycosidases, which are only active in an acidic environment. As H^+ ions accumulate in the cell wall, these glycosidases become active and begin to digest the hemicellulose and pectin "cement" that had been keeping the cell wall from stretching. The change in plasmalemma permeability also leads to increased uptake of water and ions by the cell. Auxin bound to the plasma-

Assay
Method for measuring the amount of hormone present.

Coleoptile
The sheath that surrounds a grass shoot during germination.

Nonlignified
Lacking lignin.

Plasmalemma
The outermost portion of the cytoplasm that functions as a differentially permeable membrane.

Figure 12–4. Growth curvature. Application of auxin to one side of a stem causes increased growth on that side and therefore a growth curvature.

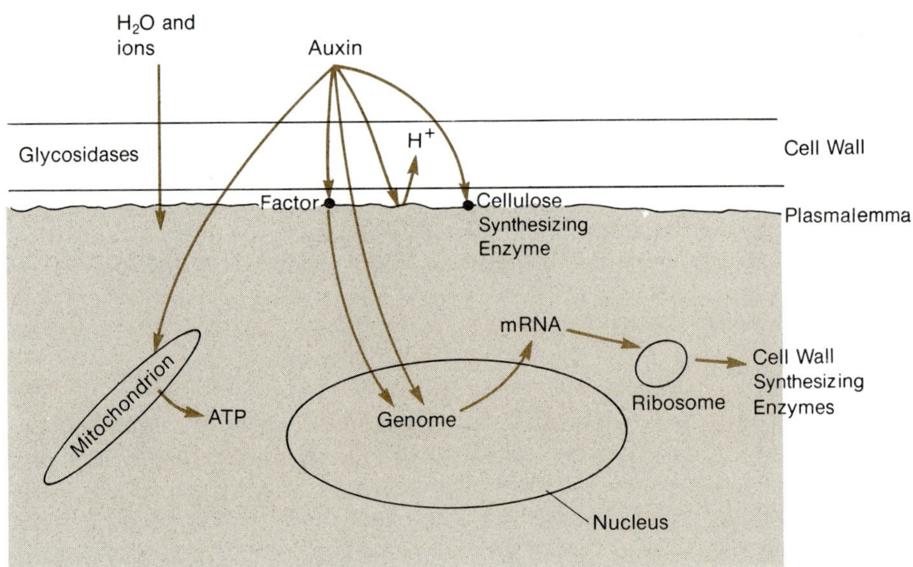

Figure 12–5. Cellular events in auxin responses. The plasmalemma contains auxin receptors that bind auxin and then initiate a number of physiological changes in the cell. A rapid response is a change in the permeability of the plasmalemma to H^+, water, and other ions so that H^+ ions leave the cell and accumulate in the cell wall while water and other ions enter the cell more readily. An increase in the number of H^+ ions in the cell wall stimulates the activity of enzymes that partially digest certain wall components making the wall more stretchable. A second response of auxin binding to the plasmalemma receptors is the release of a factor that specifically activates those portions of the nuclear genome that code for enzymes of cell wall synthesis. In addition auxin that enters the cell will activate these portions of the genome directly. Auxin binding to the plasmalemma also activates a cellulose-synthesizing enzyme bound to the membrane. The net result is the production of new cell wall material that is added to the expanding cell wall. Most of this increase in metabolic activity in the cell requires ATP, and auxin appears to act directly on the mitochondria stimulating the rate of respiration.

lemma produces two other important effects. One is the stimulation of the activity of a cellulose-synthesizing enzyme that is attached to the plasmalemma. The other is the release of a factor that diffuses to the nucleus where it promotes the specific turning on of those portions of the genetic program that carry the information for the enzymes involved in cell-wall synthesis. Some auxin also enters the cell and diffuses to the nucleus where it binds to the chromosomes. At present it is not known if the binding of auxin to the chromosome has any specific effect on mRNA synthesis. In addition to the above effects auxin stimulates respiration* thereby increasing the production of ATP* to support protein and cell-wall synthesis. Cell enlargement results from all these separate effects of auxin on the cell because (1) the existing cell wall becomes stretchable as the result of the partial digestion of the cement holding the cellulose fibrils in place, (2) the protoplast is exerting

Respiration
The metabolic process that results in the release of chemical energy from simple organic molecules.

ATP
A high-energy molecule that is involved in the transfer of chemical energy in cells.

pressure (turgor pressure) on the cell wall, which causes the wall to stretch, and (3) the existing cell wall is not allowed to become thin in the same manner as an expanding balloon, since the cell is now producing new cell wall material.

Under different physiological conditions a cell could respond differently to auxin. The events described above normally occur only in the region of elongation. However, the way in which auxin interacts with a cell would always be the same: affect membrane permeability and alter expression of the genome.* The condition of the membrane determines how the permeability changes, just as the existing state of differentiation determines which parts of the genome are activated by auxin treatment.

Genome
The complete genetic program in a nucleus, plastid, or mitochondrion.

Gibberellins

Gibberellins were discovered by Japanese scientists studying a rice disease called "foolish seedling disease." The cause of this disease is the fungus *Gibberella fujikuroi* that produces gibberellins in the rice seedlings, causing excessive stem elongation and eventual death of the plant. Application of gibberellins to dwarf varieties of plants often results in their growing as tall as the normal varieties, indicating that the dwarf characteristic is due to a reduced gibberellin level (Figure 12–6). Many normal plants treated with gibberellin also respond with stem elongation but the response is much less dramatic.

Gibberellins are characterized by their ability to cause the digestion of cereal seed endosperm* (see Chapter 21 for a complete discussion of seeds). Specifically this involves the release of digestive enzymes from the aleurone layer* into the endosperm (Figure 12–7). The cells of the aleurone layer respond to gibberellin treatment by releasing stored enzymes and by producing and releasing new enzymes that will digest the starch and storage

Endosperm
A non-embryo food storage tissue in a seed.

Aleurone layer
One of the outer layers of a cereal seed.

Unsprayed Tall Peas

Unsprayed Dwarf Peas

Dwarf Peas After Gibberellin Spray

Figure 12–6. Gibberellin and stem elongation. Dwarf pea plants treated with gibberellin spray grow to be as tall as untreated, normal pea plants.

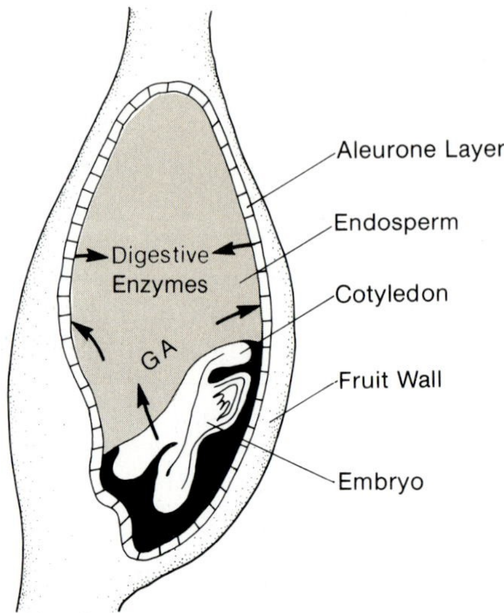

Figure 12–7. Gibberellin in the cereal seed. Germination of the cereal seed requires the mobilization of foods stored in the endosperm. Gibberellin (GA) released by the embryo diffuses to the aleurone layer cells that respond by releasing digestive enzymes into the endosperm. After digestion the simple foods are absorbed by the cotyledon and translocated throughout the embryo or seedling that develops from the embryo.

protein in the endosperm. The digested foods subsequently are absorbed by the cotyledon and are used by the embryo in its growth. To date it is not known precisely how gibberellins cause these effects. However, it is known that gibberellin treatment of a cell causes the cell to begin synthesizing specific new proteins and enzymes. Thus, like auxin, gibberellin must selectively alter the expression of the genome, which leads to the altered physiological state of the cell.

Cytokinins

Cytokinins are compounds that promote cell division in mature cells and also prevent senescence (aging) of leaves. Their mode of action is unknown except that they cause specific activation of the genome that, in turn, causes the cell to produce specific new proteins.

Abscisic Acid

Abscisic acid was actually being investigated by researchers working on two different, and supposedly unrelated, problems—bud dormancy* and leaf abscission.* Until it was purified and characterized by a chemist, it was known as either dormin or abscisin. Once the physiological properties of the purified abscisin were studied, it became apparent that the same hormone controlled both dormancy and abscission. At the cellular level, abscisic acid, like the gibberellins and cytokinins, has specific effects on the expression of the genetic program, but its mechanism of action is unknown. However, abscisic acid is like auxin in that it alters the permeability of the cell membrane.

Dormancy
A state of reduced physiological activity.

Abscission
The falling off of plant parts.

Ethylene

Ethylene is the only hormone that is a gas. Specific responses to ethylene include seed dormancy, leaf senescence, and fruit ripening. The mode of action of ethylene at the cellular level is unknown, however, in addition to effects on the expression of the genome it probably also involves changes in membrane permeability.

Interaction of Hormones

Probably no cell in a plant is exposed to just one hormone at a time. Instead cells are exposed to a complex mixture of hormones that changes over time and space. Hormone production in the different regions of the plant varies over time as a consequence of the physiological condition of the producing cells. Because the various hormones move through different tissues, there are also spatial differences in the concentrations of individual hormones. In order to produce the correct pattern of differentiation the correct balance of hormones must reach the differentiating cells at the correct time. An example that shows this clearly is the differentiation of xylem and phloem* from the vascular cambium.* If isolated stem sections are treated with gibberellins alone, the cambium begins to divide and some new phloem differentiates but no xylem. Treatment of stem segments with auxin alone also causes cambial division but this is associated with xylem and not phloem differentiation. Further study indicated that combinations of auxin and gibberellin promoted both xylem and phloem differentiation with high auxin:gibberellin ratios favoring xylem and low auxin:gibberellin ratios favoring phloem. A relationship between hormones and sucrose has been shown: for example, cytokinin in the presence of auxin increases the lignification of developing xylem cells while high sucrose:auxin ratios favor phloem differentiation. A major problem facing plant physiologists is to determine how plants manage to have such orderly development without the complex systems for the precise delivery of hormones that higher animals possess.

Phloem
Food-conducting tissue.

Vascular cambium
A secondary meristem that produces vascular tissues (xylem and phloem).

Plant Growth Responses

Stem and Root Elongation

As described in previous chapters, the growing tips of stems and roots each contain a region of elongation between the regions of cell division and maturation. This is the only region in which cell elongation normally occurs, and it is in this region where the small meristematic cells are converted into the much larger mature cells as part of their differentiation. The primary meristematic cells enlarge because auxin causes their walls to stretch as described above. At the same time each enlarging cell develops a central vacuole that is missing from a meristematic cell. Auxin-induced changes in cell membrane permeability toward water and ions help stimulate vacuolar growth and the development of sufficient turgor pressure to stretch the cell wall. The auxin that causes cell enlargement is produced by the cells of the apical meristem and is transported from cell to cell down the stem (or up the root). In the region of elongation the cells have stopped dividing and so are in the correct physiological condition to respond to the auxin by enlarging.

The cells in the region of maturation are still being exposed to auxin that is transported along the stem or root; but these cells do not respond by enlarging. The reason for the lack of response by mature cells is that there is insufficient auxin reaching these cells to cause them to enlarge. As the auxin is passed from cell to cell a small amount of it is used up and some is degraded, effectively reducing the amount of auxin below the level that will trigger enlargement of mature cells.

Apical Dominance

Once the auxin leaves the region of cell elongation in the stem, its role in regulating plant growth is not over. In many plants auxin also is involved in regulating the growth of the lateral or axillary buds. Casual inspection of the growth pattern of plants will show that some species, such as sunflower, have one main stem with no branches while others, such as flowering crabapple, have so many branches that the main stem is obscured. The difference between these two extremes is the degree to which the apical bud prevents the growth of the axillary buds—that is, the degree of **apical dominance.** Strong apical dominance results in a plant with one main stem and no branches, because the apical bud completely suppresses growth of the axillary buds. A plant with weak apical dominance will be bushy with many branches. Apical dominance is a result of auxin production by the apical meristem (Figure 12–8). This auxin is not, however, transported directly from the apical meristem to the various axillary buds that are to be controlled. Instead, apical dominance is a side effect of the auxin that is routinely being transported down the stem, the same auxin that was responsible for cell enlargement. In plants that show apical dominance the auxin that reaches the nodal region of the stem prevents the differentiation of the vascular tissue to connect the axillary bud to the vascular cylinder of the stem.

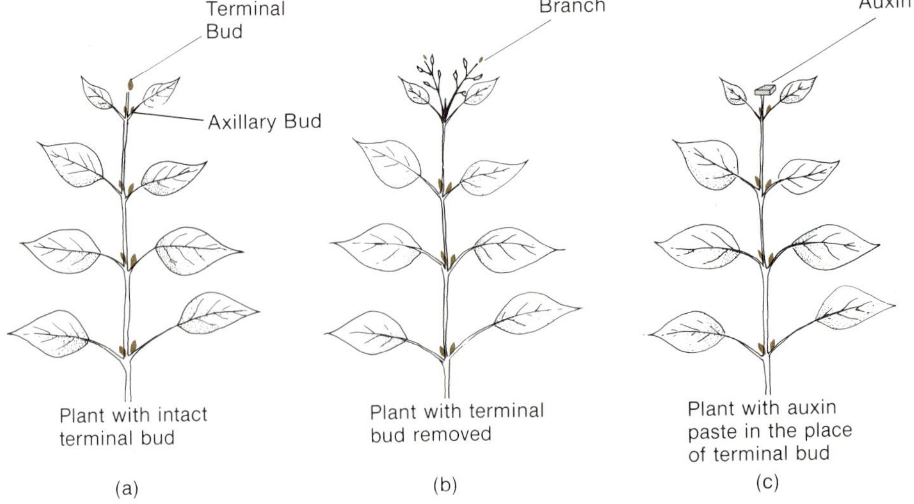

Figure 12–8. Apical dominance. Removal of the growing tip of a stem stimulates the formation of branches. The application of auxin to the decapitated plant prevents branching.

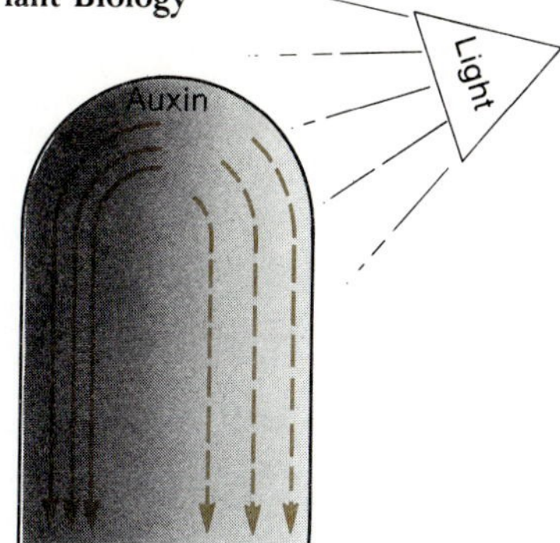

Figure 12–9. Redistribution of auxin in phototropism. In stem tips exposed to light from only one side the normal, uniform, downward movement of auxin is altered. Now each cell in the sensitive region just below the growing tip will transport at least some of the auxin laterally away from the light instead of downward. The net result is a greater concentration of auxin on the shaded side of the shoot and therefore a growth curvature toward the light.

Without a vascular connection the bud cannot begin to grow, as it would not have an adequate supply of food, water, and hormones produced by the roots. A plant that has weak apical dominance either has a low auxin level in the stems or the potential vascular cells are not as sensitive to auxin and therefore differentiate even in the presence of auxin.

Phototropism

As auxin is transported down the stem it normally is distributed equally on all sides of the stem. When the growing tip of a stem and the region of the stem immediately below it, however, are exposed to unidirectional light, the cells respond by altering their normal pattern of auxin transport. Instead of transporting auxin only to the cell below them the cells now transport some of the auxin to the cell beside them that is further away from the light (Figure 12–9). The result of this altered pattern of auxin transport is that auxin is transferred to the shaded side of the stem and moves down that side in higher amounts. This increased concentration of auxin on the shaded side is sufficient to cause greater cell elongation on that side in the region of cell elongation. Because the cell elongation occurs mainly on the shaded side, the stem appears to bend toward the light, although it actually grows toward the light. The increase in the auxin concentration on the shaded side of the stem will cause even mature cells to enlarge, and the growth curvature will extend into mature regions of the stem that have not become too rigid from the accumulation of lignin in support cells. **Phototropism** is a familiar problem to anyone who grows houseplants: it shows up as plants that grow toward the window (Figure 12–10). The solution is to rotate the plant a quarter turn once a week. This will ensure a relatively even distribution of auxin and, therefore, of elongation.

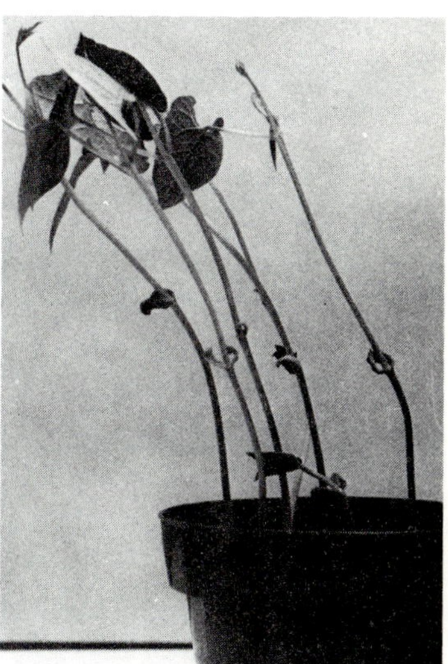

Figure 12–10. Phototropism. Bean plants exposed to light from only the side visibly begin to grow toward this light within two hours. The response pictured above was after 18 hours.

Geotropism

Another feature of the environment to which a plant responds is the gravitational field of the earth. Because plants always "know" which way is up, roots grow down into the ground and stems grow up and out of the ground when seeds germinate (Figure 12–11). This growth response to gravity is known as **geotropism.** This phenomenon is also a result of an unequal hormone distribution. The growing tips of stems and roots contain cells that respond to the gravitational field. In these cells there are amyloplasts*, containing starch granules, that tend to settle on the lower side of the cell. Since

Amyloplast
A starch-storing plastid.

(a)

(b)

Figure 12–11. Geotropism. (a) The pot containing this *Coleus* plant was placed on its side 48 hours before this picture was taken. (b) Corn seedlings showing geotropic responses of roots and shoots.

starch granules are heavy, the amyloplasts accumulate on the lower side. This causes the cells to export more of the hormone out of this side of the cell with the result that the hormone accumulates on the lower side of a horizontally oriented stem or root (Figure 12–12). In the case of stems there are actually two hormones involved in the geotropic response: auxin and gibberellin. The accumulation of these hormones on the lower side of the stem results in increased elongation of the cells on the lower side leading to the growth of stems in an upward direction. Roots grow down, not up, in response to gravity yet the mechanism of gravity perception is the same in both stem and root, that is, accumulation of amyloplasts on the lower side of the cell causes increased export of hormone out of the lower side of the cell. The difference between roots and stems is that in roots this hormone must act to inhibit the elongation of cells on the lower side in order for the growth curvature to be downward. In the root it has long been believed that auxin is the growth inhibitor, as root cell elongation is inhibited by all but extremely small amounts of externally supplied auxin; this is in contrast to stem elongation, which is stimulated by even large amounts of auxin. According to this view the accumulation of auxin on the lower side of the root causes growth inhibition. Since the amount of auxin present on the upper side of the root would be too small to be inhibitory, the upper side would elongate more than the lower side and the root would grow down. Some plant physiologists are not satisfied with this explanation of root geotropism because of the evidence which they have collected that argues against the involvement of auxin. The hormone they feel is responsible is abscisic acid that is produced in the root cap and transported up the root but which, like auxin in the stems, also is redistributed in root tips in response to gravity. Because abscisic acid is a growth inhibitor, its accumulation on the lower side would cause a downward growth curvature of roots.

This description of geotropic responses is accurate for seedlings, young plants from bulbs and tubers, and so forth, but it certainly will not agree

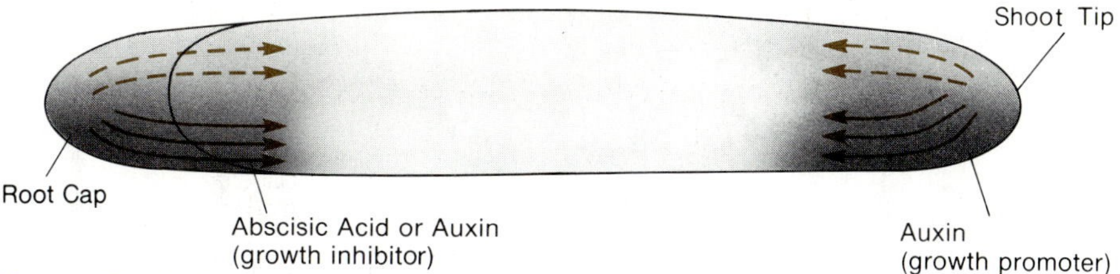

Figure 12–12. Redistribution of hormones in geotropism. In response to gravity, heavy components (statoliths) in sensitive cells sediment to the lower side of the cell. Somehow this causes an increase in the transport of the hormone responsible for geotropism through that region of the cell. Accumulation of this hormone on the lower side of a stem or root causes a growth curvature that results in stems growing up and roots growing down. The hormone responsible for geotropism in stems is auxin. In roots the sensitive region is the root cap and the hormone involved is more likely abscisic acid (ABA).

with observations of trees and bushes. Plants produce branch shoots and roots that do not grow straight up or down; instead they grow horizontally (Figure 12–13). Somehow the development of a branch shoot or root is controlled so that it does not exhibit normal geotropism. While it is of obvious survival value for the initial growth of a young plant to be vertical so that shoots grow out into the light and the roots penetrate into the soil, there would seem to be a disadvantage in all of the stems growing straight up and actually competing with each other instead of other plants for the available light for photosynthesis. Similarly a root system that spreads horizontally will come into contact with more of the nutrient-rich upper layers of the soil and also not compete with other roots of the same plant.

Dormancy

Just as environmental cues determine growth patterns for plants, they also determine periods when growth will stop. All plants in a temperate climate include a period of dormancy for one or more stages during their life cycle. For an annual angiosperm* the dormant structure would be the seed, bulb, or tuber, whereas in a perennial* the whole plant goes dormant in addition to possible dormant reproductive structures. Dormancy in plants is induced by abscisic acid. During the final stages of the development of reproductive structures such as seeds, bulbs, and tubers, abscisic acid is produced by the plant and transported to the reproductive structure, which prevents it from immediately beginning growth. As the days become shorter in late summer and early autumn, leaves of perennial plants begin to produce abscisic acid and reduce auxin synthesis thereby inducing dormancy in stem tissues and buds. At the same time abscisic acid promotes the formation of the abscission layer in the petiole* so that the leaves eventually will fall off the plant (see Chapter 8 for a discussion of autumn leaf colors). Dormancy in plant structures is broken by gibberellic acid. Gibberellic acid increases in plant tissues during the cold weather thereby overcoming the effects of abscisic acid. By the spring sufficient gibberellic acid is present in the buds, seeds, bulbs, and tubers that the warmer weather causes them to renew their growth.

Angiosperm
One of a group of plants characterized by the production of flowers.

Perennial
A plant that lives for many years.

Petiole
The stalklike region of the leaf that connects the leaf blade to the stem.

Figure 12–13. Horizontal branches. Most trees produce branches that grow horizontally instead of vertically as would be predicted from a knowledge of geotropism. The physiological mechanism for this horizontal geotropic response is unknown.

Leaf Galls

A leaf gall is a plant growth response to an insect parasite (Figure 12–14). Gall production involves the plant hormone cytokinin. A leaf gall is produced as a result of a female insect laying an egg in the mesophyll region of a leaf. The egg develops into a larval stage, which produces cytokinins from its mouthparts. Since cytokinin initiates division in mature cells, the mesophyll cells divide and produce a lump of tissue known as a leaf gall, thus providing a home for the developing insect. In addition to a home the insect needs food and, once again, a plant response to the cytokinin produced by the insect supplies this need. Plant tissues that have higher cytokinin levels receive more translocated food than other tissues. Finally, the fact that cytokinin delays senescence or aging of plant tissues provides the insect with a winter home. Leaf galls remain green and firm after the remainder of the fallen leaf has turned brown and dried up. In spring the mature adult leaves the gall, mates, and fertilized females return to new leaves to lay eggs and repeat the insect's life cycle.

All of the plant responses that produce the leaf gall are responses to a plant hormone, cytokinin, that normally is produced by the plant to regulate its own growth. Typically cytokinin is produced by roots and translocated to the shoot system where it functions to keep the leaves from aging, in addition to its other roles in plant growth and development.

Figure 12–14. Leaf gall. This gall was produced by the oak leaf in response to the secretion of cytokinins by a developing insect larva.

Flowering

Sexual reproduction in angiosperms involves the production of flowers followed by seed and fruit set (Chapter 20), all processes that involve hormones. A flower is produced from a flower bud that, in turn, is developed from a vegetative bud under the influence of a flowering hormone. A universal flowering hormone has yet to be discovered, though it has been named florigen. A few plant physiologists suggest that the different species actually use one of the various known hormones described above as their flowering hormone. After fertilization, hormones produced by the embryo developing in the ovule* (immature seed) cause the ovary* to enlarge and develop into the fruit. At the same time hormones, such as cytokinin produced by the embryo, cause the plant to translocate nutrients to the developing seed building up the food reserves required for germination. In edible fruits such as apples or strawberries the embryo produces more hormones when it nears maturity, thus initiating ripening of the fruit. Fruit ripening involves a color change from green to one such as red that strongly contrasts with the green leaves. Simultaneously there is a softening of the tissues and frequently a conversion of starch to sugar. These events have the effect of making the fruit more edible and attractive to the animals involved in seed dispersal (Chapter 21). In many edible fruits the onset of ripening triggers the production and release of ethylene gas by the fruit. Ethylene produced by one fruit speeds up ripening of adjacent fruits; it also works in conjunction with the abscisic acid produced by the embryo to cause the formation of an abscission layer* in the fruit stalk or the formation of the dehiscence zone* of a dehiscent fruit.* By the time the embryo is mature, the fruit is ripe and the seeds can be dispersed. It is important to note that the embryo regulates fruit and seed development through hormones.

Ovule
The structure within the ovary of a flower within which the egg is formed.

Ovary
The lower portion of the pistil of the flower in which ovules are formed.

Abscission layer
A layer of cells that is involved in the separation of a leaf petiole or fruit stalk from the stem.

Dehiscence zone
A region in a fruit where the fruit wall will rupture, liberating the seeds.

Dehiscent fruit
A fruit that ruptures when the seeds are liberated.

Photoperiodism*

The life of plants is regulated by the seasons in most regions of the world. As any review of weather records will indicate, the only reliable way to determine the season is to measure the day length. Even though the same day lengths occur in spring and autumn, the progression is reversed and the organism can readily distinguish between them because spring is preceded by a season of dormancy and autumn is preceded by a season of growth. If a plant is to be regulated by day length, a pigment system that functions as a growth regulator is required. **Phytochrome,** the pigment that plays this role, is a protein that exists in two interconvertible forms: one absorbs red light and the other absorbs far-red light. The absorption of red light converts phytochrome to the form that only absorbs far-red light. Absorption of far-red light converts it back to the red-absorbing form. During the day phytochrome is exposed to both red and far-red light so that both forms exist, but with the far-red form predominating. At night, however, the far-red-absorbing form spontaneously reverts to the red-absorbing form. As the result of this dark reversion, an unidentified set of metabolic events is set in motion that enables the plant to "measure" the night length and thus the day length. On the basis of this determination appropriate hormonal signals will be sent out by the leaves, the plant organs containing the majority of the phytochrome.

Photoperiodism
The regulation of plant growth and development by day length.

Plants have a number of photoperiodic responses mediated by phytochrome. For each the plant is either a long-day plant or a short-day plant. These designations do not relate literally to the number of hours of daylight with a certain number corresponding to a short day for all plants. Instead each plant species has a genetically programmed **critical day length.** A **long-day plant** is one that initiates a specific photoperiodically controlled developmental sequence in response to days longer than the critical day length. For example, a plant that is a long-day plant for flowering will begin to produce flower hormone when it is exposed to days that are longer than the critical day length. If the critical day length is 14 hours, then a day of 15 hours will trigger flowering, but one of 13 hours will not. Similarly a **short-day plant** responds to days shorter than the critical day length. A plant with a critical day length for flowering of 16 hours will produce flower hormone when the days are shorter than 16 hours. Notice the deliberate use of the critical day length of 16 hours for the short-day plant example as compared to the critical day length of 14 hours for the long-day plant. This emphasizes that for plants there is no common "definition" of short or long day lengths. In addition, it should be pointed out that in reality plants respond to night length and not day length. In nature these are irrevocably linked and all of the terminology for photoperiodism reflects the original experimenters' bias that the light period and not the dark period was controlling flowering. Despite the current awareness that night length is all important, none of the day-length terms have been replaced with their night-length equivalents, that is, short-day plant = long-night plant and long-day plant = short-night plant. In the Appendix on Green Thumb Botany is a discussion of the practical use of photoperiodism with an emphasis on the night-interruption phenomenon. A plant receiving a brief exposure to light during an otherwise long night will respond as if it were a short night. For those who are trying to induce flowering in a short-day/long-night plant, the night-interruption effect will be disastrous.

For each photoperiodic response each species has its own genetically programmed critical day length. Furthermore, a plant is not necessarily a short-day plant (or a long-day plant) for all its responses. A plant may be long-day for flowering but short-day for leaf abscission and bud dormancy. Alternately, plants can be **day-neutral** for a specific developmental sequence, which means that this sequence is not controlled by day length in this species. A plant which is day-neutral for one developmental sequence will have other responses that are photoperiodically controlled (Table 12–1). Whether a developmental sequence is photoperiodically controlled is determined by whether the development is linked necessarily to a particular season. If, for reasons of species survival, seeds must be produced only in the late summer, then flowering will be under photoperiodic control. If, on the other hand, seeds can be produced at any time during the growing season, the plant would likely be day-neutral for flowering.

Other Effects of Light on Plant Development

Plants grow toward the light, they detect the seasons by reference to day length, and they produce food as a result of photosynthesis. In addition light regulates the normal development of the shoot system. Most of us have had

Table 12–1.
Photoperiodic Responses of Plants*

Species	Short Day	Long Day	Day Neutral
Coneflower		Stem elongation; induction of flowering	
Cucumber	Increase in # of pistillate flowers	Increase in # of staminate flowers	Induction of flowering
Dahlia	Storage-root development	Fibrous-root development	
Jerusalem artichoke		Tuber formation	Induction of flowering
Kalenchoe	Induction of flowering; Crassulacean acid metabolism	Plantlets on leaves	
Potato	Tuber formation		Induction of flowering
Radish	Root thickening	Induction of flowering	
Strawberry	Induction of flowering	Runner development	
Tomato		Stem elongation	Induction of flowering

*This is only a partial list of plant species and of the photoperiodic responses of an individual species.

Internode
The section of stem between adjacent nodes.

Node
The region of a stem to which a leaf is, or was, attached.

the experience of moving a board that had been lying on the lawn or garden for a week or two. The plants that were under the board have elongated, spindly stems with small, whitish-yellow leaves. These plants are **etiolated.** Etiolated growth is desirable for celery and cauliflower and the gardener will blanch (etiolate) these plants to obtain the crop in the correct condition. For plants in nature etiolated growth appears to be an adaptation for survival because the shoot system must be in the light to carry out photosynthesis. The etiolated stem is long and spindly with long internodes.* Under natural conditions a stem that is in complete darkness will be either underground or under an object such that the stem will be prostrate. In either circumstance the plant would be living off stored food reserves such as starch in root and stem parenchyma cells. Consequently there would be a survival advantage in not wasting food reserves by producing unneeded support tissues or a compact stem with many nodes.* The leaves that form on etiolated stems are unexpanded and contain no chlorophyll; however, once they are exposed to light they expand and begin to produce chlorophyll. Plants grown under low-light conditions can also show signs of etiolated growth, most noticeably "leggy" growth that is characterized by long internodes, smaller leaves, and reduced chlorophyll content. Etiolated growth is another phenomenon controlled by phytochrome although there is no photoperiodic influence on etiolation. In fact, the three major aspects of etiolated growth are each controlled in different ways. Stem growth in the dark appears to be a result of gibberellins in the stem. Absorption of light by phytochrome either reduces gibberellin synthesis or it causes the production of inhibitors of gibberellin action. The net result is a light-induced reduction in stem growth. Leaf expansion and chlorophyll production are also under separate phytochrome control but do not involve gibberellins.

Plant ecologists (Chapter 24) have noticed another effect of light on plant development: trees tend to have two kinds of leaves, sun leaves and shade leaves. Sun leaves are smaller and thicker with more than one palisade mesophyll layer. Shade leaves are larger and thinner with more chloroplasts and chlorophyll to increase the light-gathering capacity of these leaves.

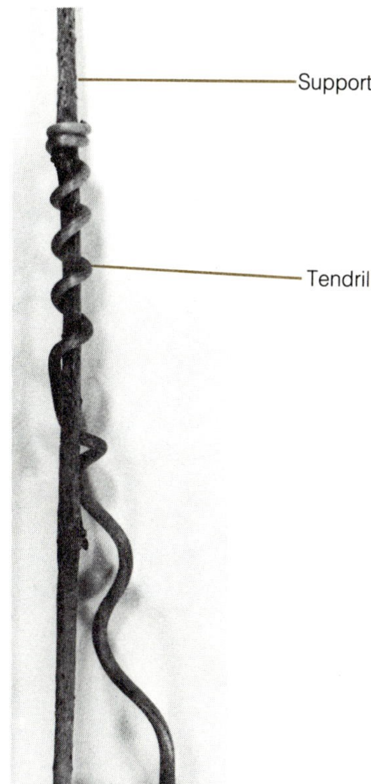

Figure 12–15. Vining stem. Contact with an object causes a pattern of differential growth in stems of this type. Cells on the side of the stem away from the object grow more rapidly than those closer to the object and this causes the stem to wrap itself around the support.

Other Plant Movements

Plants are not the "inanimate" objects they seem to be. The response of an animal to an environmental stimulus is unquestioned because it is immediate and usually accompanied by visible movement. Cause and effect are clear. Plants, on the other hand, respond slowly to environmental stimuli because they lack muscles. Thus plant responses are restricted to growth movements or a more rapid type of response due to turgor* changes.

Certain flowering plants have evolved adaptations that allow them to "invest" less food reserves in stem production and yet still receive the benefits of a sturdy, upright stem. Examples of these adaptations are the tendrils of peas and grape ivy and the twining stem of morning glory both of which wrap around the stems of other plants or on any available support (Figure 12–15). Once a tendril or twining stem touches an object, a growth response is initiated that results in the tendril or twining stem wrapping around the object, a response called **thigmotropism.** Another interesting response of certain plants is the so-called **sleep movement,** which is under phytochrome control. Sleep movement is obvious in *Oxalis* (shamrock) and *Maranta* (prayer plant) as well as in nonhouseplants such as beans. In this response the plant leaves change their position at night as the result of turgor changes in cells at the base of the petiole or leaf blade depending on the species (Figure 12–16). Sleep movement is often under the control of an internal biological clock indicating a response of some significance to the organism. Recent reports have indicated that the change in leaf position may have the effect of keeping the leaves warm at night. Measurements of heat loss by horizontal and vertical leaves while in the dark showed that vertical leaves (sleep movement) had a slightly lower rate of heat loss and therefore a warmer nighttime temperature. These plants also grew more rapidly.

Sensitive plant *(Mimosa pudica)* demonstrates well-defined sleep movements. This plant owes its common name and its popularity to the fact that it responds rapidly to touch. At night its leaflets fold up by themselves but during the day the leaflets are sensitive to the slightest touch and also respond by folding up. The basis of this response is turgor changes in the **pulvinus*** at the base of each leaflet. The response to touch is rapid, almost instantaneous; however, recovery takes about 15–30 minutes. A sensitive plant that is handled too roughly not only will have the leaflets fold up but the petiole will bend down as well. The physiological value of these responses is unknown. One suggestion is that this offers protection against predators. However, since a strong wind will also close the leaflets, the plant would be at a competitive disadvantage because this would reduce the opportunity for photosynthesis to take place.

Turgor
Pressure within a cell due to the amount of water contained in the cell.

Pulvinus
A small swelling at the base of a petiole or leaflet.

Figure 12–16. Sleep movement. (*a*) Leaflets of an *Oxalis* plant in a horizontal position in daylight. (*b*) Leaflets of an *Oxalis* plant folded down at night.

Venus flytrap is probably the most popular novelty plant, and it has been the inspiration for science fiction stories of man-eating plants due to the ability of its leaves to fold up and trap an insect. Each half of a leaf trap has three trigger hairs on it (Plate). If an insect touches one hair twice or two hairs within a few seconds of each other, this causes the turgor changes that close the trap (fold the leaf). Once the leaf folds, it releases enzymes which digest the insect. After the insect is digested, the leaf unfolds and allows the insect exoskeleton to be removed by the wind or the rain.

Summary

Growth results in the increase in mass of an organism due to increase in cell number and increase in cell size. Newly divided cells in the meristems enlarge as the result of pressure from the developing large, central vacuole and because the cell wall is stretchable. Cell wall stretching occurs as part of a cellular response to the hormone auxin. In addition auxin causes the cell to produce more cell wall material for cell enlargement. Since growth alone is insufficient to produce a functional organism, there must be a division of labor among the cells—that is, differentiation. Differentiation results from the different cell types expressing different portions of the genome with the resulting variation in proteins and enzyme activities producing the variation in cellular structure and function. Plant growth regulators (hormones) such as auxin, gibberellin, cytokinin, abscisic acid, and ethylene control the differentiation process in addition to controlling the other aspects of plant growth and development. Auxin has a major role in controlling stem and root elongation, apical dominance, phototropism, and shoot geotropism. Gibberellins are also involved in shoot elongation as well as in breaking dormancy. Cytokinins cause cell division and prevent senescence of leaves. Leaf galls are produced by insects that release cytokinins into leaves. Dormancy of buds, tubers, seeds, and bulbs is induced by abscisic acid, often as the result of photoperiodic control. Photoperiodism, the control of plant growth and development by day length, also controls flowering and leaf abscission. Short-day plants are ones that produce the particular physiological response when the days are shorter than the critical day length for the species. A long-day plant is one that responds to days longer than the critical day length. The measurement of day length requires the presence of the pigment phytochrome. Plants exposed to little or no light for several days or weeks will show signs of etiolated growth. At night certain plants show sleep movements of uncertain value; however, these are due to turgor changes in the petioles and are not growth movements. The venus flytrap shows a similar but more rapid response to turgor changes and as a result can trap insects.

Review Questions

1. What are growth regulating factors? Identify two internal and two external growth regulating factors and explain how they work.
2. Explain completely how a meristematic cell enlarges to become a mature cell.
3. Describe what would happen to the auxins in the stem of a plant which was placed horizontally on a glass plate and then illuminated from the bottom only.
4. Define photoperiodism and explain how some physiological phenomena can be regulated by short day conditions and some by long day conditions in the same plant.
5. Assume that plants you are growing in a greenhouse begin to grow toward a magnet. How would you demonstrate (a) that this is a growth response and not just a physical bending, (b) where the sensitive region of the plant is, (c) whether the internal signal from the sensitive region to the region that bends is chemical or electrical (like a nerve impulse).

Charles Darwin and the Discovery of Auxin

One of the best kept secrets of biology is that Charles Darwin did more than take a trip around the world and "discover" evolution. Charles Darwin was an important plant physiologist as well. After Darwin settled in the English countryside to let others fight over evolutionary theory, he, along with his son Francis, did many experiments on phototropism. Using orchard-grass seedlings he tried to determine why the coleoptiles grew toward unidirectional light. In simple but elegant experiments (Figure 12–17) he demonstrated that only the tip of the coleoptile was sensitive to light. Darwin concluded that some influence is transmitted from the tip to the lower portion of the coleoptile causing the growth curvature in these lower regions.

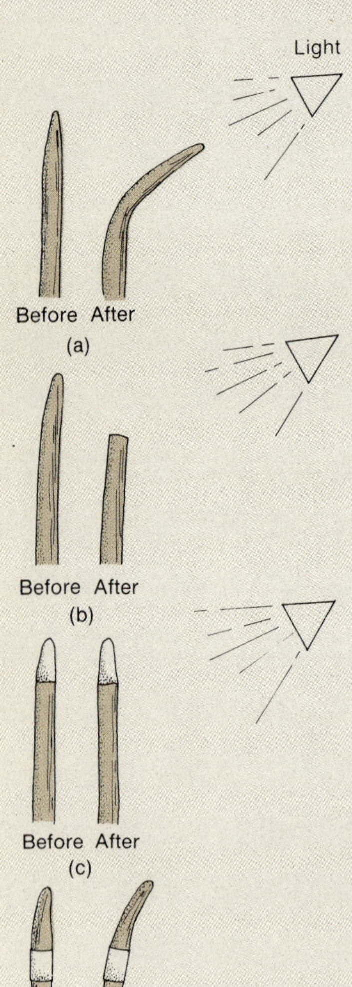

Figure 12–17. The Darwins' experiments on phototropism. After demonstrating the normal phototropic response (*a*) the Darwins attempted to determine where within a coleoptile the stimulus of unidirectional light is perceived. Removal of the terminal centimeter or so prevented bending toward the light (*b*). Similarly, covering the tip with an opaque cap prevented phototropism (c) but placing an opaque collar around the region that would bend did not (*d*). Therefore they concluded that the growing tip of the coleoptile is responsible for the phototropic response.

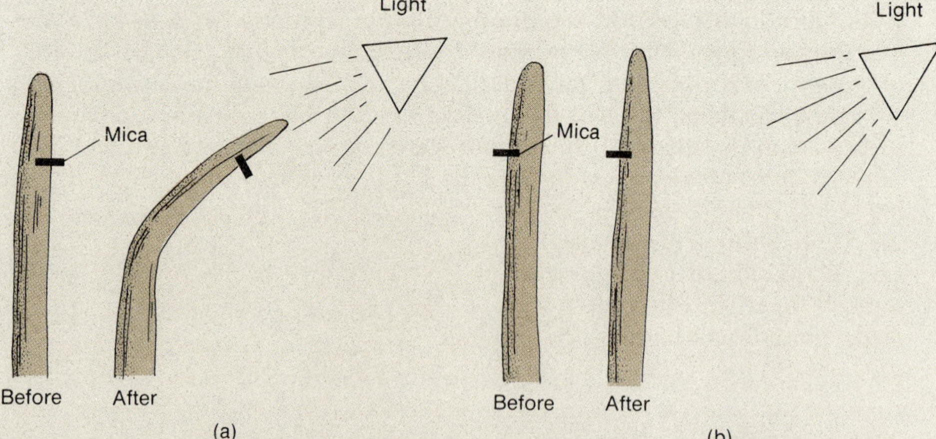

Figure 12–18. Further experiments on phototropism. A sheet of mica (a physical block to the movement of chemicals) in the coleoptile will prevent movement of chemical factors. Since mica on the illuminated side had no effect and mica on the shaded side prevented phototropism, the chemical factor responsible for phototropism had to be a growth promotor and not a growth inhibitor.

Later workers showed that this was a chemical influence and not a physical one, and that the chemical was a growth promoter and not a growth inhibitor (Figure 12–18). In the 1920s Frits Went, a Dutch plant physiologist, named this chemical auxin though no auxin was isolated or characterized from plants for another twenty years owing to its presence in such low concentrations. In fact the purification of as little as 1 gram of auxin from coleoptiles, by the limited isolation procedures available to these early workers, would have required many tons of coleoptiles. Interestingly though, a chemically pure auxin was isolated from biological material in the 1930s. However, the source was completely improbable—human urine. Apparently Went had some reason to believe that the female sex hormone might have plant growth activity so a preparation from the urine of pregnant women was tested on plants. The preparation was very active but the sex hormone was not; the active substance was indoleacetic acid, the most common plant auxin.

Commercial Applications of Plant Growth Regulators

Do you have a woody stem cutting to root? Are your potatoes sprouting in storage? Or maybe you do not like seeds in your tomatoes. The solution to these and other horticultural and agricultural problems is to use a plant growth regulator. Any garden-supply store will sell you a rooting compound that will "encourage" even difficult cuttings to root. This compound is a powder containing a small amount of a synthetic auxin that stimulates cell division in the stem with subsequent organization of this new growth into adventitious roots. The same process occurs without our intervention when we root philodendron or coleus stems by placing them in water. In this case the natural auxins are sufficient to initiate root formation. Many house plants are propagated commercially from stem cuttings that were treated with rooting compound.

The home gardener who grows enough potatoes to last into the winter and the person who buys potatoes from the supermarket will have different ideas about how easy it is to store them. Home-grown potatoes tend to sprout more than supermarket potatoes. The difference is that the commercial potatoes are sprayed with a synthetic auxin to prevent sprouting. The potato tuber, an asexual reproductive structure, is naturally dormant for a few months at the end of the growing season, but under home storage conditions it eventually breaks dormancy and sprouts. Dormancy of Jerusalem artichoke tubers, however, is broken by auxins.

If you had an apple orchard one of your problems would be that apples obey the law of gravity once the abscission layer forms in the fruit stalk. The trees can be sprayed with a synthetic auxin such as naphthalene acetic acid to prevent preharvest fruit drop. Pineapple growers have found that they can use synthetic auxins to induce flowering and fruit production out of season. This is an economic benefit because pineapple fruits can be harvested year round. Certain flowers (tomato, strawberry, or squash, for example) can be sprayed with a synthetic auxin to produce seedless fruits. The fact that the sprayed auxin substitutes for the developing seed in causing fruit development indicates that the embryo plays a major role in directing fruit growth. The commercial production of seedless fruits, however, is not done via auxin treatment. Instead cultivars that develop **parthenocarpic*** fruit have been selected and are the source of seedless grapes, oranges, and bananas. Growth regulators such as gibberellins are sold for spraying on developing strawberries to increase their size but not to induce seedless fruits.

Parthenocarpic
Having been produced without pollination having occurred.

Another major use of synthetic auxins is as herbicides. Homeowners are likely to use these on their lawns, and farmers on fields containing resistant crops. Herbicides are sold as weed killers when in reality they are plant killers. A commonly used auxin herbicide is 2,4-D, which kills broad-leaf plants (dicots) but not narrow-leaf plants (monocots). This specificity is not absolute as some monocots are sensitive to 2,4-D and some dicots are resistant. A more important consideration, however, is that 2,4-D does not discriminate between dicot weeds and dicot crops, but used carefully 2,4-D will rid a lawn of dandelions and a corn field of dicot weeds. The exact mode of action of 2,4-D is unknown but it does appear that 2,4-D–induced ethylene production is at least partially responsible for the death of sensitive plants. Monocots are resistant to 2,4-D because they metabolize it more rapidly, inhibiting ethylene production.

Ethylene is useful in agriculture for more than killing weeds. By accident it was discovered that ethylene causes the ripening of citrus fruits. This discovery came about because railroads changed from kerosene heaters, which produce ethylene, to steam heat in the boxcars of fruit being shipped from Florida to the North. Prior to this change, unripe fruit was packed in the cars in Florida and then two or three days later ripe fruit was unpacked in New York. With the advent of steam heat the fruit was still unripe when it was unloaded. Now, to induce ripening of oranges (and other fruit such as bananas), ethylene is added to the air during storage. If fruit is to be stored for a considerable time, the ethylene produced by the ripening fruits must be removed continually. The apples that are sold in March were harvested in September or October and then stored in a cool, climate-controlled warehouse where the ethylene can be removed from the air, preventing the apples from aging rapidly and becoming soft.

As more becomes known about plant growth and plant growth regulators, agriculture and human nutrition will benefit. The breakthrough that would be the most obvious to the general public, however, would be the discovery of a universal, synthetic flowering hormone. What home would be without flowers all year round once a spray bottle of flowering hormone were available? But for the agricultural scientist this would open up all manner of opportunities for accelerated plant breeding programs. Therefore this hormone is of great importance and the object of considerable research.

13

Reproduction and Meiosis

Outline

Introduction

The preceding chapters have outlined the various organs, tissues, and cell types that go together to form a flowering plant, as well as the physiological processes that allow the plant body to function. Yet the "purpose" of that plant body has not even been hinted at. Why *do* plants exist? If the philosophical aspects of this question are ignored, the answer becomes clear from a consideration of the ultimate result of the life of a plant—plants exist in order to reproduce. The only lasting effect of the existence of a plant is a new generation of plants of the same type. In Chapter 23 the point is made that organisms have increased in complexity from their one-celled origins only because those increases in complexity provided increased potential for survival and thus for reproduction. Therefore, all of the plant organs, all of the metabolic and physiological processes, have as their main function the survival of the plant body so that reproduction can take place. In fact, reproduction is so basic an aspect of being alive that the ability to reproduce is a fundamental characteristic used to distinguish between the living and the nonliving.

There are two basically different forms of reproduction—**sexual** and **asexual.** Unlike the higher forms of animal life, plants at all levels of evolutionary development are able to reproduce both sexually and asexually. Individual species of plants, however, may be capable of only one or the other type of reproduction. **Asexual reproduction,** also known as **vegetative reproduction,** usually involves a simple fragmentation of the whole plant or parts of the plant into two or more units, while **sexual reproduction** requires fusion between two specialized sex cells. It is generally accepted by botanists that the first organisms which evolved were able to reproduce only asexually (Chapter 23). Cyanobacteria* are examples of the present-day descendants of one group of these organisms that appear to lack true sexual reproduction even now. When and how sexual reproduction evolved is not known. However, certain characteristic structures associated with sexual reproduction strongly suggest that one reason reproduction evolved was as a means of survival of the organism during adverse growing conditions such as extremes of temperature, drought, or unavailability of nutrients. It is commonly observed that green algae, primitive plantlike organisms, reproduce by asexual methods during the active growing period and switch to sexual methods only at the end of the growing season prior to winter. The structures resulting from sexual reproduction often remain dormant* during winter and germinate to produce new plants in the spring. Another factor that would have favored the evolution of sexual reproduction is that sexual reproduction results in genetic variation among individuals of a species* (Chapter 22), thus increasing the likelihood of that species surviving when its environment undergoes change. On the other hand, asexual reproduction results in the production of many genetically identical individuals within a very short period of time, an ideal method to increase the population size and maintain stability of type in a static environment.

Cyanobacteria
A group of primitive, photosynthetic organisms that lack membrane-bounded organelles; also known as blue-green algae.

Dormant
In a resting state with reduced physiological activity.

Species
A group of closely related organisms that are capable of interbreeding.

Asexual Reproduction

The most primitive and the simplest form of asexual reproduction is found among unicellular prokaryotic* organisms where the single cell divides by a process described as **binary fission.** During this process a constriction is formed that pinches the cell into two, each developing subsequently as an independent organism. Among filamentous organisms, a small segment of the filament* may break off and then regenerate by repeated cell division (Chapter 4) to the size of the original filament, as observed among algae and fungi.* Another form of asexual reproduction commonly found in algae and fungi is the production of **spores*,** each of which can germinate directly into a new plant. Most algae and a few fungi produce spores in a cell called a **sporangium.** If the organism that produces these spores is aquatic, as are most of the algae, the spores usually have **flagella*** that help them to swim; such motile spores are called **zoospores** (Figure 13–1). In most algae the cells that become sporangia look like any other cell. However, prior to the production of spores the nucleus of the cell undergoes successive mitotic divisions* to produce a few to many genetically identical nuclei*, each of which is surrounded by a small amount of the cytoplasm. With the deposition of a cell wall around the cytoplasm, each nucleus with its surrounding cytoplasm and cell wall becomes a spore. The number of spores produced in a sporangium varies from one to many, the precise number being a characteristic of the species. In many cases, all cells of a filamentous alga become sporangia and synchronously* release their spores, resulting in an empty filamentous shell of cell walls (Figure 13–1). The zoospores that are released into the water swim around for a while, lose their flagella, and germinate into new filaments, giving rise to hundreds of genetically identical new organisms from a single parent.

Asexual reproduction generally is not considered to be a significant form of reproduction among the vascular plants, yet closer inspection reveals that many species do indeed reproduce asexually. One fairly common form occurs in plant species with stems that run horizontally along the surface of the ground **(stolons)** or just below the ground **(rhizomes)** (Figure 13–2). Such plants frequently form adventitious* roots at each node; thus when the stem is broken in two as the result of animal activities or other environmental effects, each part is rooted and is capable of an independent existence. Upright plants can reproduce asexually by means of **suckers,** which are shoots that form from adventitious buds on the roots. Death of the original plant or the severing of the roots again results in independent plants. Probably the forms of asexual reproduction by vascular plants that are most familiar to nonbotanists are those involving bulbs, tubers, and corms (Chapter 6). Each of these is a plant structure whose main function is asexual reproduction. Typically these are produced during the growing season, after which the remainder of the plant dies. At the start of the next growing season this structure begins to grow again and gives rise to a plant genetically identical to that of the previous year.

Among a relatively few vascular plants, asexual reproduction involves the leaves—walking fern produces new plants from the tips of leaves which

Prokaryote
(*adj.*, prokaryotic) An organism that lacks membrane-bounded organelles in the cell.

Filament
A chain of cells linked end to end.

Fungi
A group of predominately filamentous organisms that lack the ability to synthesize their own food.

Spore
A unicellular reproductive structure other than a gamete.

Flagellum
(*pl.*, flagella) Long, hairlike cell structures that function in cell motility.

Mitotic division
Division of the nucleus into two genetically identical daughter nuclei.

Nucleus
(*pl.*, nuclei) A spherical structure, found in cells, that contains the chromosomes and the majority of the genetic information.

Synchronously
At the same time.

Adventitious
Produced from a structure or region other than the usual one.

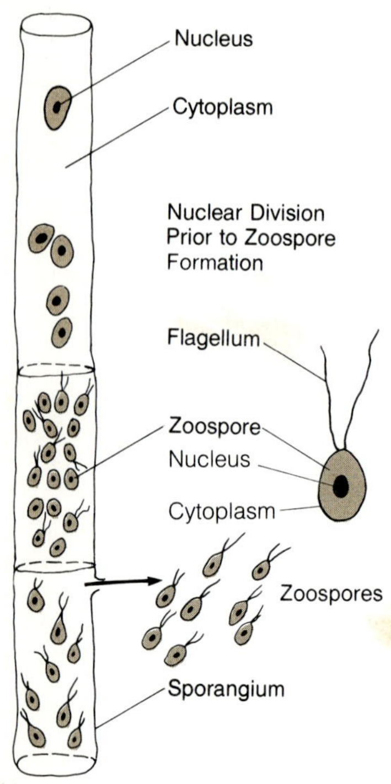

Figure 13–1. Zoospore formation. Diagrammatic representation of zoospore formation in a filamentous alga. Although this drawing shows the different stages of zoospore development on the same filament, usually all cells of a multicellular filament become sporangia and release their zoospores synchronously.

touch the soil; kalenchoe develops plantlets along the notched leaf margin (Figure 13–3); many succulents such as jade plant have leaves that produce new plants if the leaves are knocked off the plant; and piggyback plant produces plantlets at the base of each leaf blade. Another more restricted form of asexual reproduction is through the formation of stolons that form plantlets at their ends, as is the case with strawberries or spider plants.

Sexual Reproduction

Fertilization, the fusion between two **gametes** (sex cells) is a characteristic feature of sexual reproduction and results in the formation of a one-celled **zygote.** Usually gametes are produced in a specialized structure called a **gametangium.** A gametangium in which male gametes are produced is an **antheridium,** and one in which female gametes are produced can be either an **oogonium** or **archegonium,** the former being a unicellular structure with one or more eggs present, while the latter is a multicellular structure with only one egg present.

There are two methods of sexual reproduction recognized among organisms discussed in this book—**isogamy** and **heterogamy.** Those that reproduce by isogamy have morphologically similar gametes that cannot be distinguished as male or female. Such gametes are called **isogametes** (iso; "same") and are common among many algae (Chapter 17). Organisms that have a heterogamous type of reproduction produce gametes that are different in size and are called **heterogametes** (hetero, "different"). In the more primitive form of heterogamy, **anisogamy,** the gametes are similar in shape and

Figure 13–2. Asexual reproduction. A stolon, such as a strawberry, produces new plants by asexual means at successive nodes.

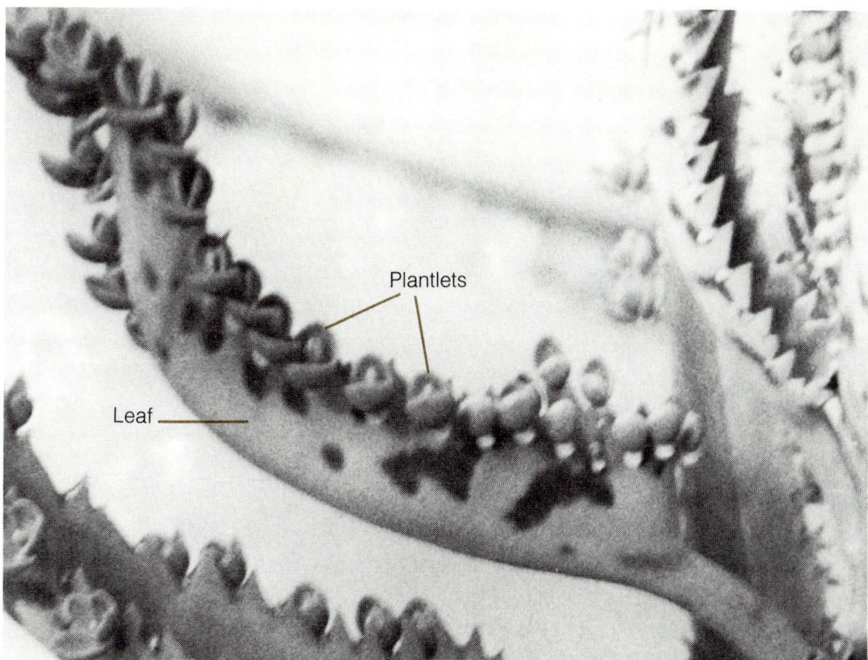

Figure 13–3. Asexual reproduction. *Kalenchoe* is an example of a plant that reproduces asexually by the production of plantlets along the leaf margin.

differ only in size. Traditionally the larger gamete has been referred to as the female gamete by analogy with the egg of more advanced organisms. **Oogamy,** the method of sexual reproduction exhibited by some fungi, many advanced algal species, and all of the bryophytes* and vascular plants*, is also a form of heterogamy. Oogamous species produce a nonmotile **egg** different in shape and larger than the **sperm,** which may or may not be motile.

Bryophyte
A group of primitive terrestrial plants, typified by mosses and liverworts.

Vascular plant
A plant that has specialized tissues (xylem and phloem) for conducting water and food.

Of the three types of sexual reproduction (Figure 13–4), the oogamous type is considered to be evolutionarily more advanced than isogamy or anisogamy. In oogamous plants, the egg is usually retained within the female gametangium and the sperm swims to the egg or is transported to the egg by other means. The fusion of male and female gametes is called **fertilization.**

Meiosis

It is recognized that fertilization is a basic requirement for the completion of sexual reproduction. However, an equally important process called **meiosis** is required, in addition to fertilization, to complete the life cycle of a sexually reproducing organism. Meiosis is a type of nuclear division during which the chromosome number is reduced to one-half the original number. Meiotic division was first observed in 1883 by Eduard van Beneden, a Belgian cytologist, in the horseworm *Ascaris.* Since then it has been recognized

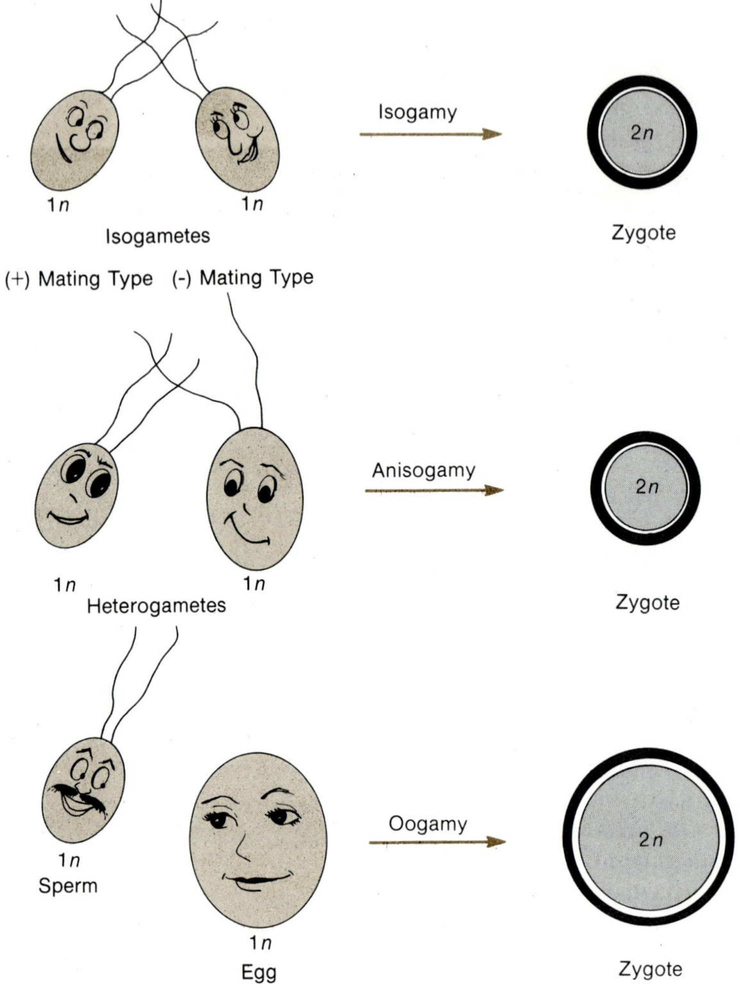

Figure 13–4. Sexual reproduction. The three methods of sexual reproduction among plants shown as a cartoon.

in organisms of all types except prokaryotes, as a process in which the chromosome number is reduced in sexually reproducing organisms. The term *maiosis* first proposed in 1905 to describe this process was subsequently changed to *meiosis*. Without meiosis there would be a doubling of the chromosome number every successive generation. It is well known that each species maintains a fixed chromosome number from generation to generation, yet a zygote always contains two sets* of chromosomes—one set received from one gamete and the second set from the other—resulting in the **diploid condition,** often called the **2n condition.** The gametes that fuse to form a zygote must therefore carry only one-half that number of chromosomes in order for the diploid number to remain constant. This chromosome number of gametes represents the **haploid condition** or the **1n condition.**

Set
In this context a set refers to a group consisting of one chromosome each from all homologous pairs.

In the life cycles of sexually reproducing plants there are both haploid and diploid cells, although the relative importance of each in the life cycle varies from one plant group to another. In vascular plants the **somatic cells—**cells that do not take part in sexual reproduction—are diploid and have two sets of chromosomes in their nuclei. For example, each living cell in the stems, leaves, and roots of a corn plant has 20 chromosomes. Ten of these chromosomes are derived from those that originally came from the male parent via the sperm nucleus and the other ten came from the female parent via the egg nucleus. In other words, there are ten pairs of chromosomes in each somatic cell. The two members of each pair of chromosomes are physically similar and each carries genes that govern the same genetic characteristics at the same location (locus) on the chromosome (Figure 13–5). Such pairs are called **homologous chromosomes;** thus corn has ten pairs of homologous chromosomes. When this corn plant produces gametes, each gamete will have only one set of ten chromosomes—half the original number. In other words, each gamete will have one member of each pair of homologous chromosomes present in the diploid cell. The reduction of the diploid number to the haploid number takes place through the process of meiosis at sometime prior to the production of gametes. The specific time at which meiosis occurs is characteristic of the organism under consideration and should be determined by reference to the life cycles in the appropriate chapters.

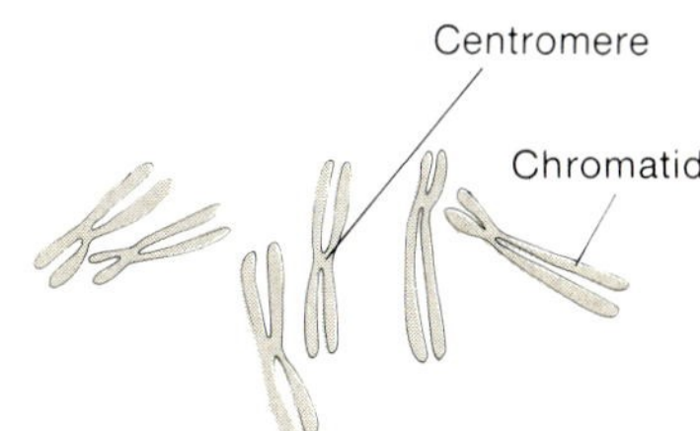

Figure 13–5. Three pairs of homologous chromosomes. Each pair is similar in shape, length, and the position of the centromere. The members of each pair originate from male and female parents.

Meiosis consists of two separate divisions—the first one called **reductional division** and the second one referred to as **equational division.** However, DNA replication takes place only once, prior to the reductional division. The stages of reductional division are prophase I, metaphase I, anaphase I, and telophase I. The stages of equational division are prophase II, metaphase II, anaphase II, and telophase II. At the end of the reductional division there typically will be cytokinesis* but this may be delayed until the end of the equational division. Similarly, the nuclei may or may not return to interphase* between the reductional and equational divisions. Usually the reductional division and equational division take place without any delay in between.

In the discussion of mitosis (Chapter 4) it was pointed out that during the interphase preceding nuclear division the DNA molecules are duplicated so that each chromosome comes to consist of two chromatids* joined at the centromere* (Figure 13–5). Similarly, chromosomes enter meiosis in the duplicated condition, even though the chromatids will not separate during the reductional division.

Figures 13–6 and 13–7 show the major steps of meiosis. During prophase I the chromatin* strands, which are usually long and intertwined, condense and produce visible chromosomes; in corn, for example, there would be 20 chromosomes. As prophase I progresses homologous chromosomes begin to associate with each other forming ten pairs of chromosomes. This pairing is called **synapsis,** the significance of which is discussed in a later section of this chapter. Later in prophase I the homologous chromosomes begin to unpair but will not become completely free. By the end of prophase I the nucleolus and nuclear envelope will have disappeared and the partially paired chromosomes moved to the equator of the cell. During metaphase I, these partially paired homologous chromosomes are found at the equatorial region of the cell, while spindle fibers* form and attach to the centromere region of each chromosome. The orientation of the paired chromosomes at the equator will be random, so that all chromosomes originally derived from a given parent will not be on the same side of the equator. The subsequent

Cytokinesis
The division of the cytoplasm that occurs during cell division.

Interphase
The phase of the cell cycle between nuclear divisions.

Chromatid
One of the two longitudinal halves of a chromosome.

Centromere
A constricted region of the chromosome that joins the two chromatids and to which spindle fibres attach.

Chromatin
Fine, threadlike genetic material composed of DNA and protein from which chromosomes are formed.

Spindle fibers
Threadlike structures composed of microtubules that originate from polar regions of a dividing cell.

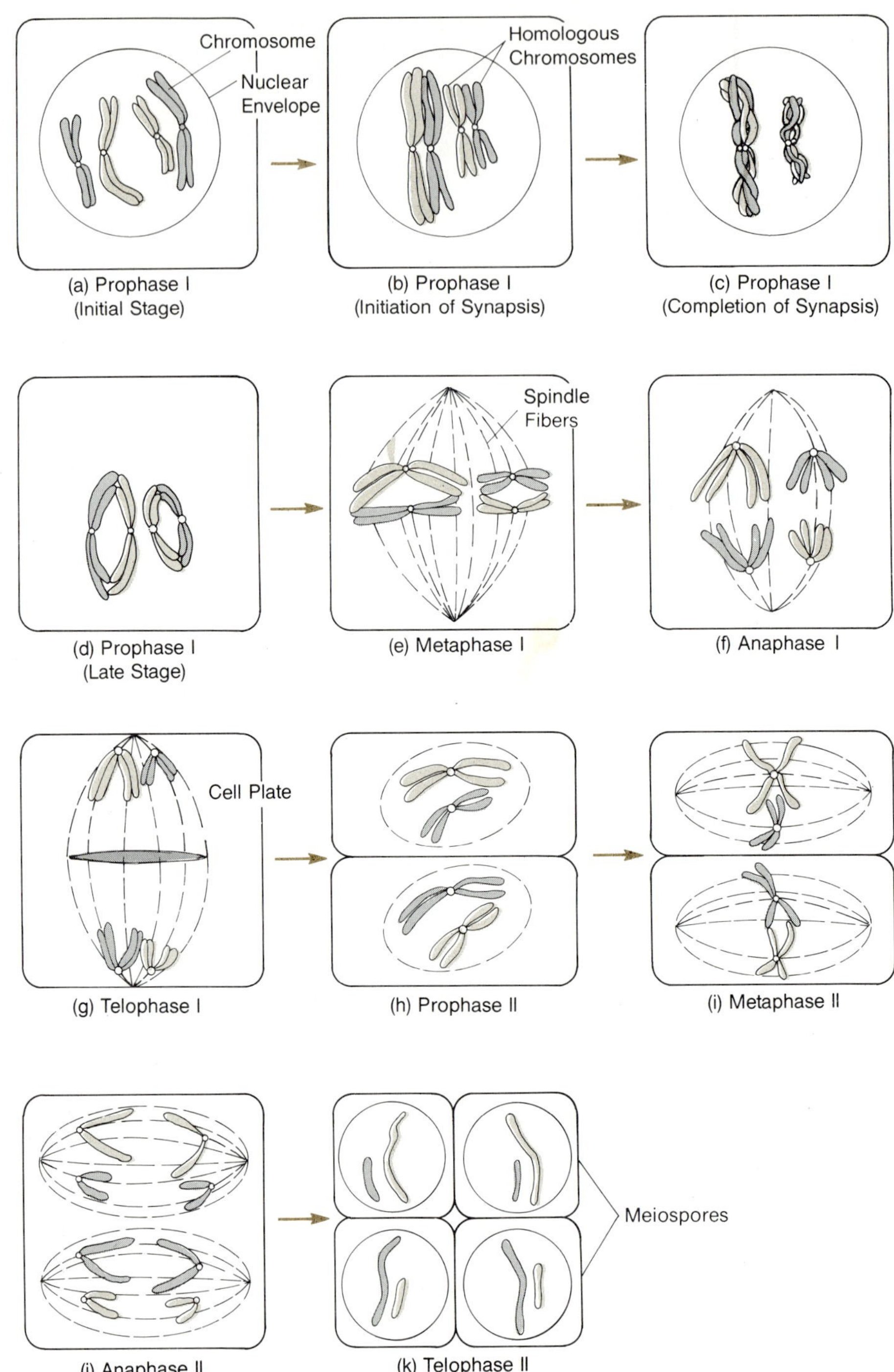

Figure 13–6. Stages of meiosis. Diagrammatic scheme of meiosis in a diploid cell with four chromosomes (two homologous pairs). Chromosomes of one color are from one parent and the other color from the other parent. Meiotic steps designated with Roman numeral I (*a–g*) are stages of reductional division and those designated with II (*h–k*) are stages of equational division. The planes of reductional and equational divisions are at right angles to each other. Note that the resulting meiospores are of two genetic types because of the random distribution of the homologous pairs during metaphase I. This shows that chromosomes originating from the same parent normally segregate randomly to different meiospores during reductional division. Crossing over, which normally occurs during prophase I, is described in Figure 13–8.

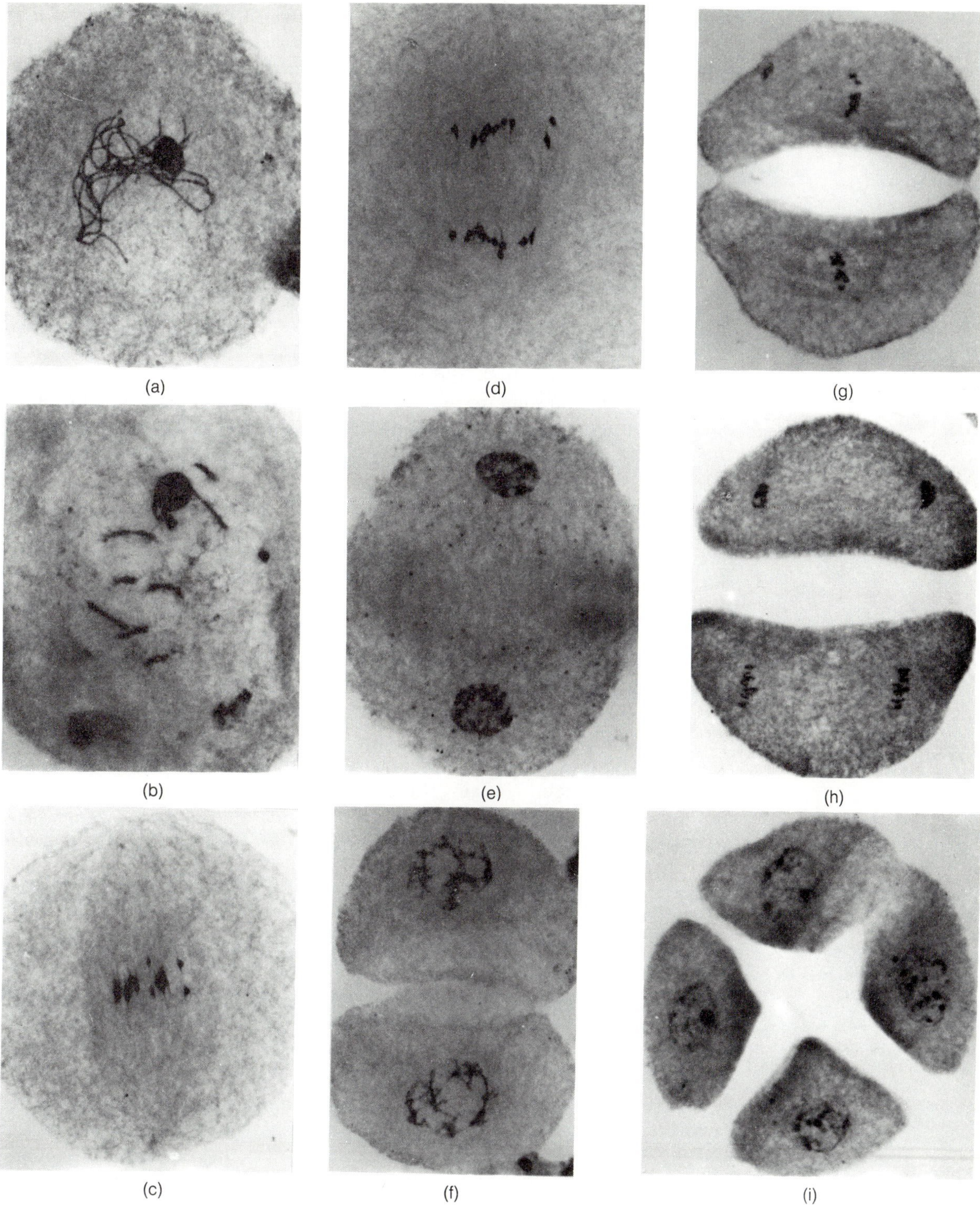

Figure 13–7. Meiosis. These photomicrographs show the different steps in sequence of meiosis in corn (*Zea mays*), which has a diploid chromosome number of 20. (*a*) Early prophase I. (*b*) Pairing of homologous chromosomes during prophase I. (*c*) Metaphase I. (*d*) Anaphase I. (*e*) Telophase I. (*f*) Prophase II. (*g*) Metaphase II. (*h*) Anaphase II. (*i*) Telophase II. (Microscope slides of meiotic preparations were provided by Dr. Jeff Shadley)

separation of chromosomes to the two poles will result in a random distribution of homologous chromosomes into the newly forming nuclei or, in other words, each newly formed nucleus will receive some chromosomes that originally came from the male parent of that individual and some others that came from its female parent.

Anaphase I constitutes the separation of the homologous chromosomes and their movement to the opposite poles of the nucleus. Thus at the end of anaphase I there will be 10 chromosomes at each pole. During telophase I the individual chromosomes in these two groups of chromosomes lose their identity while a nuclear envelope forms around both groups of chromosomes, resulting in the formation of two haploid nuclei. At the same time, cytokinesis divides the cytoplasm producing two new haploid cells. This completes the reductional division of meiosis. In some plants other than corn telophase I may lead directly into prophase II with no cytokinesis or loss of chromosome identity. It should be noted that at the end of the reductional division the chromosome number in the newly formed nuclei is reduced to one-half, although each chromosome that is represented still consists of two chromatids.

Immediately following reductional division, equational division takes place simultaneously in both newly formed cells. For all practical purposes, equational division is similar to mitosis (Chapter 4). The interphase stage between the two divisions is usually very short or absent, with no duplication of DNA, since this occurred during interphase prior to reduction division.

During prophase II the 10 chromosomes in corn reappear and the nucleoli and nuclear envelopes disappear, after which the spindle fibers become visible in both newly formed haploid cells. During metaphase II the chromosomes are found at the equatorial region of the cell with attached spindle fibers. Anaphase II begins with the separation of the two chromatids of each chromosome from each other. The newly separated chromatids move to the opposite poles and anaphase II ends when they reach the poles. Since each chromatid contains the complete genetic information of a chromosome, each of the newly separated chromatids, therefore, is a functional chromosome. Furthermore, since there were 10 chromosomes at the start of this equational division, there are 10 chromosomes at both of the poles. Hence there is no reduction in the number of chromosomes during the equational division.

During telophase II these chromosomes lose their identity, nucleoli reappear, and finally nuclear envelopes are re-formed. While these changes are taking place, cytokinesis divides the cytoplasm in both cells, resulting in the formation of 4 cells, each with a haploid nucleus. Table 13–1 and Figure 13–8 describe the differences between mitosis and meiosis.

Starting from a diploid corn cell containing 20 chromosomes, meiosis results in 4 haploid cells, each with 10 chromosomes. In an animal these haploid cells would mature and funtion directly as gametes. Most algal, fungal, and higher-plant life cycles are not that simple (see below) and the haploid cells produced by meiosis function as spores. Therefore, they are called **meiospores** and the diploid cell that produces them is a **meiocyte,** or a **meiospore mother cell.** Spores are reproductive structures that serve to produce a new organism—if the spores are produced by mitosis they are involved in asexual reproduction (see above); if the spores are produced by meiosis they are part of the sexual reproduction process.

Table 13–1
Differences between Mitosis and Meiosis

Mitosis	Meiosis
May take place in any living tissue.	Takes place only in cells or specialized tissues associated with sexual reproduction.
Both diploid and haploid nuclei can undergo mitosis.	Only diploid nuclei can undergo meiosis.
Mitosis constitutes one division.	Meiosis constitutes two divisions—reductional division and equational division.
Two new cells are produced as a result of mitosis.	Four new cells are produced as a result of meiosis.
The new cells have the same number of chromosomes as the parent cell.	The new cells have only half the number of chromosomes as the parent cell.
Homologous chromosomes do not pair or synapse during mitosis.	Homologous chromosomes pair or synapse during prophase I of meiosis.
There is no exchange of DNA between chromosomes during mitosis.	There is exchange of DNA between homologous chromosomes during synapsis in prophase I.
Chromatids separate during anaphase.	Chromosomes separate during anaphase I and chromatids separate during anaphase II.
New cells resulting from mitosis are genetically identical to the parent cell.	New cells resulting from meiosis are genetically different from the parent cell as well as from each other.

Although this discussion of meiosis is based on what takes place in corn, it can be applied equally to meiosis among most other organisms, both primitive and advanced as variations in the process of meiosis are extremely uncommon. In general, meiosis is known to take place in diploid cells ranging from single-celled zygotes to highly specialized tissues. The timing of meiosis within a life cycle is the basis for variations in life cycles among plants and plantlike organisms.

Significance of Meiosis

One of the important aspects of meiosis is that a constant chromosome number is maintained in a sexually reproducing species. If sex cells were produced without meiosis occurring prior to their production, every new generation would have twice the number of chromosomes present in the

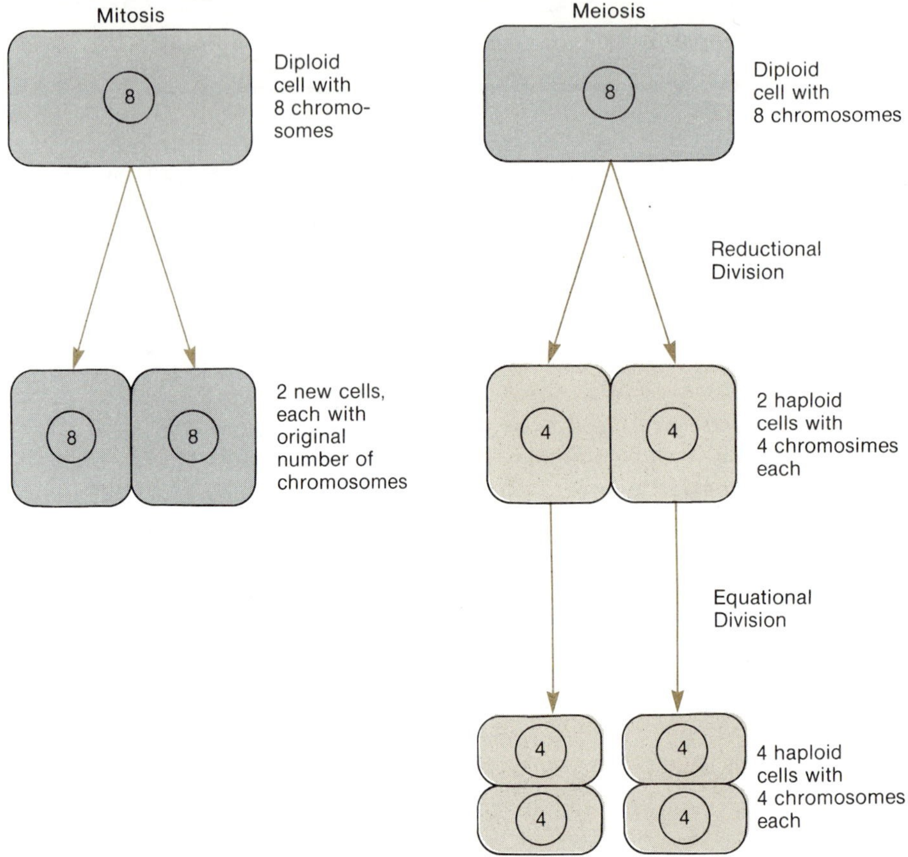

Figure 13–8. **A comparison of mitosis (left) and meiosis (right).**

parents. Another significant feature of meiosis is its ability to introduce genetic variation among the members of a given species.

One source of genetic variation introduced by meiosis is through the random distribution of homologous chromosomes into the newly formed nuclei during the reduction phase of meiosis. When the number of chromosomes is reduced from 20 to 10, as in corn, the 10 chromosomes in a newly formed cell need not all have been derived from the same parent. This makes it possible to have two genetically different types of meiospores produced from each diploid cell which undergoes meiosis. As indicated by the colored chromosomes in the diagram (Figure 13–6), the haploid cells have chromosomes of both colors instead of one color only. This is due to the fact that the orientation of paired homologous chromosomes at the equatorial region during prophase I is by chance and their subsequent separation results in chromosomes of different origin being present in the haploid cells. It is assumed that all chromosomes of the same color in the diagram originated from one parent and the others from the second parent.

A second source of genetic variation due to meiosis results from **crossing over.** Although chromatids appear thick, when highly magnified under a microscope they are in fact extremely thin and can easily get tangled, much like strands of thread. During the pairing of homologous chromosomes, or synapsis, in prophase I, the phenomenon of crossing over takes place, in which a chromatid of one chromosome becomes intertwined with a chro-

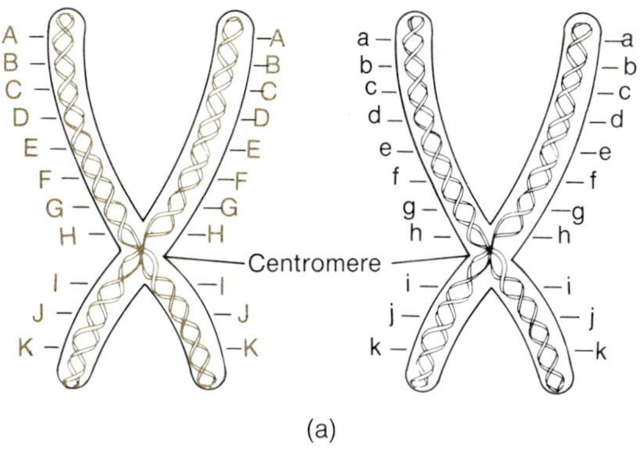

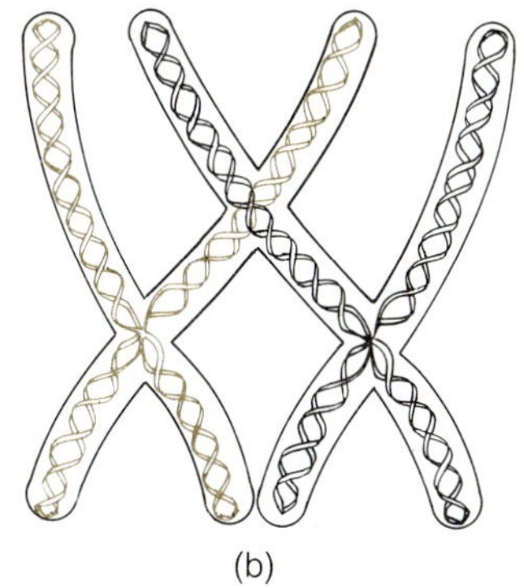

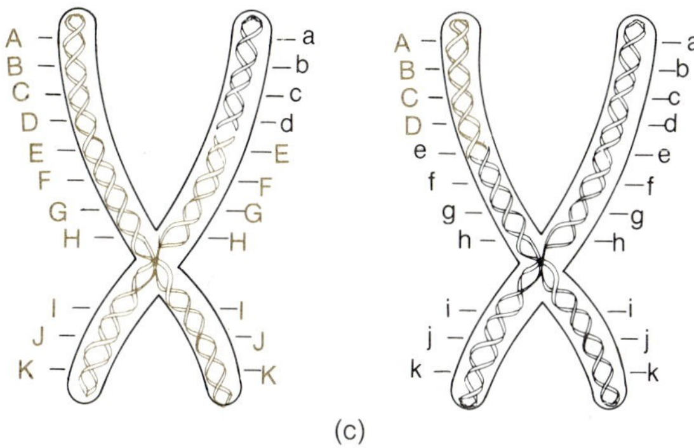

Figure 13–9. Crossing over. (*a*) Homologous pair of chromosomes, each carrying the same number of genes. (*b*) Crossing over of homologous chromosomes during synapsis in prophase I. (*c*) After separation of the homologous pair of chromosomes, segments of the chromatids have been exchanged. Note that four genes from each chromosome have been exchanged.

matid of the second chromosome of the pair (Figure 13–9). As these paired chromosomes separate, segments of the crossed-over chromatids break and are exchanged, resulting in the relocation of the chromatid segments and a corresponding reshuffling of genes between homologous pairs. As a result of this, each of the four chromatids of an homologous pair become genetically different and the resulting meiospores will also be of four different kinds in terms of their genetic makeup. Thus, through crossing over and the random distribution of homologous chromosomes during meiosis, genetic variation is introduced into progeny resulting from sexual reproduction. Variation, in turn, allows individuals within a population to adapt to changing environmental conditions, thus increasing the chances for the survival of the species.

Alternation of Generations

The lifecycle of plants and algae that reproduce sexually has two phases with respect to the chromosome condition and the function of the structures produced during the life cycle. All of the haploid structures represent one phase of the life cycle, the **gametophyte generation,** whose ultimate function is the production of gametes. The gametophyte can be an independent, free-living, haploid organism that eventually produces gametes or it can be as inconspicuous as a pollen grain, which is the male gametophyte of gymnosperms and angiosperms. Similarly, all of the diploid structures represent a second phase, the **sporophyte generation,** that produces the meiospores. The sporophyte can be an independent, free-living, diploid organism or simply a zygote.

The alternation between these two phases to complete a life cycle is described as the **alternation of generations.** The two processes that are responsible for this alternation are fertilization and meiosis, the former initiating the sporophyte generation and the latter initiating the gametophyte generation.

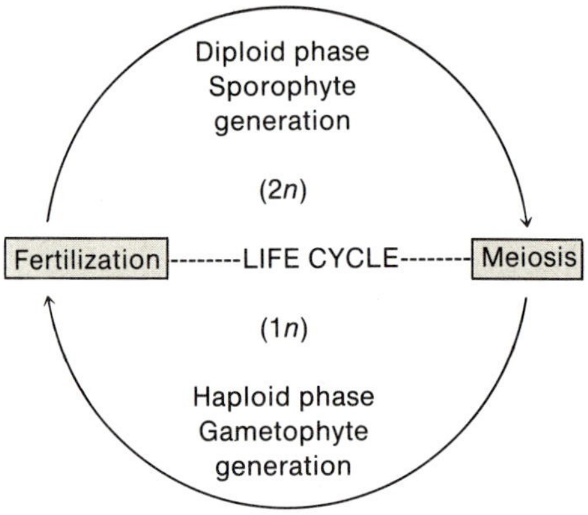

The prominence and duration of the two phases varies from one plant group to another. In general, the gametophyte generation is the prominent phase in the life cycle of most algae. Although the sporophyte of certain bryophytes achieves some conspicuousness, it never is free-living or dominant in the life cycle. On the other hand, the sporophyte generation is considerably more prominent than the gametophyte generation in the life cycle of the vascular plants.

Life cycles

Although many organisms are able to reproduce asexually and sexually, only the prokaryotic organisms, certain algae, and certain fungi use the asexual mode as the predominant means of reproduction. In addition, the asexual mode is the only known method of reproduction among some groups of fungi and algae, which generate large quantities of motile or nonmotile asexual spores to multiply their numbers. Asexual methods of reproduction are of lesser importance among most of the vascular plants and these plants do not produce any specialized asexual spores. An asexual cycle is completed in a relatively short time and may be repeated many times during the growing season. A significant characteristic of an asexual cycle is the absence of an alternation of generations because of the lack of fertilization and meiosis. Both haploid and diploid organisms are able to reproduce asexually as the following diagrams represent:

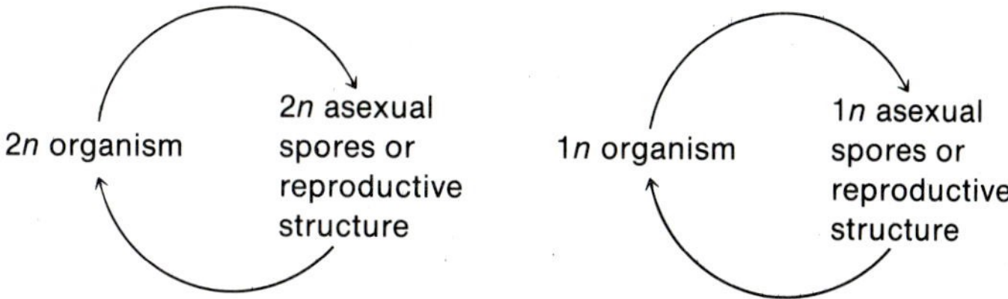

Organisms that reproduce sexually are characterized by an alternation of generations resulting from both fertilization and meiosis taking place during the life cycle. However, the occurrence of free-growing, independent haploid and diploid organisms within a life cycle is known only among certain groups of algae and primitive vascular plants. Three generalized types of life cycles are recognized among the sexually reproducing plants and plantlike organisms based on the prominence of either the gametophyte generation or the sporophyte generation or on the presence of two independent generations as represented by the following diagrams:

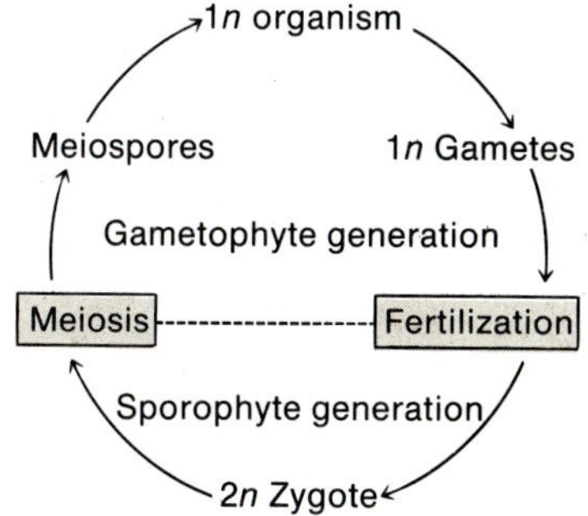

This life cycle is typical of most algae, where the gametophyte generation is dominant over the sporophyte generation, represented only by the zygote.

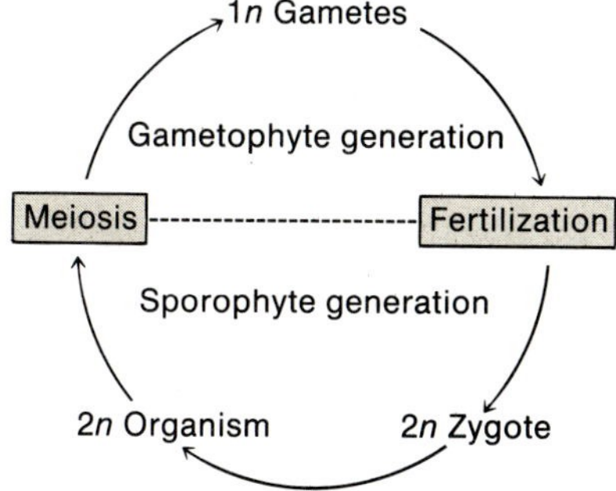

This life cycle is characteristic of animals, aquatic fungi, and certain brown algae such as rockweed *(Fucus)*, where meiosis results directly in the formation of gametes. The dominant phase is the sporophyte generation; the gametophyte generation is represented only by the gametes. This cycle does not occur in any of the plant groups.

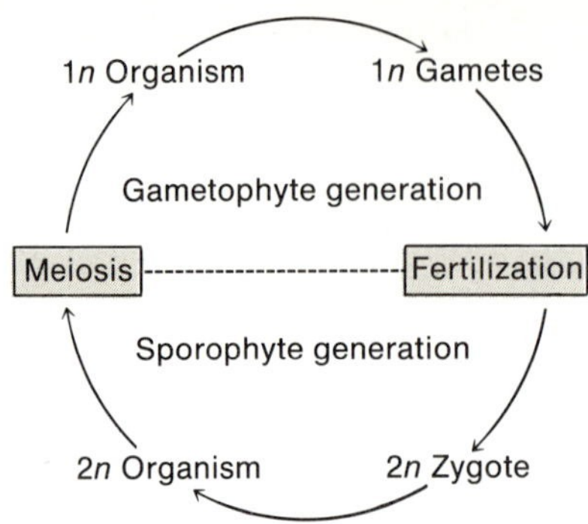

This basic cycle can be applied with modification to a number of different groups. Among certain algae *(Ulva, Ectocarpus, Cladophora)* the 1*n* and 2*n* organisms are morphologically similar and free-living. Among the bryophytes (mosses and liverworts) the 1*n* plant is free-living, with the 2*n* plant dependent on it. Among the non-seed-producing vascular plants (whisk ferns, club mosses, horsetails, true ferns) the 1*n* and 2*n* plants are free-living and morphologically different, the sporophyte phase being dominant. Finally, among the seed plants (gymnosperms and angiosperms) the dominant 2*n* phase is free-living, with the highly reduced 1*n* phase being dependent on it.

Summary

Reproduction is a primary characteristic of all living organisms and is intimately linked with the survival of the species. Asexual reproduction results in the production of genetically identical individuals, while sexual reproduction introduces genetic variation among the progeny, which facilitates adaptation of the species to a changing environment.

The most primitive form of asexual reproduction is binary fission. Regeneration of a broken piece of an algal or fungal filament is also a primitive form of asexual reproduction. Yet another form of asexual reproduction among algae and fungi is the production of motile or nonmotile spores in a sporangium. Examples of asexual reproduction among vascular plants include production of new plants from leaf margins or tips and the formation of runners and suckers.

The most primitive type of sexual reproduction is the fusion between two individual unicellular organisms. Male gametes are produced in an antheridium and the female gametes in an oogonium (unicellular) or archegonium (multicellular). Isogamy and heterogamy are the two methods of sexual reproduction among plants, algae, and fungi.

The fusion between gametes is called fertilization and results in a cell called the zygote that contains two sets of chromosomes—one set received from each parent via the gametes. Therefore, the zygote is said to be diploid (2*n*) and each of the gametes is haploid (1*n*). While fertilization results in the diploid condition, a process called meiosis initiates the haploid condition by reducing the number of chromosomes to one-half. Fertilization and meiosis are two processes that are essential for the completion of the life cycle of a sexually reproducing organism.

Meiosis takes place only in diploid cells. In a diploid nucleus, chromosomes are present in morphologically identical pairs called homologous chromosomes. During meiosis each member of a homologous pair is distributed to a newly forming nucleus. A corn plant has 20 chromosomes or 10 pairs of homologous chromosomes in its diploid cells. The eggs and sperm produced by this plant will have just 10 chromosomes each due to meiosis that occurs sometime prior to gamete production. The diploid number of 20 is restored by fertilization.

The reduction in chromosome number from diploid to haploid is achieved during reductional division. The homologous chromosomes become paired in a process called synapsis during prophase I and separate to the opposite poles during anaphase I, the two newly forming nuclei each receiving only half the original number. The equational division is similar to mitosis and thus no further reduction in chromosome number takes place. Since the orientation of the homologous pairs of chromosomes at the equatorial region during metaphase I is completely at random, the newly forming nuclei will receive a mixture of chromosomes, some originating from one parent and some from the other, a factor which is a source of genetic variation resulting from meiosis. A phenomenon called crossing over during synapsis is a second source of genetic variation, resulting from a relocation of chromatid segments and a corresponding rearrangement of genes between homologous pairs.

All of the haploid structures produced during the life cycle of a sexually reproducing plant belong to the gametophyte generation and all of the diploid structures to the sporophyte generation. The alternation between these two generations to complete a life cycle is called the alternation of generations. Three generalized types of life cycles are recognized among the sexually reproducing plants, based on the prominence of either the gametophyte phase or the sporophyte phase or on the presence of an independent gametophyte and an independent sporophyte plant. A significant characteristic of an asexual cycle is the absence of an alternation of generations because of the lack of fertilization and meiosis during asexual reproduction.

Review Questions

1. What are the basic differences between asexual reproduction and sexual reproduction?
2. What is the basic mechanism of asexual reproduction?
3. What is the basic mechanism of sexual reproduction?
4. What are the three types of sexual reproduction?
5. What are the advantages of sexual reproduction to an organism?
6. What are the major steps of meiosis?
7. How is meiosis different from mitosis?
8. What is the significance of meiosis to an organism?
9. Explain crossing over.
10. Explain alternation of generations.

Vegetative propagation

Sexual reproduction is essential for introducing variation into a plant population, variation that may permit a species to adapt to changing environmental conditions and so ensure its survival. However, from a commercial horticulturist's point of view, the primary concern is to produce a very large number of genetically identical plants in as short a time as possible. For this purpose, asexual reproduction—often termed **vegetative reproduction**—is the means favored by commercial growers for increasing the number of individuals. There are several common methods of vegetative propagation that can be easily practiced by anyone with access to clippings from a friend's plant and such simple supplies as a sharp knife, pot of damp sand or vermiculite*, and a small amount of rooting hormone, available at most garden shops. One can save a great deal of money and experience the pleasure of growing a plentiful supply of a wide variety of houseplants by utilizing one or more of the following methods:

Vermiculite
A soil substitute made from expanded mica.

Stem Cuttings

Stem cuttings are among the easiest ways of reproducing a large variety of house and garden plants such as philodendron, coleus, Swedish ivy, impatiens, peperomia, aluminum plant, purple passion vine, and many others. For best results, fill a plastic flowerpot with damp rooting medium (coarse sand is good, perlite* or vermiculite are better, and a commercial mixture of coarse sand and peat moss is best). Most cuttings can be rooted in plain water, but the root system that develops is not always as sturdy as it would be in an earth-based medium. Insert a stick or pencil into the medium to make a hole for the stem. Stem cuttings themselves should be between 3 to 6 inches long, taken from the tips of the plant. Include at least 2 nodes* and remove all but 3 or 4 of the top leaves, as well as any flowers that might be present. After moistening the cut end of the stem, dip it in the rooting hormone, then set it in the hole and firm the soil around it. If placing several cuttings in the same container, arrange them so they do not touch each other. The pot should then be placed where conditions are warm and bright, but never in direct sunlight. Keep the potting medium damp (not soggy!) at all times. An easy way of doing this is to make a miniature "greenhouse" by placing the pot containing the cuttings in a clear plastic bag, being sure that the bag does not touch any of the leaves. Using this method, it may not be necessary to water at all during the rooting period. Different species vary considerably in the time they take to form roots, coleus rooting in perhaps as short as a week while others like grape ivy typically require over a month. A hint that roots are forming is the production of new growth at the tip of the plant; when in doubt, carefully dig up the cutting and check on root size. Cuttings should be transplanted when the roots are about 1 inch long.

Perlite
A soil substitute made from expanded volcanic rock.

Node
The region of a stem to which a leaf is, or was, attached.

Leaf Cuttings

Several common houseplants such as African violets, gloxinias, and begonias are propagated in much the same manner as the above, using a leaf instead of a stem. With African violet, for example, a middle-aged leaf is removed from the plant, leaving approximately 1-½ inch of petiole extending from the blade. The cut end is inserted into the potting medium at an angle, covering about 1 inch of the petiole. New leaves will develop from the base of the cutting and when these are about one-third the size of the parent leaf, the new plant should be repotted.

Division

Undoubtedly the simplest method of vegetative reproduction, division is an effective way for propagating Boston fern, asparagus fern (actually not a fern at all, but a member of the asparagus family), or virtually any plant that sends up more than one stem. Ideally, division should be done in the early spring when the plants are beginning to produce new growth. With plants such as ferns, the method is first to cut off all foliage at soil level, then remove the plant from the pot and knock off as much soil as possible so you can see the roots. Cut the mass of roots into 2 or 3 approximately equal sections and repot into containers about 3 or 4 inches larger than the root mass. The repotted section should sit at the same level in the soil as it was before division. Firm the soil around the plant, water well, and place in a bright spot out of direct sunlight until the foliage is well developed.

Stolons

Species such as spider plants and strawberry begonias that produce "babies" at the end of trailing specialized stems called stolons or "runners" can be propagated simply by placing a pot of soil next to the parent plant and fastening the "baby" to the soil with a hairpin or paperclip. Water this plantlet like any other plant and when it appears to be well established, cut off the runner connecting it to the parent plant.

14

Classification

Outline

Introduction

To some people the mere mention of plant classification is sufficient to conjure up images of shrivelled old botanists in musty museum basements poring over their collections of dried plants. Plant classification today obviously does not enjoy the reputation of being a key aspect of botany. Yet all people classify plants—a walk in the woods or fields will bring questions such as "What kind of tree is that?" or "What is that green stuff on the rock?" Answers to these questions are examples of plant classification. For those wishing to live off the land, the most important aspect of plant classification is that which places plants in the edible or poisonous categories. For the landscape gardener, plants are classified on the basis of growth habit—tree, shrub, evergreen, deciduous, slow-growing, fast-growing, shade-tolerant, and so on. Each of these plant classification schemes is an attempt to place plants in categories that are useful to the individual doing the classification. These are valid schemes of plant classification that aid in organizing our knowledge about plants. Because there is considerable diversity in the way in which plants are used, as well as in the appearance of plants, there are many different plant classification systems. However, for botanists the only universally accepted scheme of plant classification is one that attempts to show genetic relationships among plants.

The earliest documented evidence for attempts at plant classification comes from *Historia Plantarum,* the oldest botanical writing known. In this volume Theophrastus (ca. 372–287 B.C.), a student of Aristotle, describes his classification of the plants known to him into four broad categories based on their size. This and subsequent attempts at plant classification over the next 2000 years resulted in a number of artificial systems where many unrelated plants were grouped together on the basis of a few morphological features such as size of plant, type of leaves, and number of stamens.* The classification of plants on the basis of how similar they appear might seem at first to be the most logical approach to classification. Such a system, however, does not show the relationships betwen plants. During the 18th century, botanists became more and more aware of the fact that there were indeed natural relationships among plants and attempts were made to classify plants on this basis. However, they were only partially successful in their attempts because of a lack of understanding of the basis for the relationships.

Stamen
The structure within the flower in which the pollen is produced.

In the *Origin of Species,* published in 1859, Charles Darwin explained his theory that all living organisms have evolved from ancestral forms and that a "family tree" can be derived for any species. Incorporation of this concept into classification schemes led to the development of various **natural systems** of classification, which take into account the evolutionary development of the various groups of plants, thus indicating their relationship by descent from common ancestors. The natural systems attempt to place closely rlated plants in a prescribed category called a **taxon*** and at the same time to show interrelationships among taxa with the ultimate aim of providing a complete picture of plant evolution. A natural system based solely on existing organ-

Taxon
(*pl.,* taxa) A general term to describe any unit of classification.

isms will contain informational gaps because so many ancestral species are now extinct. Incorporation of the fossil evidence as it is gathered can be used to clear up some of the uncertainty surrounding the true line of evolution, or **phylogeny,** of a given species. Present-day classification systems include both fossil and current organisms and are referred to as **phylogenetic systems.** Each classification system represents the efforts of one or more investigators to organize all of the available evidence into a coherent picture that comes the closest to describing the true phylogeny of the various types of organisms. As can be expected, this is a tremendous task that will remain incomplete until all evidence is obtained. In addition, the human bias introduced by the investigators in this field of study has resulted in the present situation of more than one classification system for plants, each based on the phylogenetic system. Since there is no single official classification scheme, different ones are used by different investigators, with the basis for selection being one's preference for the way in which the more controversial plant groups are classified.

The field of study dealing with plant classification is called **systematic botany** or **plant taxonomy.** Although in the past plant taxonomy was mainly a descriptive study, highly sophisticated biochemical analyses such as protein analysis and DNA hybridization studies*, along with computer analyses of significant characteristics, have given systematic botanists other tools with which to increase the accuracy of their classification systems.

DNA hybridization studies
Analysis of the extent to which isolated, single strands of DNA from different organisms will join together (hybridize) to form a double strand.

Kingdom Concept

Until the 1960s most biologists were satisfied to place living organisms into one of two kingdoms—the **plant kingdom,** if they either were autotrophic* or had cell walls or both, and the **animal kingdom,** if they either were heterotrophic* or lacked a cell wall or both. Organisms such as *Euglena,* which possess both plant and animal characteristics, were claimed as plants by botanists and as animals by zoologists. With an increased understanding of the structural and chemical nature of living organisms, it became evident that the two-kingdom system was not a sufficiently accurate representation of the phylogeny of organisms. Over the last twenty years there have been different attempts to assign organisms into three, five, seven, or even a larger number of kingdoms. These systems are by no means universally accepted and there are advantages and disadvantages to all of those proposed so far. However, a large number of biologists currently think that some form of a five-kingdom system comes closest to describing the true relationships among organisms. According to one form of this system, all known organisms other than the viruses can be classified under one of five kingdoms as described below:

Autotroph
An organism that can produce its own food from inorganic molecules.

Heterotroph
An organism that cannot produce its own food from inorganic molecules.

Monera: All organisms that have a prokaryotic* cellular organization—bacteria and cyanobacteria—are included in this kingdom.

Protista: Included in this kingdom is a diverse group of organisms that has been traditionally considered to belong to either the animal kingdom or

Prokaryote
(*adj.*, prokaryotic) An organism that lacks membrane-bounded organelles in the cell.

Eukaryote
(*adj.*, eukaryotic) An organism that has membrane-bounded organelles in the cell.

Mycologist
A botanist who specializes in the study of fungi.

Phycologist
A botanist who specializes in the study of algae.

the plant kingdom. It appears to be a "catch-all" kingdom. Most authors using the five-kingdom system classify all unicellular eukaryotic* heterotrophs and all eukaryotic algae under this kingdom.

Fungi: This kingdom includes all eukaryotic, cell wall-containing heterotrophs. In addition, the slime molds are also classified under this kingdom by many mycologists* though some would place them in the Kingdom Protista.

Animalia: All eukaryotic, multicellular heterotrophs lacking cell walls—the true animals—are included in this kingdom.

Plantae: Eukaryotic, multicellular autotrophs with cell walls—the true plants—are included in this kingdom. The placement of algae is still a very controversial issue. Although algae are often classified under the Kingdom Protista, many phycologists* still consider them as plants.

One of the drawbacks of this system is that the algae that have distinctively plantlike characteristics, such as a cellulose cell wall or photosynthetic pigments identical to those of true plants, are classified under the Kingdom Protista along with animal-like organisms. Therefore, many of the diverse organisms that are placed in the Kingdom Protista have no phylogenetic relationship to each other. Further research into the nature and phylogenetic relationships of organisms may help to place them in kingdoms that more closely reflect their true relationships. This may mean a rearrangement of organisms into different kingdoms and also a possible increase in the number of kingdoms beyond those presently in use.

Although only the organisms in the Kingdom Plantae presently are considered to be true plants, botany as a field of study developed under the two-kingdom classification system so that by tradition organisms such as the bacteria, cyanobacteria, algae, fungi, and slime molds are also included in botany texts. The different groups of organisms that are included in this book are presented in Table 14–1.

Units of Classification

The broadest unit of classification to which an organism belongs is the **kingdom.** All kingdoms except the animal kingdom are then subdivided into a number of **divisions.** The term *division* is of relatively new usage and is a taxon restricted to botanical classification, replacing the term *phylum,* which is still in use in zoological classification. The terms *division* and *phylum* are equivalent in status. The number of divisions included under a kingdom varies according to the particular classification system being used.

Each division is subdivided into **classes,** each class into **orders,** each order into **families,** each family into **genera,** and each genus into **species.** Each of these seven major taxa is usually divided into subunits that will not be discussed here. The classification of corn and bean as well as the characteristic endings for the major taxa names are presented in Table 14–2.

Table 14–1
Divisions of Plants and Plantlike Organisms Presented in this Text*

Kingdom	Division	Common Name	
Monera	1. Schizophyta	Bacteria	Prokaryotes
	2. Cyanophyta	Cyanobacteria	
Protista	3. Chlorophyta	Green Algae	Algae
	4. Euglenophyta	Euglenoids	
	5. Chrysophyta	Golden-brown Algae	
	6. Pyrrophyta	Dinoflagellates	
	7. Phaeophyta	Brown Algae	
	8. Rhodophyta	Red Algae	
Fungi	9. Eumycota	Fungi	
	10. Myxomycota	Slime molds	
Plantae	11. Bryophyta	Mosses and Liverworts	Bryophytes
	12. Psilophyta	Whisk ferns	
	13. Microphyllophyta	Club mosses	
	14. Arthrophyta	Horsetails	
	15. Pterophyta	Ferns	
	16. Cycadophyta	Cycads	Gymnosperms
	17. Ginkgophyta	Ginkgos	
	18. Gnetophyta	Gnetales	
	19. Coniferophyta	Conifers	
	20. Anthophyta	Flowering Plants	Angiosperms

*Divisions 1 through 11 are nonvascular organisms, while divisions 12 through 20 are vascular plants. Vascular plants are also known as Tracheophytes.

The Species Concept

The species is the basic unit of classification and probably the most important in terms of determining the identity of an organism and its relationship to others similar to it. The term *species* is used for both singular and plural forms. A species is defined as a group of morphologically similar individuals that resemble each other more than they resemble other groups. For species capable of sexual reproduction the species concept is usually extended to include only those individuals that interbreed, or could interbreed, freely in nature. In fact the ability to interbreed with other members of the same species is the basis most commonly used to determine if a particular organism belongs to the same species as the others. When members of a species are restricted from interbreeding, either by artificial means that are practiced by plant breeders or by natural barriers such as oceans, mountain ranges, or deserts, **varieties** result. The varieties created by plant breeders are often called **cultivars** (*culti*vated *var*ieties) and can be maintained only by

Table 14–2
Classification of Corn and Bean

Corn

Taxon	Old Name	Current Name
Kingdom	Planta	Plantae
Division	Spermatophyta	Antho*phyta*
Class	Angiospermae	Monocotyledon*ae*
Order	Graminales	Po*ales*
Family	Gramineae*	Po*aceae*
Genus	*Zea*	*Zea*
Species	*mays*	*mays*

Bean

Taxon	Old Name	Current Name
Kingdom	Planta	Plantae
Division	Spermatophyta	Antho*phyta*
Class	Angiospermae	Dicotyledon*ae*
Order	Rosales	Fab*ales*
Family	Leguminosae*	Fab*aceae*
Genus	*Phaseolus*	*Phaseolus*
Species	*vulgaris*	*vulgaris*

*Certain long-standing old names for families have been conserved and are used interchangeably with current family names in the botanical literature.
**The characteristic endings of taxa names are italicized.

Mutation
A change in a portion of a DNA molecule corresponding to a gene, which results in a change in the protein product of that gene.

Genetic recombination
The recombining of genetic information that occurs during sexual reproduction.

Speciation
The evolution of new species.

breeding practices. On the other hand, the natural varieties under proper conditions of extended reproductive isolation may evolve into new species (Chapter 23). Mutations*, genetic recombination*, and a variety of environmental factors are also responsible for the initiation of new species. Speciation* is an ongoing natural phenomenon and is the basis for the evolution of all organisms. There are approximately a half-million species of plants that are known and undoubtedly many more that have not yet been identified. The conclusion should not be drawn, however, that the lumping of organisms into one species represents absolute truth. Instead this represents the conclusions of the scientists who have studied the group extensively. Later new evidence may force the splitting of the species or its combining with another previously separate species. This kind of change goes on continuously as scientists seek the classification for an organism that most closely approximates its true relationship to other organisms.

There are different varieties of corn commercially grown in the midwestern states of Indiana, Illinois, and Iowa. These are produced by different seed companies, but are varieties of the same species—*Zea mays.* Similarly, there are over a thousand varieties of the common garden bean, *Phaseolus vulgaris,* grown in different regions of the world. Anyone who has browsed through a seed catalog trying to decide which type of bean seeds to order realizes that while "Topcrop," "Greensleeves," "Tenderpod," "Tendergreen," and so on, are all green beans, each one differs slightly in such characteristics as disease resistance, date of maturation, flavor, stringiness, and so forth. Nevertheless, these are varieties, not separate species, because if the distinct cultivars were not maintained by a careful plant-breeding program, they would readily interbreed.

Nomenclature

Every living organism that is identified is given a scientific name consisting of two latinized words, a generic and a specific epithet. This two-name system of nomenclature was popularized in the mid-18th century by the noted Swedish botanist and naturalist Carolus Linnaeus and is known as the **binomial system.** Presently this is the universally accepted system for naming all living organisms. Linnaeus used this system in his two-volume work, *Species Plantarum,* published in 1753, in which he described all plant species known at that time. Prior to the introduction of the binomial system, each plant species was described by a single name, followed by a description consisting of several words or a paragraph. The advantage of the binomial system, especially in scientific correspondence between individuals in different countries, is self-evident. It is also a rule of nomenclature to place the last name (or abbreviation of the last name) of the person describing and naming a new plant species at the end of the binomial. The initial *L.,* standing for Linnaeus, is seen after the names of many plant and animal species (e.g., *Zea mays* L., the scientific name for corn). In the case of a particular variety of a species such as bean, the varietal name is placed after the binomial with the authority for the variety indicated after the varietal name; for example, *Phaseolus vulgaris* L. var. *humulis* Alefeld is the bush bean.

In the binomial system the first letter of the generic name is always capitalized, whereas the first letter of the specific epithet is not capitalized. Latin words are used in the majority of scientific names. There are at least two reasons for the use of Latin in nomenclature—Latin was the language of scholars during Linnaeus' time, while continued use of Latin is due to the fact that meanings of Latin words do not change with time. Although Linnaeus' name was Carl von Linné, he Latinized it to Carolus Linnaeus, a practice common among scholars at that time.

Although the official name of a plant is the scientific name, common names are often used by most laypeople. However, common names for a particular plant may be different in different countries, as is the case of *Zea mays* L. This is commonly called corn in the United States and Canada, while in England and parts of Europe it is known as maize and in parts of India as sorghum. The term *corn* is used in England to refer to any grain, such as wheat, oats, or barley. The same plant species can also have many common names within the same country, as is the case for *Equisetum* in the United States where it is known as horsetail, scouring-rush, scouring-reed, Dutch-rush, and shave-grass. Common names can also be misleading as to the exact nature of a plant. For example, the term *moss* is used as an ending in many plant names such as peat moss, Spanish moss, club moss, reindeer moss, and

Lichen
An organism that results from the symbiotic association of certain algae with fungi.

so on. However, only one of these plants is a true moss—peat moss—whereas Spanish moss is a flowering plant, club moss is a nonflowering vascular plant, and reindeer moss is a lichen.*

Plant Collections

Plant taxonomists often take extensive field trips for the purpose of collecting plants. As a result of classifying these plants, it is occasionally necessary to name a species never before described. According to the rules of botanical nomenclature governed by the International Botanical Congress, the first description and names applied to a new species have priority over any subsequent names given the same species, even if the subsequent authors are unaware of the first author's work. For this reason, anyone collecting and naming a new species is required to preserve and maintain specimens of a newly found plant species for purposes of verification by other investigators. Each plant is dried and mounted on a sheet of paper of specific quality and size with information regarding the name of the plant, geographical location of collection, ecological characteristics of the location, the date of collection, and the name of the collector (Figure 14–1). Each preserved plant is known

Figure 14–1. Herbarium sheet. Dried, preserved plant specimens are mounted on 16.5″ × 11.5″ (42 cm × 29 cm) herbarium sheets for future reference or study. Information regarding the location of collection, name of plant, name of collector, and the date are recorded on the sheet.

as a **herbarium** specimen and the collection itself is called a herbarium. Many institutions of higher learning, research centers, and even individuals have herbarium collections.

Botanical Gardens

Botanical gardens (Figure 14–2), representing collections of living plants assembled for purposes of exhibition as well as study, have a history extending back to the groves and gardens planted around temples such as that at Karnak in Egypt nearly 3500 years ago. The Greek philosopher-naturalist Aristotle developed one of the earliest gardens devoted primarily to botanical research. His pupil Theophrastus conducted studies and made observations in these gardens which resulted in a nine-volume work, *Historia Plantarum,* which earned Theophrastus acclaim as the "Father of Botany."

After the collapse of Greek and Roman civilizations, botanical gardens were maintained primarily as collections of medicinal plants, but following the Renaissance they assumed an educational role once again. In Pisa, Italy, the Grand Duke de Medici in 1543 established one of the first botanical gardens open to the general public. During the following centuries there was a proliferation of botanical gardens in many of the countries of Western Europe, as well as in Japan, India, and eventually throughout the world. As European exploration of the little-known parts of the world accelerated, samples of exotic flora* were brought back and planted in these botanical

Flora
Plants; generally used as the opposite of fauna (animals).

Figure 14–2. Botanical garden. This photograph shows the Botanical Garden at Uppsala, Sweden, originally laid out by Carolus Linnaeus. (Photograph taken by Jan Hoh)

gardens where they were studied for their medicinal or commercial value. In a number of instances, tropical plants from one part of the world were brought to Europe and subsequently re-exported to other tropical regions. In this way quinine and rubber were introduced to southern and southeastern Asia from South America via England, while coffee, tea, cocoa, bananas, and others were similarly introduced to the New World from the Old World via European botanical gardens.

Today there are over 600 major botanical gardens and arboretums (gardens specializing in trees and shrubs) located world wide, almost half of them in the United States. The largest and most famous of all botanical gardens is the Royal Botanic Garden in Kew, England, just outside London. The second largest is the New York Botanical Garden. Excellent collections are also found at the Missouri Botanical Garden in St. Louis, the Brooklyn Botanic Garden, and Harvard University's Arnold Arboretum in Boston.

Modern botanical gardens include not only outdoor plantings but also plants inside greenhouses where temperature, light, and humidity can be regulated, thus allowing living plants from all over the world to be exhibited. In addition, many major botanical gardens also have extensive libraries of plant-related books and journals, as well as herbariums (preserved plant specimens) for research.

Summary

Because of the diversity of plants and the dependence of humans on plants for many of their necessities, attempts must have been made to catgorize plants throughout human history. Documented evidence for such attempts dates back to the 3rd century B.C. Theophrastus (372–287 B.C.) described his classification of plants in *Historia Plantarum,* the oldest botanical writing known. Classification systems until the 18th century were mostly artificial, based only on a few morphological features and showing no natural relationships. Presently the phylogenetic system attempts to classify plants according to their relationship by descent with the ultimate aim of understanding plant evolution. Unfortunately, this system will remain incomplete until all evidence is collected by taxonomists, those who study plant classification. Even then human bias may make the system somewhat subjective.

Our increased understanding of the nature of living organisms has made it difficult to place them clearly into either plant or animal kingdoms. A five-kingdom system currently in use places all prokaryotic organisms in Monera, all eukaryotic unicellular heterotrophs and algae in Protista, all eukaryotic cell-wall containing heterotrophs in fungi, all eukaryotic multicellular autotrophs with cell walls in Plantae, and all eukaryotic multicellular heterotrophs lacking cell walls in Animalia. However, in an introductory botany course organisms that either have cell walls during any phase of their life cycle or are able to synthesize their own food or both are often considered to be plants.

A unit of classification is the taxon. The most commonly used taxa for botanical classification from the broadest to the narrowest are: kingdom, division, class, order, family, genus, and species. The species is the basic taxon and consists of a group of morphologically similar organisms capable of interbreeding. Varieties result when members of a species are restricted from interbreeding by artificial means, as is often done by plant breeders. New species may result from extended reproductive isolation as the result of natural barriers.

The binomial system popularized by Carolus Linnaeus in the mid-18th century resulted in giving all living organisms a scientific name consisting of two words—the generic name and the specific epithet (e.g., *Zea mays* for corn). Common names are often used by laypersons but can be misleading because a species can have different common names or different species can be called by the same common name.

Newly identified plants are usually preserved, mounted, and deposited in a herbarium along with useful data related to their collection for verification by future investigators. Collections of living plants for exhibition and research are maintained in an outdoor setting in botanical gardens and arboretums.

Review Questions

1. What is the basis for the modern classification systems?
2. Why are organisms grouped under several different kingdoms rather than the original two-kingdom system?
3. What are the five kingdoms that are generally recognized today? What are their characteristics?
4. List the different taxa in their proper order.
5. Why is the species considered to be the basic taxon in a classification system?
6. What is the binomial system? What is its advantage?
7. Why is Carolus Linnaeus usually given credit for the binomial system?
8. What is the significance of maintaining a herbarium?

Carolus Linnaeus—The Father of Taxonomy

The study of living organisms during the 17th and 18th centuries focused primarily on the need to classify into some useful system the information about plants and animals, which was steadily accumulating as an outgrowth of both the Renaissance and the Age of Exploration. Numerous attempts had been made by various individuals to organize this information in a usable manner, but all were relatively cumbersome, requiring lengthy identifying descriptions. The publication in 1735 of the first edition of *Systema Naturae* by Carolus Linnaeus, based on a numerical method of classifying plants, introduced a far more convenient method of classification than any used previously and was rapidly and enthusiastically adopted by naturalists of the day.

Carolus Linnaeus was born near Uppsala, Sweden, in 1707 as Carl von Linné, the son of a country parson (Figure 14–3). As a child, he spent much time rambling through woods and fields with his father, observing plants and wildlife and acquiring a keen appreciation of nature. While a university student in his early twenties, Linnaeus carried out a study on plant reproductive structures and described plant sexuality. Recognizing that the stamens and pistils of flowers were not simply ornamental, Linnaeus correctly described them as male and female reproductive structures. These observations formed the basis for his first publication at the age of 23, titled *Preliminaries to the Nuptials of Plants.* These observations on plant sexuality became the foundation of Linnaeus' system of classification in which he divided plants into groups depending on the number of stamens in the flower (e.g., one stamen—Monandria; two stamens—Diandria; etc.). Subdivisions within each main group were then made on the basis of number of pistils: Monogynia, Digynia, and so on. Such a system would today be considered an artificial one, lumping together many plants that have no evolutionary relationship to each other. In Linnaeus' day, however, it had the great advantage of being simple and unambiguous; any newly discovered plant easily could be fitted into this framework. The system also had certain obvious shortcomings: organisms such as algae, fungi, mosses, and ferns—none of which produce flowers—could not be classified on this basis and were simply lumped together into a heterogeneous group which Linnaeus called Cryptogamia, or "secret marriers." Nevertheless, Linnaeus' method met with great success and within a few years he became famous throughout Europe.

Linnaeus spent three years studying and writing in Holland, where he earned a doctor of medicine degree and published several more books on classification. He then visited France and England and received much acclaim. He finally returned home to Sweden, married, and set up medical practice in Stockholm. He did not find the life of a physician entirely satisfying, and, several years later, he accepted an academic position at the University of Uppsala, first in physics and medicine and ultimately in botany, a position he held until his death. During his years at Uppsala, Linnaeus greatly upgraded the university's botanical garden, enlarging its collection

Figure 14–3. Carolus Linnaeus. Statue of Carolus Linnaeus as a young man, presently located in the Botanical Garden at Uppsala, Sweden. (Photograph taken by Jan Hoh)

from 200 to many thousands of specimens and rearranging the physical layout of the gardens to fit his classification scheme. He was a stimulating lecturer who attracted many students, whom he referred to as his "disciples." A number of them later made trips to many far-flung regions of the world to collect and identify new species of plants and animals.

Throughout his career, Linnaeus continued to publish refined and expanded editions of his *Systema Naturae,* revising his classification scheme to include not only genus and species designations but also the categories of class and order (the categories family and phylum were introduced by later taxonomists). Linnaeus ultimately recognized 24 classes of plants, describing in his *Species Plantarum* all the plants known at that time and applying the binomial system to each. This work, published in 1753, remains the starting point for modern botanical nomenclature. It was during this period that he Latinized his Swedish name Carl von Linné to Carolus Linnaeus.

Linnaeus died at the age of 71 in his home next to Uppsala's botanical garden. Today his house is maintained as a museum by the Swedish government and visitors to Uppsala can tour the gardens laid out just as Linnaeus planned them, each plant identified with the name given to it by the Father of Taxonomy.

15

Prokaryotes: Bacteria and Cyanobacteria

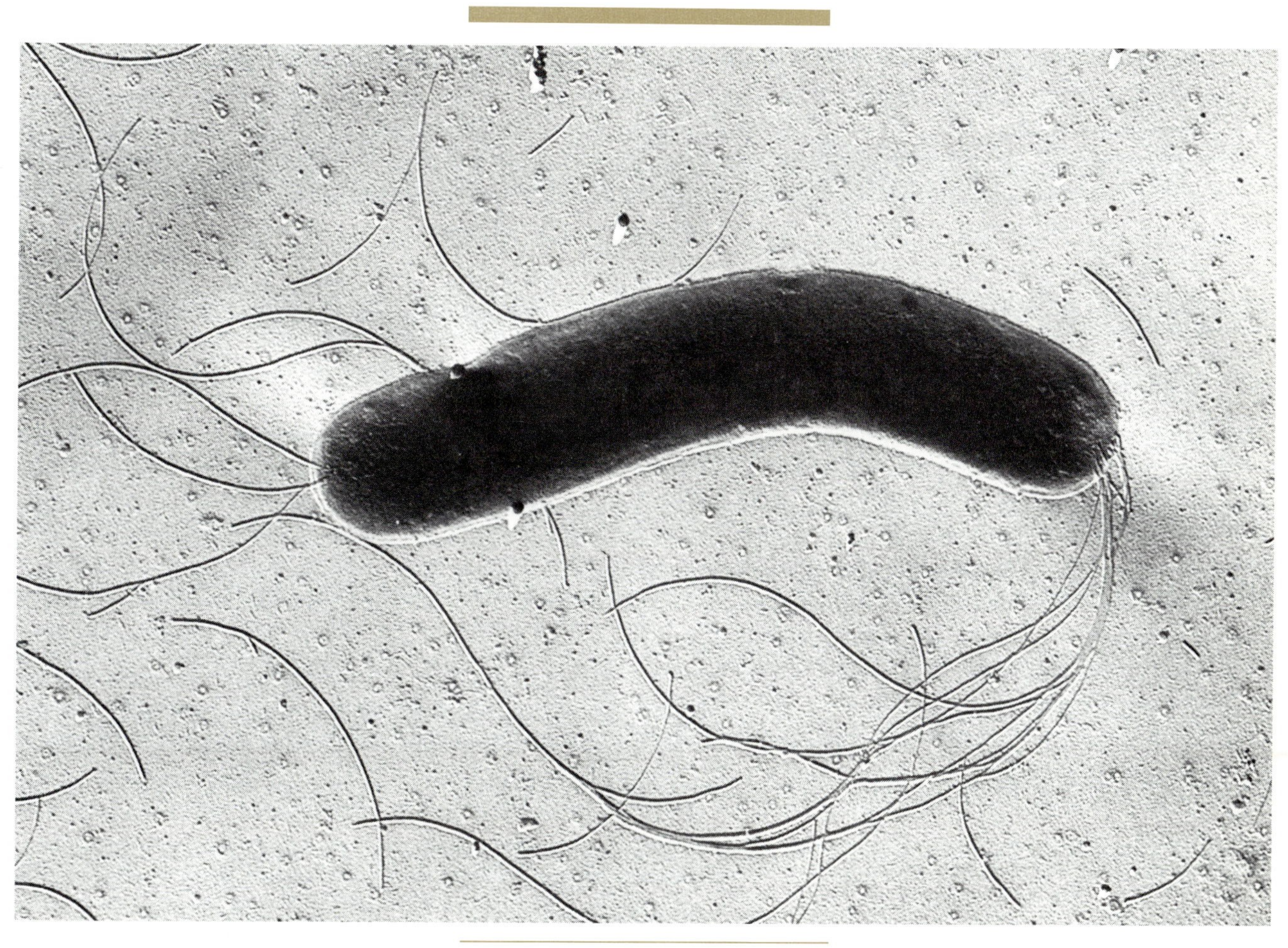

Outline

Biosphere
That portion of the earth's surface and atmosphere occupied by living organisms.

Organelle
A structure in the cell with a specific function, e.g., plastid or mitochondrion.

Nucleus
(*pl.,* nuclei) A spherical structure, found in cells, that contains the chromosomes and the majority of the genetic information.

Mitochondrion
(*pl.,* mitochondria) A membrane-bounded organelle that functions in respiration.

Plastid
A class of membrane-bounded organelles characterized by the ability to make starch.

Golgi
A cytoplasmic organelle composed of stacks of flattened, circular, membrane-bounded sacs.

Ribosome
A structure within the cell that is involved in protein synthesis.

Mitosis
The process of nuclear division that results in the production of two genetically identical daughter nuclei.

Introduction

Plague, cholera, typhoid fever, gangrene, leprosy, tetanus—these diseases were more feared by the population of Europe in historic times than were the wars and feuds of kings and princes. Bubonic plague, or "Black Death," wiped out whole villages, destroyed social organization, and in general had more influence on European history than did many of the wars that were fought so routinely. Typhoid fever and cholera were as likely to route an army as was the human enemy, while those wounded in battle died not from their injuries but from gangrene or tetanus. Yet all of these efficient "predators" on man, the most highly evolved animal, are organisms too tiny to be seen with the unaided eye—bacteria, one of the most primitive life forms. For the bacteria that cause these diseases, humans are the host to which they have become adapted. The specific disease symptoms are the response of the body to the damage caused by the growth of the bacteria. Although many bacteria are pathogenic to humans, the majority are benign or beneficial; in fact, there are bacteria that are indispensable to human life. Vitamin K must be supplied in the human diet because people are incapable of synthesizing this vitamin. However, vitamin K is not eaten directly but is a product of bacteria that live in the human intestine. Similarly, decay and decomposition are a particular benefit of bacterial growth and metabolism, essential to the recycling of minerals within the biosphere.* Life on earth without bacteria is impossible to imagine because these organisms play such a central role in most ecological cycles.

The bacteria and cyanobacteria (formerly known as blue-green algae) are probably the most primitive organisms living today. Most scientists agree that the present-day forms of bacteria and cyanobacteria are morphologically very similar to their ancestral forms, having maintained a structural identity to the first-formed cells for approximately 3 billion years (Chapter 23). Although any close phylogenetic relationship between bacteria and cyanobacteria is debatable, certain marked structural similarities suggest a possible common ancestor from which those two groups may have evolved along separate lines early during the biotic history of the earth. These common structural features and certain other characteristics are the bases for including these organisms under the Kingdom Monera.

Bacteria and cyanobacteria are **prokaryotes.** The designation prokaryote refers to the most primitive organization of cell structure in which membrane-bounded organelles*, such as nuclei*, mitochondria*, plastids*, Golgi*, and so on, that are normally found in a **eukaryote** cell are absent from the protoplast of a prokaryote. Furthermore, the ribosomes* of a prokaryotic cell are smaller than those of a eukaryote and the genetic information is contained in a single, circular chromosome, as opposed to the linear chromosomes of eukaryotes. Because there is only one chromosome per cell, no mechanism as elaborate as mitosis* has evolved to provide each daughter cell with a chromosome during cell division. In addition, both bacteria and cyanobacteria have cell walls different from those of all other plant and plantlike organisms. While the basic constituent of most plant cell walls is cellulose compound of glucose, the basic component of the cell walls of the prokaryotes is peptidoglycan, made of amino sugars and amino acids. True sexual reproduction is unknown among the prokaryotes, resulting in the absence of meiosis* and alternation of generations* from their life cycle. Reproduction among the bacteria and cyanobacteria is by asexual means such as **binary fission,** spore* formation, or fragmentation of the filaments. During binary fission (Figure 15–1) a constriction is formed that will separate the cell into two parts. Prior to the division, the single DNA* molecule is replicated so that each cell will have a copy of the hereditary material. Because of their reproduction by binary fission, the bacteria and cyanobacteria are sometimes known as "fission plants."

In addition to the prokaryotic nature of all bacteria and cyanobacteria, certain bacteria and cyanobacteria have other features in common. These include the ability to survive under high temperatures. Recent reports have indicated that there are bacterial species that not only survive, but also grow at temperatures over 250° C (482° F) and pressure of 265 atmospheres. In nature these bacteria are found in the water around deep sea vents in the Pacific Ocean. However, most of the thermophilic* bacteria and cyanobacteria are found in hot springs where the temperatures of 75° C (167° F) are close to the growth limit. Some bacteria and cyanobacteria also share the ability to utilize atmospheric nitrogen to synthesize nitrogen-containing molecules, a process described as **nitrogen fixation.** Although there is an abundance of nitrogen in the atmosphere and nitrogen is an essential component of many cellular molecules, only a small number of bacterial and cyanobacterial species are able to utilize gaseous nitrogen, whereas all other kinds of organisms require nitrogen in the form of nitrates, ammonium compounds, or organic nitrogen-containing compounds. The nitrogen-fixing bacteria and cyanobacteria play a significant role in making nitrogen-containing compounds available to other forms of life (see Highlight on nitrogen cycle in this chapter).

Meiosis
The process of nuclear division that results in the production of daughter nuclei, each containing one-half the original number of chromosomes.

Alternation of generations
The alternation between the diploid and haploid phase in the life cycle.

Spore
A unicellular reproductive structure other than a gamete.

DNA
Deoxyribonucleic acid, the genetic material of the cell.

Thermophile
(*adj.*, thermophilic) An organism that grows well at high temperatures; literally "heat-loving."

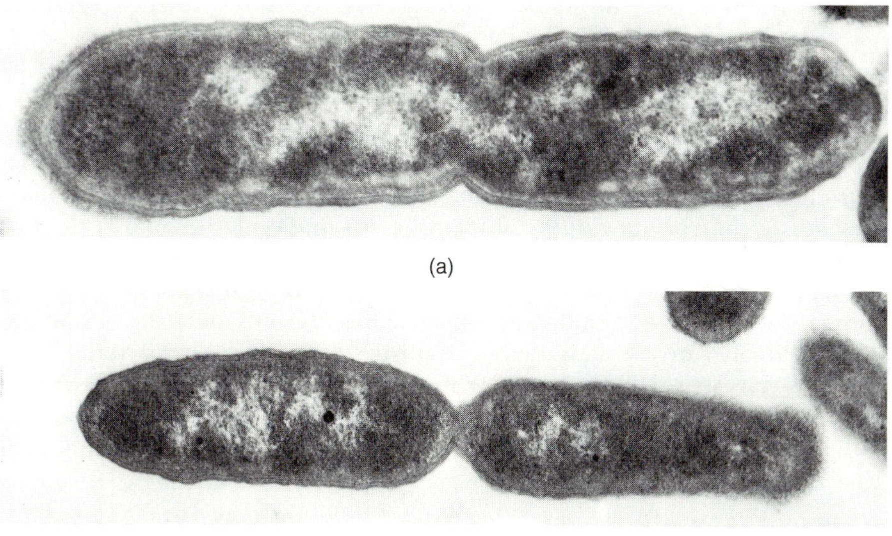

Figure 15–1. Binary fission. (*a*) Electron micrograph showing the initial stages of division in a bacterial section. (*b*) This electron micrograph shows a section of a bacterium in which the division is almost complete.

Although both bacteria and cyanobacteria are prokaryotes, there are significant differences between them to justify their classification under separate divisions, **Schizophyta** and **Cyanophyta,** respectively. The great majority of bacteria are **heterotrophs,** unable to synthesize their own food, while cyanobacteria are photosynthetic **autotrophs.** The small number of bacteria that are photosynthetic autotrophs have a chlorophyll pigment that is chemically different from chlorophyll *a,* the universal pigment present in all other photosynthetic organisms, including the cyanobacteria. The photosynthetic bacteria are also different from the cyanobacteria in that there is no oxygen production during bacterial photosynthesis. Finally, the cyanobacteria are larger and morphologically* more complex than the bacteria, as is described below.

Morphology
(*adv.,* morphologically) The study of the external appearance of organisms.

Division Schizophyta—The Bacteria

Bacteria, present everywhere, are the smallest organisms that can be seen with a light microscope, with dimensions ranging between 0.2 μm to 2 μm. The majority of bacteria are unicellular, having one of three morphological forms—the spherical types called **coccus,** the rod-shaped **bacillus,** and the spiral-shaped **spirillum** (Figure 15–2). A few species of soil bacteria called **actinomycetes** grow as branched filaments.

Many of the rod-shaped and spiral-shaped bacteria are propelled by whiplike structures called **flagella,** which differ from eukaryotic flagella in their anatomy (Figure 15–3). The number of flagella and their arrangement on the cell vary among species. In addition to flagella, structures called **pili,** short needlelike projections, are present on certain rod-shaped bacteria and play a role in the attachment of bacteria to a surface. Specialized pili are used during bacterial conjugation*, a process explained elsewhere in this chapter. Many of the rod-shaped bacteria produce **endospores*** when there is a shortage of nutrients. The endospores are extremely resistant to desiccation* and high temperatures; thus spore production is considered to be an adaptation for survival.

The individual bacterial cell consists of a **cell wall** surrounding the **protoplast.** Some species of bacteria also have a gelatinous sheath, the **capsule,** surrounding the cell wall. The outer boundary of the cytoplasm, the **plasmalemma,** controls the influx and efflux of molecules. In addition this membrane is active in aerobic respiration* because the enzymes* that are in the mitochondrial membrane of eukaryotic cells are in the bacterial plasmalemma. Although a large number of **ribosomes** are present in the cytoplasm, they are smaller in size than those of the eukaryotic cells. The bacterial chromosome consists solely of a circular molecule of double-stranded DNA and forms the **nuclear body** which, in thin sections of bacteria, can be recognized with an electron microscope as fine fibrils within the cytoplasm (Figure 15–4).

Different bacterial species produce a number of antibiotics*, such as streptomycin, chloramphenicol, tetracycline, bacitracin, and polymyxin-B,

Conjugation
Contact between two cells by means of a tubular projection during sexual reproduction.

Endospore
An internally formed, resistant spore produced by bacteria and cyanobacteria.

Desiccation
Drying out.

Respiration
The metabolic process that results in the release of chemical energy from simple organic molecules.

Enzyme
A protein that functions as a biological catalyst.

Antibiotic
A class of chemical substances produced by certain fungi and bacteria and capable of killing other microorganisms.

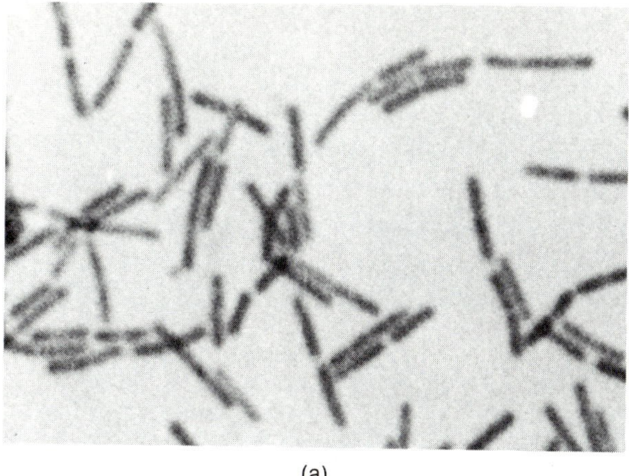

(a)

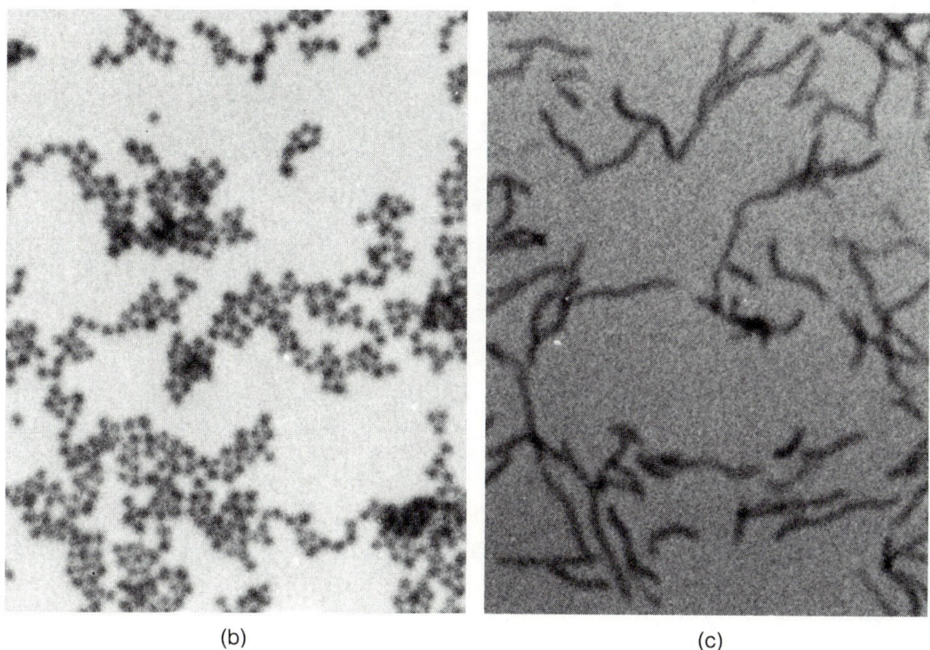

(b) (c)

Figure 15–2. Common morphological forms of bacteria. (*a*) Bacillus—rod-shaped. (*b*) Coccus—spherical. (*c*) Spirillum—spiral-shaped.

with a wide range of antibacterial properties. The first three are produced by filamentous soil bacteria of the genus *Streptomyces,* and are broad-spectrum antibiotics with considerable clinical application. However, bacitracin and polymyxin-B, produced by *Bacillus* species, are too toxic to be used internally and are therefore limited to topical (skin) applications.

A few bacterial species are **autotrophs** and synthesize their own food. Of these, some are **photosynthetic** and others **chemosynthetic.** The photosyn-

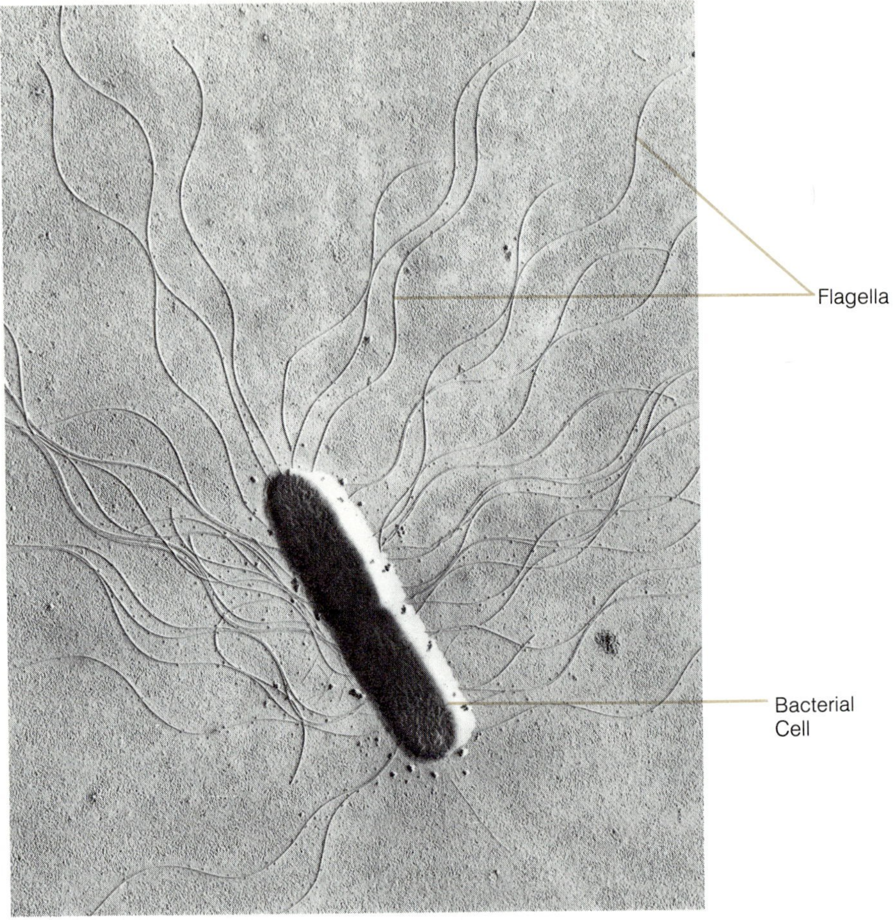

Figure 15–3. **Bacterial flagella.** Flagella are long, wavy, threadlike structures found on motile bacteria. The number and position of flagella vary depending on the bacterial species. Bacterial flagella are structurally different from the flagella of eukaryotic organisms.

thetic bacteria utilize light energy to synthesize food from atmospheric carbon dioxide, but unlike all other photosynthetic organisms they do not produce oxygen as a by-product of photosynthesis (Chapter 9). Furthermore, the photosynthetic bacteria are **anaerobes*.** The chemosynthetic bacteria, on the other hand, utilize chemical energy to synthesize their food from carbon dioxide. The recent discovery of chemosynthetic bacteria within the tissues of the tube worms of deep-sea hydrothermal vents strongly suggests the role of these autotrophic bacteria as the primary producers around the hydrothermal vents. They use hydrogen sulfide emitted from the vents as their source of energy for the incorporation of carbon dioxide into organic molecules.

Anaerobe
An organism that only can live in the absence of oxygen.

The majority of bacteria are **heterotrophs** and have to depend on an outside source for their food. These bacteria obtain energy either from nonliving organic substrates, such as dead and waste materials, or from other living organisms, including humans. The bacteria that obtain their food from dead and waste materials are called **saprophytes** and are responsible for bringing about decay and decomposition in nature, a phenomenon of great ecological significance. The **parasites**—those requiring a living source for food—cause plant diseases and varying degrees of discomfort or illness in

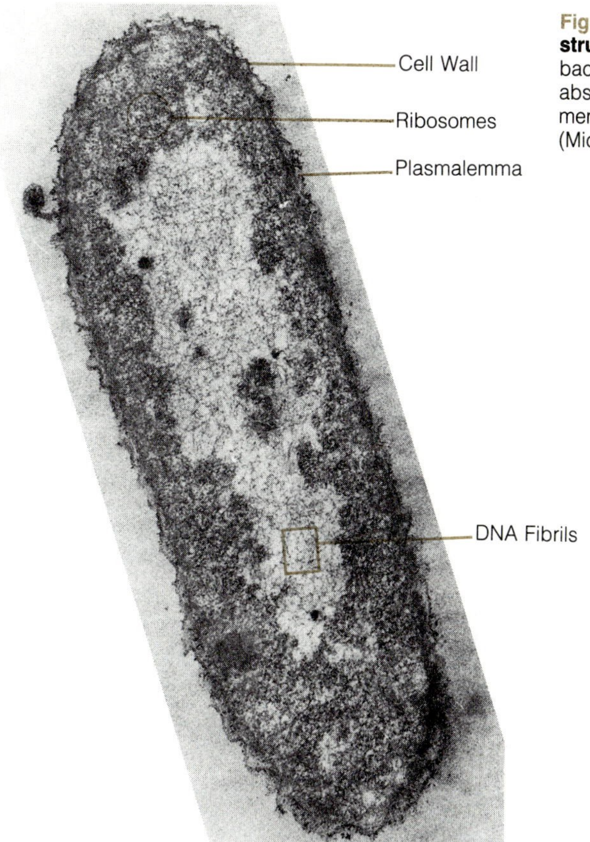

Figure 15–4. Bacterial cell structure. Electron micrograph of a bacterial section showing the absence of a nucleus and other membrane-bounded organelles. (Micrograph by Dr. Gütter)

animals and humans. Human diseases such as syphilis *(Treponema pallidum),* tetanus *(Clostridium tetani),* diphtheria *(Corynebacterium diphtheriae),* typhoid fever *(Salmonella typhi),* tuberculosis *(Mycobacterium tuberculosis),* leprosy *(Mycobacterium leprae),* cholera *(Vibrio cholerae),* and bubonic plague *(Yersinia pestils)* are caused by bacterial pathogens.* In addition, different bacterial pathogens also cause diseases such as galls, spots, rots, scabs, cankers, and wilts in many economically important crop plants, resulting in losses amounting to millions of dollars annually.

Pathogen
A disease-causing organism.

However, not all bacterial species that grow in a living host are pathogenic; rather, the symbiotic associations are mutually beneficial to both partners. The classic example of bacterial **symbiosis** is the production of **root nodules** in leguminous plants by species of *Rhizobium,* which convert atmospheric nitrogen to usable nitrogen compounds (see Highlight on nitrogen fixation). Another example is the production of vitamin K by the intestinal bacteria in humans.

Given proper growing conditions, most bacterial cells divide approximately every 20 minutes, leading to the production of huge numbers of bacteria from a single cell within a few hours. Because of this short generation time and the universal presence of bacteria, measures are usually taken either to kill them or to reduce their numbers in wounds as well as in our food and drinking water. The entry of even a relatively small number of pathogenic bacteria will quickly lead to a wound becoming infected. For this reason wounds are treated with an antiseptic solution that is really an antibacterial solution. Although simple boiling can kill most actively growing bacteria in food and water, only **sterilization** can kill bacterial spores. Sterilization is carried out by exposing food and utensils to steam under pressure, by chemical treatment, or by ultraviolet, gamma rays, and microwave irradiation. Bacterial numbers can be reduced by the process of **pasteurization,** during which food products are heated under controlled conditions to 63° C for 30 minutes.

Physiological
Pertaining to metabolic activity.

Because of the small size of their cells, morphological features are of limited use in the classification of bacteria. There is considerable variation among the bacteria in physiological* characteristics, however, and such differences form the basis of bacterial classification. All bacteria are first divided into two broad groups—Gram-positive and Gram-negative—based on their staining reaction to Gram's stain. Due to a difference in the cell wall composition of these two groups, the Gram-positive bacteria appear purple at the end of the multistep staining procedure, in contrast to the red color of Gram-negative organisms. Further classification is based on their physiological activity when the bacteria are grown on specific substrates. Bacteria, like fungi, can be cultured, under controlled environmental conditions in the laboratory, on a sterilized solid medium or in liquid broth of known chemical composition. Motility or lack of it, optimum pH and temperature for growth, ability to degrade specific molecules, production of gases, pigmentation, colony* morphology, and so on are the factors used in classification of bacteria.

Colony
An aggregation of organisms of the same species all derived from one cell or reproductive structure.

Division Cyanophyta—The Cyanobacteria

In earlier literature cyanobacteria were known as the **blue-green algae** and were classified along with all other algae. Because of their prokaryotic nature and other special features closely resembling bacteria, however, they are more correctly called cyanobacteria and have been placed in the Kingdom Monera in recent literature. As already pointed out in the introduction to this chapter, the cyanobacteria have a closer affinity to the bacteria than does any other group of organisms known today.

Cyanobacteria are primarily aquatic in habitat, although many inhabit a wide range of environmental niches. Both fresh-water and marine forms of cyanobacteria are known. Several species of cyanobacteria grow in hot springs where the water temperature ranges up to 73° C (163° F) and no other life forms can survive. Many such examples can be seen in Yellowstone National Park where the cyanobacteria impart a variety of colors to the hot springs. The terrestrial forms are usually found in soil, on rock, or on the exterior walls of buildings.

With the possible exception of a cyanobacterium species that lives in the roots of cycads*, all cyanobacteria are photosynthetic autotrophs containing several different photosynthetic pigments including chlorophyll *a,* xanthophyll, and carotene. The pigments most unique to the cyanobacteria are the water-soluble **c-phycocyanin** and **allophycocyanin,** blue pigments, and **c-phycoerythrin,** a red pigment. The difference in the amounts of these pigments present in different species is mainly responsible for the wide range of colors exhibited by cyanobacteria—black, blue-green, green, purple, and red. The Red Sea, a part of the Indian Ocean, owes its name to the periodic blooms* of a species of *Trichodesmium,* a cyanobacterium that appears red.

Cycad
A palmlike gymnosperm native to the tropics.

Most cyanobacteria are morphologically more complex than bacteria. The different morphological forms (Figure 15–5) present among the cyanobacteria include unicells, colonies, and unbranched and branched filaments, the latter being rare. The majority of cyanobacterial species are unbranched filaments. A few live as single cells or remain in colonies where the identity of individual cells is still maintained. A gelatinous sheath around the cell wall enables unicells as well as filaments to remain together in masses that often consist of thousands of individuals of a single species. Without exception, cyanobacteria lack flagella and other structures associated with motility and are considered to be nonmotile. A few species, however, demonstrate a type of gliding or oscillatory movement of unknown mechanism.

Bloom
An extremely high density of an algal species.

The individual cell of a cyanobacterium, whether from a single organism or as part of a filament, has the typical basic prokaryotic organization characterized by the absence of a nucleus and other membrane-bounded organelles. Many ribosomes are present in the cytoplasm. The most apparent structures of a cyanobacterial cell are the photosynthetic membranes known as **thylakoids** arranged along the periphery of the cytoplasm (Figure 15–6). The outer boundary of the cytoplasm is the plasmalemma. A double-stranded circular DNA molecule represents the single chromosome and can be recognized as intertwined fibrils in an electron micrograph* of a section through a cyanobacterial cell. The cell wall composition and the arrangement of wall layers are similar to those of a bacterial cell. Some species of cyanobacteria produce colorless cells called **heterocysts,** which occur along the filament (Figure 15–7). In some species the heterocysts are larger than the other cells on a filament. Available experimental evidence suggests that the heterocysts are the sites of nitrogen fixation. With our present understanding, it appears that in nature only nitrogen-fixing cyanobacteria produce heterocysts.

Electron micrograph
A picture taken through an electron microscope.

Reproduction among the cyanobacteria is by binary fission in the unicellular forms and by simple fragmentation in the filamentous forms. It has been observed that filaments with heterocysts may easily fragment adjacent to the heterocysts; the latter thus serve as points of breakage. A fragmented segment then grows to the original length by binary fission of all cells of the filamentous segment. A few species of cyanobacteria produce spores (Figure 15–8) called **akinetes,** which are larger and have thicker cell walls than the other cells of the filament. In addition, the akinetes are highly resistant to desiccation and other adverse environmental conditions and can survive as dormant spores for many years. Upon germination, the akinetes give rise to new filaments.

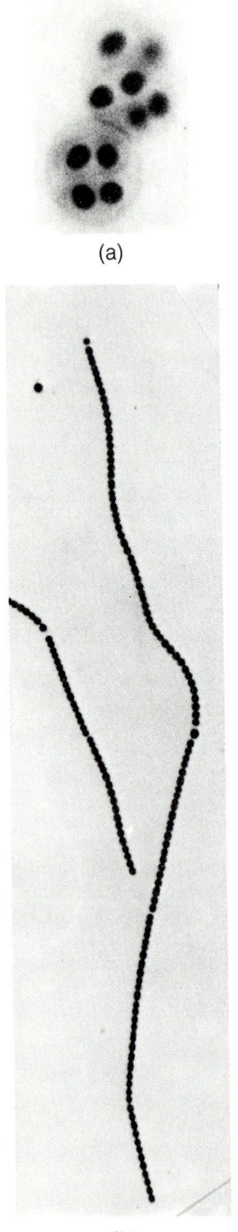

Figure 15–5. Morphological forms of cyanobacteria. (*a*) Unicellular form. Usually four cells stay together forming a colony. Unlike a true colonial form, these cells are not connected to each other. (*b*) A filament composed of cells shaped like beads.

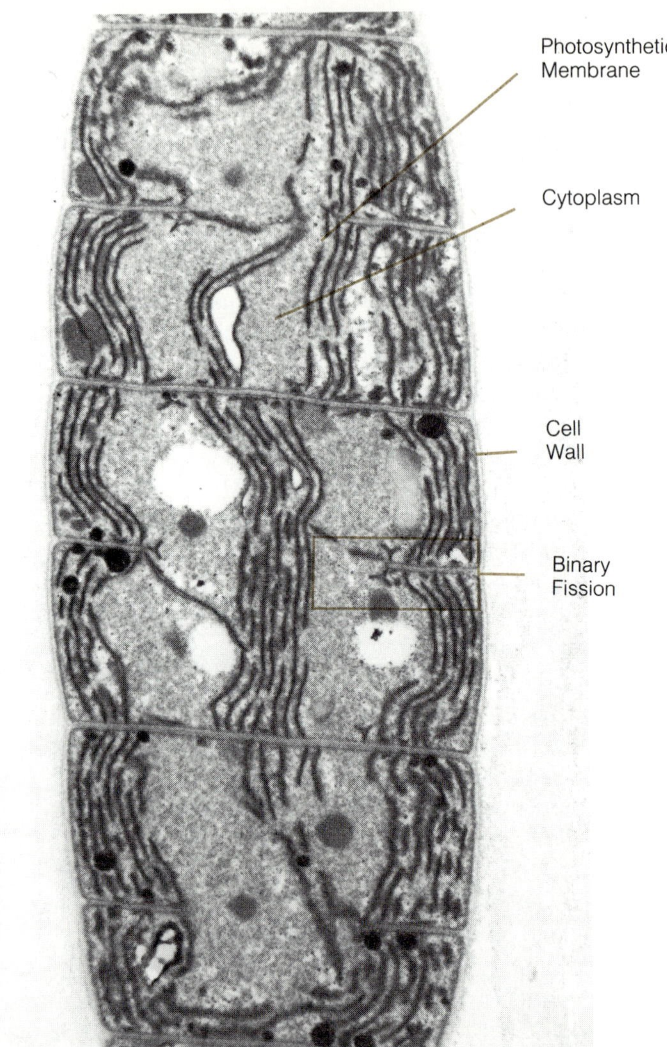

Figure 15–6. Cyanobacterial cell structure. Electron micrograph of a cyanobacterial section showing the typical prokaryotic cellular organization. The photosynthetic membranes are not enclosed by an envelope. New cells are produced by binary fission, which is visible in the micrograph.

Cyanobacteria are usually the first organisms to colonize a bare rock or soil that is devoid of all life—for example, filamentous cyanobacteria were the first organisms to colonize the sterile volcanic rocks on the island of Krakatoa in Indonesia after the volcanic eruption of 1883. The cyanobacteria, in addition to providing a rich organic substrate, contributed to the formation of soil by breaking down rock through the action of organic acids secreted by the cells. The combination of dead organic matter from decaying cells plus decomposed rock material formed a thin soil layer sufficient for the establishment of small sun-tolerant plants that continued the breakdown of rock to produce more soil. Many lichens are symbiotic associations between ascomycetous fungi (Chapter 16) and species of cyanobacteria and, like the cyanobacteria, lichens are capable of colonizing bare rock and contributing to the formation of soil and the subsequent establishment of bryophytes* and vascular plants.*

Detailed biochemical and ultrastructural* investigations of cyanobacteria and other photosynthetic organisms led to the realization that there were many similarities betwen cyanobacteria and chloroplasts*—similarities that did not exist between the chloroplast and the remainder of the eukaryotic protoplast (except for the mitochondrion). Chief among the similarities between chloroplasts and cyanobacteria are: the circular chromosome, the small ribosomes, and the factors controlling protein synthesis. Based on these observations, the **endosymbiotic hypothesis** has been proposed to explain the origin of the present-day higher-plant chloroplast. According to this theory, the chloroplast evolved from a unicellular cyanobacterial ancestor that was engulfed by a primitive unicellular heterotroph, establishing a subsequent symbiotic relationship. The cyanobacterial partner presumably lost its cell wall through evolution and became the chloroplast. Chapter 23 provides further details of this hypothesis.

In terms of economic importance, the cyanobacteria are not particularly significant. A species of *Nostoc* that grows in spherical masses is used as food in parts of China and South America because of its high protein and oil content. However, the use of cyanobacteria as food is very limited. Recently another filamentous species, *Spirulina platensis*, has been marketed in cap-

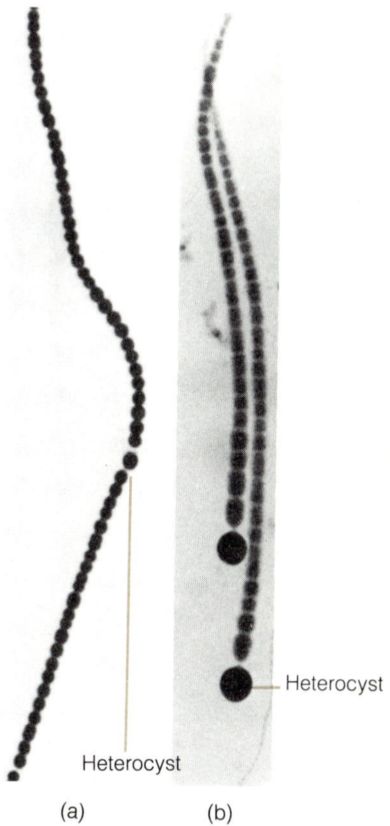

Figure 15–7. **Heterocyst.** Heterocysts are specialized cells in certain filamentous cyanobacteria. The position of the heterocyst varies and is a characteristic of the species. (*a*) *Nostoc* with heterocyst in the middle. (*b*) *Rivularia* with heterocyst at the base.

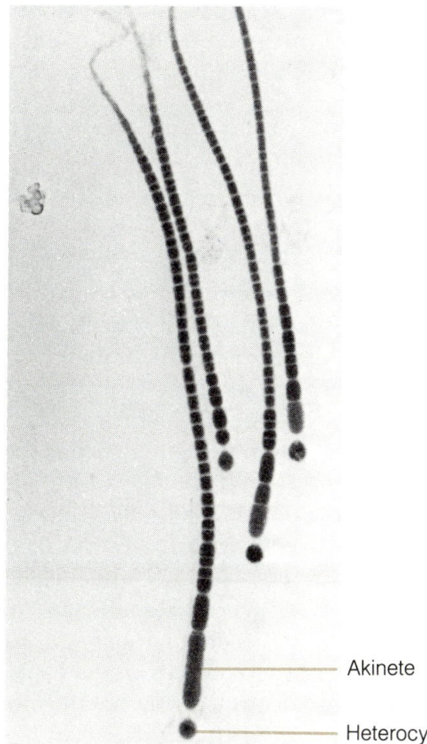

Figure 15–8. **Akinete.** *Gloeotrichia* with a large akinete next to the basal heterocyst. Akinetes are asexual spores produced in certain cyanobacteria and are resistant to adverse environmental conditions.

Bryophyte
A group of primitive terrestrial plants, typified by mosses and liverworts.

Vascular plant
A plant that has specialized tissues (xylem and phloem) for conducting water and food.

Ultrastructural
Pertaining to the fine structure of cells; only visible with the electron microscope.

Chloroplast
The type of plastid in which photosynthesis occurs.

sules in the United States and Canada as a source of vitamins and protein. The most important deliberate use of cyanobacteria, however, is in the paddy fields of Asia where the major source of nitrogen fertilizer consists of nitrogen-fixing cyanobacteria that grow in the water before the rice plants are well established.

Although most cyanobacteria are nonpathogenic, there are a number of reports where an alkaloid toxin produced by *Microcystis* and *Anabaena* species has been the cause of death of cattle and other domestic animals. These harmful effects are invariably associated with consuming large quantities of these organisms through the drinking water. In addition, reports from Australia have documented the death of large numbers of honey bees, killed by drinking water containing blooms of *Microcystis.* Often a species of *Microcystis* is also the culprit in causing taste and odor problems in many municipal water supplies.

Sexuality Among the Prokaryotes

True sexual reproduction involves the fusion of two sex cells and a resulting zygote. In addition, meiosis and a subsequent alternation of generations are also important characteristics of sexual reproduction in plants and many algae and fungi. All of those features are absent in the life cycle of prokaryotic organisms.

The most significant aspect of sexual reproduction is the genetic recombination that takes place as a result of the fusion of nuclei—one from each parent. Until about 40 years ago prokaryotes were not known to have any mechanism for genetic recombination. Between 1944 and 1952 different mechanisms for genetic recombination were discovered among the bacteria, none of which required fusion between cells and which, therefore, were unlike sexual reproduction among the eukarytic organisms. Because of the lack of cellular and nuclear fusion, zygotes are not produced by the prokaryotes.

Presently three general mechanisms for the transfer of genetic materials between bacteria are known—**conjugation, transformation,** and **transduction** (Figure 15–9). All three have certain basic similarities in that there is no cell fusion, only a small amount of the donor DNA is transferred to the recipient, and the donor DNA is incorporated into the recipient chromosome without actually increasing the size of the chromosome. The latter is achieved by the removal of a corresponding segment from the recipient DNA.

During conjugation (Figure 15–9a) in a population of a bacterial species, cells that are positive for the sex factor (F^+) produce short projections called sex pili. A cell that is negative for the sex factor (F^-) becomes attached to the sex pilus of an (F^+) bacterium, presenting a superficial resemblance to sexual reproduction. The sex factor is actually a small, circular segment of DNA called a **plasmid** that is found in the protoplast separate from the bacterial chromosome and that duplicates independently of the bacterial chromosome. During conjugation the sex factor plasmid from the F^+ bacterium is

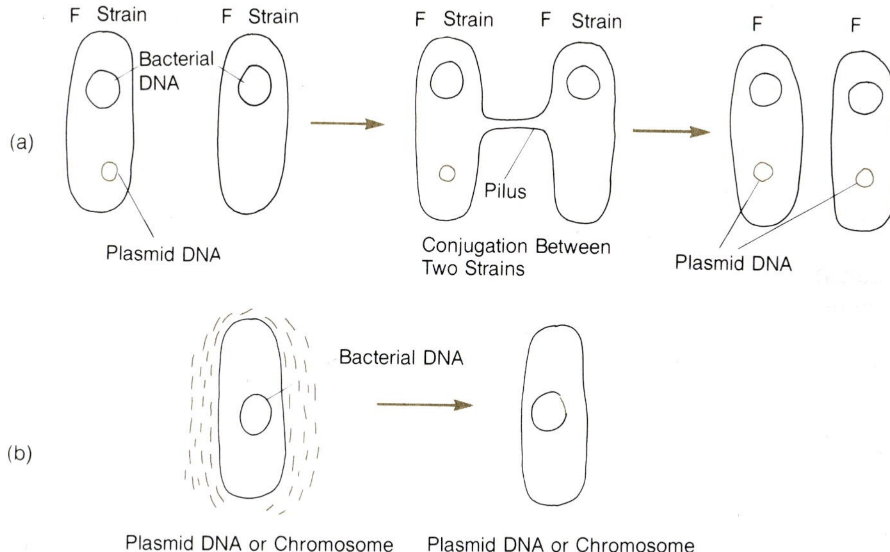

Figure 15–9. Bacterial sexuality. (*a*) Conjugation. During conjugation plasmid DNA is transferred through the sex pilus from a cell that contains a plasmid to one lacking the plasmid. Duplication of the plasmid DNA takes place prior to transfer. (*b*) Transformation. During transformation plasmid DNA or a chromosomal fragment in the environment is incorporated into the DNA of the bacterium.
(*c*) Transduction. The process of transduction results in the transfer of a DNA segment from one bacterial cell to another via bacterial viruses.

transferred to the F^- bacterium through the tubelike pilus, making both bacteria positive for the sex factor. Under certain conditions, a segment of the F^+ bacterial chromosome may also be transferred to the F^- bacterium.

Transformation (Figure 15–9b) is widespread among bacterial species. In this process free DNA is taken into the bacterium and incorporated into its genetic information. The source of this DNA is not another living cell as it would be in sexual reproduction. Instead, the DNA is in the form of either plasmids or chromosome fragments from bacterial cells that have died and broken up releasing their contents into the environment. Thus it is possible to transfer certain characteristics such as virulence* or resistance to antibiot-

Virulence
The degree to which an organism will cause disease.

ics if both donor bacteria and recipients are growing in the same culture medium or organism.

Replicated Copied.

Transduction (Figure 15–9c) is the mechanism of genetic exchange when a segment of DNA is transferred from one bacterium to another by bacteriophages, which are viruses that infect bacteria. When a phage infects a bacterium, the DNA of the phage is replicated* inside the host bacterium and new phage particles are formed. During this process segments of the host DNA can be incorporated inadvertently into the phage DNA that is then transferred into a new bacterial cell when it is infected by the phage.

Genetic recombination in the cyanobacteria was first reported only in 1968. Since then additional reports have supported the original findings that certain species of cyanobacteria have the ability for genetic recombination by transformation similar to that taking place in the bacteria—incorporation of free DNA taken up by the recipient from the culture medium. So far neither transduction nor conjugation has been observed among the cyanobacteria. The amount of research being conducted related to genetic recombination among cyanobacteria is considerably less than the amount for bacteria because of the difficulty in growing cyanobacteria in pure culture in the laboratory. Therefore, our understanding of sexuality among the cyanobacteria is limited.

The fact that there are mechanisms by which genetic recombinations can be achieved by the prokaryotes is of great survival value to these organisms. In the total absence of genetic recombination, the genetic variation necessary for a species to adapt to changing environmental conditions would only come about through mutations. Mutations are inheritable changes in the DNA that occur spontaneously or are induced by environmental factors such as radiation and chemicals, but are infrequent events under natural conditions. However, because of the large numbers of bacteria and the extremely short time needed for the completion of their life cycle, mutations play an important role in bringing about genetic variations among species of bacteria in the absence of true sexual reproduction.

Summary

Bacteria and cyanobacteria are prokaryotic organisms, lacking organized nuclei and other membrane-bounded organelles. True sexual reproduction is unknown among them, the predominant method of asexual reproduction being binary fission or simple fragmentation. A few species of both groups can utilize atmospheric nitrogen in a process called nitrogen fixation.

The majority of bacteria are heterotrophs. Most species are unicellular, being spherical, rod-shaped, or spiral-shaped, although the actinomycetes are filamentous. Many of the bacilli and spirilli have flagella for locomotion. Pili are short needle-shaped projections on some of the bacilli that help to attach the bacteria to a surface. A bacterial cell consists of an outer gelatinous capsule, a peptidoglycan cell wall, a cytoplasm containing ribosomes, and a nuclear body composed of a single DNA molecule. Most bacteria are saprophytes, but parasitic forms cause many human and plant diseases. All bacteria are divided into two broad groups—Gram-positive and Gram-negative. Further classification is based on morphological and physiological characteristics.

Cyanobacteria are primarily aquatic, and with rare exception, are photosynthetic autotrophs. The characteristic photosynthetic pigments of the cyanobacteria are c-phycocyanin and c-phycoerythrin, which are responsible for the range of colors demonstrated by this group. Most cyanobacteria are morphologically more complex than bacteria. The majority of cyanobacteria are unbranched filaments, while a few are unicells or form colonies. Cyanobacteria lack organs of locomotion, although gliding or oscillatory movements are known.

The cyanobacterial cell has the basic prokaryotic structure. The photosynthetic thylakoids are arranged along the periphery of the cytoplasm without an envelope. A peptidoglycan cell wall and a gelatinous sheath surround the cell. Nitrogen fixation takes place in specialized cells called heterocysts present along the filaments in some species. Unicellular cyanobacteria reproduce by binary fission, filamentous forms by fragmentation. Cyanobacteria often remain in gelatinous masses consisting of many individuals.

Cyanobacteria and lichens (a symbiotic association of a fungus and a green alga or a cyanobacterium) are able to colonize bare rock and contribute to the production of soil for the subsequent growth of higher forms of plant life. According to the endosymbiotic theory, the present-day chloroplast evolved from a unicellular cyanobacterium that established a symbiotic relationship within a unicellular heterotroph.

The harmful effects of certain cyanobacteria is due to their being ingested in large quantities in drinking water. They can also cause taste and odor problems in water reservoirs.

Although true sexual reproduction is unknown among the prokaryotes, conjugation, transformation, and transduction, all of which are mechanisms for genetic recombination, are known primarily among the bacteria.

Review Questions

1. List the different characteristics that can be applied equally to all prokaryotic organisms.
2. What are the different ways in which genetic recombination is achieved by the bacteria?
3. How are bacteria and cyanobacteria different from each other?
4. Discuss the role of bacteria as beneficial organisms.
5. Discuss the role of the different organisms in the nitrogen cycle.
6. Describe the different modes of nutrition among the bacteria.
7. What is the endosymbiotic hypothesis of chloroplast evolution?

Nitrogen Cycle

Nitrogen is by far the most abundant component of air, making up 78% by volume. In plants and animals nitrogen is found in important organic molecules such as amino acids, proteins, and DNA and is therefore an essential element. Most living organisms, however, are unable to utilize atmospheric nitrogen for the synthesis of organic molecules. Most autotrophs obtain nitrogen from the soil or water in the form of ammonia or nitrates. Animals obtain nitrogen by consuming plants or other animals. Certain prokaryotes such as the nitrogen-fixing bacteria and nitrogen-fixing cyanobacteria have the ability to incorporate atmospheric nitrogen directly into usable organic molecules. Species of nitrogen-fixing bacteria and cyanobacteria are found either in symbiotic association with plants, as in root nodules of leguminous plants, or are present free in the soil. The symbiotic bacteria primarily belong to the genus *Rhizobium,* while the free-living forms are found in diverse genera such as *Azotobacter, Enterobacter, Chlorobium,* and *Rhodospirillum.* The different nitrogen-fixing cyanobacteria are distributed among several genera such as *Gloeocapsa, Anabaenopsis, Anabaena, Nostoc,* and so forth. Large quantities of usable nitrogen compounds are synthesized by these nitrogen-fixing bacteria and cyanobacteria and are thus made available to plants. These nitrogen compounds are passed on to animals, including humans, by way of the food chain.

The nitrogen cycle (Figure 15–10) refers to the way in which atmospheric nitrogen moves through plants and animals and is ultimately released back into the atmosphere. Dead organisms and animal waste are decomposed by the action of saprophytic bacteria and fungi in the soil. When nitrogen-containing molecules are decomposed, ammonia is released into the soil as a byproduct. The ammonia is first converted into nitrites by a group of soil bacteria belonging to the genus *Nitrosomonas* and then into nitrates by another group of soil bacteria from the genus *Nitrobacter.* These two groups of bacteria are generally called nitrifying bacteria and the process is called nitrification. These bacteria are chemosynthetic autotrophs and utilize the energy released during nitrification to incorporate atmospheric carbon dioxide into organic molecules. Since both groups of nitrifying bacteria are usually present in the same soil, nitrites do not accumulate in the soil but are instead converted to nitrates. The nitrates are either used by plants as a nitrogen source or are converted back to atmospheric nitrogen in a process called denitrification by the denitrifying bacteria, thus completing the cycle.

Although large amounts of nitrogen-containing compounds are made available by the different groups of soil bacteria and the nitrogen-fixing cyanobacteria, intensive crop cultivation depletes the available nitrogen, making it necessary to add nitrogen in the form of fertilizers, without which the crop plants would be deficient in nitrogen.

(a)

(b)

(c)

(d)

Photos: (a) Horsetail (*Equisetum* sp.). (b) *Equisetum* strobilus. (c) Fern (*Polypodium* sp.). (d) Cycad (*Cycas* sp.). (e) Ponderosa pine (*Pinus ponderosa*.). (f) Male cones of Scots pine.

(f)

Dr. John Nadakavukaren

(e)

Color Plate 10

Photos: (a-c) Pollinating agents. **(a)** Humming bird. **(b)** Bumble bee. **(c)** Butterfly. **(d)** Carrion flower (*Stapelia* sp.) attracts flies as pollinators by its odor. **(e)** Head inflorescence of daisy. **(f)** *Anthurium* inflorescence with bright red bract.

Dr. John Nadakavukaren

(a) (b) (c) (d) (e) (f)

(a)

(b)

(c)

(d)

(e)

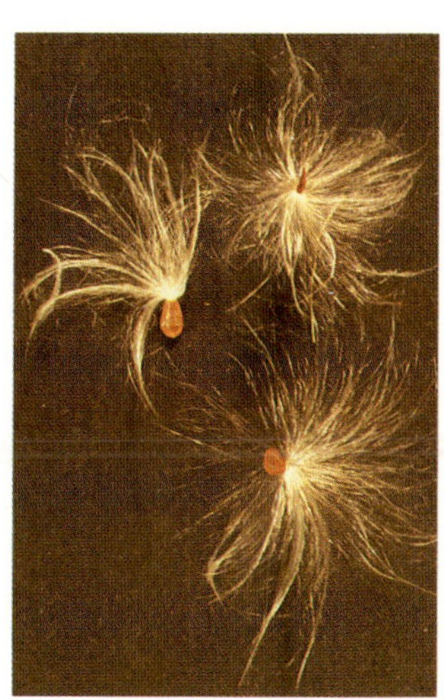

(f)

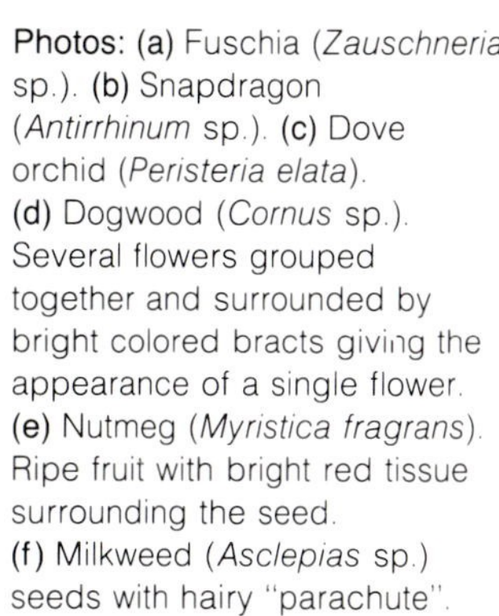

Photos: (a) Fuschia (*Zauschneria* sp.). **(b)** Snapdragon (*Antirrhinum* sp.). **(c)** Dove orchid (*Peristeria elata*). **(d)** Dogwood (*Cornus* sp.). Several flowers grouped together and surrounded by bright colored bracts giving the appearance of a single flower. **(e)** Nutmeg (*Myristica fragrans*). Ripe fruit with bright red tissue surrounding the seed. **(f)** Milkweed (*Asclepias* sp.) seeds with hairy "parachute".

Color Plate 12

Photos: (a) Tundra. (b) Northern coniferous forest. (c) Desert. (d) Grassland. (e) Tropical rain forest.

(b)

(a)

(d)

(c)

(e)

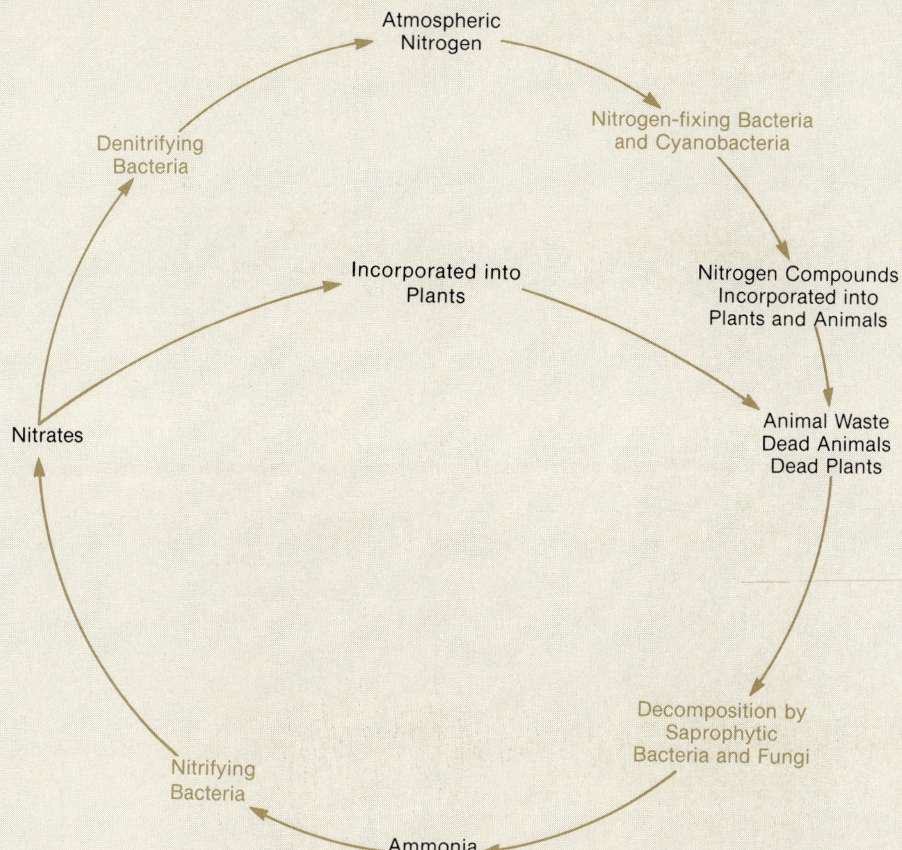

Figure 15–10. **Nitrogen cycle.**

16

Fungi

Outline

Introduction

In the previous chapter it was pointed out that bacteria cause many human diseases. Modern medicine is now able to use fungi to fight these bacteria and alleviate human suffering. In the past all a doctor could do was to treat the symptoms while letting the disease run its course, but in 1928 Alexander Fleming made the chance observation that changed medical practice. He was working with cultures of the bacterium *Staphylococcus* when he observed that one of the cultures had a fungal contaminant that was preventing bacterial growth. Instead of discarding the contaminated plate as other scientists probably had done before him, Fleming investigated this phenomenon and discovered penicillin, a substance produced by members of the fungal genus *Penicillium,* species of which grow on moldy fruit. Acting as an antibiotic penicillin is effective against certain types of bacteria. Some scientists suggest that penicillin is produced naturally as a "chemical warfare" agent—penicillin gives *Penicillium* an advantage in the competition for food by preventing the growth of bacteria that could use the same substrate. Efforts to produce penicillin commercially led to a massive search for high-yielding strains of *Penicillium,* with investigators taking samples of moldy fruit from all over. The prize-winning strain was discovered by workers from the USDA laboratory in Peoria, Illinois, from a sample collected from a cantaloupe in a local market.

Yet despite the present and historical impact of fungi, most people would have difficulty naming more than several examples—"They cause athlete's

Figure 16–1. Growth of mold. Any food left exposed will be colonized by fungi. (*a*) Moldy bread. Note the fungal filaments (hyphae) and the sporangia (spore-producing structures). (*b*) Moldy fruit. The dark coloration is due to the dark color of the spores.

foot, don't they?" Mushrooms, puffballs, yeast, bread mold, and a variety of other common organisms, on closer inspection, also turn out to be fungi. In fact, they are a very diverse group that many biologists consider to be in a kingdom by themselves (see the classification schemes in Chapter 14).

Most fungi are filamentous, but unicellular forms such as yeast also occur. The filamentous forms appear so similar that they can only be distinguished on the basis of their sexual reproductive structures. In the vegetative* stage of the fungal life cycle the fungal body is merely a nondescript group of threads or filaments called **hyphae** (singular, hypha) that grows through the substrate from which the fungus derives its nutrition. This mass of hyphae constituting the vegetative body of the fungus is called a **mycelium** (plural, mycelia). Among the more highly evolved fungi these vegetative mycelia give rise to the most conspicuous part of any fungus—the **fruiting body.** This specialized structure, produced by the organized growth of fungal hyphae, is involved in sexual reproduction. A familiar example of a fruiting body is a mushroom.

Vegetative
Not involved with sexual reproduction.

Fungal Nutrition

Fungi differ from most true plants in that they are incapable of making their own food through photosynthesis.* Therefore, like animals and most bacteria, the fungi have to obtain their food ready-made from the environment (Figure 16–1). This means that fungi have to decompose other organisms or waste materials. If the food organisms were dead, the fungus would be a **saprophyte*;** if they were living, the fungus would be a **parasite.** Some fungi exist either as saprophytes or as parasites whereas other fungi can function as both. Parasitic fungi that cause disease in humans and plants are called **pathogens.** In practice, the boundary between saprophyte and parasite is hard to draw for types of fungi such as the wood-rotter, since wood is a tissue found in both living and dead plants.

Fungi absorb already digested food from their environment (substrate) rather than ingesting undigested food as animals do. In order for fungi to do this, they must first secrete digestive enzymes* into their food supply (substrate), in a process analogous to that of digestion which occurs in animal intestines. Fungal enzymes, more than animal ones, digest a wide range of organic material including starch, cellulose, lignin, protein, leather, and chitin, as well as any other material that living organisms produce. It is this nutritional versatility of the fungi that has made them so important. Steak without mushrooms is hard for some people to imagine. More difficult to imagine is a world where plant and animal remains accumulate for generation after generation because there are no decay organisms to decompose these remains and return their elements to the soil and air. Without fungi, the decay of dead organisms and waste materials would be much slower and might not be completed before this material was buried under sediments or more layers of organic material. With plants, this may result in the production of more coal (see Chapter 18, Highlight), which might be a desirable

Photosynthesis
The metabolic process by which carbohydrates are produced from CO_2 and a hydrogen source such as water using light as the source of energy.

Saprophyte
An organism that obtains its nutrition by decomposing nonliving organic material.

Enzyme
A protein that functions as a biological catalyst.

outcome. During its lifetime every organism accumulates minerals that are required for growth and metabolism. When the organism dies these minerals remain in its body and so are unavailable to other organisms, which also need the minerals for their survival. Only through decomposition are these minerals released and made available to other living organisms. Therefore, saprophytic fungi are important to plant growth because of their role in recycling minerals from the dead to the living.

Economic Aspects

Both parasitic and saprophytic fungi can have a major economic impact. Parasitic fungi mainly attack plants. Human fungal diseases are limited primarily to skin infections such as athlete's foot or ringworm, though a serious lung infection, histoplasmosis, is of fungal origin. Most plant diseases are caused by fungi. This adds countless dollars to our food budget as the result of crop damage and food spoilage, both of which require preventative measures. The lumber industry annually loses billions of board feet of lumber both as trees and as finished lumber products. In addition, treatment of lumber to prevent fungal attack adds considerably to the cost of lumber products.

On the other hand, certain fungi are of considerable economic benefit. One obvious use for fungi is as food. Commercial production of mushrooms is big business in North America; similar commercial enterprises exist throughout the world. The actual number of fungal species cultivated for food is small, however. Some or the more desirable species, such as morels and truffles, are not produced commercially because the proper conditions for their cultivation have not been worked out. The fruiting bodies of selected fungal species are field collected by mushroom hunters. Unless you are an expert in fungal taxonomy, however, you should act as if the only edible fungi are those sold in the supermarket. Many fungi are poisonous and consumption of only a small portion of some of these is enough to cause death. Furthermore there is no reliable rule of thumb to distinguish the edible from the poisonous forms. Even someone eating a piece first, with no ill effects, is no guarantee of edibility because people differ in their responses to some of these fungal toxins, or symptoms of poisoning may be delayed for several hours.

The economic benefits of fungi extend far beyond mushroom soup. Breadmaking requires that the dough contains yeast (a unicellular fungus) to cause the bread to rise. Using sugar in the dough mix, the yeast respires and produces CO_2 gas bubbles that cannot escape from the dough, thus causing the dough to rise. Unleavened Passover bread (matzo) is made without yeast, which means that it is flatter and heavier than regular bread. In addition to bread, people of many cultures produce and consume alcoholic beverages (wine, beer, whiskey, rum, etc.), none of which would exist without fungal fermentation. Fermentation is a form of anaerobic respiration (Chapter 10) that produces alcohol as a by-product. Since many people like cheese with their bread and wine, it seems somehow appropriate that certain types of cheese (Roquefort and Camembert, for example) are produced using fungi. Even the riboflavin used as a nutritional additive in bread is produced by fungi. Fungi contribute further to human health as a source of certain antibiotics and drugs, for example, penicillin, produced by *Penicillium,* discussed above.

Finally, a major contribution of soil fungi to plant growth has been through the association of fungal hyphae and plant roots forming a **mycorrhiza,** a **symbiotic*** association benefiting both fungus and plant (Chapter 7). The fungus obtains carbohydrates from the plant roots while the plant obtains water and mineral nutrients from the fungus. The advantage of this association for the plant is that the fungus is in contact with much more soil than the plant root system giving the plant greater access to soil water and minerals. Furthermore, many of the minerals required for plant growth are locked up in the **humus***; mycorrhizal fungi along with other soil fungi contribute to the release of these minerals as they decompose the humus. The importance of the contribution of mycorrhizal fungi to most plants went unrecognized until fairly recently (see Chapter 7). Now their importance in agriculture and silviculture* is well established, as is their importance in strip mine reclamation.

Symbiosis
(*adj.,* symbiotic) A close association between two different organisms.

Humus
The partially decomposed remains of plants and animals.

Silviculture
The commercial growth of trees, as in treefarming.

Taxonomy

Fungal taxonomy has undergone major revisions in the past few years, the most significant of which has been the removal of the fungi from the plant kingdom by many botanists. Even now that the fungi are considered to constitute a separate kingdom (Kingdom Fungi), problems arise because early botanists classified organisms as fungi without a complete understanding of their life cycles. As a result, the study of fungi has come to include organisms (the slime molds) that seem to be more closely related to the protozoa than they are to the fungi. This artificial grouping reflects the interests of the botanists studying the fungi rather than any necessary evolutionary relationships. Because of this, in this book the fungi will be divided into two groups for discussion purposes—the slime molds and the "true" fungi (Table 16–1).

Slime Molds

The slime molds differ from the "true" fungi in several fundamental aspects. Slime molds lack hyphae and cell walls, except around their spores. Rather than obtaining their food by external digestion, they engulf bacteria and small particles for digestion within the cell. In fact, if it were not for

Sporophore
The structure on which the sporangium is produced.

Meiosis
The process of nuclear division that results in the production of daughter nuclei, each containing one-half the original number of chromosomes.

Table 16–1.
Classification of the Major Groups of Fungi

Kingdom Fungi
- Division Myxomycota (slime molds)
- Division Eumycota ("true" fungi)
  - Flagellate fungi
    - Class Oomycetes
  - Nonflagellate fungi
    - Class Zygomycetes
    - Class Ascomycetes
    - Class Basidiomycetes
    - Class Deuteromycetes

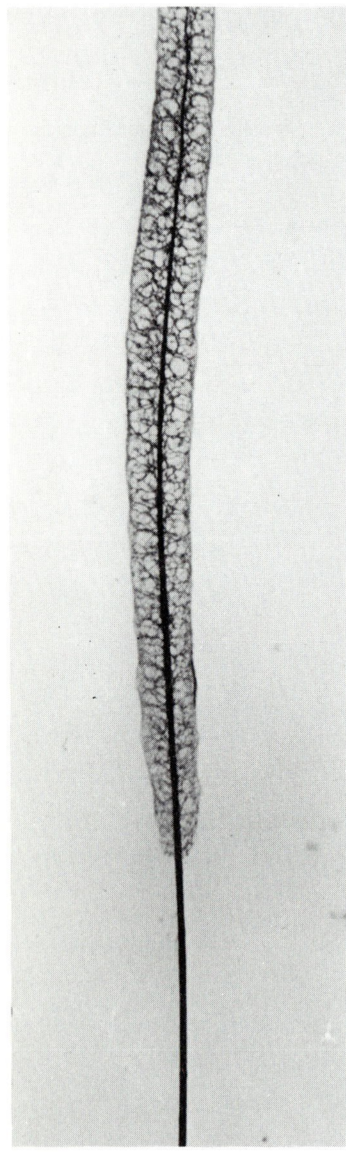

Figure 16–2. Slime mold sporophore.

their spores and spore-producing structures, the slime molds would not have been classified as fungi. There are two principal groups of slime molds, the cellular slime molds and the plasmodial slime molds.

Cellular slime molds are colonial organisms in much the same way that ants and bees are colonial. The slime mold consists of a group of individual amoebalike cells called **myxamoebae** to distinguish them from the true amoebae. These myxamoebae move and engulf bacteria in the same manner as true amoebae. When the supply of bacteria is depleted the myxamoebae start to aggregate in a mass of individual cells known as the **pseudoplasmodium.** Ultimately the cells of the pseudoplasmodium differentiate into a **sporophore*** and **spores**. After their release the spores germinate, producing new myxamoebae. This represents asexual reproduction. Sexual reproduction involving two different mating types has been observed in a number of species. During sexual reproduction a zygote is formed by the fusion of two cells that were part of a large aggregate of cells. Subsequently meiosis* and mitosis* occur so that when the zygote germinates haploid* myxamoebae are released.

Plasmodial slime molds have a distinct body made of a multinucleate mass of protoplasm called the **plasmodium**. The plasmodium moves in amoeboid fashion across the substrate on which it is growing, engulfing bacteria and other small food particles that it encounters. Eventually the diploid* plasmodium produces a sporophore with haploid spores. These spores germinate, releasing motile cells that will function as gametes under the right conditions. The zygote produced by gamete fusion divides by mitosis forming a new plasmodium. The plasmodium and associated sporophores (Figure 16–2) sometimes appear in large numbers on blades of grass in a lawn. Although they alarm the homeowner, they should be of no concern because the slime mold is feeding on bacteria and not grass.

"True" Fungi

The organization of "true" fungi is fundamentally different from that of slime molds. The fungus is typically composed of hyphae which have a rigid cell wall that usually contains chitin (basically the same material that makes up the hard outer shell, or exoskeleton, of insects). This group of organisms is classified under the division **Eumycota** and further divided into the flagellate fungi and the nonflagellate fungi.

Mitosis
The process of nuclear division that results in the production of two genetically identical daughter nuclei.

Haploid
Possessing only one of each pair of chromosomes.

Diploid
Possessing two of each pair of chromosomes.

Flagellate Fungi

Flagellate fungi encompass all of the "true" fungi that produce flagellate cells at any time during their life cycle and come under the class **Oomycetes.** These fungi are usually regarded as being primitive and they do not produce conspicuous reproductive structures such as are found in certain of the evolutionarily advanced fungal groups. Typically the members of this division produce **coenocytic*** hyphae with the only crosswalls being those that separate the reproductive structures from the rest of the mycelium. This group includes both parasites and saprophytes with many of its members being aquatic rather than terrestrial. One member of this group that is often studied in General Botany laboratory exercises is *Saprolegnia,* a common pest on fish eggs and injured fish (Figure 16–3).

Coenocytic hyphae
Hyphae that lack crosswalls resulting in a situation where many nuclei share a common cytoplasm.

Rather than attempt to describe a variety of life cycles for the diverse organisms that compose this group, certain general features of the life cycles of flagellate fungi will be considered. All members of this group produce flagellate reproductive cells that are either sexual or asexual or both. The asexual flagellate cells, **zoospores,** are released from **sporangia*** produced by the vegetative hyphae and swim around until they contact the appropriate host, which may be either living or dead organic material. Once the food supply is contacted the zoospore becomes attached, loses its flagella, and produces vegetative hyphae. Sexual reproduction among these fungi shows diversity similar to that observed in the algae (Chapter 17). In the simplest case two different hyphae fuse producing the zygotic stage. In another simple version two flagellate gametes fuse forming the zygote. The most elaborate sexual reproduction in these fungi involves separate male and female **gametangia***: male gamete nuclei are transferred to the female gametangium where they fuse with the eggs to produce the zygotes. Meiosis followed by germination of the zygotes gives rise to vegetative hyphae thus completing the life cycle.

Sporangium
(*pl.,* sporangia) A spore-producing structure.

Gametangium
(*pl.,* gametangia) A gamete-producing structure.

Figure 16–3. Parasitic growth of fungus. This fish has become infected by *Saprolegnia,* an oomycete fungus. From Cronquist *Introductory Botany* (1961) p. 276 Harper & Row.

Nonflagellate Fungi

None of the nonflagellate fungi, even if they are aquatic, produce flagellate cells at any time during their life cycle. This group contains the more conspicuous and familiar fungi—bread mold, yeasts, mushrooms, and *Penicillium,* for example. Within this category there are four major classes: **Zygomycetes, Ascomycetes, Basidiomycetes,** and **Deuteromycetes.** Since this group is mainly terrestrial and does not produce flagellate spores, spore dispersal usually is by air currents.

Zygomycetes

The organisms within this class usually have coenocytic hyphae with walls containing chitin. Frequently the vegetative hyphae grow extensively beneath the surface of their organic substrate, meaning, of course, that most of the fungus is hidden from view. Eventually aerial reproductive hyphae are produced that extend along the surface of the substrate. The colored spores that they produce are highly visible. Since black bread mold (*Rhizopus*) is in this group, this growth pattern has interesting ramifications. For instance, when you notice that a piece of bread is moldy you throw it away. Consider the other piece that you ate because you saw no mold on it. Obviously since *Rhizopus* proliferates beneath the surface of the bread before it ever sends up aerial hyphae, even bread that outwardly does not appear moldy can have hyphae growing in it. However, this is not a cause for alarm because *Rhizopus* hyphae (plain or toasted) are edible with no taste. (This should not be taken as a recommendation for the wholesale consumption of moldy bread, but is merely a description of the vagaries of fungus and human food consumption.)

As indicated above, the reproductive hyphae are produced at the surface of the substrate. Reproduction can be either sexual or asexual. Asexual reproduction involves the production of short, upright hyphae that terminate in sporangia (Figure 16–4). Within a sporangium the protoplast divides into uninucleated black spores. Rupture of the outer wall of the sporangium releases the spores into the air to be dispersed by air currents to other possible food sources. If the two mating types (involved in sexual reproduction) of a species are growing on the same substrate, they will induce each other to produce gametangia (Figure 16–5), structures that are produced when the hyphal tip becomes walled off from the rest of the hypha. Gametangia of opposite mating types are formed near each other, grow toward each other, and fuse. Fusion of the gametangia produces a compartment, the **zygospore,** connected to both hyphae. Nuclear fusion occurs within the zygospore resulting in the formation of the diploid stage of the life cycle. Preceded by meiosis, the germination of the zygospore results in the production of a short hypha that in turn produces the typical asexual sporangium which contains spores.

Ascomycetes

The Ascomycetes have played a significant role in human affairs through the centuries. This group contains the yeasts, which were domesticated as long ago as 7000 B.C. for brewing beer and 4000 B.C. for leavening bread. Other Ascomycetes recently have taken on considerable importance in biol-

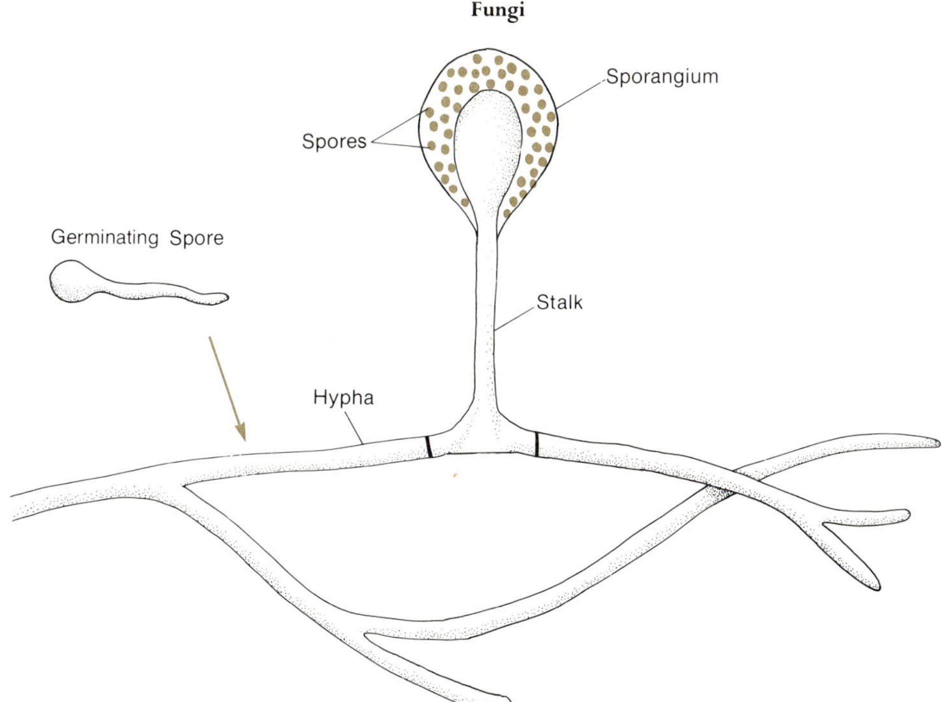

Figure 16–4. Vegetative growth of *Rhizopus*. After a spore lands on a favorable substrate (bread, for example), it grows to produce a mass of vegetative hyphae. Subsequent growth of hyphae on the surface of the substrate leads to the production of sporangia containing asexual spores.

ogy and medicine—*Neurospora* is one that is fundamental to modern genetics and molecular biology, with much of our knowledge of gene action being based on studies of various *Neurospora* mutants. Ergot, a fungal disease of cereal grains, has given modern medicine various alkaloid drugs that are used in childbirth. In the past ergot has caused considerable human suffering when, as a contaminant, it has been ground along with cereal grain to make flour. People who consumed ergot-contaminated flour suffered violent hallucinations, spasms, and often death. Some historians have suggested that the witches of Salem, Massachusetts, were in fact individuals who had consumed ergot-containing flour. Many of the Ascomycetes cause plant diseases such as ergot, chestnut blight (which has nearly eliminated the American chestnut), Dutch elm disease (which is eliminating the American elm), powdery mildew, various fruit rots, and many others. A smaller number of Ascomycetes are agents of human diseases, most notably ringworm and histoplasmosis. Two Ascomycetes, in particular the morel (Plate 8B) and the truffle, are prized as food by gourmets and mushroom hunters. The part that is eaten is the fruiting body, or sexual reproductive structure.

Unlike the Zygomycetes, the Ascomycetes have vegetative hyphae that are divided by crosswalls, or **septa,** into compartments. Each of these compartments is similar to a plant cell in that it contains one or more nuclei and cytoplasm. However, the compartments are not equivalent to plant cells because their septa have pores that allow the migration of cytoplasm, and even nuclei, between compartments and along the length of the hypha. Asexual reproduction among the Ascomycetes involves spores called **conidia** (Figure

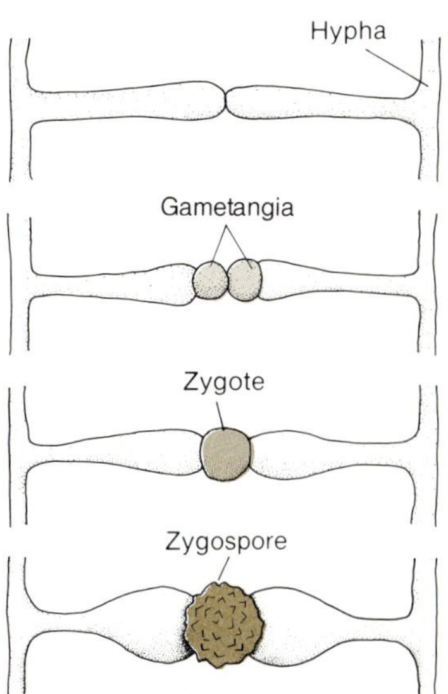

Figure 16–5. Sexual reproduction in *Rhizopus*. When hyphae of the two mating types are growing near each other, they produce branches that grow toward each other. These branches touch, produce gametangia, and fuse forming the zygospore.

16–6). The essential difference between conidia and other asexual spores is that conidia are not produced in a structure such as a sporangium; rather, they are formed directly from a hyphal tip. What really distinguishes the Ascomycetes from the previous groups of fungi is the form of sexual reproduction, which, among Ascomycetes, involves the production of a conspicuous structure called the **fruiting body** (Figure 16–7).

The vegetative growth of an Ascomycete results in the formation of a typical mycelium from either a germinated conidium or from a sexually produced spore. This mycelium may reproduce asexually many times by formation of conidia, but under proper environmental conditions sexual reproduction will occur. Sexual reproduction can be either **homothallic*** or **heterothallic***, but requires the formation of separate male and female sex organs in close proximity. The gametangia are formed by the differentiation of the apical segments of vegetative hyphae that become multinucleate. The female sex organ, or **ascogonium,** produces a slender hypha, the **trichogyne,** that makes contact with the male sex organ, the **antheridium** (Figure 16–8). After the trichogyne fuses with the antheridium, the male nuclei move through the trichogyne and pair up with the female nuclei. The resulting paired nuclear condition is referred to as the **dikaryotic** (from the Greek *di,* two, + *karyon,* a nut) phase of the life cycle. During this stage the vegetative hyphae near the sex organs grow and differentiate into the distinct mass of hyphae called the fruiting body, or **ascocarp,** that encloses the sex organs and the dikaryotic hyphae that develop from the ascogonia (Figure 16–8). These dikaryotic hyphae are usually short, with the tip of each eventually forming a hook. The two nuclei in this hook undergo mitosis to produce four nuclei. After the formation of new cell walls, two of these nuclei (one from each parent) are in the terminal region, with the other two being in two separate compartments (Figure 16–8). The two uninucleate compartments fuse, restoring the dikaryotic condition, and afterwards the cycle is repeated. Ultimately, in the dikaryotic terminal section the two nuclei fuse (fertilization), resulting in a diploid nucleus. After fertilization this terminal compartment containing the diploid nucleus is referred to as the **ascus.** Within the ascus meiosis takes place, forming four haploid nuclei; in most species a subsequent mitosis increases the number of nuclei to eight. These nuclei are separated into individual spores called **ascospores** (Figure 16–9). The ascospores are released from the ascus and germinate, forming new vegetative hyphae.

Homothallic
Producing both male and female gametes on the same organism.

Heterothallic
Producing male and female gametes on separate organisms.

While the majority of Ascomycetes follow the general life cycle outlined above, there is one important exception: the yeasts. These are unicellular, nonfilamentous organisms. Sexual reproduction in yeasts does not involve specialized cells or sex organs. In response to environmental conditions and the presence of cells of the opposite mating type, vegetative cells function as reproductive cells. After fusion of two individual cells, a diploid zygote is formed that undergoes meiosis and mitosis to produce eight ascospores. The solitary ascus in yeasts is derived directly from the zygote; no complicated fruiting body is formed.

Basidiomycetes

This group of fungi, like the Ascomycetes, produces septate hyphae with walls containing chitin, and shows considerable variety of form. The consis-

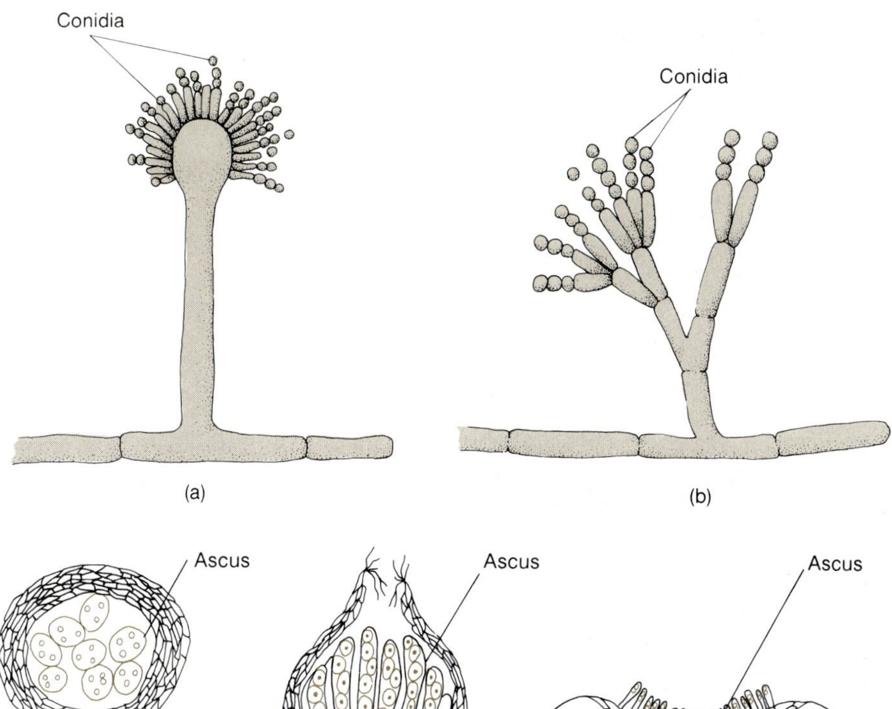

Figure 16–6. Conidia production. Conidia are asexual reproductive cells produced externally from a hyphal tip. (*a*) *Aspergillus niger.* (*b*) *Penicillium notatum.*

Figure 16–7. Ascocarps. Three different morphological types of ascocarps.

tent difference that is used to separate these two groups relates to spore formation. In Basidiomycetes the spores produced as a result of sexual reproduction form as outgrowths of a specialized hyphal segment, the **basidium,** instead of within an ascus. As saprophytes and parasites of plants, the Basidiomycetes have considerable economic impact. Most mushrooms of commerce are Basidiomycetes, as are many of the wood-rotting fungi (Plate 8C). The rusts and smuts are parasitic on cereal grasses such as wheat, oats, barley, and corn.

The vegetative phase of the life cycle begins with the germination of a spore, producing hyphae in the appropriate substrate (food supply). When two hyphae of opposite mating types come into contact they fuse, creating a dikaryotic stage that forms an extensive dikaryotic mycelium (Figure 16–10). Under the appropriate environmental conditions the dikaryotic mycelium gives rise to new hyphae whose growth is organized so as to form the distinctive fruiting body, or **basidiocarp** (Figure 16–11), of that species. Even though all hyphae of a fruiting body are dikaryotic, only a relatively few terminal segments are involved in fertilization and meiosis. Within the fruiting body the ends of certain hyphae enlarge slightly forming the clublike basidium (plural, *basidia*), the compartment in which nuclear fusion (fertilization) takes place producing the zygote nucleus. Meiosis follows and each of the four haploid nuclei which result migrates into a separate **basidiospore** that buds from the surface of the basidium (Figure 16–12). Typically each basidium produces only four basidiospores, although some can produce fewer or more. After their release from the fruiting body, the basidiospores

Figure 16–8. Sexual reproduction in Ascomycetes. Vegetative growth from an Ascomycetes spore produces a mass of monokaryotic hyphae. Production of antheridia and ascogonia in close proximity leads to the transfer of nuclei from the antheridia to the ascogonia through the trichogynes. The dikaryotic hyphae that result grow to form the ascocarp. In certain hyphae the two nuclei fuse and this region of the hypha becomes the ascus. Meiosis in the ascus produces the ascospores.

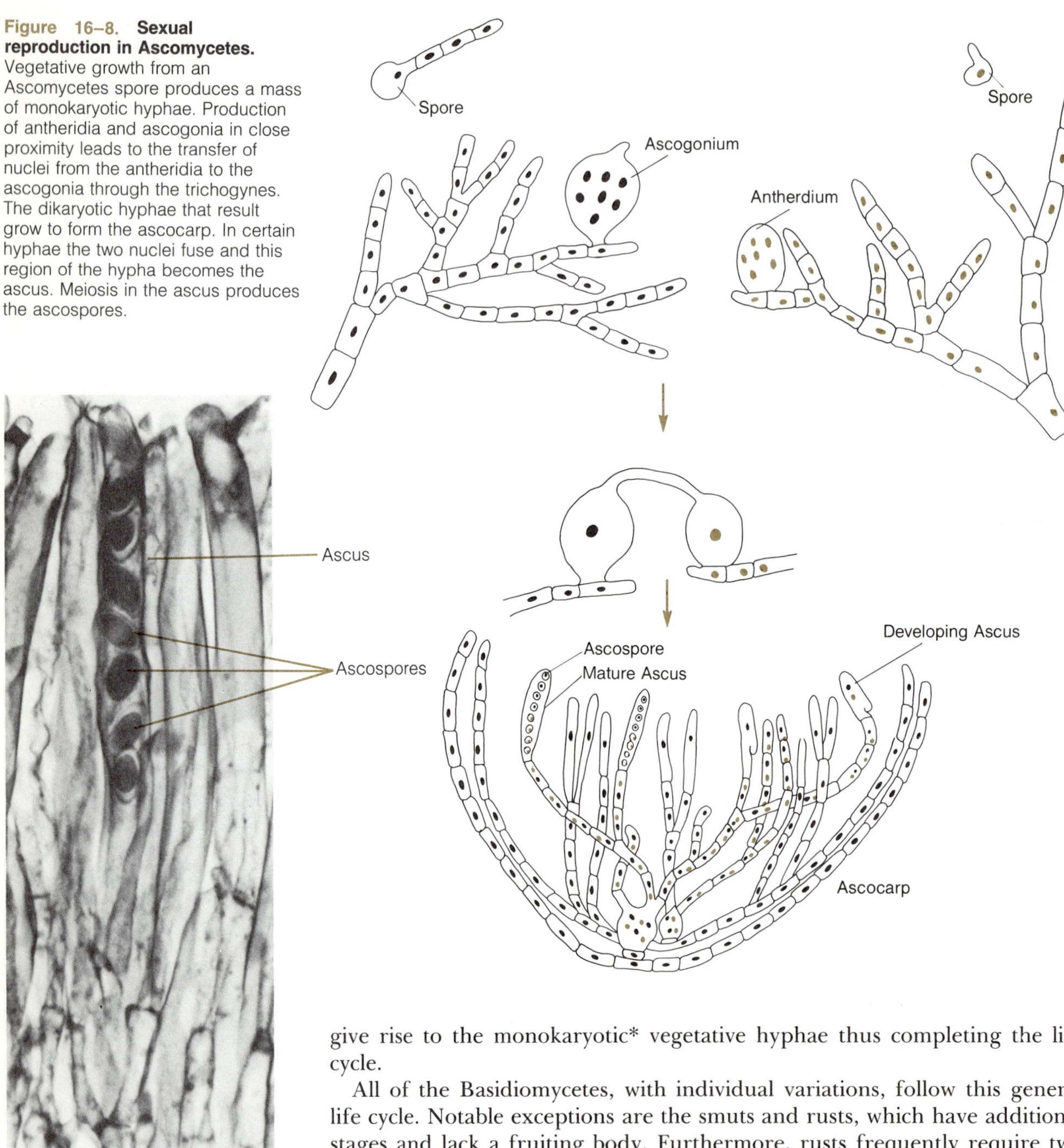

Figure 16–9. Ascospores.

give rise to the monokaryotic* vegetative hyphae thus completing the life cycle.

All of the Basidiomycetes, with individual variations, follow this general life cycle. Notable exceptions are the smuts and rusts, which have additional stages and lack a fruiting body. Furthermore, rusts frequently require two hosts in order to complete their life cycle. As a classic example, wheat rust produces a monokaryotic mycelium on the barberry plant, but it cannot complete sexual reproduction on this host (Figure 16–13). Once the dikaryotic stage has been reached, dikaryotic spores that can infect the wheat plant are produced. On wheat the dikaryotic mycelium reproduces asexually until later in the season when the wheat is maturing. At this time the fungus produces resting spores that drop to the soil or remain attached to the wheat plant all winter. In the spring these resting spores germinate directly into a

Monokaryotic
Possessing only one nucleus per cell.

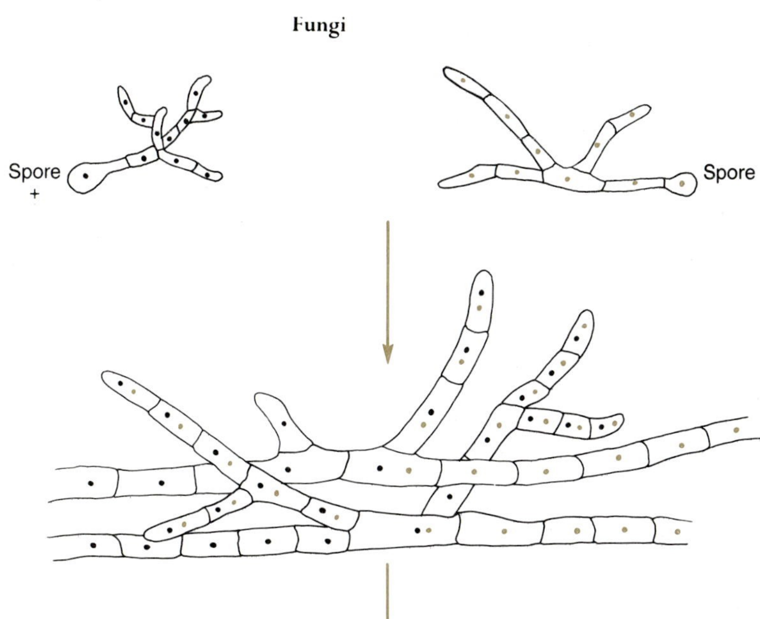

Figure 16–10. Sexual reproduction in Basidiomycetes. Vegetative growth from a spore produces a mass of monokaryotic hyphae. When hyphae of the two mating strains are growing in close proximity, the hyphae fuse and give rise to a mass of dikaryotic hyphae. Organized growth of the dikaryotic hyphae produces the basidiocarp. Within the basidiocarp the ends of certain hyphae enlarge and nuclear fusion takes place within this segment (basidium). Meiosis produces four haploid nuclei that migrate into the four basidiospores produced on each basidium.

basidium that releases basidiospores. These, in turn, can infect the first host, barberry, thus completing the life cycle. Smuts do not need two hosts but, like the rusts, they utilize a resting spore for overwintering purposes with basidia and basidiospores being produced in the spring.

Deuteromycetes

Each of the previous classes of the nonflagellate fungi represents what taxonomists refer to as a *natural group.* In other words, the organisms were placed in a particular group because they were apparently related to the other members of the group. As with plants, a major deciding factor in determining relationships is the form of sexual reproduction. In the absence of sexual stages the fungi are impossible to classify completely because they do not produce distinctive vegetative structures. Consequently, fungal tax-

(a)

(b)

Figure 16–11. Basidiocarps. The mushroom (*a*) is the most commonly recognized fruiting body of Basidiomycetes but there is a variety of other forms such as the bird's nest fungus (*b*).

(a)

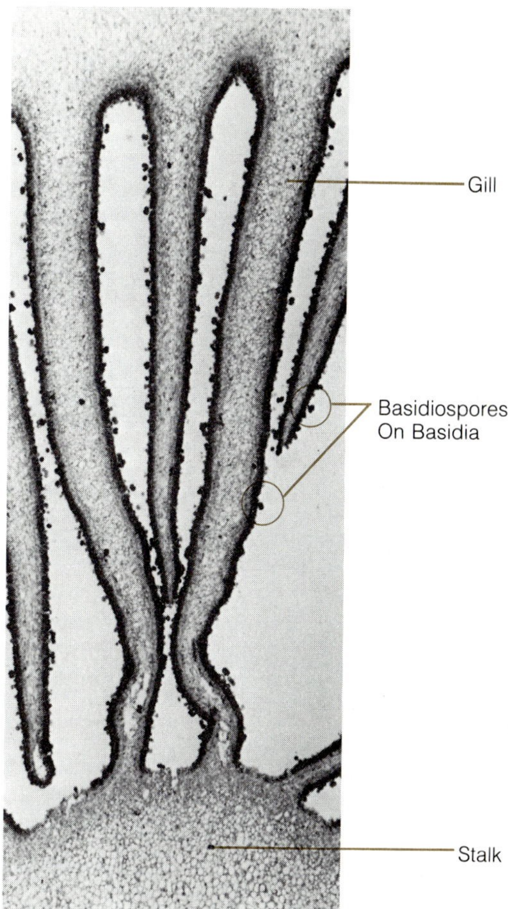

(b)

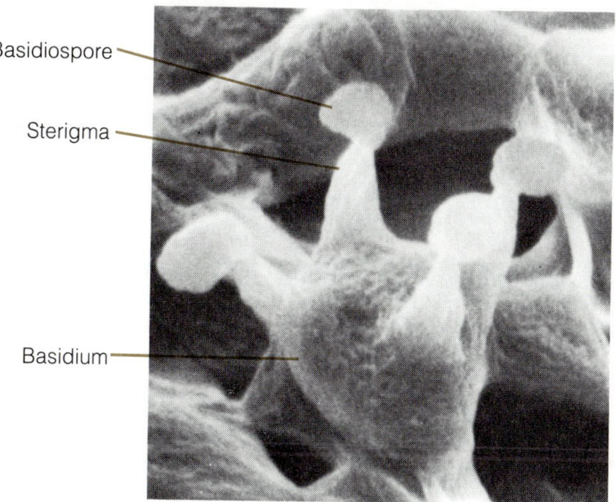

(c)

Figure 16–12. Basidiospore production. (*a*) Underside of a mushroom showing the gills. (*b*) Cross section of a mushroom through the gills. (*c*) Scanning electron micrograph of a basidium with immature basidiospores.

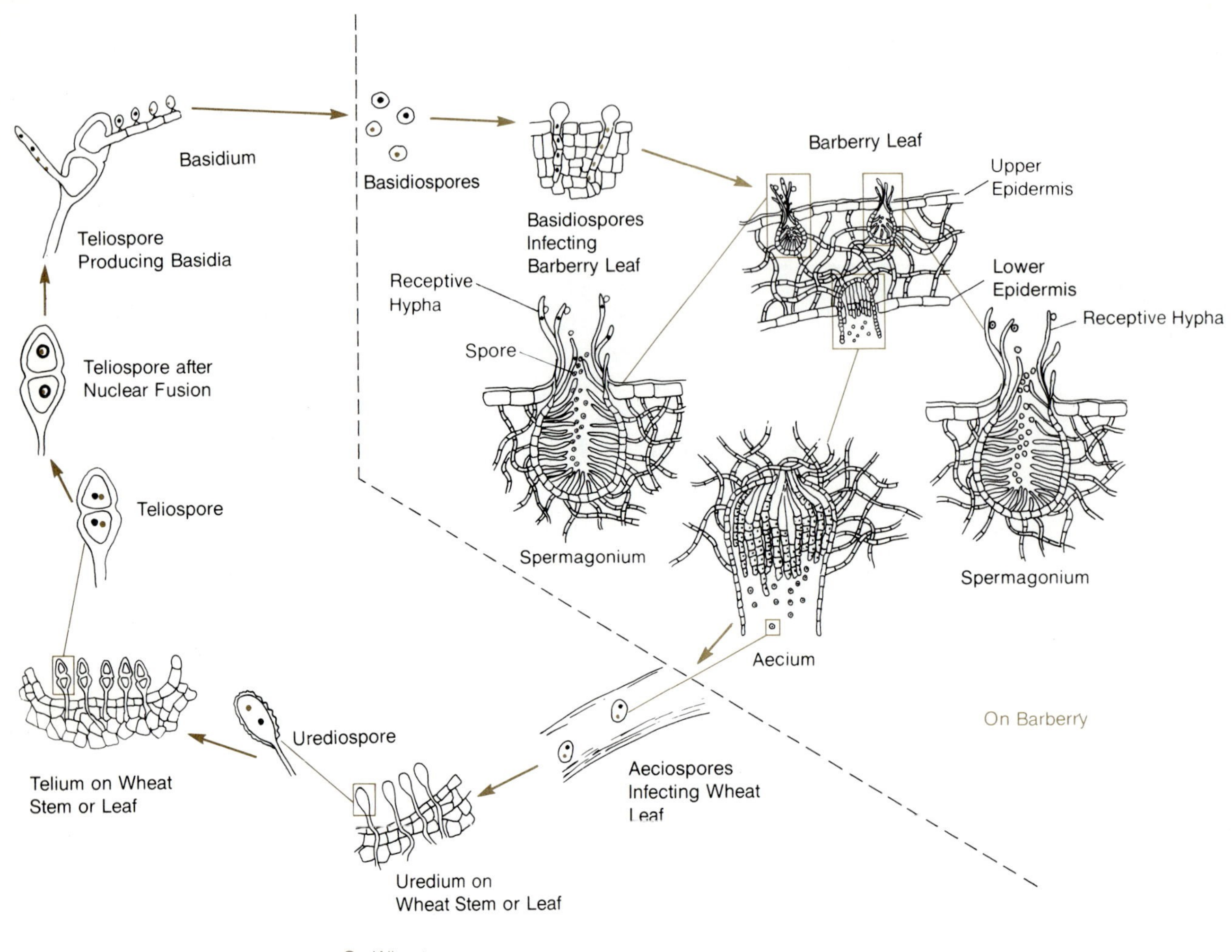

Figure 16–13. Wheat rust life cycle. Germination of a rust spore on a barberry leaf leads to the production of monokaryotic hyphae in the upper surface of the leaf. If the hyphae of the two mating strains are growing in the leaf, then hyphal fusion occurs producing dikaryotic hyphae. The dikaryotic hyphae produce dikaryotic spores on the lower side of the barberry leaf. Infection of a wheat plant with a dikaryotic rust spore leads to the growth of dikaryotic hyphae in wheat leaves or stems. Basidia are produced on wheat plants from resting spores that are the overwintering stage of the rust.

onomists have created a special class, the Deuteromycetes, in which they place all of the nonflagellate fungi for which no sexual stage is known. This group is often called the Fungi Imperfecti owing to the lack of the perfect, or sexual stage. Once the sexual stage is discovered, the organism is removed from the Deuteromycetes, given a new genus name, and placed in the proper class. In the past most of the Deuteromycetes whose sexual stage subsequently was discovered turned out to be Ascomycetes, while a few have been found to produce basidia and have been classified as Basidiomycetes. Species of *Penicillium* and *Aspergillus* are probably the most familiar examples of such change of classification. Species that have been transferred from *Penicillium* have been placed in two closely related genera of the Ascomycetes; *Aspergillus* species have also been placed in related Ascomycete genera. On the other hand, two species of *Rhizoctonia,* both of which are parasitic on plants, have been placed in two completely unrelated Basidiomycete genera.

It is expected that the sexual cycle of more species of Deuteromycetes will be discovered ultimately and the organisms subsequently transferred to the appropriate class. However, it is likely that a sexual stage will never be discovered for most Deuteromycetes either because the proper culture conditions are never achieved or because the organisms have lost their sexual stage through evolution. A loss of the sexual stage has also occurred in some plants and animals, but in these groups the general characteristics are distinctive enough to allow complete classification. The classification of fungi within the Deuteromycetes is on the basis of asexual reproductive structures; however, this does not imply any degree of evolutionary relation among the members of this class.

Lichens

Lichens (Figure 16–14, Plate 8D) are unique organisms in that each consists of a fungus and an alga living together symbiotically. Most often the fungus is an Ascomycete though a few Basidiomycetes are also involved; the partner is either a green alga or a cyanobacterium (blue-green alga). Both organisms benefit by this association, which for many lichens permits both the fungus and the alga to extend their range.

A fungus that is part of a lichen has a more organized growth of vegetative hyphae than do other fungi. Three basic body forms exist among the lichens: leaflike (**foliose**), crustlike (**crustose**), and branching and cylindrical (**fruticose**). The fungal hyphae form the upper and lower surfaces of foliose and crustose lichens, and the outer layer of fruticose lichens. The upper (and outer) surface is functionally analogous to the epidermis of a leaf in

Figure 16–14. **Lichens**

that it provides a waterproof cover. Fungal hyphae in contact with the substrate absorb water and minerals. In the middle section, particularly toward the upper surface, the fungal hyphae are interspersed with photosynthetic algal cells that provide food for the partnership. In some cases the algal partner is under the physiological control of the fungus. This causes the alga to release sugars and amino acids, which the fungus absorbs. When the alga is grown in pure culture away from the fungus, it is much less "leaky." In some lichens, the fungi "harvest" algal cells by digesting them.

Because lichens are stable associations, the rate of fungal growth must be regulated carefully so that the algal partner is never threatened with extinction. In fact, as the lichen grows, the numbers of both fungal and algal cells increase. It might appear as if the fungus has "enslaved" an alga, that receives little or no benefit. Actually the association is mutually beneficial. The fungus obtains a ready supply of food and the alga obtains protection from environmental extremes. As a result lichens are often found in environments that would be inhospitable to either partner alone. Lichens are colonizers of bare rock—a fungus would find no food there and an alga would not be able to survive the dryness. The incorporation of the alga into the lichen structure created by the fungus provides the alga with protection from the desiccation as well as from the temperature extremes encountered in such an open situation.

Lichen reproduction is complicated by the fact that they are composed of two different organisms. For a lichen to reproduce itself both the alga and the fungus must be present in the reproductive structure. Asexual reproduction, other than by fragmentation, involves the production of clusters composed of fungal and algal cells, known as **soredia,** at the surface of the lichen. The soredia are dispersed by the wind; their germination on the appropriate substrate gives rise to new lichens. Sexual reproduction involves only the fungal partner and results in the production of fruiting bodies typical of that fungus. Because the spores released by the fruiting body consist only of fungal material, they can only give rise to fungal hyphae on germination, not lichens. A new lichen can only be formed if the fungal hyphae were to begin growth in close proximity to the correct algal species. Since this is unlikely, lichen reproduction primarily occurs as the result of asexual reproduction.

Lichens are not organisms of major economic importance in most areas. They have, however, contributed to soil formation. In arctic regions the lichen called reindeer moss is crucial to the survival of grazing mammals during winter months when other food plants are unavailable. For the Eskimo and Laplander, peoples that depend on these grazing mammals, lichens are of considerable economic importance. In urban areas lichens are also taking on more significance owing to their usefulness as indicators of air quality. Since lichens are sensitive to low levels of certain types of air pollution, their presence or absence provides long-term air quality information. Finally, until chemists came up with a synthetic substitute, the acid-base indicator *litmus,* so familiar to generations of chemistry students, was derived from lichens.

Summary

The fungi are a diverse group of mainly filamentous organisms that are either parasites or saprophytes (Table 16–2). Because the majority of plant diseases are caused by fungi, the fungi have had considerable negative economic impact. However, beneficial fungi play indispensable roles in baking, fermentation, and drug production. A further beneficial role of fungi is in

Table 16–2.
Summary of the Characteristics of the Major Groups of Fungi

Group	Characteristics
DivisionMyxomycota (slime molds)	No hyphae or cell walls Engulf food particles
Cellular slime molds	Colony of amoeboid cells Aggregate into a pseudoplasmodium for asexual spore production
Plasmodial slime molds	Multinucleate mass of protoplasm
Division Eumycota ("true" fungi)	Composed of hyphae and have a rigid cell wall External digestion of food
Flagellate fungi (Class Oomycetes)	Coenocytic hyphae Flagellate reproductive cells
Nonflagellate fungi	No flagellate cells
Class Zygomycetes	Coenocytic hyphae No conspicuous reproductive structures
Class Ascomycetes	Hyphae with crosswalls Conspicuous structure (ascocarp) associated with sexual reproduction Meiosis to produce ascospores occurs in the ascus
Class Basidiomycetes	Hyphae with crosswalls Conspicuous structure (basidiocarp) associated with sexual reproduction Meiosis to produce basidiospores occurs in the basidium
Class Deuteromycetes	Hyphae with crosswalls No known sexual stage

mycorrhizal associations where the fungal hyphae increase water and mineral absorption by acting as an extension of the plant root system.

For historical reasons the taxonomic designation *fungi* includes both slime molds and "true" fungi, groups that most likely are not related. In the colonial cellular slime molds the individual amoeboid cells (myxamoebae) come together in a large aggregate only to reproduce. Plasmodial slime molds exist for most of their life cycle as a multinucleate mass of protoplasm (plasmodium) that creeps across the substrate; periodic spore production reproduces the plasmodium. The true fungi are classified under the division Eumycota. Members of the class Oomycetes are primitive and have flagellate cells some time during their life cycle. The evolutionarily more advanced nonflagellate fungi belong to the classes Zygomycetes, Ascomycetes, Basidiomycetes, and Deuteromycetes. The flagellate fungi typically have vegetative hyphae that produce flagellate reproductive cells, either zoospores or gametes. While the Ascomycetes include unicellular yeasts, more typical Ascomycetes have extensive vegetative hyphae, which produce male sex organs (antheridia) and female sex organs (ascogonia). Transfer of male gamete nuclei to the ascogonium results in the production of dikaryotic hyphae containing paired male and female nuclei. Organized growth of the vegetative hyphae and the dikaryotic hyphae results in the production of a conspicuous fruiting body, the ascocarp. In the terminal section of a dikaryotic hypha the paired nuclei fuse forming a zygote in what is now called the ascus. Meiosis and mitosis within the ascus usually produce 8 haploid ascospores. Within the Basidiomycete fruiting body, nuclear fusion occurs in terminal sections of hyphae (basidia). After meiosis occurs, spores form as specialized outgrowths of the basidium instead of within an ascus. Since placement of fungi in the correct taxonomic group depends on the observation of the sexual structures, any fungus with no known sexual stage must be placed in the Deuteromycetes, a class created to accomodate fungi of uncertain affiliations. As sexual stages are discovered, the fungi are transferred to the appropriate group.

Lichens are symbiotic associations of fungi and algae. The fungal partner provides an outer protective layer; the algal partner functions in photosynthesis in the middle portion of the organism; the fungal hyphae in contact with the substrate function in water and mineral absorption. Lichens reproduce asexually by fragmentation or by producing specialized clusters of fungal and algal cells. Sexual reproduction of a lichen is not possible.

Review Questions

1. Compare and contrast sexual reproduction in typical Ascomycetes and Basidiomycetes.
2. Describe the vegetative body of a typical flagellate fungus including a description of how it obtains its nutrition.
3. List the major groups of the fungi and their distinguishing characteristics.
4. Outline the economic impact of the fungi.
5. Describe the various modes of asexual reproduction among the fungi.

Fungi and Race Relations

One of the major social problems in the United States is caused, in part, by a fungal plant disease. Black-white relations are strained today in many large cities of the Northeast because of events in Europe nearly 150 years ago.

Ireland is basically an agricultural land noted primarily as an exporter of lace, whiskey, and people. During the 1700s a growing population placed increasing demands on food production capabilities. In order to solve the problem of insufficient food, Irish farmers turned to a newly introduced vegetable, the potato. Potatoes, native to South America, were at first an exotic novelty in Europe, but varieties were soon discovered that would grow in northern Europe. With the climatic conditions of Ireland being favorable to potato production, Irish farmers were successful at potato cultivation. This resulted in more and more acreage being planted in potatoes until Irish agriculture was almost exclusively involved with potatoes. Potatoes became the staple of the lower socioeconomic classes, in addition to being a cash crop. Unfortunately the fungus *Phytophthora infestans* (a flagellate fungus) became established in Ireland and completely devastated the potato crop in 1845. Without their main source of nourishment as many as a million people may have died in Ireland, and many more emigrated. Because of traditions, hostilities, and lack of opportunity, most of the Irish emigrants chose not to go to England. Rather they came to the United States, not least because the United States had fought two wars with England. During the late 1840s there was a major influx of Irish immigrants to the industrial cities of the Northeast.

As newcomers the Irish, like all new immigrant groups, ended up with a preponderance of the menial, poorly paid jobs. However, before they could be integrated into American society and move up the socioeconomic scale, the Civil War was fought. Victory for the North freed the black slaves who left the South in large numbers to move North to what had been the Promised Land of all escaping slaves. Once in the North, the former slaves found that the jobs they sought were already occupied by the recent Irish immigrants. For the Irish the influx of cheap black labor was an economic threat that they had to resist. Since the Irish had had a chance to establish a power base, they were able to band together and look after their own. This left the black laborers with only those jobs that even the Irish did not want.

It is interesting to speculate how black-white relations in the United States would have turned out if blacks in the 1860s had found economic opportunity in the North instead of closed doors. An interesting closing note on this issue is that John F. Kennedy as President of the United States is associated with the civil rights movement. What is sometimes overlooked is that he is of Irish descent and his ancestors were probably among those who displaced the blacks in the 1860s.

17

Algae

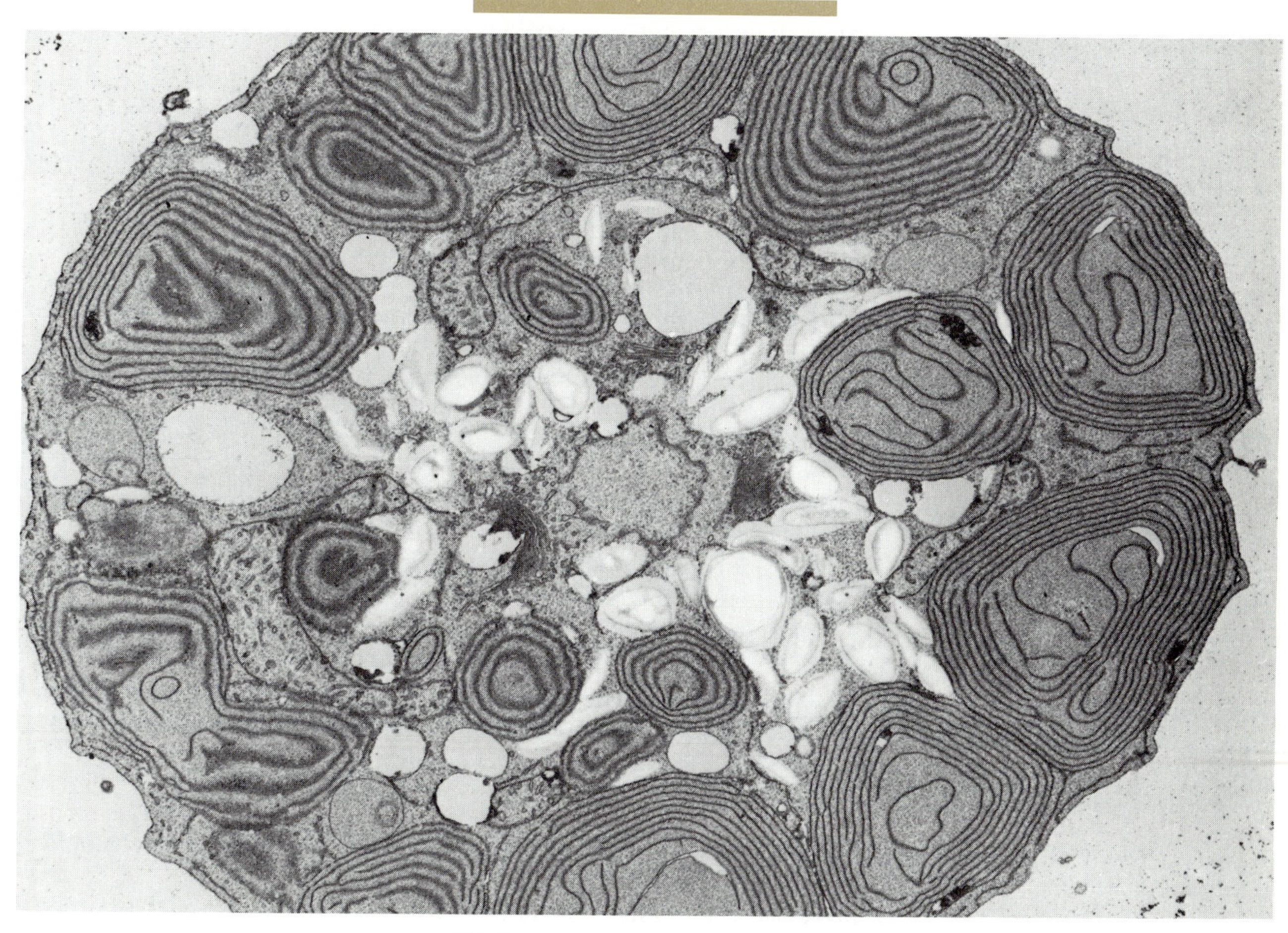

Outline

Introduction

The term ***alga*** (plural, ***algae***) was first used by Linnaeus in 1754 to describe the lower forms of plants. A number of organisms that were described as algae by Linnaeus, however, are no longer classified as algae. Nevertheless, the algae still represent a large, highly heterogeneous group of primitive photosynthetic organisms, some of which must have evolved as early as 2 billion years ago. There is at present no general agreement on whether all divisions of algae evolved from a common ancestor or from several different ancestors. Superficial structural similarities between divisions suggest a certain degree of parallel evolution*—not surprising in view of the fact that they were evolving in response to very similar environmental influences. For the most part, botanists view most divisions of algae as evolutionary dead ends; with the exception of the Chlorophyta, the various divisions of algae are not thought to have given rise to any of the present-day plants.

Although the six algal divisions are not related phylogenetically* in any apparent way, the collective designation *algae* continues to be widely used because these organisms share a number of characteristics. Among such characteristics are the following:

1. *Absence of xylem* and phloem.** Algae lack vascular tissues for conduction of food or water, absorbing such nutrients directly through their cell walls from the surrounding environment. Although some algal species may possess anchoring appendages (rhizoids or "hold-fasts") that superficially resemble roots, stemlike "stipes," or leaflike fronds, these structures lack conductive tissues and hence cannot be regarded as true roots, stems, or leaves. The only exception to this rule is found in certain of the brown algae. These organisms produce cells that, in their structure and function, resemble the sieve tube members of phloem of higher plants, but they do not produce xylem.
2. *Photosynthetic pigments.* Unlike the fungi, which also lack true stems, roots, or leaves, algae possess chloroplasts containing photosynthetic pigments, mainly chlorophyll, which enable them to manufacture their own food. Thus algae, like the plants, are said to be autotrophic. The different types and colors of pigments in various divisions of algae constitute one of the major criteria for algal classification.
3. *Reproductive structures usually unicellular.* The gametangia (structures which produce gametes for sexual reproduction) and the sporangia (structures which produce asexual spores) lack the differentiation into sterile and fertile layers that the reproductive structures of higher plants exhibit.
4. *Cell walls of cellulose.* The cell walls of most algae contain cellulose, although other components such as silica may be found in certain groups. The major exception to this rule is the division Euglenophyta, whose members lack a cell wall.
5. *No embryo formation.* The zygote* either divides meiotically* to produce

Parallel evolution
The evolution of similar adaptations to the same environmental variables by unrelated organisms

Phylogenetically
Pertaining to the evolutionary history of an organism or group of organisms.

Xylem
Water-conducting tissue.

Phloem
Food-conducting tissue.

Zygote
The cell that results from the fusion of sex cells.

the haploid* stage of the life cycle or germinates directly to produce a diploid* stage without an intervening embryonic stage.

6. *No epidermal tissue.* Since algae are either aquatic or live in moist habitats, there was no selective advantage in evolving the type of protective epidermal layer characteristic of most land plants.

Meiosis
(*adv.*, meiotically) The process of nuclear division that results in the production of daughter nuclei, each containing one-half the original number of chromosomes.

Haploid
Possessing only one of each pair of chromosomes.

Diploid
Possessing two of each pair of chromosomes.

Distribution

Algae occur widely throughout all types of aquatic habitats (the majority of the 30,000 species identified are saltwater forms) and in moist terrestrial environments as well. They can be found floating free in oceans, lakes, hot springs, and in temporary desert lakes. Some species occur anchored to the bottom in shallow water at the edge of seas, ponds, and rivers. Others find a suitable environment in damp soil, on tree trunks, rocks, and flowerpots, where they mistakenly may be referred to as "moss." More unusual adaptations include species which grow on snow, giving it a pinkish tinge; a species that grows on the fur of the jungle-dwelling three-toed sloths, giving these naturally gray-haired animals their greenish hue; and types that exist symbiotically* inside certain types of protozoa, worms, sponges, hydras, and others, manufacturing food that is transferred to the animal host while the alga receives carbon dioxide and nitrogen-rich wastes in return. Lichens, the organisms frequently observed on tree trunks, fence posts, rocks, and other substrates, represent successful symbiotic associations between an alga or cyanobacterium and a fungus (Chapter 16), where the alga and cyanobacterium produce the food while the fungal partner provides water and mechanical protection.

Symbiosis
(*adv.*, symbiotically) A close association between two different organisms.

Diversity of Form

The various species of algae exhibit a vast range of sizes and forms (Figure 17–1). Unicellular types such as *Chlorella* consist of individuals as small as 3 or 4 micrometers (it is estimated that 1 quart of a relatively dilute suspension of *Chlorella* may contain 20 billion individuals), while the largest of algal species, the Pacific kelps, frequently attain lengths of more than 30 meters (100 feet). In color, algae also exhibit considerable diversity, ranging from green to yellow-green, golden-brown, brown, and reddish-purple. In body form, most algae fall into one of five basic groups: unicellular, colonial, tubular, filamentous, or parenchymatous.

Unicellular Forms

Among unicellular algal species can be found both motile (propelled by whiplike projections called flagella) and nonmotile types, the former being

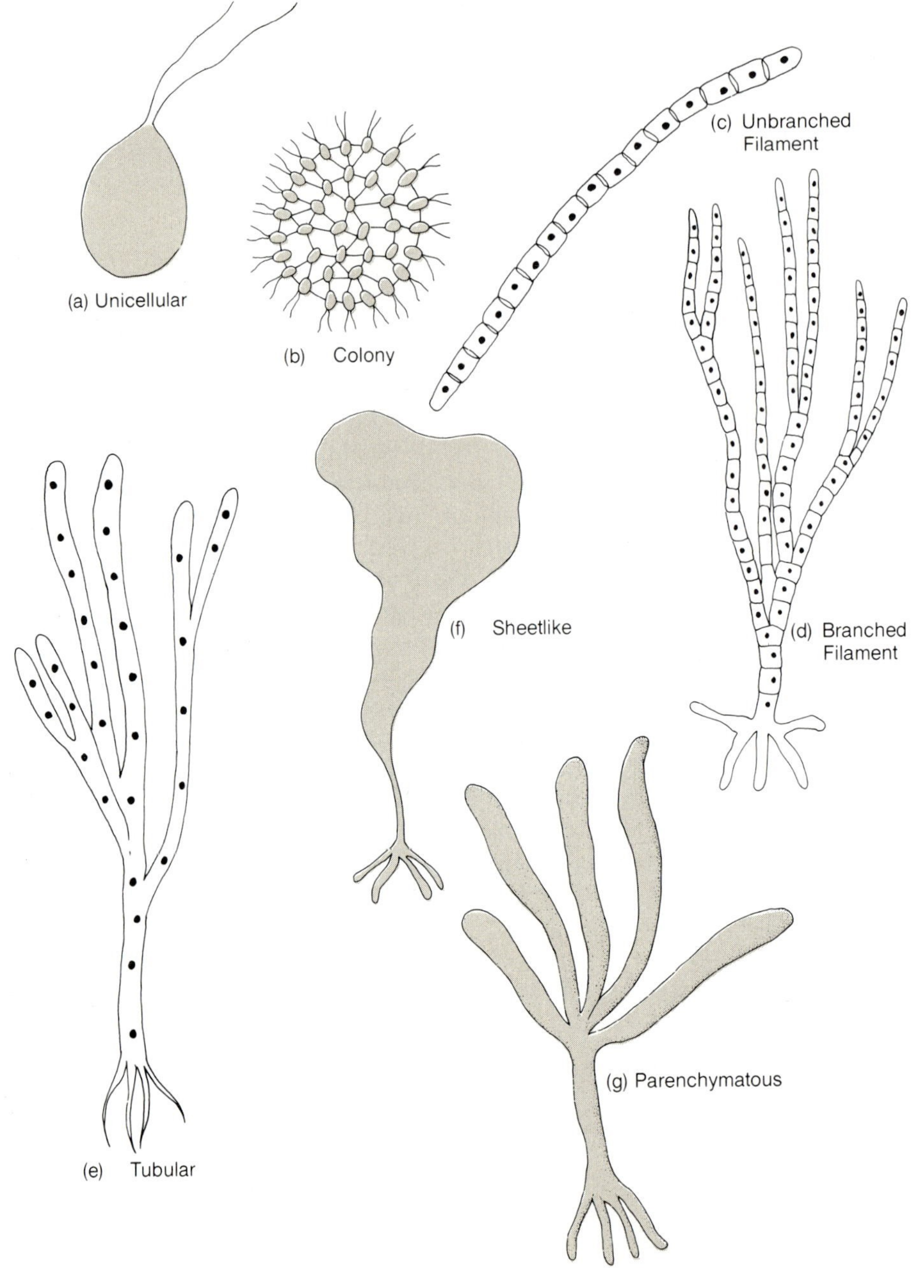

Figure 17–1. Morphological forms of algae. (*a*) Unicellular. (*b*) Colonial. Here a number of unicellular algae form an interconnected group. (*c*) and (*d*) Filamentous. Both branched and unbranched forms are shown. (*e*) Tubular. Note the absence of cross walls in this form. (*f*) Sheetlike. Cells are arranged in a single layer forming sheetlike structures. (*g*) Parenchymatous. Cells are arranged three dimensionally, forming complex branching structures.

considered more primitive. Although they may exhibit a wide variety of shapes, most unicellular algae are very small and can be seen only under a microscope.

Colonial Forms

Both motile and nonmotile unicellular algal species have given rise to a number of colonial algal species where individual cells are arranged in spherical, platelike, branched, or other configurations either within a gelatinous matrix or attached side by side. Certain species of colonial algae possess flagella and are motile, while others are free floating. Colonial forms range in size from species such as the common, nonmotile, freshwater form, *Scenedesmus,* a colony which usually consists of 4 cells arranged side by side in a single plane, to *Volvox,* a motile colony containing as many as 40,000 cells arranged on the surface of a hollow sphere, held together by a gelatinous substance.

Tubular Forms

A few algal species such as *Vaucheria,* the green velvety growth often observed on the surface of damp soil, exhibit a body consisting of extensive tubular structures with few branches and no crosswalls. Such filaments may grow to a length of several centimeters, elongation resulting from repeated nuclear divisions, unaccompanied by crosswall formation. Within such species the cytoplasm and its constituent organelles occur in a thin layer against the cell wall, while the central portion of the filament consists of a large vacuole.*

Vacuole
A membrane-bound structure that stores water and metabolic wastes.

Filamentous Forms

Probably the most familiar algal form as far as the casual observor is concerned is the filamentous type, consisting of many cells. Such species can be large in comparison with the previous three forms and can easily be observed without magnification. Filamentous algal species can be either branched or unbranched and can exhibit a certain degree of differentiation among interdependent cells.

Parenchymatous Forms

Some of the larger algal species, such as kelps, rockweeds, sea lettuce, Irish moss, and others, display three-dimensional or sheetlike arrangements of cells, with tissues bearing some similarity to those of the more evolutionarily advanced land plants.

Reproduction

Methods of reproduction among the different species of algae are extremely varied, with most groups exhibiting both asexual and sexual processes. In general, asexual reproduction predominates over sexual in terms of frequency of occurrence and contribution to population size. Some species appear to have no means of sexual reproduction at all, and in many other species it occurs only infrequently. Algae can reproduce by the formation of various types of asexual spores, which are most commonly flagellate, motile **zoospores** or, in some species, nonmotile **aplanospores,** both of which subsequently divide by mitosis* (Chapter 13), thus forming a new individual. In addition to the production of spores, algae may reproduce asexually by fragmentation when portions of the body, broken off by wave action or by the feeding activities of other aquatic creatures, regenerate new individuals.

Mitosis
The process of nuclear division that results in the production of two genetically identical daughter nuclei.

Sexual reproduction among the algae differs greatly in detail from one species to the next. Three general types—isogamous, anisogamous, or oogamous—are recognized, depending on the morphological characteristics of the gametes involved (Chapter 13). In isogamy the gametes appear morphologically identical. Isogametes usually have flagella, and upon release in water, they fuse, forming diploid zygotes. Anisogamy involves the fusion of two flagellate gametes that are identical in shape but different in size. The larger of the two is arbitrarily referred to as the female, the smaller as the male. As in isogamy, free-swimming anisogamous gametes are released into the water where they subsequently fuse, resulting in zygote formation. Oogamy occurs in species with large, nonflagellate female gametes and small, usually motile male gametes. Oogamous species produce male and female gametes within specialized structures called antheridia and oogonia, respectively. In contrast to isogamous and anisogamous species, most oogamous forms do not release both kinds of gametes into the water prior to fertilization. Instead, the female gamete (egg) in most cases remains within the oogonium and the male gamete (sperm) swims through a pore in the oogonial wall, attracted by a hormone secreted into the water by the egg.

Among isogamous, anisogamous, and oogamous species alike, some members produce gametes that are capable of fusing with other gametes from the same organism (i.e., the gametes are mutually compatible); such species are **homothallic.** Other species, called **heterothallic,** produce mutually incompatible gametes and hence require gametes from two different organisms for sexual reproduction to occur.

Algal Life Cycles

Algal life cycles are even more varied than the methods of algal reproduction; however, the life cycles, or life histories, can be categorized into three basic types. Specific examples of each will be described in detail for the various algal divisions.

Vegetative
Not involved with sexual reproduction.

Many types of algae exhibit only one free-living, vegetative* form during their life cycle. Such species are **haplobiontic.** The free-living forms of haplobiontic algae can be either haploid or diploid, depending on the species. In haplobiontic haploid types (Figure 17–2a), the haploid (1*n*) vegetative body can reproduce asexually through the release of haploid zoospores or sexually through the production of haploid gametes, both kinds of reproductive cells being produced by mitosis. Such gametes fuse, resulting in a diploid (2*n*) zygote that undergoes meiosis before it germinates. Thus the diploid phase is extremely brief and the haploid condition is reestablished upon germination of the single-celled zygote. Conversely, haplobiontic diploid species (Figure 17–2b) exhibit an equally brief haploid stage. In such types the vegetative body is diploid. Asexual reproduction by mitotically produced zoospores can occur here as in the haplobiontic haploid species, the only difference being that in this case the zoospores are diploid. The diploid vegetative algae also can produce haploid gametes, meiosis occurring in meiocytes immediately prior to gamete formation. These gametes represent the only haploid phase of the life cycle because the diploid condition is reestablished as soon as fusion occurs. The haplobiontic diploid life cycle is

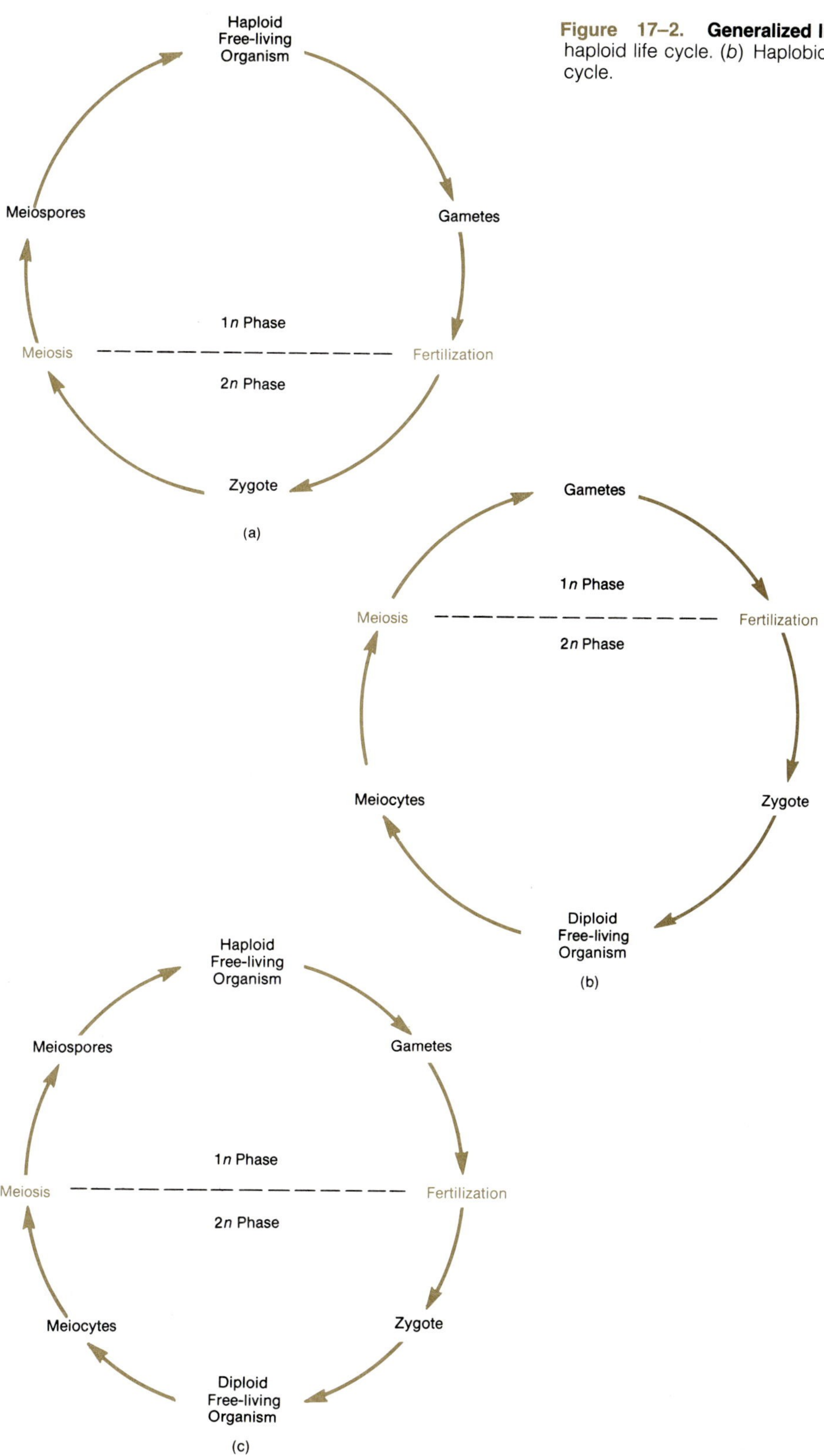

Figure 17–2. Generalized life cycles of algae. (*a*) Haplobiontic haploid life cycle. (*b*) Haplobiontic diploid life cycle. (*c*) Diplobiontic life cycle.

uncommon among the algae. It is the only type of life cycle found in the animal kingdom.

In the third type of algal life cycle, the **diplobiontic,** there occur two separate, independent, multicellular vegetative forms (Figure 17–2c). Algae exhibiting the diplobiontic life cycle have a free-living haploid form as well as a free-living diploid form. In the diploid form, meiocytes undergo meiosis to produce haploid zoospores (meiospores). These zoospores grow by mitosis into haploid organisms, which can be either unisexual* or bisexual*, depending on the species. The free-living haploid algae subsequently produce gametes by mitosis, which fuse forming a diploid zygote. This zygote undergoes mitosis producing another free-living, diploid vegetative body. The distinct haploid and diploid organisms represented by the diplobiontic life cycle may be similar in appearance (isomorphic) or may be structurally different (heteromorphic), again depending on the species. The diplobiontic type of life cycle is the same as the alternation of generations between a haploid gametophyte* phase and a diploid sporophyte* phase so characteristic of the life cycles of plants. The critical features to keep in mind when trying to distinguish among these types of algal life histories are the point at which meiosis occurs and the products of meiosis. In haplobiontic haploid species, meiosis occurs in the zygote (zygotic meiosis); in haplobiontic diploid species, meiosis occurs during gamete formation (gametic meiosis); while in diplobiontic forms meiosis occurs in a sporangium during the formation of spores (sporic meiosis).

Unisexual
Able to produce only one type of gamete.

Bisexual
Able to produce both types of gametes.

Gametophyte
The gamete-producing stage in a life cycle.

Sporophyte
The spore-producing stage in a life cycle.

In studying algal life cycles, bear in mind that in virtually all species asexual reproduction predominates and that external factors such as temperature, nutrition, and day length may affect the method of reproduction followed. Also in some species, particularly diplobiontic ones, gametes which do not fuse may germinate parthenogenetically (i.e., without fertilization) and grow into adult haploid organisms.

Ecological Importance

To human eyes algae may not appear to play a particularly important role in the biosphere* when compared to more visibly dominant forms such as the flowering plants and gymnosperms. Nevertheless, algae do play a key role in the functioning of many ecosystems.* Algae are the dominant photosynthetic organisms in aquatic habitats and, as such, are the major producers of atmospheric oxygen, along with the larger land plants. Indeed, the transformation of the earth's atmosphere from one characterized by hydrogen, methane, and ammonia, to the oxygen-rich mixture of gases found in the modern atmosphere is believed to be due to the evolution of primitive photosynthetic ancestors of present-day algae (Chapter 23). The oxygen released by algae promotes the purification of the aquatic habitat by oxidizing certain water-borne organic wastes, which might harm other organisms, and by enhancing the activities of aerobic* decomposer organisms, such as certain bacteria, which break down other organic wastes.

Biosphere
That portion of the earth's surface and atmosphere occupied by living organisms.

Ecosystem
Ecological system; the interrelationship between organisms and their physical environment in a specific location.

Aerobic
Requiring the presence of air.

Because they are capable of converting solar energy to chemical (food) energy through photosynthesis, algae are the primary producers in aquatic food chains*. All water-dwelling creatures, from the microscopic zooplankton* to the gigantic blue whale, depend directly or indirectly on algae as their source of energy. Most important in this role are the microscopic, free-floating forms of phytoplankton*, often referred to as the "grass of the sea," that occur in untold numbers in areas where upwelling currents carry the essential mineral nutrients which nourish them. The phytoplankton are food for small crustaceans, molluscs, fish, and other organisms, which, in turn, constitute the food supply of larger organisms. At the top of aquatic food chains are carnivorous fish, carnivorous birds such as eagles and pelicans, aquatic mammals such as seals, walruses, and some whales, and humans. It is well known that the number of fish present in a given body of water is directly related to the abundance of algae. Hence the most productive fishing grounds are those where mineral-laden currents support a lush algal growth. If a change in environmental conditions should adversely affect algal populations, as happened in the mid-1970s off the coast of Peru, fish stocks will drop sharply, with hardship resulting for the fishing industry.

Food chain
A series of organisms, usually starting with plants, in which members of one group consume those of the preceding group.

Zooplankton
Free-floating or motile microscopic animals.

Phytoplankton
Free-floating or motile microscopic algae.

On the other hand, in situations where an excessive amount of nutrients, such as nitrates or phosphates, are introduced into bodies of water, the resulting algal "blooms" may seriously degrade the quality of the aquatic environment leading to the depletion of dissolved oxygen and production of toxins. This process, known as **eutrophication,** results in significant changes in the species composition of the affected aquatic ecosystem.

Classification of Algae

Years ago algae were classified primarily on the basis of which pigments they possessed and were accordingly divided into groups—the blue-green algae, green algae, brown algae, and red algae. Today phycologists* recognize that other characteristics, such as the type of food stored, the nature of the cell wall, the number and arrangement of flagella, and so forth, are equally important in determining the evolutionary relationships that form the basis of our modern system of classification. The current system of algal classification has arisen as a result of the present emphasis on a variety of characteristics rather than on pigmentation alone. As indicated in Chapter 15, this approach has led to a reclassification of the blue-green algae as Division Cyanophyta in Kingdom Monera rather than as algae in Kingdom Protista. The algae are now generally classified into six divisions of Kingdom Protista: Chlorophyta (green algae), Euglenophyta (euglenoids), Pyrrophyta (dinoflagellates), Chrysophyta (golden-brown algae), Phaeophyta (brown algae), and Rhodophyta (red algae). Since the Chlorophyta are considered by most researchers to be the group that gave rise to the evolutionary line leading to higher plants, some specific examples of this diverse assemblage will be examined in detail, followed by a briefer look at the other divisions that make up the varied and phylogenetically unrelated group of organisms collectively termed *algae.*

Phycologist
A botanist who specializes in the study of algae.

Division Chlorophyta (Green Algae)

The widespread and morphologically diverse members of the Chlorophyta include approximately 6000 species in which the dominant pigments chlorophyll *a* and *b*, the same as those present in higher plants, give these algae their characteristic grass-green color. As in more evolutionarily advanced plant groups, the chlorophyll present in green algae is localized in chloroplasts, and the considerable range of sizes and shapes of chloroplasts constitutes an important characteristic in the classification of the various species. Reserve food is stored in the form of starch, again an indication of the evolutionary relationship between green algae and the land plants. In many species of Chlorophyta the formation of starch is associated with structures called **pyrenoids,** one or more of which may be located within each chloroplast. The walls of most vegetative green algal cells are composed of cellulose, although a few of the more primitive species lack a true wall. Most members of the group produce either motile vegetative or reproductive cells or both, which usually bear 2 or 4 whiplike flagella of equal length on the front end of the cell. In body form, the Chlorophyta exhibit all of the types described in the preceding section of diversity of form; methods of reproduction and life history patterns are equally varied.

Green algae can be found in all aquatic habitats. About 90% of the species are freshwater and terrestrial types, while the remainder are marine. Some aquatic species are anchored in sediments at the bottom, while others float or swim near the surface. The depth at which algae can grow is limited by the ability of the proper wavelength of light to penetrate the water. Certain species grow as epiphytes* on other aquatic plants and algae, on the backs of turtles or shellfish, and on rocks or logs. Some live symbiotically within the cells of sponges, coelenterates, or protozoans, or as the algal partner in some lichens.

Epiphyte
A plant that grows upon another plant without receiving any water or nutrients from it.

The green algae provide excellent examples of the major evolutionary trends exhibited by the algae as a group. This will be seen from the descriptions of several representative species.

Chlamydomonas

With a motile, unicellular form, *Chlamydomonas* is perhaps the most primitive species of the green algae (Figure 17–3). Commonly found on damp soil, swimming in fresh-water ponds, or even in public swimming pools, *Chlamydomonas* can occur so abundantly that it may give the water a greenish tinge. An individual consists of an egg-shaped cell containing just one large, cup-shaped chloroplast that fills most of the cell and in which are embedded one or two pyrenoids and a light-sensitive eyespot. The nucleus is located in the open area near the anterior* end of the cell, as are two contractile vacuoles that appear to function in the excretion of wastes. Two flagella extending from the anterior end of the cell propel the organism through the water.

Anterior
At the front.

Chlamydomonas reproduces asexually through the formation of either 2, 4, or 8 zoospores within the parent cell. These are identical to, though smaller than, the parent. With the breakdown of the parent cell wall, the zoospores swim out and grow into mature individuals. Sexual reproduction is isoga-

Figure 17–3. **Chlamydomonas.** *Chlamydomonas* is a unicellular, motile, extremely small green alga. (*a*) Drawing of *Chlamydomonas* showing its parts. (*b*) An electron micrograph of a *Chlamydomonas* section.

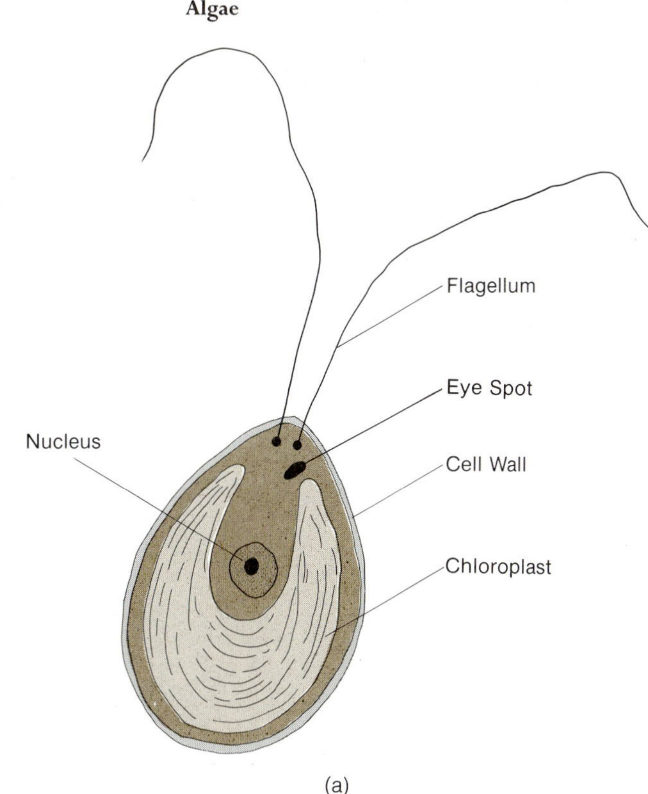

(a)

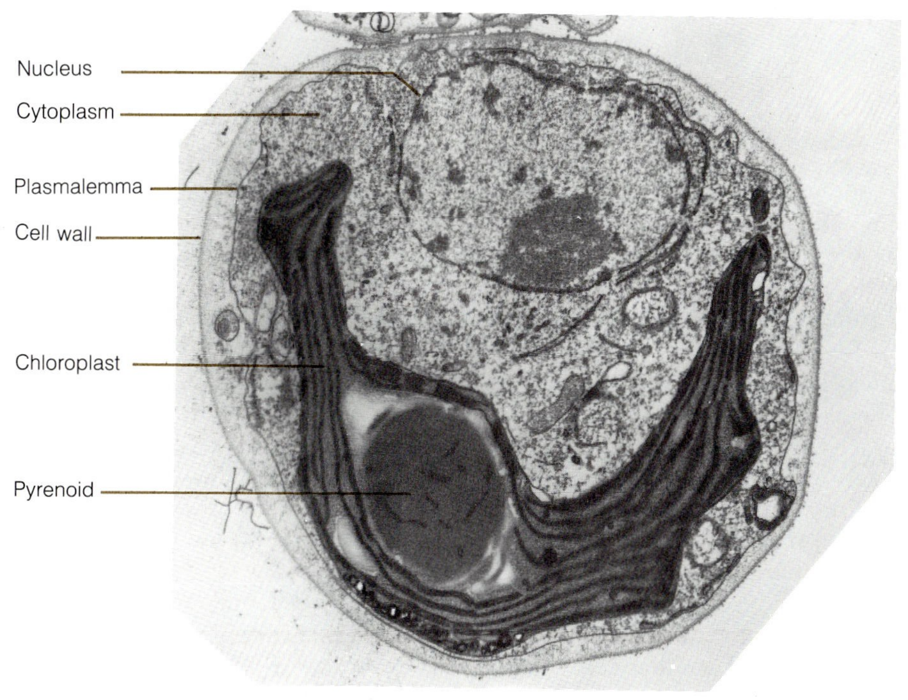

(b)

Figure 17–4. ***Chlamydomonas* life cycle.**

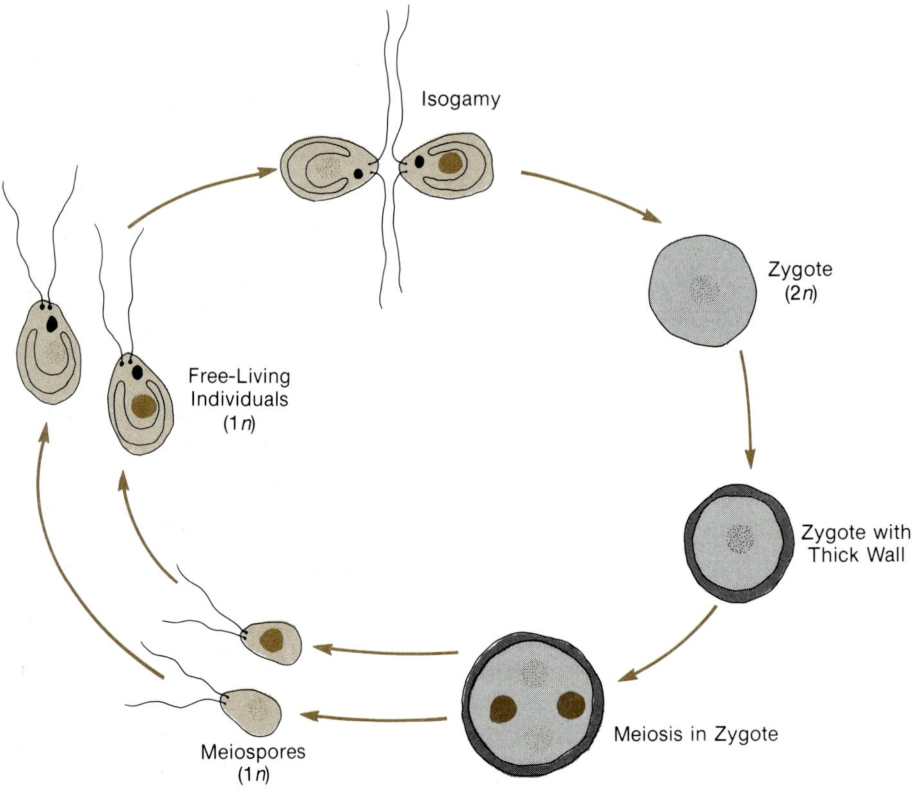

Chlamydomonas Life Cycle.

mous. As many as 32, but more commonly 8 isogametes, each resembling a miniature *Chlamydomonas,* form within the parent. After release, isogametes may fuse, producing a zygote. Meiosis accompanies germination of the zygote; depending on the species, this results in the release of 4 or 8 haploid zoospores, which develop into new individuals. *Chlamydomonas* is an example of the haplobiontic haploid life cycle (Figure 17–4) characterized by zygotic meiosis.

Volvox

Representing an evolutionary line of development leading from motile unicells to motile colonies, *Volvox* is a large, colonial, freshwater alga made up of anywhere from a few hundred to 40,000 *Chlamydomonas*-like cells that are arranged in a single layer over the periphery of a gelatinous sphere (Figure 17–5). Because of its size, *Volvox* easily can be seen without the aid of a microscope. Somewhat resembling a tiny green ball rolling through the water, it has been recognized by naturalists for several hundreds years. Most of the cells of *Volvox* are solely vegetative, incapable of any type of reproduction. Asexual reproduction occurs when certain cells enlarge and move slightly below the surface of the sphere. Here they undergo repeated mitotic

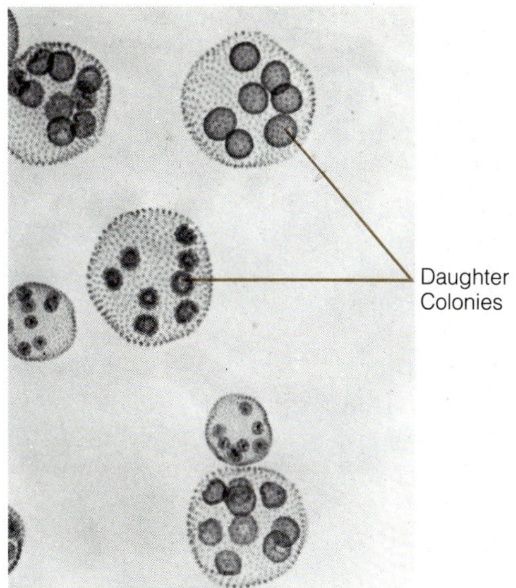

Figure 17–5. Photomicrograph of a *Volvox* colony. Individual members of a colony are interconnected with protoplasmic strands. Dense bodies in the colonies are newly produced colonies that will be released when the parent colony breaks open.

divisions and form a mass of cells that subsequently becomes oriented in the typical spherical arrangement of the parent. Eventually the daughter colony escapes from the parent and forms an independent colony.

Sexual reproduction in *Volvox* is oogamous; some species have both antheridia and oogonia within the same colony, but in most species the individual colonies are either one sex or the other. Within the antheridium numerous biflagellate sperm are produced. Upon release, the sperm swim to the oogonia, presumably guided by a chemical attractant, and there penetrate the single, large, nonmotile egg, thus accomplishing fertilization. The resulting zygote develops a thick wall and undergoes a resting period. With the disintegration of the parent colony, the dormant zygote is released and undergoes meiosis just prior to germination. Like *Chlamydomonas* to which it is closely related, *Volvox* has a haplobiontic haploid life cycle with zygotic meiosis.

Oedogonium

Another common freshwater alga, *Oedogonium* exhibits a multicellular, unbranched, filamentous type of growth with a basal holdfast. Each cell possesses a single nucleus and one netlike chloroplast containing many pyrenoids. *Oedogonium* can reproduce asexually either by fragmentation or by the production of zoospores, which in this species are characterized by a ring of flagella around the anterior end, leaving a little "bald spot" in the center. One of the most distinctive features of this alga, however, is its method of sexual reproduction (Figure 17–6). *Oedogonium* is oogamous, producing a large egg within an oogonium, and two sperm, resembling miniature zoospores, within each of several antheridia. The oogonia, rounded cells that are larger than the vegetative cells along the filament, have a small opening

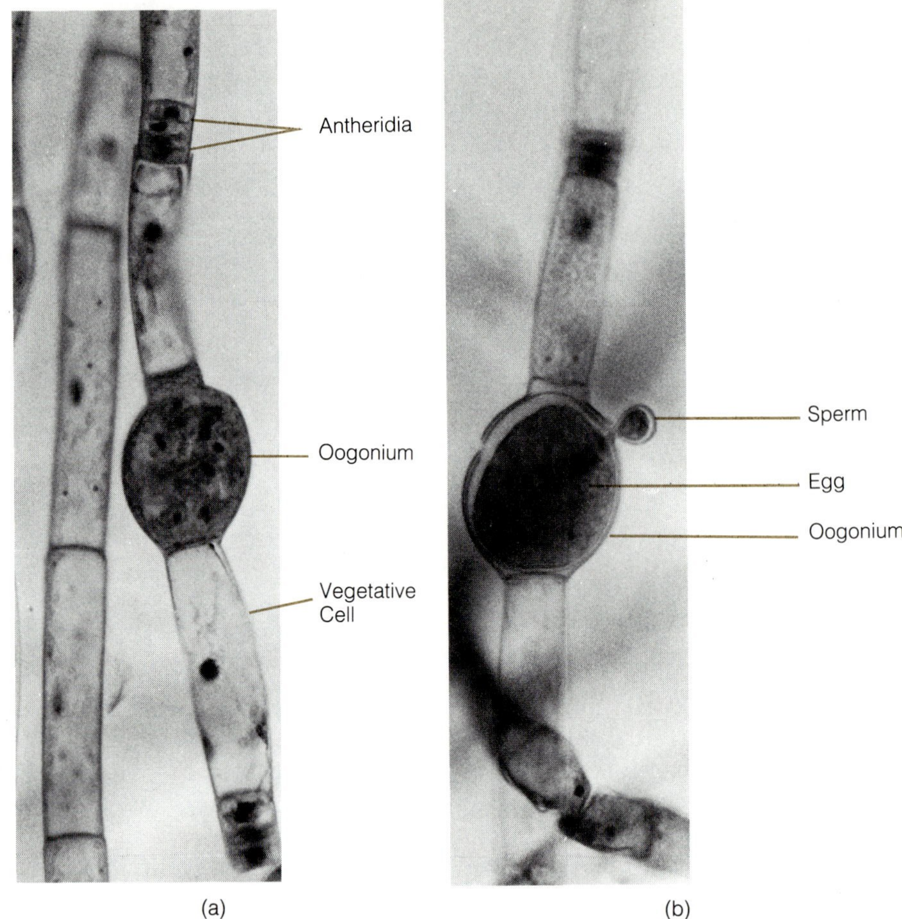

Figure 17–6. Photomicrograph of *Oedogonium.* (*a*) Segments of filament containing small, rectangular antheridia and large, spherical oogonium. Vegetative cells are cylindrically shaped. (*b*) A sperm is seen entering the oogonium. The dense body filling the oogonium is the egg.

in their wall through which sperm can swim to fertilize the nonmotile egg. The zygote, subsequently released from the filament, can remain dormant for some time before germinating to release 4 haploid zoospores. As in *Chlamydomonas, Oedogonium* exhibits zygotic meiosis, the vegetative free-living stage being entirely haploid (Figure 17–7).

Spirogyra

Spirogyra is an unbranched, filamentous, freshwater-pond alga sometimes called "water silk" because of the slippery texture resulting from the mucilaginous sheath covering its cells. Its most distinctive feature is its elongated helical chloroplast. *Spirogyra* has an unusual method of isogamous sexual reproduction in that neither zoospores nor flagellate gametes are produced. Instead, cells of two filaments that are in contact with each other form projections which develop into a **conjugation tube** (Figure 17–8). The contents of one of the cells moves into the other where they fuse, forming the diploid

Figure 17–7. ***Oedogonium* life cycle.**

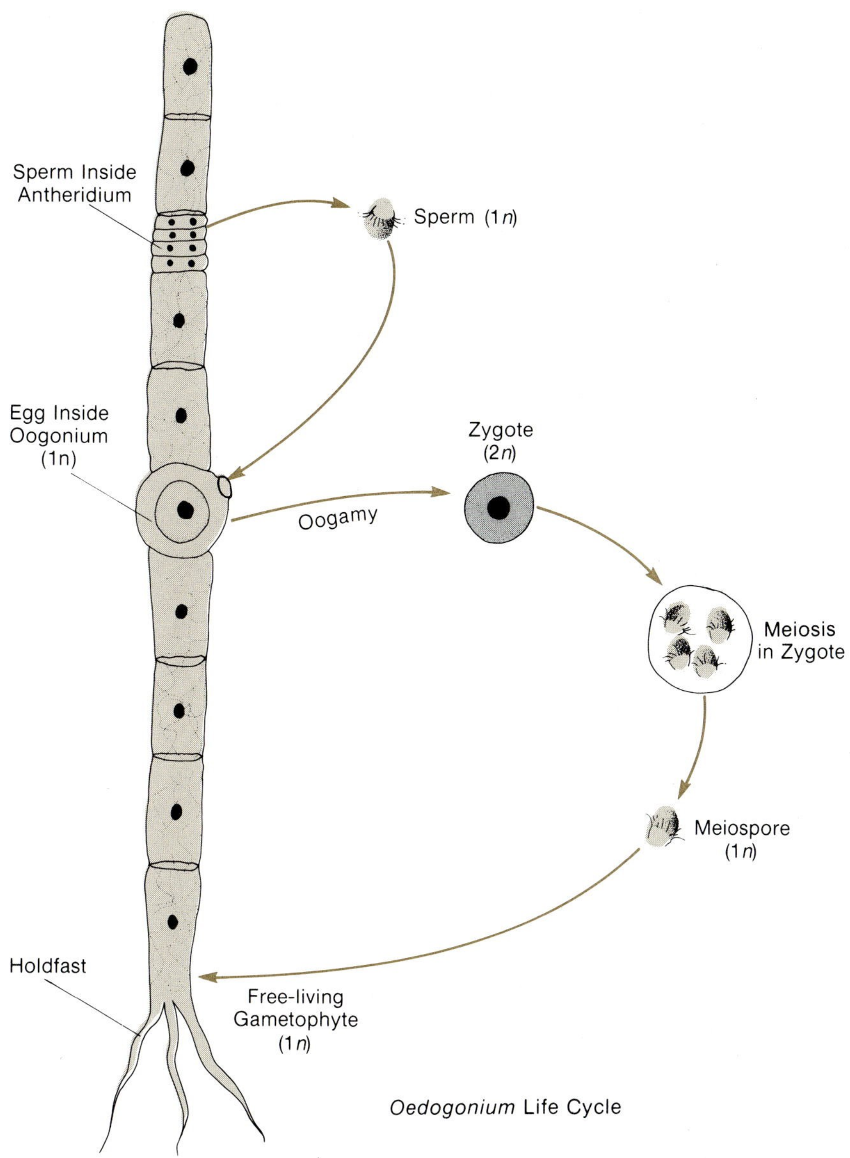

zygote. Since the contents of each cell are morphologically similar, conjugation is regarded as a specialized type of isogamy. As in the preceding examples, meiosis in *Spirogyra* is zygotic.

Acetabularia

Unlike *Spirogyra,* the marine alga *Acetabularia,* or "mermaid's wineglass," is diploid during most of its life and thus represents the haplobiontic diploid type of life cycle (Figure 17–9). Although the algal body, which resembles a small umbrella, can grow to a height of 4 or 5 inches, it consists of only one cell with the single nucleus located at the base of the stalk. At the time of

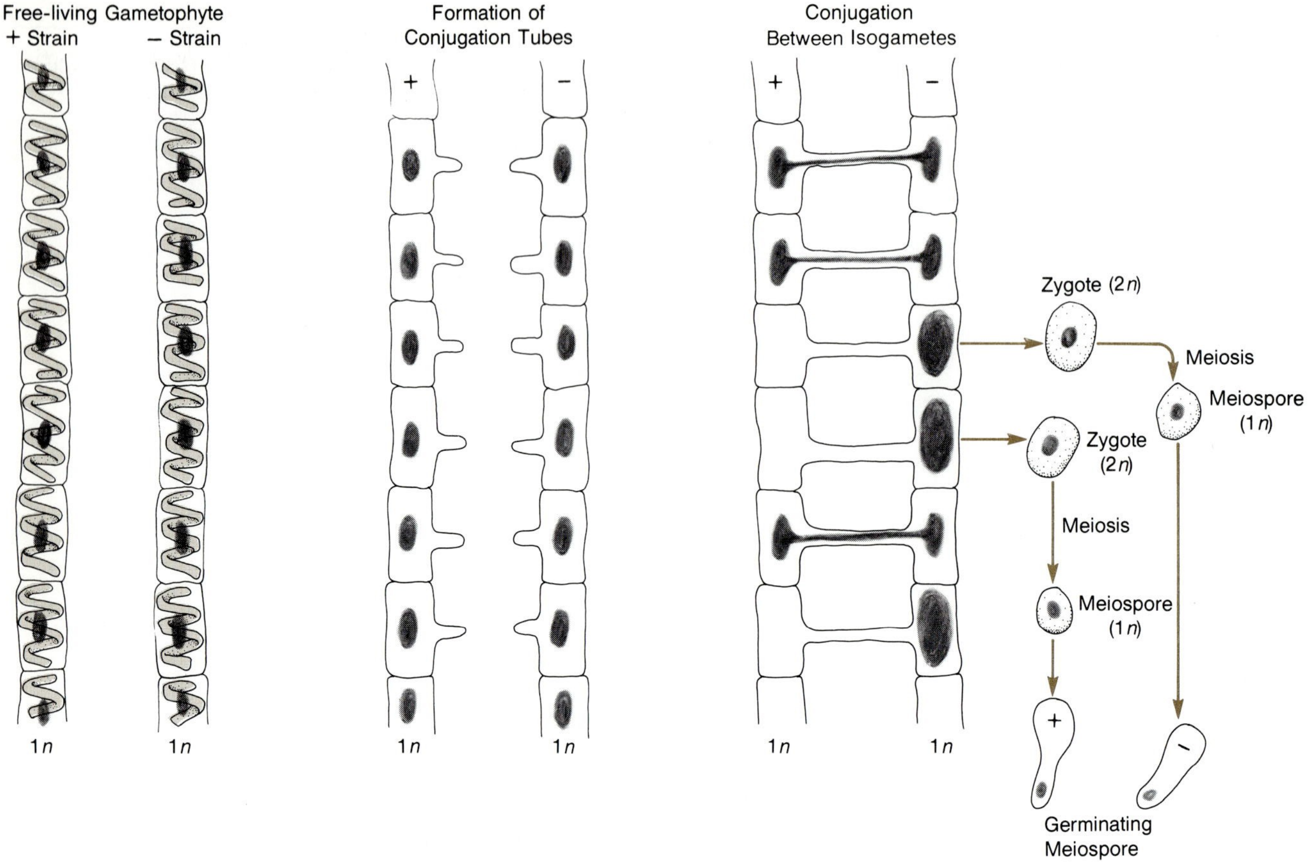

Free-living Gametophyte
+ Strain — Strain

Formation of
Conjugation Tubes

Conjugation
Between Isogametes

+ − + − + −

1n 1n 1n 1n 1n 1n

Zygote (2n)

Meiosis

Meiospore
(1n)

Zygote
(2n)

Meiosis

Meiospore
(1n)

Germinating
Meiospore

(a)

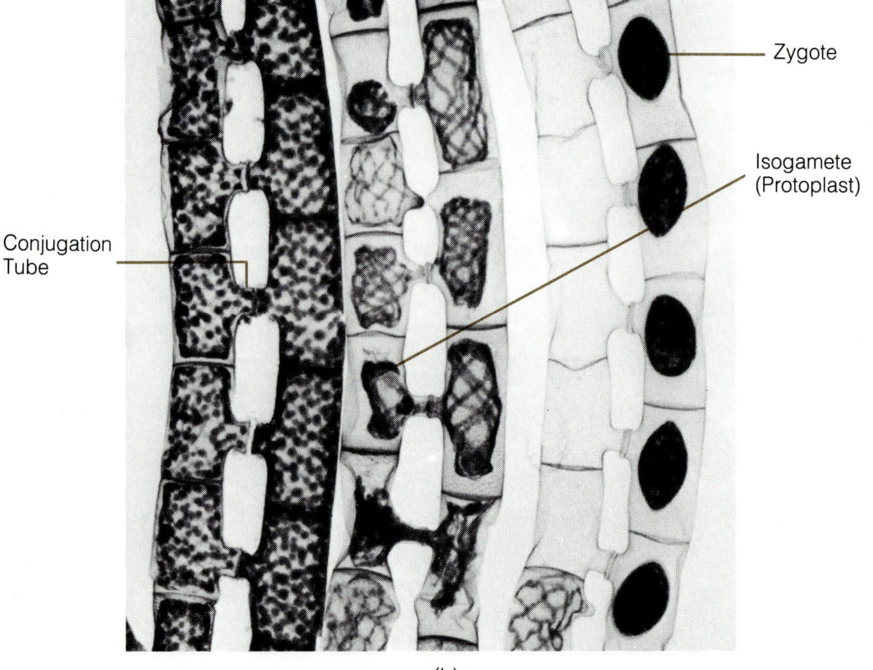

(b)

Figure 17–8. ***Spirogyra*** (*a*) Life cycle drawing. (*b*) Photomicrograph of three successive stages in the life cycle of *Spirogyra*.

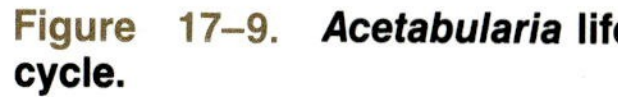

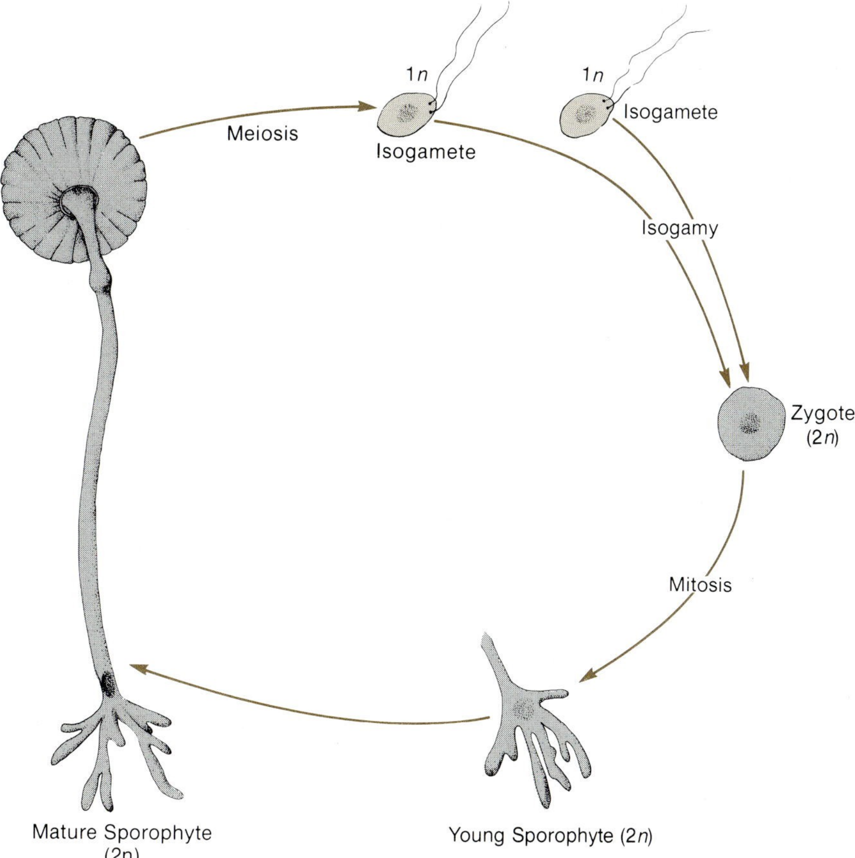

Figure 17–9. ***Acetabularia*** **life cycle.**

sexual reproduction, the nucleus enlarges considerably and undergoes numerous mitotic divisions. The resulting nuclei are carried, by cytoplasmic streaming, up the stalk to the gametangial rays, which comprise the "umbrella" portion of the alga. After a series of intermediate steps, meiosis occurs and many thousands of isogametes are released into the water. The zygotes formed by the fusion of these gametes develop into new diploid individuals.

Ulva

Called "sea lettuce" because its thin, expanded, two-cell-thick body gives it a superficial resemblance to common lettuce, *Ulva* is a marine alga found growing attached to rocks in the intertidal zone. *Ulva*, a heterothallic alga, exhibits a distinctive isomorphic alternation of generations and a diplobiontic life cycle (Figure 17–10). Haploid individuals produce gametes by mitosis (isogametes in some species, anisogametes in others) which fuse, forming the diploid zygote that develops into the diploid individual, identical in size and

Figure 17–10. *Ulva* life cycle.

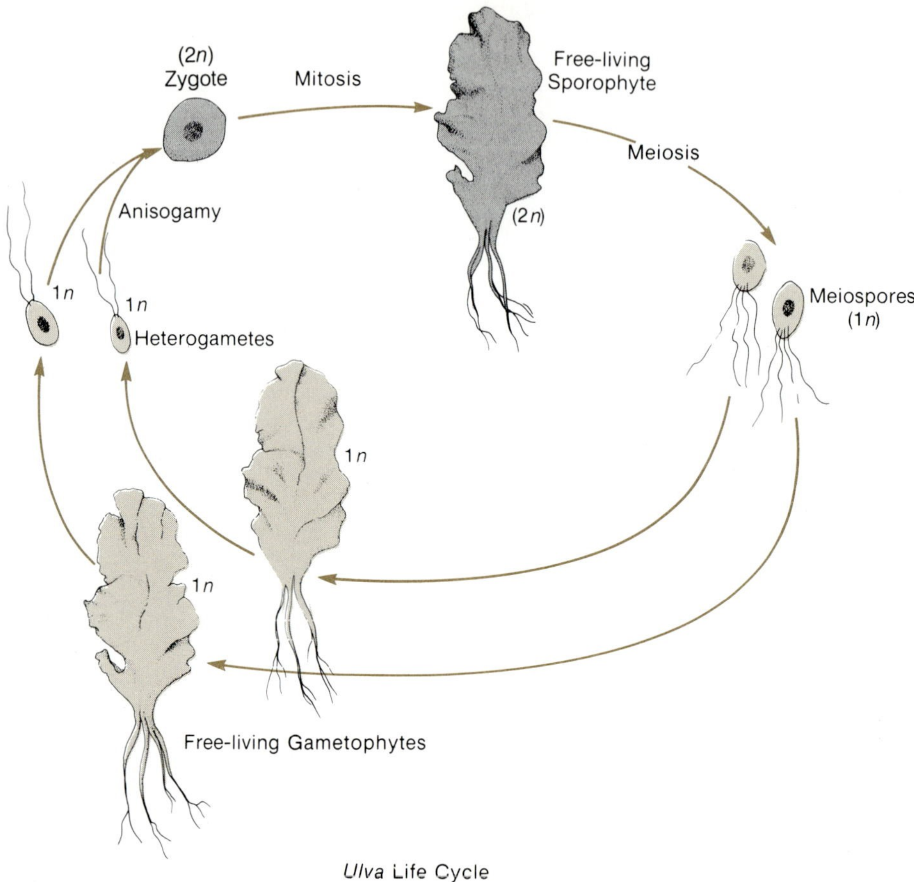

appearance to the haploid one. Sporic meiosis occurs in the diploid individual, with haploid quadriflagellate (with 4 flagella) zoospores (meiospores) being released to reestablish the haploid plant.

Economic Importance of Green Algae

In comparison with some of the other divisions of algae, members of the Chlorophyta have not been utilized commercially to any extent. A few of the filamentous species, such as *Oedogonium* and *Spirogyra,* occasionally have been used as ingredients in soup, but the only green alga to achieve much acceptance as a human food is *Ulva,* added to both soups and salads. In recent years there has been considerable interest in the possibility of mass culturing *Chlorella,* a tiny, nonmotile, unicellular green alga, for use as a protein supplement in human diets. Research has also been conducted on the feasibility of using cultures of microscopic algae such as *Chlorella* to regulate the amount of oxygen and carbon dioxide in space ships and atomic submarines. On the negative side, the algal blooms that occur when a body of water undergoes eutrophication may result in fish kills, unpleasant odors of decomposition, and general degradation of water quality which can have an adverse economic impact.

Division Euglenophyta (Euglenoids)

The euglenoids provide a good example of the type of organism that stimulates debate over the validity of the two-kingdom system of classification. These unicellular organisms (Figure 17–11) lack the cellulose cell wall characteristic of plants. Rather, their protoplast is surrounded by a tough membrane called a **periplast,** or **pellicle.** The lack of a rigid cell wall permits most euglenoids to change shape freely, although some species have such rigid pellicles that their shape remains constant. Animal characteristics of euglenoids in addition to naked cells are the presence of a flagellum in the vegetative cell, the occurrence of an eyespot, and the fact that many species lack any photosynthetic pigment, and thus they must absorb or ingest their food. On the other hand, those euglenoids that manufacture their own food display identical chlorophylls to higher plants. They store reserve food in the form of **paramylum,** an insoluble carbohydrate similar to starch, or as fats. Euglenoids reproduce only asexually, undergoing mitosis followed by a longitudinal cytoplasmic division that begins at the anterior end of the cell. Since meiosis has never been observed in the Euglenophyta, it is not known if the cells are haploid or diploid.

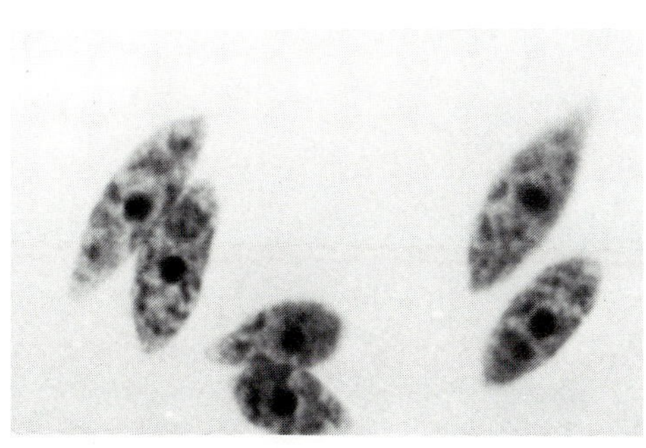

(a)

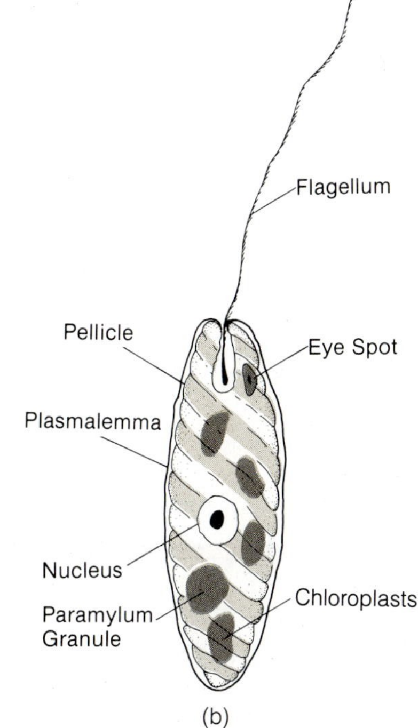

(b)

Figure 17–11. *Euglena.* (*a*) Photomicrograph of *Euglena* cells. (*b*) Drawing of a *Euglena* cell. These motile cells have a pellicle that is arranged helically forming a flexible framework enclosed by the plasmalemma.

Euglena is found almost exclusively in fresh water, particularly in small ponds rich in organic matter; occasionally it is observed on muddy banks of streams, in saltwater marshes, or even on snow. At certain times, particularly when chemical fertilizers or manures are washed into farm ponds, *Euglena* "blooms"* may occur, coloring the water. Thus the presence of large numbers of these organisms is used as an indicication of polluted conditions in lakes or ponds.

Bloom
An extremely high density of an algal species.

Division Pyrrophyta (Dinoflagellates)

The dinoflagellates, as most members of the Pyrrophyta are called, are another group of organisms that were at one time considered animals because of the presence of flagella and because of their ability to ingest particles of food. Nevertheless, because the majority of dinoflagellates possess a cellulose cell wall and are photosynthetic, botanists considered them to be plants. Now they are placed in the Kingdom Protista.

Dinoflagellates include both fresh- and saltwater species, most of which are unicellular (Figure 17–12). They are characterized by the presence of two flagella of unequal length, both of which are anchored in a central groove that divides the cell into anterior and posterior portions. The conspicuous pigment in these algae is a type of xanthophyll that imparts a brownish-yellow color. Starch and oils comprise the reserve food. Reproduc-

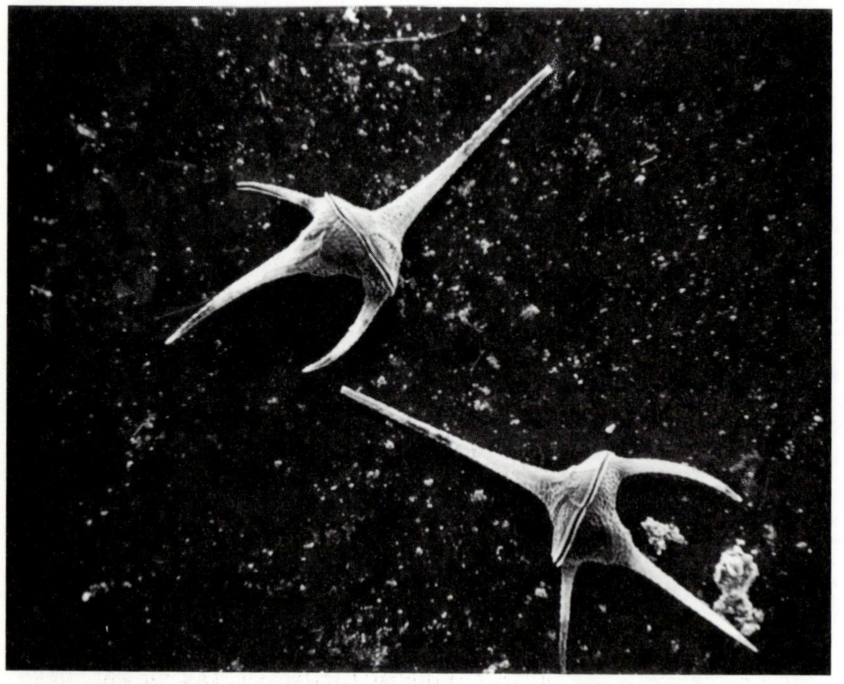

(a)

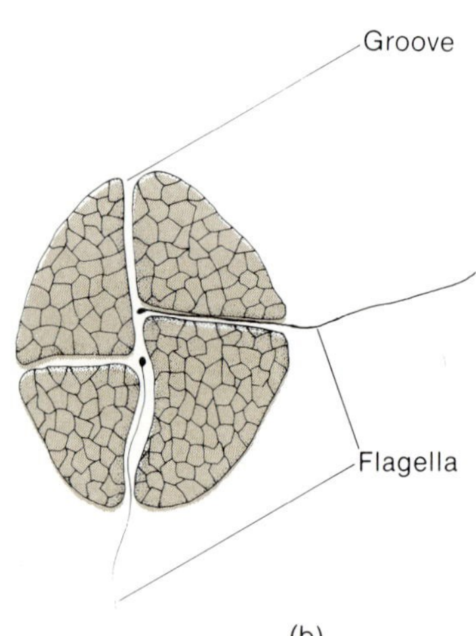

(b)

Figure 17–12. Dinoflagellates. (*a*) Scanning electron micrograph of a dinoflagellate. (*b*) Drawing of a dinoflagellate showing cellulose plates that are enclosed by the plasmalemma. One groove girdles the cell completely while the other runs along one side of the cell only. Each groove contains a single flagellum.

tion among the Pyrrophyta is almost entirely asexual by means of cell division, zoospores, or aplanospores.

At sporadic intervals certain species of dinoflagellates experience "blooms" that can result in serious economic loss, inconvenience, or illness to humans. The genus *Gymnodinium* includes species that occasionally develop a reddish color and reproduce in great numbers off the coasts of Florida and southern California. These "red tides," as the *Gymnodinium*-infested waters are called, can result in the deaths of thousands of fish owing to the release of a neurotoxin from the living dinoflagellates. Humans swimming in affected waters can suffer eye, throat, and mouth irritations. *Gonyaulax,* another genus, is responsible for outbreaks of paralytic shellfish poisoning (PSP). This ailment, which can be fatal, occurs when people eat clams, scallops, mussels, or other filter feeders, which have accumulated *Gonyaulax* toxins in their tissues. Though the most common symptom of PSP is a sensation of numbness in the lips, tongue, and fingertips soon after consuming the contaminated shellfish, in severe cases paralysis of the diaphragm can result in respiratory failure and death.

Dinoflagellate genera such as *Pyrocystis,* and some species of *Gonyaulax* and *Noctiluca* (meaning "night light") are luminescent planktonic organisms, responsible for the "glowing wake" frequently observed when ships cruise through some tropical waters. When a jar of water containing myriads of these dinoflagellates is agitated, its glow is almost as bright as that of a fluorescent tube. This light-emitting property of dinoflagellates, as well as of certain animals and fungi, is termed *bioluminescence.*

Division Chrysophyta (Golden-Brown Algae)

Often called the "jewels of the sea," diatoms are the most numerous and most economically important of the golden-brown algae. Over 5000 species of diatoms have been described, representing every conceivable moist habitat, both marine and freshwater. Although diatoms include some filamentous and colonial species, the majority are single-celled organisms (Figure 17–13) that are constructed of two overlapping parts, with the top half slightly larger than the bottom, much like a laboratory petri dish*. Most marine diatoms are radially symmetrical,* while freshwater species are for the most part bilaterally symmetrical.* The cell walls of diatoms are composed primarily of silica, which accounts for the persistence of the skeleton-like remains long after the protoplast has died. These cell walls frequently exhibit intricate and beautiful patterns in the form of ridges, striations, or small pores, which are spaced at such precise intervals that the silica skeletons can be used for testing the resolving power* of microscopes.

The conspicuous pigments are xanthophylls and carotenes, which mask the chlorophyll and impart a golden-brown or yellow-brown color to these algae. Reserve food is in the form of oils or the carbohydrate **leucosin** rather than starch.

Petri dish
A shallow, circular glass or plastic container and its overlapping cover, used for growing microorganisms.

Radially symmetrical
Able to be divided into two equal halves by any line along a radius.

Bilaterally symmetrical
Able to be divided into two equal halves only along one line.

Resolving power
The ability of an optical system to produce detailed images.

(a)

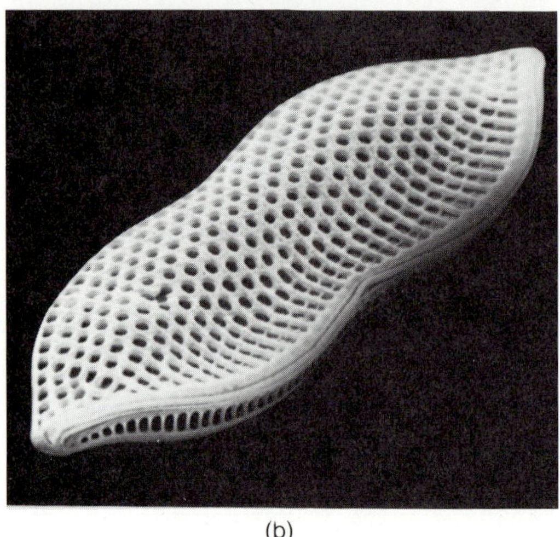
(b)

Figure 17–13. Scanning electron micrograph of diatoms. The markings are symmetrically arranged on both parts of the cell wall. The size, shape, and markings of diatoms vary. (Micrograph (a) by Dennis Banks and (b) by Laurie Overton)

Asexual reproduction among the diatoms is primarily by simple cell division. Sexual reproduction is oogamous in one group of diatoms, with vegetative cells producing gametes by meiosis. In the other diatoms naked protoplasts that function as isogametes are released after meiosis in the vegetative cells. In both cases the zygote gives rise directly to the diploid vegetative cell (haplobiontic diploid life cycle).

Diatoms are of considerable economic importance, both for their role as primary producers in aquatic food chains and for their industrial uses. Fossil deposits of diatoms, forming 100-meter-thick beds of grayish-white **diatomaceous earth,** are being mined and processed for use in many kinds of filtration processes. The chemically inert siliceous cell walls are useful in silver polish and tooth powder, in paint for removing the gloss in making flat enamels, in fertilizer to prevent caking, and for a variety of other purposes.

The division Chrysophyta includes two other classes in addition to the group represented by the diatoms. Two representative genera, *Dinobryon* and *Synura,* are of some interest because they are the source of taste and odor problems in some municipal water supplies.

Division Phaeophyta (Brown Algae)

For people living along northern coastlines, the brown algae are undoubtedly the most familiar algal species. Members of the Phaeophyta are almost entirely marine and commonly grow in the intertidal zone* of colder sea waters. Most grow attached to rocks (e.g. *Fucus* spp., Plate 8A), muddy bottoms, or shells, but a few (e.g., *Sargassum* spp.) are free floating. Brown algae are among the most conspicuous forms of algae since all are multicellular and most are large in comparison with members of the preceding divisions. The larger forms exhibit a three-dimension cell arrangement, resulting in the formation of a parenchymatous like tissue. The kelp *Macrocystis,* the largest algal species, can attain a length of more than 50 meters (160 feet). As in the other algal divisions, chlorophyll is the photosynthetic pigment, but in the Phaeophyta it is masked by the presence of the brown xanthophyll pigment, **fucoxanthin,** a characteristic of the Phaeophyta and the Chrysophyta. Reserve food is stored mainly as the carbohydrate **laminarin.**

Intertidal zone
The shoreline region between the high and low tide marks.

The body of many species of brown algae has a definite shape which bears a superficial resemblance to that of higher plants. Generally there is a rootlike holdfast that attaches the alga to a substrate, a stemlike stipe, and one or more leaflike blades. Some of the larger species display balloonlike air bladders that cause the blades or stipes to float near the surface of the water where photosynthetic activity will be maximized.

Among the Phaeophyta, reproduction may be asexual by fragmentation or zoospores, or sexual by isogamy, anisogamy, or oogamy, depending on the species. Almost all brown algae exhibit a distinct alternation of generations (i.e., they are diplobiontic) in which the diploid stage is most conspicuous.

Brown algae are of considerable economic importance largely due to their production of **algin,** a gelatinous substance secreted on their outer surface that gives these algae a characteristic slimy texture. Algin is widely used commercially as a stabilizer in foods and as a means to retain moisture in such products as ice cream, mayonnaise, marshmallows, frosting, candy, and more. Algin is employed extensively in the cosmetics industry in skin lotions and hair creams. In parts of the Orient, especially in Japan, brown algae are eaten directly as human food, while in parts of Europe they are utilized as livestock fodder. In past years brown algae were extremely important sources of both iodine and potassium and were often harvested by farmers for use as fertilizers.

Division Rhodophyta (Red Algae)

Red algae, like the browns, are predominantly saltwater species, more numerous in warm waters than in cold. They are most commonly found growing attached to rocks, shells, or to other algae either near the shore or in deep waters. Red algae generally die if detached from the substrate on which they are growing. They are so named because of the predominance of the red pigment **phycoerythrin,** in addition to chlorophyll *a* and **phycocyanin.**

Most species are filamentous, though some are membranous, superficially resembling *Ulva,* and a few are colonial or unicellular. Compared to the brown algae, most red algae (Figure 17–14) are relatively small, averaging just a few centimeters long, although a few species may attain lengths of 1 meter.

Rhodophyta are unique among the other algal divisions in that they produce no flagellate reproductive cells whatsoever. Sexual reproduction seems to be more prevalent than asexual. The nonmotile male gametes are carried by water currents to the structures containing the female gametes (i.e., oogamy). Although the life cycles of most red algae are not well known, many, though not all, are diplobiontic and exhibit a definite alternation of generations.

Figure 17–14. Red algae. Photograph of *Polysiphonia,* which grows to a length of 15–20 cm. Most red algae are much smaller than this.

Red algae such as *Porphyra* (laver) and *Rhodymenia* (dulse) are eaten by humans in both the Orient and in parts of Europe. Dulse is also fed to sheep and cattle. A species of *Gelidium* is the source of **agar,** widely used in biological laboratories as a medium for culturing bacteria or fungi. Another important derivative of the red algae, particularly from *Chondrus crispus* (Irish moss) is **carrageenan.** This product is widely used as a stabilizer in the manufacture of ice cream, chocolate milk, and other food products, as a thickener in shaving creams, and as a binder in toothpaste. Some Rhodophyta species are of ecological importance in that they have coralline walls (i.e., cell walls that contain calcium carbonate) and thus are able to contribute to the formation of "coral" reefs.

Summary

Algae represent a large, heterogeneous group of primitive organisms that are phylogenetically unrelated. These diverse groups are collectively called algae because they share the following characteristics: absence of vascular tissue, presence of unicellular gametangia, presence of photosynthetic pigments, presence of cellulose walls, absence of embryos, and absence of epidermal tissue. They are widely distributed in both freshwater and marine habitats as well as in moist terrestrial environments. The different species of algae exhibit a vast range of sizes and forms such as unicellular, colonial, tubular, filamentous, and parenchymatous.

Most groups exhibit both asexual and sexual reproduction, the former being the primary means of reproduction among most species. Three types of sexual reproduction occur: isogamy, anisogamy, and oogamy. Homothallic species are those in which the gametes produced by the same plant can fuse with each other, while in heterothallic species gametes of the same plant are mutually incompatible.

Three basic types of life cycles are recognized among algae: haplobiontic haploid, haplobiontic diploid, and diplobiontic. The algae that exhibit only one free-living form during their life cycle are haplobiontic. If the free-living form is haploid and the diploid stage is represented only by the zygote, then the life cycle is said to be haplobiontic haploid. If the free-living form is diploid and the haploid stage is represented by the gametes, the life cycle is haplobiontic diploid. The life cycle is diplobiontic when there are two independent organisms, one haploid and the other diploid, during the life cycle.

Algae are of great ecological importance. As the dominant photosynthetic organisms in aquatic habitats, they are the primary producers in aquatic food chains and thus are the organisms on which all others depend for their survival.

The classification of algae is based primarily on photosynthetic pigments; however, other factors, such as reserve storage materials, cell wall composition, and number and arrangement of flagella are also used. Presently the eukaryotic algae are classified under 6 divisions: Chlorophyta (green algae), Euglenophyta (euglenoids), Pyrrophyta (dinoflagellates), Chrysophyta (golden-brown algae), Phaeophyta (brown algae), and Rhodophyta (red algae). Of these, the Chlorophyta are recognized as having given rise to the present-day higher plants, while all the others are evolutionary dead ends.

Review Questions

1. What are the reasons for grouping together a number of phylogenetically unrelated divisions under the collective name of algae?
2. Describe the diversity of morphological forms among the algae.
3. Describe the different modes of reproduction among the algae.
4. Describe the three basic types of life cycles known among the algae.
5. List the different algal divisions and give a characteristic feature of each.
6. What is the basis for the claim that the land plants evolved from a green algal ancestor?

Saving California's Kelp Forests

Off the coast of central and southern California, extensive stands of the world's largest and fastest-growing marine alga—the giant kelp *Macrocystis pyrifera*, a brown alga—at one time formed an underwater "forest" covering more than 100 square miles. In recent decades, however, the kelp beds have been severely decimated by the onslaught of hordes of sea urchins (spiny, marine animals related to starfish), which devoured the kelp and virtually transformed the area into an undersea "desert." The cause of the urchin invasion was indirectly a result of human lack of awareness of the dangers of tampering with marine ecosystems. In the early 1900s, furhunters came close to exterminating the sea otter, which until then had preyed extensively on sea urchins, keeping their numbers under control. During recent decades, increasingly large amounts of human sewage dumped into the ocean by coastal communities polluted the kelp beds and at the same time provided an additional source of food for the scavenging sea urchins, further stimulating their explosive population growth.

The accelerating destruction of the kelp "forests" due to the combined effects of pollution and sea urchins alarmed marine biologists who recognized the vital role played by these algae. Kelp provides shelter from predators for vast schools of young fish and provides a secure habitat for crabs, abalones, clams, and other shellfish. Biologists estimate that an acre of kelp supports three times as much biomass as the same area of rocky bottom.

A major effort to restore the kelp beds was launched in the early 1970s when scientists recruited hundreds of scuba-diving volunteers around the Palos Verdes peninsula area to attack the marauding sea urchins with hammers. Within a few years after the effort began, over 2 million urchins were killed and an improvement in the kelp beds could be observed. At the same time, scientists initiated an extensive restoration effort in areas that were already denuded, bringing in algal transplants from areas as distant as Baja California. Coastal communities made their own contribution by agreeing to dispose of sewage effluent farther out at sea beyond the kelp beds. Furthermore, the reintroduction of sea otters into these areas is also aiding in the reduction of the sea urchin population.

Today the kelp beds are beginning to flourish once again and visions of controlled harvesting offer hope for future generations of their providing a source of both food and fuel, the latter in the form of methane gas from fermented kelp. Even more important, a natural ecosystem is being restored.

18

Primitive Land Plants

Outline

Introduction

Five hundred million years ago life on earth was aquatic and the first fish were evolving in the oceans, but the land was barren and devoid of life. Yet within 160 million years plants not only invaded the land but extensive forests developed. In this chapter these early groups of land plants will be described along with the environmental and ecological situations they faced. In one sense these plants were pioneers—they were the first to meet the challenges of growing in soil and exposure to air. The changes in body form and "life style" required to make this transition successfully are staggering to contemplate. An alga growing in water has far fewer demands placed on it by its environment. Plants must have water, carbon dioxide, dissolved minerals, and sunlight if they are to survive. The upper 10 or 20 feet of a lake, ocean, or river contain these in abundance, except for dissolved minerals that are often the only factors limiting algal growth. Most importantly though, the aquatic algae have a plentiful supply of available water thus permitting the maintenance of cellular water content at 90% or more. To make the transition to land, adaptations are necessary that resist the drying influence of air and allow a cell to maintain its high water content. Even desert plants have not been able to circumvent this requirement for a high cellular water content if metabolism is to occur.

As a consequence of these requirements it can be predicted that the ancestors of land plants were most likely amphibious, that is, they grew on shorelines where they were periodically submerged. Then, just as the amphibious animals gave rise to the land animals, the amphibious plants gave rise to the ancestors of all extinct and extant* land plants. The most crucial adaptation, one that had to evolve in both plants and animals, was one that permitted water retention. In plants this involved the development of an outer nonphotosynthetic layer of cells that produce a waxy, waterproof coating on their outer surface, in other words an epidermis with a cuticle. Increased competition for space in which to grow led to the development of "high rise" plants through the evolution of stems, while with increased exposure to drier habitats, roots evolved. An essential element in the evolution of stems and roots is the evolution of vascular tissue.* Land plants, such as the bryophytes, without the means of long-distance transport of nutrients and water have remained short, with the photosynthetic organs being responsible for water absorption. The end result of these evolutionary trends is a land plant, composed of stems, leaves, and roots, that is completely independant of the aquatic environment.

Extant
Still existing.

Vascular tissue
Specialized water and food conducting tissue.

In addition to changes in body form, land plants exhibit changes in sexual reproduction that distinguish them from algae. Within algal groups the type of sexual reproduction ranges from **isogamy*** to **oogamy*** (Chapter 17), and the **zygote*** does not develop into an embryo while still enclosed within the female **gametangium.*** On the other hand, all land plants are oogamous and the zygote develops into an embryo within the female gametangium. Furthermore the male and female gametangia of land plants are multicellular in contrast to the unicellular structures in most algae (Figure 18–1).

Isogamy
Sexual reproduction involving morphologically identical gametes.

Oogamy
Sexual reproduction involving distinct egg and sperm cells.

Zygote
The cell resulting from the fusion of gametes.

Gametangium
A gamete-producing structure.

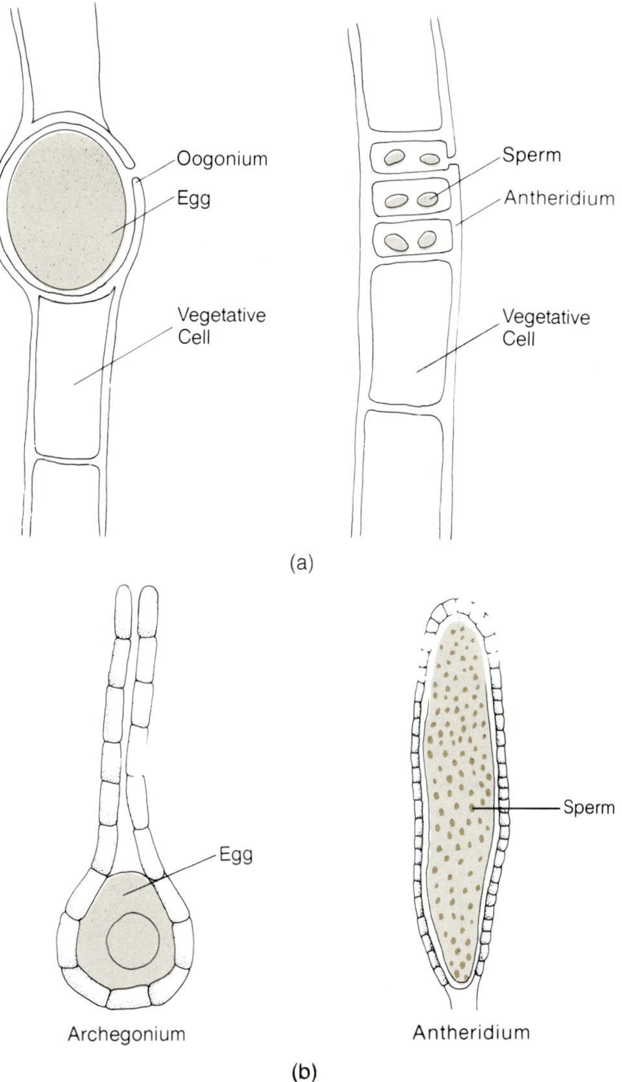

Figure 18–1. Gametangia. (*a*) Unicellular gametangia of an algae where the outer wall is a cell wall. (*b*) Multicellular gametangia of a land plant where the outer wall is composed of many cells.

Most botanists agree that the ancestors of all land plants were advanced forms of green algae (Chlorophyta), which possessed many of the features normally associated with land plants: differentiation into leaflike, stemlike, and rootlike structures, multicellular gametangia, and oogamy. Fossil records are lacking regarding the early development of the land plants so that it is impossible to state with any degree of certainty just how many times the transition to land was made. It seems reasonable to suggest that this event occurred at least twice from different ancestral algae, one line of development leading to the modern bryophytes (mosses, liverworts, and hornworts) and the other to the vascular plants. Within both of these groups a good case could be made for the separate origins of one or more subgroups. For example, liverworts and mosses, which are usually placed in the division Bryophyta, are sufficiently different from each other that they could have evolved from separate algal ancestors.

After the initial discussions of the individual groups of land plants, there is a brief discussion of their **phylogeny*** at the end of the chapter.

Phylogeny
Evolutionary history of an organism or group of organisms.

Bryophyta (Bryophytes)

Traditional classification systems place three groups of plants in the Bryophyta, or bryophytes: liverworts, hornworts, and mosses. Of these only the mosses are likely to be familiar to the nonbotanist. All of the members of the Bryophyta are small and relatively inconspicuous with most species growing in moderately moist environments, although many bryophytes can tolerate long periods of desiccation. At the time that bryophytes evolved, the substrate for plant growth consisted of bare rock, sand, or clay. Soil as such did not exist because soil is a mixture of rock particles (sand, clay) and humus (organic material derived from living organisms). Today, even though they can be found growing in soil or pure humus (decaying logs), bryophytes are still primary colonizers of bare rock in moist habitats. By their growth and metabolic activity bryophytes help break up the surface of the rock and produce a thin layer of soil.

A major characteristic of bryophytes, which traditionally has been cited as distinguishing them from other land plants, is their supposed lack of vascular tissue. In fact bryophytes traditionally are lumped with the algae as nonvascular plants. More recent investigations, however, have demonstrated that many mosses and a few liverworts do indeed contain a central strand of elongated, conducting cells in their stems. These conducting cells are of two types: water-conducting and food-conducting. Therefore they are the functional equivalent of xylem and phloem. Structurally the conducting cells of bryophytes also bear a remarkable resemblance to xylem and phloem—the water-conducting cells are nonliving at maturity, can have pitted walls, and appear similar to tracheids or vessel elements (Chapter 5), except that they lack lignin;* the food-conducting cells have end walls that are similar to the sieve plates of sieve tube members, and lose their nuclei at maturity. Consequently, the bryophytes as a group are not truly nonvascular plants although certain species may have little or no conducting tissue. Even with the recognition that vascular tissue of some sort has evolved in all groups of land plants, there is no evidence to suggest that any of the other groups of land plants evolved from the bryophytes.

Lignin
A complex organic molecule that adds mechanical strength to cell walls.

The life cycle of the bryophytes is the major characteristic that distinguishes them from all other land plants. A plant life cycle consists of an alternation of generations (Chapter 13): a haploid*, gamete-producing generation, the **gametophyte,** and a diploid*, spore*-producing generation, the **sporophyte.** Upon germination the spores develop into the gametophyte generation, which produces the egg and sperm that unite and form the zygote. From the zygote, which remains attached to the gametophyte, the sporophyte generation develops. Among the bryophytes the gametophyte is the dominant stage in the life cycle—consequently the moss or liverwort plant observed growing in the woods is a gametophyte. In all other groups of land plants the sporophyte is the conspicuous generation with the gametophyte being much reduced. The sporophyte of bryophytes grows attached to the gametophyte and in many cases is completely dependent on the gametophyte.

Haploid
Possessing only one of each pair of chromosomes.

Gamete
A sex cell such as an egg or sperm.

Diploid
Possessing two of each pair of chromosomes.

Spore
A unicellular reproductive structure other than a gamete.

Mosses

Mosses are familiar plants characterized by their small size and affinity for moist growing conditions. *Sphagnum,* a moss typically found in swamps and bogs, is a major contributor to **peat*** deposits. Other mosses can be found in lawns, on trees and rocks of moist forests, as the first colonizers of the soil of burned-over forest and grassland, and, in fact, in any environment where sufficient moisture is available. Because of their adaptability, mosses are distributed worldwide from the tropics to the Arctic.

Peat
Partially decomposed remains of *Sphagnum* moss.

The initial growth of a moss gametophyte, the conspicuous generation, is the result of spore germination in a favorable environment. Unlike the liverworts and hornworts, the mosses have two stages in the growth of the gametophyte. At first the spore produces a branching, filamentous stage, the **protonema,** similar in appearance to a filamentous green alga (Figure 18–2). The photosynthetic protonema grows on the surface of the substrate and produces branches that may penetrate the substrate. Subsequently certain cells of the protonema begin to divide giving rise to "buds" that develop into the familiar "leafy" stage of the gametophyte, the stage that produces gametes. The mature gametophyte consists of leaflike structures on a stemlike stalk, anchored to the substrate by rhizoids. Growth of the leafy shoot system is the result of division of an apical cell. The terms *stem* and *leaf* traditionally have been reserved for organs of vascular plants; however, similarities in morphology and function have led many botanists to apply these terms to structures found in bryophytes. Internally a typical moss stem is organized into three distinct regions: an outer single-layered epidermis, a large cortex,

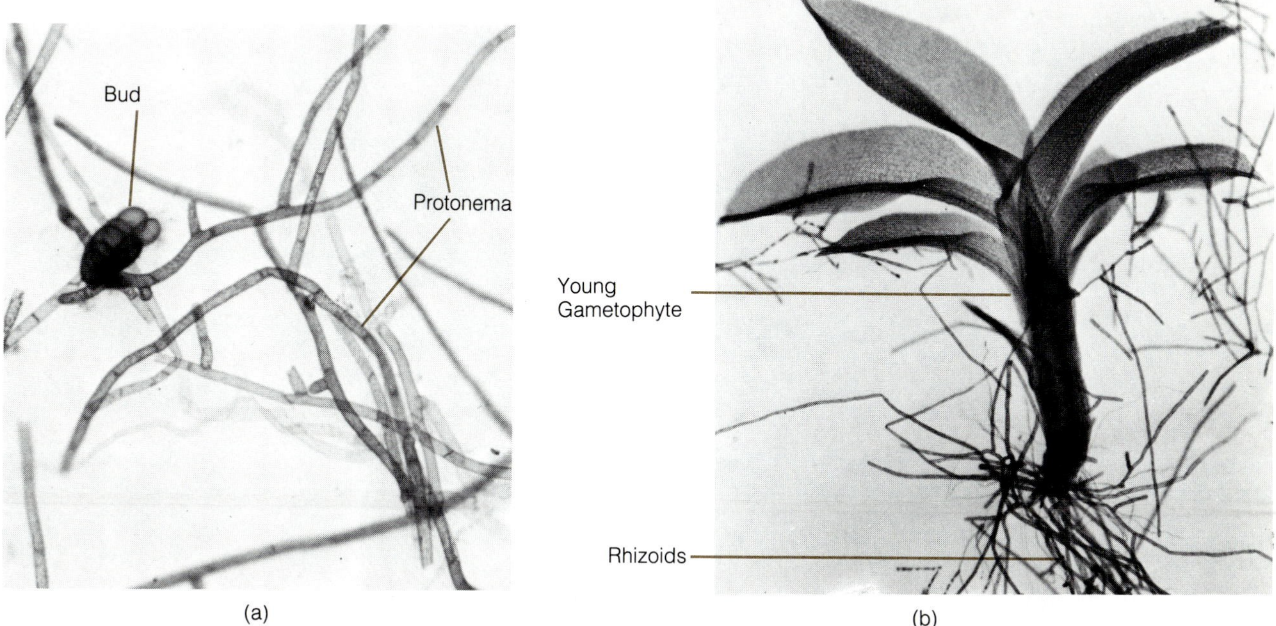

Figure 18–2. Moss protonema. *(a)* Early growth of the filamentous protonema with bud formation. *(b)* Leafy gametophyte growing from the protonema.

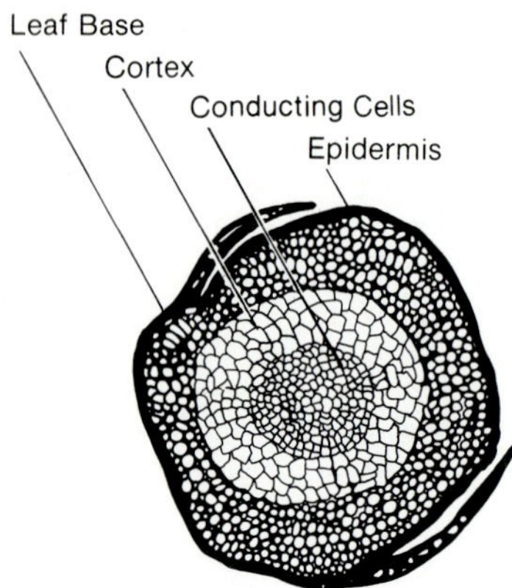

Figure 18–3. Cross section of a moss stem. The central region of this stem is composed of conducting cells.

and a central strand of conducting cells, but without xylem and phloem (Figure 18–3). Although the moss stem contains water-conducting cells there is evidence that considerable conduction of water and dissolved minerals occurs over the external surfaces of the leafy shoot, with both leaf and stem cells being able to absorb materials directly from the environment.

In response to environmental conditions, the leafy gametophyte begins to produce antheridia and archegonia. In mosses, unlike many of the liverworts, the antheridia are clustered at the end of a branch (Figure 18–4), similarly the archegonia are produced in clusters on a separate branch (Figure 18–5). Fertilization results when sperm swim from antheridia to archegonia through a layer of water resulting from rainfall, heavy dew, or fog. Embryo development commences while the zygote is still contained within the archegonium. Continued development of the embryo results in the formation of the mature diploid stage, or sporophyte (Figure 18–6). It is important to emphasize that the sporophyte generation remains attached to the gametophyte generation throughout and never leads an independent existence.

The mature sporophyte is differentiated into three regions: a **foot** that anchors the sporophyte in the gametophyte tissue, a **stalk,** and a spore-producing structure, the **capsule**. Water and nutrients are transferred from the gametophyte to the foot and then conducted through the stalk to the capsule, although the sporophyte is photosynthetic and is not totally dependent on the gametophyte. The moss capsule is the **sporangium** within which certain cells differentiate into **spore mother cells,** cells that divide by

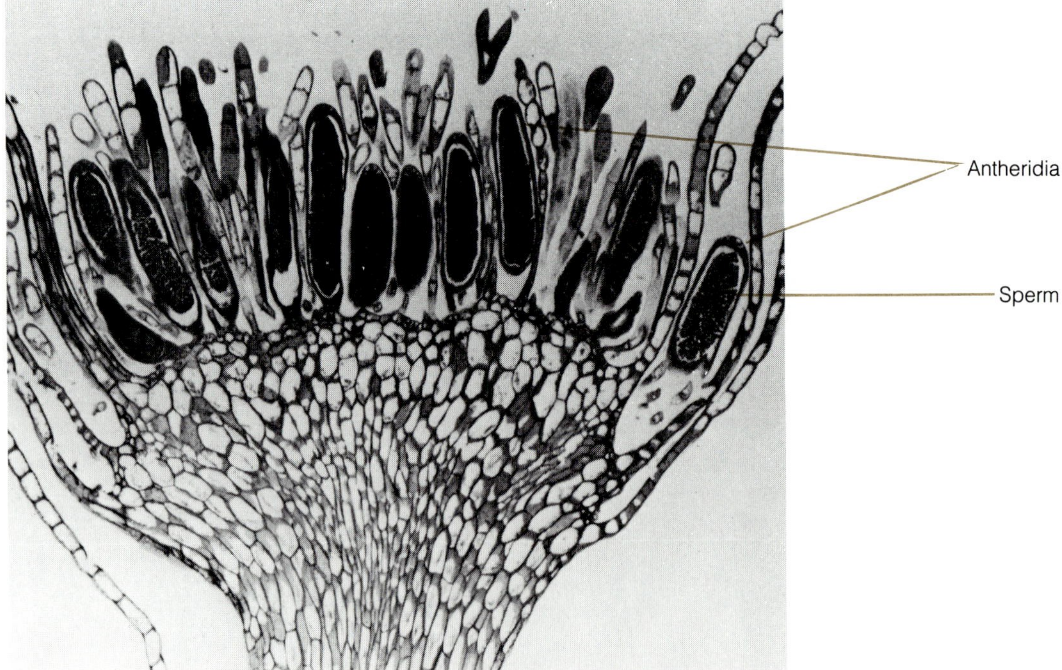

Figure 18–4. Longitudinal section of a moss branch with antheridia. Sperm formation in mosses occurs in antheridia that are produced at the end of a branch of the gametophyte plant.

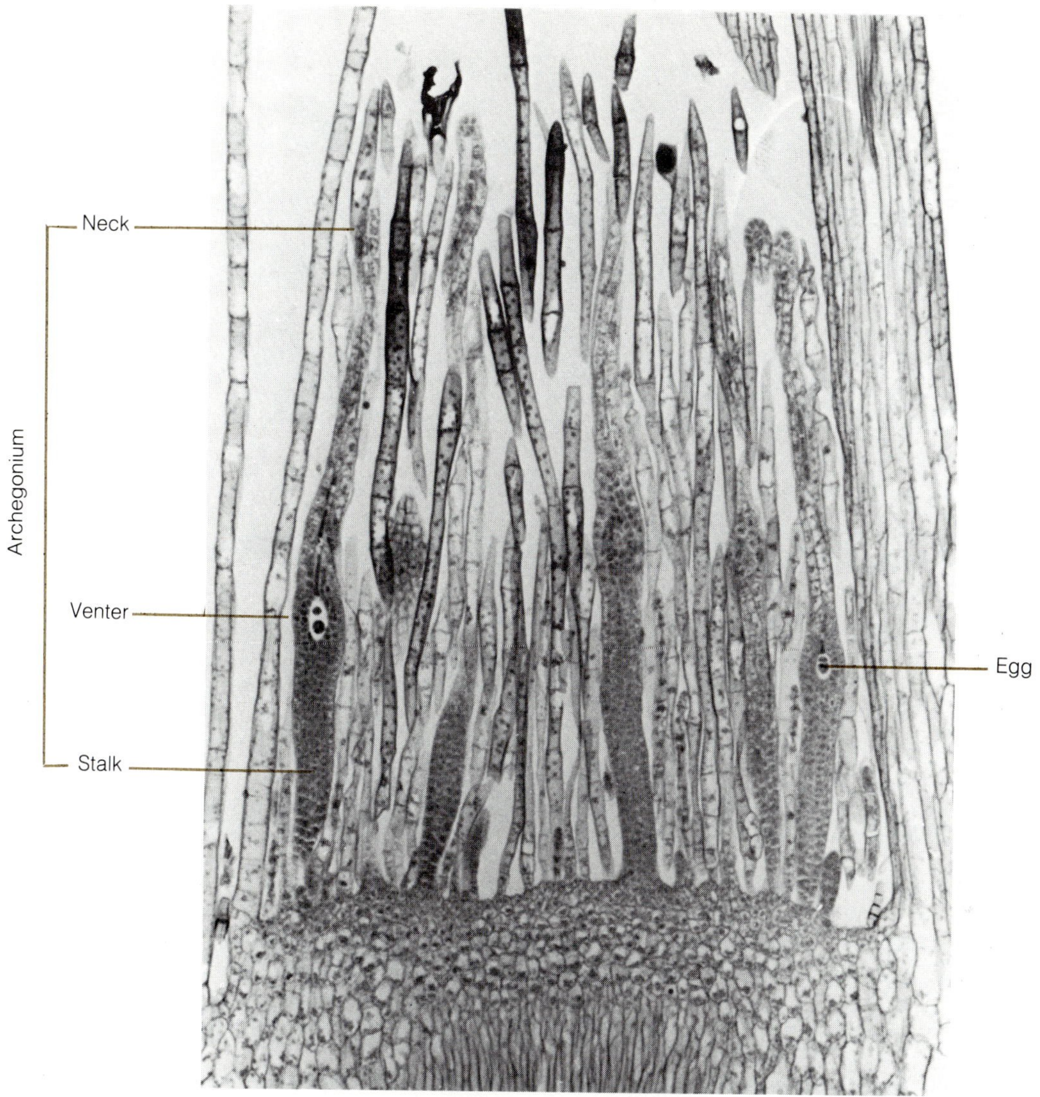

Figure 18–5. Longitudinal section of a moss branch with archegonia. The egg is produced at the base of an archegonium. Immediately above the egg is a prominent ventral canal cell that disintegrates before the sperm enters the neck canal.

meiosis producing haploid **spores.** The moss capsule is more elaborate than that of other bryophytes with respect to the mechanism of spore release. During maturation of the capsule several layers of cells in the upper portion develop specially thickened walls resulting in the formation of a ring of toothlike segments called the **peristome.** The tissue above the peristome becomes the cap. In many mosses, once the spores are mature, the capsule dries, resulting in the separation of the cap from the remainder of the capsule, and exposes the peristome. As the peristome loses moisture, it shrinks, the teeth move apart, and spores are shed. Spores that land in a favorable environment germinate and produce the haploid protonema, thus completing the life cycle (Figure 18–7).

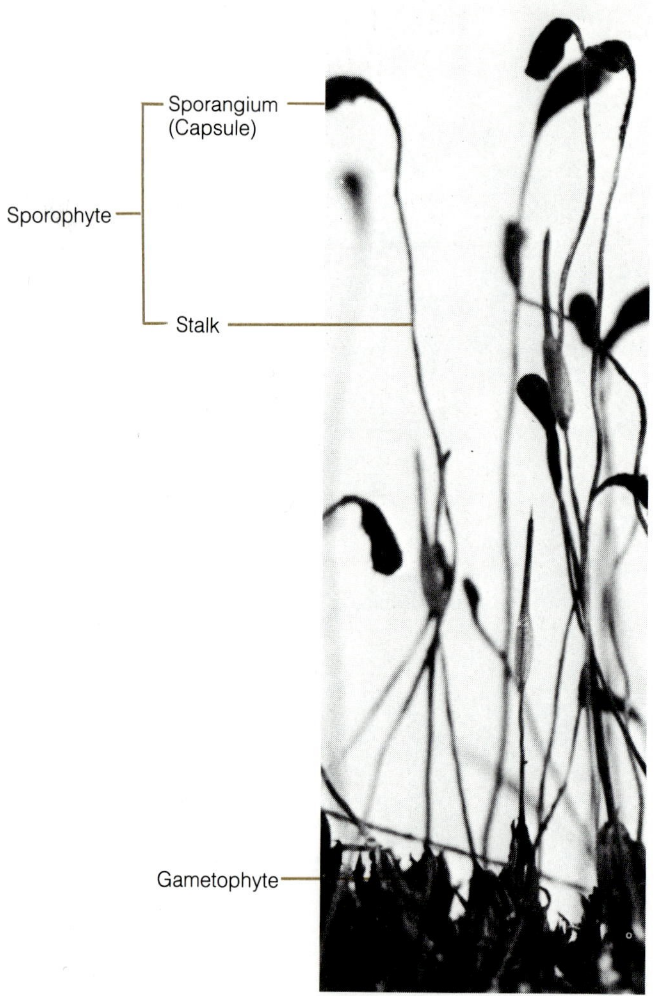

Figure 18–6. Moss sporophyte.

Liverworts

The liverworts owe their common name to a real or imagined similarity in appearance between certain genera and the human liver. According to medieval European medical beliefs, this similarity in appearance indicated that liverworts would cure diseases of the liver and, even though this belief is no longer held, the name has persisted both in the English and scientific names for the group (Hepatophyta or Hepaticae). There are two morphological* forms of liverworts, **thallose*** and **leafy,** with certain of the thallose forms being only slightly advanced morphologically and anatomically over the green alga *Ulva* (Chapter 17). With the observation that a number of genera of thallose liverworts are either aquatic or amphibious (able to grow submerged or on moist soil), it is tempting to make the suggestion that these represent the most primitive liverworts. There is considerable disagreement among plant morphologists, however, as to whether the simplicity of thallose liverworts represents a primitive condition or the secondary loss of characteristics due to evolutionary reduction.

Morphology
(*adv.*, morphologically) The study of the external appearance of organisms.

Thallose
Possessing a plant body which is relatively undifferentiated.

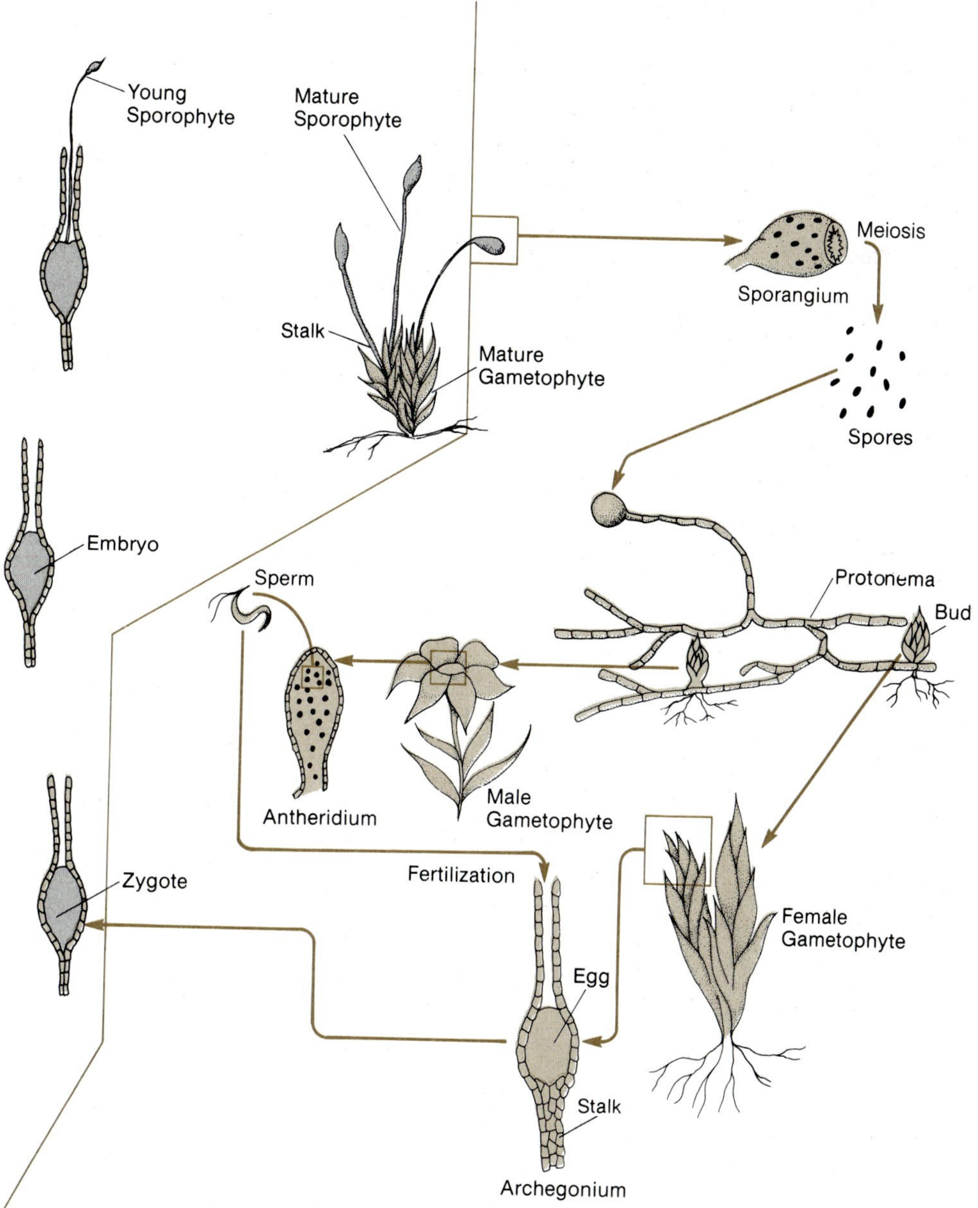

Figure 18–7. Moss life cycle. Germination of a spore produces the protonema, the first stage in the gametophyte generation. Eventually buds form on the protonema that produce the typical leafy gametophyte. Sperm are produced in antheridia and swim through water to reach an egg in an archegonium. After fertilization the zygote develops into an embryo, which is part of the sporophyte generation. The mature sporophyte consists of a foot that is attached to the gametophyte, a stalk, and a capsule in which spores are produced by meiosis. Loss of the cap from the capsule releases the spores which then develop into gametophytes.

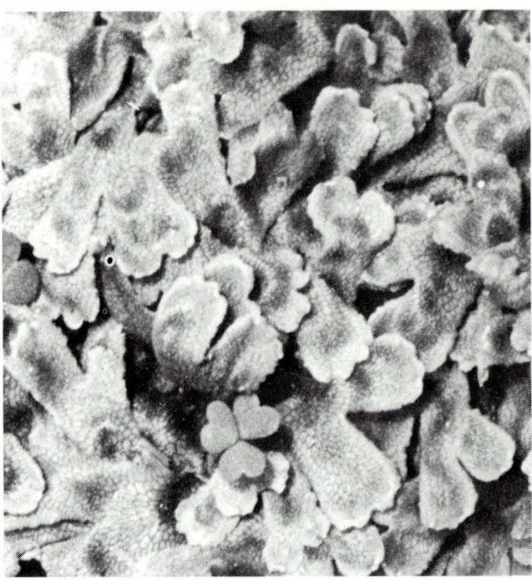

Figure 18–8. Thallose liverwort.

The gametophyte of a thallose liverwort is typically thin and flattened and varies considerably in size and shape (Figure 18–8). Its lower surface, analogous to the epidermis of a root, is involved in water and mineral absorption. To assist in this function the lower surface gives rise to multicellular filaments (scales) and unicellular extensions (rhizoids) that contact the surface of the substrate. Anchorage is by means of other rhizoids that penetrate the substrate.

Dichotomously
Dividing into two equal parts.

Growth of the gametophyte occurs by divisions of the apical cells, which in a number of genera results in the production of a dichotomously* branched plant. Since the older regions die and decay, the plant does not become very large. In branching species, the decay of older regions at the base of branches results in the fragmentation of one plant into several, representing one form of asexual reproduction. In some liverworts asexual reproduction occurs through the production of **gemmae** in **gemma cups** (Figure 18–9). The gemmae are small clusters of cells that are released and float away when the cups fill with water, as after a rain. New gametophytes will be produced if the gemmae land on a substrate favorable for their growth.

Sexual reproduction results when the gametophytes produce eggs and sperm in archegonia and antheridia respectively. Since these gametes are produced on haploid gametophytes, it might seem that a plant would, of necessity, be either male or female. While this is the case for many liverworts, there are numerous examples of bisexual species in which one gametophyte produces both antheridia and archegonia.

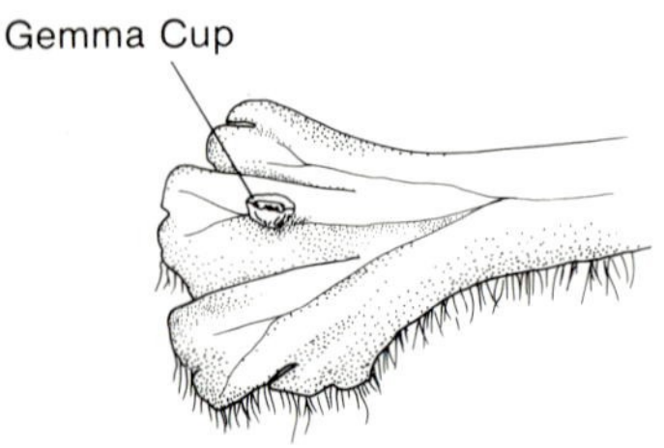

Figure 18–9. Gemma cups. Liverworts produce small clusters of cells called gemmae that function as asexual reproductive structures.

Antheridia are produced singly or in clusters on the surface of the gametophyte or, in some cases, they are produced on upright branches called **antheridiophores**. Similarly the archegonia are produced on the surface of the gametophyte or on upright branches called **archegoniophores.** Flagellate sperm swim from the antheridia to the eggs in the archegonia through a layer of water from rain or dew on the surface of the gametophyte. As in the mosses, the resulting zygote develops into a sporophyte that is attached to the gametophyte. The mature liverwort sporophyte, like that of the mosses, typically consists of a foot, stalk, and capsule. As the mature capsule

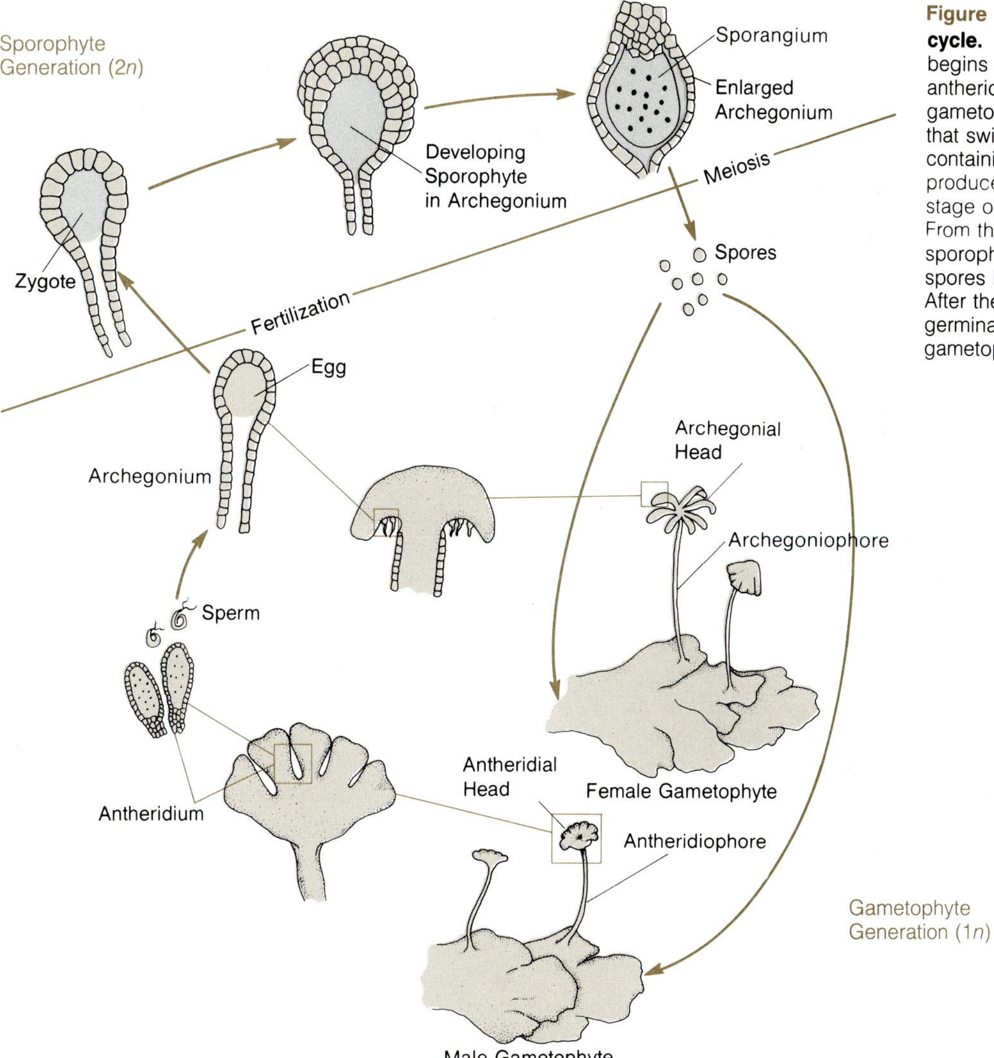

Figure 18–10. Liverwort life cycle. The gametophyte generation begins growth from a spore. After antheridia are produced by the gametophyte, sperm are released that swim to the archegonia, each containing an egg. Fertilization produces the diploid zygote, the first stage of the sporophyte generation. From the resulting embryo a mature sporophyte develops that produces spores by meiosis in the capsule. After their release the haploid spores germinate producing the gametophyte generation.

splits open, meiospores* are released. Upon reaching a favorable site for germination, they produce new haploid individuals, thus completing the life cycle (Figure 18–10).

Although the thallose liverworts represent the "typical" liverwort for many people, the largest group are the leafy liverworts, so-named because of their leaflike structures. In the leafy liverwort gametophyte, leaflike structures are attached to a prostrate stem, which is held to the substrate by rhizoids. The "leaves" are produced by the apical meristematic* region of the stem and are one cell layer thick. They have neither cuticle* nor stomata.*

Asexual reproduction in leafy liverworts occurs by means of gemmae or fragmentation. Antheridia and archegonia are produced on separate short branches. After fertilization, the zygote develops into the typical liverwort sporophyte, which remains attached to the gametophyte. Usually only

Meiospore
A haploid spore produced after meiosis.

Apical meristem
The undifferentiated, dividing cells found at the tip of a stem or a root.

Cuticle
The outer, waxy layer produced by the epidermis.

Stoma
(*pl.*, stomata) A pore in the epidermis that is opened and closed by a pair of guard cells.

one sporophyte is produced per archegonial branch with the remainder aborting.

Hornworts

The hornwort gametophyte resembles that of the thallose liverworts, but differences in both sporophyte and gametophyte characteristics have led some botanists to separate this group from the liverworts. While the gametophyte is relatively undifferentiated, the capsule of the sporophyte has more internal differentiation than that of liverworts. The hornwort sporophyte is an elongated, cylindrical structure attached to the gametophyte by the foot region. The reproductive cycle among the hornworts is similar to those of the mosses and liverworts.

Economic Importance

The bryophytes are not plants of major economic importance in most regions of the world; however, *Sphagnum* is of some economic significance. Because this moss has a large water-holding capacity, it is used by florists and horticulturists instead of soil to protect and sustain the roots of certain plants during shipping. *Sphagnum* has other horticultural applications but probably its major economic use is in the form of **peat moss.** *Sphagnum* grows prolifically in bogs, swamps, and at the margins of lakes and ponds. Since *Sphagnum* growth makes the water acidic, the growth of decay organisms such as bacteria and fungi is discouraged and dead *Sphagnum* can accumulate. The partially decomposed *Sphagnum* that results is called peat moss. In Ireland and parts of the United States and Canada, peat moss is harvested, dried, compressed, and burned as a fuel. Other peat is used as a soil conditioner for both indoor and outdoor plant growth.

Primitive Vascular Plants

The bryophytes appear to represent an evolutionary dead end although the adaptations that developed in this group were sufficiently successful that the bryophytes have survived to this day. For some reason, however, mutations* that would have led to the development of more efficient conducting tissue, roots, and cuticle did not appear in the genetic information of the bryophytes, thus limiting them in size and distribution. Modern and fossil species are very similar, indicating lack of change within the group. In fact, the habitat of the group has not really changed from that of the ancestral bryophyte: terrestrial regions with constant or periodically high moisture levels. At the time that the algal ancestors of the bryophytes were making their transition to land, other algal species were giving rise to plants with different sets of adaptations to the land environment. All of the other plants that successfully made the transition to land possess two features in common: true vascular tissue (xylem and phloem) and alternation of generations

Mutation
A change in a portion of the DNA molecule corresponding to one gene, which results in a change in the protein product of that gene.

where the sporophyte is the conspicuous generation. According to the fossil record the bryophytes and the vascular plants appear to have originated at the same time, approximately 350–400 million years ago, with some vascular plant groups apparently predating the bryophytes.

All of the fossilized remains of primitive vascular plants seem to be of sporophytes, either because the gametophytes were very short-lived or because the gametophyte tissues were too delicate to fossilize. It is also possible that the fossilized gametophytes are lying unrecognized in some paleobotanist's* speciment files. Because of this, nothing is known about the sexual cycle in these organisms, although it is likely that they produced both antheridia and archegonia. Fossil records suggest that the earliest known vascular plant sporophytes consisted of a branched stem system that had neither roots nor leaves. Anchorage and absorption were the function of a horizontal stem, a rhizome, which produced hairlike projections, or rhizoids. The upright stem system was covered by an epidermis and cuticle containing typical stomata. Since plants are typically autotrophic, it is assumed that these primitive vascular plants carried on photosynthesis in their stem cells. The ends of certain aerial branches were swollen and differentiated into sporangia, which produced haploid spores by meiosis. Germination of these spores after their release would have given rise to the gametophyte generation. Although all members of these most primitive groups are extinct, there is one living group of plants that resembles them morphologically—the members of the genus *Psilotum*. *Psilotum* is both leafless and rootless. It possesses a rhizome and aerial branches at the ends of which are produced sporangia (Figure 18–11 and Plate 8E). Morphologists and paleobotanists originally considered *Psilotum* to be a living representative of this fossil group, but now its precise taxonomic position is in doubt with some botanists claiming that *Psilotum* is more closely related to the ferns.

Paleobotanist
A botanist who studies fossil plants.

Even though the earliest vascular plants are now extinct they appear to have given rise to other groups of plants—Psilophyta, Microphyllophyta, Arthrophyta, and Pterophyta—each of which is represented by both fossil and living species.

Psilophyta

There are only two surviving genera of the Psilophyta and these occur naturally in tropical and subtropical regions. The Psilophyta are of no economic importance but they are of interest because of their primitive characteristics. As indicated above, the typical sporophyte of *Psilotum* consists of a rhizome with upright branches (Plate 8E) in which photosynthesis occurs. The spores are produced in sporangia that form on the aerial branches. After their release, the haploid spores germinate in soil to produce the gametophyte, a small, carrot-shaped plant (Figure 18–11) that lives underground. Since the gametophyte is underground and has no connection with the sporophyte, it cannot produce food photosynthetically or be dependent on the sporophyte; instead the gametophyte lives as a **saprophyte**.* In order to facilitate the decomposition and absorption of organic material from the soil, the gametophytes of the Psilophyta have evolved a **symbiotic*** association with certain fungal species. Therefore, unlike the Bryophyta, the Psilophyta produce

Saprophyte
An organism that obtains its nutrition by decomposing non-living organic material.

Symbiosis
(*adj.*, symbiotic) A close association between two different organisms.

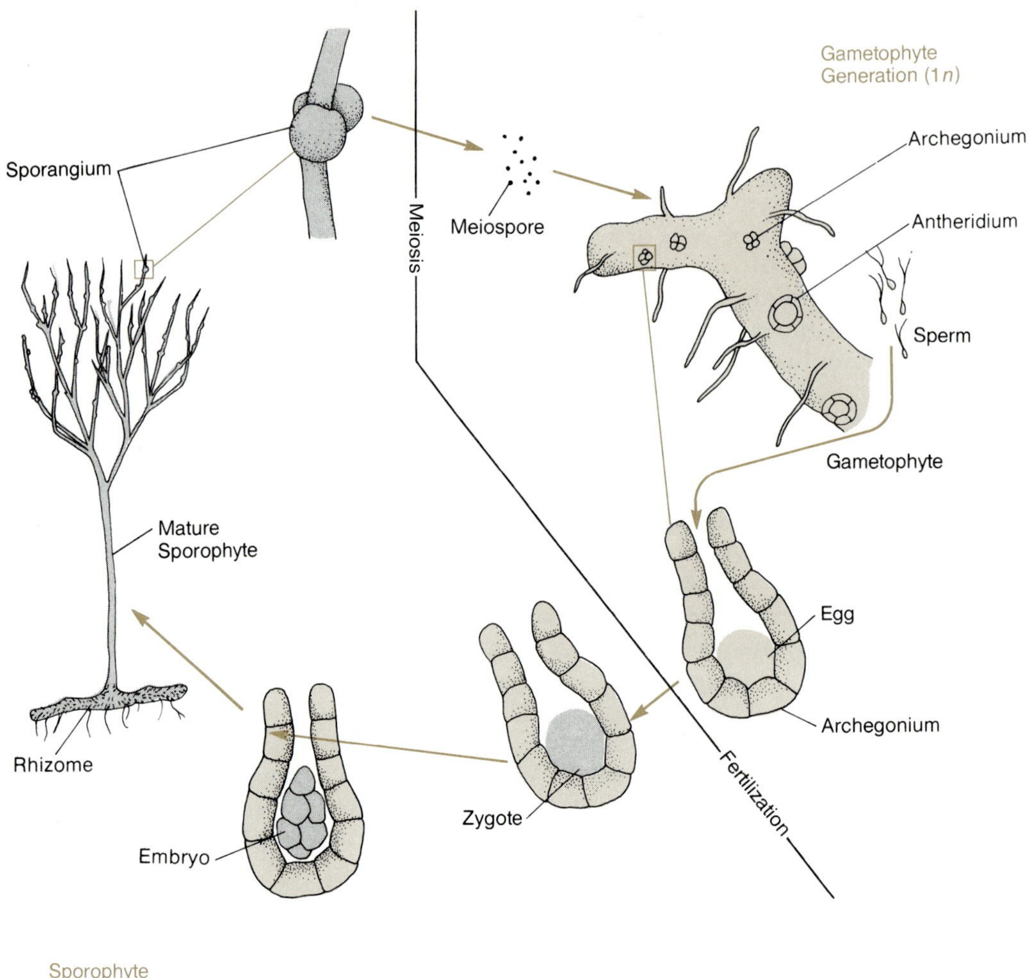

Figure 18–11. *Psilotum* life cycle. Meiosis in sporangia of a mature sporophyte produces haploid spores that germinate forming the gametophyte generation. Sperm produced in the antheridia fertilize eggs produced in archegonia. The young sporophyte develops while still attached to the gametophyte, but eventually this connection is lost and the mature sporophyte is independent of the gametophyte.

sporophyte and gametophyte generations that are completely independent of one another at maturity.

Antheridia and archegonia are borne on the same gametophyte and are similar in appearance to those produced by bryophytes. Fertilization results after sperm swim from the antheridia into the archegonia containing eggs. The resulting zygote develops into an embryo while still contained within the archegonium, so that the developing sporophyte begins its existence dependent upon the gametophyte for its nutrition. As the immature sporophyte develops further, it produces branches that come aboveground where photosynthesis will occur. Ultimately the connection with the gametophyte is lost and the sporophyte becomes physically and nutritionally independent.

Microphyllophyta

The name *Microphyllophyta* is derived from the fact that all members of this group, both living and extinct, produce characteristic small (micro) leaves (phyll). These microphylls are morphologically distinct from the more familiar megaphylls (large leaves) of ferns and seed plants. The difference is not really in size but in organization, vascular connections, and evolutionary origin. Some of the most common living members of the Microphyllophyta, as well as some of the extinct species, are commonly referred to as **club mosses,** (Plate 8F) owing to their superficial resemblance to the mosses. Other species are known as **ground pine** reflecting their similarity to miniature pines.

All of the sporophyte plants in this division have vascularized stems, roots, and leaves in contrast to the Psilophyta which have only stems. An indication of the primitive nature of this group is that branching is not due to the growth of a lateral bud*, but, rather, results from the splitting of an apical meristem into two halves, each half becoming the apical meristem for a new branch. Internal anatomy also emphasizes the primitiveness of this group. Primitive does not mean that these are not structurally complex organisms, however.

While the living plants in this division seldom grow over 30 centimeters tall, fossil members grew over 30 meters tall and more than 1 meter wide at the base. These tree forms developed **vascular** and **cork cambia*** and were capable of producing extensive **secondary growth.*** In their time the Microphyllophyta were a major component of the terrestrial vegetation and contributed to extensive forests (Figure 18–12).

In all members of the Microphyllophyta the sporophyte produces meiospores in sporangia (Figure 18–13), which are produced on special leaves

Lateral bud
A bud at the base of a leaf that will give rise to a branch shoot.

Vascular cambium
A secondary meristem that produces vascular tissues (xylem and phloem).

Cork cambium
A secondary meristem that produces cork and phelloderm.

Secondary growth
Growth in width due to the action of the cambia.

Figure 18–12. Early forest. The primitive vascular plant groups included species that grew as large trees. These trees and the associated smaller species constituted the first forests.

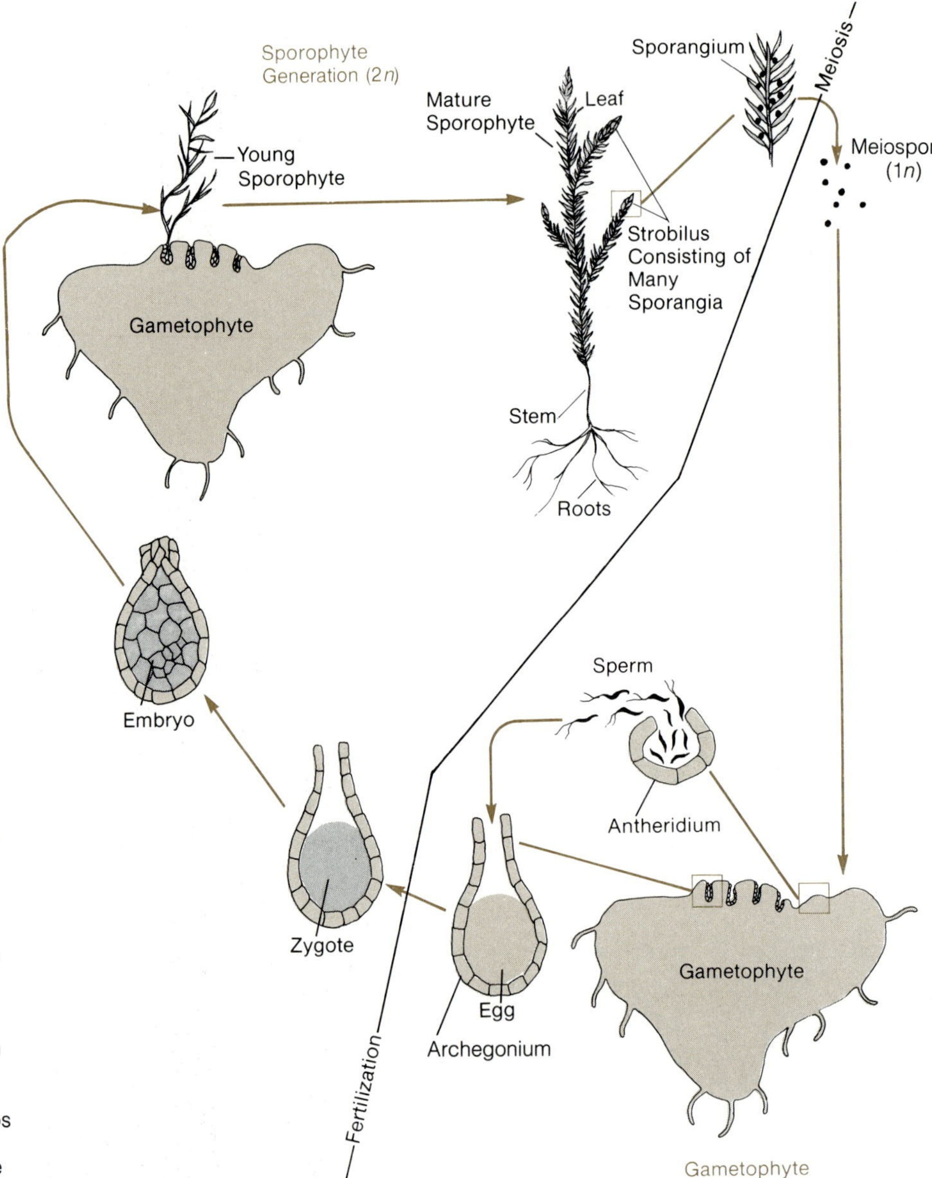

Figure 18–13. Microphyllophyta life cycle. The living genus *Lycopodium* is used to indicate a typical life cycle for this group. Sporangia are produced on sporophylls that are aggregated into a strobilus (cone). After meiosis spores are produced, released, and germinate to give rise to the gametophyte generation. Fertilization occurs after sperm from the antheridia fuse with eggs in archegonia. The sporophyte develops from the resulting zygote and eventually loses its attachment to the gametophyte.

called **sporophylls.** The sporophylls of the more primitive microphyllophytes are similar in appearance to the sterile leaves except for the sporangia at their bases. An evolutionary trend within this group is the aggregation of sporophylls at the ends of branches thus forming a structure called a **strobilus,** or **cone.**

Another important evolutionary trend within this group is the change from producing only one kind of meiospore to producing two kinds. Members of the Bryophyta and Psilophyta discussed above produce only one type of spore, a condition known as **homospory.** In the more advanced forms of the Microphyllophyta homospory gave way to **heterospory,** the condition in which two types of spores that differ in appearance and in function are produced. In a heterosporous plant one type of spore forms a male gametophyte and the other type forms a female gametophyte. Since the spores that produce male gametophytes tend to be smaller than those that produce female gametophytes, they are referred to as **microspores** and **megaspores,** respectively. The two kinds of spores are produced in different sporangia **(microsporangia** and **megasporangia)** that are borne on different sporophylls **(microsporophylls** and **megasporophylls).** Cones, or strobili, of heterosporous plants in this group contain both types of sporophylls in either a regular or a random arrangement.

The more primitive members of the Microphyllophyta have a well-developed gametophyte generation. In certain species this gametophyte plant grows aboveground and is photosynthetic, whereas in others the gametophyte grows underground and is saprophytic in association with symbiotic fungi. The independent gametophytes can persist for periods as long as a year. Within the Microphyllophyta there has been an evolutionary trend toward smaller gametophytes that are short-lived and dependent on the sporophyte for their nutrition (see below). This trend toward reduction in gametophyte size and complexity occurs in conjunction with the development of the heterosporous condition—the most advanced members of this group have well-developed heterosporous strobili and reduced gametophytes.

In the more primitive species the life cycle is similar to that of the Psilophyta. Sperm produced in the antheridia of a gametophyte are released and swim to an archegonium where the egg is fertilized. Initial growth of the developing sporophyte is at the expense of the gametophyte; ultimately the sporophyte loses all connection with the gametophyte and lives independently.

With the development of heterospory in the advanced species there was also a considerable change in the structure of the gametophyte (Figure 18–14). The male gametophyte, or **microgametophyte,** consists of a minimal number of cells—a multicellular antheridium and **prothallial cell,** which represents the rest of the gametophyte. The female gametophyte, or **megagametophyte,** also is reduced considerably. In both cases the gametophyte depends on food stored in the spore for its survival. Germination of the microspore results in the formation of a microgametophyte surrounded by the microspore wall (Figure 18–14). Similarly the megagametophyte forms completely within the confines of the megaspore wall. After spore release the microspore wall ruptures, liberating the sperm that swim through a layer of water to reach the archegonia which are contained in the part of the megagametophyte protruding through the megaspore wall (Figure 18–14). The resulting zygote develops into an embryo using food translocated from the megagametophyte. Since the food originally was derived from the preceding sporophyte generation, this situation foreshadows the later evolutionary development of the seed in other groups of plants.

Although fossil Microphyllophyta are often depicted as part of tropical forests, the surviving species are distributed from subarctic regions to the equator. In particular they are found in moist shaded locations; however, the resurrection plant is native to the arid regions of Mexico and the American southwest.

Arthrophyta

Evolution from the earliest vascular plants led to another major, but now mainly extinct, group—the Division Arthrophyta. The only living represen-

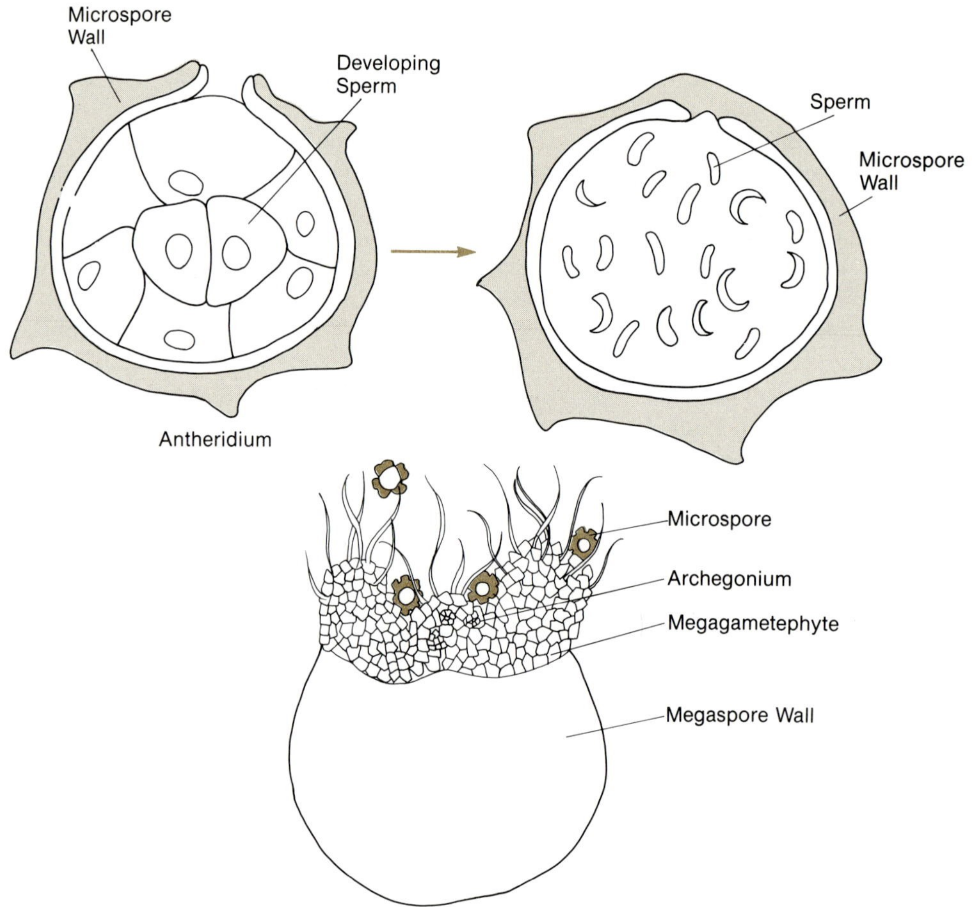

Figure 18–14. Evolutionarily advanced Microphyllophyta gametophytes.
(*a*) Microgametophyte. The microspore germinates producing a reduced microgametophyte within the microspore cell wall. Sperm are still produced within antheridia.
(*b*) Megagametophyte. The megagametophyte develops within the megaspore cell wall and projects beyond it.

tatives of this division are species of *Equisetum,* (Plate 9A) commonly called horsetails or scouring rushes because of their use by pioneers for scouring pots and pans. Although the living members of this group are herbaceous and range from 30 centimeters to 5 meters tall, some extinct genera formed large trees. Like the extinct Microphyllophyta trees, these Arthrophyta trees were important components of the forests of 300 million years ago that contributed significantly to coal formation. Members of this division are characterized by ribbed and segmented stems, cell wall impregnated with silica, whorled arrangement of leaves and branches, and an extensive system of rhizomes. As in the Microphyllophyta, sporangia are localized on sporophylls that are organized into strobili (Plate 9B) that form at the ends of branches. Within the Arthrophyta heterospory is known from fossil forms; however, the majority of species appear to have been homosporous. The germination of a spore of *Equisetum* produces a small, free-living photosynthetic gametophyte (Figure 18–15), the sex of which is determined by environmental conditions. Within a short time antheridia and archegonia are produced. As in the other groups of plants discussed thus far, motile sperm fertilize the eggs in archegonia. Initially the developing sporophytes are nurtured by the gametophyte.

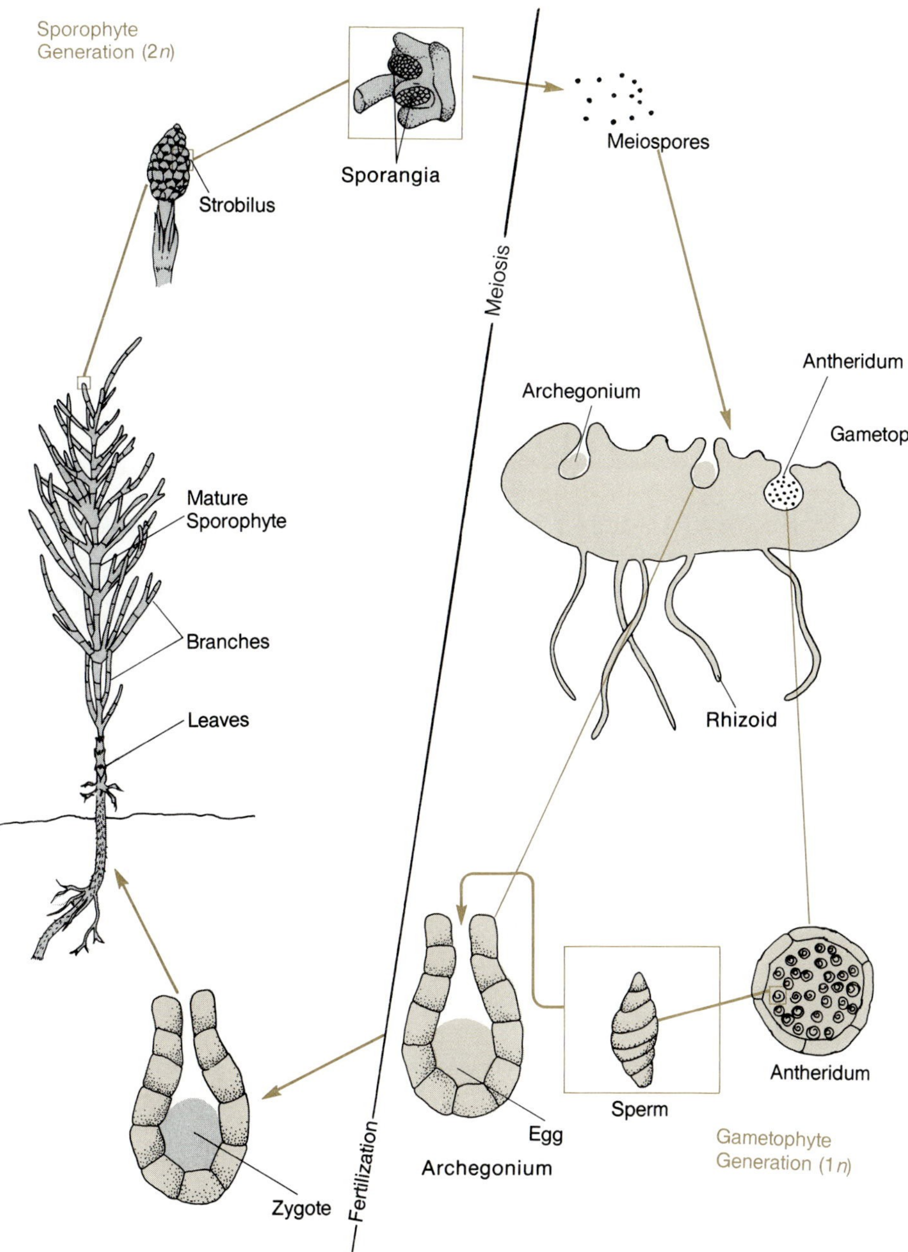

Figure 18–15. **Arthrophyta life cycle.** Since *Equisetum* is the only living genus of this group, it is used as the model for the life cycle of the group. A strobilus containing sporophylls produces the spores by meiosis. After gametophyte development, antheridia and archegonia are produced. Sperm swim to the eggs and fuse with them producing the zygotes. The young sporophyte begins growth attached to the gametophyte but eventually grows independently.

Pterophyta

The fourth division to have evolved from the earliest vascular plants is the Pterophyta or **ferns.** As a heterogeneous group of plants ferns include small, floating, aquatic forms (Figure 18–16A) and large, treelike forms (Figure 18–16B). Within this group there does not appear to have been a trend from homospory to heterospory as most living fern genera, including evolutionarily advanced forms, are homosporous. The ferns are remarkable because there has been no tendency in this group for sporophylls to be clustered into strobili. However, sporangia are typically produced in groups called **sori**

(a)

(b)

Figure 18–16. Atypical ferns. The typical fern is a familiar sight to most, but considerable variation in body form exists within this group. (*a*) An aquatic fern that consists of two leaflike fronds and one that is modified and resembles roots. (*b*) A tree fern is the largest of the ferns.

(singular, **sorus**), which are usually found on the lower surface of a sporophyll (Figure 18–17).

Compound leaf
A leaf whose blade is subdivided into a number of individual leaflets.

The typical fern sporophyte (Plate 9C) is a horizontally growing, underground rhizome at the tip of which are produced leaves, which may be **compound.*** Fern leaves, as well as the large, compound leaves of **cycads** (Chapter 19), palms, and some other plants, are commonly called **fronds.** Growth of a fern results from cell divisions at the growing tip of the rhizome. Immature leaves are rolled up in such a way that they are commonly referred to as fiddleheads, a resemblance that is lost as the leaf uncoils and flattens. Roots are produced along the length of the rhizome in much the same manner as they are on rhizome-bearing flowering plants. Upright stems as tall as 20 meters or more are produced in tree ferns. Interestingly, most of the trunk of a tree fern is root, not stem.

Sexual reproduction in the ferns, as in other land plants, begins with the production of spores by meiosis (Figure 18–18). As indicated above, the sporangia are clustered in sori. In many ferns the sori are covered with a protective structure called the **indusium,** which dries and shrivels at the time of spore dispersal. Spore release and dissemination is literally explosive in a considerable number of ferns. In one group of ferns, the sporangium

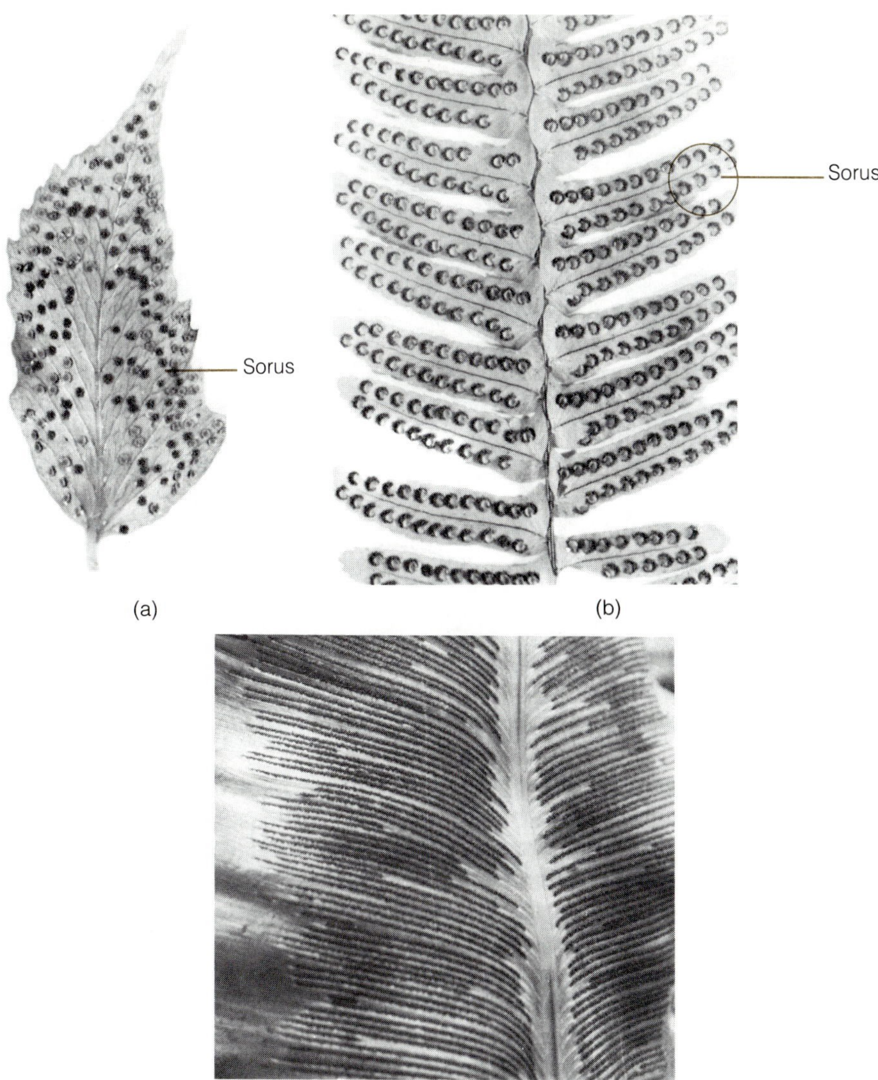

Figure 18–17. Sori. Fern sporangia are produced in clusters called sori. The sori are produced on the underside of the frond in a variety of patterns.

has a row of large, thick-walled cells called an **annulus** that forms a ridge which almost completely encircles the sporangium (Figure 18–19). This circle of cells is completed by thin-walled **lip cells.** As the spores mature the indusium is "lost" and the sporangia are exposed to the drying influence of the air. Loss of water by the sporangium causes the cells of the annulus to contract, resulting in the rupture of the lip cells and the opening of the sporangium. Further water loss causes increased contraction of the annular cells and further opening of the sporangium into two halves. Finally there is a sudden release of the tension in the annulus and the two halves of the sporangium catapult forward to the closed position. The sudden catapulting is often sufficient to eject most of the spores, their momentum carrying them out through the opening left by the ruptured lip cells (Figure 18–20).

The gametophyte produced from the spores is independent of the sporophyte and may be either saprophytic or photosynthetic depending on the species. Saprophytic fern gametophytes, like those of *Psilotum* and members

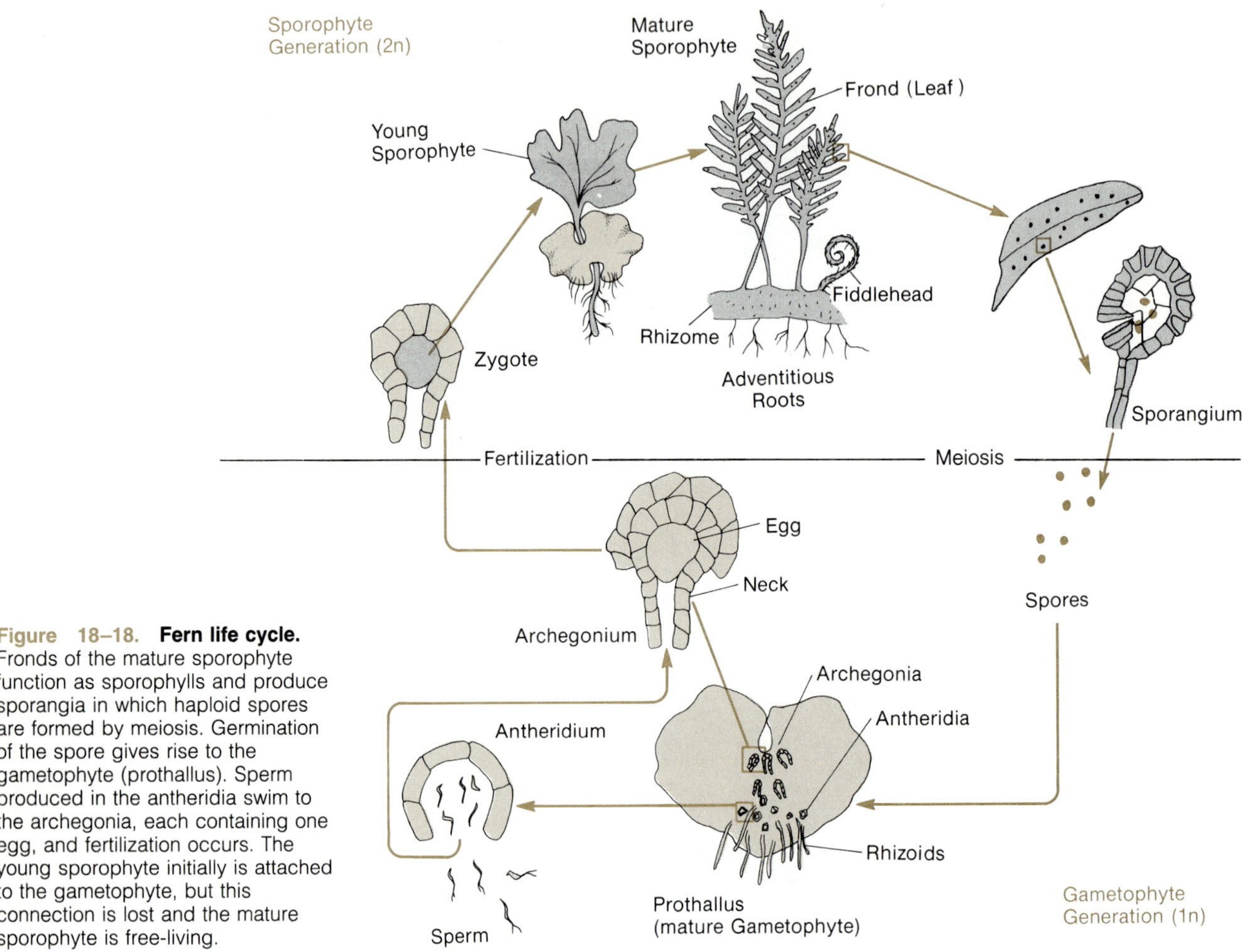

Figure 18–18. Fern life cycle. Fronds of the mature sporophyte function as sporophylls and produce sporangia in which haploid spores are formed by meiosis. Germination of the spore gives rise to the gametophyte (prothallus). Sperm produced in the antheridia swim to the archegonia, each containing one egg, and fertilization occurs. The young sporophyte initially is attached to the gametophyte, but this connection is lost and the mature sporophyte is free-living.

of other divisions, form symbiotic relationships with fungi. The photosynthetic gametophyte, or **prothallus,** is used most frequently to illustrate fern gametophytes (Figure 18–21). In either case, antheridia and archegonia are formed, motile sperm are released, and eggs are fertilized. As in the other primitive land plants, the young sporophyte begins its development dependent on the gametophyte, but it soon loses this connection as it becomes photosynthetically active (Figure 18–22).

Ferns are not agriculturally or economically important plants although the fiddleheads of certain species are considered to be a delicacy and florists use fern fronds as greenery to make floral arrangements and corsages more aesthetically pleasing. The major importance of ferns lies in their fossilized remains: some of the coal deposited between 280 and 350 million years ago represents the remains of ferns.

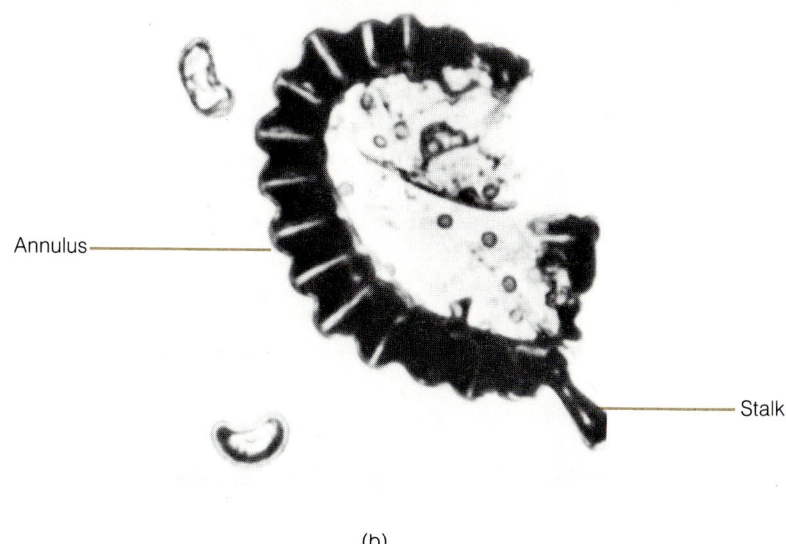

Figure 18–19. Fern sporangium. (*a*) A cluster of sporangia as would be found in a sorus. (*b*) An empty sporangium with ruptured lip cells.

Phylogenetic Relationships

The fossil record is incomplete with regard to the evolutionary relationships among these early land plants. In many cases the fossilized plants themselves are incomplete thus rendering their classification difficult, if not impossible. Despite these problems, a general overview of what must have been happening in the evolution of the various groups can be attempted with a certain

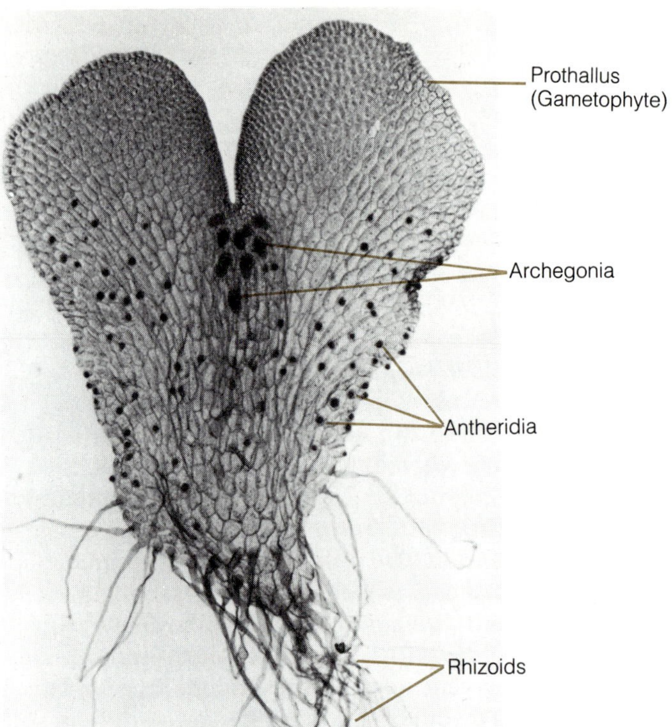

Figure 18–21. Fern gametophyte. The fern gametophyte or prothallus is a free-living plant anchored to the substrate by the rhizoids. Although the antheridia are produced in scattered fashion, the archegonia are clustered at the base of the cleft in the prothallus.

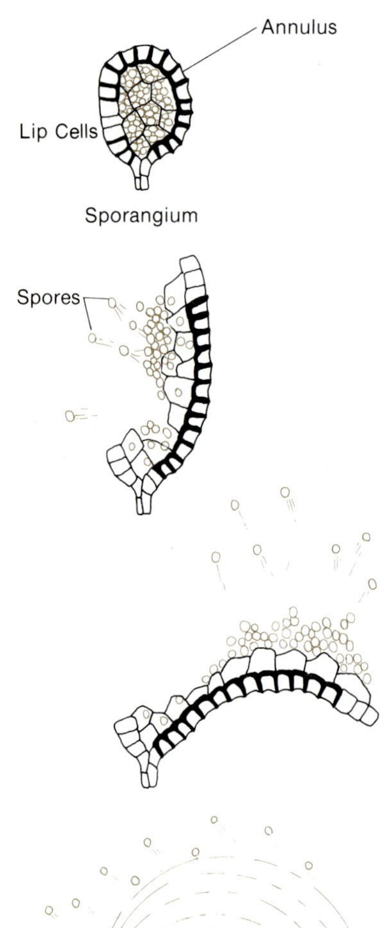

Figure 18–20. Spore release in ferns. Loss of water causes the cells of the annulus to contract, rupturing the lip cells. Further contraction of the annulus causes pronounced bending of the sporangium. Release of tension in the annulus catapults the sporangium forward, ejecting the spores.

degree of confidence. At the beginning of this chapter it was indicated that the bryophytes and vascular plants resulted from at least two different evolutionary sequences leading from the green algae (Figure 18–23). Within the Bryophyta there were either three separate lines of development leading from a common ancestor or else multiple origins of the three groups that compose the division. There is no evidence that the Bryophyta are ancestral to the vascular plants and, in fact, both groups appear to have evolved as new colonists of the terrestrial environment at the same time.

Although the fossil record is not clear on the nature of the common ancestor of the vascular plants, it is apparent that a number of primitive vascular plants had evolved about 390–400 million years ago. Within a very short period in terms of geological time, these earliest vascular plants gave rise to four major divisions of more advanced vascular plants (Figure 18–23). All of these evolutionary lines appear in the fossil record at about the same time with no apparent relationship between them other than a common ancestry in the first vascular plants.

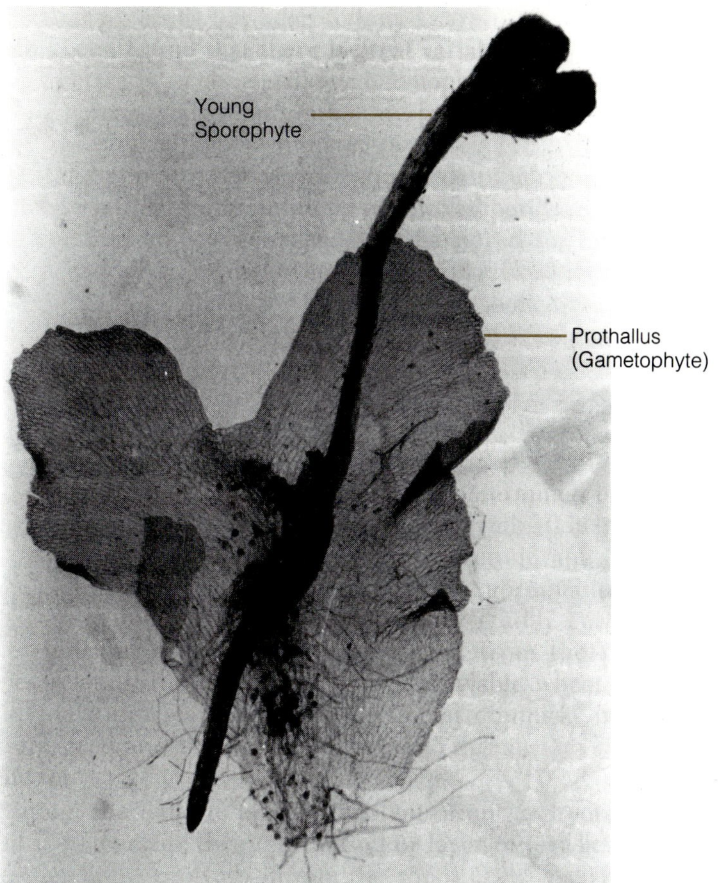

Figure 18–22. Development of the sporophyte. The fern sporophyte begins growth attached to the prothallus.

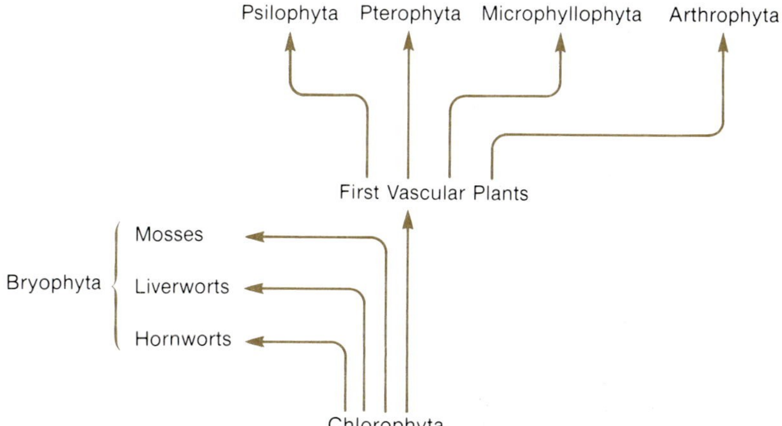

Figure 18–23. Phylogeny of land plants. Adaptation to a terrestrial environment by one or more green algal ancestors led to the evolution of the various bryophytes. The first vascular plants developed from another evolutionary line leading from the green algae. Subsequent evolution from these earliest vascular plants led to the development of more advanced groups of vascular plants.

Summary

Approximately 350–400 million years ago plants evolved that could live on land. Prior to this time all plants were aquatic algae. In order for this transition to occur the ancestors of land plants had to have evolved various means of overcoming water stress. Successful adaptation to a terrestrial environment occurred in at least two evolutionary lines leading to the development of two different groups of land plants—the bryophytes and the vascular plants. The Division Bryophyta consists of the liverworts, hornworts, and mosses, all of which are small, relatively inconspicuous plants that grow in moderately moist environments. All of the bryophytes differ from the remaining land plants in that the gametophyte generation of the bryophyte is the dominant generation in the life cycle with the sporophyte being attached to the gametophyte throughout its existence.

Moss gametophyte growth begins with the germination of a spore to produce the filamentous protonema from which is produced the typical "leafy" stage of the moss gametophyte generation. After fertilization the zygote develops into the sporophyte, which grows attached to the old archegonial tissue and relies on the gametophyte for water and nutrients. Within the sporophyte capsule diploid spore mother cells divide by meiosis producing the haploid spores. Most bryophytes are not of economic importance, although sphagnum moss is used in horticulture and peat moss serves as a fuel or a soil conditioner.

Thallose liverworts are typically thin, flattened structures with little external differentiation. Asexual reproduction is accomplished by gemmae and by fragmentation resulting from the death and decay of older regions.

The hornwort gametophyte resembles that of the thallose liverwort, but the capsule of the hornwort sporophyte has more internal differentiation.

The early vascular plants represent an independent evolution of land plants from green algae. All of the true vascular plants have a life cycle in which the sporophyte is the dominant generation. The most primitive vascular plant sporophytes very likely consisted of a branched stem system with neither roots nor leaves and may have resembled living members of the genus *Psilotum.* Sporangia produced on aerial branches form haploid spores by meiosis; these germinate, producing the gametophyte. From these earliest vascular plants four other divisions evolved—Psilophyta, Microphyllophyta, Arthrophyta, Pterophyta—each of which is represented by fossil and living species.

The Psilophyta, extremely primitive in their morphology, have photosynthetic branches produced by a subterranean rhizome. Among living Psilophyta the gametophyte developing from the haploid spore is completely independent of the sporophyte and grows underground. Association with a symbiotic fungus aids the gametophyte to survive heterotrophically.

Familiar members of the Microphyllophyta include the club mosses, or ground pines. Living forms are herbaceous; however many fossil species were trees that grew over 30 meters tall and more than 1 meter in diameter. Typically, stems, roots, and leaves are produced. Sporangia form at the bases of leaves called sporophylls. In more advanced forms the sporophylls are clustered at the end of a branch and form a structure called a strobilus, or cone. A trend toward heterospory occurs in this division whereby microsporangia are borne on microsporophylls and megasporangia are borne on megasporophylls. Both types of sporophyll can be contained within the same strobilus. In the homosporous species the gametophyte is developed and may be either photosynthetic or heterotrophic. In the evolutionarily advanced heterosporous species the male gametophytes develop within the confines of the microspore wall and are reduced to a prothallial cell (representing the gametophyte body) and a multicellular antheridium. Similarly the female gametophyte is considerably reduced and completes its development within the confines of the megaspore wall.

The Division Arthrophyta primarily contains extinct forms. The only living species belong to the genus *Equisetum.* Extinct Arthrophyta trees along with those of the Microphyllophyta and Pterophyta are responsible, in part, for extensive deposits of coal. Evolution within this division led to strobilus formation; however, heterospory appears to have been of limited occurrence. Gametophyte production and sexual reproduction is similar to that of preceding groups of vascular plants.

The fourth division to have evolved from the earliest vascular plants is the Pterophyta, or ferns. The typical fern sporophyte consists of a rhizome with attached compound leaves (fronds) and roots along its length. Fern sporangia are clustered in groups called sori that are produced on a sporophyll; however, the sporophylls are not clustered in strobili. The independent fern gametophyte formed from a spore produces antheridia and archegonia. Fertilization results in the formation of a young sporophyte.

Review Questions

1. Compare and contrast the life cycles of a moss and a fern.
2. What were the environmental conditions that led to the invasion of the land by plants and what were the major environmental challenges that had to be overcome.
3. Describe the major evolutionary advances that occurred in the early vascular plants and explain why these would be an advantage.
4. Describe the typical sporophytes of a moss, *Psilotum,* a club moss, *Equisetum,* and a fern.
5. Explain how each of the groups of early land plants is still dependent on water for the completion of their life cycle. How does this limit their distribution?

Coal

A rose is a rose is a rose, but coal is many things. Coal is the fuel that powered the Industrial Revolution and made possible our modern way of life. Coal is the source of the air pollution that killed thousands of people in the famous smog outbreaks in Donora, Pennsylvania, and London, England. Coal is formed from algae. Coal burning contributes to the greenhouse effect resulting in a warming of global climate. Coal is formed from ferns. Coal is the source of industrial chemicals. All of these statements about coal describe a different aspect of a common but complex substance, a substance that has been in daily use for centuries, but one that still provides mixed blessings.

Coal, wherever it is found, formed because plant material was buried before decay was complete. Typically, coal formed when large areas of partially decomposed vegetation were flooded and subsequently buried under large amounts of sediment. In swampy areas the source of coal was ferns, horsetails, and club mosses, whereas drier upland environments gave rise to coal from gymnosperms. Today this process can still be seen occurring in peat bogs, or swamps such as the Okefenokee, for example. Continued subsidence of the land containing buried vegetation increased both pressure and temperature, two factors that further altered the partially decomposed vegetation through nonbiological chemical reactions. As these reactions continued, the plant material became more homogeneous in physical appearance, which is why coal appears like rock rather than parts of plants. The ferns, horsetails and club mosses which resulted in vast amounts of coal deposits were treelike in appearance and size, unlike those of the present-day forms of these primitive vascular plants. Still more coal was formed from various gymnosperms (Chapter 19) that were also flourishing during this same period (about 280–345 million years ago). Another major coal-forming episode occurred 65–136 million years ago resulting in the immense deposits of coal in the western United States on either side of the Rocky Mountains. This coal, however, was formed from conifers and angiosperms (Chapter 20) because these more advanced vascular plants had replaced the primitive vascular plants as the dominant vegetation on Earth.

Although coal is usually thought of as a fuel, it is also a major potential source of industrial chemicals. Currently petroleum provides many of these chemicals, but world supplies of petroleum are variously estimated to be sufficient for only 50 to 100 years. Coal reserves are much more plentiful and will be of increasing significance as petroleum supples diminish.

Coal can be treated in a number of ways to produce usable chemicals. One of the most recent approaches is gasification in which coal is converted to gases that can then be used as fuel or for synthesizing more exotic chemicals. Coal has also been converted to coke, which is then converted to calcium carbide. The latter reacts with water and forms acetylene, which can be used to produce polyvinylchloride (PVC) or acrylic fibers. Other processes yield coal tar. This liquid can be processed to produce gasoline, road tar, pitch, creosote, plastics, dyes, insect repellents, insecticides, herbicides, terylene, nylon, polystyrene, aspirin, paint solvents, fertilizer, and so on. It is not far-fetched to state that coal alone could provide the fuel, clothing, drugs, agricultural chemicals, and plastics that are needed by our present-day society. The loss of petroleum, a fossil fuel derived from algae, does not have to spell the end of civilization as we know it—coal can fill the gap.

19

Gymnosperms

Outline

Introduction

In the natural world plants hold the record for the tallest, the largest, and the oldest of all living things. By coincidence, all three records are held by gymnosperms native to California. Although these record-setting species are native to California, gymnosperms as a group have a worldwide distribution.

A coastal redwood tree *(Sequoia sempervirens),* which grows in Humboldt Redwoods State Park in northern California, is the tallest of all trees, with a height exceeding 112.5 meters (369 feet) and still growing. Other redwood trees only slightly shorter are found along a 30-mile-wide stretch extending along the Pacific coast from the southern corner of Oregon to the Monterey Peninsula in California. These redwoods, growing in virtually pure stands, present an awesome spectacle with their massive trunks soaring upward for 30 meters (100 feet) or more before branching.

Within a few hundred miles of the coastal redwood stands is found a related species, the giant sequoia *(Sequoia gigantea),* sometimes referred to simply as "Bigtree." Although not as tall as the tallest redwoods, the giant sequoias are far more massive. The largest of all living things, which grows in Sequoia National Park, California, is a giant sequoia affectionately called "General Sherman." Its circumference at the base is 31 meters (102 feet), its height is 83 meters (273 feet), and its estimated weight is 625 tons. A close second to the General Sherman tree is another sequoia, "General Grant," with a circumference of 33 meters (108 feet), a height of 82 meters (268 feet), and an estimated weight of 565 tons. Unlike the coastal redwoods, giant sequoias seldom grow in pure stands but are usually widely spaced, intermixed with other species of smaller trees. Thus they lack the immediate visual impact of redwood forests. The range of the giant sequoia is limited to the western slopes of the Sierra Nevada Mountains of California. It grows at elevations extending between 900 meters (3000 feet) and 2700 meters (8900 feet) above sea level.

The oldest of all living things, the bristlecone pine *(Pinus longaeva),* is far less impressive in size and height than the redwoods or giant sequoias (Figure 19–1). Found only in the American Southwest, the largest stands of bristlecone pine can be seen in the Ancient Bristlecone Pine Forest, a part of Inyo National Forest, California. At least 17 bristlecone pine trees exceed an age of 4000 years and one specimen has been dated to be more than 4600 years old. Comparatively the redwoods and giant sequoias are young—the largest redwoods are estimated to be 2500 years old and the oldest giant sequoia is approximately 3500 years old. The age of a tree is determined by counting the growth rings (Chapter 6) on a thin core of wood extracted with a tubelike borer from the base of the trunk. Because of the harsh climatic conditions of their habitat, bristlecone pines grow extremely slowly. One tree estimated to be 700 years old is only 1 meter (3 feet) tall and has a trunk diameter of only 7 centimeters (3 inches). The older bristlecone pines, seldom taller than 11 meters (35 feet), usually assume a gnarled, twisted shape, and frequently exhibit only a narrow strip of bark and a few living branches, the remainder of the tree being deadwood.

Figure 19–1. Bristlecone pine. *Pinus longaeva,* is one of the oldest living things known. Several of the bristlecone pines are estimated to be older than 4000 years.

The term *gymnosperm* is derived from the Greek *gymnos,* "naked," and *sperma,* "seed." This term has no taxonomic* status, however, but rather is descriptive of a heterogeneous group of plants characterized by the production of naked seeds. Estimates from fossil records indicate that gymnosperms must have evolved approximately 300 million years ago from nonseed-producing ancestors of the extinct division Progymnospermophyta, which were fernlike in appearance. From the progymnospermous ancestors several different lines of evolution led to the various groups of gymno-

Taxonomy
(*adj.,* taxonomic) The science of naming and classification of organisms.

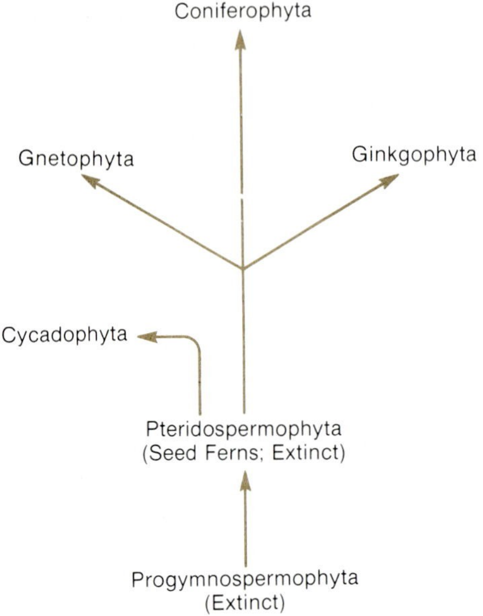

Figure 19–2. Gymnosperm evolution. Possible evolution of present-day gymnosperms from the non-seed-producing progymnosperms. The so-called seed ferns are presently considered to be the first gymnosperms. The term *seed fern* was given incorrectly to the fossilized members of this group because of their close resemblance to ferns.

sperms shown in Figure 19–2. Seed production is an adaptation of great significance for the survival and dispersal of plants. In fact, this was part of the competitive advantage that allowed the gymnosperms to supercede the other vascular plants as the dominant type of vegetation on land. Only the later evolution of the flower and the fruit allowed another group of seed plants (the angiosperms, see Chapter 20) to displace the gymnosperms from their preeminent position.

Evolution of the Seed

It is generally accepted at present that seed production must have evolved initially among now extinct, primitive gymnosperms. Although the seed represents a significant evolutionary advance in plant reproduction, it developed from existing structures. In all vascular plants the spores* that give rise to gametophytes* are formed in sporangia. Among certain primitive vascular plants, as noted in the previous chapter, there was a trend to the production of different-sized spores—microspores and megaspores—which would produce separate male gametophytes (microgametophytes) and female gametophytes (megagametophytes), respectively. This trend to **heterospory** is believed to have made the evolution of the seed possible because it resulted in separate microsporangia and megasporangia. Ultimately the heterosporous life cycle became established as the sole method of sexual reproduction only among the seed-producing plants.

Spore
A unicellular reproductive structure other than a gamete.

Gametophyte
The gamete-producing stage in a life cycle.

In the evolutionary lines leading to the various gymnosperm groups, the megasporangium became surrounded by a protective layer of tissue, the **integument.** This new structure (megasporangium plus integument) is called an **ovule.** In the seed plants the megaspores are not released from the megasporangium, so the megagametophyte develops inside the megasporangium and thus within the ovule. Therefore, when the egg is formed it is enclosed within the ovule. As will be described later, once the microgametophyte is physically transported to the ovule, the sperm will be released in the vicinity of the egg contained within. After fertilization, growth of the embryonic sporophyte* occurs within the ovule, the entire structure now referred to as a **seed.** In most gymnosperms, seeds are produced in a cone-like structure, the strobilus, which is an evolutionarily modified branch. The strobilus, or cone, as a reproductive structure is known to have evolved independently among the club mosses and horsetails as described in Chapter 18.

Sporophyte
The spore-producing stage in a life cycle.

It should be obvious now that the seed is the result of two different evolutionary developments: the enclosure of the megasporangium within protective tissue and the retention of the megagametophyte and the developing embryo* within the megasporangium. Another change that accompanied the evolution of the ovule was the elimination of the necessity of water to transport the sperm from the microgametophyte to the megagametophyte for fertilization to occur. This development made it possible for seed plants to complete their life cycles in a relatively dry environment, compared to those of the non-seed-producing vascular plants.

Embryo
Rudimentary plant formed after fertilization of the egg by a sperm.

It is reasonable to assume that gradual changes toward the direction of seed production had been taking place among the primitive vascular plants. Nonetheless, there is no known fossil evidence to support a theory that seeds may have evolved independently in more than one group of plants prior to the evolution of the gymnosperms. As in all vascular plants, the conspicuous phase in the life cycle of seed-producing plants is the diploid* sporophyte, or the meiospore* producing phase. The haploid* gametophyte, or gamete-producing phase, is much reduced in size and is completely dependent on the sporophyte in the gymnosperms and flowering plants, in contrast to the situation in the primitive vascular plants.

Diploid
Possessing two of each pair of chromosomes.

Meiospore
A haploid spore produced after meiotic division.

Haploid
Possessing only one of each pair of chromosomes.

Division Coniferophyta

The most familiar and most abundant gymnosperms, particularly in the temperate regions of the world, including vast areas of the United States and Canada, are members of the Division Coniferophyta, known commonly as conifers. This division includes types such as pine, fir, spruce, juniper, larch, cypress, redwood, hemlock, and arbor vitae. The conifers, with few exceptions, are trees. The branches are arranged at regular intervals in whorls or scattered on the trunk. The simple leaves of conifers are typically needle-shaped and are produced singly or clustered in definite numbers, as in the different pines, or they can be scalelike as in arbor vitae. With the exception of bald cypress and larch, the conifers, unlike the typical temperate-zone angiosperms, are **evergreen** plants, retaining their leaves for more than one season. Another characteristic of this group of plants is the presence of resin ducts in the wood and bark.

Within this division there are species that produce both microsporangia and megasporangia on the same plant, a condition referred to as **monoecious** (Greek *mon,* one + *oikos,* house); other species are **dioecious** (Greek *di,* two) and produce microsporangia and megasporangia on separate plants.

Pine Life Cycle

Since pine (*Pinus* sp.) is the commonest of all gymnosperms and also because the major events related to the reproduction of pine are similar in a general way to those of all divisions of gymnosperms, pine is used here as an example of the gymnosperm life cycle (Figure 19–3).

A pine tree represents the **sporophyte generation** in the life cycle (Plate 9E). As a heterosporous plant, two types of meiospores, microspores and megaspores, are produced. The **gametophyte generation,** produced from the development of these spores, develops on the sporophyte and remains completely dependent upon it.

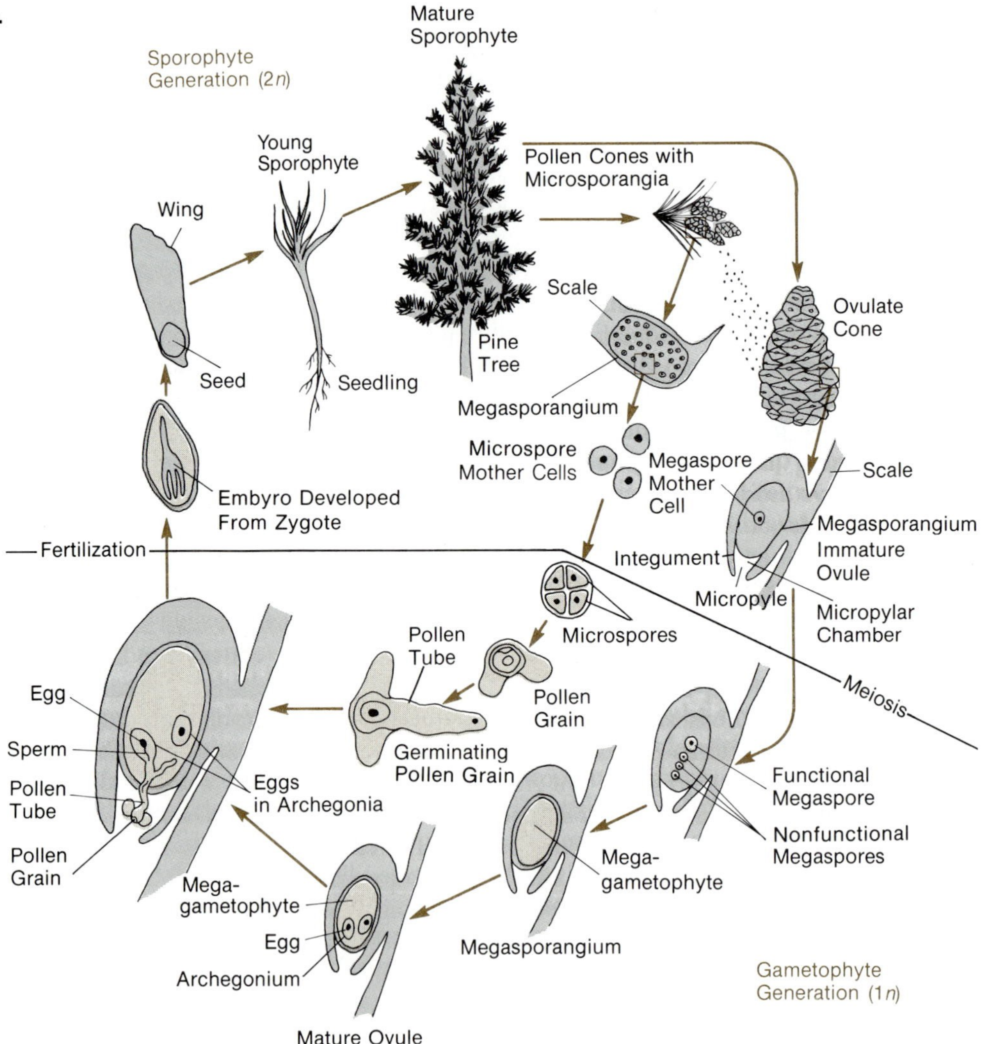

Figure 19–3. Pine life cycle.

(a)

(b)

Figure 19–4. Pine cones. (*a*) A cluster of microsporangiate (pollen) cones. (*b*) Three megasporangiate (ovulate) cones. Both types are produced on a tree, but the ovulate cones are always larger than the pollen cones of the same tree.

All pines are monoecious; however, the sporangia are produced in separate cones (Figure 19–4), the place where all of the events of sexual reproduction occur. Both microsporangial and megasporangial cones consist of **scales** arranged spirally on a central axis. Scales bearing microsporangia remain small and soft, while ovule-bearing scales become large and woody after pollination. In addition to paleobotanical evidence, anatomical and de-

Microsporophylls
Modified leaves that produce microsporangia.

velopmental studies have shown that the microsporangial scales are probably derived from microsporophylls*, whereas the ovule-bearing scales are modified branches. Microsporangial cones are usually produced in clusters near the tips of lower branches and the ovulate cones are produced either singly or in pairs on the younger branches. Within the same species the mature microsporangial cones are always small, commonly less than 1 inch long, whereas the ovulate cones are large, measuring up to 2 feet long in sugar pines.

Microgametogenesis

The microsporangial cones (Plate 9F) develop during early spring; the production of two **microsporangia** (Figure 19–5) on the lower surface of each cone scale is the first in a series of events leading to the development of the pine **microgametophyte.** As the cones mature, the diploid **microspore mother cells** within each microsporangium undergo meiosis* to produce haploid **microspores,** each of which divides mitotically* producing the **microgametophyte** (Figure 19–6). This entire process is usually completed in less than two weeks. The pine microgametophyte, or **pollen grain,** is much reduced in size and complexity compared to gametophytes of other terrestrial plants.

Meiosis
The process of nuclear division that results in the production of daughter nuclei, each containing one-half the original number of chromosomes.

Mitosis
The process of nuclear division that results in the production of two genetically identical daughter nuclei.

At this stage in its development the microgametophyte, or pollen grain, consists of only four cells: a **tube cell,** a **generative cell,** and two **prothallial cells** (Figure 19–7). The prothallial cells are considered to represent the vegetative cells of the gametophyte and take no apparent part in the further development of the microgametophyte. When the pollen matures, the mi-

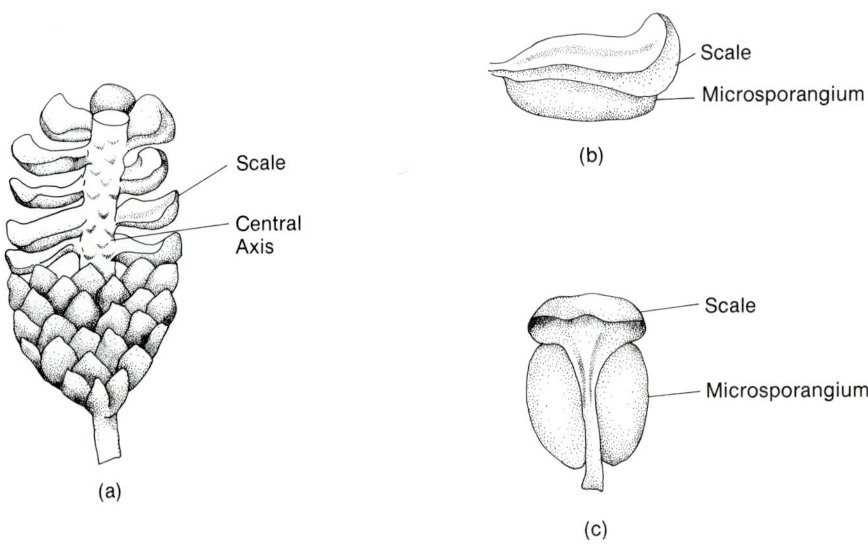

Figure 19–5. Microsporangiate structures of pine. (*a*) Microsporangial cone showing the spiral arrangement of scales on a central axis. (*b* and *c*) Lateral and top views of a single scale showing the attachment of microsporangia on the lower surface.

crosporangia break open releasing it into the air. Each pollen grain possess two wings that develop from the microspore cell wall. It is believed by some botanists that the wings keep the pollen bouyant during its transport via air currents to the female cones. This transfer of pollen from the microsporangial cone to the ovulate cone is called **pollination.** After the release of pollen grains, the microsporangial cones dry up and fall off. A pine tree produces massive amounts of pollen, most of which never reaches the female cones, but instead settles on any stationary object as a fine yellow powder. The yellow film often seen on lakes near dense pine stands during early spring is invariably pollen. After reaching the ovulate cone, the pollen grain will slowly develop further, as described in the next selection.

Megagametogenesis

Unlike the short-lived microsporangial cones, the ovulate cones usually remain on the tree for at least two years. Paired ovules develop on the upper surface of each scale of a newly forming ovulate cone (Figures 19–8, 19–9). As the cone matures, each ovule undergoes a series of developmental changes, culminating in the formation of a megagametophyte.

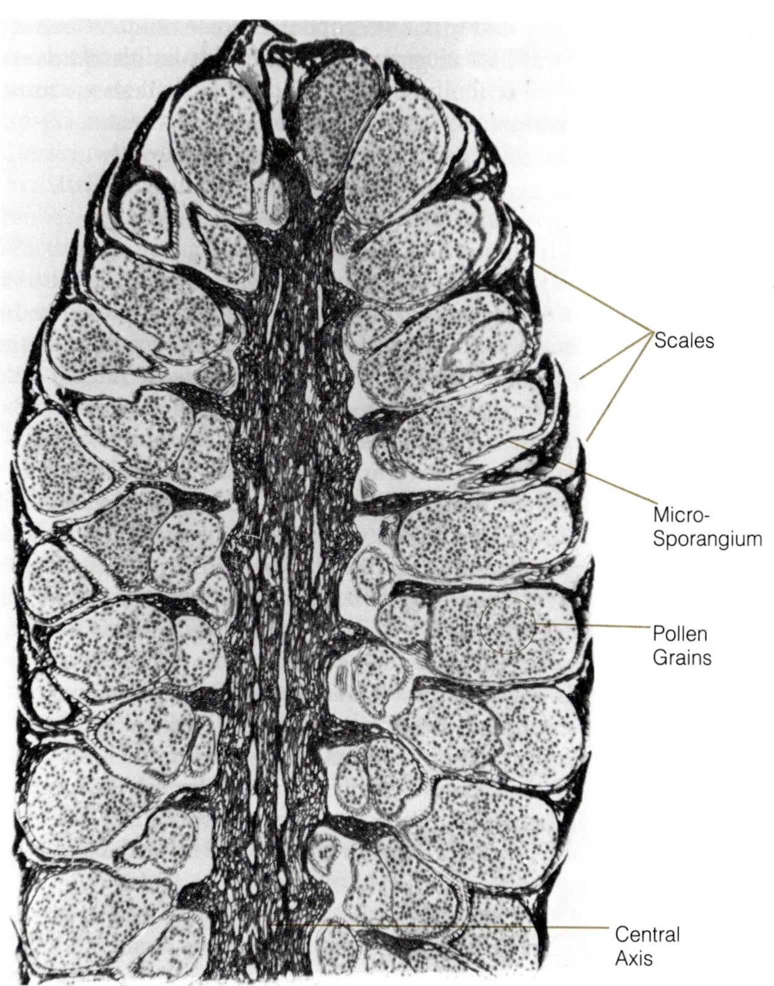

Figure 19–6. Longitudinal section of a microsporangial cone. Note the arrangement of scales on a central axis and the microsporangia on the lower surface of each scale.

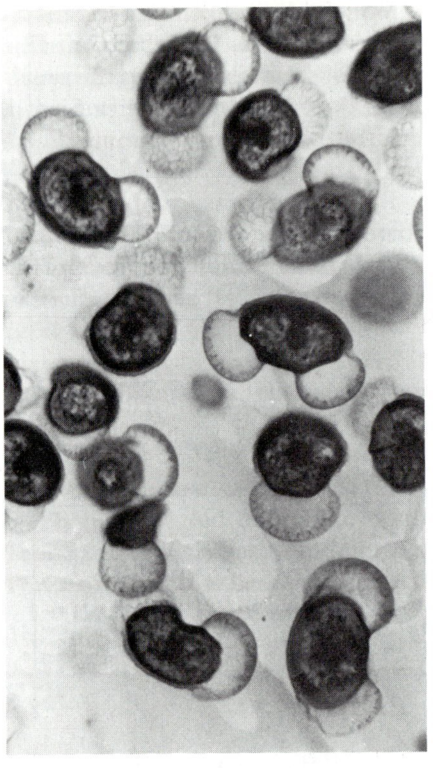

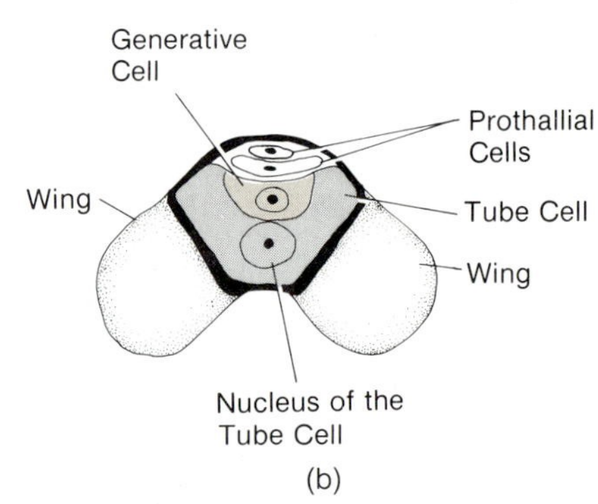

(a) (b)

Figure 19–7. Pine pollen grain. (*a*) Photograph of sectioned pollen grains. (*b*) Diagram of pollen grain showing details.

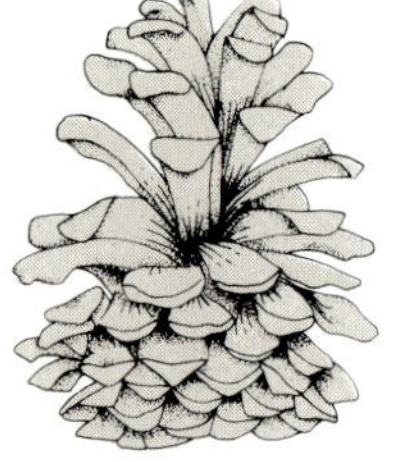

(a)

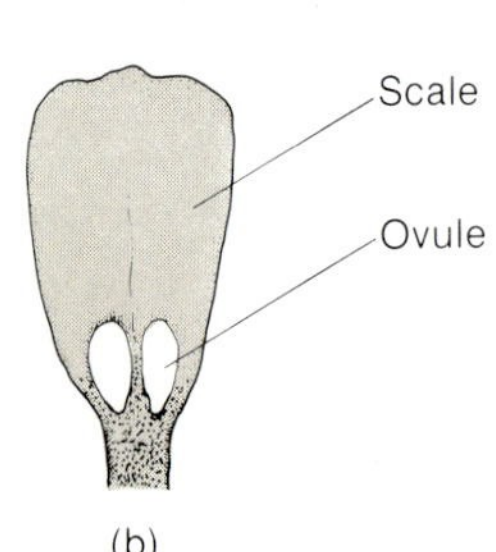

(b)

Figure 19–8. Megasporangial structures of pine. (*a*) Ovulate cone showing the spiral arrangement of scales on a central axis. (*b*) Upper view of scale showing two ovules.

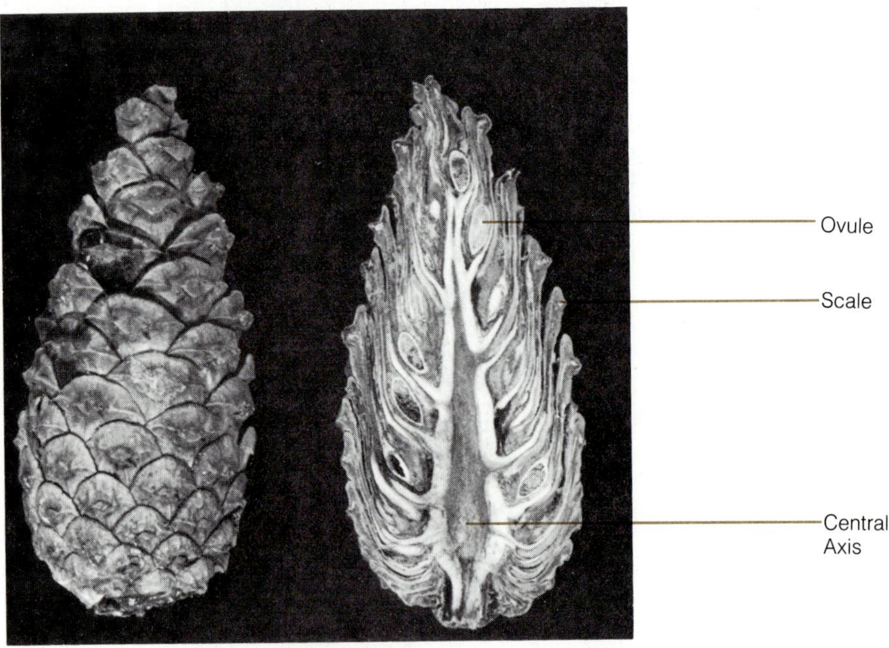

Figure 19–9. Ovulate cone. Photograph of ovulate cone split through the center. Scales with ovules on their upper surface are attached to a central axis.

A young ovule consists of a multicellular **megasporangium** surrounded by an integument. The opening through the integument, toward the base of the scale nearest the cone axis, is called the **micropyle.** Within the megasporangium a single, diploid **megaspore mother cell** is produced that will eventually divide by meiosis, forming four haploid **megaspores,** three of which degenerate. The functional megaspore that remains divides by mitosis and becomes the **megagametophyte.** Two or more **archegonia*** are produced within the megagametophyte near the micropylar end, followed by the development of an **egg cell** in each archegonium, thus completing the process of **megagametogenesis*** (Figure 19–10).

When the ovule is at the megaspore mother cell stage of development, the cone scales spread apart temporarily, facilitating pollination. The pollen grains that are trapped in the sticky fluid exuded from the ovule near the base of the scale are drawn, as this fluid dries, through the micropyle to a space between the integument and the megasporangium. Further development of both the ovule and the pollen grains continues slowly over the next 12 to 14 months prior to gamete production and fertilization.

Archegonia
(*sing.*, archegonium) Gametangia in which eggs are produced.

Megagametogenesis
The process that results in the production of the egg.

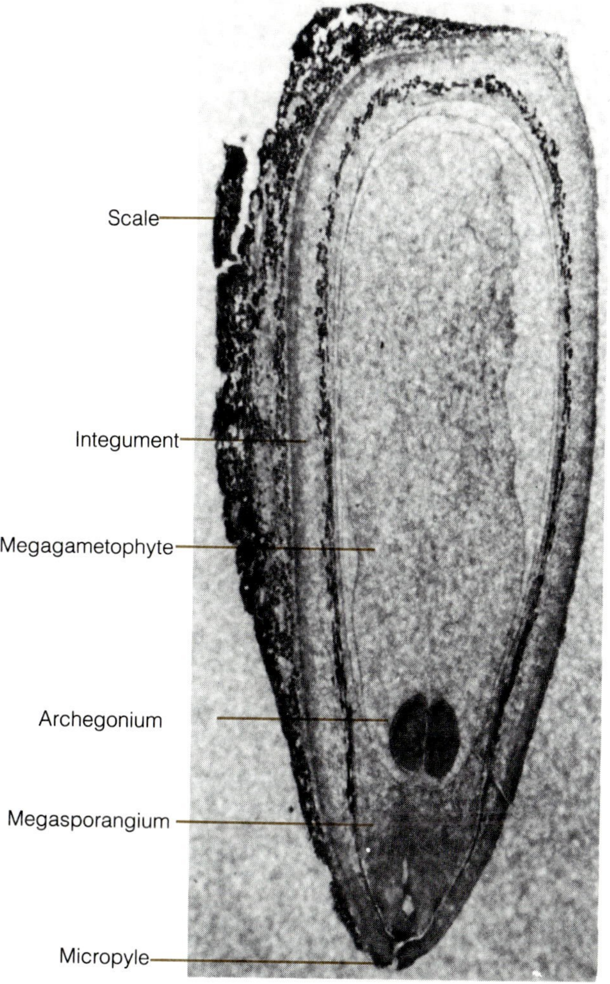

Figure 19–10. Pine ovule. Photomicrograph of a mature ovule sectioned longitudinally.

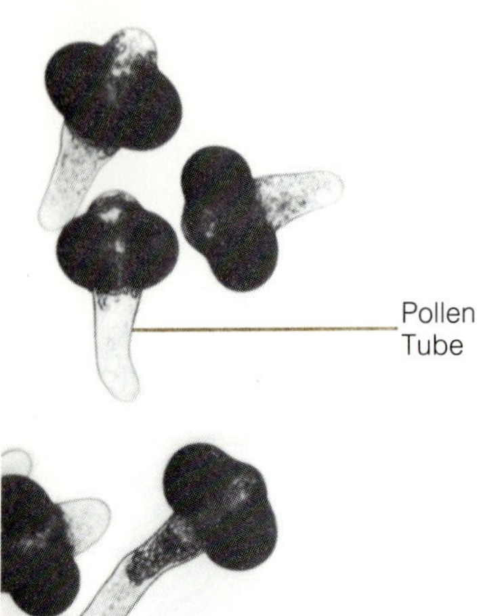

Figure 19–11. Germinating pollen grains of pine. In nature the pollen tubes often branch.

Upon release from the microsporangium, the pollen grain is an immature microgametophyte because it does not contain sperm. Furthermore, there is no way for sperm to reach an egg contained within an archegonium inside the megagametophyte. During the approximately one year that the microgametophyte is enclosed within the ovule, it begins to produce branched tubular extensions, known as **pollen tubes,** that digest their way into the megasporangium. These **pollen tubes** (Figure 19–11) seem to function both as an absorptive surface for nutrients and a transport system for the nonmotile sperm to reach the egg cell. A pollen grain with its tubular extensions represents the mature **microgametophyte,** which in nature is completely dependent on the ovule for its survival.

Sometime during the continued growth of the pollen tube toward the egg cell, two nonmotile sperm are derived by mitosis from the generative cell of the pollen grain, which by then has migrated into the tube. (It is important to note that in the gymnosperms the sperm are not produced in an antheridium* as they are in the bryophytes and the primitive vascular plants.) When the pollen tube reaches the egg cell, the tube breaks and the sperm are released into the egg cytoplasm. The first sperm to reach the egg nucleus fertilizes it, resulting in the formation of the **zygote,** while the second sperm degenerates. The zygote, by repeated cell divisions, then gives rise to a multicellular **embryo.** After fertilization, an additional 4 to 6 months are needed for the development of the embryo in the seed and its subsequent dispersal.

Antheridium
Gametangium in which sperm are produced.

Seed Development

Since there are several archegonia in a mature ovule, several eggs are often fertilized; however, during subsequent development only one embryo becomes established, while the others degenerate. The elongated pine embryo bears a **radicle*** at one end, a shoot apex, or **plumule,** at the other end, and 3 to 18 (typically 8) needle-shaped **cotyledons*** attached near the shoot apex. During embryo development other changes take place in the ovule that result in the transformation of the ovule into a **seed.** A **seed coat** is derived from the integument, a portion of which expands, forming a winglike appendage that facilitates wind dispersal of the seed (Figure 19–12). The megagametophyte tissue, which surrounds the embryo, contains stored food that will be used during germination of the embryo into a pine seedling. Pine seeds are dispersed during autumn when the scales of the ovulate cone separate. This complete process of embryo development and seed formation and dispersal takes approximately 16 to 20 months from the time of pollination.

Radicle
Embryonic root.

Cotyledon
A leaflike structure in the seed that may contain stored food.

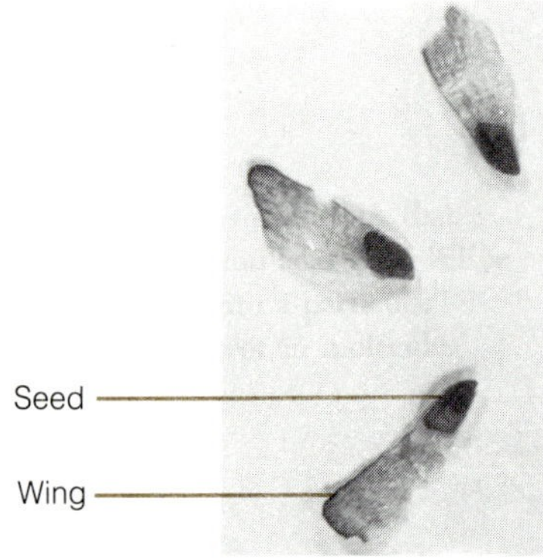

Figure 19–12. Winged pine seeds. The wings are produced by the expansion of the seed coat, which originates from the integument.

Division Cycadophyta

Members of the Division Cycadophyta are primitive gymnosperms and superficially resemble palm trees and tree ferns. These plants are commonly called cycads, from the Greek *kykas,* "a palm." The cycads (Figure 19–13 and Plate 9D) have an unbranched, nonwoody trunk, which, in a few spe-

Figure 19–13. A cycad. Cycads are palmlike gymnosperms.

cies, may reach a height of 45 feet. The trunk is topped with a crown of large, palmlike, compound leaves. Cycads are characterized by slow growth: in one case a plant reached a height of only 6 feet after almost 1000 years of growth. Their distribution is restricted to subtropical and tropical regions of the world. The only genus of cycads native to the continental United States is *Zamia,* which grows in Florida.

All cycads are dioecious, producing either microsporangial cones or ovulate cones at the terminal portion of the trunk. The sperm of the cycads are flagellate*, a feature considered to be an ancestral carryover; however, the motile sperm are transported to the egg through the pollen tube without the aid of water. The cycad life cycle generally resembles that of the typical pine described above.

Flagellated
Possessing hairlike structures for motility.

Division Ginkgophyta

Ginkgo, commonly called maidenhair tree because of the resemblance of its leaves to those of maidenhair fern, is the only extant genus of the Division Ginkgophyta. Owing to its fossil record, it is considered to be the oldest genus of extant seed plants. *Ginkgo biloba* is a large, highly branched tree that resembles a woody angiosperm* because of its laminar leaves and deciduous nature. However, the fleshy seed coat of *Ginkgo* can easily be mistaken for a fruit.

Angiosperm
One of a group of plants characterized by the production of flowers.

The *Ginkgo,* found during historic times only in the central part of mainland China, was introduced to the American continent, Europe, and other parts of Asia as an ornamental tree, a fact credited with saving it from extinction. In addition to its stately shape, the popularity of *Ginkgo* as an ornamental tree is growing due to its resistance to many types of air pollution common around industrial areas.

Microsporophyte
A sporophyte that produces microspores.

Megasporophyte
A sporophyte that produces megaspores.

Ginkgo is dioecious, with the microsporangia arranged in paired cones at the end of short branches on the microsporophyte* trees and paired ovules (not in cones) produced at the tips of short branches on the megasporophyte* trees (Figure 19–14). After pollination and subsequent fertilization, each ovule develops into a seed with a fleshy seed coat (Figure 19–15). Because of the high butyric acid content of the fleshy seed coat, mature seeds have a foul odor. As in the case of cycads, the motile sperm of *Ginkgo* are delivered through a pollen tube, and water is not needed for fertilization. *Ginkgo biloba* is a modern-day survivor of an ancient evolutionary line extending back 150 million years. It appears that this species has undergone very little change from its ancestors.

Division Gnetophyta

Vessel element
One type of water-conducting cell in xylem, characterized by its relatively large diameter and lack of end walls.

The Division Gnetophyta is comprised of three morphologically dissimilar genera of gymnosperms: *Ephedra, Gnetum,* and *Welwitschia.* One significant feature they all share, in addition to being gymnosperms, is the presence of vessel elements* in their wood, a cell type absent in other gymnosperms. The sperm produced by the members of this division are nonmotile. All three genera are dioecious.

Ephedra (Figure 19–16) is a shrublike plant with jointed stems and highly reduced leaves. Although species of *Ephedra* have a wide distribution, they are usually restricted to desert or semi-desert habitats. In the United States, they are found in the arid regions of the Southwest. *Ephedra* is the source of ephedrine, a drug used for the relief of allergic symptoms because of its ability to constrict blood vessels.

Netted venation
Arrangement of veins in a leaf in a branching network.

Species of *Gnetum* (Figure 19–17) are mostly woody vines native to tropical India, southeast Asia, tropical West Africa, and Brazil. The broad leaves of *Gnetum* have netted venation*, giving these plants a close resemblance to angiosperms. Even the naked seed of *Gnetum* can be mistaken for a fruit. As in most other gymnosperms, the reproductive structures of these plants are produced in cones.

Xerophyte
(*adj.,* xerophytic) A plant capable of growing under dry conditions.

Welwitschia (Figure 19–18) is a strange-looking gymnosperm and is probably the most unusual of all seed-producing plants. The only species known is restricted to the coastal region of Angola and southwest Africa where it grows under extreme xerophytic* conditions. A mature plant has a highly compressed, distinctly V-shaped stem that is mostly underground. The upper portion of this woody stem is known to reach a diameter of 4–6 feet. The two large leaves—up to 9 feet long—produced by each plant persist throughout the life of the plant, growing continuously by the activity of mer-

Figure 19–14. *Ginkgo.* (*a*) Microsporangiate cones produced on "male" trees. (*b*) Young ovules in pairs produced on "female" trees.

Figure 19–15. A branch of a female *Ginkgo* tree. Note the characteristic fan-shaped leaves and the seeds with fleshy covering, often mistaken for fruits.

Figure 19–16. *Ephedra.* This shrubby gymnosperm has scale leaves and appears to have jointed stems.

Figure 19–17. *Gnetum.* This is a woody vine with leaves that look like those of dicotyledonous plants.

Figure 19–18. ***Welwitschia.*** This unusual looking gymnosperm has two large leaves that split longitudinally as the plant grows older, giving the appearance of having many leaves. The stem is mostly underground.

istems* present at the leaf base. As the plant ages, these leaves periodically split lengthwise giving the plant the appearance of having many leaves. The male and female cones originate from the axils* of leaves. Growth of these plants is unusually slow due to their dry environment.

Meristem
A tissue whose cells are capable of repeated division.

Axil
The angle formed between the upper side of the petiole and the stem.

Summary

Gymnosperm, meaning "naked seed," refers to a group of vascular plants that produce seeds on a cone instead of inside a fruit. Seeds are considered to have evolved first among an extinct group of gymnosperms, the progymnosperms, from which the present-day gymnosperms are thought to have evolved approximately 300 milllion years ago. The gymnosperms are classified under four separate divisions—Coniferophyta, Cycadophyta, Ginkgophyta, and Gnetophyta.

Heterospory and the production of spores on cones are considered to be forerunners of seed production; both features are present among the primitive vascular plants. The enclosure of the female sporangium within a protective layer to produce an ovule must have preceded seed evolution since the ovule becomes the seed after fertilization. Unlike that of the primitive vascular plants, the gametophyte phase of all seed-producing plants is completely dependent on the sporophyte phase because it is retained within the sporangium.

The Division Coniferophyta includes such familiar trees as pine, fir, spruce, redwood, cypress, and others. The conifers, with few exceptions, are evergreens with needle-shaped or scalelike leaves.

The most familiar conifer is *Pinus,* a monoecious tree. Both male and female cones consist of scales attached to a central stemlike structure. The two microsporangia on each male scale produce many microspre mother cells that undergo meiosis forming haploid microspores. The microspores develop into winged, four-celled pollen grains, which are carried by wind to the female cones.

Two ovules are present on each female scale. A young ovule contains a megasporangium surrounded by an integument with its micropyle directed toward the base of the scale. The single megaspore mother cell within the megasporangium divides meiotically producing four haploid megaspores, three of which degenerate. The remaining megaspore enlarges by cell division to become the megagametophyte, inside of which are produced two or more archegonia, each containing an egg.

As the ovule undergoes these changes, the pollen grain, which is trapped in the micropylar chamber, produces a branched pollen tube that slowly works its way into the egg cell. One of the sperm derived from the generative cell of the pollen grain fertilizes the egg, resulting in a zygote that eventually develops into an embryo. After fertilization the ovule is transformed into the seed with an enclosed embryo. It takes approximately 12 to 14 months after pollination for fertilization to take place and an additional four to six months before maturation of the embryo and seed dispersal.

The cycads, Division Cycadophyta, are palmlike, dioecious plants that are native to tropical and subtropical regions. *Ginkgo* (maidenhair tree), the only extant genus of the Ginkgophyta, resembles a woody angiosperm tree. *Ephedra, Gnetum,* and *Welwitschia,* three morphologically distinct genera, make up the Division Gnetophyta, characterized by the presence of vessel elements in their wood, unlike those of other gymnosperms.

Review Questions

1. What are the apparent evolutionary trends that culminated in seed production?
2. Describe characteristic features of the different gymnosperm divisions.
3. What are the major events in the life cycle of a pine?
4. Discuss the economic importance of the gymnosperms.

Economic Importance of Gymnosperms

Many gymnosperms are of considerable economic importance, especially in the wood products industry where the major areas of wood utilization are lumber, plywood, and wood pulp. Much of the wood used for construction comes from different species of pine, as well as from redwood, Douglas fir, spruce, cedar, cypress, and larch. The wood of redwood and of bald cypress, among others, is resistant to attack by insects and fungi. The harvest of redwood lumber was so extensive in the United States at one time that the number of redwood stands fell to a dangerously low level. Similarly, the unrestricted logging of eastern white pine during the last century resulted in the sparse stands of these trees, which at one time covered vast areas of the northeastern United States and Canada. Indiscriminate logging, a serious problem until recent years, is prevented by widely practiced conservation measures, which, in turn, have improved the outlook for tree species used in the wood products industry.

Three-fourths of all plywood is made from conifers. Plywood is made by glueing together several thin sheets of wood called veneer, which are sliced from a log rotating against a veneer knife. The direction of grain of adjacent sheets is at right angles to each other, giving plywood its dimensional stabiilty and flexibility.

Most of the wood pulp used in the paper industry in this country is made from white spruce. Because of the use of paper as newsprint and for numerous other purposes, it is no surprise that billions of trees are needed annually to meet the heavy demand for paper in the United States alone.

Another product of commercial value that is obtained from different conifers is **resin,** from which is made **turpentine** and **rosin. Turpentine** is a volatile liquid used as a solvent and thinner for paints and varnishes; rosin is used in the manufacture of floor waxes and paper coatings, on the bows of stringed musical instruments, and even baseball pitchers use rosin—the only legal foreign substance that they can put on their fingers. **Amber,** a hard yellowish to brownish translucent substance, is fossilized resin, used today chiefly for jewelry.

There are also other products of commercial value obtained from gymnosperm wood by distillation* and hydrolysis.* One of the advantages of these processes is that waste products from saw mills and pulp manufacture, such as sawdust, trimmings, branches, and stumps, can be utilized. Methanol, an organic solvent used extensively in industry and for isolating fatty or fat-soluble materials from biological systems, and acetic acid, a solvent used in the production of a variety of substances, are two of the distillation products of wood. Some of the typical products of wood hydrolysis include ethyl alcohol, glycerol, lignin, carbolic acid, and synthetic rubber.

Distillation
Process of separating a mixture by boiling a liquid and condensing the resulting vapor.

Hydrolysis
Chemical reaction in which a molecule breaks apart after combining with water.

The seeds of certain species of pine are eaten plain or are ground up and used in cooking or baking. These seeds have a relatively high protein and oil content. In addition, other parts, particularly the phloem-containing inner bark, have been utilized as a source of food by various native American

Indians and the early settlers. The reserve food materials in the seeds of cycads and ginkgo are also utilized in certain Asian countries.

In addition to the uses described above, many gymnosperms are used extensively for ornamental plantings. The seasonal Christmas tree industry in temperate regions of the world also uses conifers.

20

Angiosperms: Flowers and Life Cycle

Outline

Introduction

Water-meal, an aquatic plant barely 1 millimeter long that grows in freshwater ponds, and *Eucalyptus,* growing to heights of 92 meters (300 feet) and native to the Australian continent, have one thing in common—they both produce flowers. Water-meal *(Wolffia)* produces the smallest flowers known. The largest flowers, however, are not produced by the giant *Eucalyptus,* but by a parasitic plant, stinking-corpse lily *(Rafflesia arnoldi),* which grows in the jungles of Sumatra, one of the Indonesian islands. Whereas water-meal flowers are barely visible to the naked eye, the giant flowers of stinking-corpse lily are 1 meter (3 feet) across when fully open and weigh approximately 7 kilograms (15 pounds). In spite of this tremendous difference in the size of their flowers, both carry out the same basic function in the life cycle of these plants—sexual reproduction. Flowers are the specialized sexual reproductive structures of plants belonging to the Division **Anthophyta.** Flower production is the most significant feature of this division.

The Anthophyta is comprised of plants that differ greatly in their morphological* features. Included are both small and large plants that grow as herbs, shrubs, trees, woody and nonwoody vines, and trailing plants. The Anthophyta are the most widely distributed vascular plants*, ranging from arid habitats to completely aquatic habitats, thus showing a high degree of adaptation.

Morphology
(*adj.,* morphological) The study of the external appearance of organisms.

Vascular plant
A plant that has specialized tissues (xylem and phloem) for conducting water and food.

Plants of the Anthophyta are commonly referred to as **angiosperms,** from the Greek *angeion,* a vessel, and *sperma,* a seed. This indicates the nature of seed production in this division—seeds are enclosed within a vessel, or fruit. The possession of fruit has been of considerable value in terms of the widespread distribution and survival of the flowering plants. This will be considered in the next chapter.

The angiosperms are the most abundant and familiar plants around us. Included are grasses, roses, orchids, dandelions, lilies, daisies, chrysanthemums, peas, beans, pumpkins, squashes, tomatoes, maples, oaks, walnuts, beeches, birches, ashes, and cacti, just to name a few. The angiosperms in many ways, directly and indirectly, provide us with food, a source of building material, fiber for clothing, products of medicinal value, and many other useful substances.

Evolution of Angiosperms

The earliest known fossils of angiosperms are approximately 135 million years old. However, these fossils are similar to some of the present-day forms, suggesting that the evolution of angiosperms as a group must have taken place prior to this time. Although fossil records are not available to determine the time of evolution of the angiosperms, estimates place this period to be around 150 million years ago. One suggestion is that angiosperms evolved from an extinct gymnosperm group known as the seed ferns. Without doubt, the angiosperms are the most highly evolved group of plants among the present flora* and have been the most abundant group of plants in terms of number of species for the past 100 million years. The production of vessel elements* as the major cell type in xylem*, the different modes of pollination*, and the efficient mechanisms for seed dispersal are all features that contributed to the successful establishment of angiosperms as the predominant vegetation on land.

Flora
Plants, generally used as the opposite of fauna.

Vessel element
Large water-conducting cell of xylem.

Xylem
Water-conducting tissue.

Pollination
Transfer of pollen from anther to stigma.

Classification

The angiosperms fall naturally into two taxonomic* classes based on the number of **cotyledons*** present in their seeds. Those containing one cotyledon are classified as **Monocotyledonae** (monocots) while those with two cotyledons are classified as **Dicotyledonae** (dicots). Other morphological and anatomical* differences between monocots and dicots are presented in Table 20–1. Approximately 70% of all angiosperm species are dicots (about 170,000 species) and 30% are monocots (about 65,000 species). It is believed that the monocots are of more recent origin than the dicots. Examples of monocots include all of the cereal grains and other grasses, lilies, daffodils, iris, orchids, bananas, and palms. With the exception of the palms and bamboos, almost all of the monocots are nonwoody herbaceous plants. Common dicots include tomato, rose, sunflower, pea, bean, cucumber, mulberry, honeysuckle, apple, peach, cherry, willow, chestnut, maple, and oak. Almost all of the flowering woody shrubs and trees are dicots.

Taxonomy
(*adj.*, taxonomic) The science of classification and naming of organisms.

Cotyledon
A leaflike structure in the seed that may contain stored food.

Anatomy
(*adj.*, anatomical) The study of the internal arrangement of tissues.

Table 20–1.
General differences between monocots and dicots

Characteristic	Monocots	Dicots
Number of cotyledons	One	Two
Number of individual members within each floral whorl	Three or multiple of three	Two, four, five, or their multiples
Leaf venation	Parallel	Pinnately or palmately netted
Root system	Fibrous	Tap
Vascular cambium	Absent	Present
Arrangement of vascular bundles	Scattered	Circular
Secondary tissues	Absent	Present
Seed storage tissue	Endosperm	Cotyledons

The most economically important family of Monocotyledonae is the Poaceae, the grass family, which includes all of the cereal grains, such as corn, wheat, rice, oats, barley, rye, and sorghum, as well as sugarcane and bamboo. The largest monocot family is the Orchidaceae, the orchid family. There are over 290 families of dicot plants, of which Fabacaea (the bean family), Rosaceae (the rose family), and Asteraceae (the sunflower family) are among the most common.

Evolution of the Flower

Homologous
Having the same origin.

Determinate growth
Growth that halts after a specific size has been reached.

The flower as a reproductive structure must have evolved from structures serving a similar purpose in the more primitive vascular plants. Unfortunately, fossil flowers are scarce, and even when they are found, they are similar to modern types. Therefore, current assessments of the evolution of the flower must be based on comparative studies of present-day flowers and of reproductive organs of primitive vascular plants. A survey of the reproductive organs of such plants shows that the sporangia in primitive vascular plant species are produced on leaves or on structures homologous* to leaves (Chapter 18). Sporangia among some ferns are not restricted to any group of leaves; instead, all leaves are capable of producing sporangia. In the strobilus of club mosses the sporophylls (spore-bearing leaves) are restricted to the terminus of the upright shoot. Among the advanced vascular plants such as the conifers, the reproductive structures are modified branches of determinate growth* in which the sporangia are attached to scales similar in origin to leaves or modified branches. Analysis of the development of a flower in a bud has shown that the flower is a compressed shoot system of determinate growth and each floral part represents a leaf on this modified shoot. In most flowers, the resemblance to leaves can be easily recognized in the petals and sepals, which have only a secondary role in reproduction. The resemblance to leaves is less apparent in the reproductive structures (stamen and pistil) within a flower. Although stamens of most flowers bear no resemblance to leaves, there are leaflike stamens among a few of the primitive angiosperms, such as the magnolias, strongly suggesting possible origin of stamens from a sporophyll. Stamens are believed to have evolved through a folding and fusion of leaves (sporophylls). The pistil is also believed to have evolved from a sporophyll or leaflike organ by folding, as seen in Figure 20–1. Differences in the folding and fusion of floral parts can be recognized among flowers from evolutionarily primitive and advanced families.

Parts of a Flower

Most flowers are composed of four groups of floral parts arranged in successive whorls (Figure 20–2) on a **receptacle,** which is the enlarged tip of

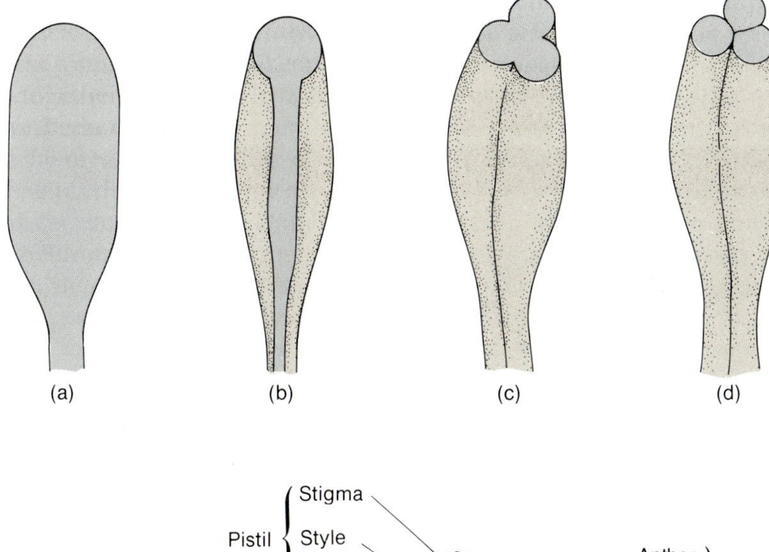

Figure 20–1. Evolution of a pistil. Diagrammatic representation of the evolution of a pistil by the folding of a leaf. (*a*) Simple leaf. (*b*) Simple pistil with marginal placentation composed of a single carpel formed from one leaf such as in *a*. (*c*) Fusion of three leaves producing a compound pistil with three carpels. The fusion between the carpels in this case is only along the plane where the margins touch each other, resulting in parietal placentation. (*d*) Fusion of the margins and part of the sides, producing a compound pistil with three carpels, resulting in axile placentation.

Figure 20–2. Parts of a typical flower. (*a*) Longitudinal view showing four successive whorls of floral parts. (*b*) Entire view.

the flower stalk, or **peduncle.** The number of parts within each group varies among species and can be used to distinguish monocots from dicots (refer to Table 20–1).

The outermost whorl is the **calyx,** which is made up of **sepals.** The sepals are usually green and enclose the entire flower before it opens. Inside the calyx is the next whorl, the **corolla,** the individual members of which are the **petals.** In many flowers, the petals are colorful, showy, and scented—all traits that attract insects and other pollinators. The petals may be fused together forming a **corolla tube,** characteristic of such flowers as morning glory. Although the sepals and petals make the flower attractive to pollinating agents, they are not essential for reproduction and, hence, are often considered as **accessory parts.**

A group of **stamens** make up the next whorl. Each stamen consists of a **filament** and an **anther.** The anther contains four sporangia within which are produced the microspores* and ultimately the pollen grains. Since the pollen grain represents the male gametophyte*, the stamen has been traditionally (but not necessarily correctly) referred to as a male reproductive

Microspore
A spore that gives rise to the male gametophyte or microgametophyte.

Gametophyte
The gamete-producing stage in a life cycle.

structure. The collective term for all the stamens of one flower is **androecium,** which literally means "male household."

The innermost floral whorl is the **pistil** which consists of a basal **ovary,** a terminal **stigma,** and a **style** that connects these two. The basic unit of a pistil is the **carpel.** Along the margin of a carpel is a tissue called the **placenta,** from which ovules are produced. Pistils composed of only one carpel are called **simple pistils,** those made up of two or more carpels are **compound pistils.** The number of lobes or branches of a stigma, or the number of lobes of the ovary, or both usually correspond to the number of carpels in a compound pistil. The eggs are produced within the ovules of an ovary; therefore the pistil is often called the female reproductive structure, even though only a small portion of each ovule is actually female. The pistil, or more commonly the several pistils of a multipistillate flower, is referred to as the **gynoecium,** meaning "female household."

Placentation, the position of the placenta in the ovary, varies among species depending on whether the pistil is simple or compound and on the degree of fusion of the carpels comprising a compound pistil (Figure 20–3). In a simple pistil containing only a single ovule, the placenta is usually produced at the lower end of the carpel, a situation known as **basal** placentation. When several ovules are produced in a simple pistil, the placentation is **marginal,** the placenta extending the length of the carpel along the margin. The edges of the carpels of a compound pistil may or may not fuse forming individual chambers within the ovary. If separate chambers are not formed, placentas are produced along the margins where the adjacent carpels meet, a condition called **parietal** placentation. If separate chambers are formed, the placentation is **axile** as the placenta forms in the center. In some species the walls dividing the chambers of the compound pistil were lost during the course of evolution, leaving only a central column of tissue projecting from the base of the ovary. Thus the axile placentation in such species was re-

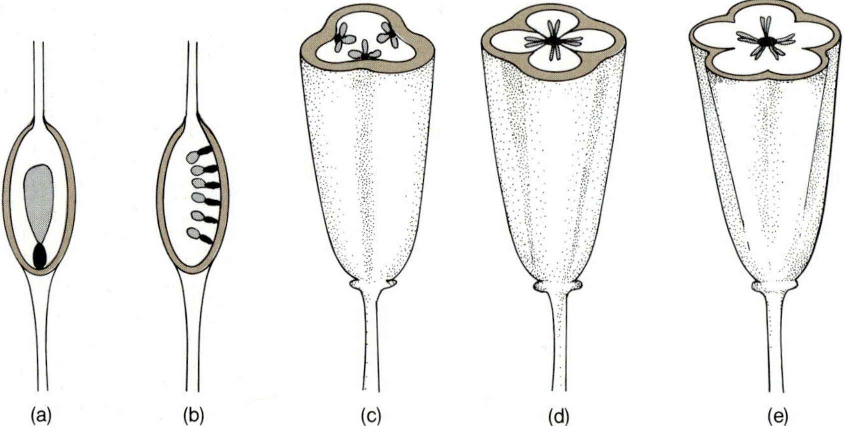

Figure 20–3. Types of placentation. (*a*) Basal. (*b*) Marginal. (*c*) Parietal. (*d*) Axile. (*e*) Free central. Figures *a* and *b* represent simple ovaries; *c*, *d*, and *e* represent compound ovaries.

placed by **free central** placentation, where the placenta is formed around this central column.

In its simplest form the flower consists of concentric whorls of floral parts, each whorl being fractionally higher on the receptacle than the previous one. The pistil, or ovary, of such a flower originates higher than all of the other floral parts; therefore the placement of the ovary is said to be **superior** (Figure 20–4). In many flowers, however, the bases of the other floral parts are fused to the ovary wall, which gives the appearance that the ovary originates below these other structures. The placement of this ovary is said to be **inferior** (Figure 20–4 and Plate 11A).

A flower that has all floral whorls is a **complete** flower. When any one of the floral whorls is missing, the flower is considered to be **incomplete** (Figure 20–5). If either the pistil or stamens is missing, the flower is **imperfect;** a **perfect** flower has both of these structures present. An imperfect flower is always incomplete, but an incomplete flower need not be imperfect.

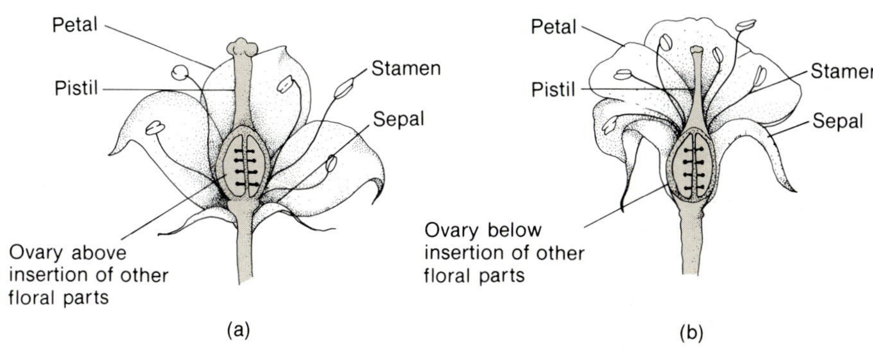

Figure 20–4. Flower with superior ovary (*a*) and inferior ovary (*b*).

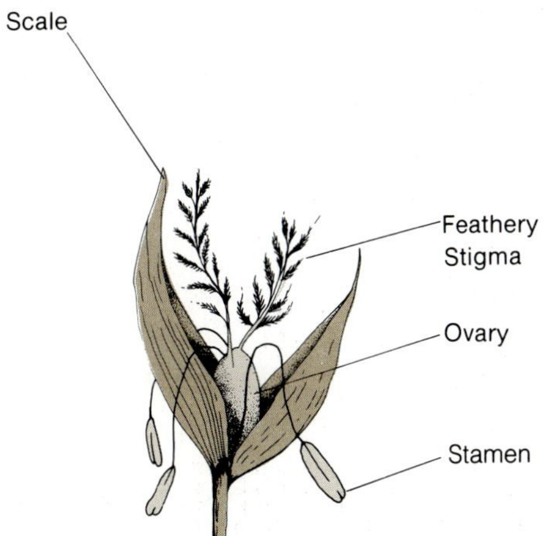

Figure 20–5. Incomplete flower (grass).

Where imperfect flowers occur on separate staminate or pistillate plants, the plant is said to be dioecious; where both staminate and pistillate flowers occur separately on the same individual, the plant is said to be monoecious. The majority of flowering plants have perfect flowers.

Flowers with parts arranged in a radially symmetrical manner are called **regular** flowers. A regular flower, such as a morning glory, can be cut into two identical parts along several planes, as long as the plane of section passes through the radius (Figure 20–6). An **irregular** flower, such as a snapdragon, has its floral parts arranged in a bilaterally symmetrical fashion and can be divided into two identical parts only along one plane.

Arrangement of Flowers

Flowers can be arranged differently on plants, depending on the species; the collective term **inflorescence** refers to the pattern of arrangement. On some plants the flowers are arranged singly, whereas in others they are arranged in groups on a stalk that produces only flowers. The following are common types of inflorescences (Figure 20–7):

Spike	sessile* flowers arranged alternately on an unbranched stalk
Raceme	flowers with short stalks arranged alternately on a main unbranched stalk
Catkin	staminate* or pistillate* flowers arranged on a pendulous stalk; found on many kinds of trees
Panicle	flowers arranged alternately on a branched stalk

Sessile
Not attached to the stem by a stalk.

Staminate flower
A flower that lacks a pistil.

Pistillate flower
A flower that lacks stamens.

(a)

(b)

Figure 20–6. Floral symmetry. (*a*) The floral parts of a regular flower are arranged symmetrically. This flower can be divided into two equal parts along many planes, as indicated by the lines. (*b*) An irregular flower can be divided equally only along one plane, as indicated by the line.

Corymb	flowers arranged alternately on a stalk, yet all are at the same height owing to the differences in pedicel* length, giving the cluster a flat-topped appearance; blooms from the edge toward the center
Cyme	flowers arranged in basically the same manner as a corymb, the major difference being that the flower cluster blooms from the center toward the edges and the main (center) stalk is always terminated by a flower
Simple Umbel	the pedicels of all flowers originate from the terminal portion of the stalk
Compound Umbel	similar to a simple umbel except that the stalk is branched
Head	a cluster of sessile flowers on a single receptacle. In the Asteraceae there are usually two types of flowers, ray flowers and disc flowers, arranged on the flattened receptacle. The ray flowers are located around the periphery, while the disc flowers are located in the center and give the head the appearance of a single flower (Plate 10E). Some heads consist only of ray flowers or only of disc flowers.

Pedicel
The stalk on which flowers are borne.

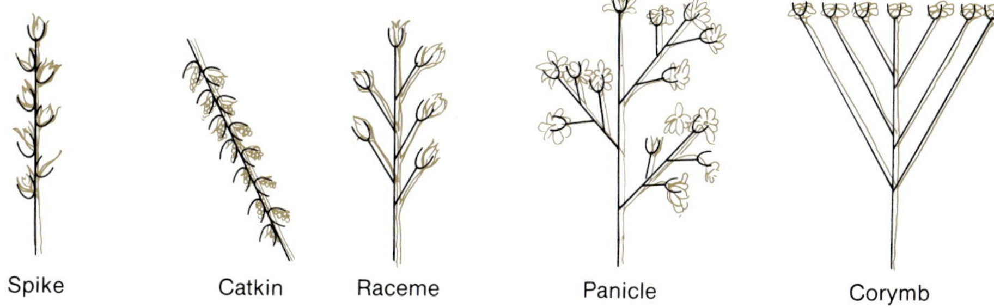

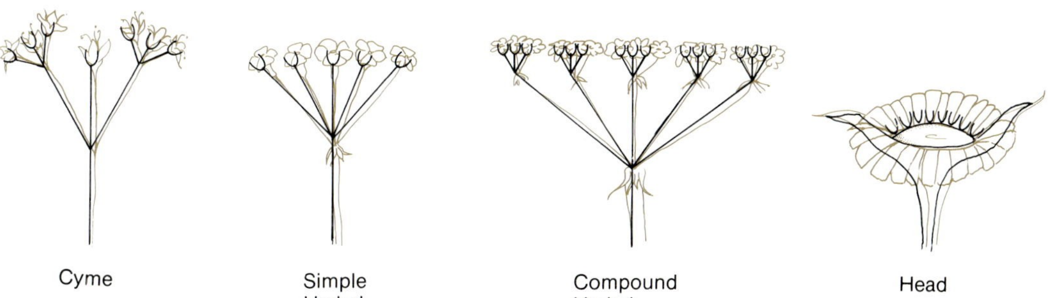

Figure 20–7. **Diagrams of some common inflorescence types.**

Angiosperm Life Cycle

Sexual reproduction among the angiosperms is the most advanced of all the vascular plants. Their unique production of seeds inside a fruit increases the chances for survival and distribution of the species. Like the gymnosperms, the angiosperms have a **heterosporous** life cycle (Figure 20–8) during which two types of meiospores*—microspores and megaspores*—are produced. However, several significant differences between the life cycles of the two groups are known, the most important one being the phenomenon of **double fertilization** in the angiosperms, which is unknown in any other plant group and is explained later in this section. A second difference is the absence of archegonia among the angiosperms.

Meiospore
A haploid spore produced after meiosis.

Megaspore
A spore that gives rise to the female gametophyte or megagametophyte.

The **sporophyte generation***, represented by the flower-producing plant, is the dominant phase of the life cycle, with the highly reduced **gametophyte generation*** being short-lived and completely dependent on the sporophyte plant. Once flowers are produced, the reproductive process is completed usually within a few days.

Sporophyte generation
The diploid stage in the life cycle of a plant.

Gametophyte generation
The haploid stage in the life cycle of a plant.

Microgametogenesis

The anther of a stamen contains four **microsporangia** within which are produced a number of **microspore mother cells** (Figure 20–9). As the anther matures, **meiosis** takes place in each microspore mother cell, resulting in the formation of a tetrad* of unicellular, haploid* **microspores.** Each microspore undergoes **mitosis***, producing a two-celled **pollen grain,** which consists of a large **tube cell** and a small **generative cell.** The pollen grain is surrounded by a thick, sculptured pollen wall (Figure 20–10). It is at this stage that the pollen grain, which is an immature microgametophyte, is released from the anther and is transferred to the stigma by means of wind, water, or other agents discussed later in this chapter. This process is known as **pollination.** Almost immediately after pollination, the tube cell of the pollen grain begins to elongate and grows through the style toward the ovary, forming the **pollen tube.** In the meantime, the nucleus of the generative cell divides once by mitosis and the resulting nuclei, along with a small amount of cytoplasm, become two nonmotile **sperm.** The mature **microgametophyte,** consisting of the germinating pollen grain and its two sperm, normally develops only in the pistil.

Tetrad
A cluster of four cells.

Haploid
Possessing only one of each kind of chromosome.

Mitosis
The process of nuclear division that results in the production of two genetically identical daughter nuclei.

Megagametogenesis

Depending on the species one or more **ovules** can be produced within an ovary. An immature ovule consists of a **megasporangium** enclosed by an outer and an inner **integument** (Figure 20–11). A small opening, the **micropyle,** occurs in the integuments at one end of the ovule. The single **megaspore mother cell** that develops within the megasporangium undergoes **meiosis** to produce four **megaspores,** three of which degenerate. During the development of the **megagametophyte** from the functional megaspore, the nucleus of the megaspore undergoes three successive mitotic divisions

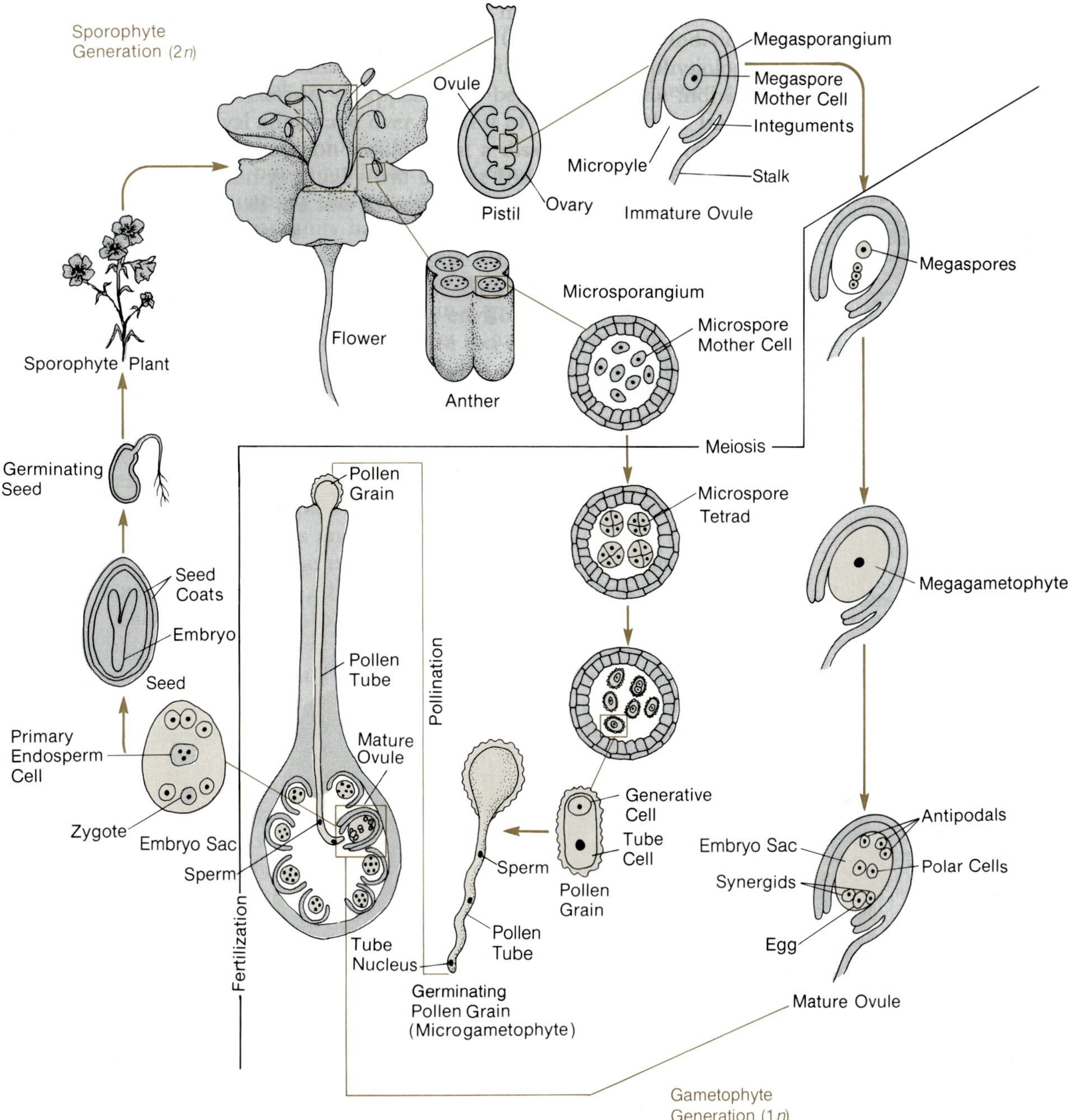

Figure 20–8. Angiosperm life cycle.

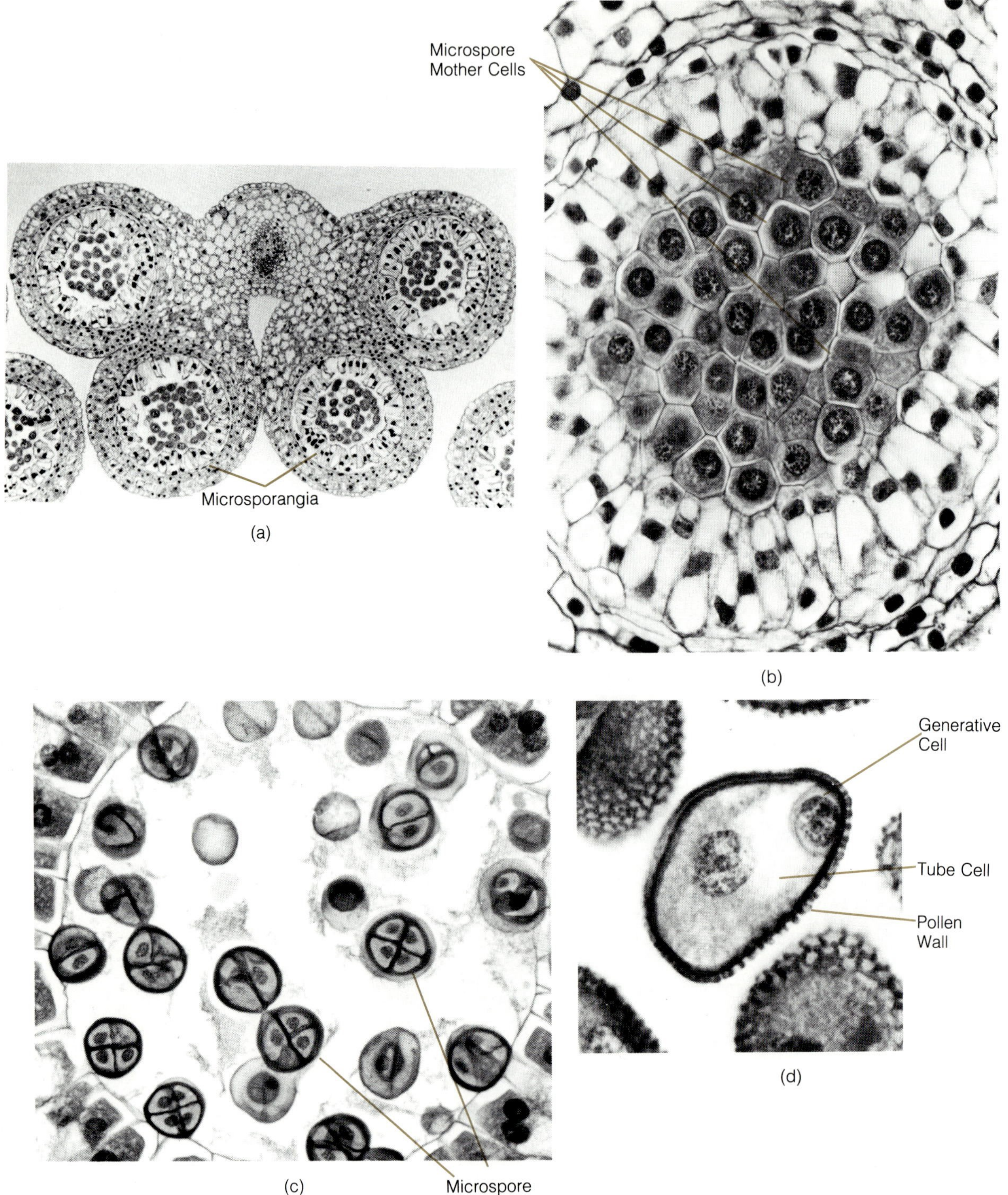

Figure 20–9. Microsporogenesis. (*a*) Transverse section of a lily anther showing four microsporangia. (*b*) A group of microspore mother cells. (*c*) Microspore tetrads resulting from meiosis in microspore mother cells. (*d*) Pollen grain with a large tube cell and a small generative cell.

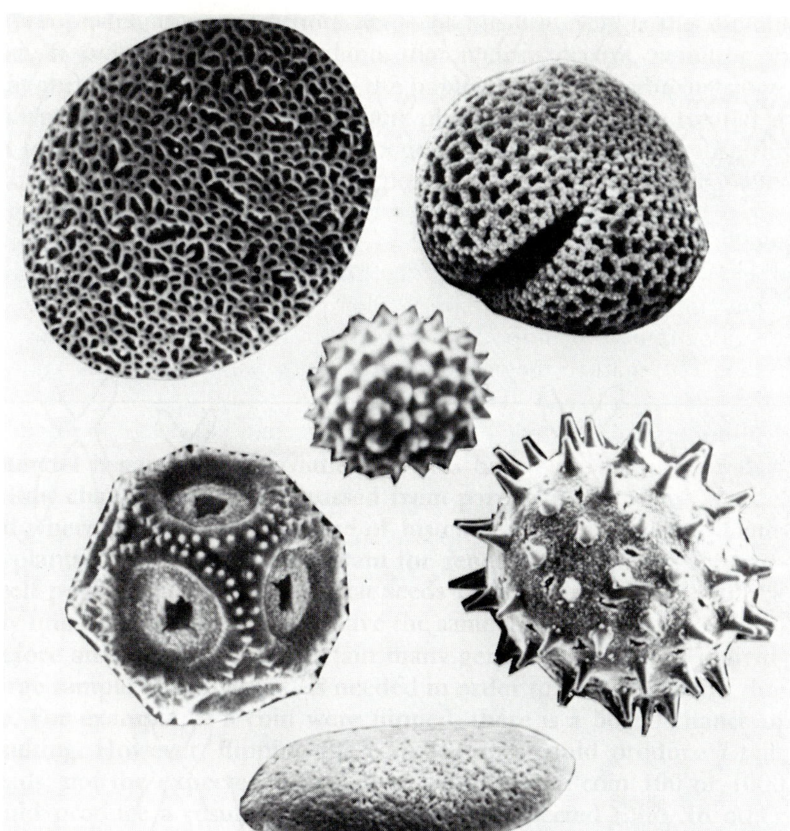

Figure 20–10. **Pollen grains.** The external pollen wall is usually highly sculptured as seen in this scanning electron micrograph. These walls are also highly resistant to decay, hence pollen grains are preserved in fossil deposits.

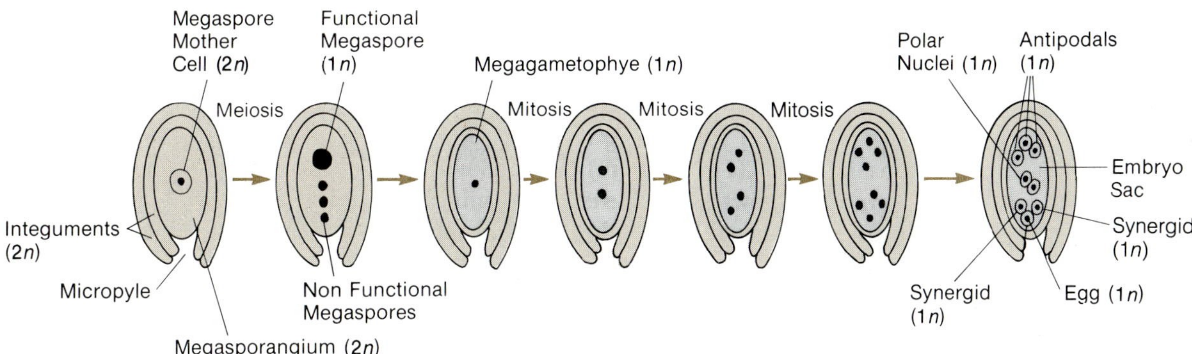

Figure 20–11. **Typical megasporogenesis and megagametogenesis.** These diagrams show the sequence of events that take place during the maturation of a typical angiosperm ovule.

resulting in eight nuclei, four at each end of the megagametophyte (Figure 20–11). One nucleus from each end migrates to the center and together these become the **polar nuclei.** Each of the remaining six nuclei is surrounded by a small amount of cytoplasm and a cell membrane. Similarly, a central binucleate polar cell is also formed. Together they form seven cells within the original megaspore.

The three cells at the end opposite the micropyle are known as the **antipodals.** Of the three at the micropylar end, one becomes the **egg** and the other two are called **synergids.** The function of the synergids and antipodals is not fully understood, although they probably represent the remnants of the gametophyte plant body that has been reduced during angiosperm evolution. The mature **megagametophyte** with its seven cells is known as the **embryo sac** (Figure 20–12).

Although the life cycle described in this chapter is basic to the majority of angiosperms, including both monocots and dicots, there are several variations to this basic theme, which occur during the development of the megagametophyte. One such variation that occurs in *Lilium* (lily) is described because lily is the most common example of the angiosperm life cycle demonstrated in introductory botany laboratories (Figure 20–13).

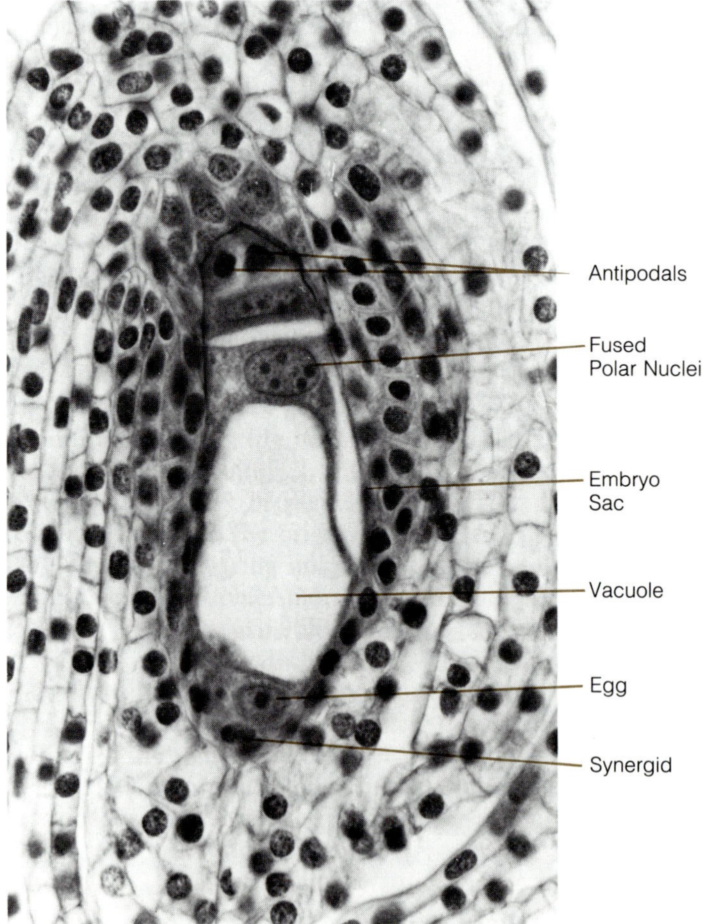

Figure 20–12. Embryo sac. This photomicrograph of a lily ovary section shows a mature ovule with the enclosed embryo sac containing seven cells and a large vacuole. The cell walls are not distinct. The polar cell was formed by the fusion of the two polar nuclei.

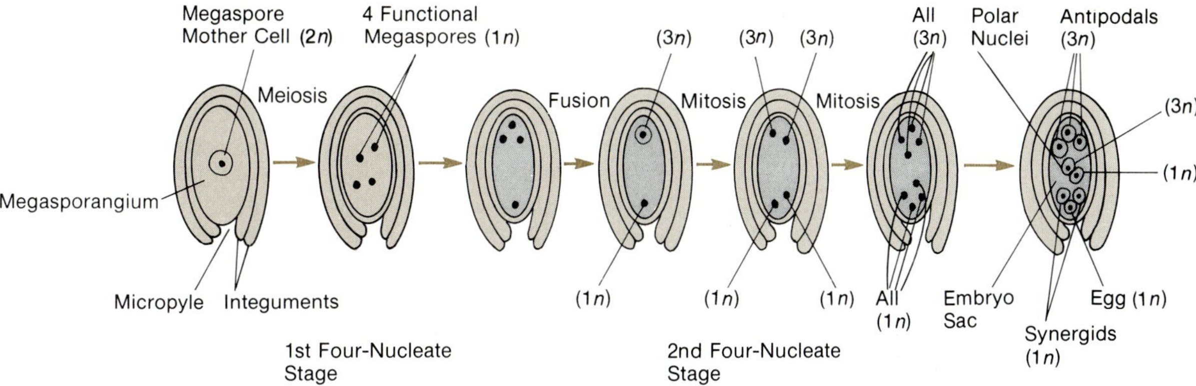

Figure 20–13. Megasporogenesis and megagametogenesis in the lily. The sequence of events during the maturation of a lily ovule is different from that of the typical angiosperm shown in Figure 20–11.

A single megaspore mother cell is produced inside the lily megasporangium. Although the nucleus of the megaspore mother cell undergoes meiosis to produce four haploid megaspores as in other angiosperms, all four megaspores of lily are functional, unlike the typical case; this stage is known as the first four-nucleate stage. Three of these megaspores migrate to the end opposite the micropyle and fuse while one remains near the micropylar end, resulting in a triploid nucleus at one end of the megasporangium and a haploid nucleus at the other end. Mitosis occurs in both of these nuclei, resulting in the second four-nucleate stage. Each of the four nuclei then undergoes another mitotic division, producing eight nuclei, four at each end of the embryo sac. The four nuclei at the micropylar end are haploid whereas the four at the opposite end are each triploid.

One nucleus from each end now migrates to the center of the megasporangium where they are referred to as the polar nuclei, one of which is triploid* and the other haploid. The remaining cells of the megagametophyte are distributed as described above. When fertilization takes place, a diploid* zygote and a pentaploid* primary endosperm cell are produced. Unlike the triploid endosperm of most angiosperms, the endosperm of lily is pentaploid.

Triploid
Possessing three of each kind of chromosome.

Diploid
Possessing two of each kind of chromosome.

Pentaploid
Possessing five of each kind of chromosome.

Pollination and Fertilization

Flowers usually open when the ovules are mature and pollination can take place immediately. The developing pollen tube grows into the embryo sac through the micropyle and releases the two sperm. One of the sperm fertilizes the egg and the other sperm fuses with the polar nuclei. This unusual process, called **double fertilization,** is a phenomenon found only among the angiosperms. The diploid **zygote,** the fusion product of the egg and sperm, develops into the **embryo,** whereas the triploid **primary endosperm cell,** the fusion product of the sperm and the two polar nuclei, undergoes repeated cell division producing the **endosperm.** At the same time, other changes taking place transform the ovule into a **seed** and the ovary into a **fruit.**

Seed and Fruit Development

Usually, ovules do not become seeds or ovaries become fruits unless fertilization occurs. However, there are a number of plants, such as bananas, pineapples, and certian varieties of grapes and melons, in which fruit development takes place even without fertilization, a phenomenon called **parthenocarpy.** Parthenocarpic fruits are characterized by the absence of seeds, a feature that is of economic value. Often commercial crop varieties are specially selected for this characteristic due to the consumer preference for seedless fruits. It is also known that parthenocarpic fruit development can be induced by treatment of flowers with naturally occurring or synthetic plant hormones (Chapter 12). Plants that produce seedless fruits naturally can be propagated only by vegetative* means.

Vegetative
Growth not associated with sexual reproduction.

Each seed has two outer coverings, the inner and outer seed coats formed from the inner and outer integuments, respectively. All seeds have an embryo that develops into the mature sporophyte plant. In addition, some seeds have an endosperm, which is an important tissue for the storage of reserve food. Although an endosperm is present at the initial stage of development of all seeds, in some the endosperm is depleted during the maturation of the embryo (Chapter 21). Such seeds store their reserve food in the cotyledons that are part of the embryo. Examples of mature seeds that have an endosperm include corn, wheat, rice, coconut, onion, castor bean, nutmeg, and coffee. Seeds that do not have an endosperm when mature include pea, bean, peanut, cashewnut, sunflower, and almond.

Mechanisms of Pollination

Pollination is a significant step during the life cycle of a flowering plant and is a prerequisite for fertilization, without which seeds cannot be produced in most species. However, pollination is a random process that frequently results in pollen of the wrong species landing on a stigma. Instead of this resulting in a cross or mating between a maple and a willow, for example, there are a variety of mechanisms that prevent the completion of the development of the microgametophyte (pollen grain) in this foreign environment. Plant species that are closely related can have complete development of the microgametophyte followed by sexual reproduction. Hybrids produced in this way may or may not be viable as is the case with hybrids between related animal species.

Within a species pollination may take place within the same flower, between flowers on the same plant, or between flowers on different plants. The former two are examples of self-pollination, while the latter represents cross-pollination. Although many plants are self-fertile, a major contributing factor for the introduction of variation and the subsequent adaptability of plants to changing environmental conditions (Chapter 23) is cross-pollination. Therefore it is advantageous to the species, in the long run, to have mechanisms that can enhance the chances for cross-pollination. There are more angiosperm species in which adaptations to bring about cross-pollination have evolved than there are species that are limited to self-pollination. These adaptations can be considered under two categories. Among the first are adaptations that prevent self-pollination by the inhibition of pollen tube formation if pollen from an individual lands on the stigma of flowers of the same individual or even of different but genetically similar individuals. This condition, called self-incompatibility, is recognized among many angiosperm species, including varieties of apples, pears, and plums. In most cases, the self-incompatibility is due to the slow growth of the pollen tubes of the pollen that has landed on the stigma of the same varieties rather than to a complete inhibition of pollen germination.

In the second category are found adaptations that prevent self-pollination by means of differential maturation of the male and female reproductive structures. An example of this is Jacob's Ladder, in which the stigma of a flower is not receptive when anthers of the same flower are mature and shedding their pollen.

In addition to these adaptations, some angiosperms have flowers in which structures have evolved that favor pollination by insects, birds, or bats (Plate 10). In many cases, coevolution of the flowers with their respective pollinators can be recognized even though the mechanism for such coevolution is not understood. Among some of these plants, pollination cannot take place in the absence of their pollinators. The pollinators transfer pollen from one plant to another as they visit flowers to collect nectar. In many cases, these flowers are very highly modified, such as in alfalfa, mint, clover, and snapdragon, which are pollinated by bees. These flowers have a bilateral symmetry, whereby the lower parts of the petals fuse and form a corolla tube while the free ends of the petals form an upper lip and a lower lip (Plate 11B). The latter provides the bee with a convenient landing platform. The stamens and style are located in the upper portion of the corolla tube in such a way that they contact the back of the bee as it enters the corolla tube to collect the nectar that is secreted by glands at the base of the pistil. Thus the pollen that is deposited on the back of the bee can be transferred to a mature stigma as the bee visits another plant. The stamens and pistils of these individual flowers mature at different times, thus preventing self-fertilization. Bees are able to perceive the ultraviolet, blue, and yellow colors that are predominant in flowers pollinated by these insects. Some flowers, such as the marsh marigold, have distinctive ultraviolet markings that are invisible to humans but visible to bees.

Some common insect pollinators are butterflies, moths, flies, beetles, and wasps. *Ophyrus,* an orchid, has flowers that so closely resemble a species of female wasp that the males try to mate with the flowers resulting in cross-pollination as the males visit flowers on different plants.

The hummingbirds, the African sunbird, and the Hawaiian honey creeper are examples of bird pollinators. Although bats as a group are not major pollinators, species of bats belonging to the genus *Leptonycteris* pollinate the organ-pipe cactus. Since bats are nocturnal mammals, they are attracted to the flowers mainly through their sense of smell. The flowers they visit bloom at night.

Pollination among the members of the grass family, many temperate trees, and other plants such as ragweed is by means of wind. Pollen grains of such plants are buoyant and are thus easily carried by air currents. The floral parts, especially the petals and sepals of many wind-pollinated flowers, are highly reduced or absent. Many of the wind-pollinated plants also produce large amounts of pollen and are the cause of most pollen allergies.

Summary

Flower production is the most significant feature common to all plants that belong to the Division Anthophyta, the angiosperms. The angiosperms include two taxonomic classes, the Monocotyledonae and the Dicotyledonae. A flower is a specialized organ for sexual reproduction. All flowering plants produce seeds within fruits. The angiosperms are estimated to have evolved between 250 and 150 million years ago from the so-called seed ferns.

It is generally accepted that the flower is a shoot system of determinate growth and that each floral part represents modified leaves. A typical flower consists of four groups of floral parts arranged in successive whorls on a receptacle. The outermost whorl is the calyx, consisting of sepals. The next whorl is the corolla, composed of petals. The sepals and petals are accessory parts and have no direct role in reproduction. The stamens that make up the next whorl each have a filament and an anther, the latter containing four sporangia within which are produced the pollen grains. The pistil consists of a basal ovary, a terminal stigma, and a style that connects these two. A simple ovary is formed from one carpel, whereas two or more fused carpels form a compound ovary. Ovules are produced within an ovary from a specialized tissue called the placenta. The common arrangements of the placental tissue are basal, marginal, parietal, axile, and free central. The ovary may be superior or inferior, depending on its apparent position with respect to the other floral whorls.

A complete flower has all four whorls; if any one of these whorls is missing, the flower is termed incomplete. A perfect flower has both stamens and pistil. Flowers with symmetrically arranged parts are called regular flowers. Some different types of inflorescences, or the arrangement of flowers on a plant, are spike, raceme, catkin, panicle, corymb, simple umbel, compound umbel, and head.

The angiosperms have a heterosporous life cycle with a dominant sporophyte phase and a highly reduced, short-lived gametophyte phase. Microspore mother cells produced within the microsporangium of the anthers undergo meiosis giving rise to the haploid microspores that, in turn, develop into pollen grains. Pollination transfers the pollen to the stigma where the elongating pollen tube transports the sperm produced by the generative cell to the egg within a mature ovule. An immature ovule consists of a megasporangium enclosed by two integuments. Within the megasporangium the single megaspore mother cell undergoes meiosis producing four haploid megaspores, only one of which divides to become the megagametophyte.

The nucleus of the functional megaspore undergoes three successive mitotic divisions to produce eight nuclei that are distributed in the mature ovule as three antipodals at one end, two polar nuclei in the middle, and one egg and two synergids at the other end close to the micropyle, the small opening in the integuments through which the pollen tube enters the ovule. The microgametophyte and megagametophyte are represented by the germinating pollen grain and the embryo sac, respectively.

Double fertilization, a characteristic process of the Anthophyta, is the fusion of one sperm with the egg forming the diploid zygote and the fusion of the second sperm with the polar nuclei forming a triploid primary endosperm cell. The zygote develops into the embryo while the primary endosperm cell develops into endosperm, which in some seeds is the food storage tissue. A new sporophyte plant develops as the seed germinates into a seedling, completing the life cycle. A slight variation from the typical maturation of the ovule described above is found in lily, where all four megaspores are functional, subsequently resulting in a pentaploid primary endosperm cell rather than the typical triploid primary endosperm cell. Other variations are also known.

Review Questions

1. Explain the concept that a flower is a compressed shoot system of determinate growth.
2. Discuss the evolution of the flower.
3. What are the sequential stages of microgametogenesis?
4. What are the sequential stages of megagametogenesis?
5. What is cross-pollination? Why is it considered more beneficial than self-pollination? How does nature ensure cross-pollination?

Pollen Grain

The pollen grain is a highly efficient structure for the long-distance transfer of the male gametes. As described elsewhere in this chapter, this is accomplished through wind, insects, or other animals. The pollen grain also provides a mechanism to transport the sperm directly to the egg by means of the pollen tube without any need for water. These features have enabled the seed plants, flowering plants and gymnosperms, to attain a much wider geographic distribution than all other vascular plants.

A pollen grain has an inner and an outer wall. The outer wall is extensively sculptured and is composed of sporopollenin, a chemical substance that is highly resistant to degradation. There are one or more openings in the outer wall through which tubes of a germinating pollen grain emerge. The wall of the pollen tube is an expansion of the inner wall.

Considerable variation in the sculpturing present on the outer pollen wall has been strikingly demonstrated with the scanning electron microscope in recent years. These markings on the pollen wall, along with size and shape variations, are being used to a limited extent as taxonomic characteristics. There are also a number of allergenic substances present on the pollen wall, which are responsible for causing hay fever in many individuals. Because of their resistance to degradation, pollen grains commonly occur as fossils. Fossil pollen has been utilized by archaeologists and plant geographers to determine the types of plant communities characteristic of a given area at different times in the past, thus giving an indication of climatic changes or human settlement patterns. In addition, fossil pollen is used by the petroleum industry as a tool to correlate sedimentary deposits, to determine the age of such deposits, and also to determine if a given area would produce oil, gas, or anything at all.

Pollen or Parathion?

Pollen grains are a major source of food for honeybees and certain kinds of beetles. Honeybees are insects of great economic importance to agriculturalists for their role as the chief pollinators of such crops as alfalfa and many types of orchard fruits, in addition to making honey. The following incident shows how precarious the balance of nature is and how modern technology can backfire even when carried out with the best of intentions.

One of the more common insecticides used in orchards, particularly in California, has been the organophosphate, methyl parathion. Unlike DDT and related compounds, parathion breaks down within two weeks after application and, hence, does not cause toxic residue problems. However, before its degradation parathion is extremely toxic to both insects and humans. In recent years, efforts to enhance the safety of farm workers resulted in the development of a technique whereby methyl parathion is encapsulated in tiny plastic beads about the size and color of pollen grains. In this form the pesticide is released slowly, thus extending its effectiveness and at the same time reducing its hazard to those who handle it. This innovation, however, has proven disastrous to the honeybees because they are unable to distinguish between the deadly beads of parathion and pollen grains, both of which they carry back to their hives. Gradual release of the pesticide inside the hives has resulted in large-scale destruction of bee populations and great economic loss to beekeepers.

21

Angiosperms: Fruits, Seeds, and Seedlings

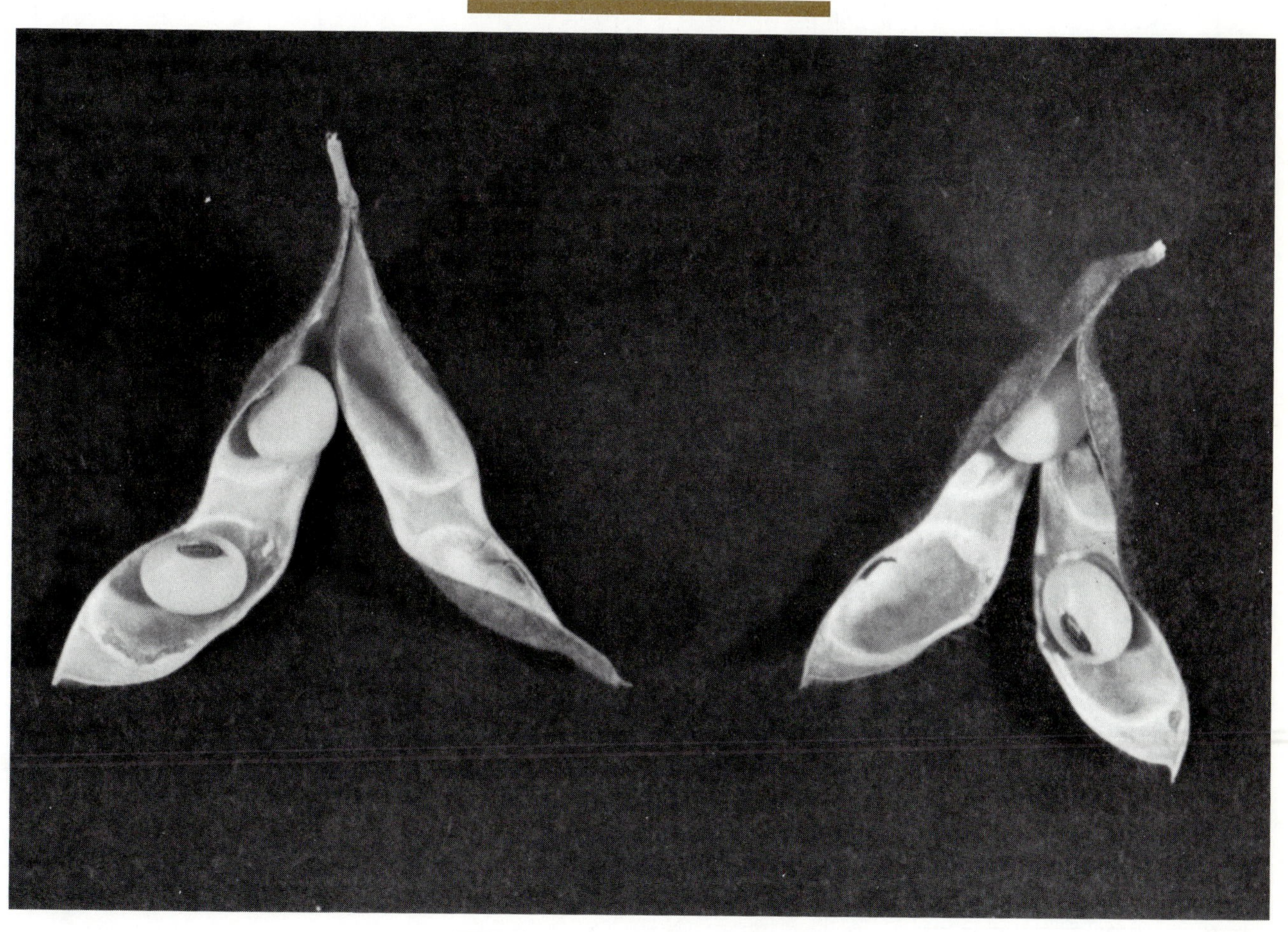

Outline

Introduction

Even though edible fruits comprise a significant part of the human diet, many people lack a comprehension of their botanical nature or their origin. This situation is further complicated by the insistence on classifying edible plant organs as either fruit or vegetable. For the layperson the distinction between these two is not botanical, but gastronomical—a fruit is sweet and eaten alone or as part of a dessert, whereas a vegetable is served with the main course. For the botanist, many vegetables are more properly called fruits, even though at home a request for fruit will more than likely be met with an apple than with a bean pod.

The fruit is a distinguishing feature of the angiosperms, a group believed to have evolved from gymnosperm ancestors more than 120 million years ago. When the gymnosperm cone evolved over 300 million years ago, there were no large terrestrial animals, as the reptiles were just evolving from the amphibians during that period. Consequently there was no evolutionary pressure on gymnosperms that would lead to the development of seed dispersal mechanisms involving animals. The angiosperm flower, like the gymnosperm cone, is derived from an aggregation of **sporophylls*** on a compact branch (Chapter 20). The major difference between the two groups is that in the angiosperms the megasporophyll* forms a closed structure, or carpel, enclosing the **ovules**.*

Sporophyll
A modified leaf that produces sporangia.

Megasporophyll
A sporophyll that produces megasporangia.

Ovule
The structure in seed plants that contains the megasporangium.

The carpel evolved at a time when both birds and mammals existed. The presence of these organisms provided evolutionary pressure that favored the development of seed dispersal mechanisms involving animals. Consequently fleshy fruits evolved that would be eaten by an animal, which unknowingly, would distribute the seeds in its droppings. In other plants structures evolved on seeds and fruits that would adhere to an animal's fur. Wind and water dispersal of fruits and seeds evolved in yet other angiosperms.

Fruit Structure

Observation of the development of any fruit confirms the fact that the fruit develops from a flower (Figure 21–1). A fruit is a mature **ovary** even though some fruits incorporate other flower parts into their structure. Being derived from the ovary, the fruit contains the **ovules**, which at their maturity are called **seeds.** As the fruit develops, the ovary wall, now known as the **pericarp,** will often become modified in such a way that seed dispersal is facilitated.

The different types of fruit that have evolved are each suited to a different seed dispersal mechanism. The pericarp can be either **fleshy** or **dry** (Table 21–1, Figure 21–2). In fleshy fruits the pericarp is frequently made up

Figure 21–1. Development of a cucumber fruit. All fruits develop from the ovary of a flower. In this picture the petals can be seen still attached to the enlarging ovary that is developing into the characteristic cucumber fruit.

of three layers—**exocarp***, **mesocarp***, and **endocarp***—which can differ in their textures. For instance, a **berry** (Table 21–1) has an exocarp that forms an outer skin, as in a grape or tomato, whereas the mesocarp and endocarp are fleshy and indistinguishable. On the other hand, a **drupe** (Table 21–1) has a skin formed by the exocarp, a fleshy mesocarp, and a stony endocarp that surrounds a seed, as in a peach or plum. The pericarp of a dry fruit may be paper-thin or very thick and heavy; however, there is no differentiation into layers. The dry fruit may open at maturity or may remain tightly closed. **Dehiscent** fruits open when seams or sutures in the pericarp split as the fruit dries; **indehiscent** fruits only open as the result of physical damage or metabolic activities of other organisms.

Exocarp
The outer layer of the fruit wall or pericarp.

Mesocarp
The middle layer of the fruit wall.

Endocarp
The inner layer of the fruit wall.

Variations in the structure of flowers lead to variations in the structure of the fruits derived from them. A **simple** fruit is derived exclusively from a single pistil, whereas an **aggregrate** fruit is formed from several separate pistils of the same flower. A fruit derived from more than one flower is a **multiple** fruit. The fruit types listed in Table 21–1 are all simple fruits; each aggregate or multiple fruit would consist of a group of simple fruits of one of the types in this list.

An aggregate fruit frequently has the appearance of a group of coalesced, small fruits. The raspberry is an aggregate fruit composed of many small drupes (not berries). The blackberry, a similar fruit, is an aggregate-accessory fruit because the receptacle* forms part of the fruit in addition to the small drupes (Figure 21–3). Although this is not as immediately obvious, strawberries are also aggregate-accessory fruits. In this case the major portion of the fruit is the much enlarged receptacle while the "seeds" on the surface of the strawberry are actually achenes (Figure 21–3).

Receptacle
The end of the flower stalk to which the floral parts are attached.

Familiar examples of multiple fruits are the pineapple and mulberry. In each case a multiple fruit develops from the individual flowers of an inflo-

Table 21–1.
Simple Fruits

Types	Description	Examples
Fleshy		
Berry	Derived from either a simple or a compound ovary; contains one to many seeds; most of the pericarp is fleshy	Tomato, grape
Hesperidium	Modified berry with a separable leathery rind composed of exocarp and mesocarp; the endocarp is fleshy	Orange, grapefruit
Pepo	Modified berry with an inseparable, rigid rind composed of the exocarp; mesocarp and endocarp form the flesh	Cucumber, watermelon
Drupe	Thin skin (exocarp) surrounds fleshy or fibrous mesocarp; endocarp at center forms a stony wall around seed.	Cherry, peach, plum, almond, olive, coconut
Dry, Dehiscent		
Follicle	Forms from a single carpel; splits open along one margin	Milkweed
Legume	Forms from a single carpel; splits open along two margins	Pea, bean; peanut as an exception does not dehisce
Capsule	Forms from two or more fused carpels; splits open along multiple seams, or forms pores, or forms a lidlike cap that falls off	Lily, poppy, horse chestnut, okra, columbine
Dry, Indehiscent		
Achene	Single-seeded fruit in which seed is attached to pericarp at only one point	Sunflower, dandelion
Samara	Winged achene	Ash, elm, maple (double samara)
Caryopsis	Single-seeded fruit in which seed is completely fused to pericarp	Corn, wheat, rice, oat, barley
Nut	Single-seeded fruit derived from compound ovary; possesses hard pericarp	Acorn, walnut, hazelnut (filbert), cashew

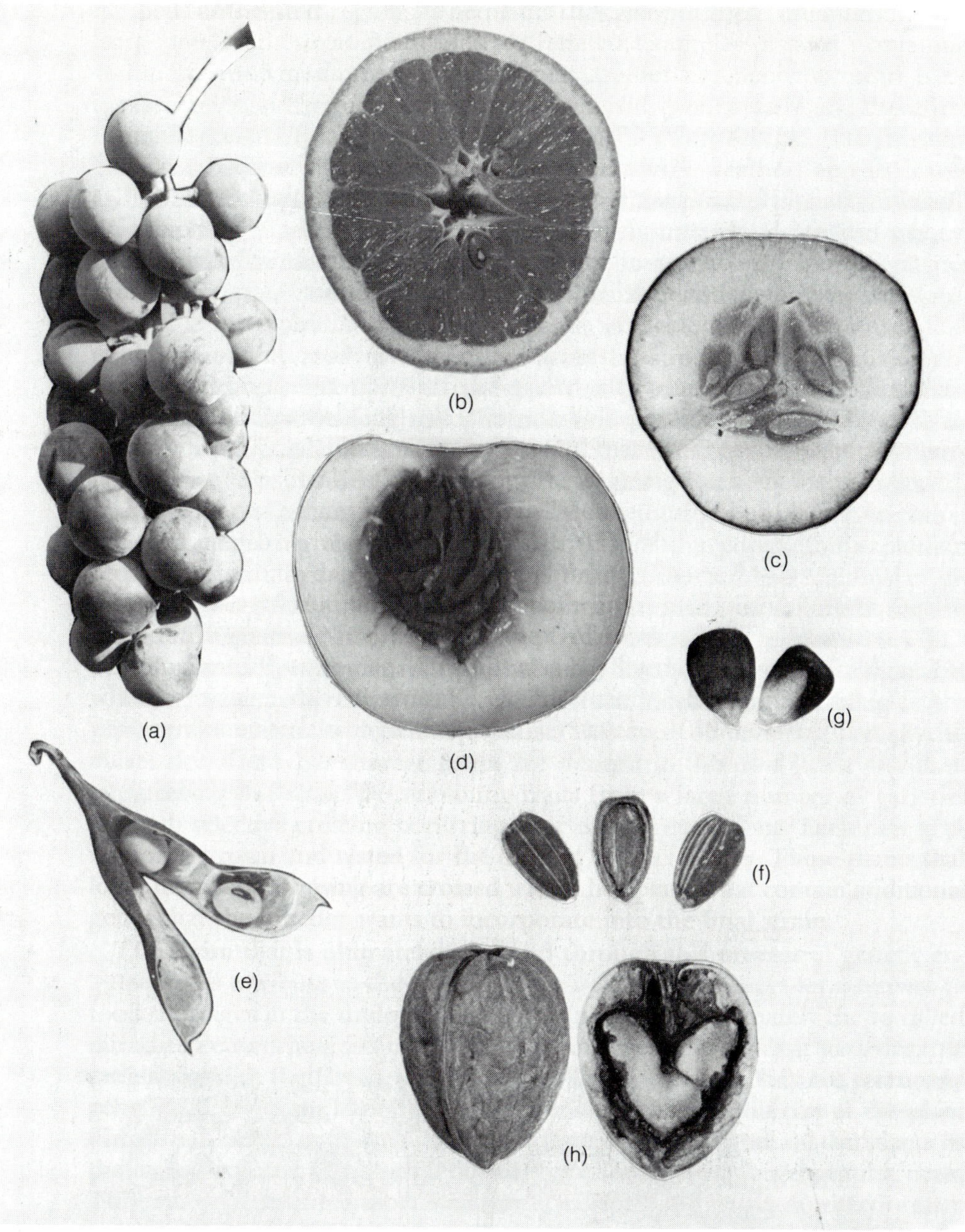

Figure 21–2. Fruit types. (*a*) Berry. (*b*) Hesperidium. (*c*) Pepo. (*d*) Drupe. (*e*) Legume. (*f*) Achene. (*g*) Caryopsis. (*h*) Nut. See Figure 21–9 for a follicle and a capsule and Figure 21–7 for a samara.

rescence.* The central core of the pineapple develops from the stem to which the flowers of the inflorescence were attached; most of the fleshy fruit is derived from the individual flowers that have fused together as they enlarged (Figure 21–4). Each individual "berry" in a mulberry fruit is developed from a single pistillate flower*. The "seed" in the center is actually an achene and the fleshy part is derived from the calyx (fused sepals).

Accessory fruits result when flower parts in addition to the pistil(s) are incorporated into the fruit structure. Typically a simple accessory fruit is derived from a flower with an inferior ovary (Figure 21–5), that is, a flower in which the lower portions of sepals*, petals*, and stamens* are fused to

Inflorescence
A group of flowers on a common stalk.

Pistillate flower
A flower that lacks stamens.

Sepal
A leaflike flower part found at the base of the petals.

Petal
A leaflike flower part that is typically brightly colored.

Stamen
A flower part consisting of a filament and an anther in which the pollen grains are produced.

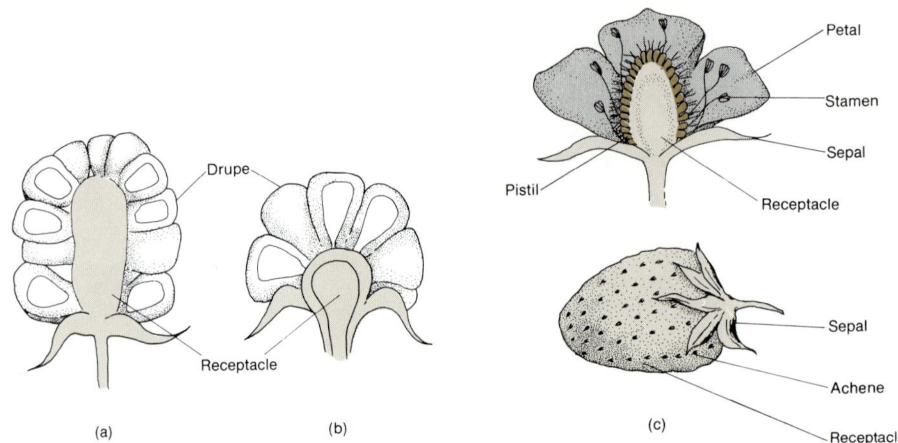

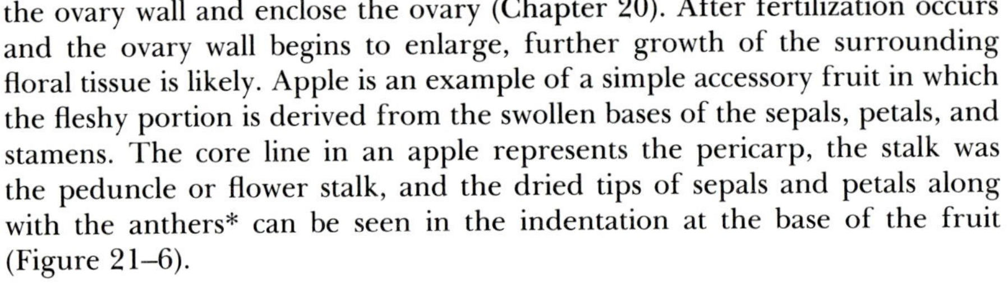

Figure 21–3. Aggregate fruits. (*a*) Raspberry. The drupes are fused and separate as a unit from the receptacle once they are ripe. (*b*) Blackberry. This is an aggregate accessory fruit because the receptacle does not separate from the drupes and is part of the fruit. (*c*) Strawberry. The strawberry flower contains a large number of pistils on an elongated receptacle. During maturation of each of the fruits (achenes) of a strawberry, the receptacle enlarges, further separating the achenes. The fleshy part of the strawberry is the receptacle.

the ovary wall and enclose the ovary (Chapter 20). After fertilization occurs and the ovary wall begins to enlarge, further growth of the surrounding floral tissue is likely. Apple is an example of a simple accessory fruit in which the fleshy portion is derived from the swollen bases of the sepals, petals, and stamens. The core line in an apple represents the pericarp, the stalk was the peduncle or flower stalk, and the dried tips of sepals and petals along with the anthers* can be seen in the indentation at the base of the fruit (Figure 21–6).

Anther
The floral structure composed of four microsporangia in which the pollen is produced.

Figure 21–4. Pineapple. A multiple fruit formed by the fusion of fruits from individual flowers of an infloresence.

Fruit and Seed Dispersal Mechanisms

Since reproductive success is a major factor driving evolution (Chapter 23), there has been selection for species that have effective seed dispersal mechanisms. Each of these, in different ways, increases the probability that a few of the many seeds produced by a plant will land on the appropriate substrate. The agents of seed dispersal are wind, water, explosive ejection, and animals. Even though it is the seed that must be dispersed, the structure of the fruit can be modified for dispersal or as in the case of tumbleweeds, the whole plant can take part in dispersal.

Wind

At some time or another, almost everyone has taken a fruiting head of dandelion and blown on it to watch the "parachutes" float away, or thrown a handful of maple fruits in the air to watch the "helicopters" spin back to the ground. In these and other cases, the fruit is modified for wind dispersal; other plants have seeds that are modified for wind dispersal. Fruits and seeds adapted for wind dispersal are relatively lightweight and have an outer coat that either is winglike or has various projections to catch the wind (Figure 21–7, Plate 11F). Seed dispersal by tumbleweeds is the result of an ad-

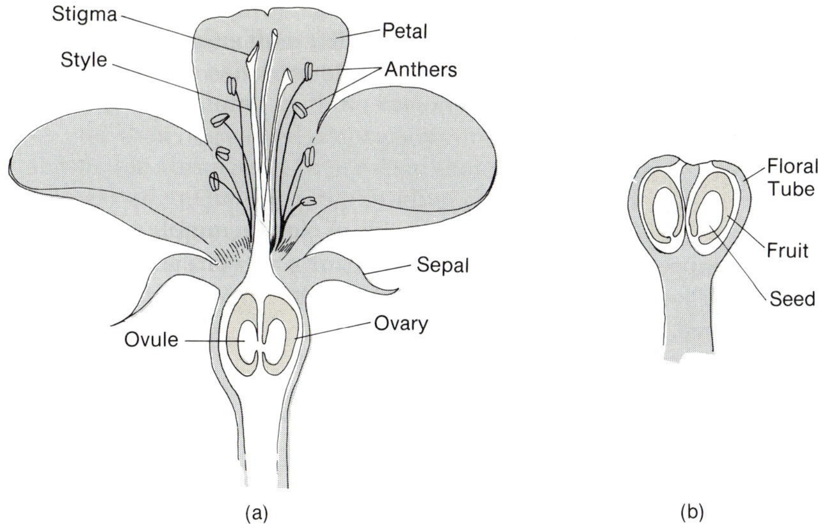

Figure 21–5. Development of a simple accessory fruit. When the bases of the other flower parts are fused to the ovary wall in the flower, they are likely to become included in the developing fruit even if their upper ends fall off.

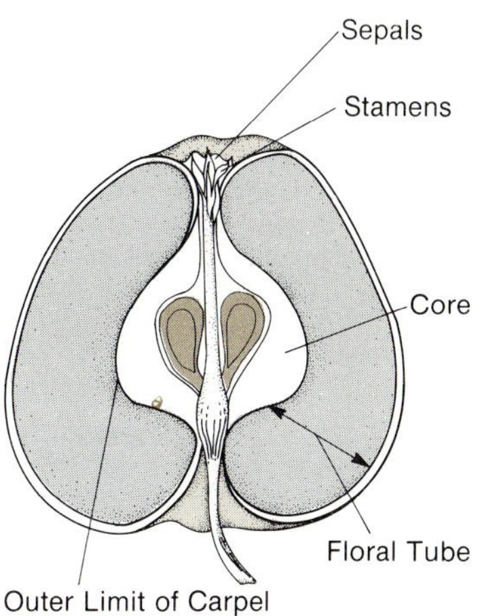

Figure 21–6. Apple. The apple is a simple accessory fruit developed from a flower with an inferior ovary. In this case the bases of flower parts fused with the ovary wall forming the fleshy edible part of the fruit. The anthers and sepals remain attached to the developing fruit and may be seen in the indentation opposite the fruit stalk.

aptation by a whole plant whereby the stem separates from the roots after the seeds are mature. As the plant is blown about by the wind, the seeds drop out (Figure 21–8). Many plant species have wind dispersal mechanisms that rely on the wind shaking the seed out of a dehiscent fruit (Figure 21–9).

Water

Evolutionary selection pressure has favored the development of adaptations to water-dispersed seeds by species that live near water. The coconut palm has become dispersed over vast areas of the Pacific region by virtue of the fact that the coconut fruit floats. Coconuts possess an outer fibrous layer that gives them buoyancy (Figure 21–10). Coconuts floating with ocean currents may be deposited on shore at the high tide mark; this is why the first coconut palms on an island will be found at the shoreline, not inland. Likewise, other island-dwelling plant species may be water distributed, although many are dispersed by birds. Freshwater rivers and lakes also function in seed dispersal, particularly for species that inhabit floodplains and wetlands. Spring floods carry floating seeds and fruits that become stranded as the floodwaters gradually recede, leaving the seeds or fruits partially or completely buried in a deposit of fertile silt.

Explosive Ejection

Seeds of certain species are forced out of the fruit when it opens (dehisces). For example, as the fruit of witch hazel (*Hamamelis*) dries, the buildup of stresses and strains bursts the pericarp along the sutures, releasing the stress suddenly. This results in the explosive ejection of the seeds. Another plant that ejects seed explosively is touch-me-not (*Impatiens*).

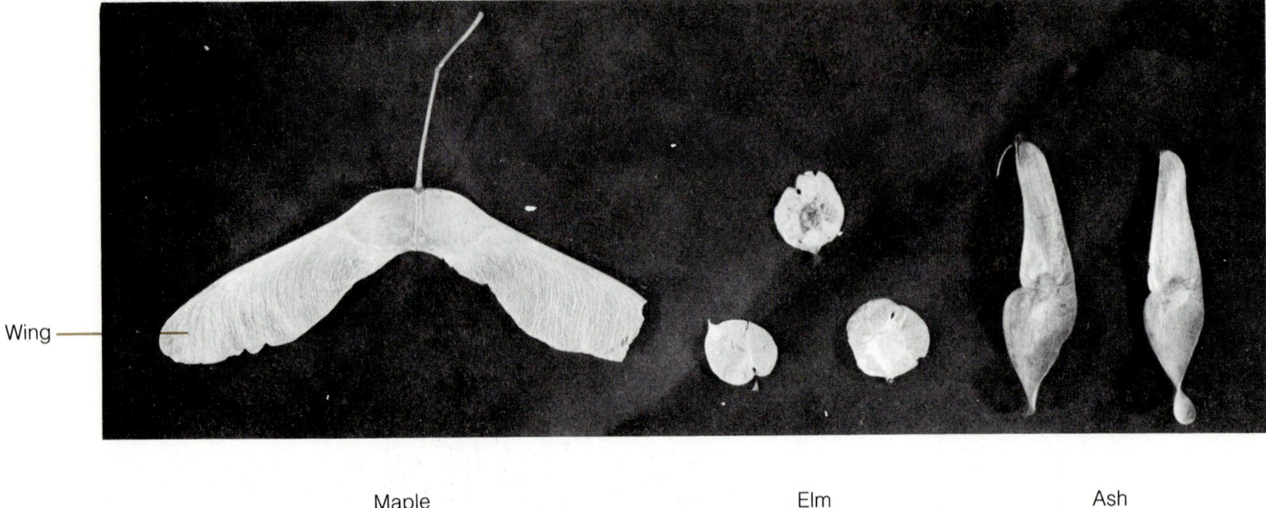

Figure 21–7. Seeds adapted for wind dispersal. Each of the seeds or fruits in this picture possesses a thin flattened region that provides buoyancy in the air.

Figure 21–8. Tumbleweed. The stems of tumbleweeds break off at ground level after the seeds have reached maturity and the plant is blown around by the wind. As it blows, seeds fall off and are dispersed in the environment.

Pericarp
Seed
(a)

Seeds
Pericarp
(b)

Figure 21–9. Dehiscent fruits. Dehiscent fruits split open along predetermined lines and release their seeds. (*a*) Follicle. (*b*) Capsule.

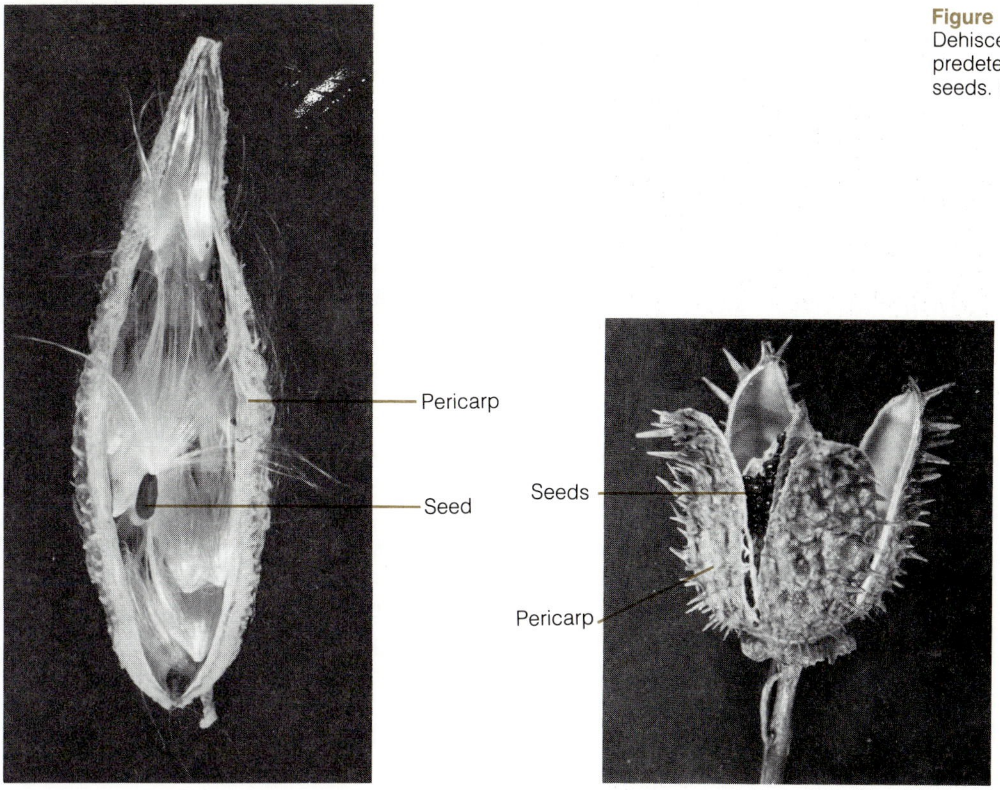

Endocarp
Mesocarp
Exocarp
(a)

Endocarp
(b)

Figure 21–10. Coconut. (*a*) The outer layers of the coconut fruit are fibrous and this makes the fruit buoyant. (*b*) The coconut seed after removal of the husk.

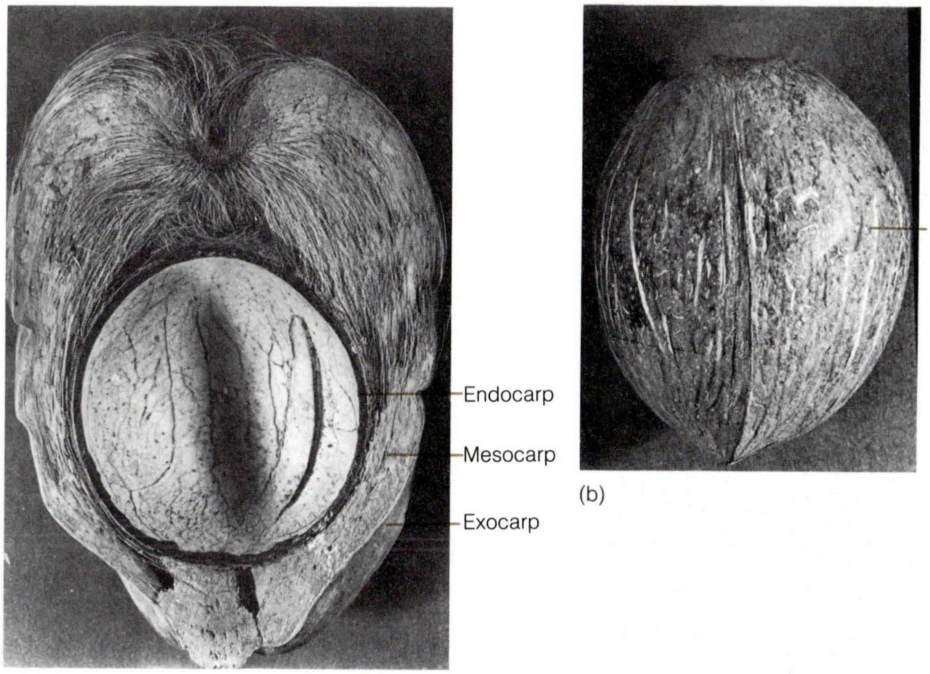

Animals

Both fleshy and dry fruits, as well as some seeds, are adapted to animal dispersal. Fleshy fruit is often gathered and eaten by humans, an experience useful in understanding the fleshy fruit as a special adaptation for seed dispersal. Whether it is a banana in the supermarket, a tomato in the garden, or raspberries in the wild, there is a distinct difference between ripe and unripe fruit. As many fruits ripen their flesh becomes softer and in fruits like apples and bananas the flesh becomes sweeter as starch is broken down to sugar. Accompanying the changes in texture and flavor of the flesh there is usually a dramatic change in the color of the skin. An unripe, fleshy fruit is usually green—the same color as the leaves. An animal searching for food will disregard the green fruit either because it has learned that the unripe fruit is unpalatable or because it cannot distinguish the fruit from the green leaves.

As the embryo in a seed matures, hormonal signals (Chapter 12) are released that initiate ripening. Ripening is a physiological process that evolved in species with fleshy fruits because it increased the efficiency of seed dispersal and therefore was favored by natural selection (Chapter 23). Consequently, the skin changes from green to a color that contrasts with the vegetation with the result that animals cannot fail to notice the fruit (Plate 11E). A characteristic aroma that indicates ripeness also is often present. Because the food contained in the flesh cannot be used by the plant embryo or the developing seedling, it appears that the fleshy fruit evolved solely as part of a seed dispersal mechanism by animals.

Seed coat
The outermost protective layer of the seed that is derived from the outer layers, or integuments, of the ovule.

Seeds contained in edible fruits must possess adaptations that allow them to survive ingestion by an animal. Typically this involves the possession of either a hard **seed coat***, or a slimy layer around the seed, or both, so that the seed escapes the grinding action of the teeth. Once in the stomach the seed receives further protection against digestion from the outer layer of the seed. After passage through the remainder of the digestive system the seed is left behind in the animal's feces. Here the seed is in a warm, moist environment that contains all of the mineral nutrients necessary for its growth. Furthermore the seed has been "planted," giving it an advantage over seeds left exposed on the surface of the ground. The tomato is a familiar example of a plant that has evolved this type of seed dispersal system.

Epiphyte
A plant that grows upon another plant without receiving any water or nutrients from it.

Not all seeds of edible fruits are adapted to pass through an animal's digestive system. Some species, such as mistletoe, have seeds that are coated with a viscid substance that sticks to a bird's beak when the fruits are eaten. The bird's response to this turn of events is to fly to another tree and wipe its beak against the bark of a branch, a process that dislodges the mistletoe seed and leaves it adhering to the bark. Since mistletoe is a parasitic **epiphyte***, this mechanism of seed dispersal ensures that at least some seeds end up in the correct environment.

A most unusual example of seed dispersal by animals has been reported from the jungles surrounding the Amazon River. The seasonal flooding of the river enables certain varieties of bottom-feeding fish to scavenge the fruits that fell from the trees and shrubs growing on the flooded lands. As with many other animal-dispersed edible fruits, the seeds pass through the digestive tract of the fish and are thereby dispersed over the floodplain.

The possession of thick seed coats offers the needed protection for seeds that must endure the grinding and acids encountered in the animal digestive tract. If such seeds do not pass through an animal, they may not germinate because the thickness of the seed coat does not permit the penetration of oxygen or water. In order to germinate such seeds artificially, it is necessary to damage the seed coats by **scarifying** them with a file or even acid. Artificial scarification mimics the natural process of passage through a digestive system.

There is one report in the scientific literature that the extinction of the dodo bird has adversely affected the reproduction of a tree species native to an island off the African coast. From the observations that no trees were younger than about 300 years old (the length of time that the dodo bird has been extinct) and that the seeds had a thick seed coat, these investigators hypothesized that seeds of this tree germinated naturally only after being scarified in the crop* of a dodo bird, the only native bird that ate these seeds. To test their hypothesis, these scientists force-fed seeds from these trees to geese, which also have a crop. The sprouting of seedlings from the goose droppings confirmed the very close relationship that had evolved between a plant and the animal that dispersed its seeds.

Crop
A portion of a bird's esophagus in which seeds are ground up.

Since seeds contain stored food, they are eaten by a number of animals, some of which store their seeds underground. The squirrel gathering acorns is a familiar sight; however, not all of the acorns and other fruits and seeds buried by a squirrel are recovered and eaten. Those not consumed will germinate in the spring under favorable conditions. In addition to these larger animals, small animals such as insects also accidentally disperse plant seeds. Seeds taken into burrows or nests by ants and other insects have the potential to germinate if they are not eaten. Although these animals reduce the number of viable seeds, plant reproduction is favored because the remaining seeds are in a location more favorable to their germination and survival as seedlings.

Because they produce hair, mammals are particularly useful agents of seed dispersal, even for plants that produce inedible fruits. Many fruits and seeds are modified with little hooks and barbs (Figure 21–11) that catch in animal hair. Under natural conditions the burs and other similar structures remain with the animal until the hair falls out or until the animal grooms itself and picks them out. Due to the mobility of animals, these seeds and fruits may be dispersed over wide areas.

Figure 21–11. **Bur fruits.** Burs such as these have barbs that catch in the hair of passing mammals. The seeds are left behind when the animal cleans its fur.

Seed Structure

The integuments, or outer layers of the ovule, upon maturation form the seed coats, the outer layers of the seed. Enclosed within the seed coats are the embryo and the stored food. Depending on where food is stored in angiosperm seeds, different seed structures can result. In one type food for germination and seedling growth is stored in the **endosperm** tissue (Chapter 20), resulting from the fusion of one of the two sperms with the polar nu-

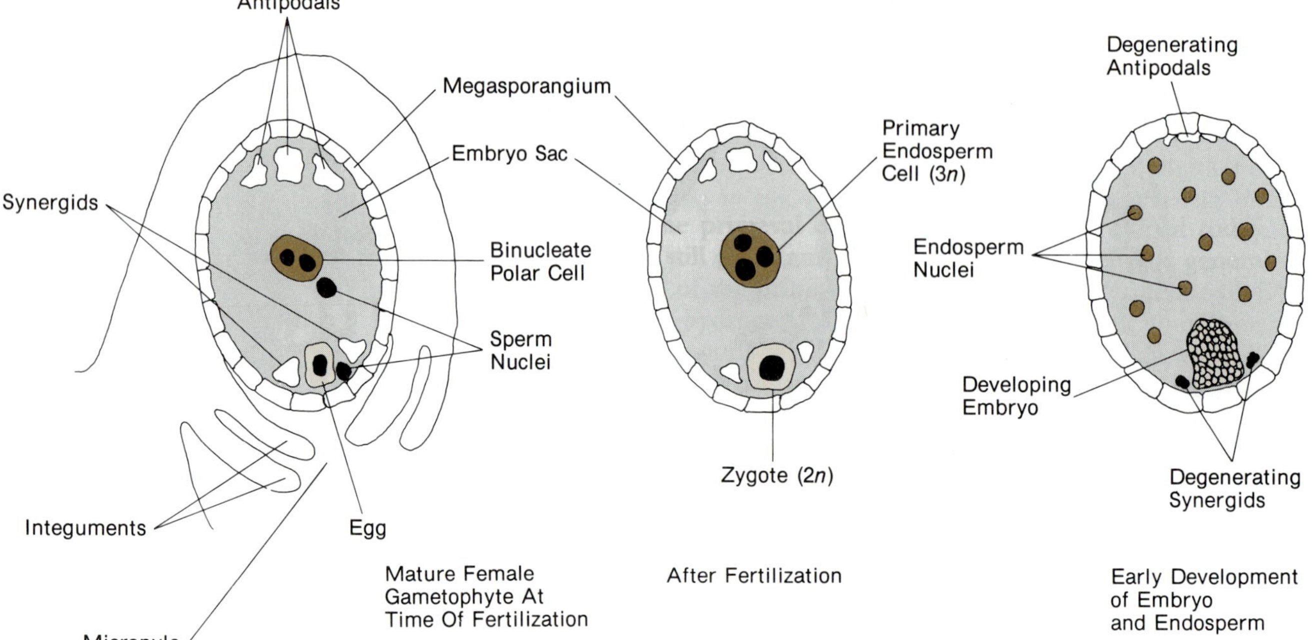

Figure 21–12. Endosperm development. During fertilization in angiosperms, one sperm fuses with the egg producing the zygote and the other fuses with the polar nuclei forming the primary endosperm nucleus. Subsequent divisions of this nucleus produces the large endosperm tissue that functions in food storage in certain angiosperm seeds.

clei in the embryo sac (Figure 21–12). Food translocated to the developing seed accumulates in the endosperm, which comprises the major component of the mature seed as seen, for example, in corn and coconut seeds (Figure 21–13). In the second type food is stored in the leaflike **cotyledons** of the embryonic plant, not in the endosperm of the seed (Figure 21–14). Even though an endosperm is formed in this second situation, food is translocated from the developing endosperm to the cotyledons of the developing embryo. The cotyledons enlarge at the expense of the endosperm, which, at maturity of the seed, completely disappears in many instances.

Recall that in the gymnosperms, discussed in Chapter 19, no endosperm forms. (Endosperm development from double fertilization is only an angiosperm characteristic.) Rather, the food-storage function is carried out by the tissue of the female gametophyte.

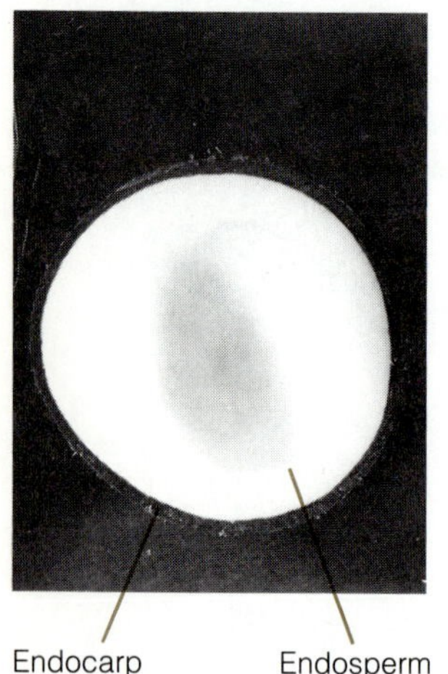

Figure 21–13. Coconut endosperm. The solid white portion of this coconut is the endosperm of the seed.

An angiosperm embryo consists of the same organs as the mature vegetative plant: root, stem, and leaves (Figure 21–15). Together the stem and root of an embryo form the **embryonic axis,** just as they will form the stem-root axis of the mature plant. Unlike the mature stem and root, however, the embryonic structures are quite simple with an essentially arbitrary boundary being drawn between the embryonic root, or **radicle,** and the embryonic stem. Attached to the embryonic stem are the cotyledons, either one or two in angiosperms, but several in gymnosperms. The region of the stem-root axis immediately below the point of attachment of the cotyledons is referred to as the **hypocotyl** while the region above is the **epicotyl.** At the very tip of the epicotyl there may be embryonic leaves, which, after germination, produce chlorophyll and begin to photosynthesize. In plants such as radish, the cotyledons, which are modified leaves, will carry out photosynthesis for a week or more, until the seedling becomes established.

Cotyledons of seeds that have endosperm are modified to function as organs of food absorption. As the seedling germinates, the food stored in the

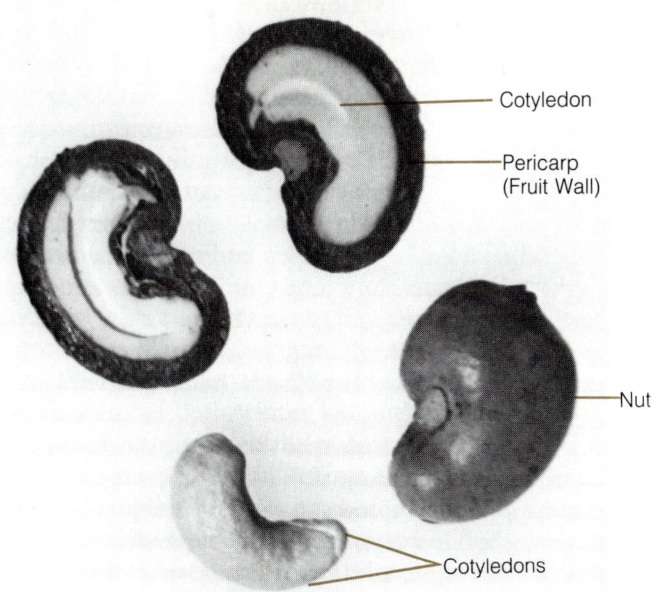

Figure 21–14. Cashew cotyledon. In many edible seeds it is the large fleshy cotyledons that are consumed.

endosperm is digested. Since the endosperm is outside of the embryo rather than being a part of the embryo, the digested food must be absorbed by the embryo instead of being translocated through vascular tissue. In many seeds that contain an endosperm, the cotyledons remain within the seed at or below ground level, while shoot growth results from epicotyl elongation (Figure 21–16). When the cotyledons are the food-storage organ of the seed, they may remain belowground as in peas (Figure 21–17) or may emerge aboveground as in beans (Figure 21–18), where they will contribute to photosynthetic food production during the first few weeks of seedling growth.

The angiosperms are divided into two groups—the monocots and dicots—which differ in many characteristics (see Chap. 20). The group names, however, reflect a difference in a characteristic that in itself is probably not particularly significant. Monocots (monocotyledonous plants) have one (mono) cotyledon and dicots (dicotyledonous plants) have two (di) cotyledons. Although both monocot and dicot seeds can have an endosperm at maturity, it is more typical for monocot seeds to contain an endosperm at maturity whereas dicot seeds are more likely to lack an endosperm.

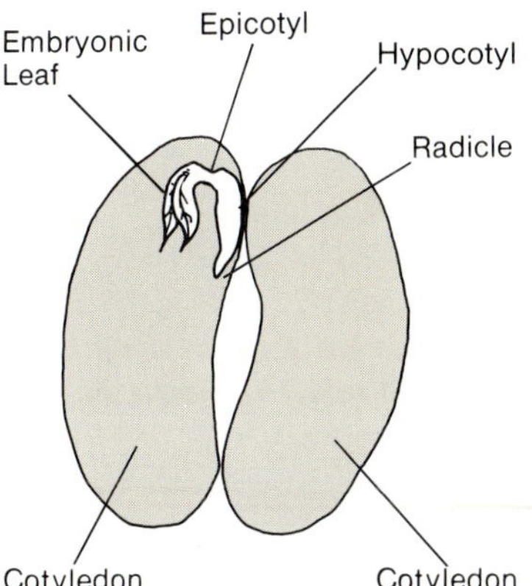

Figure 21–15. Plant embryo. The bean seed contains a large embryo that contains structures typical of many plant embryos.

Seed Germination

Germination of a seed involves a renewal of embryonic growth and requires that certain conditions be met. The first requirement for seed germination is the absorption of water allowing the rehydration of the embryo, which had become dehydrated during the final stages of seed maturation. Nor-

Figure 21–16. **Corn germination.** In the germination of corn the coleoptile emerges after the radicle. Once it has been exposed to light, the coleoptile stops growing and the shoot breaks through. The cotyledon remains within the kernel where it absorbs foods released by digestion of the endosperm.

mally, plant cells are 90% water, but the cells of the embryo are rendered dormant by a reduction in water content to levels as low as 10%–15%. Maple seeds that do not enter true dormancy contain 60% water. As a result of water uptake metabolism is activated, thus promoting the physiological events of germination such as synthesis of new proteins, mobilization of food supplies, and growth. Water uptake by the cells results in their swelling, thus accelerating the rupture of the seed coat and the emergence of the embryo. In addition to water, oxygen is needed by the germinating embryo to support the required metabolic activity through energy generation in respiration (Chapter 10). For this reason many seeds fail to germinate in waterlogged soils where water has replaced the air in the pores of the soil.

Seeds that are produced in the spring and early summer will most likely be able to germinate immediately, but seeds produced later in the season are more likely to require a period of **dormancy*** before they can germinate. Dormancy of seeds usually can be broken artificially by soaking them in water or by providing a cold treatment, which functionally mimics winter, although the coldness does not have to be of the same duration or intensity. In many temperate species seed dormancy is broken naturally by a period of cold temperatures, or by the seed spending a minimum time in dormancy. In the latter case, called after-ripening, specific physiological changes occur in the embryo that permit it to germinate when placed under favorable conditions.

Dormancy
A state of reduced physiological activity.

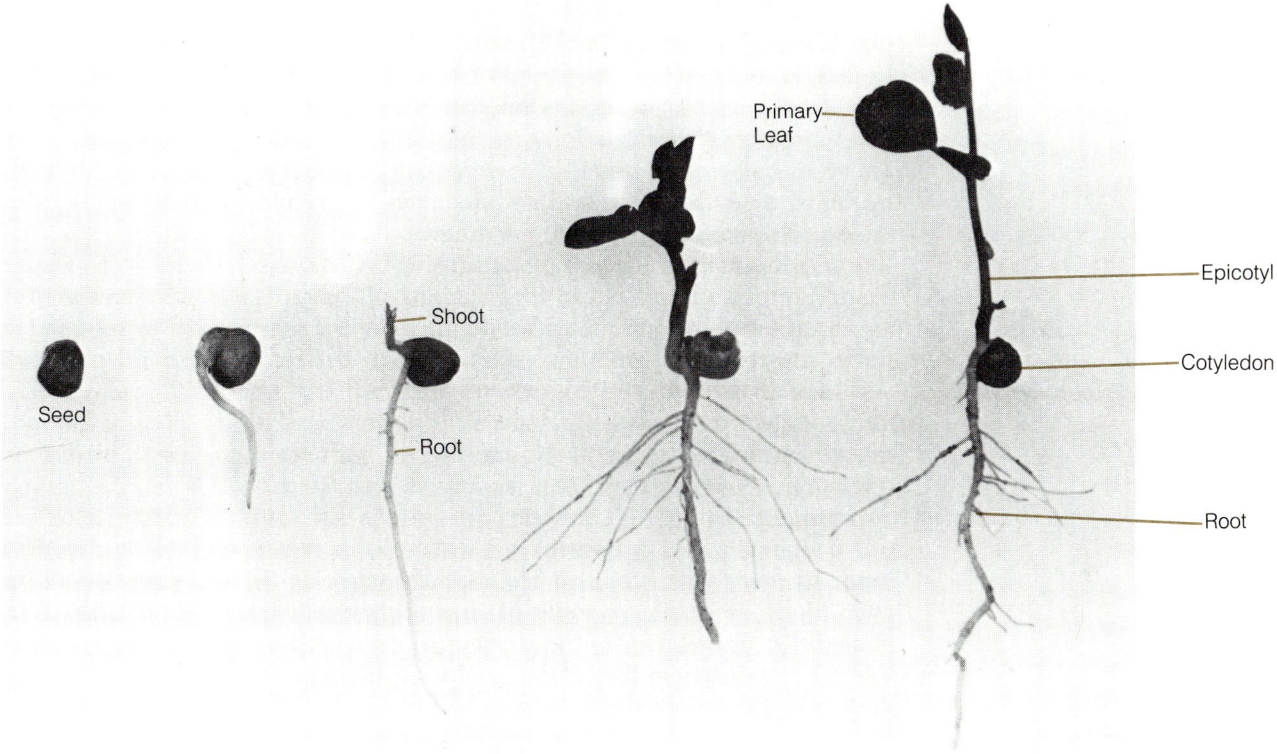

Figure 21–17. Pea germination. The radicle of the pea seedling emerges first followed by the epicotyl. The fleshy cotyledons stay below ground within the seed coat.

Desert species have their greatest possibility of seedling survival if germination occurs after the heavier rains that occur during the "rainy" season. Germination does not occur after one of the light rains, which may fall at other times of the year, because the seed coat contains water-soluble germination inhibitors that would not be removed completely by a light shower.

There are other species of plant seeds whose dormancy can only be broken by scarification of the seed coat permitting water and oxygen to enter. Scarification can be induced by weather, such as repeated freezing and thawing cycles that would crack the seed coat, or by microbial* digestion. Seeds adapted to passage through an animal's digestive tract often possess a thick seed coat that protects the embryo; however, this thick seed coat prevents germination. The digestive process scarifies the seed, either through grinding or acid treatment, so that, when the seed is eliminated along with other indigestible material, germination will occur. Table 21–2 lists the conditions that induce the germination of seeds of some common plants.

Microbial
Pertaining to microorganisms such as bacteria.

Figure 21–18. Bean germination. After emergence of the radicle, the hypocotyl begins to elongate. Because of the uneven growth in the stem, the hypocotyl forms a hook that pulls the epicotyl and cotyledons through the soil. Shortly after exposure to light, the hook begins to straighten out. The cotyledons of a young bean seedling turn green and carry out photosynthesis during the initial period of seedling growth.

Table 21–2.
Conditions required for induction of seed germination for some common plants

Common Name	Scientific Name	Conditions or Special Treatment
American Basswood	*Tilia americana*	soak in concentrated nitric acid for 30–120 minutes then cold treatment for 90–150 days
Apple	*Malus domestica*	cold treatment for 50 days
Buckeye	*Aesculus glabra*	cold treatment for 120 days
Honeylocust	*Gleditsia triacanthos*	scarify the seed coat or soak for up to 2 hours in concentrated sulfuric acid
Oaks (most species)	*Quercus* spp.	cold treatment for 60–90 days
Orange, Grapefruit, Lime, Lemon	*Citrus* spp.	none required
Peach	*Prunus persica*	cold treatment for 100 days
Sweet Cherry	*Prunus avium*	cold treatment for 100–120 days

The survival of dormant seeds has been followed carefully for a number of species, and it has been found that there is great variability in the longevity of seeds. Some tree seeds, such as willow and maple, survive for only a few days or weeks, whereas seeds of a South American legume, *Cassia,* that were collected in 1776 could still be germinated in 1934—158 years later. If no special storage conditions are employed, the seeds of most plants are viable for only a few years. Under conditions of low temperature, reduced seed-moisture content, and low oxygen, seeds can be stored for longer periods. These storage conditions seem to favor seed longevity because they reduce respiration rates that result in the buildup of metabolic wastes which could damage the embryo.

Extreme examples of natural seed longevity include Oriental lotus, a member of the waterlily family, and arctic lupine, a member of the bean family. The facts that Oriental lotus seeds are surrounded by a thick pericarp and were found buried in a peat deposit account for the reduced oxygen availability to the embryo. These lotus seeds were determined by radiocarbon dating to be about 2000 years old, yet their germination rate was extremely high. Some arctic lupine seeds found frozen in the Alaskan permafrost are the oldest seeds that have been capable of germination. Even though the seeds were dated as 10,000 years old, they readily germinated producing lupines indistinguishable from ones grown from the previous year's seed. Stories relating the discovery of ancient, viable wheat seed in Egyptian pyramids have thus far proven to be tourist gimmicks, whereby tour guides drop the seed and then proceed to "find" this ancient seed that they will sell to the tourist.

During germination, the delicate apical meristem of an embryonic shoot must be protected in its passage through the soil or layer of leaf litter. Two methods of protection have evolved: in one the shoot apex is surrounded with protective tissue; in the other the shoot apex is pulled backwards out of the soil instead of being pushed out of it. Corn is a plant that provides a protective sheath, the **coleoptile,** around the epicotyl. The epicotyl grows within the protective tube formed by the coleoptile. Once the seedling emerges aboveground and is exposed to light, the coleoptile stops growing and the epicotyl pushes through the sheath. In order for the apical meristem to be pulled out of the soil, a portion of either the epicotyl (as in peas) or the hypocotyl (as in beans) must grow more rapidly than the upper portions of the embryonic stem. This will cause the stem to grow in a hooked configuration, with the apical meristem being protected by facing down (Figure 21–18).

Given adequate water and oxygen and a reasonable temperature, a nondormant seed will germinate. The first embryonic structure to emerge from the seed coat is usually the radicle. In fact, root growth is more pronounced than shoot growth in early seedling development (Figure 21–16, 21–17, 21–18) resulting in firm anchorage of the seedling. Subsequently, the shoot emerges either as the hook (Figure 21–18) or under a protective sheath (Figure 21–16), described above. Once the cotyledons or leaves are exposed to light, they begin to turn green as the immature plastids differentiate into chloroplasts. The period following germination is one of rapid differentiation of the embryonic tissues, which produces the young seedling.

Seedlings

Following germination the seedling is a rudimentary version of the mature plant that ultimately develops. At this stage the stem is still considered as being either hypocotyl or epicotyl and the first leaves are still expanding. During this time the seedling grows by using food reserves stored in the cotyledons or endosperm. Bean seedlings, for example, are not independent of food stored in their cotyledons until about a week after the seedling emerges, even though the plant has well-formed green leaves before this time. As seedlings mature, food-storing cotyledons and endosperm shrivel and disappear leaving a seedling that, under favorable ecological conditions (Chapter 24), will reach maturity. In many plants, the embryonic leaves that become the first, or primary, leaves of the seedling are shaped differently than leaves formed from the shoot apical meristem after germination.

Summary

Fruits are plant organs derived from flowers. Simple fruits develop from the ovary of a single flower. An aggregate fruit is formed from several pistils of the same flower and a multiple fruit is formed when pistils of several flowers on the same inflorescence fuse, forming a single fruit. Accessory fruits contain other floral structures, such as petals or sepals, in addition to the ovary. Simple fruits are classified as either fleshy or dry depending on the condition of the fruit wall, or pericarp. Fleshy fruits are classified on the basis of differences in texture of the three layers of the pericarp—exocarp, mesocarp, and endocarp. Tomatoes (berries) have an exocarp that forms an outer skin whereas the mesocarp and endocarp are fleshy. Plums (drupes) have a skin formed by the exocarp, a fleshy mesocarp, and an endocarp that forms a hard layer around the seed.

Fruits are structures that contain the mature seeds and thus can be involved in seed dispersal. On some seeds hairs and wings have evolved that make them adapted for wind dispersal. Other plants have evolved buoyant seeds or fruits that allow for dispersal by water, while still other seeds are "shot" from the fruit as it dries. For many angiosperm species, animals, such as birds and mammals, serve as unwitting seed-dispersal agents. Fleshy fruits have evolved as adaptations for animal dispersal of seeds. After the fruit is consumed, the majority of the seeds pass unharmed through the digestive tract and are deposited by the animal in a fertile substrate—the feces. Certain dry fruits and seeds have barbs or hooks that adhere to an animal's fur or feathers, thus ensuring a wide dispersal.

A seed develops from an ovule and contains an embryo and a stored food supply. In many angiosperms the food is stored in a nonembryonic tissue called the endosperm, formed by the fusion of a sperm nucleus with the polar nuclei. Other angiosperms store food used in germination in modified leaves called cotyledons, which are part of the embryo. The embryo consists of root (radicle), stem (hypocotyl and epicotyl), cotyledon(s) and, in some cases, embryonic leaves. In seeds that contain an endosperm the cotyledon(s) function in absorption of food from the endosperm. During germination the apical meristem is protected either by being pulled backwards out of the soil or by being enclosed in a protective tube called the coleoptile. Germination frequently does not follow seed dispersal because of ecological factors, in which case the embryo becomes dormant. Dormancy can be broken by cold treatment in many temperate species and by heavy rain in desert species. Although most seeds remain viable for only a few years, a few exceptional cases are known where seeds have germinated after more than 100 years had passed. In order to initiate germination most seeds require water to rehydrate embryonic cells to 90% water from the 10%–20% water typically found in dormant seeds. The seedling that results from germination of a seed is a rudimentary plant. As food reserves are depleted, the endosperm or cotyledon shrivels and the plant becomes photosynthetically competent.

Review Questions

1. Draw and label a corn kernel. For each of the structures indicate the floral structure or cell from which it developed.
2. Describe with examples food and seed adaptation that aid in seed dispersal. Why is seed dispersal necessary?
3. Define seed dormancy and explain its role. What are the different ways by which dormancy in seeds can be broken?
4. What is the advantage of having coleoptile or a cotyledonary hook during seed germination?
5. Distinguish between simple and multiple fruits.
6. How does an accessory fruit develop? Give an example and show what floral structures are involved in the development of an accessory fruit.

Self-planting Seeds

Many different mechanisms have evolved in the flowering plants to ensure seed distribution and germination, from the wind-dispersed dandelion to the bur that is carried by mammals. However, some unusual adaptations have evolved in certain plants that actually result in the seed being planted without human or animal intervention. As a first example, even though peanuts actually develop underground like potatoes or radishes, the peanut is neither stem nor root; instead it is a seed inside a fruit. The peanut plant is a legume—a relative of peas and beans—in which a unique mechanism has evolved that ensures the seeds are planted properly. After fertilization has occurred the stamens and petals fall off and the flower stalk begins to elongate while bending toward the ground. Further elongation of the flower stalk pushes the ovary into the soil where the fruit and seed mature. An ovary that fails to become buried will die. Peanut farmers increase their yield by hilling the plants once the flowers are mature.

A second example of a plant that buries its own seed is needlegrass *(Stipa)*. The needlegrass fruit, containing a single seed, has a sharp point at the lower end and a long, bristlelike awn, which is an extensions from a bract*, at the other end (Figure 21-19). A needlegrass fruit lands with its point facing down ready to be planted. This is accomplished by the awn twisting around in response to changes in its water content. As the awn twists it comes into contact with plant debris that prevents the free end of the awn from moving, thus effectively transferring the twisting motion to the fruit. Because the end of the fruit is pointed, this motion drills the fruit into the ground and plants it up to 2 inches deep, an optimal depth for the germination of this species.

Bract
A modified leaf

Figure 21–19. Needlegrass. Twisting action of the long awn drills the pointed needlegrass fruit into the ground to a depth of about 2 inches.

22

Plant Genetics

Outline

Introduction

A sure sign of oncoming spring in temperate regions is the arrival of mail-order seed catalogues and the seed displays in stores. When we find the pictures and descriptions irresistable and purchase different kinds of seeds, we expect specific results. Planting a package of *Coleus* seeds results in an assortment of variegated plants. Why? A package of "Sunburst" petunia seeds gives rise to petunias that all have bright yellow flowers. Again we ask why. Why do seeds of some plants produce offspring that are all different from one another, while seeds from other plants produce offspring that are all apparently identical? The answer is not that this is a difference between species. Instead, these results reflect differences in the genetic status of the plants that produced the seeds. In the case of the *Coleus* seeds, the parents were genetically diverse and contained a variety of inheritable leaf pigmentation characteristics that were passed to their offspring. The parent petunia plants, on the other hand, were genetically homogeneous for flower color and could only pass the yellow flower characteristic to their offspring.

The observation that physical characteristics can be inherited is an ancient one. Farmers have always recognized that the best animals are produced by breeding two parents with the desired characteristics. Traditionally plant breeders also have chosen to plant seeds from the best female plants as a way to increase crop yield. Until the early 1900s, however, plant and animal breeding was more of an art than a science because the breeders had no awareness of how characteristics were inherited. In fact, there was a school of thought associated with the French zoologist Jean Baptiste de Lamarck which held that organisms acquire characteristics in response to environmental changes, an idea that was later discredited. A major event in explaining the true mechanism of heredity occurred in 1857 when Gregor Mendel, an Austrian monk, began a study of the inheritance of characteristics in peas. This study subsequently established the foundations of modern genetics.

Genetics is the study of the inheritance of specific characteristics. Initially, only the physical characteristics of an organism were studied (**classical genetics**). Increasingly geneticists have become concerned with the inheritance of specific proteins (**molecular genetics**) because these proteins, through their metabolic activities, determine the physical characteristics of the organism. To understand both classical and molecular genetics it is essential to remember that the genes* for characteristics in offspring are received from their parents through sexual reproduction and that their parents' gametes (egg and sperm) were ultimately produced by meiosis.* In the following discussion of genetic principles, continuing reference will be made to meiosis and fertilization.

Chapters 13 and 18–20 have pointed out that sexual reproduction in plants does not involve direct production of gametes by parental organisms. Instead, there is an alternation of haploid* and diploid* generations, with only the haploid generation producing gametes. Among all land plants, ex-

Gene
A portion of a DNA molecule that encodes the genetic information for a protein or protein subunit (polypeptide).

Meiosis
The process of nuclear division that results in the production of daughter nuclei, each containing one-half the original number of chromosomes.

Haploid
Possessing only one of each pair of chromosomes.

Diploid
Possessing two of each pair of chromosomes.

cept the bryophytes, the conspicuous stage of the life cycle is the diploid generation. It is in these diploid plants that meiosis occurs, resulting in spores that germinate and develop into the haploid gamete-producing stage. For the purposes of discussing genetics, the plant life cycle will be treated as if meiosis leads directly to gamete production. This is possible because the gametes are genetically identical to the spores that give rise to the gamete-producing generation.

Mendel's Experiments

Before Mendel began his work, little progress had been made in understanding how characteristics were passed from parent to offspring. Mendel succeeded where others failed because of his choice of experimental material—pea plants. The pea is an ideal plant for genetic studies because it normally is self-pollinated. This means that seeds from the same plant will be genetically uniform (i.e., they all will have the same inherited characteristics) and therefore an investigator can obtain many genetically identical individuals. A large sample of individuals is needed in order to eliminate error due to chance. For example, if a coin were flipped, there is a 50:50 chance of heads resulting. However, flipping the coin 10 times could produce 7 tails and 3 heads, not the expected distribution. Flipping the coin 100 or 1000 times would produce a result much closer to the expected ratio. In other words, the large sample size reduced the error due to chance.

Similarly in Mendel's experiments, many plants were used to eliminate chance variations. Once Mendel had the large numbers of plants he needed, he took an additional step that others before him had not: he followed the inheritance of single characteristics. In one experiment Mendel crossed tall plants with dwarf ones. This necessitated opening the flowers, removing the anthers (Chapter 20) to prevent self-pollination, and then manually transferring the appropriate pollen. The resulting seeds, when planted, gave rise to the first generation (F_1) offspring all of which were tall. Instead of stopping at this point, Mendel allowed these plants to self-pollinate and produce seeds. Among the plants from these second generation (F_2) seeds he observed both tall and dwarf plants in a 3:1 ratio (Figure 22–1). From these results Mendel concluded that in an individual there must be two hereditary units for each characteristic. This follows from the observation that dwarfness disappeared in the **F_1 Generation** and reappeared in the **F_2 generation.** If there were only one hereditary unit in each individual, then dwarfness could never have reappeared in subsequent generations. A second conclusion that Mendel drew was that one of the hereditary units must be **dominant** over the other. In this cross the dominant trait is tall whereas the dwarf trait is **recessive.** In the F_1 generation where the hereditary units for both traits are present, only the dominant trait is expressed; the recessive trait is masked and can only be expressed in an F_2 plant containing both recessive hereditary units.

Subsequent to Mendel's work it has been shown that the hereditary units

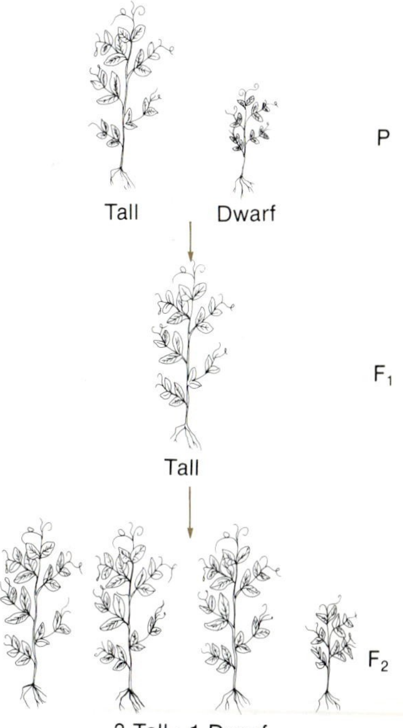

Figure 22–1. Results of a cross of tall and dwarf pea plants. The symbol *P* represents the parental generation with F_1 and F_2 representing the two generations of offspring.

are genes, which are segments of chromosomes. The different forms of expression of a particular gene are called **alleles.** As explained in Chapter 3, the expression of a gene takes the form of protein production, with many proteins being enzymes.* In practice, the difference between dominant and recessive alleles is the difference between the two forms of the same enzyme that are produced. In the example above, the pea plants quite likely differ in their ability to produce gibberellin, a hormone that stimulates stem elongation (Chapter 12). The tall allele appears to code for an active enzyme in the metabolic pathway leading to gibberellin production, whereas the recessive allele appears to code for an inactive enzyme or one much reduced in activity. Therefore the presence of the recessive dwarf allele causes less gibberellin production and thus less stem elongation. Although in this case only two alleles are involved, there are other characteristics for which there are multiple alleles. This means that a particular gene exists in several different forms within a population of the organism. For example, flowers colored red, blue, purple, and white can result from the expression of different alleles of the same gene. In such a multiple allele system there will be a hierarchy of dominance, such as red dominant over all of the others with blue recessive to red but dominant over purple and white, and so on.

Enzyme
A protein that functions as a biological catalyst.

Homologous chromosomes
Two chromosomes that carry the genes that govern all of the same traits.

Genotype
The genetic makeup of an individual; refers to the specific alleles carried on the chromosome.

The two alleles of the gene for height in peas are carried on **homologous chromosomes*** (Figure 22-2). When both homologous chromosomes of the diploid nucleus carry the same allele for a trait, the organism is said to be **homozygous** for that particular trait. An organism that is **heterozygous** for a particular trait has two different alleles for that trait. A heterozygous individual is also referred to as a **hybrid** because it represents a new combination of genetic information from two genetically different parents. The term hybrid is also used to refer to the individuals produced as the result of successful crosses between different species.

Using this information, Mendel's cross can be written with the symbol *T* to represent the chromosome carrying the tall allele and *t* to represent the chromosome with the dwarf allele. The parental generation would consist of individuals with the genetic makeup, or **genotype,*** *TT* (tall) and *tt* (dwarf). During meiosis homologous chromosomes separate, producing haploid cells that ultimately produce gametes:

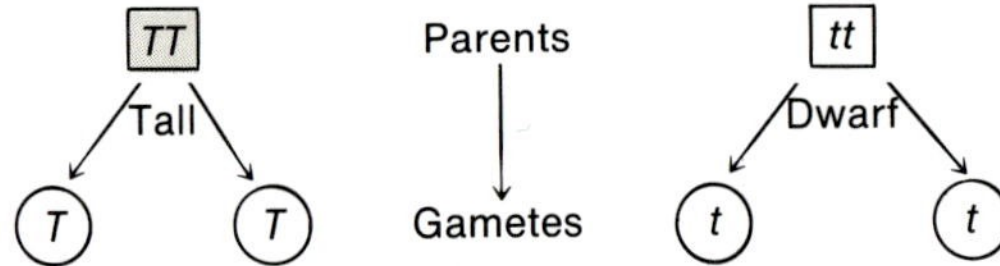

Sexual reproduction involves fusion of a gamete from the tall parent with the gamete from the dwarf parent forming an individual of the F_1 generation:

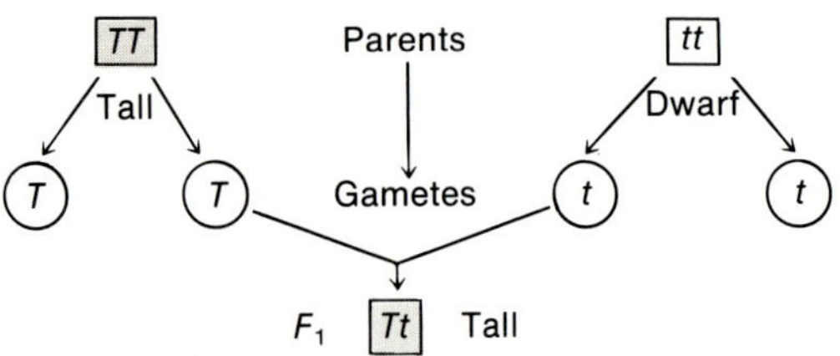

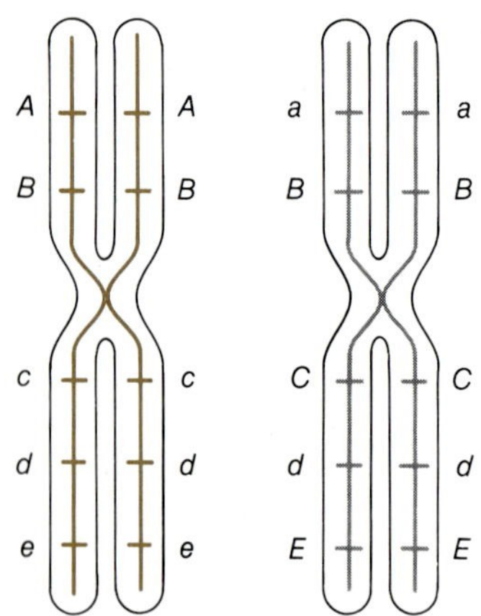

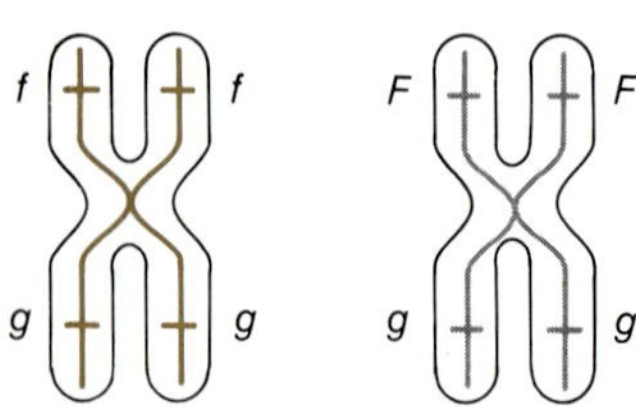

Figure 22–2. Two pairs of homologous chromosomes with alleles indicated.

Because the F_1 generation is heterozygous, when the homologous chromosomes separate in meiosis, two genetically different gamete types are produced by each individual:

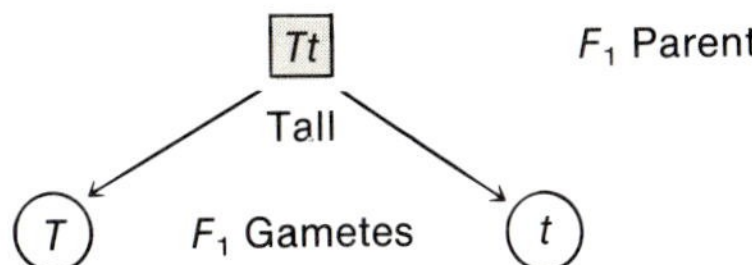

During sexual reproduction preceding the F_2 generation, random fusion of gametes will produce all possible genetic combinations:

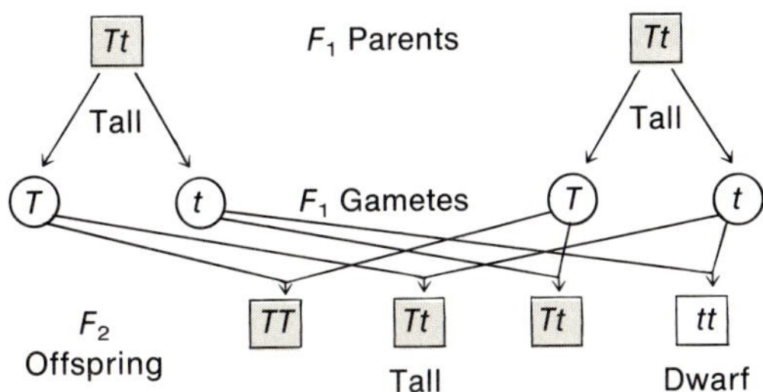

To avoid the confusion of lines connecting gamete types to show how the F_2 generation is formed, geneticists have adopted the use of a checkerboard, or Punnett square. Across the top of the square are written all of the possible gametes that could be produced by one of the parents; along one side are written all of the gamete types that the other parent could produce.

	T	*t*
T		
t		

To determine the genotypes of all possible offspring, all that remains is to combine the gamete in the first row with that in the first column, then with that in the second column and finally repeating the procedure for the gamete in the second row:

	T	*t*
T	*TT*	*Tt*
t	*Tt*	*tt*

This procedure results in the gametes being combined in all possible combinations, just as the joining of gametes by lines did, but in a fashion that is easier to read.

Regardless of the method by which the genotypes of the F_2 generation are derived, the results are the same. The theoretical ratios of F_2 genotypes are $1TT{:}2Tt{:}1tt$. Unfortunately, it is impossible to look at a plant and determine the genotype for a particular trait unless the plant is homozygous recessive *(tt)* for that trait. If even only one of the dominant alleles were present, the physical appearance, or **phenotype,** of the plant would be that of the dominant allele. Therefore, in this cross the F_2 generation will have a theoretical phenotypic ratio of 3 tall: 1 dwarf. From the calculations we see that two out of every three tall individuals will be heterozygous (*Tt*), but there is no way to distinguish these from the homozygous *(TT)* individuals by their appearance. It is possible, however, to determine the genotype of any individual for a particular trait or traits by performing a **test cross.** A test cross is a cross between the individual in question and a homozygous recessive parent. Then, if the F_2 individual were *TT,* all the offspring would have a *Tt* genotype and a tall phenotype; if the F_2 individual were *Tt,* there would be equal numbers of tall *(Tt)* and dwarf *(tt)* individuals produced (Figure 22–3).

Because the inheritance of only one trait was followed in the above example, it is known as a **monohybrid cross.** However, Mendel also performed **dihybrid crosses** in which he followed two traits simultaneously. As will be shown, following more than two traits simultaneously becomes cumbersome. Figure 22–4 shows the results of crossing tall, green-seeded individuals with dwarf, yellow-seeded individuals. Since the tall and yellow-seeded characteristics show up in the F_1 generation, these must be the dominant alleles, with dwarf and green-seeded being recessive. In the F_2 generation the following phenotypes are observed in a 9:3:3:1 ratio—tall, yellow-seeded; tall, green-seeded; dwarf, yellow-seeded; dwarf, green-seeded. As in the monohybrid cross this dihybrid cross can be diagrammed to show the gamete production (meiosis) and fusion (fertilization) involved. Yellow-seeded, the dominant al-

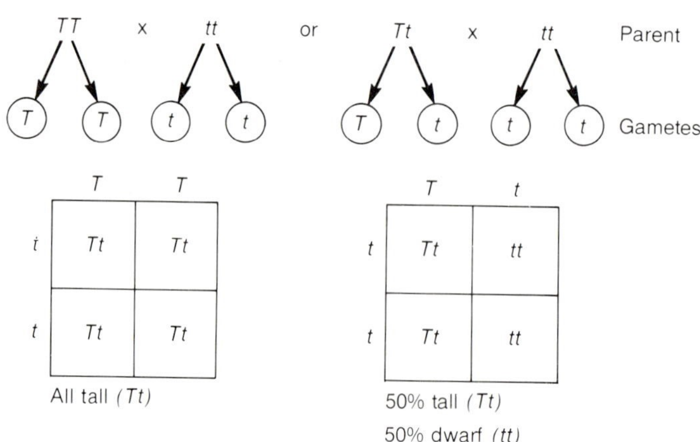

Figure 22–3. Test cross. Crossing an individual of unknown genotype with an individual homozygous recessive for the trait(s) under consideration allows the determination of the unknown genotype.

		F_1 Gametes			
		TY	Ty	tY	ty
F_1 Gametes	TY	$TT \cdot YY$	$TT \cdot Yy$	$Tt \cdot YY$	$Tt \cdot Yy$
	Ty	$TT \cdot Yy$	$TT \cdot yy$	$Tt \cdot Yy$	$Tt \cdot yy$
	tY	$Tt \cdot YY$	$Tt \cdot Yy$	$tt \cdot YY$	$tt \cdot Yy$
	ty	$Tt \cdot Yy$	$Tt \cdot yy$	$tt \cdot Yy$	$tt \cdot yy$

Figure 22–4. Dihybrid cross. F_2 genotypes can be determined using the Punnett square when all possible F_1 gamete types are placed along the top and side of the square. Sorting of the resulting individuals on the basis of phenotype produces a 9:3:3:1 ratio of tall, yellow-seeded:tall, green-seeded:dwarf, yellow-seeded:dwarf, green-seeded.

lele, is represented by *Y*, whereas *y* represents the green-seeded, recessive allele. The result of this cross is:

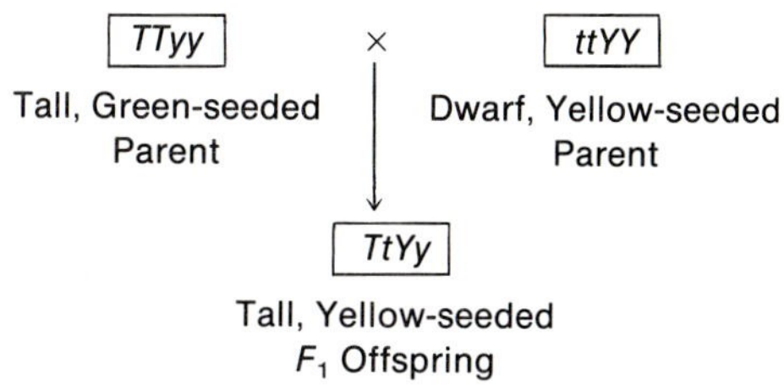

How was this F_1 genotype obtained? During meiosis the chromosomes of the *TTyy* individual line up in metaphase I as homologous pairs, then separate in anaphase I producing the haploid *Ty* genotype:

$$(Ty) < \begin{matrix} TT \\ yy \end{matrix} > (Ty)$$

Notice that homologous chromosomes separate from each other. However, each daughter nucleus *must* receive one of each chromosome type, so it is impossible for a viable gamete to be produced that contains only the *T* or the *y* chromosome. The *ttYY* individuals by a similar process produce *tY* gametes. Fertilization unites a gamete from each, establishing the diploid condition which in this case has the genotype *TtYy* (for the characteristics being considered). During meiosis in F_1 individuals, the chromosomes could segregate in either of two ways, producing equal numbers of four different gamete genotypes:

$$(TY) < \begin{matrix} Tt \\ Yy \end{matrix} > (ty) \qquad (Ty) < \begin{matrix} Tt \\ yY \end{matrix} > (tY)$$

This conclusion is reached on the basis of a knowledge of meiosis and the location of genes on chromosomes. Based on his analysis of the F_1 and F_2 generations of his crosses, Mendel arrived at the same conclusion that the hereditary units must segregate independently.

Selfing
Allowing to self-pollinate.

When the F_1 plants *(TtYy)* are **selfed*** to produce the F_2 generation, these four gamete types combine randomly without regard for genotype. Because there are four different sperm genotypes and four different egg genotypes there are sixteen possible combinations of egg and sperm involved in forming the F_2 generation (Figure 22–4). Since gamete fusion is a random process not related to gamete genotype, gametes will fuse in all possible combinations only if there are large enough numbers of individuals. An experiment producing only 100 seeds could, by chance, result in an F_2 generation with 40 tall, yellow-seeded plants, 30 tall, green-seeded plants, 20 dwarf, yellow-seeded plants, and 20 dwarf, green-seeded plants—hardly the expected 9:3:3:1 ratio. As outlined earlier, Mendel's success was due in large part to his use of experimental populations numbering in the thousands that allowed him to determine the true genetic ratios.

Once Mendel had worked out these basic laws of inheritance (Table 22-1) it was no longer always necessary to use large populations or to start with parents of known genotype. In fact, once the basic laws of inheritance became known, more was learned by applying the laws to determine the genotype of individuals that did not have homozygous parents or grandparents. The following example serves to indicate the application of these fundamental laws. A red-flowered, cut-leaf plant is crossed with a white-flowered, smooth-leaf plant, producing in the F_1 generation red-flowered, cut-leaf plants and white-flowered, cut-leaf plants. A red-flowered, cut-leaf F_1 individual was selfed and produced plants that were red-flowered and cut-leaf, red-flowered and smooth-leaf, white-flowered and cut leaf, and white-flowered and smooth-leaf (Figure 22–5). From the phenotypes of the parents and offspring it should be possible to determine the genotypes of all individuals. The first thing to do is to determine which characteristics are dominant: both red and white flowers show up in the F_1 generation but when the red-flowered, cut-leaf F_1 individual is selfed, both red and white flowers appear. Therefore red is dominant over white. Since the smooth-leaf

Table 22–1.
Mendel's laws or rules of inheritance

Law of Unit Characteristics. An organism contains two hereditary units (now known as alleles) for each inherited characteristic.

Law of Dominance. One of the alleles is dominant and the other is recessive.

Law of Segregation. The pairs of hereditary units separate from each other during the formation of gametes.

Law of Independent Assortment. Because gamete fusion is random, not determined by genotype, gametes fuse in all possible combinations independently of their genetic composition.

Red Flower, Cut Leaf x White Flower, Smooth Leaf P

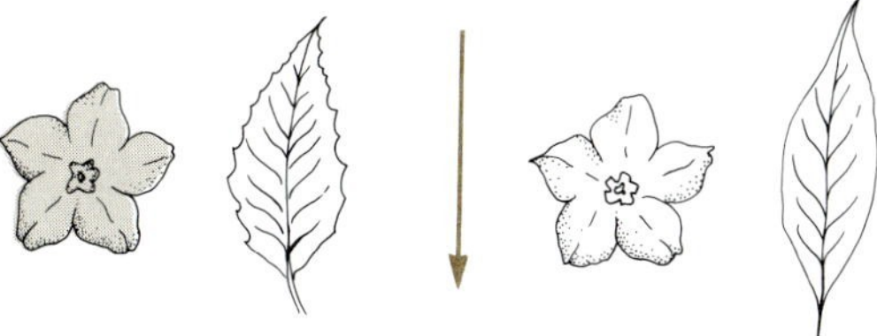

Red Flower, Cut Leaf and White Flower, Cut Leaf F_1

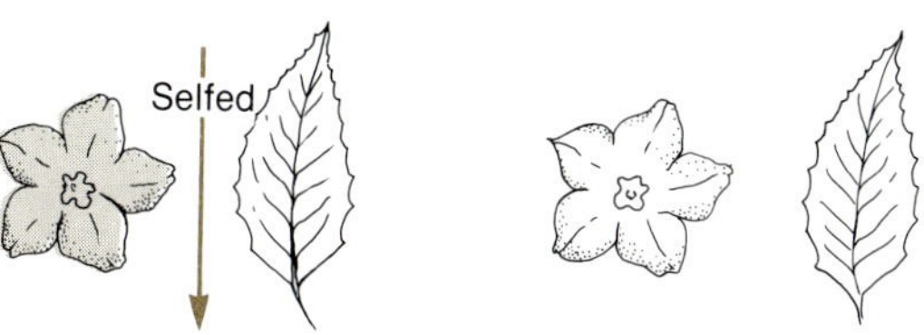

Red Flower, Cut Leaf; Red Flower, Smooth Leaf; White Flower, Cut Leaf; White Flower, Smooth Leaf. F_2

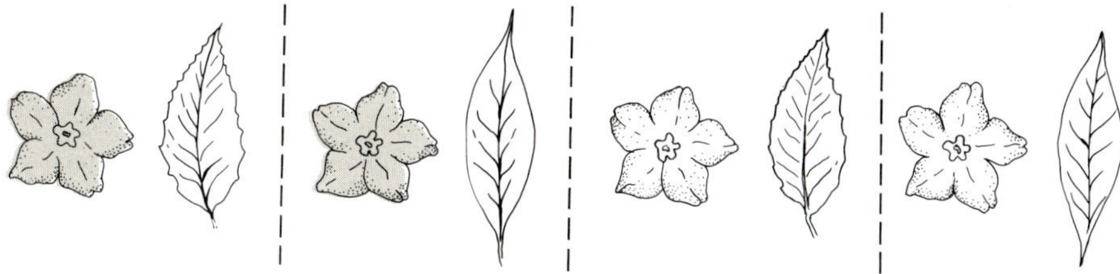

Figure 22–5. A cross between parents of unknown genotypes. From the information presented here, determine the genotypes of the parents.

characteristic disappears in the F_1 generation, it must be recessive. The genotypes of the parents must be *R__C__* (red-flower, cut-leaf) and *rrcc* (white-flower, smooth-leaf). To fill in the blanks in the parental genotype, inspect the F_1 offspring and notice that the white-flower characteristic was present. The only way that a recessive characteristic can show up in the F_1 generation is for both parents to carry the recessive allele. Therefore the parent is *RrC__*. Since the smooth-leaf did not appear in the F_1 offspring, the parent had to be *RrCC*.

Gene Linkage

Peas have a chromosome number of $n = 7$. Since peas have more than seven genes it should be obvious that each gene cannot exist on a separate chromosome. In order to accomodate the thousands of genes that are needed to code for a pea plant, each chromosome contains many separate genes. Mendel has variously been credited with being lucky or having amazing foresight

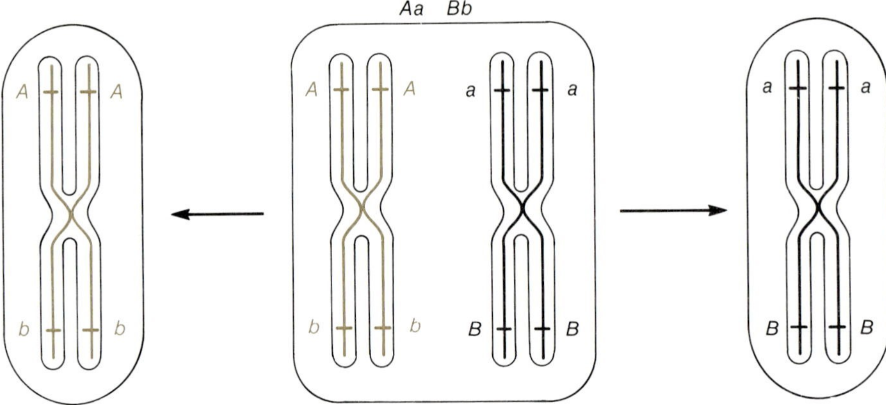

Figure 22–6. Gene linkage. During meiosis, genes on the same chromosome will move together and therefore will be linked.

because the characteristics he followed were on a different chromosome and thus were transmitted independently from generation to generation. Less charitable individuals have suggested that Mendel just disregarded those instances where a pair of characteristics did not conform to the pattern set by the other characteristics. Regardless of how he did it, Mendel did in fact report results for genes that were not linked. How does gene linkage affect the inheritance of characteristics? During meiosis, genes on the same chromosome will not segregate, but rather will stay together during metaphase I and anaphase I. Therefore, in a cross of two homozygous parents (*AAbb* × *aaBB*) the F_1 generation will have the expected genotype (*AaBb*) but the F_2 will not show the expected 9:3:3:1 phenotypic ratio. As can be seen from Figure 22–6, meiosis in the F_1 progeny will only produce two gamete genotypes (*Ab* and *aB*) since *A* and *b* are on the same chromosome and cannot segregate during meiosis. In practical situations where a plant breeder is studying the inheritance of various traits, gene linkage will show up as above, that is, two or more traits will stay together from generation to generation.

Crossing Over

One of the functions of sexual reproduction is to allow the reshuffling of the genetic information, thereby improving the chances of species survival (Chapter 23) by maintaining genetic diversity. If, however, large blocks of genetic information always stayed together because of linkage on the same chromosome, genetic diversity would be greatly reduced. This does not happen because of crossing over. During prophase I of meiosis, homologous chromosomes exchange portions of their genetic material (Figure 22–7). Consequently, in the F_1 generation of this cross there will be a small number of *AB* and *ab* gametes produced by crossing over, in addition to the large number of *Ab* and *aB* gametes produced by cells that have no crossing over

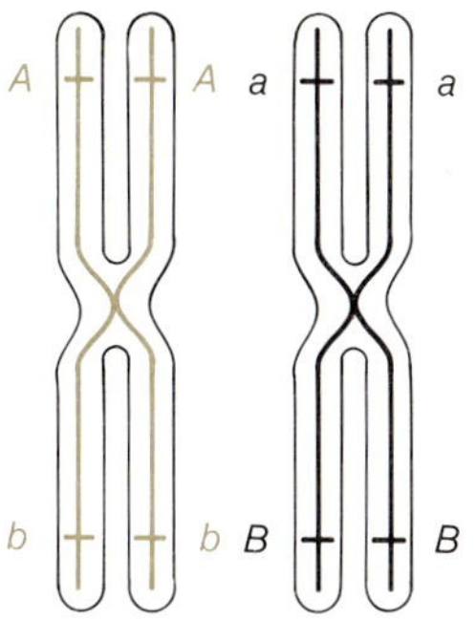

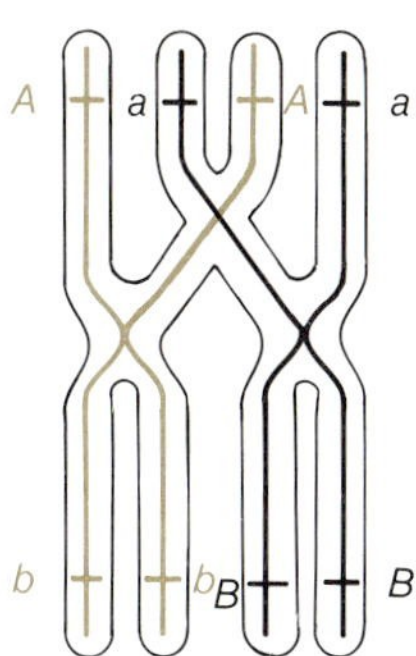

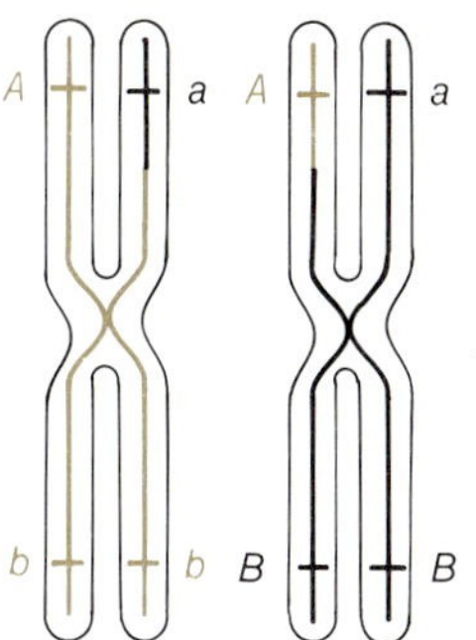

Figure 22–7. Crossing over during prophase I of meiosis. In the region where the chromosomes are in contact, the DNA molecules are cut and re-spliced enzymatically.

in that region of the chromosome. The frequency of crossing over and **recombination*** is a function of the distance between two gene **loci.*** The closer two genes are on the chromosome, the less likely it is that crossing over will separate them. Genes at opposite ends of a chromosome are likely to be separated if there is any crossing over at all in that chromosome. Information gained from crosses involving linked genes has allowed **geneticists*** to map the position of genes on a chromosome (Figure 22–8).

Recombination
The appearance of new combinations of alleles as a result of crossing over.

Locus
(*pl.*, loci) The physical location of a gene on a chromosome.

Geneticist
A scientist who studies genetics.

Incomplete Dominance

As can be seen from the preceding sections, the simple picture developed by Mendel has been complicated by subsequent discoveries. Another of these is the phenomenon of incomplete dominance.* Actually, incomplete dominance, or blending, is what most people expect when they cross two plants or animals. How many times have you heard (or said) that a puppy, kitten, or baby has characteristics of both of its parents? The implication is that the combination of genes from the two parents has resulted in a blending of characteristics—it has, but not due to the blending of individual pairs of alleles. Instead, the offspring show dominant characteristics from both parents. True blending, or incomplete dominance, occurs when neither of a pair of alleles dominates the other, so that the phenotype for this gene is a blend of the two characteristics (Figure 22–9). If a red-flowered plant were crossed with a white-flowered plant and the offspring were pink-flowered, this would be an example of incomplete dominance. In this situation the same type of symbolism is still used to represent the alleles but it is purely arbitrary which one is written in capitals. For this example the red-flowered parent could be *RR*, the white-flowered parent *rr*, and the pink-flowered F_1 progeny *Rr*. Or the red-flowered plant could be *ww*, the white-flowered *WW*, and the pink-flowered *Ww*. Since neither allele is dominant it makes no difference which set of symbols is used. In the example used, the pink color arises because there is only half as much anthocyanin* (Chapter 2) in the cells of the pink petal as there is in the red. The white allele codes for an inactive enzyme in the anthocyanin biosynthetic pathway so that in *Rr* or

Incomplete dominance
A situation in which neither allele is dominant so that in the heterozygous condition the phenotype is a blend between the two homozygous chromosomes.

Anthocyanin
A water-soluble pigment of variable color present in cell vacuoles.

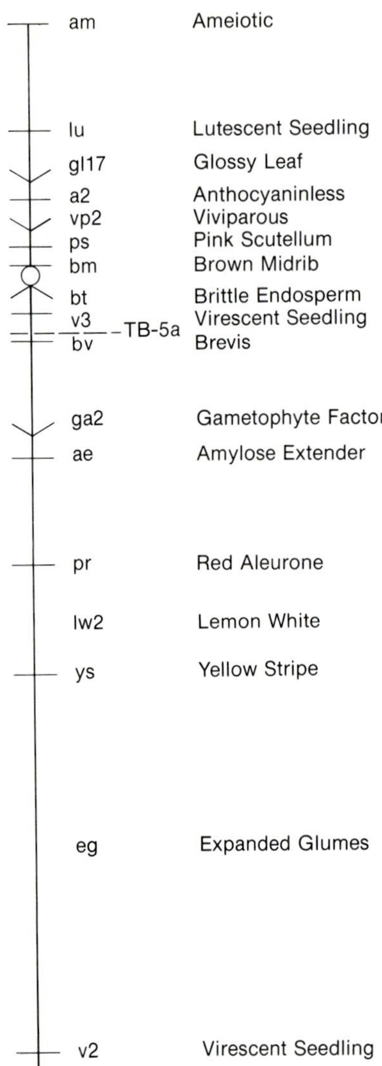

Figure 22–8. Chromosome map. This diagram shows the relative positions of the various genes that have been localized on chromosome 5 of corn. The symbols are the shorthand designation for each gene.

Ww individuals there is only one functional allele for the required enzyme instead of two. When white is dominant over red, the white allele must prevent the formation of an active enzyme from the red allele. A dominant red allele would result in the production of sufficient enzyme to produce enough anthocyanin to give the petals a red color.

Cytoplasmic Inheritance

Occasionally an enzyme or a characteristic will be inherited in a non-Mendelian way. When this happens it is invariably found that the gene for this

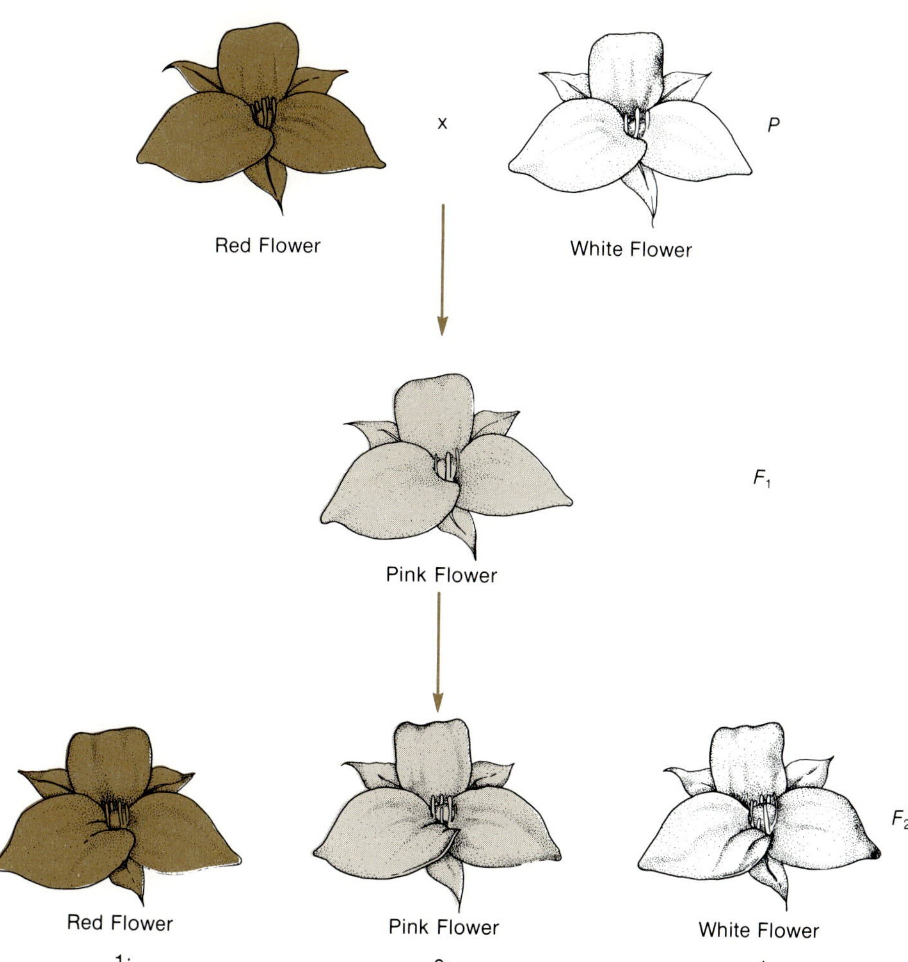

Figure 22–9. Incomplete dominance. In this cross the F_1 phenotype is a blend of the phenotypes of the parents. However, in the F_2 the parental characteristics reappear in their true form.

enzyme or characteristic is localized in the cytoplasm, not the nucleus. Since genes do not float around in the cytoplasm, they must be associated with specific structures. In fact, cytoplasmic genes are associated with plastids and mitochondria, both of which contain unique, circular chromosomes (Chapter 23). During sexual reproduction in many plants, gamete fusion involves the fusion of male and female nuclei but the male cytoplasm does not enter the egg cell and therefore is not passed on to the next generation. All of the plastid and mitochondria of the offspring are received from the mother via the egg cytoplasm, thus exhibiting **maternal inheritance.** Typically, cytoplasmic genes are not of major significance to plant breeders, although they are of considerable interest to plant physiologists in their quest to understand the workings of a cell. In a few instances, however, cytoplasmic genes are significant factors in plant breeding. The first discovery of maternal inheritance and cytoplasmic genes was in the early 1900s in *Pelargonium* (geranium). Subsequent experiments with *Oenothera* (evening primrose) showed that there was an interaction between the plastid **genome*** and the nuclear

Genome
The complete genetic program in a nuclei, plastid, or mitochondrion.

genome. In crosses between *O. hookeri* and *O. lamarckiana* it was observed that *O. hookeri* plastids developed into normal chloroplasts in the hybrid offspring. *O. lamarkiana* plastids, on the other hand, could not develop into functional chloroplasts in the presence of the nuclear genes contributed by *O. hookeri* to the hybrid offspring. Studies of this interaction between nuclear and plastid genes have continued and are an active area of plant physiology research today. It has become obvious that even under normal circumstances plastid genome expression is under nuclear control so that plastid differentiation is appropriate to the cell type. Furthermore there are a significant number of nuclear genes that code for plastid proteins: the carbon-fixing enzyme, ribulose bisphosphate carboxylase, contains protein subunits coded for in the nucleus in addition to subunits coded for in the plastid.

Mutations

Mutation
A change in a portion of the DNA molecule corresponding to one gene, which results in a change in the protein product of that gene.

Mutagen
A chemical or physical agent that causes mutations.

Codon
The three bases in mRNA that specify the amino acid that will be incorporated into a particular position in a protein molecule.

The basis of evolutionary theory is that mutations* in inheritable characteristics occur and are subject to natural selection (Chapter 23). Normally when an unexpected characteristic shows up in a cross it is the result of a recessive allele pair, not a mutation. Under natural conditions however, mutations do occur in both nuclear and cytoplasmic genomes, with the result that new forms of proteins and of characteristics appear in a population over a period of time. Once the mutation has occurred, the altered allele is inherited exactly as any other allele; no distinction is made in the organism between "normal" and mutated alleles. The rate at which mutations occur is a function of the organism and even of the gene itself. Some species seem more prone than others to mutational change in the genome, with the result that the rate of species evolution varies considerably. Plant breeders and horticulturists are always on the lookout for mutations in agricultural and horticultural plants and will induce mutations by use of chemical **mutagens*** or radiation to modify the genome. In this situation the mutagen functions to increase the rate of mutation considerably over the natural rate. In both natural and induced mutations, changes occur in the DNA coding for a particular protein or protein subunit. As a result of one type of mutation, the triplet codon* (Chapter 3) will be altered so that an incorrect amino acid is incorporated into the protein. In other mutations the loss of one base from the DNA molecule can completely alter the way in which the program is read.

If a particular mutation results in some desired characteristic, the plant breeder will maintain this change in the population through either sexual or asexual reproduction. Although mutations are sought by the plant breeder and are an important factor in evolution, most mutations are either harmful or have no obvious effect. In a natural population, then, many mutations place the organism at a selective disadvantage. This leads to reduced reproductive success and eventual loss of the mutant characteristic from the population.

Polygenic Inheritance

Finally, not every phenotypic characteristic is under the control of a single gene locus. A number of physical traits are under the control of two or more genes. An obvious human example is skin color. In plants, traits such as fruit weight or ear length in corn have been shown to be under the control of multiple genes. Traits that are controlled in this way show continuous variation* as opposed to the discontinuous variation* demonstrated by a trait controlled by a single gene. Polygenic inheritance certainly complicates the work of the plant breeder because more crosses are required in order to maximize the gene complement for a particular trait.

Continuous variation
A continuous gradation in the phenotypes of a population between the opposite extremes.

Discontinuous variation
A condition in which the phenotype for a particular trait is at one extreme or the other.

Summary

Genetics, the study of the inheritance of specific characteristics, in its modern form had its origin with the work of Gregor Mendel. Mendel studied the inheritance of single traits and was led to the conclusion that each trait is controlled by two inheritable units (alleles). One of these alleles is dominant and the other recessive. A cross between a homozygous dominant individual and a homozygous recessive individual results in a heterozygous F_1 generation with the dominant phenotype. Selfing an F_1 individual produces an F_2 generation in which 3 out of 4 individuals in the population have the dominant phenotype. The genotypic ratios in the F_2 generation would be 1 homozygous dominant to 2 heterozygous to 1 homozygous recessive. A dihybrid cross followed to the F_2 generation would produce a 9:3:3:1 phenotypic ratio. All of the characteristics reported by Mendel were on separate chromosomes. Inheritance of genes on the same chromosome is complicated by the fact that these genes are linked and cannot segregate during meiosis. As a consequence, the expected dihybrid phenotypic ratios are not obtained. Crossing over can occur between homologous chromosomes, however, allowing a certain amount of segregation of linked genes with consequent recombination of alleles in the offspring. Measurement of the frequency of crossing over between genes allows the construction of a chromosome map.

After Mendel's pioneering work, later investigators discovered the phenomenon of incomplete dominance. This results from the situation where neither allele is dominant so that the heterozygous phenotype is a blend of the two homozygous phenotypes. A further complication to Mendel's straightforward model is cytoplasmic inheritance. Certain characteristics are coded for by cytoplasmic genes carried in plastids and mitochondria. In many plants these cytoplasmic genes are inherited maternally because pollen cytoplasm is not included in the zygote. Mutations of the nuclear and cytoplasmic genomes also add to the genetic variability within a population. In addition, some phenotypic characteristics show polygenic inheritance, with multiple genes controlling the characteristic, thus resulting in continuous variation between individuals.

Review Questions

1. Outline Mendel's experiments and explain how Mendel derived his laws of inheritance from them.
2. What is gene linkage and how does crossing over affect the inheritance of linked genes.
3. Explain the crosses that would be needed to determine whether a characteristic 1. is showing incomplete dominance, 2. is inherited cytoplasmically, 3. results from a mutation or a hidden recessive allele.
4. Consider the following crosses of tomato plants:
   A (red fruited) x B (yellow fruited) several of the offspring (C) from this cross were red fruited; A (red fruited) x C (red fruited) several of the offspring from this cross were yellow fruited.
   Which gene is dominant: red fruited or yellow fruited? What are the genotypes of A, B, and C? Explain your answer.
5. Write out the F_2 genotypes that would be expected from Mendel's crosses if the tall and green seed traits were linked as were the dwarf and yellow seed traits. What would the F_2 genotypes be if there were 10% crossing over in the F_1?

Genetics and the Green Revolution

In 1798, the English clergyman Thomas Malthus pointed out to the world that the human species is potentially capable of outstripping its food supply. He feared that human population growth over the long term would be greater than the increase in food production. According to Malthus, the inevitable result would be starvation, disease, and war, until the human population size decreased to levels that could be supported. Two major approaches have been tried to prevent the predictions of Malthus from occurring. A most obvious way is to slow and eventually stop the growth of the human population. This approach is not of direct relevance in a botany text, although it should be pointed out that this is the only long-term solution. The contribution of botany to solving the food-population imbalance is through increasing the food supply. In spite of the population explosion, recent increases in per-acre crop yields have prevented the Malthusian disaster from reaching global proportions thus far. Much of the credit for this is due to intensive effort directed to crop-breeding programs in Third World countries, where introduction of new, high-yielding dwarf varieties of wheat and rice ushered in the so-called **Green Revolution,** doubling or tripling grain harvest within a few years. This type of breeding actually involves deciding what characteristics are desired in the new strain and then engineering this strain by combining traits from a large number of varieties through selective crossing occurring over many generations. Each new generation is grown and tested for the desired characteristics. Those plants that look the most promising are crossed with other plants that contain additional genes that the breeder wants to incorporate into the final strain.

The strain that is ultimately produced through this process of genetic engineering is a tribute to the plant breeder's art and the short-term answer to food shortages in the underdeveloped countries. Unfortunately the so-called miracle rice or wheat is only a short-term solution. The plant varieties that are the basis of the Green Revolution are fatally flawed. They are essentially genetically uniform. Therefore any increase or mutation in one of the plant disease organisms or any variation in climate will affect all of the plants in the same way and could wipe out the crops over a large geographic area. This was demonstrated most dramatically in the United States with the corn crop in 1970 and 1971.

The goal of the corn breeder is to produce hybrid corn seed with the desired genetic qualities at the lowest possible cost. A major component of this cost is labor. In areas where seed corn is produced, a major source of employment for junior high and high school students is detasseling. Workers on a detasseling crew walk or ride down the rows of parental corn, stripping the male flowers, or tassels, from the top of the plants that have been selected to be the females in this cross. The silks (stigmas and styles) of the female flowers are lower on the corn stem and so are unaffected by the removal of the male flowers. The pollen for the desired cross comes from the male parental line that was planted every fifth row and left untouched by the detasselers.

In order to reduce costs, corn breeders looked for other ways to prevent self-pollination. An obvious solution was to use a male-sterile line. This might produce pollen, but the pollen would be sterile and unable to fertilize the eggs. A suitable male-sterile line (Texas male-sterile) was found and incorporated by selective breeding into the parental lines prior to a hybrid cross. This particular male sterility was a cytoplasmic characteristic. Therefore all parental lines carrying male sterility had the same cytoplasm regardless of the nuclear genes that were being carried. For a number of years this form of cost cutting worked and the number of detasselers needed decreased significantly. However, a new race of Southern corn-leaf blight fungus appeared that was particularly devastating to corn carrying the Texas male-sterile cytoplasm. As the fungus spread, so did the economic disaster, because the majority of the hybrid corn contained Texas male-sterile cytoplasm. Fortunately some hybrid lines did not contain susceptible cytoplasm and the corn yield losses were not as great as they might have been. Now the hybrid seed-corn companies are again using manual labor to prevent self-pollination in seed-corn production and Southern corn-leaf blight is not a major corn pathogen anymore.

The varieties of corn, rice, and wheat that are traditional to Third World countries are wind-pollinated and genetically variable. Thus when changes in disease organisms or climate occur, some of the individuals will survive to provide food and seed for the next season. However, the Green Revolution is changing all of that. Because of the genetic uniformity introduced into the crop varieties, catastrophic famine such as the Irish potato famine of the 1840s (Chapter 16) could happen anywhere in the Third World at any time. Therefore the Green Revolution should be considered only as a stopgap measure in the effort to feed a hungry world. By means of genetic engineering, time has been bought in which population and sociological problems can be addressed; however, the plants of the Green Revolution cannot feed the millions of new mouths that unchecked reproduction will produce over the next twenty years.

Recombinant DNA

Genetic engineering, biotechnology, gene splicing, genetic surgery—these are the newest areas of biology to capture the fancy of the public. If even a few of the claims of the scientists using these techniques were realized, there would be profound changes in all areas of medicine and agriculture. The basis for all of these technologies is the revolutionary development of recombinant DNA. Simply stated, recombinant DNA is DNA taken from one organism and spliced into the genetic material of a bacterium where it is expressed (Figure 22–10).

In practice, DNA is isolated from the tissue or cells in a purified form. Depending on the experiment to be conducted, this DNA could be from any source. Typically the botanist would be interested in using DNA from the

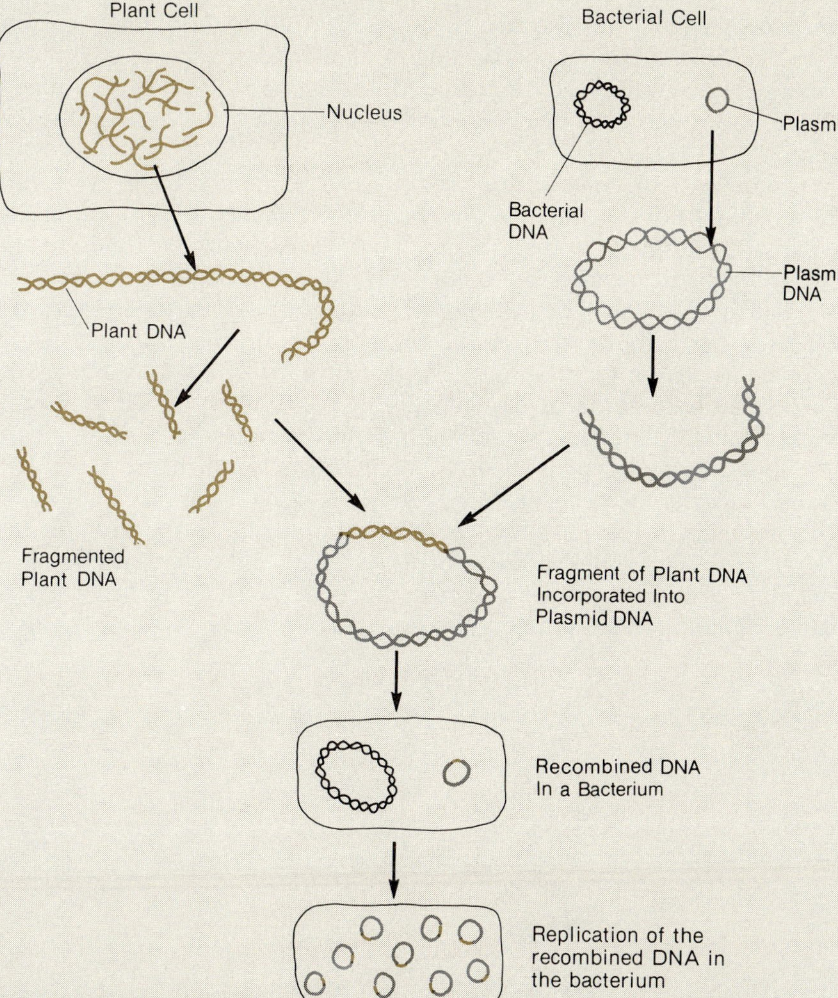

Figure 22–10. Recombinant DNA. In this technique DNA of nucleus, plastid, or mitochondrion is isolated in pure form and then cut enzymatically into fragments. These fragments are fused to plasmid DNA, which is subsequently incorporated into bacteria. Under ideal conditions the bacteria express this piece of DNA as protein.

nucleus, plastids, or mitochondria of plant tissue, or from bacterial or cyanobacterial cells. Once the DNA has been collected, it is digested enzymatically to produce smaller fragments. These fragments are combined subsequently with small, circular strands of bacterial DNA called plasmids. Each type of plasmid carries a small amount of genetic information, but is not part of the bacterial chromosome. Since plasmids can be isolated from the bacteria carrying them, the researcher is able to cut the plasmid DNA, splice the DNA fragments, and reseal the plasmid DNA—all in the test tube with the appropriate enzymes.

Once the modified plasmids have been prepared, standard bacteriological techniques are used to incorporate the plasmids in the appropriate bacteria. At this point, the DNA that has been incorporated into the bacteria is the recombinant DNA. If the technique is to be of any significance to the investigator, this foreign DNA must be expressed in the bacterial cell through the synthesis of foreign proteins. While it has become relatively simple to incorporate DNA from any source into bacteria, it is often difficult to detect the expression of this recombinant DNA as protein. In many cases this could be the result of the DNA fragment not containing a complete gene for the synthesis of a complete protein. In other cases more than one gene would be needed or an additional part of the foreign DNA that controls expression of this gene is required. Recombinant DNA success stories are accumulating however, and with each new achievement confidence in future gains is enhanced.

A major area of botanical interest in recombinant DNA is agricultural. The techniques of recombinant DNA described above would be sufficient if all that were important was the protein itself—an enzyme for industry, a protein for medicine, and so on. For crop plant improvement, however, the recombinant DNA must be incorporated into plant cells. Current research is being done with *Agrobacter tumefasciens,* a bacterium that causes crown gall tumors. The reason for this interest is that tumor formation is induced by a plasmid that becomes incorporated into the plant DNA after infection. Any recombinant DNA that is part of the plasmid should also become incorporated into the plant genome. Considerable effort is now being expended to find and incorporate genes for nitrogen fixation into crop plants in this way. Other investigators would like to improve the quality of the proteins made by some crop plants. Resistance to drought, heat, cold, salty soils, herbicides, and other plant stresses are some of the genetically controlled traits that someday might be incorporated into crop plants.

23

Plant Origin and Evolution

Outline

Introduction

A perennial question that people ask is How did life begin? Scientists attempt to answer this question in the same way that people trace their family trees to determine their roots—by studying various documents indicating who is descended from whom and when they lived. Investigation of the evolutionary origins of organisms leads to the conclusion that today's complex, multicellular plants (and animals) are descended from simple, unicellular organisms. However, in any genealogical search, a point is reached where so much time has passed that vital records are missing or incomplete then the investigator is forced into conjecture followed by searches of secondary records to confirm or deny the various postulated ancestries for a family. So it is with the origin of the unicellular ancestors of all organisms on earth. The clear trail stops with the simple cell, but logic says that the cell had to have its roots in simpler, nonliving chemical systems. Consequently scientists have hypothesized that life began as the result of a long period of chemical evolution that led up to the first cells. Once living cells had evolved, biological evolution followed, giving rise to the wondrous variety of organisms that exist today. It should be stressed from the outset that the scientific theory for the origin of life does not assume any special properties for the chemicals involved—every reaction that is hypothesized to have occurred relies on known chemical properties of the molecules involved. In fact, the reactions that are suggested as having led to the origin of life are all inevitable ones for those molecules under the postulated environmental conditions. The scientific theory of the origin of life as the result of chemical evolution is itself still evolving, with new ideas being tested before their rejection or incorporation into the current version of the theory. In its present form the theory is similar in broad outlines to that being discussed by scientists in the late 1800s but the details reflect current understanding of molecular biology, that is, DNA, RNA, and protein synthesis.

Origin of Life

In order to develop a theory that will explain how life began on Earth, it is necessary to determine fairly accurately the nature of the **primeval earth.** Obviously the primeval earth had to be devoid of all traces of life: this means a planet that had no coal, oil, coral islands, and chalk cliffs; even minerals such as iron ore, which contain oxygen, were absent. To complete this picture of the earth before life, the work of geologists and astronomers must be consulted. These scientists have determined that the earth is about 4.5 billion years old, and that during its formative stages there was tremendous heat and extensive volcanic activity. As the surface of the earth cooled, outgassing from volcanoes released steam that condensed in the atmosphere before falling as rain. In this way the lakes, streams, and oceans formed. Geologists have also determined that the land masses of the earth have changed considerably over the years, both in composition and distribution. Impossible as it might seem, North America is a continent formed from many individual land masses that have been pushed together by **continental drift.** Two hundred million years ago all of the land is believed to have formed one supercontinent, Pangaea, which gradually broke up and the fragments drifted off becoming separate continents and islands (Figure 23–1). Collisions between fragments led to mountain formation such as the Himalayas that are the result of India bumping into Asia. While continental drift

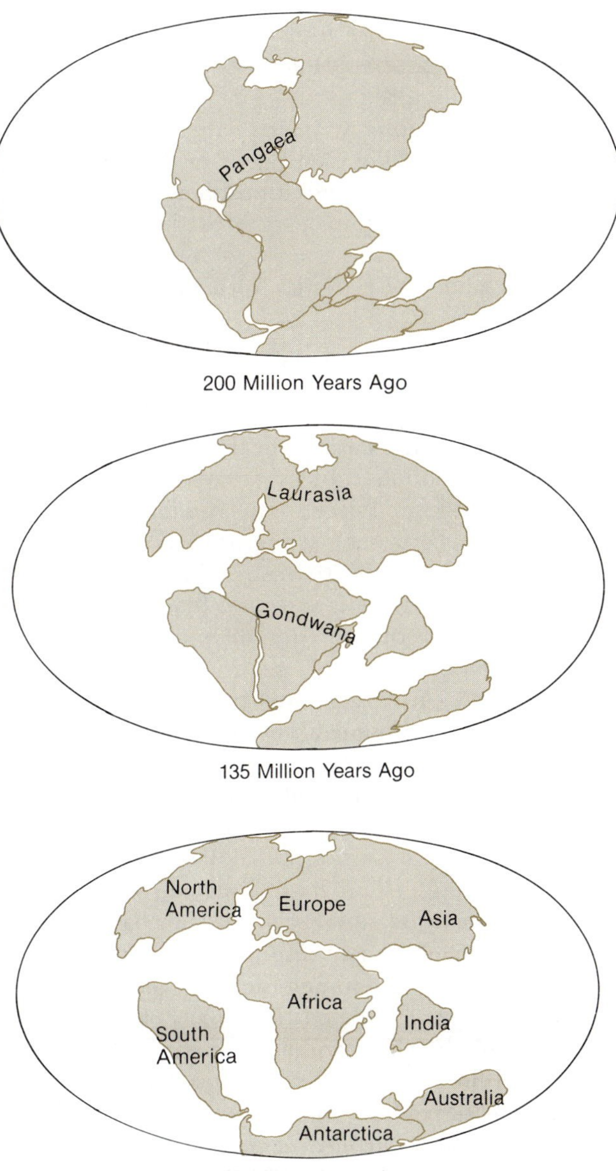

Figure 23–1. **Continental drift.** Positions of the major land masses during the last two hundred million years.

did not influence the origin of life, it is important to realize that when life did begin on earth the present distribution of land masses did not exist, so that no single present region can be investigated as being the possible cradle in which life formed. At the time that life was forming, the most feasible scenario is that most of the planet was land, with only small, shallow seas or streams, and climatic conditions that were much more uniform without cold poles and a hot equator.

The atmosphere of the primeval earth was totally unlike that of today. Although some dispute still exists among various scientists, there is general agreement that the "air" of the primeval earth contained carbon monoxide, carbon dioxide, nitrogen, hydrogen sulfide, and water vapor, but no oxygen. Owing to the geologically young state of the earth, the atmosphere was subjected to violent electrical storms and, because of the lack of a protective ozone layer, to intense ultraviolet radiation from the sun. Added to all of this was the intense volcanic heat and the decay of radioactive elements in the rocks forming the earth's surface. The complete picture of the primeval earth is one of a desolate planet with an environment and climate completely hostile to life. Yet life evolved from this "witches brew" with profound effects on the nature of the atmosphere and the climate. Without the development of living organisms the earth would have remained desolate and hostile. However, it was the very combination of conditions on the primeval earth that made the evolution of living organisms possible—even inevitable.

Chemical Evolution

Living cells are composed of organic molecules such as **proteins, lipids, nucleic acids,** and **carbohydrates** that currently are produced naturally only by the metabolism of living organisms or artificially in a chemical laboratory. Since these molecules are fundamental building blocks of a cell, they must have been present at the time of the formation of the first cell. This might lead one to ask, then, whether the chemical building blocks or the chemical-producing cells were the first to exist. Regarding the origin of life scientists propose that the organic molecules came first, followed by the formation of living cells.

During the initial stages of chemical evolution, the complexity of the organic molecules present on earth would have increased until some form of primitive cell resulted from the correct combination of organic molecules within a small compartment. Beginning with the successful experiments of Stanley Miller in 1953, chemists and biologists have found that the first steps in this chemical evolution could have occurred on the primeval earth. In order for living cells to have formed, organic molecules such as proteins, lipids, nucleic acids, and carbohydrates had to be present. Miller's contribution was to demonstrate that simple organic molecules could be formed from inorganic precursors under the conditions that likely prevailed on the primeval earth. He simulated the effects of violent electrical discharges on the hot gases of the primeval atmosphere and found that a number of simple organic molecules including amino acids were produced. Similar experiments have been done by others using ultraviolet radiation, heat, and even the intense sonic shock wave created by thunder as energy sources; in each case a mixture of organic molecules was produced. None of these conditions produce the complex molecules typical of cells, although the precursors of these molecules are produced. Subsequent experiments were based on the premise that these organic molecules would be washed out of the atmosphere by rainfall and would accumulate in shallow seas and lakes. Here they would have the opportunity to react with each other, forming new and more complex molecules. The drying up of shallow seas and lakes could have exposed the organic molecules to hot, dry conditions and strong ultraviolet radiation. The evidence that has accumulated from this experimentation suggests that prebiotic environmental conditions, indeed could have led to the formation of proteins from amino acids, nucleic acids from nucleotides, and lipids from fatty acids and glycerol (chapter 3). Thus all of the essential ingredients for a living cell could have formed **abiotically** as the result of the effects of environmental conditions on inorganic molecules present on the primeval earth.

The theory of the origin of life must now explain how the essential molecules came together forming the first cell. This question is complicated by the fact that even the most primitive cell now known is so complex biochemically that it is difficult to conceive of a functional cell without its metabolic "necessities" (enzymes or a complete protein-synthesizing system). Yet evidence is accumulating that such a cell could have existed under primeval conditions and that this cell could have grown and reproduced itself. Perhaps even more basic to this discussion is the question of why the cell would form—that is, what is the essential function of a cell? Philosophical debates on the meaning of life do not help answer this question. From the broadest biological perspective, the role of the cell is in the reproduction of the individual and, therefore, the species. It is becoming more and more apparent from molecular biology and biochemistry, however, that the ultimate function of a cell is the reproduction of its **genetic information*** stored in DNA. All of the metabolic functions of a single-celled organism have the singular effect of producing more copies of the DNA, each of which is put into a compartment (a cell) along with the metabolic machinery required to continue making more DNA. Sexual reproduction, when it occurs, is only a variation on this theme in which there is some mixing of genetic information from two cells of the same species before DNA copying occurs again. The rearrangement of DNA that occurs in sexual reproduction increases the chances for survival of the hereditary material of that species. This may have been the basis for the evolution of sexual reproduction in essentially all groups of organisms. In multicellular organisms each mature cell may not be directly responsible for cell division or sexual reproduction; however, by contributing to the survival of the organism, these cells make possible the continued copying of the DNA in **meristematic*** cells or the production of egg or sperm for sexual reproduction.

Genetic information
Coded instructions that are used in the synthesis of proteins.

Meristem
(*adj.*, meristematic) A tissue whose cells are capable of repeated division.

Seen from this perspective, the first cells are hypothesized to have formed by accident, but those that increased the chances of reproduction of a particular DNA molecule survived. These earliest **protocells** probably resulted when nucleic acids became surrounded by a lipid envelope or membrane. This could have occurred easily because the surfaces of the various bodies of water were most likely covered by a lipid layer, just as they are today. Wave action or other disturbances of the surface could have created small compartments, or **vesicles,** with an outer lipid layer surrounding a tiny drop of water containing nucleic acids (and other dissolved substances). Because nucleic acids have the property of being self-duplicating, even in the absence of enzymes, the absorption of nucleotides (precursors for nucleic acids) from the environment through the lipid membrane would have allowed the DNA to copy itself, albeit at a slow rate. Random splitting of larger protocells into smaller ones would have produced, at least some of the time, two protocells each containing identical nucleic acid molecules. It is suggested that the sequence of nucleotide bases in nucleic acid, at this point, was not the coded instructions for assembly of proteins that it became in subsequent cells. Instead, proteins and amino acids were included within the protocell by chance just because they were present in solution along with the nucleic acids, and they played no role in nucleic acid duplication. Proteins are believed to have become vital components of protocells when they facilitated the copying of nucleic acids. According to this hypothesis, abiotically produced polypeptides, which had combined with inorganic molecules that catalyzed nucleotide formation and coupling, provided a competitive edge to the protocell that happened to have accidentally incorporated them. This competitive advantage could have been lost during any division of the protocell. Therefore, any change in nucleic acid function that allowed a nucleotide sequence to become the code for a specific polypeptide would have been favored. To date not enough is known about the properties of nucleic acids and polypeptides to allow a theoretical model to be developed that would explain how this step in the evolution of the cell came about. This gap is not seen as a major problem for the complete hypothesis, however, because even more significant gaps have been closed with the discovery of essential basic facts in chemistry, geology, or molecular biology.

Biological Evolution

With the development of specific protein synthesis using nucleic acid-coded information, protocells would have become cells. These first cells must have been little more than bags of chemicals that could reproduce themselves; but as they would have represented a quantum jump in efficiency, they would have rapidly replaced the less-efficient protocells. At this stage in the development of the cell, no internal compartmentalization would have developed—the cells would have been **prokaryotic** and would have resembled primitive bacteria and cyanobacteria. In fact, microfossils that are 3 billion years old have been examined under the electron microscope, and they appear morphologically similar to bacteria and certain cyanobacterial groups. Geochemical evidence for the even earlier existence of bacteria and cyanobacteria is strong, with most investigators being confident that life existed on earth at least 3.5 billion years ago. Subsequent to the chemical evolution that led to the establishment of the first cells, there followed a period of biological evolution that has continued to the present and can be expected to continue as long as life exists.

The first primitive cells that formed were probably of many metabolic types. Fusion of cells would form new cells with new combinations of metabolic abilities, whereas fragmentation of larger cells would separate and redistribute the different nucleic acid molecules with the result that these small cells would differ metabolically. Survival of a particular nucleic acid molecule would depend on the ability of the cell to maintain itself and to duplicate the nucleic acid. Consequently, there would be **selection pressure** leading to mechanisms that would keep successful combinations of nucleic acid molecules together as a unit such as the **chromosome.** The price of failure would have been death, and the reward for success would have been survival. Also favored would be the development of an orderly process of cell division that produced identical daughter cells. These events had to be accompanied by the evolution of metabolic pathways. All of these primitive cells were **heterotrophs**—they absorbed from their environment the specific amino acids, nucleotides, and other substances that were required to duplicate cell components. During the initial stages of biological evolution, there is believed to have been a plentiful supply of abiotically produced nutrients available for a cell to absorb. Under these conditions rapid growth of all viable cells could occur. As the total number of cells increased, however, there eventually would have come a time when the more desirable nutrients were being used up faster than the various abiotic processes (ultraviolet radiation, electrical storms, volcanic eruptions, etc.) could produce them, leading to enhanced competition among the cells for the limited resources. A competitive advantage that would allow for survival in this situation is the ability to use plentiful, alternate compounds by converting them to those needed to duplicate cellular components. In other words, there would have been selection for cells with more complex types of metabolism. Ultimately some of these early heterotrophs must have developed mechanisms whereby they could use other cells as a source of needed nutrients instead of merely absorbing these nutrients from the environment.

Food shortages and the resulting competition would have selected for any adaptation that would allow a cell to obtain the nutrients required for growth and division. One of the most successful adaptations was the development of the ability to make organic molecules from inorganic ones, particularly CO_2 and H_2O. Organisms with this ability are called **autotrophs** and the process is **photosynthesis.**

Photosynthesis.

The development of photosynthesis freed organisms from the nonbiological, or abiotic, synthesis of organic molecules. Photosynthesis, as it occurs today in autotrophs, is too complex to have originated in one step. Furthermore it could not have originated from the gradual accumulation of nonfunctional bits and pieces of metabolic machinery, until finally all of the parts were assembled. Nevertheless it is difficult to conceive of earlier, less complex versions of the process.

The process of photosynthesis evolved as the result of the addition of one favorable adaptation at a time. Photosynthesis most likely began as a process that activated nucleotides by phosphorylating them, for example, ADP + inorganic phosphate → ATP. This reaction was essential to the inclusion of nucleotides into new nucleic acid molecules, and the duplication of nucleic acid was the prime "motivating" factor behind cellular evolution; consequently, improvements on the initial phosophorylation reaction would have been favored. In a proposed molecular model for this initial reaction, ultraviolet radiation is absorbed by an abiotically produced molecule (a quinone) that, in the presence of an abiotically produced iron sulfide complex, results in the transfer of inorganic phosphate to ADP or other nucleotides. Because this reaction is central to the duplication of the nucleic acid molecule, organisms with this capability would have had a selective advantage over those that had to rely on absorbing activated nucleotides from the environment. For organisms that remained heterotrophs the analogous metabolic advance would have been the earliest stages in the development of cellular respiration (Chapter 10), which also produces activated nucleotides. As photosynthesis accumulated reactions and new enzymes and pigments, it gradually became transformed into the highly organized set of reactions observed in even the most primitive autotroph today.

One of the most important additions to photosynthesis, from the point of view of biological evolution, was the reaction that results in the release of oxygen as a by-product. Oxygen production is not essential to photosynthesis and only occurs because water (H_2O) is split and used as the source of hydrogen atoms for carbohydrate synthesis (Chapter 9). The release of oxygen into the atmosphere had profound implications for biological evolution because, before this, there was no free oxygen in the atmosphere. Until oxygen evolution began 2 to 3 billion years ago, the earth's atmosphere contained nitrogen, carbon dioxide, carbon monoxide, and hydrogen sulfide. Reaction of oxygen with carbon monoxide and hydrogen sulfide removed them from the air, after which free oxygen could accumulate, thus changing the nature of the primeval atmosphere completely.

This change in the gaseous composition of the atmosphere and the presence of dissolved oxygen in the bodies of water where life existed led to two major evolutionary developments: **aerobic respiration*** and the rise of terrestrial organisms. Before oxygen became available, all respiration (Chapter 10) was of necessity **anaerobic.*** Aerobic respiration in its current form is much more efficient than anaerobic respiration; however, this increase in efficiency seems to have been a side benefit of the evolution of aerobic respiration. Surprising as it may seem, oxygen is a cellular poison, and aerobic respiration evolved out of a metabolic pathway that detoxified oxygen. In aerobic respiration the only set of reactions that need oxygen are those that involve the transfer of hydrogen atoms to oxygen, resulting in the conversion of oxygen into water. As the oxygen content of the water in lakes, streams, and oceans increased, the early organisms that were able to take the hydrogen atoms generated in anaerobic respiration and transfer them to oxygen had a pronounced survival advantage.

Aerobic respiration
The metabolic process that, in the presence of oxygen, results in the release of chemical energy from simple organic molecules.

Anaerobic
In the absense of oxygen.

The accumulation of free oxygen in the atmosphere also changed the optical properties of the atmosphere. The pre-oxygen atmosphere was very transparent to the ultraviolet radiation emitted by the sun, whereas the **ozone*** layer formed by free oxygen in the atmosphere is an effective barrier to ultraviolet radiation. Ultraviolet radiation, in the amounts reaching the surface of the earth when life evolved, is lethal to living organisms. Because water absorbs ultraviolet radiation, all organisms were restricted to living below the surface of bodies of water. As oxygen accumulated in the atmosphere, organisms could move closer to the surface of the water. Finally, about 500 million years ago, there was sufficient oxygen present in the atmosphere to prevent ultraviolet radiation damage to plants and animals invading the land. It is after this time that the bryophytes and early vascular plants began to appear in the fossil record (Chapter 18). (If the ozone layer

Ozone
A molecular form of oxygen in which three oxygen atoms are linked, forming O_3.

were reduced to one-third of its thickness today, the increase in ultraviolet radiation would kill most higher organisms.)

The presence of photosynthetically-produced free oxygen in the atmosphere did one final, irrevocable thing—it made it impossible for chemical evolution to begin again. The conditions for the abiotic production of organic molecules included ultraviolet radiation and an oxygen-free atmosphere; in their absence, no abiotic synthesis of amino acids, nucleotides, lipids, and other substances would occur. After photosynthetic organisms "polluted" the atmosphere, the only source of new organic molecules for cell growth and division was biological, that is, photosynthesis. If life had been destroyed at this point by some drastic change in environmental conditions, even the return to the environmental norms for the primeval earth would not have resulted in the reinitiation of chemical evolution because the required atmospheric gases would have been absent.

Origin of Eukaryotic Organization.

The earliest cells that formed had a primitive cell structure similar to that of bacteria and cyanobacteria—a cell structure that lacked internal compartmentalization. In these cells the metabolic pathways are carried out in the cytoplasm and even the genetic information which exists in the form of discrete bodies, or chromosomes, in higher organisms, is not separated from the remainder of the cell (Chapter 15). These cells, also lacking a true nucleus surrounded by a membrane, are referred to as **prokaryotes.** More advanced organisms, however, have compartmentalized metabolism and possess a true nucleus; these are referred to as **eukaryotes.**

The eukaryotic form of cellular organization appears to be associated with the evolution of structural complexity of organisms and the transition from **unicellular** to **multicellular** organisms. But the way in which the required compartmentalization came about is still a matter of dispute. According to some biologists, compartmentalization resulted from the evolution of internal membrane systems. These membranes segregated the metabolic reactions of photosynthesis into a chloroplast, those of respiration into a mitochondrion, and surrounded the chromosomes forming a nucleus.

Most biologists now favor the **endosymbiotic hypothesis** for the origin of the eukaryotic cell. Neither group of biologists has satisfactorily explained how the circular chromosome of the prokaryote became the linear chromosome of the eukaryote, but this step is not central to the explanation of compartmentalization. According to the endosymbiotic hypothesis, the eukaryotic cell evolved from an assemblage of prokaryotic cells that initially lived together in a **symbiotic association** (Figure 23–2). It is further hypothesized that the chloroplast evolved from a cyanobacterium and that the mitochondrion evolved from a bacterium that was capable of aerobic respiration.

The following sequence of events is postulated as leading up to the origin of the eukaryotic cell. Over 1 billion years ago heterotrophs evolved that had a true nucleus, respired only anaerobically, and engulfed and digested other cells as food (much as *Amoeba* does today). Normally the incorporation of a cell into a food vacuole would result in digestion. As the result of genetic changes in either the predator or the prey, however, one such heterotroph was unable to digest an aerobic bacterium. Retention of the living aerobic bacterium would have provided the heterotroph with protection against

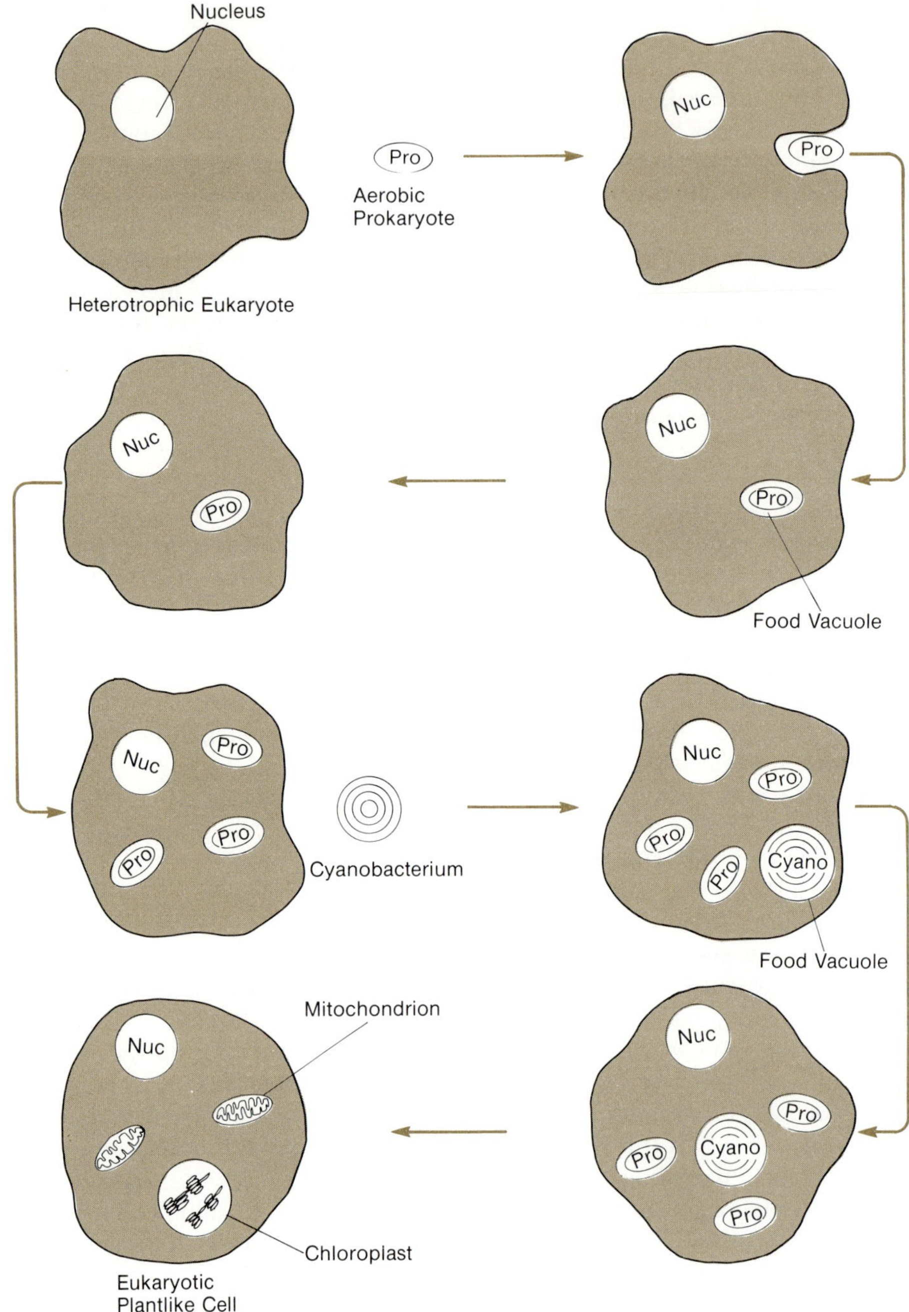

Figure 23–2. Endosymbiotic origin of plastids and mitochondria. The first step in the origin of the eukaryotic cell is believed to have been the incorporation into an anaerobic eukaryotic heterotroph of a formerly free-living aerobic prokaryote. Subsequently, an endosymbiotic organism of this type engulfed a cyanobacterial cell that also became an endosymbiont. A long period of coevolution led to the complete integration of the aerobic prokaryote and the cyanobacterium with the host cell as mitochondrion and plastid, respectively.

toxic levels of oxygen. Therefore the heterotroph with the symbiotic aerobic bacterium could exploit food that was available only in areas with high levels of dissolved oxygen, such as would be found around photosynthetic cells. Because of the survival advantage conferred on the predatory heterotroph as well as on the captured aerobic bacterium, continuance of this symbiotic association would have been favored. Similarly any mutations in the genetic material of either partner that would ensure a longer and closer association would have been selected for. The net result of this selection pressure is the evolution of the mitochondrion, now completely integrated into the eukaryotic cell and incapable of an independent existence. From symbiotic cells such as these, the animals and fungi are postulated to have evolved.

Plants and other eukaryotic autotrophs are believed to have evolved as the result of the addition of yet another symbiont to this initial eukaryotic cell. At the time that the eukaryotic cell was evolving, one of the major forces directing evolution was the food shortage resulting from the rapid increase of cells competing for the abiotically produced organic molecules. A heterotroph, with a true nucleus and mitochondria, that engulfed but did not digest a cyanobacterium would have a captive photosynthetic factory for as long as the cyanobacterium was retained. For the cyanobacterium, as with the symbiotic aerobic bacterium, this forced symbiotic association would have the advantage of reducing its likelihood of becoming prey because of the protection afforded it through its association with the larger heterotroph. Since a cyanobacterium excretes organic molecules as metabolic by-products, which the heterotroph could use, and the host heterotroph produces CO_2 and nitrogen-containing wastes, which the cyanobacterium could use, both partners would benefit from this association. It is postulated that chloroplasts arose from this situation where the cyanobacterium became an integral component of the cell.

Evidence in favor of the symbiotic hypothesis is not all direct because internal details of cellular structure do not fossilize. However, there is considerable direct biochemical evidence, as well as indirect evidence from observations of other symbiotic organisms living today, in support of this hypothesis. Probably the major direct supporting evidence is that both the chloroplast and the mitochondrion have circular chromosomes and protein-synthesizing machinery just like those found in bacteria and cyanobacteria, but different from those present in the cytoplasm of the eukaryotic cell. Indirect supporting evidence includes a symbiotic association formed between the animal *Hydra* and the alga *Chlorella* that produces a photosynthetic *Hydra*. Other supporting evidence comes from the associations between *Paramecium* and algae, and marine worms and algal chloroplasts.

Evolution

The first part of this chapter has considered evolution and the origin of life. The following discussion describes what evolution is (and is not), the processes involved, and how these relate to the evolution of the organisms around us. The modern theory of evolution has its origins in the research and writings of both Charles Darwin and Alfred Russel Wallace, with Darwin being given most of the credit now because of the influence of his book *On the Origin of Species by Means of Natural Selection, or the Preservation of Favoured Races in the Struggle for Life,* first published in 1859. Since the time of Wallace and Darwin, the theory of evolution has undergone revision as the result of new knowledge in all areas of biology; however, the basic principles of this theory remain unchanged. They are:

1. Within any species there exists considerable morphological and physiological *variability,* even when individuals are derived from the same parents.
2. All species have the *potential for rapid increase* in their numbers, yet population sizes are relatively constant.
3. There is a *struggle for existence* between individuals (and species) that are competing for limited resources.
4. This competition leads to *natural selection* whereby the best adapted individuals are successful in the struggle for existence and the least adapted die out.
5. The end result of natural selection is *survival of the fittest.*

Gene Pool
All of the genetic information of the members of a species.

Allele
One of the two or more forms of a gene that codes for a different form of the same protein or polypeptide.

The raw material for evolution is the genetic variability within the **gene pool*** of a species. This genetic variability results in the observed variability among organisms. Through the action of natural selection particular genetic characteristics either can become very common in the gene pool or can disappear completely from it. An allele* for a favorable adaption would enhance the survival of the individual possessing it, and, therefore, the individual would be more likely to reproduce and pass on the allele. Unfavorable alleles would be lost gradually from the population because their carriers would be at a disadvantage in the struggle for existence and would not be as likely to reproduce. Alleles of genes that have a neutral effect on survival in a particular habitat neither will be selected for nor selected against. Because of the random nature of gene recombination during sexual reproduction, neutral alleles can be lost, by chance, from a population over time or, again by chance, they could become part of the genome of the majority of organisms of the species.

Under altered ecological conditions a previously neutral characteristic may take on new significance for the organisms possessing it and become advantageous or disadvantageous. The more demanding the environment, the greater the selection pressure is on genes for critical characteristics. Under conditions where the struggle for existence has been eliminated, all individuals, regardless of their fitness, will live to reproduce and contribute to the gene pool of subsequent generations.

After a sufficient time the effects of natural selection acting on the gene pool will result in the evolution of new species. Oftentimes the development of new species appears to be the result of a gradual increase in the genetic variability between two reproductively isolated populations of a species. However, a number of biologists believe that there are also sudden, major increases in the number of species resulting from drastic changes in conditions regulating natural selection. For example, as a result of oxygen levels becoming high enough to effectively screen out much of the ultraviolet radiation, plants could invade the terrestrial environment. This one change in environmental conditions, it is believed, led to an explosion in the number of plant species, whereas, prior to this time, the number of algal species would increase only slowly.

Changes, or **mutations,** occasionally occur in the genetic information. These mutations alter the structure, and possibly the activity, of the protein resulting from the mutated gene. This provides more variability to the gene pool acted upon by natural selection. Mutations usually result in individuals that are less well adapted. Occasionally a mutation causes a new physiological or morphological characteristic to appear that better adapts the individual and affords it a survival advantage over the rest of the population. Individuals with a deleterious mutation would be at a disadvantage. Furthermore the deleterious mutation gradually will be lost from the population as the disadvantaged individuals are less successful in reproduction or they die as the result of lethal mutations.

A simplified model of plant evolution that will demonstrate the effect of mutation follows. In this model (Figure 23–3) a hypothetical species *A* occupies a range that is restricted by various environmental conditions: at the northern edge of the range the winters are too cold, at the southern edge the summers too hot, at the western edge it is too dry, and at the eastern edge it is too wet. Species *A* is well adapted to the climatic conditions that prevail within its range; however, it is poorly adapted to the extremes that would be found beyond the perimeter of its range. Within its range, species *A* is represented by a population of individuals that differ genetically. Individuals that are less well adapted fall victims to the struggle for existence as members of the same species and of different species compete for limited resources. Each generation there is a reshuffling of the genetic material as a result of sexual reproduction, thus renewing the competition for resources with only the fittest surviving to contribute significantly to the next generation.

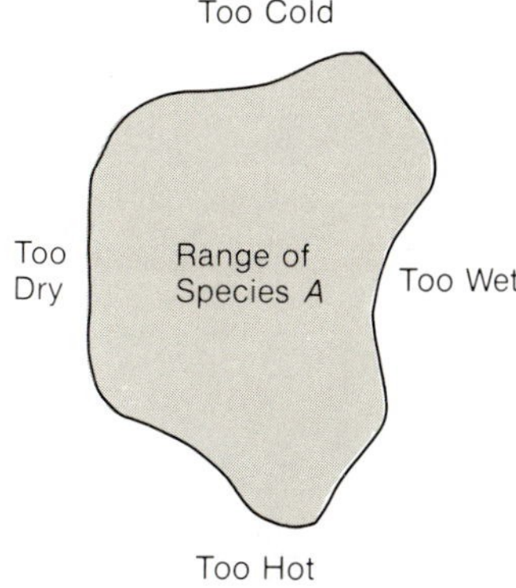

Figure 23–3. The distribution of a hypothetical plant species. In this example climatic factors limit the distribution of species to the range indicated here.

No plant engages in parental care of the young or any sort of deliberate behavior that will ensure that the young are placed only in areas where they can grow. Plants produce large numbers of offspring that are distributed randomly and only those that are fortunate enough to land in a favorable location begin growth. Subsequent survival, after germination and initial growth, is determined by the individual's adaptations to the environmental conditions indicated above. Growth of an individual of species *A* at the boundary of the range does not restrict seed distribution to only areas within the range. Many seeds will be distributed outside of the range. Seeds of species *A* that land outside of the range will not survive to reproduce the species, however, unless they happen to possess an adaptation that permits their survival outside the normal range.

Let us assume in our hypothetical example that, in a plant near the western boundary, a mutation occurs which results in the production of a thicker leaf cuticle. This change, which reduces water loss and allows the plant to survive drier conditions, is passed on through the seeds to the next generation. Seedlings carrying this adaptation will have improved survival chances in the dry areas immediately adjacent to the western edge of the range. If this characteristic were sufficient for survival, some of these individuals would survive to reproductive maturity. Of their progeny only those that carry the mutant gene would be able to survive and form the next generation; that is, there would be selection pressure for this characteristic. Ultimately the geographical region adjacent to the western edge of the original range would be occupied by a specific subgroup of species *A*—those individuals possessing a thicker cuticle.

This same mutation has an equal chance of occurring in individuals anywhere within the range of species *A*. In fact, the allele for thick cuticle will probably be found in a limited number of individuals as part of the normal genetic variation within the species unless this characteristic is lost because it significantly reduces the fitness of the individual. In all other regions this mutation would have no survival value and would not likely become a dominant characteristic of the population. Only in the western regions would this mutation have survival value and therefore be kept in the gene pool by natural selection.

Over a long time mutations and other genetic changes would occur among all segments of species *A,* including the thick-cuticled subgroup. If a change were to occur that prevented the exchange of genetic material between members of the main group and a subgroup, these two populations would be on their way to becoming separate species. A change as simple as one that allows the subgroup to be self-pollinated (Chapter 20) instead of bee pollinated would be sufficient to initiate divergence. During the subsequent **reproductive isolation,** sufficient genetic changes could occur between the two populations that individuals from the two populations, when crossed, would be incapable of producing fertile offspring therefore they would be distinct species. These two species could be similar in their physical appearance or enough genetic differences could have accumulated that they were visibly dissimilar.

The inability of different species to interbreed has been stressed here and in other chapters. However, in certain situations sexual reproduction between species is possible, particularly if the two species are closely related genetically. Through such **hybridization** a new population of individuals could arise. This can be illustrated by use of the same hypothetical model as above. A cross between individuals of species *A* and species *B* at the southern border of the range of species *A* would produce hybrid plants with characteristics of both parent species. If these characteristics allowed survival in this hotter summer climate, a hybrid population would have evolved. Depending on the genetic closeness of the two species, the new hybrid might or might not be fertile. A fertile hybrid could reproduce sexually and, because it would be different genetically from both parents, it would tend to reproduce with other hybrid individuals. Individuals of this hybrid species would then compete with individuals of both parental species. Natural selection based on fitness for the environment would determine which species would be successful. On the other hand, a nonfertile hybrid population also could give rise to a new species if the hybrid individuals were successful competitors and could reproduce asexually.*

Asexually
Without sex.

Polyploidy represents another major mechanism of evolution of plant species. In any population of plants a low frequency of individuals will arise that are polyploid, that is, they contain two or more times the normal diploid* complement of chromosomes. Any polyploid individual with an even number of chromosomes would be able to produce functional egg and sperm and, therefore, be able to reproduce sexually. Offspring from such crosses would give rise to a new species, one sexually isolated from the three immediate ancestral species (Figure 23–4). Again, using our hypothetical model, if a polyploid of species *A* arose anywhere within the normal range of species *A,* it would be isolated reproductively from the rest of species *A.* Because a polyploid is often more vigorous than the normal diploid, this

Diploid
Possessing two of each pair of chromosomes.

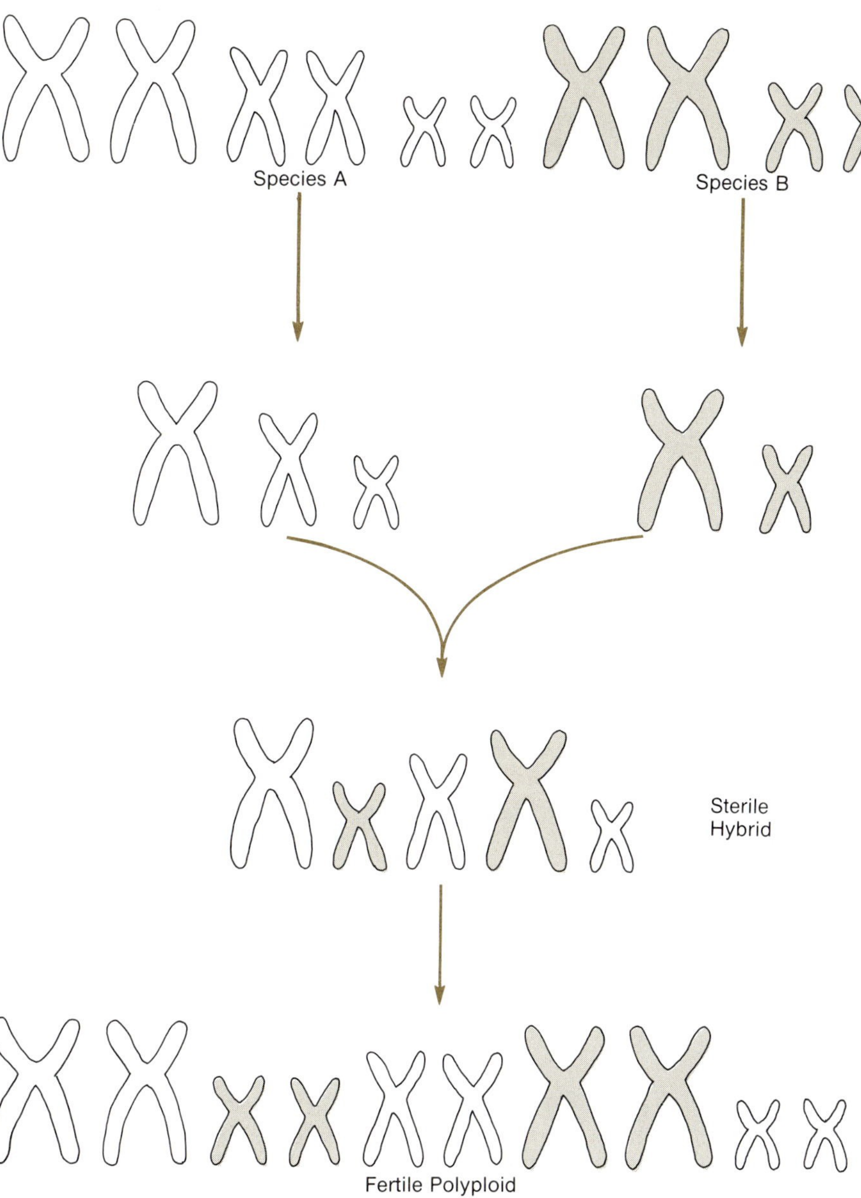

Figure 23–4. Role of polyploidy in speciation. A cross between two related species may give rise to a viable but sterile hybrid. Although this hybrid species may grow vigorously and reproduce asexually, it is incapable of sexual reproduction because all chromosomes exist singly and not in pairs. Spontaneous doubling of each chromosome would produce an individual capable of forming a viable gametophyte generation and therefore of undergoing sexual reproduction. The progeny of this individual would give rise to a fertile polyploid species.

would allow the establishment of a population of this polyploid. Eventually the enforced reproductive isolation between the diploids and the polyploids would lead to two species instead of one. Competition between the two would determine their survival rate and distribution range. The form that has a greater reproductive ability will be the better adapted in its particular environment.

These simplified examples from our hypothetical model demonstrate in a general way how new species are formed—genetic variation affects the struggle for existence which leads to natural selection and survival of the fittest. In times of rapid changes in ecological conditions there will be rapid evolution because natural selection eliminates those individuals that are not well adapted to the new conditions. When conditions change rapidly, on a geological time scale, a considerable number of species can evolve and become extinct within a short time. This results from natural selection associated with the competition for resources. Species become extinct when they are no longer the best-adapted (the fittest) species to the particular ecological conditions that prevail in the region. Ferns and other early vascular plants (Chapter 18) were replaced as the dominant vegetation on the earth because the gymnosperms, and later the flowering plants, were better adapted to use the available resources in many habitats, and were able to compete better than the existing species. Since the struggle for existence is continuing and plants continue to show genetic variation, natural selection will continue to be the driving force behind the evolution of new plant species.

There is no reason to assume that flowering plants represent the final stage in the evolution of plants. As well adapted as the flowering plants are, eventually they will be replaced by an evolving group of new plants that possess some adaptations which allow them to compete better than the flowering plants. The flower, as the organ of sexual reproduction (Chapter 20), is the major adaptation that gave the flowering plants their advantage over earlier groups. Future competitions could be decided on the basis of efficiency of sexual reproduction, on the ability to photosynthesize more efficiently, on the tolerance to pollution, or on some other major environmental factor. Whatever the basis, evolution appears to result in the production of organisms that most efficiently utilize their particular habitat, with inefficiency causing extinction if a more efficient competitor occurs in the same area. In many ways a study of evolution is a study of ecology over a longer time scale; many of the principles governing both processes are the same (Chapter 24).

Evolutionary History of Plants

As indicated earlier, life began with unicellular organisms and evolved into the vast array of complex, multicellular organisms we see around us. The evolutionary history of plants can be traced back to the green algae with certainty; however, the origins of the green algae are unclear because there are no known surviving links between the existing green algae and the first eukaryotic autotroph. Among the green algae the most primitive ancestor for the higher plants is likely to have been a unicellular flagellate species similar to *Chlamydomonas* (Figure 23–5). In the evolutionary sequence leading to the land plants certain events must have occurred. First the cells, with the exception of reproductive cells, had to lose their flagella and become nonmotile. At the same time colonies or filaments of cells developed leading to the formation of multicellular species. Subsequent developments would have included two-dimensional growth followed by three-dimensional growth as is observed in the most advanced green algae. Sexual reproduction was also evolving and changed from the fusion of identical sex cells to the fusion of an egg and a motile sperm.

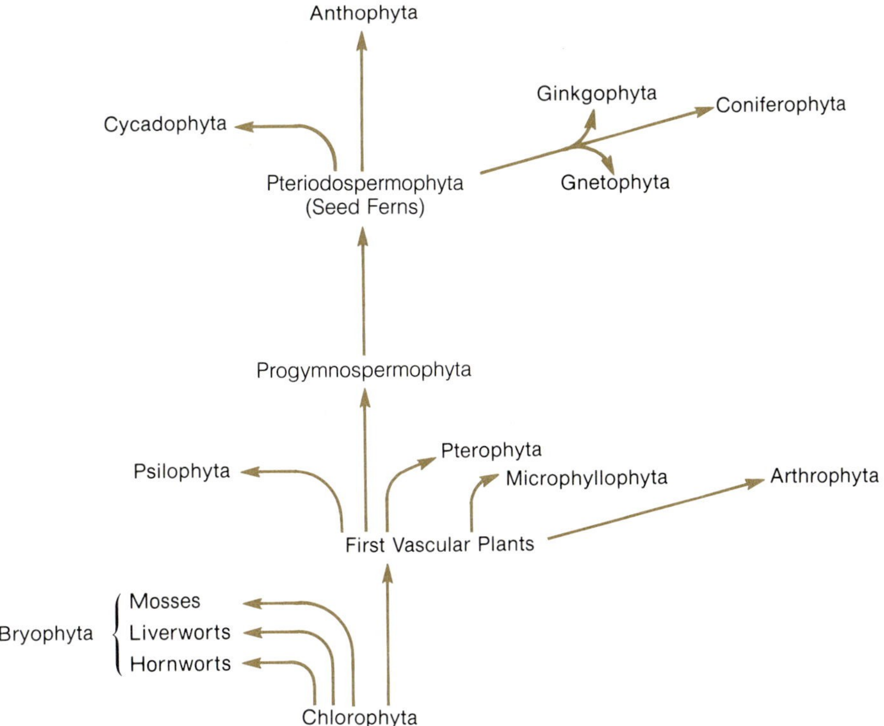

Figure 23–5. Phylogeny of land plants. Evolutionary adaptations to the terrestrial environment led to both the bryophytes and the first vascular plants. Subsequent evolution of these first vascular plants led to the evolution of the remainder of the early vascular plants. One of these lines leads directly to the seed ferns, which have given rise to all of the different kinds of gymnosperms (Cycadophyta, Ginkgophyta, Gnetophyta, and Coniferophyta) and the flowering plants (Anthophyta).

About 500 million years ago conditions on the earth changed as a result of the development of an ozone barrier to ultraviolet radiation from the sun. Now, to survive on land, it was not necessary for organisms to have adaptations to protect them against this lethal radiation. Instead, all that was needed were adaptations that resisted the drying influence of the air. Individual advanced green algae that periodically were exposed to air, because of fluctuating water levels, and, by chance, had a genetic composition that provided any form of resistance to dehydration would have given rise to more progeny than those without this resistance. Thus natural selection would have led to the evolution of a waxy cuticle, of water absorbing structures such as rhizoids, and so forth. Even the evolution of multicellular gametangia* with embryo development occurring within the female gametangium could have been a response to the periodic dry conditions. An embryo that completed its development within the protective layers of the female gametangium, with food and water supplied by the female plant, would be more likely to survive drought than an independent, free-living embryo.

Gametangium
(*pl.*, gametangia) A gamete-producing structure.

The fossil evidence indicates that 350–400 million years ago selection pressure on these amphibious green algae had led to the evolution of two different types of land plants from at least two different green algal ancestors. One group of terrestrial plants was the **Bryophyta,** consisting of **mosses, liverworts,** and **hornworts.** Bryophytes have been an evolutionary dead end to date and, while they have been successful, they have not given rise to more advanced forms. A dramatically more successful invastion of the terrestrial environment was made by the green algal ancestors of the **vascular plants.** Vascular plants show up in the fossil record at the same time as, or

Vascular tissue
Specialized water and food conducting tissue.

Xylem
Water-conducting tissue.

Arborescent
Treelike in appearance.

Gymnosperms
A plant that produces naked seeds, typically in cones.

Gametophyte
The gamete-producing stage in a life cycle.

Megasporangium
A sporangium that produces megaspores.

Sporophyte
The spore-producing stage in a life cycle.

Angiosperm
One of a group of plants characterized by the production of flowers.

Ovule
The structure in seed plants that contains the megasporangium.

slightly earlier than, the bryophytes. Because they possessed true **vascular tissue,*** in addition to adaptations to prevent dehydration, the vascular plants were able to grow upward as opposed to the relatively low-growing bryophytes. The evolution of an efficient xylem* "pipeline" to transport water within the plant made possible the evolution of roots that could penetrate the soil and reach underground water supplies. This, in turn, allowed vascular plants to colonize drier areas away from bodies of water. To support an extensive root system, plants had to have a food-translocating tissue such as the **phloem.** It is important to remember that the evolution of the root system facilitating water absorption proceeded hand in glove with the evolution of the xylem and phloem, and that the role of natural selection was to favor the plants with the best adaptations for survival.

Among the vascular plants each group developed large, arborescent* forms and proceeded to dominate the terrestrial environment (Figure 23–6), then each was gradually replaced by another group that had evolved better adaptations to the environment of that specific time (Figure 23–7). For most of these extinctions of plant groups, it is impossible to determine what crucial selective advantage the new group possessed and the old group lacked. However, the rise of the **gymnosperms*** can be traced to the evolution of the **seed** (Chapter 21).

The seed is a reproductive structure formed when the female gametophyte* is retained within the megasporangium* throughout embryo development. Survival and reproductive success of the female gametophyte is greater in this protected environment than it is for the free-living gametophytes produced by spore-releasing plants. In addition the developing embryo can receive a major input of water and nutrients from the sporophyte* generation that further increases the likelihood of survival. Furthermore, seed producers do not rely on the presence of water for fertilization; instead, sperm is delivered to the egg via pollen tubes. Among the seed producers there also was an evolutionary trend to concentrate the male and female reproductive structures in short branch systems—the **cone** and the **flower.**

The various gymnosperm groups have been replaced by the angiosperms* as the dominant plants of the world. This is most likely due to the evolution of the flower. The flower as an organ of sexual reproduction not only results in greater protection for the ovules* within an enclosed ovary but is more efficient than the cone and, therefore, favored by natural selection. One of the reasons for this greater efficiency is that insects can be attracted to the flower and function as agents of pollination. Insect pollination is more efficient than wind pollination. The flower, as a replacement for the cone, only could have evolved because insects had evolved that could be used as pollinators. At the time that the gymnosperms were evolving there were no suitable pollinating insects, and the only means of pollination was wind. The presence of insects resulted in natural selection favoring the evolution of flowers, which, in turn, favored the continued evolution of insects that could exploit flowers as a food supply while still pollinating them—an example of coevolution of plants and animals. Since the ovules in the flower were enclosed within the ovary, a **fruit** resulted that protected the seed during development and subsequently served as a means of seed dispersal (Chapter 21). Evolution of the various types of fruits with their different means of

(a)

(b)

Figure 23–6. Two forests. (*a*) A forest composed of primitive vascular plants. (*b*) A forest composed primarily of angiosperms.

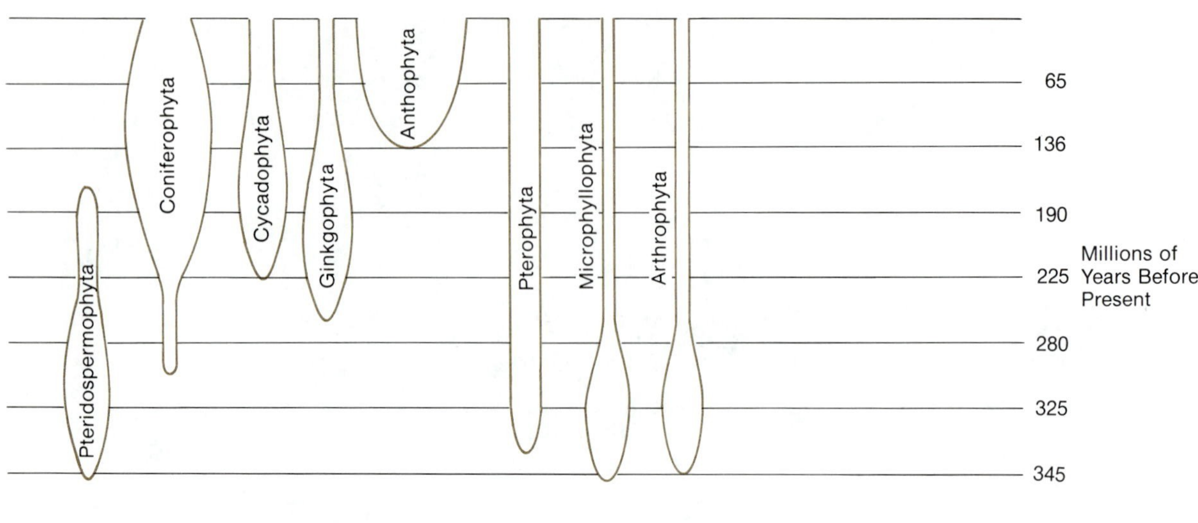

Figure 23–7. Timescale for the origin of vascular plants. Each major group is represented both in terms of time of origin and relative abundance.

seed dispersal has further increased the selective advantage of angiosperms over gymnosperms and has helped push the gymnosperms from their former place as the dominant plant group. What still-to-be evolved plant group will be dominant in 10 million years is impossible to predict, as are the selection pressures that will cause the group to evolve.

Summary

According to the scientific theory for the origin of life, the first cells formed as the end result of a long period of chemical evolution. Four billion years ago the earth's surface likely had large areas of extensive volcanic activity that contributed many gases to the atmosphere and produced water vapor, which, upon condensation, gave rise to shallow seas. The primeval earth's atmosphere most likely contained carbon dioxide, carbon monoxide, hydrogen sulfide, nitrogen, and water vapor, but no oxygen. Re-creation of these conditions in the laboratory has resulted in the formation of amino acids, nucleotides, lipids, carbohydrates, various other simple organic molecules, and ultimately proteins and nucleic acids.

The first protocell could have formed at the surface of the sea as the result of nucleic acids becoming surrounded by part of the lipid surface layer covering the water. Certain abiotically produced polypeptides could have facilitated the copying of nucleic acids, which probably led to the evolution of protein synthesis directed by the information encoded in the nucleic acid.

Microfossils that are 3 billion years old are morphologically similar to the prokaryotic bacteria and cyanobacteria, though even earlier cells most likely would have been heterotrophs similar to most bacteria. The release of oxygen as a by-product of photosynthesis altered the nature of the atmosphere by removing carbon monoxide and hydrogen sulfide as well as allowing free oxygen to accumulate in the air. Aerobic respiration probably evolved out of a mechanism to detoxify oxygen, which is a cellular poison. Terrestrial organisms also owe their origin to free oxygen in the atmosphere as it forms the ozone shield against ultraviolet radiation that otherwise would kill all exposed organisms.

The earliest true cells were prokaryotes. The origin of the eukaryotic cell from the prokaryote is conjectural. One hypothesis states that internal membrane systems just evolved to compartmentalize the cell. A majority of biologists now favors the endosymbiotic hypothesis that states that the eukaryotic cell formed as the result of several symbiotic associations. According to this hypothesis the mitochondrion originated as a free-living bacterium capable of aerobic respiration. Chloroplasts are believed to have originated as the result of a symbiotic association between a heterotroph containing mitochondria and a cyanobacterium.

Modern evolutionary theory has its origins in the writings of Charles Darwin and Alfred Russel Wallace. Darwin's evolutionary principles can be summarized as (1) the variability within a population combined with (2) the potential for the rapid increase in numbers leads to (3) a struggle for existence that results in (4) natural selection causing (5) the survival of the fittest. Rapid increases in selection pressure because of major changes in the ecological conditions speed up the rate of evolution leading either to sudden extinctions or to significant increases in the numbers of new species or to both. Periods of little or no change would favor a slower rate of evolution.

The evolutionary history of higher plants can be hypothesized to have originated from a unicellular green algal ancestor similar to *Chlamydomonas*. Evolution of filamentous, nonmotile species led to three-dimensional body forms as observed in the advanced green algae. Once the ozone layer developed in the atmosphere, advanced green algae could begin to survive on land so long as they had adaptations to prevent dehydration. Because of this selection pressure, two different lines of evolution developed—one green algal ancestor gave rise to the bryophytes and another to the vascular plants. The development of true vascular tissue, xylem and phloem, gave these plants a considerable advantage over the bryophytes. This included the ability to use more effectively the terrestrial environment because roots and vertical growth were possible with vascular tissue. The evolution of the seed in several advanced groups was a significant advance over spore release. In the seed-producing plants, the reproductive structures are concentrated in short branch systems—the cone in gymnosperms and the flower in angiosperms. The angiosperms are the most recent plant group and have replaced the gymnosperms as the dominant plant group because of the greater efficiency of the flower in reproduction.

Review Questions

1. Outline and explain the scientific theory for the origin of life.
2. Discuss the endosymbiotic theory for the origin of plastids and mitochondria.
3. Distinguish between prokaryotic and eukaryotic organisms.
4. Outline the theory of evolution.
5. "Evolution is the study of ecology over a large time span." Discuss this statement.

Evolution of Crop Plants

Corn, wheat, and potatoes are examples of widely cultivated crop plants which, in their present form, do not exist in the natural state. The wild ancestors of these species have undergone such hybridization and selection by human cultivators that they bear little resemblance to the original stock.

The course of evolution is guided by the process of natural selection, in which the organisms best adapted to a given environment or situation produce the most offspring. If a human cultivator, rather than natural forces, is guiding the selective process, the advantageous characteristics will be such qualities as superior yield, good taste, attractive color, ease of harvesting, resistance to pests, and so forth. Many centuries of such human-directed selection have resulted in the complete domestication of many of our most important crop plants. Just as evolution through natural selection can result in the formation of new species, so can human intervention result in the genetic alteration of plants to such an extent that the domesticated types bear little resemblance to their wild ancestors and constitute, in fact, distinctly different species.

The potato *(Solanum tuberosum)* originated in the Andes Mountain region of South America, where a number of species have been cultivated for thousands of years. It is quite likely that *S. tuberosum* arose, either in the wild or in the garden of an Amerindian farmer, as the offspring of an accidental cross between two earlier species of *Solanum.* Although the original hybrid was probably sterile, as the progeny of two different species frequently are, a subsequent spontaneous polyploid from this hybrid stock was fertile. (Since potatoes are generally propagated vegetatively, the sterility of the hybrid was not a problem insofar as cultivation of the new potato was concerned.) Because the polyploid had more desirable agricultural characteristics than the parent stock, Indian farmers selectively propagated the new species. European cultivators carried the selective process even further, though not to the point of creating a new species.

Both cultivated and wild species of wheat *(Triticum)* exist. Within this genus there is evidence for frequent hybridization between species and even with closely related genera. Sterile hybrids that form give rise to fertile polyploids which themselves hybridize with any of the diploid species or with other polyploids. The commonly cultivated wheat, *Triticum aestivum,* could have originated as the result of a cross between a polyploid hybrid and a diploid species. Selection by the farmers in whose fields this event occurred is responsible for the retention of this valuable genetic combination.

The origin of corn *(Zea mays)* is more open to question. After considerable investigation it would appear that modern corn is descended from teosinte, a wild plant with which corn freely interbreeds. In fact, some corn geneticists consider teosinte to belong to the same species as corn, with Amerindians developing corn by selection from natural populations of teosinte.

Modern plant breeders are deliberately creating hybrids between species in attempts to combine their food-producing qualities. The tangelo, a fruit produced as a hybrid between tangarine and grapefruit trees, is an example of a successful experiment. Unfortunately, doubters can also point to *Raphanus-Brassica,* a hybrid between radish and cabbage as an example of a hybrid where the desirable properties of both parents were lost. There is, however, one example of a man-made hybrid that has great agricultural potential. *Triticale* is a new genus of plants created from a cross between wheat *(Triticum)* and rye *(Secale).* After the resulting sterile hybrid was converted to a fertile polyploid, plant breeders have been able to produce many varieties for agricultural testing. Whether these varieties ever become major grain plants, it is interesting to note that the production of the first species of *Triticale* took less than 20 years. Since the only step that the plant breeder accelerated was that of converting the sterile hybrid to a fertile polyploid, it is likely that natural speciation by this mechanism would also occur quickly.

24

Plant Ecology

Outline

Introduction

Throughout this text plants have been treated as if they were growing under ideal conditions and in isolation from all other organisms. This is a necessary approach for understanding the structure and functioning of a plant, but it does not represent the true situation in which most plants are found. Plants live in a world filled with other organisms and are affected by external influences. **Plant ecology** is the scientific study of the relationship between plants and their environment, and the resulting patterns of plant distribution. Casual observation reveals that seeds and spores are distributed widely and randomly, yet even within a small geographic area organisms are not distributed uniformly. Human influence is a major factor determining the distribution of plants over much of the earth's surface.

The changes that would result if human civilization were to be destroyed by disease are aptly described in a science fiction novel by George Stewart, where, in the absence of humans, plant distribution becomes a matter of the survival of the fittest. The description of a cross-country trip, taken by one of the survivors, from California to New York is actually a description of how plant distribution is affected by the environment and how, in turn, the environment is affected by the distribution of plants.

> The broad roadway, unused, showed few signs of change—only roughnesses and a few cracks here and there. Where blown dust had settled into cracks and corners, a little grass was growing, and a few hardy weeds, not many. . . . That afternoon he crossed into Arkansas, and though he knew that state-lines were only imaginary, he suddenly became conscious of another change. Here all the dryness of the plains country was left far behind, and the weather was hot and humid. As a result the growth was everywhere pushing in upon the roads and buildings. Runners from vines and climbing roses already dangled across windows and hung swinging from eaves and porch roofs. The smaller houses looked as if they were shrinking back shyly and beginning to hide in the woods. Fences also were being obscured. There was no longer a sharp line between the road and the surrounding country. Grass and weeds were showing green at every little crack in the concrete; blackberry shoots were pushing in from the shoulders, breaking the clean white line. In one place the long runners of some vine reached clear to the white line in the middle of the pavement, and met others advancing from the other side. . . . The progress eastward had become more and more laborious. Floods, windstorms, and frost had transformed the once open and smooth highways into rough lines of concrete chunks strewn with gravel from washouts, overgrown with vegetation and crisscrossed with fallen tree-trunks. . . . An island within an island, the green oblong of [Central] Park will remain. It has open soil where the rain penetrates. The sun shines upon it. In the first season the grass grows tall; the seeds fall from the trees and bushes, the birds bring in more seeds. Give it two seasons, three seasons, and the eager saplings

are sprouting. Give it twenty years, and it is a jungle of second growth with each tree straining upward to gain light above its fellows, and the hardy natives, fast-growing ash and maple, crowding out the soft exotics which man once planted there. You hardly see the bridle path any more; leaf-litter lies thick on the narrow roads. Give it a hundred years, and you walk in full-grown forest, scarcely knowing that man was ever there except where the stone arch still spans the under-pass, making a strange cave. (George R. Stewart, *Earth Abides,* [New York: Random House, Inc., 1949], p. 62,75,205,261).

Ecological Factors

The **ecological factors** that determine the distribution of plants can be grouped in three categories: 1) **climatic,** 2) **edaphic,** and 3) **biotic.** Plants are exposed to various combinations of factors from each of the groups. The distribution of each plant species is determined by the interaction of these factors with the genetic potential of the plant.

Climatic Factors

The major climatic factors include temperature, precipitation, day length, and light intensity.

Temperature

All aspects of a plant's metabolism are affected by temperature: higher temperatures speed it and lower temperatures slow it. Regardless of whether a plant comes from northern Canada or from equatorial Brazil, metabolism occurs over basically the same range of temperatures, 5° C to 40° C (40° F to 105° F). Even though plants from these different environments respond differently to cold temperatures, because the metabolic pathways of all organisms are fundamentally the same (Chapter 3), they all require temperatures in this range. Plants from arctic regions grow during the warm months. In the protected **habitats*** that they occupy temperatures can approach 27°–32° C (80°–90° F). Arctic plants differ from tropical plants in that they have evolved adaptations that allow them to tolerate extreme cold. Tropical plants, on the other hand, are usually unable to withstand even a light frost. Plants from temperate regions are intermediate between these extremes in their ability to survive subfreezing temperatures. To be able to withstand the extremes of temperature a plant must have special physiological adaptations.

Habitat
An area in which conditions are favorable for the growth of a particular plant or group of plants.

Living plant cells contain 80%–90% water. Freezing temperatures (below 0° C) can damage a plant by causing ice crystals to form in the intercellular spaces surrounding the cells (Chapter 2). Severe water loss from the cells to the growing ice crystals results in the death of the cells. A more rapid drop in temperature can cause ice crystals to form inside cells, damaging cytoplasmic membranes and leading to cellular death. The evolution of physiological adaptations that prevent this damage has enabled many plants to grow in cold climates.

Dormancy
A state of reduced physiological activity.

Angiosperm
One of a group of plants characterized by the production of flowers.

Phloem
Food-conducting tissue.

Cambium
(pl., cambia) A secondary meristem.

Xylem ray
Row(s) of parenchyma cells in the xylem.

Bud
Structure that gives rise to a branch shoot or flower.

Cuticle
The outer, waxy layer produced by the epidermis.

Stoma
(pl., stomata) A pore in the epidermis that is opened and closed by a pair of guard cells.

Dormancy* is a major component of cold survival that has evolved in most plants. Annual plants overwinter as seeds that contain dormant, partially dehydrated embryos which are more resistant to cold temperatures than are the mature plants. Perennials may overwinter as types of underground stems that are protected from extreme cold by being below ground. In trees and shrubs, by contrast, adaptations have evolved that allow the retention of stem growth from previous years. Except for the leaves, much of the shoot system of a tree or shrub is composed of dead xylem and cork cells (Chapter 6); consequently cold temperatures will have little or no lasting effect on these tissues. Cold-resistant leaves have not evolved in many angiosperm* tree and shrub species of temperate and cold regions. Consequently, these are shed each autumn, while the few living cells of the stem (phloem*, cambia*, xylem rays*, and buds*) survive by becoming dormant. These living cells are further protected from cold by adaptations that either prevent freezing or prevent the formation of ice crystals within the protoplast. In many cold-acclimated plants, water in protoplasts can supercool to as low as −40° C before flash freezing occurs. Cryoprotectant chemicals (such as sugars, amino acids, and amino acid derivatives) act both to reduce intracellular ice formation and to stabilize proteins during the water stress that occurs as water is lost from a protoplast to extracellular ice crystals. Anatomical and physiological adaptations have evolved in northern evergreen conifers that allow their leaves to survive subfreezing temperatures and even to carry on photosynthesis when the temperature inside the leaf becomes high enough.

High temperatures can be just as damaging to plant cells. Heat denatures proteins and makes them nonfunctional. Since all enzymes are proteins and proteins also serve structural roles in membranes, chromosomes, microtubules, and ribosomes, high temperatures can disrupt key cellular structures and metabolism by inactivating the proteins associated with them. In addition, the lipid component of cellular membranes is temperature sensitive. At high temperatures the lipids in the membranes of many plants would be too fluid to function as a diffusion barrier. For a plant to survive at the upper extreme of the physiological temperature range, its proteins and lipids must possess modifications that make them stable. These modifications typically involve substitution of different amino acids in the proteins and different fatty acids in the lipids. Although an individual plant may produce modified lipids in response to increased temperatures, the modified proteins are produced at all temperatures. In both cases the genetic ability to produce heat-stable proteins and lipids is an evolutionary adaptation.

At elevated temperatures a plant is also susceptible to increased water loss (Chapter 11), a problem that is compounded in the hot deserts of the world. Morphological and physiological adaptations that have evolved to reduce water loss include thicker cuticles*, modified photosynthetic pathways (Chapter 9), water-storing cells, total absence of leaves and presence of photosynthetic stems, a reduction in the number of stomata*, and various kinds of root systems that either grow into deep groundwater supplies or efficiently absorb rainwater from a wide area around the plant. Some of the same characteristics that better adapt northern species to cold have also evolved in southern species, resulting in their adaptation to high temperatures—loss of leaves during temperature extremes and dormancy of seeds or plants.

In temperate regions, the highest temperatures are usually encountered in open, unforested regions. The plants that grow in these situations have at least some of these adaptations. Two familiar examples of lawn plants that routinely survive summer temperature extremes are dandelion and Kentucky bluegrass. Kentucky bluegrass has a shallow root system and therefore is unable to obtain water as the upper levels of the soil dry out. It survives, however, because it goes dormant during hot, dry weather, and its leaves can wilt and die without causing any lasting damage to the plant. Dandelion, on the other hand, stays green and grows all summer long because it has a deep root system allowing it access to water unavailable to Kentucky bluegrass. Desert plants show more extreme combinations of these adaptations. Cacti, for example, have thick cuticles, water-storing cells, and leaves that either are absent or are modified into spines.

The extremes of temperature in a particular location are controlled basically by latitude. Seasonal variations in temperature at any latitude fall within well-defined limits. These normal limits, and perhaps more importantly, the occasional record-breaking extremes govern the distribution of plants. There are many examples of well-defined ranges of plant species that are restricted by winter temperatures. In fact, the north-south distribution of plants is mainly determined by temperature (Figure 24–1).

This effect of latitude is modified both by large bodies of water and by altitude. In central North America the Great Lakes have a significant effect on temperature extremes. Anyone who has lived in or near Chicago, Toronto, or any of the cities on the Great Lakes will be familiar with weather reports that state lakefront temperatures at least 10° F cooler than those of

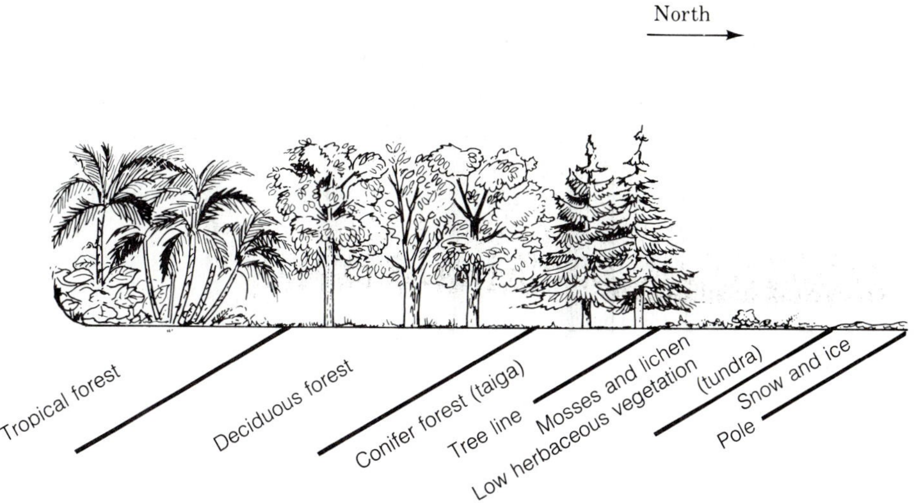

Figure 24–1. Effect of latitude on plant distribution. Seasonal fluctuations in temperature increase with latitude and (ignoring availability of water as an ecological factor) evergreen tropical rainforests give way to deciduous forests that become mixed with conifers and, then, replaced by coniferous forests further north. Close to the Arctic, even conifers cannot survive the harsh winters where they are replaced by tundra.

the inland suburbs in the summer and 10° F to 20° F warmer in the winter. The Great Lakes moderate winter extremes to such an extent that many introduced southern species are able to survive at these northerly latitudes only because of the lake effect. In fact, southern Ontario has had a prosperous peach-growing industry only because of the influence of the lakes.

Western Europe would have an entirely different flora if it were not for the Gulf Stream (Figure 24–2). Instead of winters like those of northern Canada, England and Ireland have winters comparable in severity to those of Virginia or North Carolina. Scotland, which receives less of the effects of the Gulf Stream, has a winter more typical of its latitude in terms of snowfall, but the ocean still moderates the temperatures. Consequently, the **biome*** of the British Isles is the temperate, deciduous forest, as it is for most of Western Europe, instead of the coniferous forest or tundra* found at the same latitudes in North America.

Biome
A combination of plants that is characterized by a particular vegetation type such as grasses or deciduous trees.

Tundra
An assemblage of plants that tolerate extreme winter conditions.

The effect of latitude on seasonal temperature extremes is also modified by altitude (Figure 24–3). Travel to the Rocky Mountains, for example, leads to the observation that above a certain altitude no trees will grow. This zone of demarcation is the **timberline.** Vegetation maps of North America indicate that the timberline should not be encountered below far northern Canada and Alaska (Figure 24–2). However, climbing a mountain is like travelling north: the higher one climbs, the more the climate approaches that of northern latitudes—longer and colder winters during which snow is present for more time. A general rule of thumb is that every four feet of altitude equals one mile of latitude. These alpine conditions, then, result in the presence of **tundra** which "belongs" hundreds of miles further north and has no similarity to the **plant community*** found at the base of the mountain.

Plant community
The group of plants growing in a particular region.

Precipitation

All forms of precipitation are important in regulating the distribution of plants whose cells contain 80%–90% water. Under ideal conditions water is freely available, but for terrestrial plants these ideal conditions are only typical of **rain forests.** In other regions rain comes mainly on a seasonal basis, meaning that plants have to be adapted to dry periods of varying duration.

Distribution of the major vegetation types is determined by a combination of two main factors: temperature and amount of rainfall. As described above, the effect of temperature on plant distribution is readily observed on a north-south line or over a major change in altitude. Within North America, the role of precipitation can be demonstrated on an east-west line. For example, travelling from eastern North America to the Rocky Mountains takes one from a region with precipitation sufficient to support forests to a region where rainfall is only adequate for tall-grass prairie*, then into drier areas with short-grass prairie (Figure 24–2). Within the prairies there are limited areas, particularly along stream and river beds, where sufficient water is available for growth of trees. West of the Rocky Mountains in the United States, many areas receive so little rain that they only support desert vegetation. Plants that live where water availability is a growth-limiting factor must have mechanisms which enable them to conserve the water that they have as well as to reach deep water supplies when the top soil is dry. These mechanisms were listed above under adaptations to high temperatures.

Prairie
A plant community in which grasses comprise the dominant form of vegetation.

Fog and snow are important forms of precipitation in different situations.

Fog is converted to water droplets by contact with objects such as tree limbs and leaves. These droplets then fall to the ground. In California this source of water is vital for the existence of the coastal redwoods, otherwise there is not enough summer rainfall in these regions for the continued survival of these trees and their associated plant community. When the redwoods became established in this part of their range thousands of years ago, the summer rainfall was adequate. With the gradual change to a drier climate, fog has come to play a dominant role in the survival of this community.

The significance of snow as a form of precipitation relates not only to its contribution, as it melts, to surface water and groundwater, but also to its function as an insulating blanket for seeds and small plants. Those people who live in regions with significant winters know that the thermometer alone is not an accurate measure of coldness; the wind velocity has to be considered. For unprotected seeds and plants, the combined effects of low temperature and wind can prove fatal. A major problem for home owners is winterkill that results from dry, cold winds blowing over exposed shrubs including evergreens. If, however, the seeds or plants were covered with snow, they would not experience this drying effect. Furthermore, a layer of snow prevents the ground temperature from becoming as low as the air temperature. For many plants that lack special adaptations for extreme cold, the protection afforded by snow cover is essential for their survival. In fact for annuals and low-growing perennials this insulating effect of snow is more important than its water content in determining their distribution.

Day Length

For plants, just as it is for animals, day length is an important indicator of the season of the year. The value of this information to humans is obvious; however, it is less obvious why plant survival is linked to seasonal responses. The life cycles of many plants have evolved in such a way that their reproduction is tied to a specific season. Since day length is the only climatic variable that is completely predictable and consistent from year to year, plants that reproduce seasonally have evolved a mechanism that measures day length (Chapter 12).

Day length is not only a function of season but also of latitude. More northern latitudes have longer days in the summer than southern ones. A southern plant that is induced to flower as the result of receiving 15 hours of light or less on summer days would not be exposed to days this short until later in the summer at a more northern latitude. Therefore it might have insufficient time to complete the reproductive cycle before the onset of winter. If a plant species were unable to reproduce in a particular location, its ability to maintain a population there would be limited.

Light Intensity

This is not a feature of climate that determines the major vegetation distribution patterns; however, it does determine the distribution of plants within a region. As any nursery catalogue indicates some plants grow best in the shade while others grow best in full sun. There are, in addition, intermediate plants; for other plants, such as tree seedlings, their requirements change as they grow.

Light intensity is effective in breaking up an otherwise homogeneous lo-

Figure 24–2. Major biomes of the world. The distribution of vegetation types (biomes) around the world is mainly determined by a combination of temperature and water availability. Although there are local exceptions to the broad generalizations presented here, this kind of map gives a good indication of the assemblage of plants that would be found on an undisturbed site at any location.

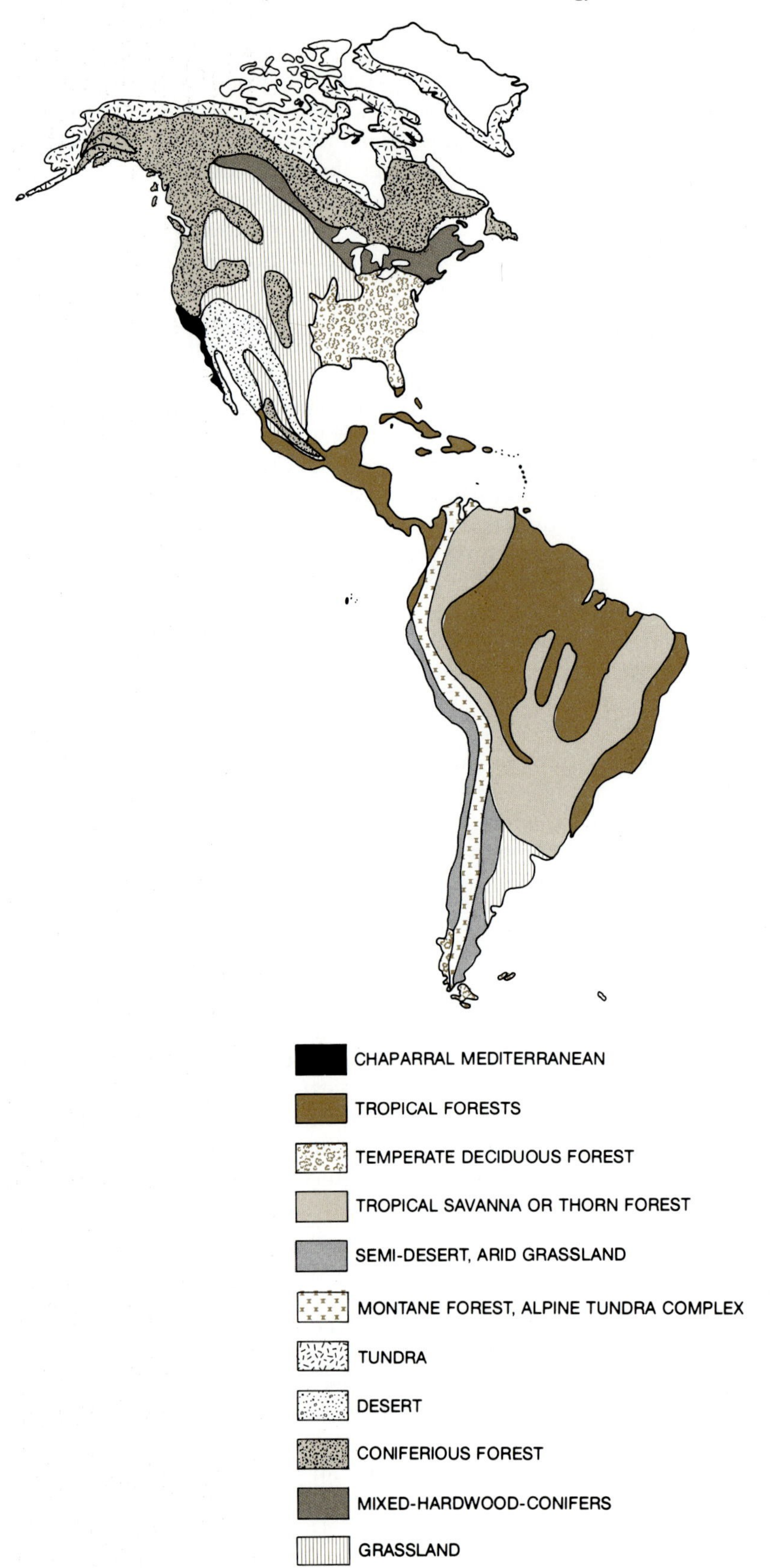

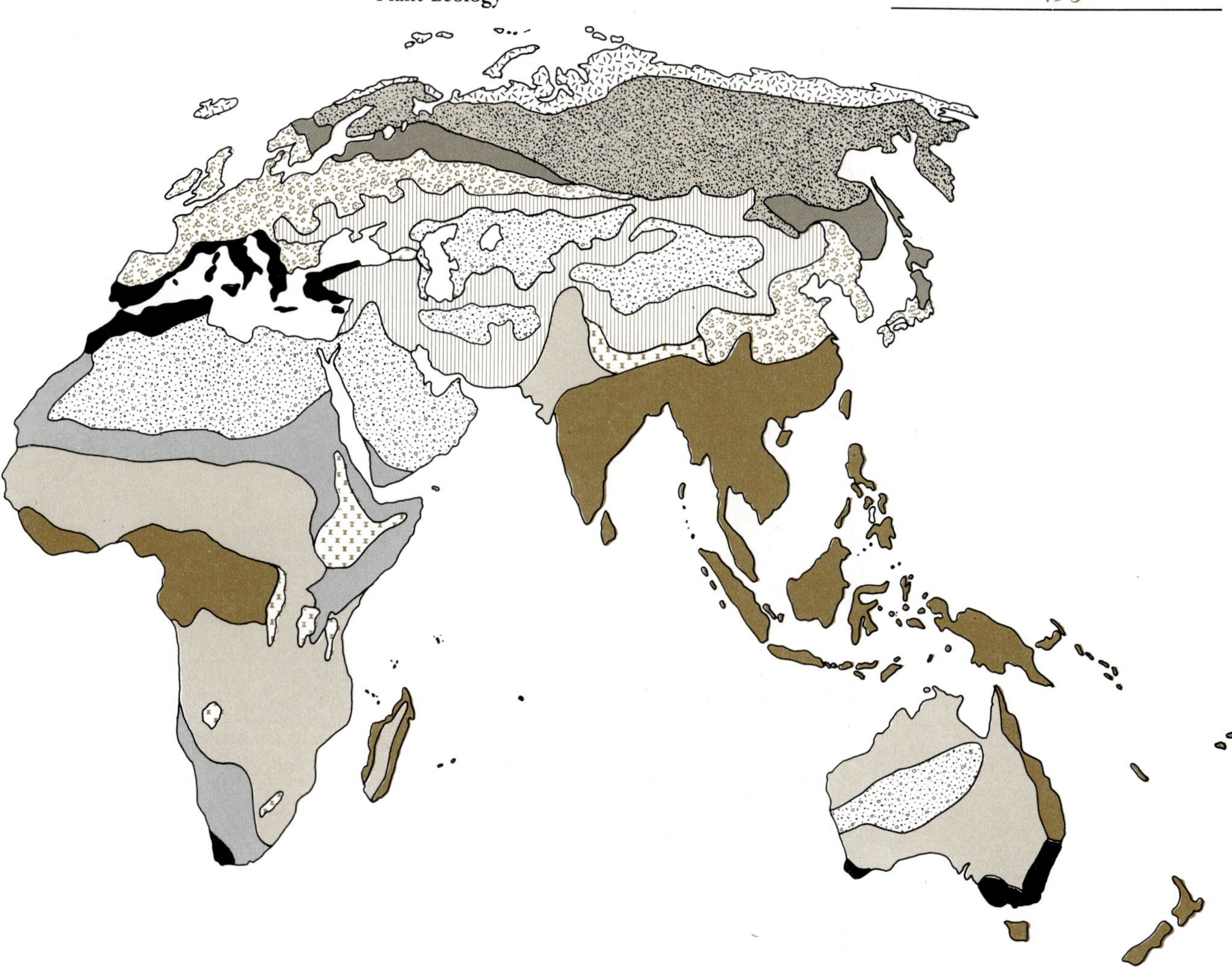

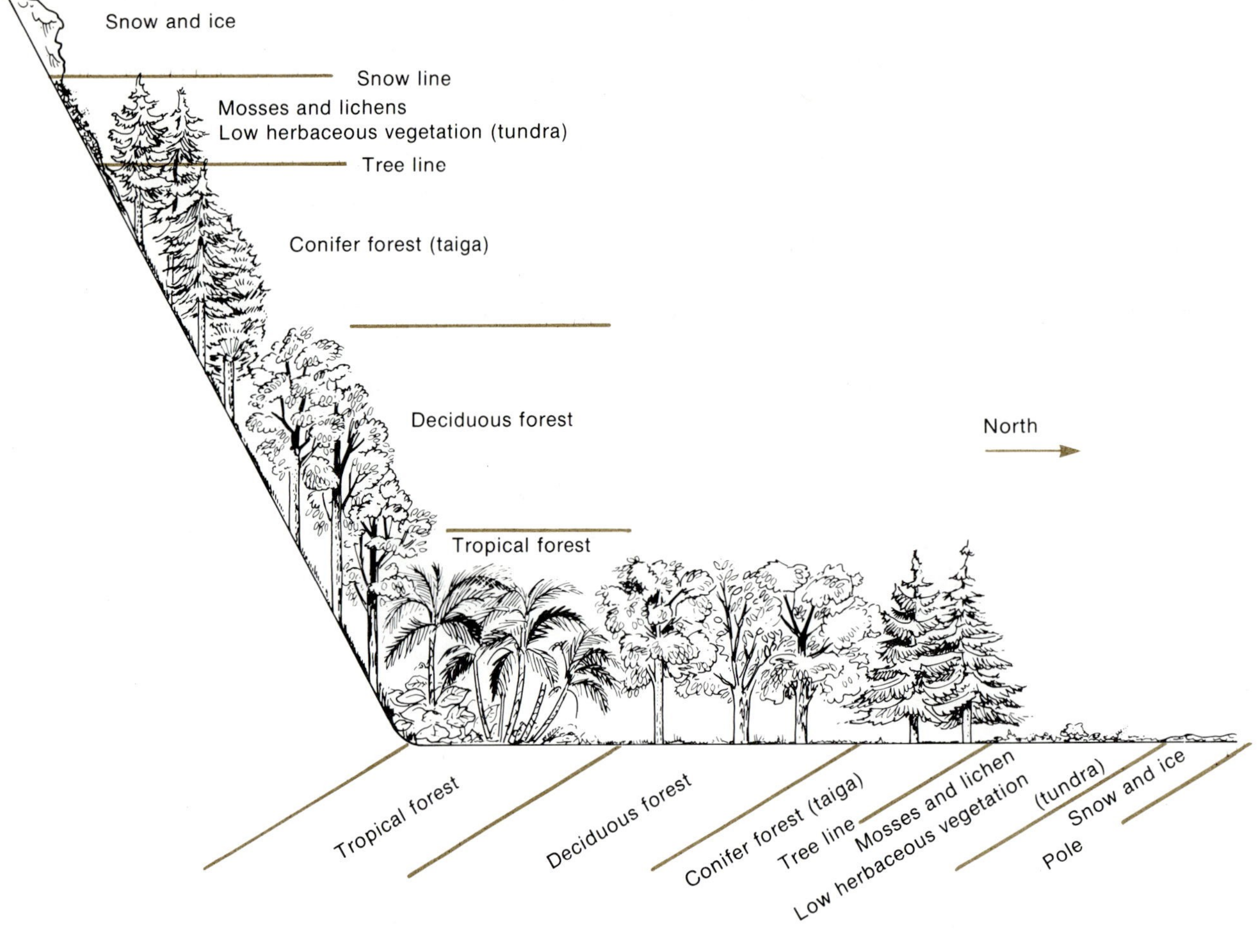

Figure 24–3. **Effect of altitude on plant distribution.** At higher altitudes the vegetation types are similar to those found at more northern latitudes.

cale into many distinct habitats: a clearing in a forest will have different plants than the forest floor (Figure 24–4); the north and south slopes of a hill will differ; the sites at the top and bottom of an overhanging cliff will not have the same plants. Light intensity will not be the only determinant in these situations, however; soil moisture and temperature will also be determining factors.

Edaphic Factors

The ecological factors relating to the soil or other substrates on which plants grow are called edaphic factors. All plants are adapted to a particular type of soil or other substrate, although most plants can grow in a wide range of soil types particularly if other ecological factors are favorable. This is evident in a greenhouse or home garden where the horticulturist often grows different plants in just one type of soil. For the optimum growth of some plants, a particular soil mixture will be needed.

Figure 24–4. A clearing in a forest. The plants found in a forest clearing are typical of those found in open, high-light-intensity situations.

Plants are usually considered to grow in soil; however, some of the more primitive plants, such as mosses and liverworts (Chapter 18), grow on tree bark and bare, wet rock. Liverworts grow well on soil in the greenhouse, but without constant attention they will be replaced by neighboring plants that are better adapted to soil. On bare, wet rock, however, liverworts possess a competitive advantage because the adaptations necessary for survival in this situation have evolved in these organisms.

Visitors to Florida immediately become aware of another group of plants that do not grow in the ground. Spanish moss, which is not a moss but an angiosperm related to pineapple, grows draped over the limbs of trees. This and plants such as mistletoe and orchids that grow on other plants are known as **epiphytes.** Some epiphytic species are parasitic*, however, most are not—they are simply well-adapted for growing in this special environment. Spanish moss is an example of an epiphyte that absorbs all of its water and mineral nutrients directly from the air. Most epiphytes, however, absorb at least some of their water and minerals through a root system that also anchors the plant to the bark of a tree or a layer of dust and **humus*** in the fork of a tree (Figure 24–5).

Parasite
An organism that obtains its food from another organism.

Humus
The partially decomposed remains of plants and animals.

For plants that grow in soil, the nature of the soil will affect seed or spore germination as well as root and rhizome growth. Differences in soil structure*, texture*, and chemical composition result in differences in acidity, water retention, availability of minerals (both toxic and beneficial ones), degree of aeration, and types of soil organisms. Different plants favor different soils: for example, prickly pear cactus grows in sandy, well-drained soils; venus-flytrap grows in water-logged, swampy soils; conifers grow in acid

Soil structure
Arrangement of soil particles.

Soil texture
Relative proportions of the different-sized soil particles.

(a)

(b)

Figure 24–5. Epiphytes. (*a*) Spanish moss. This pineapple relative grows draped over tree branches, power lines, and any other similar support. Because this plant does not grow in soil, a root system is absent. (*b*) Grass. Although this is not thought of as an epiphyte by most people, grasses, like many other plants, will grow as epiphytes. These plants, however, require the presence of at least a thin layer of soil for root growth and the absorption of water and minerals.

soils; certain maples grow in nonacidic soils; and so on. In each case, the plant is best adapted to one type of soil. Most plants, however, can grow in a variety of soil types. Some epiphytes, such as Christmas cactus grown commonly as a houseplant or greenhouse plant, are at home on soil as well as on trees in the tropical rain forest.

Biotic Factors

A plant does not live alone in a physical environment—it constantly is interacting with other organisms and this interaction influences the distribution of plants. Biotic ecological factors include not only interactions between organisms, but also the modifications of the physical environment caused by living organisms. For instance, the intensity of light at the base of a tree is reduced by the tree's many leaves. The reduced light intensity, or shading, causes a lowering of temperature at the base of the tree. Thus we can see that the mere presence of a plant, in this case a tree, is going to affect the microenvironment around itself. Consequently, the ground cover in a forest will consist of shade-loving, or shade-tolerant, plants that are adapted to cooler summer temperatures. The soil in a forest will be moist because reduced summer temperatures result in less evaporation from soils and plants. Therefore, plants associated with a forest do not have to be adapted to withstand as much water stress as a plant growing in an open field.

Winter extremes are also reduced by the presence of trees. All trees reduce wind velocity, which reduces both winter kill and drifting of snow. In addition to the snow that can build up, leaf litter covers seeds and the bases of small plants providing further protection against the cold.

Finally, leaf litter and other plant debris that becomes incorporated into the organic component of soil has profound effects on soil characteristics. Organic material in the soil increases water retention, improves soil structure, and serves as food for small animals, such as earthworms, that by their tunneling activities aerate the soil and promote root growth. In addition the organic material serves as a reservoir of the mineral nutrients required by plants in their growth. Since these nutrients are bound to the organic material, they are not leached out by water passing through the soil. Soil organisms, including bacteria, fungi, earthworms, and other animals, break down the organic material thereby releasing the minerals. Absorption of these minerals by the roots of plants recycles the minerals and promotes plant growth. Differences in the rates of mineral recycling between soils will have significant effects on plant distribution. In temperate, deciduous forests decomposition of plant debris occurs at rates that permit the growth of an extensive groundcover of shrubs and herbaceous plants. Coniferous forests, however, are characterized by the reduced amount of ground cover. One of the contributing factors is that needles from conifers decay slowly, acidify the soil, and tend to build up in thick layers. In addition conifers accumulate and retain minerals which makes the soil nutrient-poor and restricts the number of groundcover species that can become established.

The actions of animals are also able to affect plant distribution. Autotrophic plants are the only organisms capable of producing food from inorganic materials. Therefore heterotrophic organisms must feed on plants or on other heterotrophs that have fed on plants. This observation led ecologists to the concept of **food chains** (Figure 24–6) with plants at the bottom being eaten by **herbivores;** further up the chain the herbivores are eaten by **carnivores.** Of concern to the botanist is the influence that the distribution of carnivores has on the distribution and population size of the herbivores because plant distribution is profoundly affected by herbivore population size and distribution, as any observation of a cow pasture will demonstrate. Cows eat grasses and certain other plants such as tree seedlings, but avoid thistle, milkweed, and other noxious plants. Herbivore pressure thereby determines the plant community found in the pasture. Similarly rabbits, caterpillars, birds, and humans influence the distribution of plants by their se-

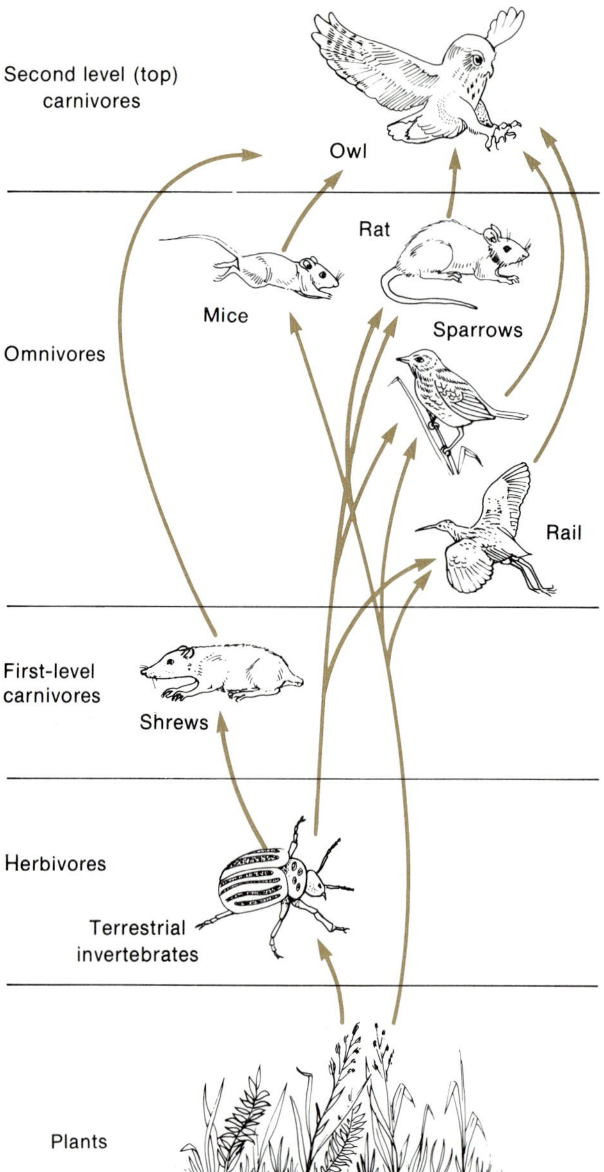

Figure 24–6. Food chain. Plants are the ultimate source of food for all animals. Herbivores eat plants and they, in turn, are eaten by carnivores that may be eaten by larger carnivores. Omnivores, such as mice or humans, eat both vegetation and other animals.

lective feeding habits. In the absence of carnivores, the herbivores could completely alter the type of plant community found in a particular location, just as goats have done when released on islands where there were no predators. Under natural conditions the herbivore population usually does not expand unchecked: predator populations increase as the prey populations increase. This, in turn, reduces the ability of the herbivore to completely alter the plant community.

Animals further influence plant distribution through their role in plant reproduction. Pollination of many angiosperm species is accomplished by birds and insects (Chapter 20). Sometimes the relationship is so close that pollination will not occur in the absence of a specific pollinator. In these cases, plant distribution is restricted to locations within the range of the animal pollinator. Seed dispersal is also accomplished through animal mechanisms in many cases (Chapter 21). Here also the absence of the animal agent could either restrict the rate of spread of the plant (for example, burs would fall only at the base of the parent plant and would not be carried away) or else prevent reproduction entirely (for example, seeds that need to be scarified* by passage through an animal's gut would not germinate in the absence of the animal).

Scarification
(adj., scarified) Mechanical or chemical damage to a seed coat.

Many other examples of biotic factors and their influence on plants and the physical environment could be elaborated; however, further discussion goes beyond the scope of this book.

Genetic Factors

The nature of a plant and its response to its environment is programmed into the inherited information contained in the chromosomes. The actual genetic makeup of a species is the result of evolution (Chapter 23). Through evolution, adaptations become incorporated into the genetic program if they convey survival advantage on the individuals possessing them. However, the genetic program of the species only lays down broad outlines within which individuals may develop. Individual variations in genetic information as the result of sexual reproduction (Chapter 13) produce a range of individuals in terms of environmental responses. Furthermore, two genetically identical individuals can differ significantly in appearance as a result of growing in different microenvironments within the species range.

Plant Distribution

From the preceding discussion of the ecological factors determining the distribution of plants it can be concluded that (1) there is a very complex interaction of factors, (2) there are a variety of special **niches*** within each habitat, and (3) plant species have a natural range encompassing the ecological factors to which the species is best adapted. For each set of environmental conditions (ecological factors) there is one plant community that is the best adapted and which will be found in this habitat (Plate 12). All members of this community will not be responding to exactly the same ecological factors. By their very presence, some members of the community will have modified the physical environment in such a way as to create a series of microenvironments that will be advantageous to other species. Each individual species in a community is the best adapted to its particular set of environmental conditions. Its place in the community was achieved as the result of a long process of **competition** and **natural selection** (evolution). If another plant species with similar requirements were introduced from another continent or other geographically distant area, the two species would compete with one another until the better adapted species replaced the less well adapted (survival of the fittest). This competition involves all stages of the life cycle, not

Niche
The unique set of ecological factors to which an organism is best adapted.

just vegetative growth, so that the "winner" does not have to be more efficient in photosynthesis or mineral uptake, for example. In an extreme case, the competitive edge could be that one species produces more seeds that germinate earlier. Thus seeds of the second species are not as likely to become established, with the result that after a few years the second species would have died out in that particular region.

Australia is an ideal place to document this competition between native and introduced species because this continent has been separated from the other continents for so long that there are many niches occupied by native Australian species that are occupied by different species elsewhere. When *Opuntia* (prickly-pear cactus) (Figure 24–7) was introduced to Australia, it soon escaped from cultivation by animal dispersal of its seeds. Unfortunately for the Australian sheep herder, *Opuntia* turned out to be better adapted than many of the native forage plants to the dry, short-grass areas where

Figure 24–7. *Opuntia.* This cactus spread extensively in Australia until the introduction of a specific herbivore that reduced its reproductive efficiency.

sheep were grazed and soon replaced them. Since sheep will not eat *Opuntia,* the range land was rendered useless for grazing and *Opuntia* became a serious pest. However, its competitive advantage was almost completely eliminated by the introduction of a new biotic factor—a moth whose larvae feed specifically on *Opuntia* flowers, destroying them before seeds can be set. Now *Opuntia* is found but rarely and the native plant community has recovered.

Dynamics of Plant Communities

The group of plant species that occupies a particular habitat constitutes a plant community. This does not mean that the plant community you see is a stable feature of that habitat or is necessarily the best adapted group of species for that area over the long term. The plant community is a dynamic feature of a region and changes in response to changes in the ecological factors. Past major climatic changes, associated with changes in global weather patterns, glaciation and continental drift, have certainly caused changes in the distribution of plants as well as causing extinction of species and evolution of new ones. Since the major climatic features of a region undergo only slow change and since the physical nature of soil also changes slowly in the absence of climatic changes, the major influence that causes changes in the plant communities must be biotic.

Plant ecologists have observed that, over time, a given habitat is frequently occupied by a sequence of different plant communities. The explanation for this process of **succession** is that the members of each community by their presence alter the local environment in such a way that the new conditions favor the growth and survival of a different set of species. This is not a deliberate setting of the stage for the new plant community and there is not a rapid and complete change of plant species as one community succeeds another. The transition from one community to the next is accomplished as the result of competition among the members of each community for space, light, water, and so forth. The plants that survive constitute the community. It is possible for a species to be a member of several plant communities in the succession sequence as the result of being the best adapted species for the particular niche that it occupies. Furthermore, chance plays a considerable role in determining the species composition of a given plant community. Chance, and not climatic or edaphic conditions, determines which seeds and spores will actually land in a particular habitat. Many plant species can grow under a variety of environmental conditions in the absence of competition. Therefore the chance establishment of one species before a better-adapted competitor could preclude the growth of what would normally be considered the best adapted species for that niche.

The following example of succession in a plowed field in eastern North America (Figure 24–8) should clarify these principles. After a plowed field or other disturbed site is abandoned, it will be invaded by pioneer plant species which are adapted to open habitats subject to the extremes of the weather. These pioneer species are typically annuals or short-lived perenni-

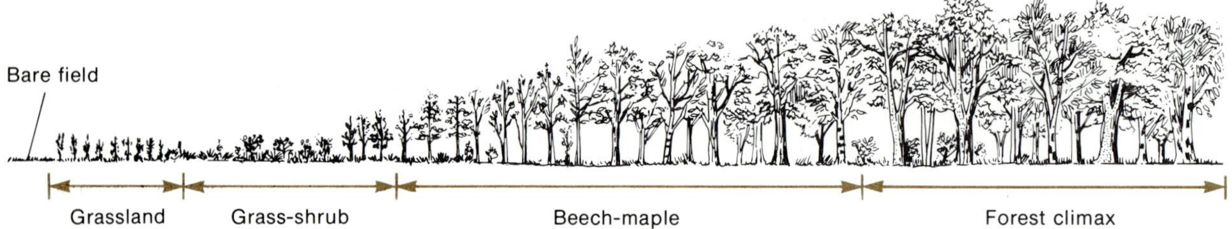

Figure 24–8. Succession. A field in an area with abundant precipitation will be home to a changing variety of plant species and communities over the years. First to dominate are various colonizers that tolerate the harsh open conditions (mainly grasses and weeds). Gradually, shrubs and small trees become established followed by larger trees. Eventually a mature forest results with the accompanying undergrowth.

als that frequently occur around human habitation sites where they are called weeds. Many weeds are merely plant species that are particularly well adapted to the open, disturbed habitats, such as lawns, gardens, building sites, and cropland, created by humans. After the weeds have grown on the site for one or two years, the ecological factors become changed. Instead of open ground, there is now a layer of dead plant material covering the soil and keeping the surface of the soil moister and cooler than before. These conditions permit the germination of seeds and spores that were unable to tolerate the wide fluctuations in temperature and moisture availability that are typical of exposed soil. Now as seeds and spores germinate there will be competition between plants of the original pioneer species and these newcomers. Because of the differences in numbers of seeds and spores in the habitat, it will take a few years to complete the transition, but ultimately many of the pioneer species will have vanished from the community on this site. They will disappear not because their seeds or spores are absent but because the competitive advantage of pioneer species is their ability to survive in an open, disturbed environment.

Once the environment is no longer as open, then the advantage is lost and the pioneer species are replaced by species better adapted to the new conditions. The cover provided by the growing plants and their litter further changes the microclimate and the soil and produces conditions conducive to the growth of trees and shrubs that are shade-intolerant. After the establishment of more extensive cover, the environment near the ground becomes shady. This favors the growth of different ground cover species than in the original open situation as well as the growth of tree seedlings that are shade-tolerant instead of the existing shade-intolerant species. This leads to a gradual change in the tree species composition of the community. As trees die, the only seedlings that are available to replace them are the shade-tolerant ones that have survived for several years. By this time they have grown five to ten feet tall under shady conditions and are immediately able to grow rapidly to fill the space left by the fallen tree. Shade-intolerant tree seedlings would not even be able to begin growth until the space becomes available by which time shade-tolerant seedlings would have the competitive advantage. This transition from open field to forest can take 100 years or more to complete, with each successive plant community altering the ecological factors. Since each spring brings renewed growth and therefore renewed competition, the continual changes in ecological factors drive the process of succession. During the successional process, which species become established depends upon what species are available in the surrounding areas. Furthermore, a plant does not wait on the outer boundary of the area undergoing succession until conditions favor the survival of this species. Instead, seeds and spores are broadcast indiscriminately, some being able to germinate and others never finding conditions that favor germination. As a result of this type of seed dispersal, trees of the final forest community could become established as members of any of the prior communities.

Ecologists have observed that succession seems to end with the establishment of a stable plant community whose members are those species best adapted to the environment created by the community. Since this community is the culmination of the process of succession, it is referred to as the **climax community.** The climax community, then, by its presence (not by intent), establishes and maintains a balance of ecological factors that favors the growth and reproduction of members of that community. The species composition of a climax community will only change as the result of the evolution or introduction of better adapted species, a major catastrophe such as climatic changes and fire, evolution or introduction of new pests, or human activities.

The concepts of succession and climax communities can be applied to familiar situations that do not require hundreds of years for their completion. The dream of North American suburbanites is to have a lawn as green and lush as any golf course (in ecological terms the intent is to establish a "climax" community composed exclusively of lawn grasses). To achieve this goal, lawns are regularly mowed. This not only keeps the grass short but it is also an ecological factor that limits the establishment of tree seedlings and other tall plants. Additionally, herbicides are used to kill broad-leaf (dicot) weeds because these possess a competitive advantage over lawn grasses in many lawns. Fertilizer also is applied frequently because the soils do not contain enough of the essential mineral elements (Chapter 7) for maximal growth of grass. Finally, many areas require irrigation because natural rainfall is inadequate to support season-long growth by the introduced grasses that are found in lawns. In the presence of these biotic modifications to the natural environment, a lawn grass "climax" community can be maintained. In the absence of any of these modifications a different climax community will exist as can be seen by examination of different lawns in a neighborhood.

Summary

The distribution of plants is determined by the interaction of three groups of ecological factors: climatic, edaphic, and biotic. Climatic factors include temperature, precipitation, day length, and light intensity. Although plants inhabit a variety of environments that differ considerably in upper and lower temperature extremes, all plant metabolism and growth occurs over basically the same range of temperatures (5°–40° C). Plants from different habitats have different adaptations to deal with the extremes. A major adaptation to extremes of both heat and cold is dormancy. Similarly the deciduous habit evolved in species adapted to both desert and cold climates because it reduces damage due to the extremes of temperature. Hot, dry climates require further adaptations to decrease water stress with various species evolving thicker cuticles, modified photosynthetic pathways, water-storing cells, leafless photosynthetic stems, reduced number of stomata, and more efficient root systems in response to this harsh environment.

The effects of latitude on seasonal temperature extremes are modified by large bodies of water such as the Great Lakes and the oceans. Increases in altitude mimic the effects of increases in latitude.

Rainfall determines plant distribution with lesser amounts requiring increasing adaptation to water stress. Snow is a source of groundwater and an insulating blanket against the extremes of winter cold and windchill.

Plants that reproduce in specific seasons usually do so in response to day length, the only reliable indicator of season. A species that is adapted to a particular day length might be restricted in its geographic range because these conditions are never met or occur at the wrong time. Because of the requirement of light for photosynthesis, light intensity determines plant distribution with specific species having evolved adaptations to shade, full sun, or intermediate intensities.

The substrate available to a plant for root growth is important because most plants absorb their mineral nutrients and water from the soil. The difference in soil structure, texture, and chemical composition causes differences in water and mineral retention, degree of aeration, and types of soil organisms.

Plants do not live in isolation; therefore their environment includes not only the physical world but also the living organisms in the same area. All aspects of the physical environment are modified to some degree by living organisms which creates many subsets of ecological factors instead of one homogeneous habitat in each location. Because plants act as competitors for light, water, minerals, and so on, a plant community consists of the individuals that are best adapted to the existing sets of ecological factors.

Plant adaptations are a reflection of the genetic program of the species that in turn is a result of evolutionary selection. The distribution of plants within a geographic area is a function of the above ecological factors with the best adapted plants surviving to form the different plant communities. If the major physical ecological factors remain constant in a region, there will be a succession of plant communities, the members of which act as biotic factors to modify the environment. Most frequently this results in gradual changes in species composition of the plant community as species better adapted to the changed conditions compete better than earlier species. Ultimately a climax community is reached that, by chance, creates the environmental modifications that favor the growth and reproduction of the species that form this community. The climax community only undergoes changes as a result of major changes in the ecological factors such as climatic shifts, fires, new pests, new competitor species, or human activities.

Review Questions

1. List the major ecological factors and explain how these influence plant distribution.
2. What is succession and why does it occur?
3. Why does succession not occur in a climax community? Give two examples of ecological situations that would lead to the resumption of succession after a climax community has become established.
4. Explain the role of climatic factors in determining plant distribution and why all plants are not restricted to the same set of climatic conditions.
5. The majority of plants may be grown in a greenhouse under the same growing conditions (ecological factors) yet in nature each would be found only in its own habitat. Explain what factor is responsible for these plants growing equally well under greenhouse conditions which may be quite different from those of their native habitats.

Fire

Anthropologists have long been aware of the significance of fire in the development of human culture. The ability to control fire gave early man a competitive advantage: fire chased away large predatory carnivores, hardened wooden weapons, provided warmth in winter, and cooked meat. Therefore fire was an important factor in human ecology.

Fire is also an important factor in plant ecology in many environments. Tall-grass prairie constituted a climax community in Illinois before modern agriculture only because periodic fires killed tree seedlings before they could become established. Even today pines remain dominant members of plant communities along the Atlantic coast of the United States because they are more resistant to fire than other trees. Grass fires or forest fires usually convert most of the vegetation and litter into an ash layer and at the same time convert the habitat to an open site with altered climatic, edaphic, and biotic factors. In regions where fires have been of frequent occurrence, adaptations have evolved in some plants that permit them to use fire to their advantage. Various plants of the chaparral and certain legumes, such as partridge pea, produce fire-resistant seeds that lie dormant until fire stimulates their germination. Jack pine cones will only open after they have been through a fire. In both of these cases the plants with these types of adaptations have a competitive advantage as their seeds will be the first to begin growth in this newly exposed site. Still other plants, including some grasses, are stimulated by fire and produce large quantities of seed which allows them an advantage over species that have to be reintroduced from unburned areas. Not all fire-tolerant species rely on new seed to provide the advantage. Forested regions that have frequent fires will not burn as hot as regions with infrequent burns. Ponderosa pine is able to survive fires because it possesses a thick bark that can protect the living tissues of the stem from the extreme heat. Similarly underground stems often escape the killing effects of fire and can give rise to new plants rapidly.

Smokey the Bear has convinced Americans that forest fires are bad and should be "stamped out." However, controlled burning in many habitats is absolutely essential for the maintenance of certain plant communities. Species that require fire for seed release or germination would gradually die out in their natural habitat if fire were eliminated. Furthermore, frequent, light burns, which reduce litter levels in the drier forests, will prevent the catastrophic hot fires that kill all of the trees and destroy the forest environment for centuries. Prairies that have been reestablished also require periodic burning to maintain the climax community. Western sagebrush-grass communities used for grazing purposes need burning at regular intervals to reduce the number of sagebrush plants. In the absence of fire the sagebrush will dominate this community because grazing animals do not eat it.

Indiscriminate fire setting is not to be encouraged; however, in our controlled "natural areas" there should be a reevaluation of the usefulness of fire in maintaining the desired types of plant communities. Forest fires, brush fires, and grass fires are local (frequently short-term) disasters that are natural ecological factors that plant species have been exposed to since plants became terrestrial. The human concern with fires is more for the economic damage that results than for any long-term ecological effects. In fact, except for higher altitudes and more northern latitudes, even forest fire damage results in only a momentary disruption on the geological time scale.

25

Plants of Special Importance

Outline

Introduction

Since ancient times and in virtually every culture, beliefs persisted until fairly recently that plants which bear a resemblance to parts of the human body possess curative powers for ailments of those parts. Among American Indians, squash tendrils were eaten to purge the body of worms; plants with milky sap were recommended to promote lactation in nursing mothers; and phallus-shaped roots were fed to newly married women to promote fertility. Ginseng plants, with forked roots fancifully resembling the human torso (Figure 25–1), have been revered for thousands of years in China as plants that can cure all sorts of human ills. The mandrake plant, like ginseng, has a forked root and, in addition, sometimes has a protuberance suggesting male genitals. People of the eastern Mediterranean region have regarded mandrake as having a positive effect on male reproductive powers, as well as the ability to help would-be mothers to conceive and to promote sexual ardor. In India the snakeroot plant *Rauwolfia*, so named because the twisted root system looks like a coiled snake, has long been used to prepare an antidote for snakebite.

Such persistent folk beliefs were most fully formulated in 16th century Europe by the alchemist-physician Paracelsus who set forth what he called "The Doctrine of Signatures." According to this concept, God placed a mark on every plant to indicate the purpose for which it was intended. Any person seeking cures for human ills need only look closely to see which herbal remedies should be effective for specific diseases. According to this doctrine, walnuts, with their convoluted nutmeats resembling the human brain, can make one more intelligent or cure madness and migraine headaches; kidney beans are effective against urinary disorders; Dutchman's-breeches can be used to treat venereal disease; liverwort is good for liver ailments; the yellow spice turmeric can be used to treat jaundice; bloodroot sap is effective for blood diseases; the Jews's-ear fungus for ear inflammation, and so on.

In a very few cases medicinal uses prescribed by the Doctrine of Signatures may have had some validity. For the most part any connection between plant and medicinal cure was purely imaginary, offering handsome profits to the herbalists who traded in such plants and providing little help to the afflicted who utilized them. Today science and modern medicine have relegated the Doctrine of Signatures to the position of historical curiosity that is still reflected in some plant names and in fairy tales but is otherwise largely forgotten.

Despite the demise of the Doctrine of Signatures, modern humans would do well to remember something of which our primitive hunter-gatherer ancestors were well aware—the survival of the human species depends directly and indirectly on plants which provide food, fiber, fuel, building materials, medicines, paper, and a host of other products. It would be impossible in a single chapter to describe all the contributions to human well-being made by plants, but the following brief survey of a few of the major food and medicinal plants will provide some idea of the extent to which we are

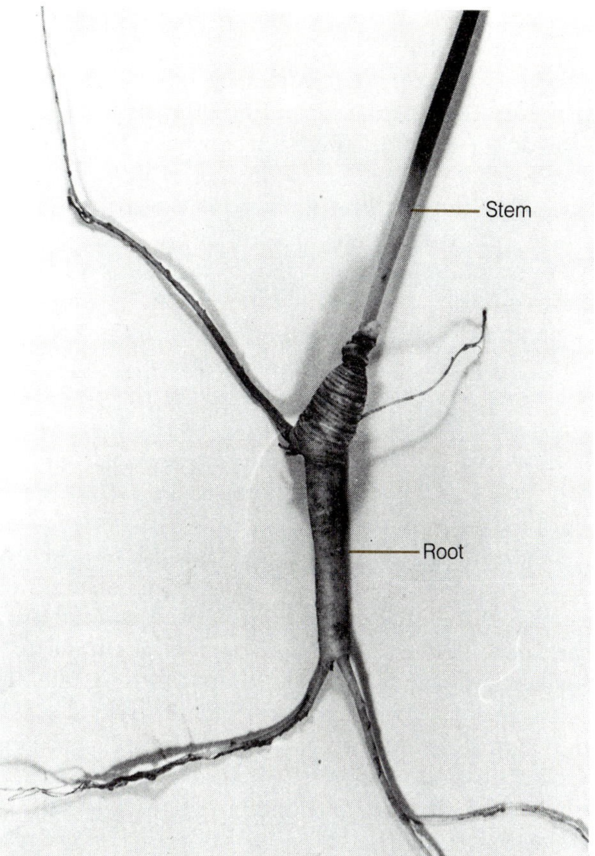

Figure 25–1. Ginseng. The forked root of ginseng resembles the human torso.

dependent on botanical products. In recent years a great deal of attention has been given to hallucinogenic plants. Because their use has become a controversial issue and is of concern to the general public, a short discussion of the more important hallucinogens is included. Finally some of the more common plants that are poisonous to eat or that cause skin irritations or allergies are considered.

Plants as Food

The discovery, made independently in several different parts of the world approximately 10,000 years ago, that wild grain seeds, deliberately sown in prepared soil, could provide a much more reliable food source than was available through random gathering, marked the beginning of agriculture. The rise of an agricultural way of life, resulting in the domestication of many species of both plants and animals, permitted a significant increase in human populations and provided a sort of stable, settled lifestyle that eventually gave rise to modern civilization.

Interestingly, however, plant domestication concentrated human attention on a relative handful of the species that potentially could serve as food for people. It is estimated currently that of the more than 350,000 species of plants in the world, close to 80,000 are edible. Throughout human history only about 3,000 of these species have been utilized as food and at present a mere 30 supply 95% of all our protein and calories. Indeed, more than half of all human food energy is derived from just 3 plants, all grains—rice, wheat, and corn (Plate 13–a, b, c).

Almost all plants important as human food are flowering plants. Of these, the group that accounts for more than half the world's caloric intake are the members of the grass family, known as cereal grains. Wheat *(Triticum aestivum),* rice *(Oryza sativa),* corn *(Zea mays),* oats *(Avena sativa),* barley *(Hordeum vulgare),* rye *(Secale cereale),* the millets—all are characterized by a single-seeded fruit, called a caryopsis, that contains an abundant starchy endosperm* but only a moderate to small amount of protein. Because of their low water content, grains can be stored for relatively long periods and can be transported easily.

Endosperm
A non-embryonic food storage tissue in a seed.

Other major food crops, in terms of acreage devoted to their cultivation, include various types of beans *(Phaseolus spp.),* the primary source of high-quality protein for populations lacking animal products in their diet. Within the past 50 years there has been a phenomenal increase in cultivation of soybeans *(Glycine max),* particularly in the United States which now leads the world in soybean production. Characterized by a dry-weight protein content of about 45%, as well as by a desirable oil, soybeans today are used in a wide variety of food products, from cooking oil, margarine, and soy sauce to synthetic steaks, bacon, and hamburger extenders.

Several starchy root crops—cassava *(Manihot esculenta),* (Plate 14–d, e) sweet potato *(Ipomoea batatas),* and the white, or "Irish," potato *(Solanum tuberosum)* (which is actually a tuber*, not a root)—are major sources of energy for millions of poor people throughout the world. Although tasty and filling, such root crops contain very little protein; heavy dependence on such foods is a frequent cause of malnutrition among people who are consuming an adequate number of calories but lack certain essential nutrients.

Tuber
An underground stem modified for vegetative reproduction.

In tropical regions, both bananas *(Musa spp.)* (Plate 15d) and coconuts *(Cocos nucifera)* are major cash crops and are important items in the local diet. The different varieties of bananas include types that are eaten cooked—either boiled, steamed, or deep fried and salted to make "banana chips"—as well as the sweet varieties consumed uncooked. The coconut is surely one of the most valuable and versatile of fruits: the fibrous outer portion surrounding the "nut" is used in making rope, doormats, and a variety of other products. The seed consists of a solid white "meat" and a refreshing liquid, both of which are types of endosperm. The coconut meat can be dried to form "copra" from which coconut oil is extracted, or it can be grated fresh and used in desserts, curries, vegetable dishes, or in dozens of other culinary preparations.

Sugar cane *(Saccharum officinarum)* and sugar beets *(Beta vulgaris)* constitute two other food crops of great commercial importance. Though their product is identical, the two plants are totally unrelated. Sugar cane (Plate 13d), a tropical member of the grass family, stores high concentrations of sucrose in its stems. Sugar beet is a temperate zone plant in which the enlarged root is the site of sucrose accumulation.

Although grown in substantially lesser amounts than the above, plants such as the green leafy vegetables, orchard fruits, nuts, tomatoes *(Lycopersicon esculentum),* melons *(Cucumis spp.),* and so forth, make an important contribution to many local economies and are invaluable in providing essential vitamins and minerals necessary for good nutrition. Indeed, one of the ironies in the great boost per/acre wheat and rice production which accompanied the Green Revolution* in several of the developing nations is that the enhanced profitability of growing grain has caused farmers to substitute acreage formerly planted to high-protein legumes and vitamin-rich leafy vegetables to less-nutritious cereal grains. In effect, quality has been sacrificed for quantity, with a deleterious effect on the health of local populations.

Green revolution
The replacement of low-yielding indigenous crop plants with high-yielding, genetically improved varieties.

Today agricultural experts are becoming increasingly concerned about the worldwide trend toward decreased crop variability, both in the number of species on which we depend and in genetic diversity within a species. The rise of agribusiness, among other factors, has in recent years resulted in increasing concentration on just a few crop varieties for reasons of profitability and efficiency, with the result that today more than ever before we rely on only a handful of crops for our food security. Many scientists feel that this state of affairs is both unwise and dangerous, stating that a wider variety of foods would improve nutritional levels and provide a buffer against such inevitable natural calamities as bad weather, crop diseases, insect pests, or climatic change.

Today there is an increasing interest in "new" food crops—plants that currently are being raised only in isolated local areas but which offer great possibilities for large-scale cultivation. Included among the little-known plants that could make a valuable contribution to the dinner tables of the future are the following. Amaranths *(Amaranthus spp.),* relatives of common pigweed, were first cultivated by Mexican Indians. Today they can be grown for their tasty leaves and nutritious seeds. Of significance is that amaranth grows well even in poor soils. Marama beans *(Tylosema esculentum)* are gathered wild in the Kalahari Desert of southern Africa by wandering tribesmen. The beans have a high protein content; their taste is similar to cashew nuts. The seeds of buffalo gourd *(Cucurbita foetidissima)* have been a traditional food among western Amerindians. The seeds are high in both oil and protein while the large root contains abundant starch reserves. Buffalo gourd is a tough perennial that does well under semidesert conditions and could constitute an important protein source in many semiarid lands where malnutrition is now common. Finally, for a discussion of the winged bean, see the Highlight at the end of this chapter.

NonFood Cash Crops

Many other plants are of great economic value for nonnutritive uses. So important are several of these so-called "cash crops" to some national economies, that farmlands which could produce food are used to grow more profitable nonfood commodities.

Among the world's leading cash crops must be included tobacco *(Nicotiana tabacum)*. It is grown virtually worldwide for its leaves that are used primarily in the production of cigarettes and cigars. At present China and the United States are by far the largest tobacco-producing countries, followed by India, the Soviet Union, Turkey, and Brazil.

Two fiber-yielding plants, cotton *(Gossypium spp.)* and jute *(Corchorus capsularis)* have been grown since ancient times in both tropical and subtropical areas. High tensile strength, pliability, durability, and ability to take dyes well have made cotton a premium fiber for the weaving of textiles. Cotton is raised extensively in the southern United States, the world's leading producer, as well as in India, Egypt, China, and elsewhere. The cotton fibers are actually seed hairs. The fruit is commonly called the cotton "boll" (Figure 25–2). Unlike the soft cotton fibers, the coarse fibers of jute are obtained from the fiber bundles of the tall, thick stalks of the jute plant. The fibers, which may range between 5 to 10 feet long, are produced mainly in India, Bangladesh, and Brazil. Jute is used almost exclusively for the manufacture of rope, twine, and gunny sacks and is second only to cotton as a fiber of commercial importance.

Rubber is produced from the white liquid latex obtained by tapping the native Brazilian tree, *Hevea brasiliensis* (Plate 15b). Rubber trees were introduced into Malaysia, Indonesia, India, the Philippines, Liberia, and Ghana, all of which developed very profitable rubber cultivation. Synthetic rubbers developed during World War II were not quite as satisfactory as the natural product and today most manufactured rubber is a mixture of natural and synthetic materials. Another latex-containing plant, the desert shrub guayule *(Parthenium argentatum)*, (Plate 15a) was cultivated in the American Southwest earlier in this century and supplied about half the rubber used in the

Figure 25–2. Cotton boll. The white hairs on the seed are the fibers from which cloth is woven.

Table 25–1.
Commercially Important Spices

Spice	Scientific Name	Plant Part Used
cayenne pepper	*Capsicum frutescens*	fruit
black pepper	*Piper nigrum*	dried berry
cloves	*Eugenia caryophyllata*	dried flower bud
cinnamon	*Cinnamomum zeylanicum*	bark
nutmeg	*Myristica fragrans*	seed
mace	*Myristica fragrans*	aril (red membrane around nutmeg seed)
ginger	*Zingiber officinale*	rhizome
turmeric	*Curcuma longa*	rhizome
saffron	*Crocus sativus*	stamens
allspice	*Pimenta dioica*	dried unripe berry
cardamom	*Elettaria cardamomum*	seed
coriander	*Coriandrum sativum*	seed
cumin	*Cuminum cyminum*	seed

United States between 1900 and 1910. Cheap Asian rubber and the development of synthetic forms made guayule cultivation uneconomical. Recently, however, there has been renewed interest in guayule as a cash crop that can be grown on otherwise useless arid lands, thus ensuring a domestic rubber supply.

Tea *(Camellia sinensis)*, coffee *(Coffea arabica)*, and cocoa *(Theobromia cacao)* are the world's most widely consumed nonalcoholic beverages (Plate 14–a, b, c). In the case of tea, leaves are used to make the drink, whereas ground seeds are used in making coffee and cocoa. Tea is grown primarily in China, India, and Sri Lanka. Central and South America and eastern Africa are the world's leading coffee-producing areas, while Ghana and Brazil dominate world cocoa production.

Tropical spices such as black pepper (Plate 14f), cinnamon, cloves, ginger, turmeric, nutmeg (Plate 11e), mace, cardamom, and so forth, have been important articles in world trade for millennia. As in the past, the most important suppliers of spices today remain India, Indonesia, and the West Indies. Table 25–1 lists some of the common commercially important spices.

Medicinal Plants

Since prehistoric times people have recognized that certain plants possess curative powers for many types of human ailments. Although the precise nature of how and why these herbal remedies were effective was not understood (or in some cases was misunderstood—e.g., The Doctrine of Signatures), their practical nature was appreciated and widely applied. A noted succession of medical writers, such as Hippocrates, Dioscorides, Avicenna, and the eccentric Paracelsus, compiled lists of drug plants that constituted the primary guides to medical practice until just the last century. There was a great emphasis on plants as curative agents for human illness until the development of organic chemistry and the creation of synthetic products at the end of the 19th century. Today we tend to overlook the fact that one-fourth of the medicinal drugs dispensed in the United States are derived from plants. Many other drugs are synthetic copies of plant extracts, identical in every way to the natural substance which served as the blueprint, but manufactured in the laboratory for cost-efficiency reasons. For many widely used medicines, laboratory synthesis has proved either too difficult or too expensive and the world's needs for such drugs are still supplied by plants, fungi, and bacteria. Some of these are described on the pages that follow.

Opium poppy *(Papaver somniferum).*

The milky juice obtained by slitting the capsule of the opium poppy (Figure 25–3) (not the red or orange oriental poppy, *Papaver orientalis,* commonly found in gardens) has been known since ancient times as a superb painkiller. Unfortunately this beneficial aspect is marred by the fact that opium use is

(a)

(b)

Figure 25–3. Poppy. (*a*) Flower. (*b*) Capsule from which raw opium is collected.

addictive. The major active ingredient in poppy extract is the alkaloid* **morphine,** which acts on the nervous system first as a stimulant and then as a depressant and induces a peaceful, dreamless sleep. Codeine, another alkaloid from poppy, is a common ingredient in cough syrups and cold medicines. Its pain killing properties are considerably less than those of morphine; however, it is significantly less habit-forming. Heroin, a prime component of the illicit narcotics trade, is several times more addictive than morphine. Heroin is made in the laboratory as a derivative of morphine.

Alkaloid
Member of a large group of nitrogen-containing organic chemicals, many of which are bitter in taste and frequently poisonous to animals and humans.

Cinchona (Feverbark) tree *(Cinchona officinalis).*

The bark of *Cinchona,* a small, evergreen tree native to the eastern slopes of the Andes Mountains, was known by Indians in the region to be an effective cure for malaria. Word of the miraculous medicine, called "quinine" (after *quina,* the native name for bark), eventually reached Europe where malaria had long been a major cause of debilitation and death. The trade in quinine bark was a lucrative one, and the producing countries of South America jealously guarded their monopoly. By the mid-19th century, however, seeds smuggled out of Peru became the basis for extensive quinine plantations on Java in the Dutch East Indies (today a part of Indonesia). By the mid-20th century synthetic substitutes for quinine, such as atabrine, chloroquine, and primaquine, were developed. In addition to being more expensive, some of the substitutes are less effective than quinine and also have undesirable side effects.

Belladonna *(Atropa belladonna).*

Native to Europe but grown as a garden ornamental in North America, this member of the tomato family is both useful and toxic, as may be implied from its common name, "deadly nightshade." All parts of the plant contain the alkaloid **atropine,** which can cause dilation of the pupils, rapid heartbeat, fever, or death. In controlled amounts, however, atropine has great value as a painkiller, and in treatment of hay fever (because of its tendency to dry out tissues), colds, Parkinson's disease, and many other ailments.

Snakeroot *(Rauwolfia serpentina).*

A low evergreen shrub native to southern and Southeast Asia, snakeroot (Figure 25–4) has been recognized for over 3000 years in India as an effective remedy for cases of mental disturbance and high blood pressure. Unfortunately, not until the 1950s was the value of this plant recognized in the West, when the active ingredient, the alkaloid **reserpine,** was isolated. Reserpine is obtained from the bark of the roots, which are ground to a powder. This substance, which acts on the brain and central nervous system and produces a sedative effect, has now largely replaced the use of shock therapy for the treatment of schizophrenia. It is also widely used as a tranquilizer, to some extent replacing barbiturates. Reserpine is now equally important in the treatment of hypertension, effecting a reduction in blood pressure in as little as three weeks. Although reserpine has been synthesized in the laboratory, the process is not economically feasible for commercial production and reserpine continues to be derived from snakeroot plantations in India.

Figure 25–4. Snakeroot. The taproot of this plant has the appearance of a coiled snake.

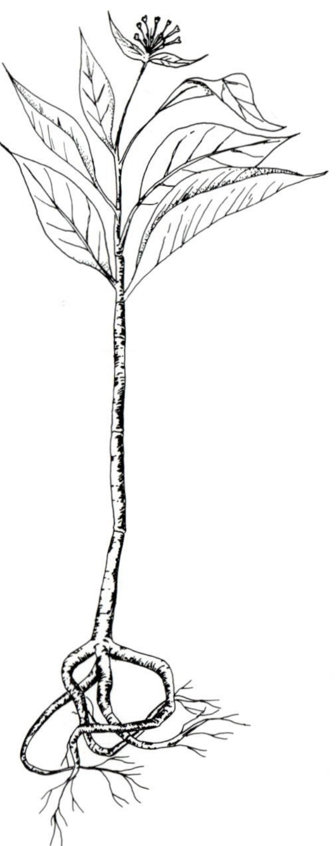

Snakeroot (*Rauwolfia serpentina*).

Foxglove *(Digitalis purpurea).*

This common garden flower (Plate 16d), native to Europe and widely planted in North America, has been employed as a medicinal herb since ancient times. It was used to cure a wide variety of ailments ranging from leg ulcers and cuts to pimples and sore ears. In fact, the active ingredient of foxglove, **digitalis,** composed of different glycosides*, functions primarily as a heart stimulant. Digitalis, localized mainly in the leaves, was first used by the medical profession in the mid-18th century for the treatment of dropsy (edema), a condition caused when irregular heartbeat inhibits proper kidney function, resulting in the accumulation of fluids in the arms, legs, and abdominal regions. Digitalis regulates the heart so that it can pump more normally, thus promoting proper kidney functioning and the excretion of excessive fluids. In addition to its usefulness in treating heart disease, digitalis more recently has been shown to be effective in easing glaucoma, asthma, and neuralgia. Digitalis continues to be derived solely from plants, since it has not yet proven possible to synthesize the drug in the laboratory.

Glycoside
A chemical derived from a sugar reacting with a substance such as an alcohol.

Penicillin *(Penicillium notatum).*
The discovery by Alexander Fleming in 1928 that an extract of the common bluish green mold often seen on decaying fruits and bread could kill bacteria ushered in the new era of antibiotics. Penicillin, as Fleming named his new discovery, proved to be just one of a whole family of antibiotics—streptomycin, aureomycin, terramycin, chloromycetin. These invaluable drugs are all derived from fungi and bacteria—penicillin from an ascomycete fungus, the others from a group of soil-inhabiting, filamentous bacteria known as actinomycetes. Use of antibiotics against a broad range of bacterial diseases came into widespread use at the end of World War II.

The Future for Medicinal Plants

Although plant products are vital ingredients in many of our most effective medicinal drugs, scientists believe that the potential of plants for alleviating human ailments has been barely tapped. Only a small proportion of the world's plants have been tested for their medicinal properties and the presumption is that thousands more valuable drugs still await discovery. This conviction has led both pharmaceutical companies and university researchers to launch expeditions into many obscure regions. There they interview so-called "witch doctors" and practitioners of folk medicine for the purpose of obtaining clues that will narrow the scope of their search. Results to date have been encouraging, with the identification of many hundreds of plants around the world having previously unknown medical properties. Unfortunately large areas of the tropics are being cleared to provide lumber or agricultural land. As these last remaining natural areas disappear, so do many plant species that ultimately could have produced the cure for some disease or have been a new source of food for human populations. People will never know what plant resources were lost to expanding human populations in Europe, China, and North America, but the potential exists now to prevent any futher loss of our natural heritage. Even if there were no economically significant plants left in any of the natural areas of the world, humanity would be better served in the long run if these areas and their species diversity were preserved.

Hallucinogenic Plants

A discussion of economically important plants would not be complete without at least mention of several species, known for a long time but attaining greatly enhanced notoriety within the past two decades. Not generally thought of as medicinal plants, although they are occasionally used for medical purposes, hallucinogenic plants are those that, when ingested or smoked, bring about temporary changes in mood and perception of reality. In the past most hallucinogenic drugs were employed almost exclusively by medicine men or shamans during ceremonies; only recently have they become widely used for nonreligious purposes.

Angiosperm
One of a group of plants characterized by the production of flowers.

Only about 10 species of hallucinogenic plants are cultivated, although many more species, mostly angiosperms* and a few fungi, are known. The following are among the more important ones today.

Marijuana *(Cannabis sativa).*
A native of central Asia, marijuana (Plate 16e) is a weedy annual with a worldwide distribution. It has been used for hundreds, perhaps thousands, of years in many parts of the world as a medicine effective in dulling pain and as a muscle relaxant. It has been used even longer as an intoxicant; archeological records indicate that in a number of different Asian civilizations marijuana seeds were burned and the smoke inhaled to produce a stimulating effect. It was widely used in India by mystics and peasants alike, being called by such epithets as "joy giver" and "sky flyer." The constituent in marijuana responsible for producing hallucinations is tetrahydrocannabinol (THC), produced in special glands located on the leaves and in female flower parts. The amount of THC produced varies from one strain of marijuana to another and is influenced by growing conditions as well. In general, marijuana from dry, hot regions is higher in THC than is that grown in cooler regions. Most of the varieties of marijuana growing wild in the United States are low in THC; such strains were introduced originally for their fiber-producing qualities. In fact, until about 40 years ago marijuana was a cash crop in the United States because the fibers from the plant were used to make hemp rope. Only the introduction of nylon rope caused farmers to forsake cultivation of marijuana.

Peyote *(Lophophora williamsii).*
For thousands of years the Indian tribes of central and northern Mexico used this small, spineless cactus (Figure 25–5) as an integral part of their

Figure 25–5. Peyote. A mixture of psychoactive alkaloids is present in these spineless cacti.

(a)

Photos: **(a)** Corn (*Zea mays*). **(b)** Rice (*Oriza sativa*). **(c)** Wheat (*Triticum* sp.). **(d)** Sugar cane (*Saccharum officinarum*).

(b)

(c)

(d)

Color Plate 14

(a)

Photos: (a) Cocoa (*Theobroma cacao*). **(b)** Coffee (*Coffea arabica*). **(c)** Tea (*Camelia sinensis*). **(d)** The brown roots in the basket are of cassava (*Manihot* sp.). **(e)** Cassava plants. **(f)** Black pepper (*Piper nigrum*) growing on vines.

(b)

(c)

(d)

(e)

(f)

Photos: (a) Guayule (*Parthernium argentatum*). **(b)** Rubber (*Hevea* sp.). **(c)** Cashew (*Anacardium occidentale*) nuts attached to the swollen receptacle. **(d)** Banana (*Musa* sp.) plant with fruits.

(a)

(b)

(c)

(d)

Color Plate 16

Photos: (a) Trumpet creeper (*Campsis radicans*). (b) Wild parsnip (*Pastinaca sativa*). (c) Poison hemlock (*Conium maculantum*). (d) Foxglove (*Digitalis purpurea*).

Dr. Jacob Varkey

(a)

(b)

(c)

(d)

religious ceremonies. They ate the dried, disc-shaped tops ("mescal buttons") to induce brightly colored visions and the perception of communicating directly with the gods. More recently the peyote cult, or "peyotism," has spread widely among North American Indian tribes, resulting in the establishment of the Native American Church, which incorporates some Christian elements into a religion centering around peyote.

The physiological effects of peyote are due to at least 15 psychoactive alkaloids, of which **mescaline** seems to be the most active. Mescaline has now been synthesized for use in psychological experiments; in its pure form it is much more powerful than the complex of substances found in peyote.

Psilocybe mexicana.

Figure 25–6. *Psilocybe.* Two or three of these mushrooms, when consumed, can cause hallucinations.

Psilocybe *(Psilocybe spp.).*

Called by the Aztecs "flesh of the gods," just two or three of these small mushrooms (Figure 25–6) can produce hallucinations that feature visions, laughing spells, escapism, altered time and space perceptions, and a general relaxed feeling. The alkaloid **psilocybin** has been identified as the main hallucinogenic compound in these fungi.

Ergot *(Claviceps purpurea).*

This fungus, which frequently infects rye and other cereal grasses (Figure 25–7), is responsible for the serious type of food poisoning known as ergotism or "St. Anthony's Fire." Ergot is also the source of **lysergic acid,** which, when combined with the synthetic chemical diethylamide, produces LSD, the most powerful hallucinogen known. LSD may be of some benefit for psychiatric researchers, but its widespread misuse has resulted in a proliferation of drug-induced mental disorders.

Immune reaction
Production of antibodies as a response to the presence of a foreign substance in the blood.

Figure 25–7. Ergot. The black sclerotium, or resting body, of the ergot fungus is the cause of ergotism poisoning. Lysergic acid obtained from ergot, when combined with diethylamide, produces LSD.

Plants That Cause Hay Fever

A sizeable percentage of any given population is known to be allergic to pollen from a variety of plants such as ragweed (Figure 25–8), corn and other grasses, many flowering trees, and to the spores of fungi. Sensitive individuals react to certain antigenic proteins present on the walls of the pollen and spores. A consequence of this immune reaction* is an increased synthesis of the chemical **histamine,** which is responsible for the swollen membranes, runny nose, itchy eyes, and sneezing—all symptoms of hay fever. The reactions to pollen and spore allergens vary considerably from one individual to another. To the hay fever sufferer it is probably obvious that this allergy recurs throughout the months when plants are growing. What is probably not obvious is that there are three different hay fever seasons caused by three different groups of plants: in the spring the pollen comes mainly from trees; in the summer it comes from grasses and late-blooming trees; and in autumn ragweed and grasses contribute the most pollen. For those suffering from hay fever it makes good sense to avoid ragweed. They would also do well not to plant oak, hickory, elm, or box elder as shade trees on their property because these are the trees that contribute most to hay

(a)

Figure 25–8. **Ragweed.**
(a) Branch of a ragweed plant.
(b) An antigenic protein present on the walls of these pollen grains is the cause of ragweed hay fever.

(b)

fever. A normally nonsensitive person may become sensitized to any of these plants by constant exposure to the pollen or spores. Exposure to these plants is difficult to avoid, short of moving to a place where they do not grow.

Poisonous Plants

There are many common house, garden, and roadside plants that can cause illness or even death. Although the fact is not widely recognized, about 3.5% of all poisonings in recent years have been caused by plants. With the current interest in camping, backpacking, and "returning to nature," many people are coming into contact with plants of whose characteristics they have no awareness. Of particular concern are the natural food faddists who assume that anything "natural" is nutritious and are unaware that many plants, both wild and domestic, contain poisonous substances. For the sake of one's own health, and particularly for the safety of small children, by far the most frequent poisoning victims, it is important to be able to recognize which plants are hazardous.

The types of poisonings caused by plants fall primarily into the categories of skin irritations due to bodily contact with a plant and internal poisoning when poisonous plant parts are ingested.

Plants Causing Dermatitis

Each year millions of campers, picnickers, nature lovers, farmers, and others are afflicted with irritating and sometimes painful skin rashes or blisters originating from contact with certain poisonous plants. Such eruptions may cause only temporary discomfort, but others may persist for weeks and give rise to secondary infections, or even, in severe cases, require hospitalization. The plants most frequently causing dermatitis in the United States and Canada are discussed below.

Poison ivy and its relatives.

Poison ivy *(Toxicodendron radicans)* formerly classified as *Rhus radicans* (Figure 25–9) is a climbing vine or woody shrub found throughout the United States and southern Canada, with the exception of the Pacific coast. Another, less common, species, *T. rydbergii* is found only west of North Dakota in the northern tier of states bordering Canada. This grows as an upright herba-

Figure 25–9. **Poison ivy.** This plant can be recognized by its trifoliate compound leaves.

ceous plant rather than as a vine. Poison ivy prefers the moist conditions along lake shores, stream banks, and floodplains and also is frequently found at the edges of woods, often growing on the trunks of large trees, on fence posts, and around buildings. The characteristic three leaflets ("leaflets three, leave them be") and the longer stalked middle leaflet aid in identification and help to distinguish poison ivy from Virginia creeper, a 5-leaflet vine with which it is often confused. In autumn, poison ivy turns a brilliant red, making it a tempting target for leaf collectors who learn to their regret that the plant is as poisonous then as it is during the season of active growth. The substance causing the irritation is an aromatic oil in the sap. When cutting poison ivy one should be careful not to touch the cutting blade, since the sap can be indirectly transferred to the skin. Because the irritant can be carried on dust or ash particles, it is inadvisable to stand in the smoke generated by the burning of poison ivy. Even wind-blown pollen can result in dermatitis among extremely sensitive individuals. Fifty to seventy percent of the population is sensitive in some degree to poison ivy and its relatives.

Some other closely related species, poison oak *(T. diversilobum* and *T. toxicarium)* and poison sumac *(T. vernix),* are similar in toxicity but are somewhat less common and more restricted geographically. Poison oak is generally found in areas of sandy soil or in pine forests in the eastern and south-central portions of the United States and along the West Coast from Washington to Mexico. Poison sumac, growing as a shrub or small tree, principally inhabits bogs, swamps, and river bottoms and is most common along the Atlantic and Gulf coastal plains and in the Great Lakes region.

Stinging nettle *(Urtica dioica).*
This plant, as well as several others in related genera, is a tall, perennial herb that grows in damp woodlands and along roadsides. Both leaves and stem are covered with conspicuous stinging hairs that function like miniature hypodermic needles, injecting an irritating chemical into the skin upon contact. This substance causes a painful burning and itching sensation that can last for several hours.

Trumpet creeper *(Campsis radicans).*
Throughout the eastern United States this attractive woody vine with red or yellow trumpet-shaped flowers and opposite, pinnately compound leaves grows up fence posts, trees, and utility poles (Plate 16a). Skin contact with either leaves or flowers can result in a reddening of the skin and the formation of blisters that can last for several days.

Wild parsnip *(Pastinaca sativa).*
Extensive stands of this showy weed, which resembles a yellow version of Queen Anne's Lace, are common in fields and along roadsides throughout the United States in late spring and early summer (Plate 16b). Wild parsnip causes a dermatitis (inflammation of the skin, followed by the formation of blisters and a persistent redness that can last for weeks or even months)–through the action of the photosensitive compond **furocoumarin.** Exposure will only result in dermatitis when skin, moist from perspiration or water, is exposed to bright sunlight.

Ingested Plant and Fungal Poisons

To some people the term *poisonous plant* immediately evokes visions of deadly mushrooms or the Socratic "cup of hemlock." Indeed, these images are accurate, but there are many other more common plants and fungi (Table 25–2) that can, if eaten, cause illness or even death. The degree of hazard presented by such organisms depends on a number of factors: the part eaten (organisms can contain deadly toxins in some tissues yet none in others); the stage of development of the organism or the season of the year; the amount eaten (in some cases a bite or two can be fatal, while in others large quantities must be consumed to invoke a response); and the age or physical condition of the victim (in general, a given amount of toxin will be more harmful to a small child than to a healthy adult). A knowledge of which plants and fungi pose danger is helpful to anyone, but is particularly important for parents of small children who are the most frequent poisoning victims. The following discussion is limited to those plants and fungi found in the United States and Canada that either present the greatest danger or are exceedingly common.

Table 25–2.
Some Commonly Encountered Poisonous Plants*

Plant: Common Name	Scientific Name	Poisonous Parts
Yew	*Taxus* spp.	most parts except fleshy red covering around the seed
Lily of the valley	*Convallaria majalis*	all parts of the plant
Hyacinth	*Hyacinthus orientalis*	bulb
Daffodil	*Narcissus* spp.	bulb
Dumbcane	*Dieffenbachia* spp.	leaves and stem
Philodendron	*Philodendron* spp.	leaves and stem
Jack-in-the-pulpit	*Arisaema triphyllum*	corm
Nutmeg	*Myristica fragrans*	seeds (not harmful in minute amounts as a spice)
Delphinium	*Delphinium* spp.	seeds and young plants
Mayapple	*Podophyllum peltatum*	roots, stem, flower, leaves, and unripe fruit
Bleeding heart	*Dicentra eximia*	all parts of the plant
Rhubarb	*Rheum rhaponticum*	leaf blades
Wild cherry	*Prunus serotina*	bark, leaves, seeds
Sweet pea	*Lathyrus* spp.	seeds

*For additional information on poisonous plants see John M. Kingsbury, *Poisonous Plants of the United States and Canada.* (Englewood Cliffs, NJ: Prentice-Hall Inc., 1964) or James W. Hardin and Jay M. Arena, *Human Poisoning from Native and Cultivated Plants,* 2nd ed. (Durham, NC: Duke University Press, 1974)

Mushrooms

Mushrooms are a gourmet's delight, provided of course that the ones in question are nonpoisonous. There is no simple rule of thumb for distinguishing between those wild forms that are safe and those that are not. Even today we still hear of occasional deaths among people who have eaten mushrooms they picked from woods, fields, or lawns. Although only a relatively few of the thousands of species of mushrooms found in North America are poisonous, they may look very much like nonpoisonous species and frequently even grow together. Members of one of the most poisonous genera, *Amanita* (one species of which is called the "death angel"), grow commonly in "fairy rings" in woodlands and on lawns. Just one or two bites of these alkaloid-containing fungi can be fatal to an adult, and, in fact, the majority of deaths due to mushroom poisoning are caused by species of *Amanita.*

Castor bean *(Ricinus communis).*

This attractive shrublike plant is grown as an ornamental foundation planting, as well as commercially for its oil. The leaves of the plant are only slightly toxic; however, the attractive mottled seeds can be deadly, containing the toxin **ricin.** Children who chew on the seeds experience intense irritation of the mouth and throat, gastroenteritis, extreme thirst, dullness of vision, convulsions, uremia, and death. One to three seeds are sufficient to kill a child; as few as four to eight may be fatal to adults.

Jimsonweed *(Datura stramonium).*

This common weed, found in gardens, pastures, and along roadsides, contains toxic alkaloids in all parts of the plant (Figure 25–10). The seeds and leaves are particularly dangerous. Children have been poisoned by sucking nectar from the flowers, eating the seeds, or drinking liquid in which the leaves have been steeped. A very small amount can be fatal to a child. Symptoms of jimsonweed poisoning include dilation of the pupils, hallucinations, extreme thirst, redness of the skin, dry sensation in the mouth, delirium, convulsions, and death.

Poison hemlock *(Conium maculatum).*

Poison hemlock, an extract of which the Greek philosopher Socrates purportedly drank to commit suicide, is a tall, weedy member of the parsley family and bears a resemblance to Queen Anne's lace. An import from Eurasia, poison hemlock commonly grows along roadsides and railway tracks, particularly in the Midwest, and blooms from early summer to midsummer (Plate 16c). The most toxic parts of the plant are the fleshy taproot and the seeds. Because of their unpleasant taste, fatal quantities are seldom accidentally consumed; however, lesser amounts can produce gastroenteritis, muscular weakness, trembling, dilation of pupils, convulsions, or coma.

Water hemlock *(Cicuta maculata).*

Also called spotted cowbane, this native North American plant resembles poison hemlock except that the pinnately compound leaves are not so finely divided and it has clustered, short, thickened tuberous roots instead of a single taproot. The hollow stems are spotted or striped with purple (poison hemlock may also have a mottled stem, however). As the name suggests, water hemlock is found most frequently in moist soil, growing along stream

Figure 25–10. Jimsonweed. A common weed found in fields and pastures produces beautiful white flowers that bloom at night. All parts of this plant contain toxic alkaloids.

banks, roadside ditches, and marshes throughout most of the United States and Canada. Its active toxin **cicutoxin** is localized primarily in the root, one mouthful of which can kill an adult. Death generally occurs when the plant is misidentified as wild artichoke or wild parsnip and eaten. Poisonings among children have occurred when the hollow stems were used for whistles or peashooters.

Pokeweed *(Phytolacca americana).*
In the summer of 1980, 21 children at a day camp in New Jersey developed nausea, stomach cramps, and vomiting a few hours after eating a salad made from the young leaves of pokeweed. Pokeweed (Figure 25–11) is a tall, up to 8 feet, shrublike, perennial herb, common in open fields, wastelands, and along roadsides throughout the eastern United States and southern Canada. It produces clusters of purplish-black juicy berries ("inkberries"), which, if cooked in pies, are edible but which eaten raw can cause serious poisoning, especially if consumed in any quantity. The toxic substance phytolaccine is most concentrated in the root but is present also in lesser amounts in leaves,

Figure 25–11. Pokeweed. This plant produces clusters of purplish-black, juicy berries that are safe to eat only if they are properly cooked.

stems, and berries. Supposedly the young leaves and stems can be eaten safely if thoroughly cooked in at least one change of water; however, in the New Jersey situation, even though the greens were boiled, serious gastroenteritis nevertheless resulted. Pokeweed presents a special danger because many people either do not cook the leaves thoroughly or they pull up some of the root along with the stem and leaves. In small amounts, pokeweed's most obvious poisoning symptoms are vomiting and severe stomach cramps; consumption of large amounts can result in convulsions and death. Even handling the plant can be damaging to human health because phytolaccine can be absorbed through breaks in the skin. Once absorbed, it acts on the blood, causing a variety of ill effects: in particular, lymphoid cells are stimulated to divide and mature, thus altering the ability of the immune system to respond to infection.

Summary

Human well-being is heavily dependent on the plants which provide food, fiber, fuel, building materials, medicines, paper, and many other products.

Hallucinogenic plants and poisonous plants also exert a direct impact on human welfare, although often in a negative manner.

Of the approximately 80,000 species of edible plants, only about 3,000 have ever been utilized as human food, and of these a mere 30 species supply 95% of our protein and caloric needs. Most of the plants important as human food are angiosperms; in terms of acreage devoted to their cultivation, the world's major food crops are rice, wheat, corn, oats, barley, rye, the millets, several species of beans, cassava, sweet potato, white potato, coconuts, bananas, sugar cane, and sugar beets. Leafy vegetables and fruits, although grown in lesser amounts, supply vitamins and minerals necessary for good health. Nonfood crops such as tobacco, cotton, jute, rubber, coffee, tea, and spices are also of great economic importance.

Concern about the worldwide trend to decreased crop variability has stimulated interest among scientists in "new" food crops, which are now being raised in localized areas but offer potential for widescale cultivation. Of special interest in this context are the winged bean, amaranths, marama beans, and buffalo gourd.

The value of certain plants for medicinal purposes has been recognized since prehistoric times. Even today one-fourth of all medicinal drugs in the United States are derived from plants and many others are synthetic copies of plant extracts. Some of the common medicinal plants and their products include: opium poppy (morphine, codeine), cinchona tree (quinine), belladonna (atropine), snakeroot (reserpine), foxglove (digitalis), *Penicillium* (penicillin).

Plants and fungi with hallucinogenic properties formerly were used for religious, ceremonial, and medicinal purposes by aboriginal peoples. Such hallucinogenic organisms include marijuana, peyote cactus, psilocybe mushrooms, and the fungus ergot.

Among plants causing human discomfort are ragweed and some trees and grasses whose pollen induces the allergic reaction known as hay fever. Other plants such as poison ivy, stinging nettles, trumpet creeper, and wild parsnip can cause irritating rashes. Still other plants such as castor bean, jimsonweed, poison hemlock, water hemlock, and pokeweed, in addition to some fungi, can induce illness or death if consumed.

Review Questions

1. Discuss some of the reasons why many scientists feel a sense of urgency in promoting the development of "new" food crops such as winged beans or amaranths. What obstacles must be overcome before these plants are widely cultivated?
2. In what ways are over-population and the need for economic development in tropical countries endangering the earth's biological diversity? How can such trends have an impact on the welfare of temperate-zone peoples?
3. Describe several plants which have medicinal value if properly used but which, under certain conditions or taken in excess, constitute a serious health threat.
4. List some specific words of advice you might offer a friend who was planning to spend a summer backpacking and "living off the land."

The Winged Bean

The future "wonder crop of the tropics" may well be a plant long known to peasant farmers in Southeast Asia and New Guinea, but one which has come to scientific attention only within the last few years. The winged bean *(Psophocarpus tetragonolobus)* (Figure 25–12), so called because a cross section of the pod is square or oblong with a winglike projection extending from each corner, has been dubbed by some writers as a "supermarket on a stalk." Every part of the plant, which grows vigorously up to 15 feet when staked to a pole, is edible. The long pods can be eaten either cooked or raw; the leaves, rich in vitamin A, can be prepared and eaten much like spinach; the root is a flavorsome, potatolike, white tuber with a protein content 10 times that of the common tropical energy source, cassava; the delicate tendrils can be eaten, and the cooked flowers have the appearance and texture of mushrooms. Of greatest potential, however, are the dried seeds. Similar to soybeans, winged bean seeds have an edible oil content of about 17% and a protein content as high as 40%; they are also rich in vitamin E and iron. Like other members of the bean family, winged beans live in symbiotic association with nitrogen-fixing bacteria, which enable them to thrive without large inputs of chemical fertilizers. Thus the plant offers great promise in the effort to improve nutritional levels in poor agricultural societies throughout the tropics. In many parts of the world where winged beans were unknown just a few years ago, the plant is rapidly gaining in popularity, especially in small private or village plots and market gardens. Among agronomists there is great optimism that the winged bean may be as revolutionary an innovation in tropical agriculture as the soybean was to the temperate regions only 50 years ago. The major hurdle to be overcome, however, may not be agricultural, but sociological, as it takes time to introduce new foods and customs into traditional societies.

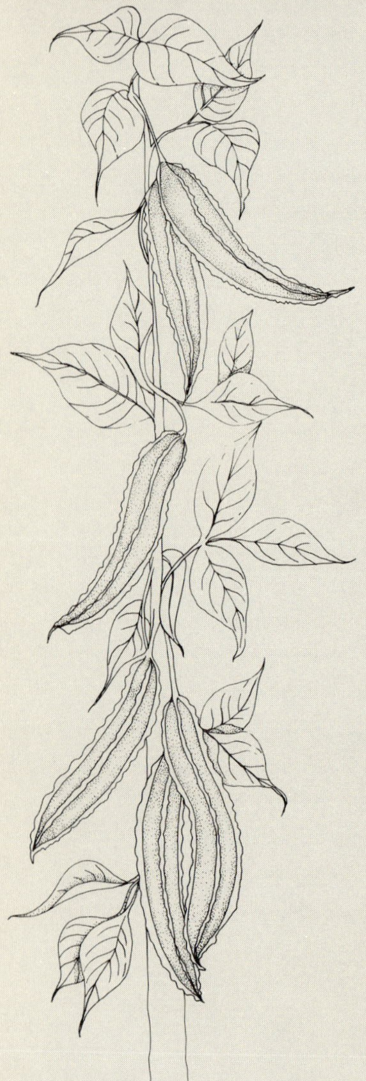

Figure 25–12. **Winged bean.** Every part of this plant is edible. The four-angled beans produced on this plant have small winglike extensions from which it gets its name.

A

Appendix A:

Chemical Principles

In order to understand how plants function an understanding of basic chemical principles is essential because all living organisms are composed of chemicals and all metabolic reactions are chemical reactions. For many individuals there is an artificial distinction between living organisms and the nonliving world such that the living organism is seen as being composed of special substances that differ from the chemicals found elsewhere. Although this is not true and all organisms consist of chemicals, chemists in the nineteenth century drew a distinction between the chemical substances produced by living organisms, *organic chemicals,* and all other chemicals, *inorganic chemicals.* This distinction is still useful and in this book chemical substances produced by living organisms are referred to as organic molecules even though the basic principles of chemistry are the same for organic and inorganic molecules.

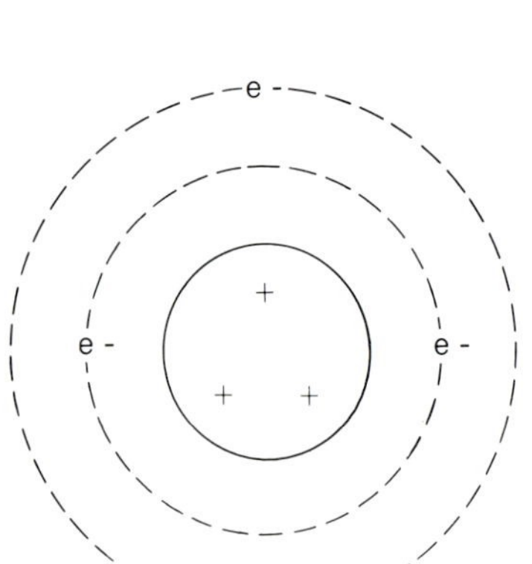

Figure A–1. Atomic model. Positively charged protons are in the atomic nucleus, with an equal number of electrons distributed in shells around the nucleus.

Atoms

Chemical elements are the fundamental substances that comprise the building blocks of all chemicals. Among the 92 naturally occurring elements are included oxygen, nitrogen, hydrogen, carbon, gold, silver, uranium, and aluminum, to name just a few; however, chemists and physicists have synthesized additional elements bringing the total to 109. Elements, in turn, are composed of *atoms,* the smallest stable unit of matter; each element has its own unique atomic composition. Further, atoms are composed of *subatomic particles—protons, neutrons,* and *electrons.*

Positively charged protons and neutrons with no electrical charge together form the positively charged *nucleus* of an atom, while the negatively charged electrons are in orbit and form an electron cloud around the nucleus. The negative charges of the electrons cancel out the positive charges of the protons; because both protons and electrons occur in equal numbers, the atom is electrically neutral.

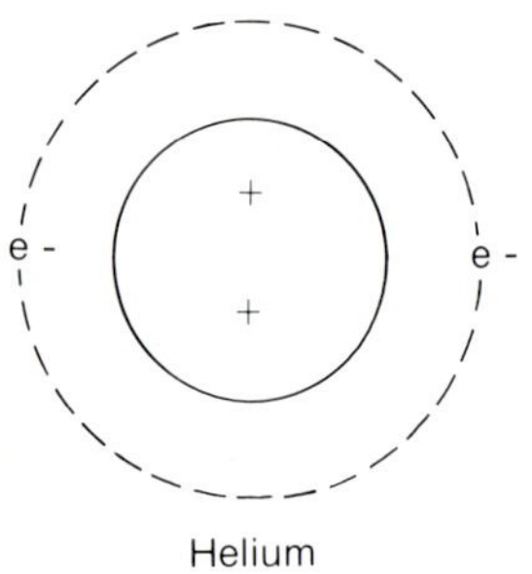

Figure A–2. Helium. This element has 2 protons in the nucleus and 2 electrons in the first shell.

Considerable research has been done on the organization of the subatomic particles within the atom with emphasis on the distribution of electrons within the electron cloud. Although this particular model is outdated for the purposes of advanced chemistry, the planetary model of the atom formulated by Niels Böhr is adequate to explain the basic principles of chemistry. According to this model the electrons move in orbits, called *shells,* around the nucleus in much the same way as planets move around the sun (Figure A–1). Based on the results of studies of atomic properties we know that each shell can contain a limited number of electrons and each shell must be filled before electrons can be added to a new shell. The first shell can contain only two electrons as in atoms of the element helium (Figure A–2). Atoms with more than two electrons will have a first shell containing two electrons with the remaining electrons added to subsequent shells, each of which takes more than two electrons. An atom of carbon has six protons in its nucleus which means that it must have six electrons to balance the electrical charge;

two of these electrons are in the first shell and the remaining four are in the second shell (Figure A–3). Oxygen with eight electrons has two in the first shell and six in the second (Figure A–3); but neither carbon nor oxygen fills this second shell because this shell can hold eight electrons. For reasons beyond the scope of this botany book, an atom "needs" its outermost electron shell completed even though having a filled outer shell would violate the proton-electron equivalency rule. Consequently atoms combine with each other in what are called *chemical reactions* with the result that each atom ends up with a filled outer shell.

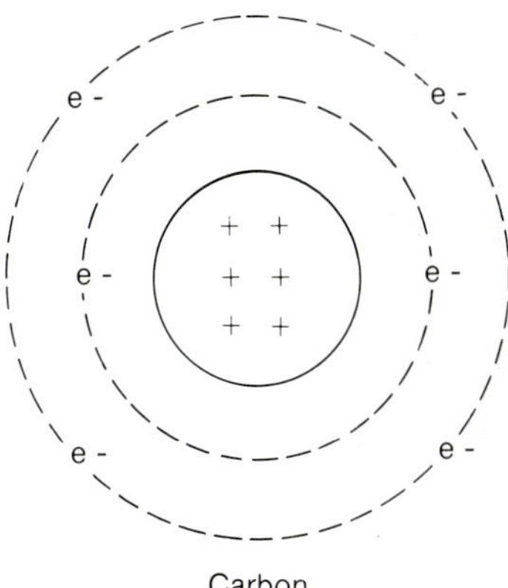

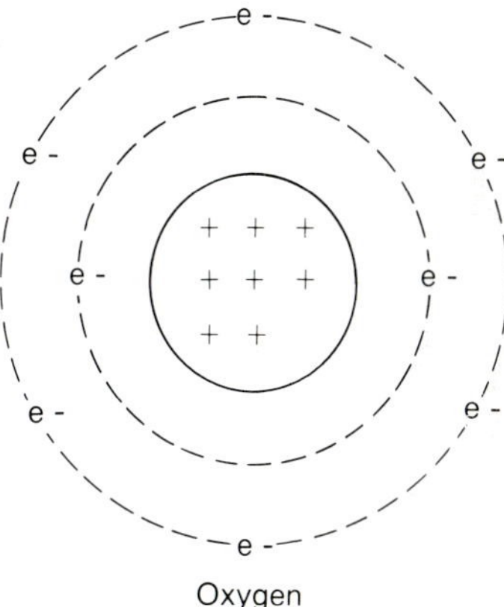

Figure A–3. Carbon and oxygen. The first electron shell can contain only 2 electrons, but the second shell can hold up to 8.

Chemical Bonds

In some of these chemical reactions atoms share electrons with each other which results in each atom having access to enough electrons to fill its outermost shell. This sharing of electrons between atoms creates a link or chemical bond, a *covalent bond,* that holds the two atoms together. Each covalent bond involves two electrons, one from each atom. For example, a hydrogen (H) atom has one electron (e^-) but since its outer shell will hold two electrons, a hydrogen atom will react with another atom forming a covalent bond. Two hydrogen atoms can react and share their electrons in a covalent bond thus satisfying each atom's "need" for two electrons in its outer shell since both electrons orbit both nuclei instead of orbiting just their own nucleus (Figure A–4). The result of this bonding between two hydrogen atoms is the formation of a hydrogen *molecule:* H-H or H_2. Oxygen has six electrons in its outer shell, therefore two more must be added to complete this shell. One way an oxygen (O) atom can obtain the additional electrons is to share with two hydrogen atoms (Figure A–5). In this molecule each hydrogen atom forms a covalent bond with the oxygen atom: H-O-H or H_2O, better known as water. A few other examples of molecules formed as the result of covalent bonding are shown in Figure A–6. In fact, most of the molecules that make up both living organisms and inanimate objects are held together by covalent bonds.

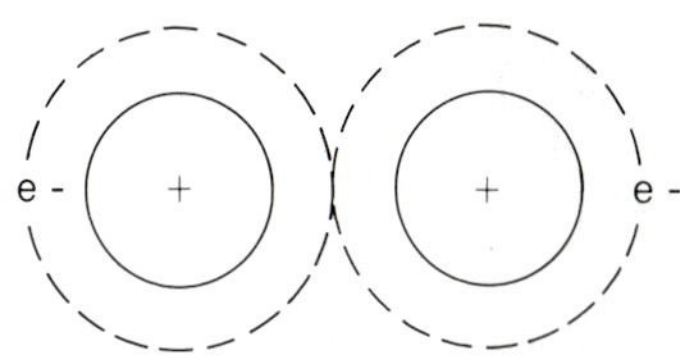

Figure A–4. Covalent bonding. Two hydrogen atoms can each fill their outermost electron shell if they share their electrons.

In some chemical reactions instead of electrons being shared between atoms, electrons from one atom are captured by the other atom. The atom which loses electrons is now positively charged because it has too few electrons to balance the positive charge in the nucleus while the atom that gained electrons is negatively charged. Since a positive and a negative attract, the two atoms are bound together in a chemical bond called an *ionic bond* because charged atoms (or molecules) are called *ions.* Table salt is an example of a molecule held together by an ionic bond. In this case sodium (Na) has one electron in its outer shell and must either gain seven more to fill that shell or lose the lone electron in the outermost shell which would leave the atom positively charged but with a filled outer shell (Figure A–7). Chlorine (Cl) has seven electrons in its outermost shell and must have one more to fill it. Chlorine could form a covalent bond; however, in a reaction

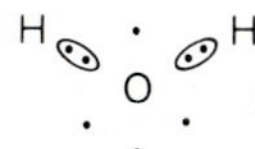

Figure A–5. Water. Water (H_2O) results from the formation of covalent bonds ( ) between 2 hydrogens and 1 oxygen.

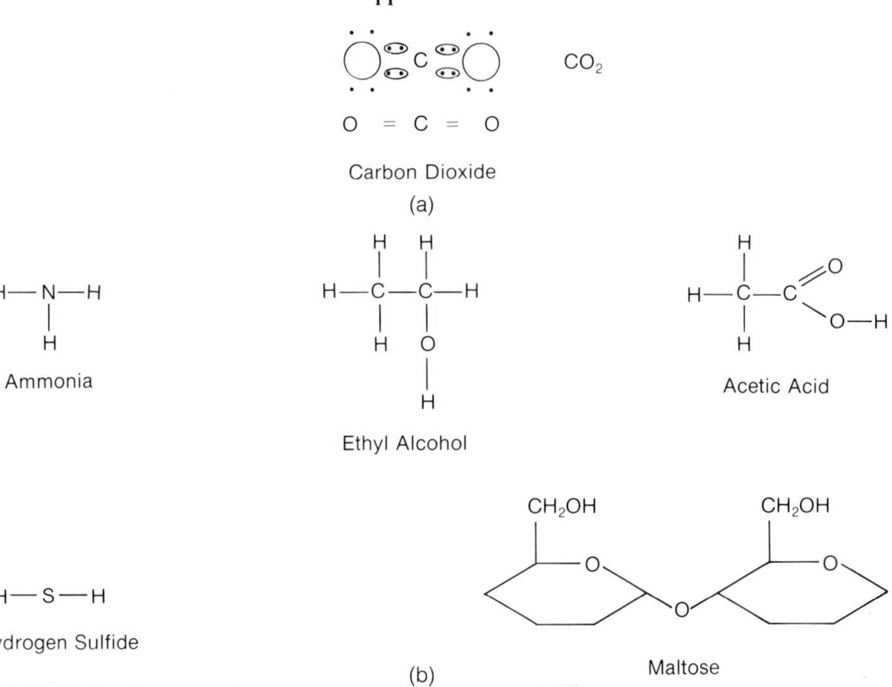

Figure A–6. Covalent bonding involving carbon. (*a*) Carbon dioxide. Covalent bonding between one carbon atom and two oxygen atoms forms CO_2, or carbon dioxide. In this case two covalent bonds form between each oxygen and the carbon. This provides the four extra electrons needed by the carbon atom and the two needed by each oxygen. (*b*) Many different kinds of compound are formed by covalent bonding. For the more complex molecules with a carbon backbone the individual carbon atoms and their associated –H and –OH groups are frequently omitted from structural representations. In the representation of maltose each line represents a covalent bond and each point (except for the one with an oxygen atom) in the hexagon is occupied by a carbon atom and associated –H and –OH groups.

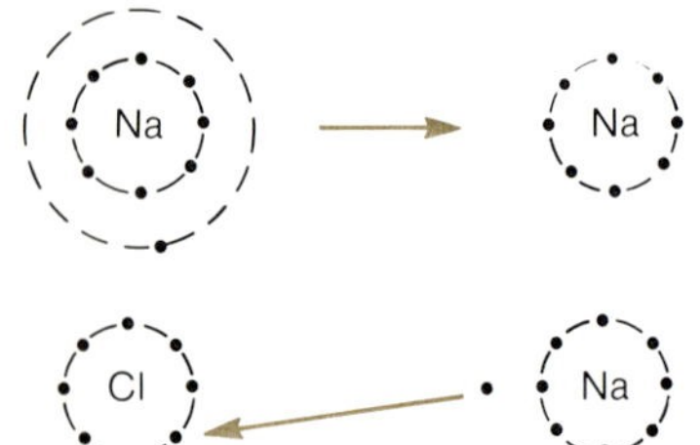

Figure A–7. Sodium chloride. Sodium (Na) has only one electron in its outermost shell and therefore is unable to form enough covalent bonds to fill that outer shell. Loss of that outermost electron leaves sodium with a filled outer shell and a positive charge. The capture of an electron by chlorine (Cl) leaves it with a negative charge. Since opposites attract, the Na^+ and Cl^- ions are held together by an ionic bond.

with sodium, chlorine is much more likely to just "grab" the electron it needs instead of sharing it (Figure A–7). Na^+ and Cl^- ions attract each other and stick together forming NaCl or table salt.

The chemical reactions described above result in the formation of chemical bonds between atoms that produce molecules. Most chemical reactions of interest in biological systems involve molecules, not free atoms. In these chemical reactions chemical bonds are also formed; however, in order to form new bonds, old bonds must first be broken. Therefore metabolism involves the selective breaking and remaking of covalent bonds (Figures A–8 and 13).

Hydrogen Bonds

The covalent and ionic bonds are strong bonds that form between atoms. There are other, weaker bonds that develop between atoms and molecules as the result of weak forces of attraction, for example, the hydrogen bond formed between weakly charged atoms. Within a molecule the different atoms differ in the extent to which they monopolize the electrons that are being shared in a covalent bond. If an electron spends more time with one member of the pair of atoms in a covalent bond, this atom will have a slight negative charge whereas the other atom will have a slight positive charge. A good example of this phenomenon is the water molecule, H_2O. Since oxygen exerts a greater pull on the bond electrons, they spend more time with the

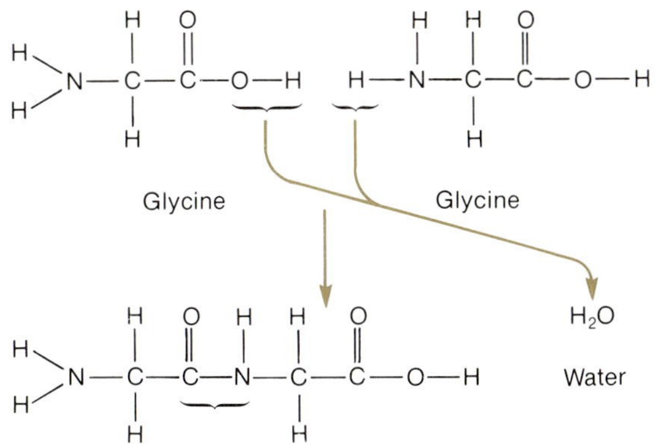

Figure A–8. Example of a metabolic reaction. In this reaction the covalent bond between C and OH of one glycine is broken. At the same time the covalent bond between N and H of the other glycine molecule is broken. The OH and H bond forming water and the C and N form a covalent bond that links the two molecules of glycine forming glycylglycine.

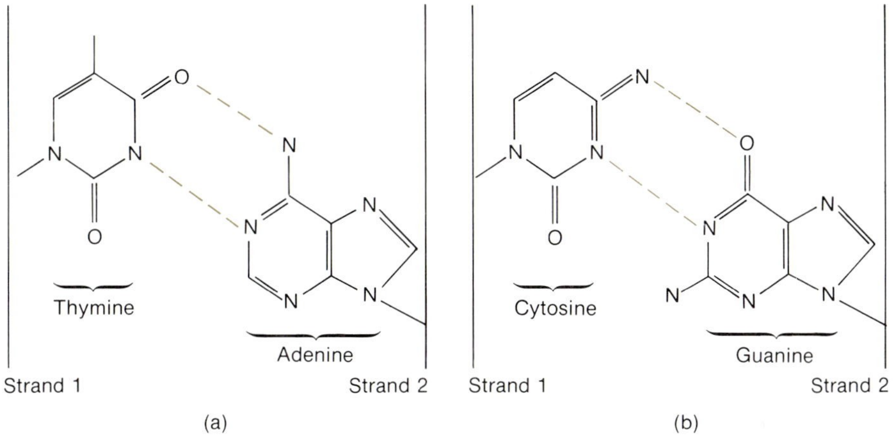

Figure A–9. Hydrogen bonding. Weak ionic attractions between atoms in adjacent molecules will link molecules together without uniting them.

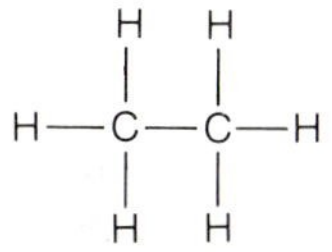

Figure A–10. Carbon–carbon bonding.

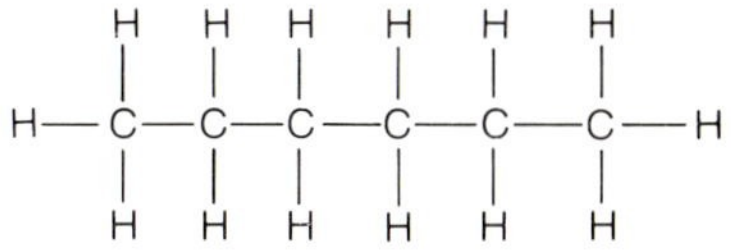

Figure A–11. Carbon backbone. Any number of carbons can be linked in a linear arrangement with additional covalent bonds to other atoms such as hydrogen.

oxygen atom giving it a slight negative charge and the two hydrogen atoms a slight positive charge.

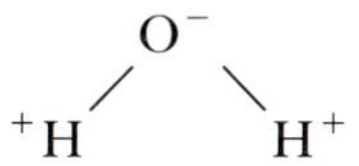

When two water molecules come together they bond to each other by hydrogen bonds.

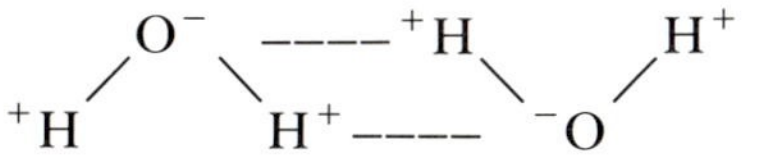

The hydrogen bonds are not strong enough to cause the formation of new molecules but large numbers of such bonds, as formed between bases in nucleic acids (Figure A–9), result in strong molecular associations.

Organic Molecules

The chemistry of life on earth is based on the element carbon (Chapter 3) and the molecules containing carbon are referred to as *organic molecules*. One reason for carbon being essential to the chemistry of life is that each carbon

Figure A–12. Carbon compounds. A wide variety of carbon-containing compounds can be formed with different atoms bonded to the carbon. These can be linear or ring combinations. By convention, the ring forms are written without the carbon atoms and the hydrogen atoms bonded to them.

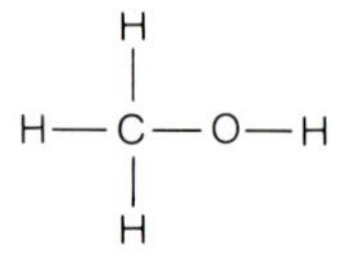

Methyl Alcohol

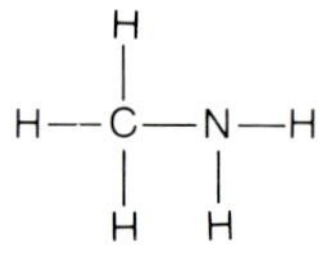

Methyl Amine

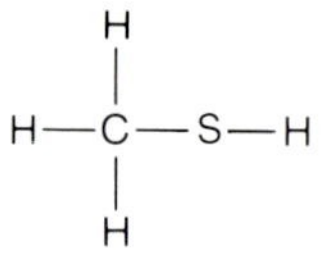

Methyl Mercaptan

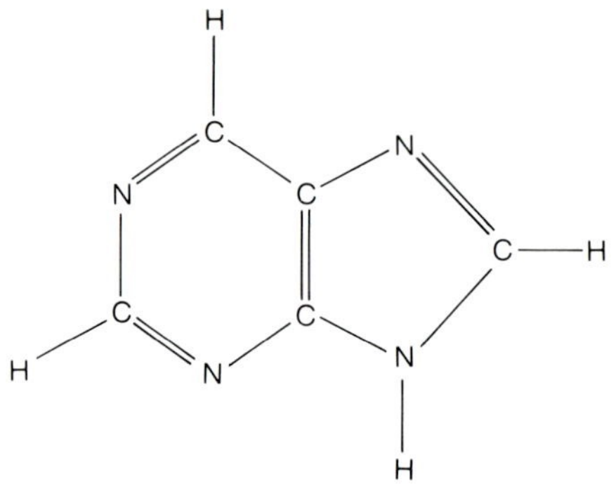

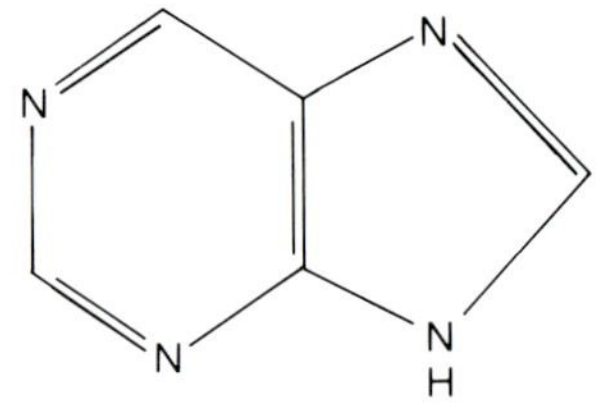

Purine

atom can form four covalent bonds because the outermost shell of a carbon atom contains four electrons. Consequently carbon could react with four hydrogen atoms forming H—C—H (with H above and below) or CH_4 (methane, also known as swamp gas), or one carbon atom could form a covalent bond with another carbon atom (C—C) and each would still be able to form three more covalent bonds (Figure A–10). Long chains of carbon atoms can be formed, yet each atom in the chain would be able to form at least two additional covalent bonds (Figure A–11). In this way many different molecules can be formed with a carbon backbone to which are attached any atoms that can form covalent bonds (Figure A–12). Metabolic reactions involving organic molecules require the breaking of existing bonds so that new bonds can form (Figure A–13) because each carbon atom can only form a total of four covalent bonds. The breaking and making of specific covalent bonds are facilitated by the action of enzymes, which are biological catalysts (Figure A–14).

PGA: $HOOC-HCOP-H_2COH$ ⇌ ($-H_2O$) PEP: $HOOC-COP=CH_2$ ⇌ (ADP → ATP) Pyruvic Acid: $HOOC-C(=O)-CH_3$

Pyruvic Acid → ($NADH + H^+$ → NAD^+) Lactic Acid: $HOOC-HCOH-CH_3$

Figure A–13. Metabolic pathway. The removal of H and OH in the form of water from phosphoglyceric acid (PGA) allows the formation of a carbon–carbon double bond. Removal of the phosphate group (symbolized by P) to form ATP allows another rearrangement of covalent bonds, and pyruvic acid results. Addition of hydrogen atoms to pyruvic converts it to lactic acid.

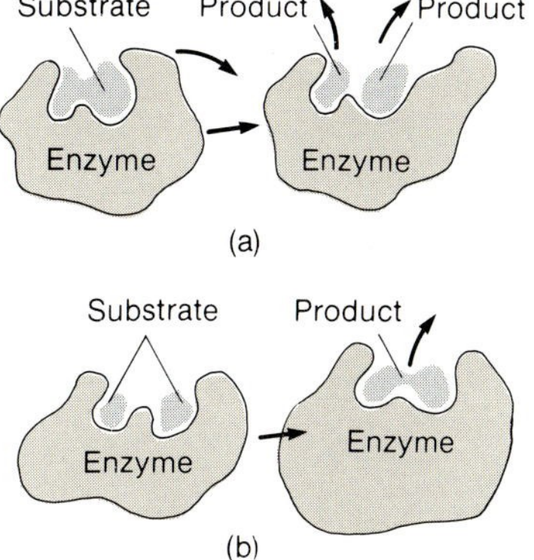

Figure A–14. Mechanism of enzyme action. As the result of binding substrate an enzyme changes its shape. (a) A conformational change in the enzyme that puts a strain on a bond within the substrate can result in the breaking of the bond and the separation of the substrate into two molecules of product. (b) This conformational change can result in two substrate molecules being brought closer together thus overcoming the energy barrier to bond formation between them.

B

Appendix B:

Microscope

Microscopes are commonly used to observe the details of tissue and cell structure of plants and other living organisms. Many of the illustrations in this book are photographs originally taken through a light microscope or an electron microscope. Light microscopes are much more common and are the only type which students in an introductory botany course will have an opportunity to use in their laboratory.

Light Microscopes

The light microscope was invented in the late 16th century in the Netherlands by J. and Z. Janssen. Since that time the light microscope has been refined and is considered to be an instrument that has reached its theoretical limits. Much of the refinement of the light microscope took place in Germany during the 19th century under the leadership of Ernst Abbe.

Presently there are two types of light microscopes used in a general botany laboratory—the dissecting microscope and the compound microscope. The former is used to examine large, opaque objects such as leaf surfaces, flower parts, and seeds, whereas the compound microscope is used to study the structure and arrangement of cells. Unless objects are very small and transparent they must be thin-sectioned before they can be examined with a compound microscope.

Microscopes, like our eyes, are optical instruments, all of which have one thing in common—the ability to show details of an object. This ability is known as the resolving power of an optical instrument. The resolving power is usually expressed as the shortest distance that two objects can be separated and still be recognized as two objects. For most human eyes this value is 0.1 mm, which means that if two points are separated by a distance shorter than 0.1 mm, most individuals will not be able to distinguish two points, but will see the two as one. Under these conditions, one will need a microscope to help resolve the two points.

The amount of detail that can be seen with a light microscope depends on a number of factors. The most fundamental factor and the one that limits the resolving power is the wave nature of light. Photons, or light particles, travel in a wave form, with the distance between two crests or two troughs being the wavelength (Figure B–1). Each color within the visible spectrum has its characteristic wavelength. Wavelengths are measured in nanometers (nm), one nanometer being 1/1,000,000 of a millimeter (mm). The shorter the wavelength, the better the resolving power will be.

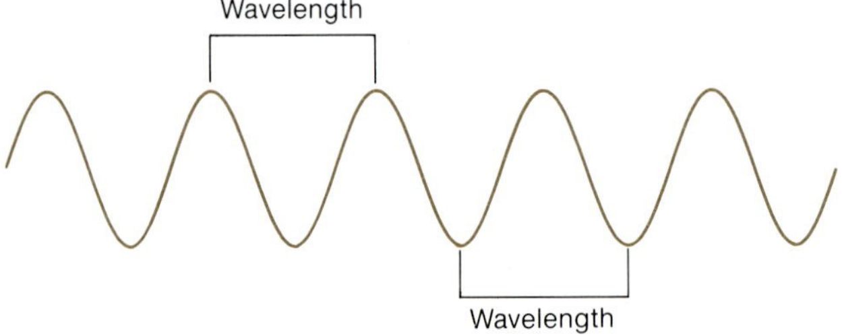

Figure B–1. Diagram representing the wave nature of light. The distance between two crests or two troughs is the wavelength. The wavelength for each color of the visible spectrum is different, as shown in Figure B–2.

The fact that the resolving power of light microscopes is limited by the wavelength of light becomes all the more evident when it is understood that the shortest wavelength usable from the visible spectrum is 400 nm, which is perceived as violet (Figure B–2). As a rule of thumb, light microscopes are able to detect details approximately half the size of the wavelength of light used in illuminating the object. Therefore, considering wavelength as the sole factor, the smallest structures that can be detected with the best light microscope is approximately 200 nm, which is half the wavelength of violet light. The resolving power of most laboratory microscopes used by undergraduate students is considerably poorer–more in the range of 600–1000 nm. Since many of the organelles in a plant cell are smaller than the resolving power of light microscopes, not all of them are visible through a compound microscope.

The theoretical resolving power of a microscope can be calculated by using the equation

$$d = \frac{0.61 \times \lambda}{N.A.}$$

where d = the minimum distance of separation needed for resolving the two points that are next to each other; 0.61 = a constant; and λ (lambda) = wave length. The value of λ can be obtained from the color of light used for illumination. In the case of white light, the mean value of the shortest (violet, 400 nm) and the longest (red, 700 nm) wavelengths, which is 550 nm, can be used. Numerical aperture *(N.A.)*, or the light-gathering power of a lens, is marked on the body of the objective lens of a microscope as a number without any unit after it. Keep in mind that the theoretical resolving power can be attained only if all other conditions of operation are ideal, a situation which is seldom met.

Although it is not apparent from the equation, there are a number of factors in addition to the wavelength that can influence the resolution of an image seen through a microscope. These are mainly related to the physical nature of the lenses and the quality of glass used in making the microscope lenses. This is the reason why inexpensive, department store microscopes are not able to show the details one can achieve with a laboratory microscope. The average laboratory microscope may cost around $500, while good quality research microscopes can cost between $5,000 and $15,000, or even more.

Another feature that all microscopes have in common is the ability to magnify. The total magnification that can be obtained with a microscope is the product of the magnifications of the individual lenses taking part in the formation of an image. In a light microscope these include the objective lens and the ocular lens (viewing lens). The magnifying power of each lens is usually marked on the body of the lens. For compound microscopes the total

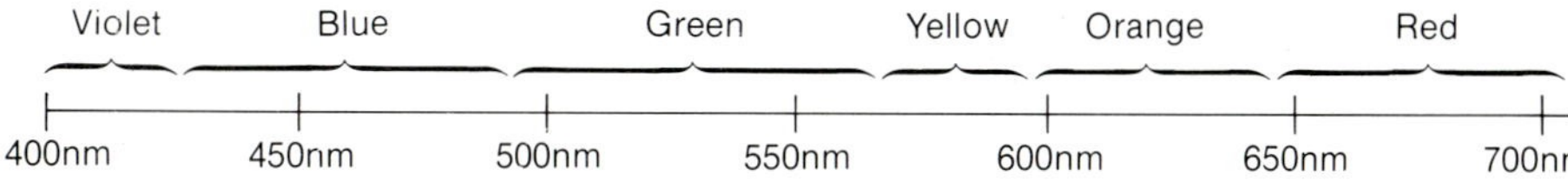

Figure B–2. Spectrum of visible light. This shows the range of wavelengths corresponding to the different colors that comprise white light.

magnification ranges between 40× and 1000×, depending on the objective lens in use.

Unlike the resolving power, magnification can be increased by projecting the image produced by a microscope onto a screen or by taking a picture of the image and then photographically enlarging the image when a print is made. What most people do not realize, however, is that beyond a certain magnification, the image will become less and less sharp. In other words, the resolution of the image decreases with increasing magnification beyond a certain limit. The useful magnification to which an image can be enlarged is directly proportional to the resolving power of the optical system—that is, the better the resolving power, the greater the possible enlargement without losing resolution.

Perhaps many of us have encountered a related situation where a photograph that seems to have a reasonable amount of detail on a wallet-size print appears fuzzy when enlarged to an 8 × 10 inch print. This is because the resolution on the negative was not good enough for the extra enlargement. This restriction of enlargement is a common problem when negatives are made with inexpensive cameras. All other features being equal, the cost of a camera increases depending on the quality of the lens.

Electron Microscopes

Electron microscopes were first built by two teams of researchers working independently in Germany and Canada in the early 1930s. These instruments, called transmission electron microscopes (TEM), did not become available commercially until the late 1940s. In terms of resolution, a TEM is at least a thousand times better than a light microscopes. Present-day TEMs (Fig. B-3) have resolving powers of 0.25 nm or better and are able to show very fine details of even the smallest cell organelle.

How is such incredible resolution achieved through a TEM? The basis for this is again the wavelength, except that in the case of TEM it is the wavelength of electrons rather than of photons (light). Moving electrons, like moving photons, have wavelengths. In addition, moving electrons can be focused using a magnetic or electric field, just as photons are focused by a glass lens. (Fig. B-4) Therefore, a TEM has electrons as a source of illumination instead of light and magnetic lenses in the place of glass lenses. The wavelengths of electrons in a conventional TEM are in the range of 0.004 nm to 0.006 nm, which is approximately 10,000 times smaller than the range of wavelengths for light. Although the numerical apertures of the magnetic lenses of a TEM are much smaller than those of glass lenses, the resolving power of a TEM is still a thousand times better than that of a light microscope. Since the resolving power of a TEM is so much better, the image in a TEM can be enlarged 400,000 times without losing any resolution, compared to the maximum enlargement of 5,000 times for the best light microscope.

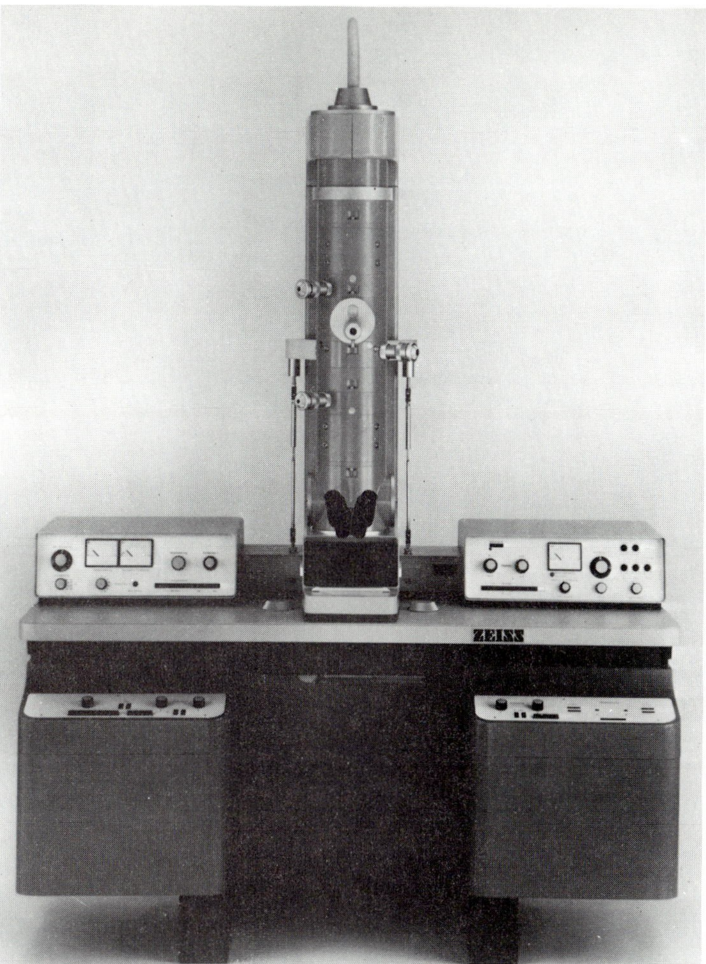

Figure B–3. A transmission electron microscope. Conventional electron microscopes operate in the range of 20,000 to 100,000 volts, which accelerate electrons from the electron gun at the top of the column to the fluorescent screen at the bottom. A high vacuum is maintained in the column, which consists of several electromagnetic lenses in addition to the electron gun. The specimen is inserted into the focal plane of the objective lens. The primary image produced by the objective lens is magnified stepwise by the other lenses. (Photograph courtesy of Carl Zeiss, Inc.)

The scanning electron microscope (SEM) is different from a conventional TEM. Commercially available since the 1960s, the SEM (Fig. B-5) has become very useful in the study of minute organisms and other plant and animal structures. The SEM is able to produce a three-dimensional image of an object, as seen in several photographs included in this book. Although both the TEM and SEM have an electron beam and magnetic lenses and operate at very high voltages under vacuum, the basic principles of image formation in these two microscopes are very different. It is beyond the scope of this appendix to describe image formation in electron microscopes except to state that in the case of a TEM it is similar to that of a light microscope, whereas in the case of an SEM it is closely related to the production of an image on a television screen. The resolving power of the SEM is based on the diameter of the electron beam that scans the surface of the specimen, rather than the wavelength of the electrons as in the case of the TEM. The resolving power of an SEM increases with decreasing diameter of the electron beam. The resolving power of the newer SEMs is approximately 30 nm

Figure B–4. Comparison of the sources of illumination and the lens systems in a light microscope (*a*) and a transmission electron microscope (*b*). All lenses in a light microscope are glass, whereas all lenses in an electron microscope are electromagnets. The image in an electron microscope is viewed on a fluorescent screen.

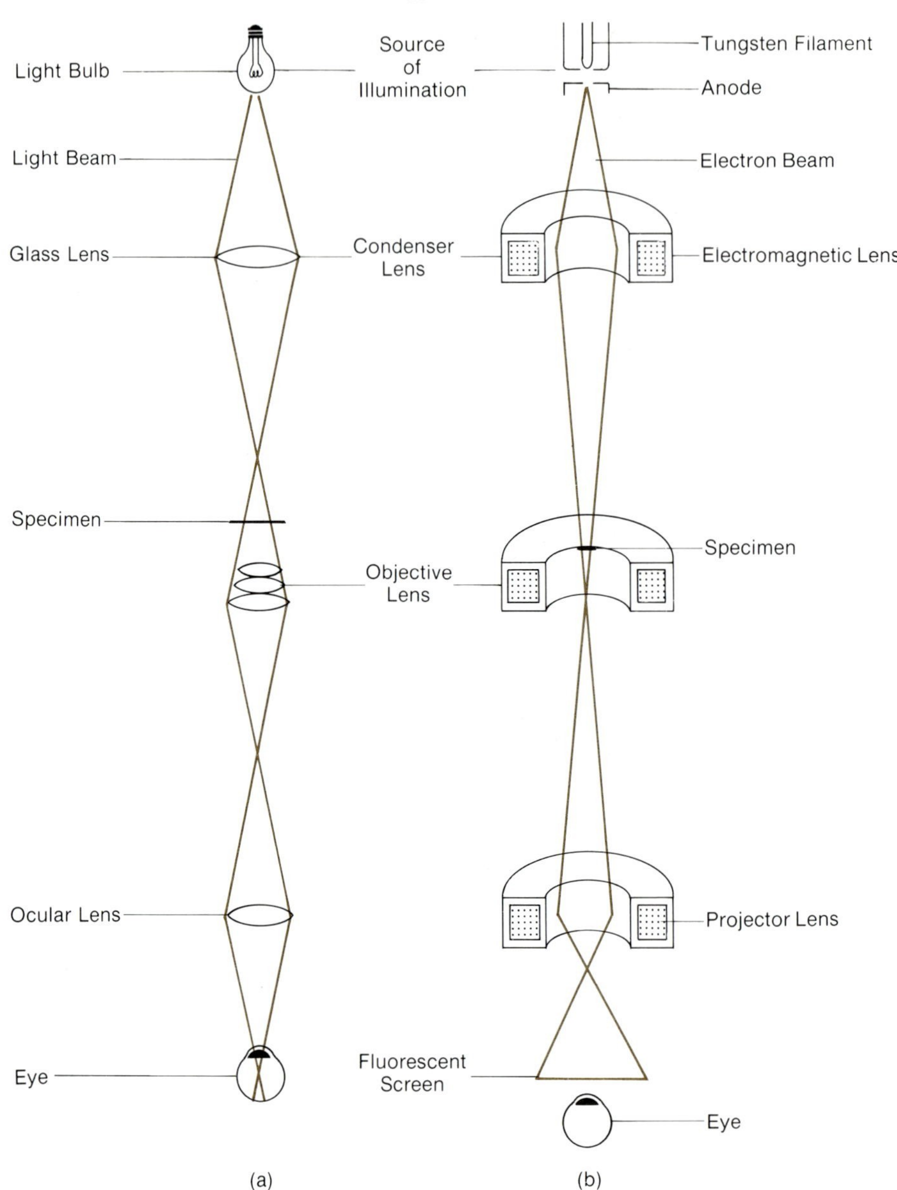

to 40 nm, not as good as a TEM but much better than a light microscope. Magnitications from 5 times to 100,000 times can be routinely obtained with the SEM.

A new kind of electron microscope that has become available in recent years is a combination of a TEM and an SEM known as the scanning transmission electron microscope (STEM). As the name indicates, it has the features of both an SEM and a TEM, and therefore can reveal details that neither an SEM nor a TEM alone can show. This kind of a microscope was first built in 1970 at the University of Chicago in Professor Crewe's laboratory and was used to visualize single atoms for the first time.

Figure B–5. A scanning electron microscope. Scanning electron microscopes operate in the range of 5,000 to 30,000 volts. The electromagnetic lenses in the column condense the electrons to a tiny probe, which scans the surface of the specimen situated at the base of the column. Usually the image in a scanning electron microscope is produced from the electrons that are emitted from the specimen surface as a result of the scanning by this probe.

Specimen Preparation

Most specimens for examination in the compound (light) microscope and the TEM are specially prepared by a rather time-consuming procedure. The reason for this is that since most plant materials are opaque to a beam of light or electrons, they have to be sliced very thin (approximately 0.1—0.01 mm thick for light microscopy and 50–80 nm thick for TEM) before they can be examined. It is not possible to make such thin slices from a living tissue. Therefore small pieces of tissues are first killed and preserved using chemical fixatives, then gradually dehydrated by passing them through increasing concentrations of alcohol, and finally embedded in paraffin, in the case of light microscopy, or plastic resins in the case of TEM work. After the embedding material hardens, thin slices of the embedded tissue are made with instruments called microtomes (Fig. B-6). Very sharp cutting edges, like those of razor blades, are used for making the sections to be examined in the light microscope, while specially broken glass or diamond edges are needed for making the extremely thin sections required for TEM. The sections for light microscopy are mounted on glass microscope slides, soaked in xylene to dissolve the paraffin, stained with dyes, and finally made permanent by gluing a cover glass on the specimen. The thin sections for examination in a TEM are mounted on copper discs with microscopic holes and treated with electron opaque stains before examination.

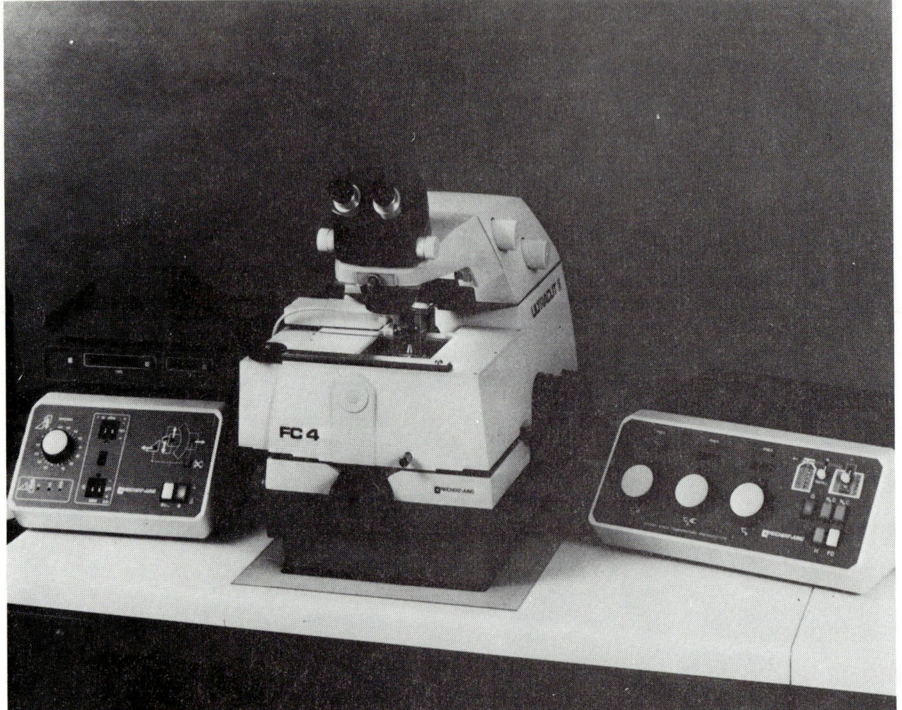

Figure B–6. An ultramicrotome. These instruments are capable of making sections of specially prepared biological specimens embedded in plastic. Extremely sharp glass or diamond knives are used in making sections in the range of 50 to 80 nanometers thick.

Specimens to be examined in an SEM are fixed and dehydrated in acetone, after which the acetone is removed under special conditions to minimize distortions that would occur as the result of air-drying. Samples also can be prepared by freeze-drying (a process similar to the preparation of frozen-dried food) or they simply can be air-dried if the material is noncollapsible. These are then coated with a thin layer (about 20 nm thick) of gold by evaporation of gold in a vacuum chamber before they are examined in the SEM.

Although the sectioning method of specimen preparation is the most commonly used in the light and electron microscopic study of cell structure, there are other methods of specimen preparation, especially for TEM work. Because a discussion of these is beyond the scope of a botany text, the reader is referred to the many excellent books on theory and techniques of light or electron microscopy found in a library.

C

Appendix C:

Green Thumb Botany

For many of us, general knowledge of what plants are and how they grow is interesting but what we really want is a green thumb. To be able to grow plants and not have them die or look abused seems to be an art and not a science when, in fact, all that is needed is the application of general botanical knowledge to the growth of the plants that have been adopted as "pets." Actually, a major part of good plant care is realizing that plants, like animal pets, are living organisms that must have food and water and the proper environment if they are to survive. In addition to the how and why of plant culture, the botanical information in this text provides the basis for understanding the nature of the plants that we grow both indoors and out.

Ornamental Foliage

Probably the major reason that people grow plants in their homes is that they are attractive. The attractiveness of many houseplants is determined by the appearance of their leaves, not their flowers, because flowers, if present, usually last but a short time. Features that contribute to the popularity of foliage plants—plants grown for their leaves—include leaf color, leaf shape, and a striking surface of the leaves.

Leaf Color

Ornamental plants come in a wide variety of colors but they are predominantly green. Green leaves contain the photosynthetic pigment chlorophyll, which is localized exclusively in chloroplasts. The different shades of green leaves is related to different concentrations of chlorophyll, with darker leaves having a higher concentration of chlorophyll. Some leaves with a low concentration of chlorophyll may appear yellow. This can occur when the leaves are dying or, on the other hand, this may be the normal, healthy leaf color of a particular species or variety. The presence of carotenoids in special plastids, or chromoplasts, accounts for the yellow color leaves. Green leaves also contain carotenoids in their chloroplasts but the greater concentration of chlorophyll masks the yellow carotenoids.

Although yellow in leaves is due to carotenoids, yellow petals in a flower could either contain carotenoids or another pigment called a flavone. Flavones are found only in the vacuole of a plant cell and usually only in epidermal cells. Leaves usually do not contain significant amounts of yellow flavones but they do contain other related vacuolar pigments, the anthocyanins, that produce purple, red, and blue leaves or parts of leaves. Like the flavones, the anthocyanins are usually restricted to the vacuoles of epidermal cells. Since anthocyanins are produced when the cells have excess sugar, these colors are most prominent in plants which have been grown under good light conditions. The leaves of some plants appear brown but are obviously not dead or dying. The color of these leaves is due to the presence of just the right amount of red anthocyanin in the epidermal layer overlying a green mesophyll layer.

Both white and silver regions on leaves are due to the absence of all pigments. The white regions reflect all of the colors of sunlight back to the eye and the brain perceives this mixture of colors as white. The silvery regions on the leaves of aluminum plant result from a blister created by the separation of the colorless epidermis from the mesophyll layer. Light passing through the water of the epidermal cells to the air of the blister is reflected resulting in a silvery-gray appearance. If the air were squeezed out of the blister, the epidermis would collapse onto the mesophyll and the region would appear green.

The value to the plant of any of these pigments or pigment distributions depends on the specific situations. Obviously chlorophyll is valuable as it is required in photosynthesis. The advantages of leaf colors other than green, however, are not so obvious. Before searching for functions for these other colors it is important to realize that many ornamental plants are cultivars that were selected from the natural population of that species by humans precisely because they had unusual leaf color patterns. Therefore the frequency and overall importance of these unusual patterns of leaf coloration tend to be over exaggerated by nonbotanists. Naturally occurring species and populations of species that have nongreen leaves, however, may derive some advantage from the presence of these additional leaf pigments. Because these pigments absorb light, they may serve to protect a plant from high light intensities. In fact, the amount of anthocyanin in the leaves of plants such as the purple wandering Jew is directly related to the amount of sunlight they have received—low light intensity results in plants with green leaves, but direct sunlight causes the entire leaf to appear purple. This ability to produce anthocyanins in response to light intensity would appear to be an adaptation that permits certain shade-tolerant plant species to extend their habitat into sunny environments. A purple or blue leaf also may be less appetizing to a herbivorous animal than would a normal green leaf—just as blue mashed potatoes and red fried eggs would be difficult for most humans to eat. Because green is the predominant leaf color, however, the other adaptations of plants to high light intensity and herbivory must be most effective.

Leaf Surface

The presence or absence of hairs is a distinctive feature of the leaf surface. Most leaves possess some epidermal hairs but usually a microscope is needed to see them. These hairs serve a variety of purposes including protection against insects and reduction of transpiration. Larger hairs, visible to the naked eye, contribute to the aesthetics of the plant. In the case of velvet leaf (purple passion plant) the epidermal hairs contain anthocyanin, which gives the leaf its purple color.

Plants such as rubber tree and certain peperomias have shiny leaves devoid of all hairs. The cuticle of these leaves is invested with an extra thick layer of wax. Rubber tree leaves, in particular, suffer from "wax buildup" and will appear dull unless buffed regularly with a soft cloth to raise a gloss.

Leaf Shape

In chapter 8 the considerable variation in the shape of leaves and leaf margins was indicated. For certain ornamental plants, growing conditions determine the shape of the leaves. Boston ferns, for example, need adequate light if they are to produce the doubly compound leaves typical of this species. Certain types of split-leaf philodendrons climb a support into which aerial roots can grow. In the absence of this support, the leaves will not develop their characteristic split appearance.

Cultural Requirements of Houseplants

While it is necessary to know that plants need light and water to grow, it is essential to know how much of these and other plant growth factors must be provided. The ideal situation would be to have a list of the cultural requirements for each plant in a collection; however, this is not usually the case. A rule of thumb might be to let the plants show you what they need: for instance, it is better not to water any more often than is necessary to prevent wilting; house plants should be kept out of the strongest sunlight unless they begin to show straggly growth, or fail to develop an expected anthocyanin. These and other cultural requirements of houseplants will be explained below.

Light

Plants require light for the photosynthetic production of food. In the dark, plants live off their stored reserves of starch. After several days in the dark new plant growth will be spindly and yellow, that is, etiolated. The yellow color comes about because light is necessary for chlorophyll synthesis, but not for carotenoid synthesis. Spindly growth with unusually elongated internodes occurs because, in the absence of light, sufficient auxins are available that excessive stem elongation will occur. Even when a plant is in the light it can still exhibit spindly growth if there is insufficient light to prevent this excessive internode elongation.

Despite the requirement of sufficient light for plant growth and survival, people insist on using plants to decorate offices and shopping centers where the level of light may be inadequate. The plants used in these areas are typically slow-growing, foliage plants that can tolerate low-light conditions. Since they grow slowly, etiolated growth appears more slowly and the aesthetic value of the plant takes longer to deteriorate.

Although the plants used in these environments are shade-tolerant, they still need sufficient light to produce enough food by photosynthesis that they are not on a "starvation diet." As food reserves become depleted, older leaves are cannibalized and their carbohydrates and amino acids are translocated to the younger leaves and growing tip. This process is obvious to the observer because the older leaves turn yellow and fall prematurely. A common complaint of people trying to grow rubber trees is that every time a new leaf forms, an old leaf drops. Leaf drop could be prevented by increasing the amount of available light. The problem with trying to determine whether there is adequate light for plant growth is that the human eye is a poor light meter because it functions equally well over such a wide range of light intensities. Even such shade-tolerant species as *Dieffenbachia* (dumbcane) or *Aspidistra* (cast iron plant) need at least 100 footcandles of light if they are to survive long, yet well-lighted interior areas often have no more than 60–80 footcandles of illumination. To the human eye there is ample light, whereas to the plant the light is insufficient. Despite this warning all plants should not be grown in the window providing the highest light intensity. All species have adapted to a particular range of light intensities through evolution; consequently they will not do well under other lighting conditions. Shade-loving plants either will develop bronze-colored areas of "sun burn" on their leaves or else the leaf tips will die and turn brown. This is where knowing your plants is important so that the correct lighting conditions can be provided before plant damage occurs. In the absence of this information a medium light intensity is a good starting point. The plant can be moved about as its growth dictates.

If the average home does not have enough areas with sufficient light intensities, plants can be grown under lights. However, in an office or a shopping center or even a home an alternate solution is to rotate plants between areas with insufficient light and those with adequate light. In many commercial settings plants can be rented from an interior landscaper who returns them to the greenhouse every several months to recuperate.

Water

Although houseplants require water, different species require different amounts, best determined by trial and error. Most of the water taken up by the plant is lost in the process of transpiration. If water were lost from a leaf faster than it could be replaced, the leaf cells would become plasmolyzed and the leaf would wilt. Misting a wilted leaf will not reverse this process. Instead, water must be added to the soil or the plant should be taken out of direct sunlight. The increase in relative humidity from misting the plant can be sufficient to reduce the transpiration rate and to prevent wilting. The dry, heated, indoor air found in our homes in wintertime can mimic desert conditions and can pose a greater problem to the indoor horticulturist than just wilting. Under the conditions of extreme low humidity delicate plants more likely will develop brown leaf tips, a result of water stress in cells furthest removed from the central xylem, causing these cells to die. Brown leaf tips and wilting also can be caused by over- and underwatering: the underwatered plant has insufficient available water; excessive watering damages the roots, thereby preventing adequate water uptake and resulting in water stress in the leaves.

Fertilizer

As indicated in Chapter 7, plants need certain minerals if they are to grow well. A good potting soil provides an ample supply of these minerals for the initial growth of plants; however, these are used up by the plant and are leached out of the soil during watering. Fertilizer applied to the soil will replace these minerals and promote healthy plant growth.

Fertilizers come in forms such as soluble powders, liquid concentrate, slow-release pellets, and slow-release sticks; however, the form of the fertilizer is not important to the plants. In each case the manufacturer's instructions on frequency and amount of fertilizer to be used should be followed. During periods of rapid growth, liquid fertilizer should be applied about once a month and then, as growth slows, the frequency of fertilization should be reduced. Dormant plants should not be ferilized at all, whereas plants that have reached the maximum desired size should be fertilized only as frequently as necessary to prevent leaf yellowing or leaf drop.

Plant Growth Responses

We often observe houseplants growing toward a window, a phototropic response that can result in deformed plants. To prevent deformity rotate the plants a quarter turn on a regular basis. Fast growers will require more frequent turning.

Some plants are normally branched while others are not. Houseplant branching is controlled by apical dominance (Chapter 12). Plants, such as *Coleus,* with low apical dominance branch readily and produce bushy plants. *Dieffenbachia* and split-leaf philodendron normally produce but a single stem, yet if the top part of the plant were removed, branches would form. Typically only the two or three lateral buds nearest the top of the cut stem will begin to develop into branches. With some plants the loss of the leaves is sufficient to induce branching even though the apical region is undamaged. Since the control of branching is due to the auxin concentration reaching the lateral buds, the loss of leaves most likely stimulates branching because the leaf is a major source of auxin.

Propagation

Plants can be propagated either sexually or asexually. Sexual reproduction of most houseplants results in seed production because most houseplants are seed-producing plants. However, not all flowering plants flower regularly in the house and even those that do seldom produce seed either because the proper pollinating mechanism is not functioning or because they flower only once in the year. Therefore the most feasible method of houseplant propagation is by asexual reproduction.

The method used most successfully with houseplants involves the stem cutting—a terminal portion of a stem bearing leaves and lacking roots. Production of a new plant from a stem cutting only requires conditions that will induce root formation on the cutting. The usual practice for herbaceous stems is to place the lower part of the cutting in water, and within two or three weeks roots will form. When roots do not form under these conditions, perhaps darkness is required in the root-forming region or the water contains insufficient dissolved oxygen. Rooting of certain species is inhibited by light, so it should be a routine practice to use an opaque container for rooting. The requirement for oxygen is more difficult to overcome when water is the rooting medium, although bubbling air through the water with an aquarium pump would be beneficial in most cases. Oxygen is necessary for root formation in all species because cell division and growth are energy-requiring processes, the energy being produced in aerobic respiration. Species that root in water are able to obtain enough oxygen from the water to support aerobic respiration at the required rates. A species that will not root in water is best rooted in sand or a specialized commercial rooting mixture of sand and peat moss.

The greatest success rate comes in propagating cuttings that are 3–6 inches long and have most of the leaves removed. For a plant with large leaves only one or two should be left on the cutting because water lost by transpiration is not replaced as readily in the absence of roots. To increase the probability that roots will form, the cut should be made about 1/2 inch below a node because cells in the nodal region are more likely to initiate roots. Root initiation requires auxins, often present in sufficient quantity in herbaceous cuttings that rooting will occur naturally. To initiate rooting in woody cuttings, often containing insufficient auxin levels, the cut end should be treated with a rooting compound containing auxin.

Other methods of asexual propagation include leaf cuttings, "eye" cuttings, and runners. Species such as African violet, which do not have an elongated stem, often can be propagated by taking whole leaves and partially burying them in sand or rooting mixture until they form new plants. *Dieffenbachia* are prepared by cutting the stem into short segments, each containing two nodes. The segments are then buried horizontally half in sand or rooting mixture. After about a month a new shoot forms from a lateral bud at one node while roots form at the other node. Spider plants and strawberry begonias are known for the plantlet-bearing runners that they produce. These plantlets either can be separated from the runner and placed on soil to root, or the runner can be rested on soil until the plantlet roots and then be severed from the parent plant.

Forcing and Induction of Flowering

It is very easy to have hyacinths, tulips, crocuses, and daffodils bloom in your home in January and February. Bulbs purchased in October or early November should be planted immediately in soil in pots so that only their tips show. After watering, the pots must be placed in a cool, dark place. The temperature should be below 50° F and above freezing (a refrigerator is ideal for several pots). Low temperatures, which mimic winter conditions, are required to break the dormancy of the bulbs. The bulbs should be kept watered and under cool dry conditions until etiolated shoot growth appears. At this time they can be transferred to warm, well-lighted conditions after which flower buds will develop. Forced bulbs can be planted in the garden later, but they should not be forced a second time.

Other more typical houseplants can also be induced to flower. Many cacti will flower if exposed to cool, dry conditions during the winter months. In fact temperatures close to the freezing point at night are ideal when combined with dryer than normal growing conditions. Other houseplants flower in response to the photoperiod (chapter 12) and a knowledge of their photoperiodic requirements will pay off in flowers instead of foliage. Two easily induced plants are the poinsettia and the Christmas cactus. To induce flowering in a poinsettia it must be kept in complete darkness for 14 continuous hours out of every 24-hour day for at least 1 month starting in October. Regular interruptions of the dark period will produce leafy poinsettias with no flowers. The Christmas cactus is also a short-day plant. It can be induced by keeping it in darkness from sundown to sunup during the autumn, at a time when the soil is dryer than usual. Under these conditions it should flower during the Christmas season.

Plant Diseases

Plant diseases can be either infectious or noninfectious. Infectious diseases are caused by living organisms such as bacteria, viruses, fungi, or insects. Noninfectious diseases are caused by environmental factors such as mineral deficiency, improper temperature, water stress, improper light, and so on. The way to treat both kinds of plant diseases is to identify the causative agent and eliminate it. The problem is that the identification of the causative agent is not always easy. Plants have a limited range of visible symptoms: wilting, chlorosis (breakdown of chlorophyll), and necrosis (tissue death) are the most obvious. As indicated above the same symptom can have different causes—wilting can result from both over-and underwatering. Therefore it is important to observe plants regularly so that the symptoms can be seen in the context of the total care that the plant has received.

Most infectious houseplant diseases are caused by insects and other small invertebrate organisms that possess either sucking or biting mouth parts and use the plant as a source of food. The most successful cure for these diseases is to catch them early before they can spoil the appearance of the plant or spread to the rest of the collection.

A few of the more common organisms are mealy bugs, scale insects, and spider mites. Mealy bugs are oval-shaped, white organisms with many legs. Both the mealy bug and their white, cottony egg masses are found at the base of a leaf or on the underside of a leaf. The mealy bug is slow-moving, and scale insects never appear to move. A plant infested with scale will have hard, oval-shaped, brown lumps along young stems and on the undersides of leaves, and, in addition, the leaves feel sticky. Fortunately scale insects prefer leathery-leaved plants like cycads, some ferns, and rubber trees, and not the majority of houseplants. Like scale, spider mites cause sticky leaves and they also cause small, yellow spots to appear on the upper surface of a leaf. Although spider mites are barely visible to the naked eye, they can be detected by the presence of these symptoms or the presence of their webs on the undersides of leaves.

The simplest way to get rid of plant pests is to discard the infected plant. Alternatively, mealy bugs and scale insects can be removed manually, or if they are too numerous, the whole plant could be dipped in warm, soapy water (which would also affect spider mites). Another alternative is to kill the organisms by dabbing each with rubbing alcohol or nail polish remover. If these treatments are neither practical nor effective, a last resort would be a pesticide, although sprays should not be used indoors. A pesticide treatment that works with minimum hazard to the user is to place the plants in a plastic bag with a pest strip for 24 hours.

Outdoor plants are susceptible to a number of fungal infections, but in houseplants fungi usually are not a major problem. Sometimes a gray-black layer will form on leaves due to the growth of powdery mildew. African violets are particularly susceptible to *Botrytis,* which also appears as a grayish mold. When a plant shows signs of a fungal disease, it is probably best to discard the plant because fungicides are not that effective and they are hazardous to humans. Bacterial and viral diseases are uncommon in houseplants. There is essentially no way of controlling them short of throwing away the plant.

With the correct choice of plants to adopt and the application of the botanical principles outlined in this text, you, too, can have a green thumb.

Glossary

ADP adenosine diphosphate, a low-energy molecule that is a precursor to ATP.

ATP adenosine triphosphate, a high-energy molecule that is involved in the transfer of chemical energy in cells.

Abiotically in the absence of living organisms.

Abscisic acid a plant hormone that induces bud dormancy and leaf abscission.

Abscission layer a layer of cells that is involved in the separation of a leaf petiole or fruit stalk from the stem.

Abscission zone (see abscission layer).

Abscission the falling off of plant parts.

Actinomycete a filamentous form of soil bacteria.

Active transport the transport of molecules across a differentially permeable membrane that requires the expenditure of energy.

Adventitious produced from a structure other than the usual one as in roots produced from a stem.

Adventitious buds buds produced from structures other than apical meristems.

Adventitious root a root from an organ other than a root.

Aerial refers to the aboveground parts.

Aerial root an aboveground root that absorbs moisture from the air.

Aerobic requiring the presence of oxygen.

Aerobic respiration the breakdown of organic molecules in the presence of oxygen to release stored energy that can be used to make ATP.

Agar a substance, obtained from red algae, that forms a gel on which microorganisms can be grown.

Aggregate fruit a fruit derived from several separate pistils of the same flower.

Agronomy the study of agricultural crop production practices.

Akinete a thick-walled, nonmotile spore produced by certain cyanobacteria.

Alcoholic fermentation the process of anaerobic respiration in which glucose is converted to ethyl alcohol.

Aleurone layer one of the outer layers of a cereal seed.

Alga a photosynthetic organism characterized by the absence of vascular tissue and the presence of unicellular gametangia.

Algin a jellylike substance, obtained from the cell walls of certain brown algae, used as a stabilizer in dairy products.

Alkaloid member of a large group of nitrogen-containing organic chemicals, many of which taste bitter and are frequently poisonous to animals and humans.

Allele one of the two or more forms of a gene that codes for a different form of the same protein or polypeptide.

Allergenic capable of causing an allergic reaction.

Allophycocyanin a blue, water-soluble pigment found in cyanobacteria and red algae.

Alternate leaf arrangement arrangement of leaves along a stem in which there is only one leaf at a node.

Alternation of generations the alternation between the diploid and the haploid phase of the life cycle.

Amber fossilized resin.

Amino acid a simple organic molecule that is the building block of proteins.

Amyloplast a starch-storing plastid.

Anaerobe an organism that lives only in the absence of oxygen.

Anaerobic in the absence of oxygen.

Anaerobic respiration the breakdown of organic molecules in the absence of oxygen to release stored energy that can be used to make ATP.

Anaphase the stage in nuclear division during which the chromatids separate and move to opposite poles.

Anatomy the structure and organization of tissues within an organism.

Androecium a collective term for all of the "male" reproductive structures of a flower.

Angiosperm one of a group of plants characterized by the production of flowers.

Anisogamy a form of sexual reproduction involving the fusion of motile gametes of unequal size.

Annual a plant that completes its life cycle in one growing season.

Annual ring one year's growth of secondary xylem in a woody plant.

Annulus the row of unevenly thickened cells around a fern sporangium.

Anterior at the front.

Anther the portion of the stamen in a flower in which pollen is produced.

Antheridiophore a stalk that bears antheridia.

Antheridium a gametangium in which sperm are produced.

Anthocyanin a water-soluble pigment of variable color present in cell vacuoles.

Anthophyta the division to which all flowering plants belong.

Antibiotic a class of chemical substances produced by certain fungi and bacteria and capable of killing other microorganisms.

Anticodon a sequence of three bases on the tRNA complementary to the triplet codon on the mRNA.

Antipodals the group of nuclei in the angiosperm megagametophyte at the end opposite the micropyle.

Apical bud the growing tip of a stem.

Apical meristem a small group of undifferentiated, dividing cells present at the tip of a stem or a root.

Aplanospore among algae, a zoospore that has not developed flagella; among fungi, a nonmotile spore.

Arborescent treelike in appearance.

Arbuscle highly branched fungal structure within the root cell that functions for the exchange of nutrients.

Archegoniophore a stalk that bears archegonia.

Archegonium a gametangium in which an egg is produced.

Arthrophyta the division to which the horsetails belong.

Ascocarp the fruiting body of an ascomycete fungus.

Ascogonium the female gametangium of an ascomycete fungus.

Ascomycetes a class of fungi in which nuclear fusion and meiosis occur in an ascus.

Ascospore the haploid spore produced meiotically in the ascus.

Ascus the terminal portion of a dikaryotic hypha in which nuclear fusion and meiosis occur producing ascospores.

Asexual reproduction reproduction of organisms by nonsexual means.

Asexually without sex.

Assay method for measuring the amount of hormone present.

Atropine an alkaloid produced in all parts of belladonna and used in small doses as a pain killer and in the treatment of hay fever.

Autotroph an organism that can produce its own food from inorganic molecules.

Auxin a plant hormone that is responsible for cell enlargement.

Axil the angle formed between the stem and the leaf.

Axile placentation the arrangement of placenta in the center of a compound pistil with two or more separate chambers.

Axillary bud (see lateral bud).

Bacillus a rod-shaped bacterium.

Bacteria prokaryotic organisms that typically are nonphotosynthetic.

Bark all tissues located outside the vascular cambium.

Basal placentation placenta produced at the lower end of the carpel in a simple pistil containing only a single ovule.

Basidiocarp the fruiting body of a basidiomycete fungus.

Basidiomycetes a group of fungi in which nuclear fusion and meiosis occur in a basidium.

Basidiospore the haploid spore produced meiotically on the basidium.

Basidium the terminal portion of a dikaryotic hypha in which nuclear fusion and meiosis occur producing basidiospores.

Berry a simple fruit whose exocarp forms an outer skin surrounding the fleshy mesocarp and endocarp.

Betacyanin a water-soluble, red-purple pigment found in the vacuoles of cells of a limited number of flowering plants.

Biennial a plant that requires two growing seasons to complete its life cycle.

Binary fission the process of cell division in prokaryotic cells in which the protoplast is divided in two by an ingrowth of the cell wall.

Binomial system the system for naming organisms in which each organism receives a unique name consisting of two parts, the genus and the species.

Bioluminescence the production of light by living organisms.

Biome a combination of plants that is characterized by a particular vegetation type such as grasses or deciduous trees.

Biosphere that portion of the earth's surface and atmosphere occupied by living organisms.

Bipinnately compound a compound leaf in which the leaflets are arranged along secondary petioles that, in turn, are arranged along the length of the primary petiole.

Bisexual able to produce both male and female gametes.

Blade the broad, flat, photosynthetic portion of a leaf.

Bloom an extremely high density of an algal species.

Blue-green algae (see cyanobacteria).

Botanical garden a garden that functions as a living herbarium in which is grown a wide variety of plant species from many parts of the world.

Botany the scientific study of plants (including algae, fungi, and bacteria).

Branch a stem that develops from a lateral (axillary) bud.

Branch root a root formed as an outgrowth of an existing root.

Breathing root an exposed root of a swamp tree (such as mangrove) that serves as an avenue of gas exchange for the submerged root system.

Bryology the study of mosses, liverworts, and hornworts.

Bryophyta the plant division containing the mosses, liverworts, and hornworts.

Bryophyte a member of a group of primitive terrestrial plants (mosses, liverworts, and hornworts).

Bud primordium (primordia, plural) a group of meristematic cells that gives rise to a bud.

Bud scale a modified leaf that protects a bud.

Bud a structure that gives rise to a branch shoot or flower.

Bulb an asexual reproductive structure composed of a small, flattened stem to which is attached a tightly packed mass of swollen, storage leaves.

Bulliform cell a large epidermal cell that is believed to be involved in the curling of grass leaves during water stress.

Bundle cap the fiber cells that form a protective layer just to the outside of the primary phloem.

Bundle sheath a single layer of parenchyma cells that surrounds the vascular tissues in a leaf; also a group of fiber cells that surrounds the vascular tissues in a monocotyledonous stem.

C_3 pathway the pathway of carbon fixation in which the first product to contain the newly incorporated carbon atom is a three-carbon compound.

C_4 pathway the pathway of carbon fixation in which the first product to contain the newly incorporated carbon atom is a four-carbon compound.

Calyx the collective term for all of the sepals of a flower.

Cambium (cambia, plural) a secondary meristem.

Capillary action the upward movement of water through narrow tubes as a result of the surface tension of water.

Capsule a dry, dehiscent fruit formed from two or more fused carpels.

Carbohydrate an organic molecule consisting of carbon, hydrogen, and oxygen in an approximate ratio of 1:2:1, the simplest carbohydrate is known as a sugar.

Carnivore an animal that eats other animals.

Carnivorous consuming animal flesh.

Carotenoid one of a group of yellow, orange, or red pigments some of which function as accessory pigments in photosynthesis.

Carpel a simple pistil or one of the units of a compound pistil.

Carrageenan a jellylike substance obtained from certain red algae used as a stabilizer in dairy products.

Carrier molecule a membrane-associated molecule that functions to transport specific molecules or atoms across the membrane.

Casparian strip a continuous band of suberin that is located in the walls of an endodermal cell.

Catkin a type of inflorescence in which the unisexual flowers are arranged on a pendulous stalk.

Cell the smallest unit of a living organism that can remain alive by itself; typically composed of cytoplasm, nucleus, and cell wall in plants.

Cell cycle a sequence of events that cells undergo before, during, and after cell division.

Cell division the process of nuclear and cytoplasmic division that forms two new cells.

Cell membrane (see plasmalemma).

Cell plate the first cell-wall layer laid down during cytoplasmic division.

Cell theory the statement that all living organisms are composed of cells.

Cell wall the outer, rigid boundary of a cell that is composed primarily of cellulose, hemicellulose, and pectin in plants.

Cell/tissue culture a technique for growing isolated cells or tissue in a sterile medium of known composition.

Cellulose a polymer of glucose that forms the main structural component of plant cell walls.

Centriole a cytoplasmic organelle present in cells that are motile; characteristic structure of all animal cells. Centrioles are associated with the bases of flagella as well as with the formation of spindle fibers in such cells.

Centromere a constricted region of the chromosome which joins the two chromatids and to which the spindle fibers attach.

Chemosynthesis the metabolic process in which CO_2 is incorporated into organic molecules using chemical energy as the energy source.

Chlorophyll a green pigment that functions in photosynthesis.

Chlorophyta the division to which the green algae belong.

Chloroplast the type of plastid in which photosynthesis occurs.

Chromatid one of the two longitudinal halves of a chromosome

Chromatin fine, threadlike genetic material composed of DNA and protein from which chromosomes form.

Chromatography the separation of molecules on the basis of their relative affinities for the stationary and mobile phases of the system.

Chromoplast a plastid containing carotenoid pigments.

Chromosome a structure composed of DNA that contains all or part of the genetic information of a cell.

Chrysophyta the division to which the golden-brown algae belong.

Cicutoxin a highly toxic substance produced in the roots of water hemlock.

Cilia (cilium, singular) small, hairlike cell structures that function in cell motility.

Cisterna a flattened, circular, membrane-bounded sac.

Cladode a modified stem that serves as the photosynthetic organ of a plant in place of the leaves.

Class a major taxonomic category between division and order.

Classical genetics the study of the inheritance of observable physical characteristics (contrast with molecular genetics).

Climax community the community of organisms that by its presence establishes and maintains the balance of ecological factors that favors the growth and reproduction of members of that community.

Climbing root an adventitious root produced from the stem that can attach a plant to a vertical surface.

Club moss common name for certain surviving species of the Microphyllophyta.

Cocaine an alkaloid obtained from coca leaves and used as a narcotic.

Coccus a spherical bacterium.

Codeine an alkaloid obtained from extracts of opium poppy; a common ingredient in cough syrups.

Codon a sequence of three bases in mRNA that codes for one amino acid.

Coenocytic hyphae hyphae that lack crosswalls resulting in a multinucleate condition.

Coleoptile the sheath that surrounds a grass shoot during germination.

Collenchyma a cell with cell walls thickened at the corners.

Colloid (colloidal, adj.) very fine particles that will remain in suspension in a liquid.

Colony (colonial, adj.) an aggregation of organisms of the same species all derived from one cell or reproductive structure.

Companion cell a cell type of phloem tissue; always associated with sieve tube members.

Complete flower a flower that contains all floral parts (sepals, petals, stamens, and pistil).

Complex tissue a tissue composed of more than one cell type.

Compound leaf a leaf whose blade is divided into leaflets.

Compound pistil a pistil formed by the fusion of two or more carpels.

Concentration gradient an unequal distribution of molecules in solution.

Cone a short branch along which sporophylls or ovuliferous scales are arranged; a strobilus.

Conidium a unicellular, asexual reproductive structure produced from the terminal portion of a fungal hypha.

Conifer the major group of gymnosperms; e.g., pines, firs, spruces.

Coniferophyta the plant division composed of conifers.

Conjugation tube a connection between mating cells through which nuclei (or protoplasts) pass.

Conjugation contact between two cells by means of a tubular projection during sexual reproduction; a form of isogamy.

Continental drift movement of continents as consequence of the spreading of the ocean floor.

Coral reef a reef produced by the secretion of the skeleton that houses the coral polyps.

Cork cambium a secondary meristem that produces cork and phelloderm.

Cork a secondary tissue containing suberized cells that forms the outer layer of woody stems and roots.

Corm an upright underground stem modification that functions in asexual reproduction.

Corolla a collective term for all of the petals of a flower.

Corolla tube a tubular structure produced by the fusion of all of the petals of the flower.

Cortex a region of the stem or root consisting mainly of parenchyma cells, located between the epidermis or cork and the vascular cylinder.

Corymb an inflorescence in which the flower stalks arranged on the peduncle are of unequal length thus giving a flat-topped appearance to the inflorescence.

Cotyledon a leaflike structure in the seed that may contain stored food.

Covalent bond a chemical bond that results from two atoms sharing a pair of electrons.

Crassulacean acid metabolism a variation of the C_3 pathway of photosynthetic carbon fixation in which all of the CO_2 incorporated comes from the breakdown of organic acids that had been formed at night when the stomata are open for gas exchange.

Crista (cristae, plural) the tubular projection of the inner mitochondrial membrane where electron transport reactions take place.

Critical daylength a characteristic daylength associated with photoperiodically-controlled phenomena. Short-day plants respond to daylengths shorter than their critical daylength and long-day plants respond to daylengths longer than their critical daylength.

Cross pollination transfer of pollen from a flower of one plant to a flower on a different plant.

Cross-section a plane of section at right angles to the long axis.

Crossing over the exchange of chromatid segments between homologous chromosomes during meiosis.

Cultivar a cultivated plant variety.

Cuticle (cuticular, adj.) a waxy layer produced by the epidermis.

Cutin the waxy component of the cuticle.

Cyanobacteria a group of primitive, photosynthetic organisms that lack membrane-bounded organelles; also known as blue-green algae.

Cyanophyta the division to which the cyanobacteria belong.

Cycad a palmlike gymnosperm native to the tropics.

Cycadophyta the division to which the cycads belong.

Cyclic photophosphorylation the generation of ATP as the result of cyclic flow of electrons in photosystem I.

Cyme an inflorescence in which the youngest flowers are at the base and are thus furthest from the apex which ends in a terminal flower.

Cytokinesis the division of the cytoplasm that occurs during cell division.

Cytokinin a class of plant hormone the characteristic function of which is to induce cell division.

Cytology the study of cell structure and function.

Cytoplasm the portion of the protoplast exclusive of the nucleus.

Cytoskeleton the substructure of the cytoplasm consisting of microtubules and microfilaments.

DDT dichlorodiphenyltrichloroethane, an insecticide.

DNA deoxyribonucleic acid, the genetic material of the cell.

DNA hybridization studies analysis of the extent to which isolated, single strands of DNA will join together (hybridize) to form a double strand.

DNA replication the synthesis of new DNA strands that takes place during the S-phase of the cell cycle.

Dark reactions of photosynthesis the second major step of photosynthesis during which CO_2 is incorporated into a carbohydrate, also called carbon fixation.

Day-neutral not regulated by daylength.

Deciduous refers to a plant that loses its leaves before the onset of a period of dormancy.

Defoliation loss of leaves due to natural or artificial cause.

Dehiscence zone a region in a fruit where the fruit wall will rupture, liberating the seeds.

Dehiscent fruit a fruit that ruptures when the seeds are liberated.

Dendochronology historical and climatological study based on tree ring analysis.

Derivative a cell derived from a meristematic cell.

Dermal tissue a tissue present on the outside of an organ, e.g.; cork and epidermis.

Dermatitis a skin condition caused by an allergic reaction or by an infection.

Desiccation drying out.

Determinate of limited extent; a type of growth pattern.

Deuteromycetes the taxonomic class to which all fungi of unknown sexual reproduction belong.

Diatomaceous earth deposits of diatoms.

Dicotyledonous plant a flower-producing plant whose seeds have two cotyledons.

Differential permeability the selective passage of molecules across a membrane.

Differential centrifugation a technique whereby mixed components of differing sizes and densities are separated by centrifugation.

Differentially permeable possessing the property of selectively controlling the movement of molecules.

Differentiated possessing a specialized function.

Differentiation the development of specialized structure and function by cells, tissues, or organs.

Diffuse root system a root system characteristic of monocotyledonous plants in which the primary root is replaced by several roots of the same size; also called a fibrous root system.

Diffusion the net movement of molecules along a concentration gradient.

Digestion the enzyme-mediated breakdown of complex organic molecules.

Digitalis a glycoside, produced mainly in the leaves of foxglove plants, that functions primarily as a heart stimulant in man.

Dihybrid cross a genetic cross in which two traits are considered at the same time.

Dikaryotic a condition in certain stages of life cycles in some fungi (Ascomycetes and Basidiomycetes) where two nuclei are present in each cell.

Dilated ray the V-shaped areas composed of parenchyma cells in the secondary phloem of woody plants.

Dioecious the condition in which the male and female reproductive structures are present on separate organisms.

Diplobiontic algal species that produce two separate, independent, multicellular forms during their life cycle.

Diploid possessing two of each pair of chromosomes in a cell, typical of the sporophyte phase in a life cycle; also referred to as the $2n$ condition.

Division a taxonomic unit used for plant classification equivalent to the animal phylum.

Dominant trait the genetic trait that is always expressed when present in an organism.

Dormancy a state of reduced physiological activity.

Dormant being in a resting state with reduced metabolic activity.

Double fertilization the fusion between the egg and sperm and a second fusion between the polar nuclei and a sperm; characteristic of all flower-producing plants.

Drupe a fleshy fruit such as a peach in which the fruit wall can be differentiated into a thin exocarp, a fleshy mesocarp, and a stony exocarp.

ER (see endoplasmic reticulum).

Ecology the field of study related to the interaction of organisms with their environment as well as other organisms.

Ecosystem ecological system; the interrelationship between organisms and their physical environment in a specific location.

Ectomycorrhiza a symbiotic relationship between fungi and the roots of trees resulting in only an external growth of fungal filaments around the roots.

Egg the nonmotile female gamete.

Electron micrograph a picture taken through an electron microscope.

Electron transport chain a series of molecules arranged on a membrane, such as in a chloroplast or a mitochondrion, capable of transferring electrons and at the same time releasing energy that is used for ATP synthesis.

Electrophoresis a technique by which charged molecules are separated in an electric field.

Embryo a multicellular structure developed from the zygote and from which the adult will develop.

Embryo sac the eight-nucleate structure representing the female gametophyte of a flowering plant.

Embryonic axis the short, vertically elongated structure of the embryo in a seed to which the cotyledons are attached.

Endodermis a single-layered tissue composed of thick-walled and thin-walled cells encircling the vascular cylinder in a root.

Endomycorrhiza a symbiotic relationship between fungi and roots that results in the growth of fungal filaments within the root tissue.

Endoplasmic reticulum a network of membranes within the cytoplasm.

Endosperm a food-storage tissue present in certain angiosperm seeds but not part of the embryo.

Endospore an internally formed, resistant spore produced by certain bacteria.

Endosymbiotic a symbiotic association in which one organism lives inside the cells of the other.

Endosymbiotic hypothesis the hypothesis that plastids and mitochondria originated as endosymbiotic organisms within a host cell.

Enzyme a protein that functions as a biological catalyst.

Epicotyl the portion of the embryonic axis above the point of attachment of the cotyledon(s).

Epidermis the one-layered tissue present on the outside of stems and roots of nonwoody plants and all leaves.

Epiphyte a plant that grows on another plant without obtaining any water or nutrients from it.

Equational division the second set of divisions during meiosis that does not involve reduction in the number of chromosomes.

Erosion loss of soil due to the action of wind or water.

Ethylene a hormone produced by plants that regulates the ripening of fruits and the sex of flowers in certain plants. Ethylene is also a product of incomplete combustion of hydrocarbons.

Etiolation a condition resulting when plants are grown in the dark, characterized by yellowish, spindly stems and unexpanded leaves.

Euglenophyta the division to which euglenoid algae belong.

Eukaryote (eukaryotic, adj.) an organism that has membrane-bounded organelles in its cell(s).

Eutrophication the overenrichment of a body of water that results in an overabundance of aquatic plant life.

Evergreen a plant that does not lose all of its leaves at the same time.

Evolution the origin of new types of organisms from ancestral forms as the result of natural selection.

Facultative anaerobe an organism that normally grows in the presence of oxygen, but also can survive in the absence of oxygen.

Family a unit of classification larger than a genus but smaller than an order.

Fascicular limited to the bundle; with reference to cambium it means the vascular cambium within a vascular bundle.

Fat an organic molecule made up of one glycerol molecule and three fatty acid molecules.

Fatty acid a molecule composed of a hydrocarbon chain with a carboxyl group at one end.

Fenestrated possessing a lacelike structure.

Fern a non-seed-producing primitive vascular plant.

Fertilization the fusion between an egg and a sperm.

Fiber a dead, vertically elongated cell with a thick secondary wall and tapering ends.

Fibrous root system (see diffuse root system).

Filament a chain of cells linked end-to-end.

Fixing the act of killing and preserving cells.

Flaccid the condition of being wilted due to loss of water from cells.

Flagellum (flagella, plural) long, whiplike cell structures used for motility.

Flavone a major class of water-soluble plant pigments responsible for flower color; e.g., anthocyanins.

Fleshy scale leaf a modified leaf that functions in storage of water and food as in onions.

Flora plants, generally used as the opposite of fauna (animals).

Fluid mosaic model a membrane model that shows a discontinuous distribution of proteins in a double layer of lipids.

Food chain a series of organisms, usually starting with plants, in which members of one group consume those of the preceding group.

Forestry the field of study dealing with forest trees, their products, and their management.

Fossil preserved remains or impressions of organisms of past geological periods.

Free central placentation the arrangement of placenta around a free standing central column of tissue projecting from the base of a compound ovary.

Free space the cell walls and intercellular spaces of a tissue.

Free water water molecules not tightly associated with solute molecules.

Frond the leaf of a fern.

Fruit the transformed ovary, with or without any associated floral parts, within which are produced the seeds of flowering plants.

Fruiting body a characteristic reproductive structure of an ascomyete or basidiomycete fungus within which are produced the meiospores.

Fucoxanthin a form of xanthophyll present in golden-brown algae and brown algae.

Fungi a group of predominately filamentous organisms that lack the ability to synthesize their own food.

Furocoumarin a photosensitive compound present in wild parsnip that can cause skin inflammation and blisters.

G_1 phase a stage in the cell cycle immediately following cell division and characterized by active RNA and protein synthesis and cell growth.

G_2 phase a stage in the cell cycle just prior to cell division characterized by synthesis of spindle fiber proteins and storage of energy.

Gametangium (gametangia, plural) a gamete-producing structure.

Gamete a sex cell.

Gametogenesis the steps leading to the production of gametes.

Gametophyte the gamete-producing phase in a life cycle.

Gemma (gemmae, plural) a mass of tissue specialized for asexual reproduction in liverworts and capable of producing a new organism.

Gemma cup a cavity in which gemmae are produced.

Gene the basic unit of heredity; a segment of a DNA molecule to which a specific function can be assigned; e.g., the code for the synthesis of a particular protein.

Generative cell one of the cells in a pollen grain, responsible for the generation of sperm.

Genetic information coded instructions on the DNA molecule that are used in the synthesis of proteins.

Genetic material the DNA molecules that contain coded instructions for the synthesis of proteins.

Genetic recombination the recombining of genetic information that occurs during sexual reproduction.

Genetics the field of study dealing with the transmission of traits from parents to progeny, including the study of gene structure and function.

Genome the complete genetic program in a nucleus, plastid, or mitochondrion.

Genus (genera, plural) a unit of classification larger than the species and smaller than the family; includes one or more species with common characteristics.

Geotropism the growth response of a plant organ to the unidirectional stimulus of gravity resulting in the upward growth of stems and downward growth of roots.

Germination the process of renewed growth by a reproductive structure.

Gibberellin a type of plant hormone involved in stem elongation and breaking dormancy.

Ginkgophyta the division of gymnosperms to which ginkgos belong.

Glycolysis the first stage in the respiration of glucose resulting in the production of two molecules of pyruvic acid from each glucose molecule.

Glycoside a chemical derived from a sugar reacting with a substance such as an alcohol.

Glyoxysome a type of microbody functioning in lipid metabolism.

Gnetophyta the gymnosperm division to which *Ephedra, Gnetum,* and *Welwitschia* belong.

Golgi body a cellular organelle composed of flattened, disclike sacs with fenestrated margins functioning in packaging and secretion of cell wall materials.

Golgi cisternae flattened, circular, membrane-bounded sacs that make up the Golgi body.

Grain a type of dry fruit characteristic of grasses in which the single seed is fused with the fruit wall.

Gram-negative bacteria that do not stain with crystal violet.

Gram-positive bacteria that stain with crystal violet.

Granum a structural component of the chloroplast; each granum is composed of a stack of two or more flattened, circular membranous sacs called thylakoids.

Green Revolution the replacement of low-yielding indigenous crop plants with high-yielding, genetically improved varieties.

Ground meristem one of the primary meristems from which the ground tissues, such as parenchyma, collenchyma, and so forth, are derived.

Ground pine a common name for certain plants belonging to the division Microphyllophyta; also known as club mosses.

Ground tissue a primary tissue derived from the ground meristem, e.g., parenchyma.

Growth ring secondary xylem produced during one growing season.

Guard cell a chloroplast-containing, kidney-shaped, modified epidermal cell functioning in the regulation of stomatal opening and closing.

Gymnosperm a plant that produces seeds in cones.

Gynoecium a collective term for all of the pistils in a flower.

Habitat an area in which conditions are favorable for the growth of a particular plant or group of plants.

Hallucinogen a substance that causes hallucinations.

Haplobiontic algal species that produce only one free-living form during their life cycle.

Haploid possessing only one of each pair of homologous chromosomes.

Haustorium a root of parasitic plants specialized for the absorption of water and nutrients.

Head a type of inflorescence in which a cluster of sessile flowers is produced on a single receptacle.

Heartwood the dark, inner part of xylem in a woody stem or root that cannot transport water and minerals.

Hemicellulose a type of polysaccharide that is present in the cell walls of plants.

Herbaceous plant a plant composed mainly of primary tissue.

Herbarium a place for the deposition of preserved plants for scientific study.

Herbivore an organism that feeds on plants.

Heroin a narcotic derived from morphine.

Heterocyst a specialized cell of certain cyanobacteria functioning in nitrogen fixation and as an asexual spore.

Heterogamy the fusion between two morphologically different gametes.

Heteromorphic a description of a life cycle in which the haploid and diploid phases have different shapes.

Heterospory a condition in which two functionally different types of meiospores are produced in the life cycle.

Heterothallic producing male and female gametes on separate organisms.

Heterotroph an organism that cannot produce its own food from inorganic molecules.

Heterozygous a hybrid; a condition in which both dominant and recessive alleles that control a single trait are present.

Hexose a six-carbon sugar.

Histamine a derivative of the amino acid histidine; plays an important role in causing the typical symptoms of hay fever.

Homologous chromosome one of a pair of morphologically identical chromosomes that carry the genes representing the same characteristics, one derived from the male parent and one from the female parent.

Homospory a condition in which only one type of meiospore is produced.

Homothallic producing both male and female gametes on the same organism.

Homozygous a genetically pure line; a condition in which both alleles are of the same type, either dominant or recessive.

Hormone a molecule produced by one group of cells within an organism that regulates the growth and development of other cells.

Horsetail a primitive vascular plant with jointed stems and terminal strobilus.

Horticulture the study of orchard and garden plants.

Humus the organic component of soil; the partially decomposed remains of plants and animals.

Hybrid (see heterozygous).

Hybridization the mixing of different gene pools during sexual reproduction.

Hydrocarbon a molecule that consists exclusively of a carbon skeleton with attached hydrogen atoms.

Hydrogen carrier a molecule that can accept and donate hydrogen atoms or hydrogen ions.

Hydrophilic molecules that will associate with the aqueous phase; literally "water-loving."

Hydrophobic molecules that associate with the lipid phase but not the aqueous phase; literally "water-fearing."

Hydroponics the method of growing plants without soil.

Hypha (hyphae, plural) a fungal filament.

Hypocotyl the portion of the embryonic axis below the point of attachment of the cotyledons.

Hypothesis an unproved explanation for a scientific observation.

Immune reaction production of antibodies as a response to the presence of a foreign substance in the blood.

Imperfect flower flower in which only one sex is represented.

Incomplete flower a flower in which one or more floral parts is missing.

Indehiscent fruit a dry fruit that does not split open even when mature.

Indusium the covering of a sorus (a cluster of sporangia) on a fern leaf.

Inflorescence a group of flowers on a common stalk.

Initial a cell that divides producing a new cell.

Inoculum a small amount of a culture of microorganisms that is used to start a new culture.

Inorganic molecules molecules that lack carbon atoms.

Integument the outer covering of an ovule.

Interfascicular present in the space between two vascular bundles.

Internode (internodal, adj.) segment of stem between two nodes.

Interphase the phase of the cell cycle between nuclear divisions.

Ion an atom or molecule with a positive or negative charge.

Irregular flower a flower that has nonsymmetrically arranged floral parts.

Isodiametric an object having a shape with approximately the same dimensions in all directions.

Isogamy the fusion of morphologically similar gametes.

Kingdom the largest unit of classification.

Knee a modified root produced by some plants that grow in swampy areas, e.g., cyprus.

Knot a structural feature resulting from the base of a branch becoming embedded in the wood of the main trunk.

Krebs cycle a series of reactions of respiration taking place in the mitochondrial matrix resulting in the breakdown of acetic acid to CO_2 and H atoms that are transferred to hydrogen carriers.

LSD lysergic acid diethylamide, a combination of lysergic acid obtained from ergot fungus and diethylamide; LSD is the most powerful hallucinogen known.

Lactic acid fermentation the respiration of glucose to lactic acid in the absence of oxygen.

Lacuna an air space in a tissue.

Laminarin the carbohydrate-storage material of brown algae.

Lateral bud a bud present in the axil (angle between the stem and leaf) of a leaf; gives rise to a branch. Also known as an axillary bud.

Leaf primordium (primordia, plural) a group of meristematic cells that gives rise to a leaf.

Leaf gall a tumorous growth on a leaf induced by certain insects that lay eggs in leaves.

Leaf gap a small break in the vascular cylinder of the stem above the point of departure of a leaf trace at a node.

Leaf scar a scar on a branch left after a leaf falls.

Leaf sheath the sheathlike petiole or base of a leaf that encircles a portion of the stem as in grasses.

Leaf trace the branch of the vascular tissues of the stem that enters a leaf.

Leaflet a unit of a compound leaf.

Legume a type of dry fruit that breaks open along two margins releasing the seeds; also called a pod.

Lenticel a small opening in the cork layer of woody stems where loosely arranged parenchyma cells replace cork cells facilitating gas exchange between the internal tissues and the atmosphere.

Leucoplast a colorless plastid.

Leucosin the characteristic carbohydrate storage material of golden-brown algae.

Lichen an organism that results from the symbiotic association of certain algae with fungi.

Light reactions of photosynthesis the first steps in photosynthesis during which light energy is converted to chemical energy.

Lignin a complex organic molecule that adds mechanical strength to cell walls.

Lipid (see fat).

Lipid bilayer a double layer of lipid molecules formed when their adjacent nonpolar tails associate with each other and the polar heads are directed outward.

Liverwort lobed liverlike, flattened nonvascular succulent plant that grows along the soil surface; a member of the division bryophyta.

Lobed leaf a leaf whose blade has a deeply indented margin.

Long-day plant a plant that requires light periods longer than the critical daylength in order to initiate a physiological response such as flowering.

Longitudinal section a section made along the long axis of an object.

Longitudinal a plane of section that passes through the long axis.

Lumen the space within a dead cell.

Lysergic acid an alkaloid obtained from the ergot fungus which, when combined with the synthetic chemical diethylamide, produces LSD, a potent hallucinogen.

Lysosome a single-membrane-bounded organelle that contains digestive enzymes; a type of microbody.

Macronutrient an essential element that is needed in large quantities.

Marginal placentation the condition in which the tissue to which the ovules are attached (the placenta) is found along the edge where the two margins of a carpel meet.

Mass/pressure flow hypothesis an explanation for the movement of food based on pressure differences in the phloem between two regions of a plant—food-synthesizing region (source) and food-using region (sink).

Maternal inheritance inheritance of genetic characteristics through the maternal parent only; associated with the transmission of genes in plastid and mitochondrial genomes.

Mating type sexually compatible strains of organisms.

Matrix the fluid part of the mitochondrion where Krebs cycle reactions take place.

Megagametophyte the egg-producing phase in the life cycle.

Megasporangium a sporangium that produces megaspores.

Megaspore the meiospore ($1n$) that initiates the megagametophyte phase in a heterosporous life cycle.

Megasporophyll a spore-bearing leaf that gives rise to megaspores.

Meiocyte the diploid ($2n$) cell that undergoes meiosis. Also known as a meiospore mother cell.

Meiosis (meiotically, adv.) the process of nuclear division that results in the production of daughter nuclei each containing one-half the original number of chromosomes.

Meiospore mother cell the diploid ($2n$) cell that undergoes meiosis producing meiospores. Also known as a meiocyte.

Membrane a differentially permeable bilayer structure composed of proteins and phospholipids.

Meristem (meristematic, adj.) a tissue whose cells are capable of repeated cell division.

Mescaline a hallucinogenic alkaloid, present in the spineless cactus peyote, that induces brightly colored visions.

Mesophyll the photosynthetic parenchyma present in leaves.

Messenger RNA a copy of a DNA segment that contains the information for the synthesis of a particular polypeptide or protein.

Metabolism (metabolic, adj.) all of the chemical reactions that occur in a living cell.

Metaphase a stage of nuclear division (meiosis or mitosis) during which chromosomes line up along the equatorial plane of the cell.

Microbody a single-membrane-bounded organelle such as lysosome, peroxisome, and glyoxysome.

Microfibril a structure made up of strands of cellulose molecules.

Microfilament a cytoskeleton component made up of globular protein subunits. Also called an actin filament.

Microgametophyte the sperm-producing phase in the life cycle.

Micrometer one-millionth of a meter or one-thousandth of a millimeter.

Micronutrient an essential element that is needed in small quantities.

Microorganism an organism too small to be seen with unaided eyes, e.g.; bacteria and certain algae and fungi.

Microphyll a small leaf.

Microphyllophyta the division to which the club mosses belong.

Micropyle a small opening in the integument(s) of an ovule for the entry of the pollen tube or the pollen grain.

Microsporangium a sporangium in which microspores are produced.

Microspore the meiospore ($1n$) that initiates the microgametophyte phase in a heterosporous life cycle.

Microsporophyll a microspore-producing leaf.

Microtubule a tubular organelle of indeterminate length composed of protein subunits; component of spindle fibers and flagella.

Middle lamella (lamellae, plural) the structure that glues the primary walls of two adjacent cells together.

Midrib the major vein in a leaf that divides the blade into two halves.

Millimeter one-thousandth of a meter.

Mitochondrion (mitochondria, plural) a membrane-bounded organelle that functions in respiration.

Mitosis the process of nuclear division that results in the production of two genetically identical daughter nuclei.

Mitotic division division of the nucleus into two genetically identical daughter nuclei.

Mitotic toxins chemicals that block or disrupt mitosis.

Molecular genetics the field of genetics related to the study of gene structure and function; contrast with classical genetics.

Molecule a chemical compound formed when two or more atoms are held together by chemical bonds.

Monera a kingdom to which all prokaryotic organisms (bacteria and cyanobacteria) are assigned.

Monocotyledonous plant a flower-producing plant whose seeds have only one cotyledon.

Monoecious an organism that has both male and female reproductive structures on it.

Monohybrid cross a genetic cross in which only one trait is considered at a time.

Monokaryotic possessing only one nucleus per cell.

Morphine an alkaloid present in the extracts from opium poppy and used as an effective painkiller.

Morphology (morphologically, adv.) the study of the external appearance of organisms.

Moss a type of bryophyte characterized by its typical leafy appearance.

Multicellular consisting of many cells.

Multiple fruit a fruit derived from ovaries of different flowers.

Mutation a change in a portion of the DNA molecule corresponding to one gene, which results in a change in the protein product of that gene; a detectable change in the genetic information.

Mycelium a collection of fungal hyphae (filaments).

Mycologist a botanist who specializes in the study of fungi.

Mycology the study of fungi.

Mycorrhiza a mutually beneficial association between fungi and plant roots.

Myxomycota the division to which the slime molds belong.

Nanometer one-billionth of a meter.

Natural selection the selective influence of the environment on living organisms leading to the survival of the fittest.

Netted venation the arrangement of veins in the form of a network; characteristic of leaves of dicotyledonous plants.

Niche the unique set of ecological factors to which an organism is best adapted.

Nitrogen fixation the conversion of atmospheric nitrogen to nitrogen-containing compounds by nitrogen fixing organisms such as certain bacteria and cyanobacteria.

Node the region of a stem to which a leaf is, or was, attached.

Nomenclature the naming of organisms.

Nonlignified lacking lignin.

Nonsuberized lacking the wax suberin.

Nonvascular plant a plant lacking specialized food and water conducting tissues (xylem and phloem).

Noncyclic photophosphorylation synthesis of ATP as a consequence of unidirectional electron flow from water to NADP during the light reactions of photosynthesis.

Nuclear body the DNA-containing region in the cytoplasm of a bacterium or cyanobacterium.

Nuclear envelope the double membrane surrounding a nucleus.

Nucleation center a particle around which ice crystals are formed.

Nucleic acid a complex organic molecule involved in the storage and expression of genetic information.

Nucleolar organizing region specific portions of certain chromosomes associated with the formation of nucleoli.

Nucleolus a dense region in a nucleus composed of RNA and protein where ribosomal subunits are synthesized.

Nucleoplasm that part of the protoplast enclosed by the nuclear envelope.

Nucleotide the basic unit of nucleic acid containing a phosphate, a five-carbon sugar, and a nitrogen-containing base.

Nucleus (nuclei, plural) a spherical structure in a cell that contains the chromosomes and the majority of the genetic information of the cell.

Oligosaccharide a carbohydrate consisting of a small number of linked sugar molecules.

Oogamy sexual reproduction involving the fusion between an egg and a sperm.

Oogonium a unicellular gametangium in which one or more eggs are produced.

Opposite leaf arrangement the form of leaf arrangement where two leaves are present opposite each other at a nodal region.

Order a unit of classification composed of one or more families.

Organelle a structure in the cell with a specific function, e.g., plastid or mitochondrion.

Organic molecules molecules containing carbon atoms.

Osmosis the diffusion of water through a differentially permeable membrane.

Osmotic equilibrium the condition in which the movement of water due to osmosis has ceased.

Ovary the lower portion of the pistil of the flower in which ovules are formed.

Ovule the structure within the ovary of a flower within which the egg is formed.

Ozone a molecular form of oxygen in which three oxygen atoms are linked forming O_3.

Paleobotany the study of fossil plants.

Palisade parenchyma the photosynthetic parenchyma cells arranged in columns toward the upper epidermis in a leaf.

Palmately compound a type of compound leaf in which all of the leaflets are attached to the tip of the primary petiole.

Palmately netted a type of venation in which two or more veins of equal size originate from the base of a leaf blade.

Panicle an inflorescence in which the flowers are arranged alternately on a branched stalk.

Parallel evolution the evolution of similar adaptations to the same environmental variables by unrelated organisms.

Parallel venation an arrangement in which several veins of equal size run parallel to each other over the length of a leaf except at the tip; typical of monocotyledonous plants.

Paramylum a carbohydrate stored as a reserve food by euglenoids.

Parasite (parasitic, adj.) an organism that obtains its food from another organism.

Parasitic plant a plant that derives part or all of its food from another organism.

Parathion an organophosphate compound that is widely used as an insecticide.

Parenchyma relatively undifferentiated cells that frequently function in food synthesis and food storage.

Parietal placentation arrangement of placenta along the margins where adjacent carpels are fused.

Parthenocarpy (parthenocarpic adj.) fruit production without fertilization; such fruits lack seeds.

Passage cell an endodermal cell that allows passage of water into the xylem of a root.

Pasteurization partial sterilization of food items by raising the temperature to 62°C for 30 minutes or 72°C for 15 seconds.

Pathogen (pathogenic, adj.) a disease-causing organism.

Peat partially decomposed remains of *Sphagnum* moss.

Pectin a chemical component of the middle lamella which glues adjacent cells together.

Pedicel the stalk of a single flower in an inflorescence.

Peduncle the stalk of a solitary flower or of an inflorescence.

Pellicle a nonrigid covering present under the plasmalemma of euglenoids and certain other algae.

Pentaploid possessing five of each type of chromosome.

Pentose a five-carbon sugar such as ribose and deoxyribose.

Perennial a plant that lives for many years.

Perfect flower a flower that contains both stamens and pistil.

Pericarp the fruit wall (derived from the ovary wall).

Pericycle a one-layered tissue that gives rise to branch roots.

Peripheral toward the outside.

Periplast (see pellicle).

Peristome a ring of triangular cells around the opening of a moss sporangium.

Perlite a soil substitute made from expanded volcanic rock.

Permeability the ability of molecules to penetrate a membrane or cell wall.

Peroxisome a single-membrane-bounded organelle associated with photorespiration; one type of microbody.

Petal one of the floral parts that is usually showy; a unit of the corolla.

Petiole the stalklike region of the leaf that connects the leaf blade to the stem.

Petri dish a shallow, circular glass or plastic container used for growing microorganisms.

pH a quantitative measure of the acidity of a solution.

Phaeophyta the division to which all brown algae belong.

Phage a bacterial virus.

Phelloderm parenchyma cells produced by the cork cambium.

Phenotype the physical appearance of an organism.

Phloem specialized food-conducting tissue.

Phospholipid a fatty substance containing a phosphate group, a component of all membranes.

Photoperiodism the regulation of plant growth and development by daylength.

Photophosphorylation the synthesis of ATP during photosynthetic light reactions.

Photorespiration light-induced consumption of O_2 and release of CO_2 that does not result in ATP production.

Photosynthesis the metabolic process by which carbohydrates are produced from CO_2 and a hydrogen source, such as water, using light as the source of energy.

Photosystem a group of chlorophyll molecules that function to trap light energy and funnel it to a strategically located chlorophyll molecule that is involved in electron transfer reactions.

Phototropism a growth response to a unidirectional light stimulus, e.g., the bending of a stem toward a light source.

Phycocyanin a blue, water-soluble, photosynthetic pigment present in cyanobacteria and red algae.

Phycoerythrin a red, water-soluble, photosynthetic pigment present in cyanobacteria and red algae.

Phycologist a botanist who specializes in the study of algae.

Phycology the study of algae.

Phylogeny the evolutionary history of a group of organisms.

Physiology (physiological, adj.) the study of structure and function of organisms.

Phytochrome a blue pigment responsible for the control of a variety of light-mediated growth responses in plants, e.g., flower production.

Phytogeography the study of plant distribution.

Phytoplankton free-floating or motile microscopic autotrophs.

Pigment a molecule that can absorb light.

Pili short rigid extensions used in bacterial conjugation.

Pinnately compound a type of compound leaf in which the leaflets are attached along the primary petiole.

Pinnately netted a type of venation in which one main vein divides the leaf blade into two. Secondary veins of a smaller size branch off from the main vein.

Pistil the "female" reproductive organ in a flower; it consists of a basal ovary, a terminal stigma, and a style connecting the two.

Pit a small opening in the secondary wall of a cell.

Pit membrane that portion of the primary wall present between a pair of pits on adjacent cells.

Pitch a dark resin.

Pith a region composed of parenchyma cells at the center of the vascular cylinder.

Pith ray the strip of parenchyma tissue connecting the pith and cortex and present between two vascular bundles in a herbaceous stem.

Placenta the tissue from which ovules are produced in an ovary.

Placentation the arrangement of placenta in an ovary.

Plane-sawed lumber cut tangentially.

Plant an organism that belongs to the Kingdom Plantae.

Plant community the group of plants growing in a particular region.

Plant ecology the study of the interactions of plants with other organisms and the environment.

Plant food the various mineral salts that are necessary for plant growth (fertilizer).

Plant pathology the study and control of pathogens causing plant diseases. Common plant pathogens are fungi, bacteria, viruses, and nematodes.

Plant physiologist a botanist who studies plant structure and function.

Plant physiology the study of plant structure and function.

Plantae the Kingdom to which all true plants belong—bryophytes, whisk ferns, club mosses, horsetails, ferns, gymnosperms, and angiosperms.

Plasmalemma the outermost portion of the cytoplasm that functions as a differentially permeable membrane.

Plasmid a small DNA fragment usually found in the cytoplasm of a bacterium. Plasmids are replicated and integrated into the bacterial DNA through genetic recombination.

Plasmodesmata cytoplasmic connections between adjacent cells that result in intercellular continuity.

Plasmodium a multinucleate mass of protoplasm without a wall.

Plasmolysis the loss of water from the vacuole due to a difference in free-water concentration between the inside and outside of a cell resulting in the shrinkage of the protoplast.

Plastid a class of membrane-bounded organelles characterized by the ability to make starch.

Polar nuclei two centrally located nuclei in the female gametophyte of angiosperms, one derived from each pole. After fusion with a sperm they become the primary endosperm nucleus.

Polar molecules that have a positive and/or a negative charge.

Pollen grain a microspore-derived structure containing the haploid microgametophyte of seed-producing plants.

Pollen tube a tubular extension, produced by the tube cell during germination of the pollen grain, that transports the sperm to the ovule.

Pollination the transfer of pollen from the anther to the stigma in angiosperms or from the microsporangium to the ovule in gymnosperms.

Polyhedron a many-sided structure.

Polymer a molecule made up of a large number of repeating subunits.

Polypeptide a molecule composed of amino acids.

Polyploid a cell or an organism that has more than two sets of chromosomes.

Polysaccharide a carbohydrate consisting of a large number of simple sugars.

Prairie a plant community in which grasses comprise the dominant vegetation.

Primary endosperm cell a cell resulting from the fusion of a sperm with two polar nuclei; it subsequently gives rise to the endosperm.

Primary meristem a meristem responsible for the production of primary tissues.

Primary phloem the phloem tissue derived from the procambium.

Primary tissue a tissue derived from a primary meristem.

Primary wall the first wall layer produced by a newly formed cell.

Primary xylem the xylem tissue derived from the procambium.

Primary tissues produced by primary meristems such as protoderm, ground meristem, and procambium.

Primordium a precursor that gives rise to a structure.

Procambium the primary meristem from which the primary vascular tissues and pericycle are formed.

Prokaryote (prokaryotic, adj.) an organism that lacks membrane-bounded organelles in the cell.

Prop root a modified root that functions in support.

Propagation the production of new plants by sexual or asexual means.

Prophase the first stage of nuclear division during which chromosomes form from chromatin material.

Proplastid an immature plastid that is the precursor of all mature plastid types.

Protein a large molecule consisting of a long chain or chains of amino acids.

Prothallial cell one of the cells of a gymnosperm pollen grain the function of which is not known.

Prothallus a multicellular, autotrophic, independant structure that represents the gametophyte phase in the life cycle of ferns and other primitive vascular plants.

Protista the Kingdom to which all algae and all protozoa belong.

Protocell the hypothetical precursor of the cell that formed on the primeval earth.

Protoderm the primary meristem from which the epidermis is derived.

Protonema the branched filamentous structure produced by the meiospore during the life cycle of a moss.

Protoplasm a term to describe the total amount of living substance in an organism.

Protoplast the living portion of a cell, composed of cytoplasm and nucleoplasm.

Pseudoplasmodium a mass of cells formed by the aggregation of the myxamoebae in cellular slime molds.

Psilocybin a hallucinogenic compound present in mushrooms of the genus *Psilocybe.*

Psilophyta the division to which the whisk ferns belong.

Pteridology the study of ferns.

Pterophyta the division to which the ferns belong.

Pulvinus a small swelling at the base of a petiole or leaflet.

Pyrenoid a chloroplast structure associated with starch deposition in many algae.

Pyrrophyta the division to which dinoflagellates belong.

Quarter-sawed lumber cut radially, passing through the center.

Quinine an alkaloid obtained from the bark of *Cinchona,* a small evergreen, used for controlling malaria.

RNA a type of nucleic acid, ribonucleic acid, synthesized as copies of DNA segments and used in protein synthesis.

Raceme a type of inflorescence in which flowers are arranged alternately on an unbranched stalk.

Radial a plane parallel to the long axis that passes through the center.

Radial section a lengthwise section passing through the center of an object.

Radially symmetrical able to be divided into two equal halves by any line along a radius.

Radicle the lowermost portion of the embryonic axis of a seed, responsible for the production of the primary root.

Rain forest an evergreen forest that develops when annual rainfall exceeds 100 inches.

Receptacle the upper end of a flower stalk from which the floral parts are produced.

Recessive trait a trait governed by a recessive allele and not expressed in the presence of the dominant allele.

Red tide a condition caused by large populations of *Gymnodinium,* a dinoflagellate, resulting in the death of fish feeding on these neurotoxin-producing algae.

Reductional division the first nuclear division of meiosis during which the number of chromosomes is reduced by half.

Regular flower a flower in which the parts are arranged symmetrically around a central axis.

Replicated copied.

Reproductive isolation the presence of a physical or biological barrier that prevents sexual reproduction.

Reserpine an alkaloid isolated from the snakeroot plant; used in the treatment of high blood pressure, schizophrenia, and snake bite.

Resin duct tubular structures in plants, mostly gymnosperms, for the transport of resin; also known as resin canals.

Resolving power the ability to distinguish details of an image. The minimum distance two particles can be apart for them to be recognized as two separate objects.

Respiration the metabolic process that results in the release of chemical energy from simple organic molecules.

Rhizoid a rootlike structure lacking vascular tissue, present in primitive vascular plants.

Rhizome a modified underground stem that may store food.

Rhodophyta the division to which the red algae belong.

Ribosomal RNA the ribonucleic acid component of ribosomes responsible for the translation of the code on messenger RNA during protein synthesis; also called rRNA.

Ribosome an organelle that is involved in protein synthesis.

Ricin a toxin produced mostly in the seeds of castor bean.

Root the primarily underground organ of a plant derived from the radicle of the seed; it serves to anchor the plant and absorb water and minerals.

Root cap a group of cells that forms a protective layer around the growing tip of the root.

Root hair a tubular projection of an epidermal cell of a root that serves to increase the surface area for absorption.

Root hair zone that segment of a root corresponding to the region of maturation where root hairs are present.

Root nodule a spherical outgrowth on roots composed of nitrogen-fixing bacteria and root tissue.

Root pressure the sum total of turgor pressure developed in root cells due to osmosis.

Rosin a translucent, amber to dark brown substance obtained from resin; sometimes used as a synonym for resin.

Runner a modified stem that grows along the soil surface; also called a stolon.

S-phase this stands for the synthetic phase of the cell cycle during which DNA is synthesized.

Saprophyte an organism that obtains its nutrition by decomposing nonliving organic material.

Sapwood the outer part of xylem in a woody stem or root characterized by its light color and ability to transport water.

Saturated fatty acid a fatty acid in which each carbon atom in the hydrocarbon chain forms only one bond with each of its carbon neighbors and the remaining bonds are with hydrogen atoms.

Scarification (scarified, adj.) mechanical or chemical damage to a seed coat.

Schizophyta the division to which bacteria belong.

Science knowledge of natural laws obtained through experimentation.

Sclereid a type of dead cell with a thick secondary wall, small lumen, and pit canals; a group of these make up sclerenchyma tissue.

Sclerenchyma a primary tissue that is composed of either fiber cells or sclereids.

Sclerenchymatous containing sclerenchyma fiber cells or sclereids.

Secondary tissues produced by secondary meristems such as the vascular cambium and cork cambium.

Secondary meristem a meristem that gives rise to secondary tissues; vascular cambium and cork cambium.

Secondary tissue a tissue derived from a secondary meristem.

Secondary wall a wall layer laid down on top of the primary wall of a mature cell, usually resulting in the death of that cell.

Seed a reproductive structure containing an embryo and derived from an ovule.

Selection pressure the influence of the environment and competition on the survival and reproduction of an organism.

Self-pollination the transfer of pollen from the anther of a flower to the stigma of the same flower or the stigma of a different flower on the same plant.

Senescence aging.

Sepal the outermost flower part, usually green in color; a unit of the calyx.

Septum the crosswall produced during division that separates a cell into two.

Sessile a stalkless structure as in a leaf without a petiole.

Set in this context a set refers to a group consisting of one of each of the homologous chromosomes.

Sexual reproduction the form of reproduction involving the fusion of two sex cells.

Shoot the aerial portion of a plant consisting of stems, leaves, buds, and reproductive organs.

Short-day plant a plant that requires light periods shorter than the critical daylength in order to initiate a physiological response such as flowering.

Shrub a woody plant with two or more stems arising near the ground.

Sieve cells the food conducting cells of the phloem in a gymnosperm.

Sieve plate the perforated end wall of a sieve tube member.

Sieve tube several sieve tube members connected end-to-end.

Sieve tube member a vertically elongated cell type in the phloem tissue with sievelike perforations on the end walls; also called sieve tube element.

Silica silicon dioxide, a component of sand.

Silviculture the commercial growth of trees, as in tree farming.

Simple fruit a fruit type derived from a simple or compound pistil of a single flower.

Simple leaf a leaf whose blade is not divided into leaflets.

Simple pistil a pistil consisting of one carpel.

Simple tissue a tissue composed of only one cell type.

Sink an area in a plant where food is used.

Sleep movement a type of nastic movement that is reversible and regulated by changes in turgor pressure as in the folding of leaves at night.

Smooth ER endoplasmic reticulum without attached ribosomes.

Soil structure arrangement of soil particles

Soil texture relative proportions of the different sized soil particles.

Solute a substance that can go into solution.

Soredium an asexual reproductive structure of a lichen composed of algal cells surrounded by fungal hyphae.

Sorus a cluster of sporangia on the lower surface of a fern leaf.

Source an area in a plant where food is available for translocation.

Speciation the evolution of new species.

Species a group of closely related organisms that are capable of interbreeding.

Spectrophotometry the technique by which the absorption or transmission of light of specific wavelength by a substance is measured.

Sperm the male sex cell.

Spike an inflorescence in which sessile flowers are arranged alternately on an unbranched stalk.

Spindle fibers threadlike structures composed of microtubules functioning in chromosome movement during cell division.

Spine a needlelike structure that is usually a modified leaf or part of a leaf.

Spirillum a spiral-shaped bacterium.

Spongy parenchyma the loosely arranged photosynthetic parenchyma cells in a leaf.

Sporangium (sporangia, plural) a spore-producing structure.

Spore mother cell a diploid cell in which meiosis takes place.

Spore a unicellular reproductive structure other than a gamete.

Sporogenesis the steps leading to the formation of spores.

Sporophore the structure on which the sporangium is produced.

Sporophyll a leaf bearing sporangia.

Sporophyte the spore-producing stage in a life cycle.

Sporopollenin the complex, highly resistant substance that makes up the outer walls of pollen grains.

Spring wood xylem tissue composed of large cells that are produced during early spring.

Stamen the structure in the flower within which pollen is produced.

Staminate flower a flower that lacks pistils.

Starch a polysaccharide composed of a long chain of glucose molecules; the reserve food in all plants, many algae, and certain fungi.

Stele that portion of a stem or root surrounded by the cortex; includes the vascular tissues.

Sterilization the process by which living organisms are killed.

Stigma the uppermost part of a pistil that forms a receptive surface for pollen grains.

Stipule a leaflike structure at the base of the leaf stalk found on certain plants, e.g., rose.

Stolon (see runner).

Stoma (stomata, plural) pore in the epidermis through which gas exchange occurs.

Stomatal chamber large air space next to the epidermis associated with a stoma.

Strobilus a conelike structure made up of sporangia-bearing leaves.

Stroma the nonmembranous part of a chloroplast where dark reactions of photosynthesis take place.

Style the central portion that connects the ovary and stigma of a pistil.

Suberin a wax produced as a cell wall component by cork and endodermal cells.

Suberized containing the wax suberin.

Substrate a molecule that can be acted upon by an enzyme.

Succession the gradual change in the composition of organisms of a community between the time of initial colonization of a bare area and the establishment of a climax community.

Succulent a plant that stores water in its leaves or stem.

Sucker a root modification in which new plants are produced from the root as in cherry and crab apple.

Sugar a simple carbohydrate molecule; produced in photosynthesis.

Summer wood relatively small cells of xylem produced during summer.

Superior ovary refers to the higher position of an ovary in a flower with respect to other floral parts.

Symbiosis (symbiotic, adj.) an association between two dissimilar organisms that is advantageous to one or both.

Synapsis the pairing of homologous chromosomes during prophase I of meiosis.

Synchronously at the same time.

Synergid one of the cells associated with the egg in an angiosperm female gametophyte.

Synthesis the production of a substance.

Systematic botany the field of plant science dealing with classification.

THC tetrahydrocannabinol, the chemical component of marijuana responsible for producing hallucinations.

Tangential section a lengthwise section that does not pass through the center.

Tap root system a root system characteristic of dicotyledonous plants in which successively smaller secondary roots are produced from a single primary root called the tap root.

Target cell the cell whose development is controlled by the hormone under consideration.

Taxon (taxa, plural) a general term to describe any unit of classification.

Taxonomy the field of science dealing with the classification of organisms.

Telophase the last stage of nuclear division during which the newly formed nuclei revert to interphase.

Temperate having a moderate climate with reference to regions of the world.

Tendril a coillike stem or leaf modification that attaches a weak-stemmed plant to a support.

Terminal-bud-scale scar an encircling scar indicating the position of the terminal bud during previous years.

Test cross a cross between an unknown genotype and a known recessive homozygous to determine the genotype of the unknown.

Thermophile (thermophilic, adj.) an organism that grows well at high temperature.

Thigmotropism a growth reaction to the stimulus of touch as in the twining of a morning glory on a support.

Thorn a hard, spiny, stem modification.

Thylakoid a flattened membranous sac in the chloroplast, two or more of which make up a granum; photosynthetic pigments are localized in these membranes.

Timberline the narrow zone beyond which trees will not grow because of harsh winter conditions.

Tissue a functional unit composed of a group of cells.

Tissue culture a technique for growing isolated cells or tissues in a sterile medium of known composition.

Tonoplast the membrane surrounding the vacuole.

Toothed possessing teethlike indentations along the margin of the leaf blade.

Totipotent possessing the potential to develop into any mature cell type.

Toxin a poisonous substance.

Trace element an essential mineral element needed only in small quantities; also known as a microelement.

Tracheid one type of water-conducting cell in xylem, characterized by its relatively narrow diameter and the presence of end walls.

Transcription the copying of specific DNA segments in the form of RNA; the first step in the process of protein synthesis.

Transduction genetic exchange during which a DNA fragment is transferred from one bacterium to another via a bacterial virus.

Transfer RNA the ribonucleic acid responsible for the transfer of amino acids to the growing peptide chain during protein synthesis.

Transformation genetic exchange during which a DNA segment from one cell is incorporated into the DNA of a second cell.

Translation the second step in protein synthesis during which amino acids are assembled forming a protein molecule.

Translocation in physiology, the transport of substances such as food from one part to another.

Transpiration the evaporation of water from leaf and stem surfaces through stomatal openings.

Transpirational-pull cohesion hypothesis the explanation that water movement in the xylem is due to a pull created by transpiration on a cohesive water column.

Transverse a plane at right angles to the long axis.

Tree a woody plant with one main stem, the trunk, from which branches are produced some distance above ground level.

Trichogyne a tubular extension from the female gametangia of red algae and ascomycetous fungi through which sperm pass.

Triose a three-carbon sugar molecule.

Trunk the main stem of a tree.

Tube cell one of the cells of a pollen grain from which the pollen tube is produced.

Tuber an underground stem modified for vegetative reproduction.

Tubulin the protein subunit of microtubules.

Tundra the biome typical of polar and alpine regions, characterized by permanently frozen subsoil and by plants that can tolerate extreme cold and aridity.

Turgid the condition in which the cell vacuoles are filled with water.

Turgor pressure the pressure developed inside a cell resulting from osmosis of water into the cell.

Turgor pressure within a cell due to the water content.

Turpentine a distillation product of resin and used as a thinner for paints and varnishes.

Twiner a weak-stemmed plant that needs support to grow upright.

Tyloses the blockage of a vessel element or tracheid resulting from the growth of a parenchyma cell through a pit.

Ultrastructural pertaining to the fine structure of cells; only visible with an electron microscope.

Umbel an inflorescence in which all flowers are borne at the same height with all stalks originating from the same point.

Unicellular made up of one cell.

Unisexual able to produce only one type of gamete.

Unit membrane a model for a membrane as described from cross-sections of membranes examined in an electron microscope that show a lightly stained layer of lipids in the center with a continuous dense layer of proteins on either side.

Unsaturated fatty acid a fatty acid in which one or more pairs of carbon atoms form double bonds with each other.

Vacuole a membrane-bounded structure that stores water and metabolic by-products.

Variety a genetically distinct subgroup within a species that results from restricted interbreeding.

Vascular bundle scar a scar representing the position of a vascular bundle within a leaf scar on the stem.

Vascular bundle an aggregation of vascular tissues (xylem and phloem).

Vascular cambium a secondary meristem that produces vascular tissues (secondary xylem and secondary phloem).

Vascular plant a plant that has xylem and phloem for conducting water and food.

Vascular tissue conducting tissue such as xylem and phloem.

Vegetative propagation reproduction of organisms by nonsexual means; also called asexual reproduction.

Vegetative reproduction (see vegetative propagation).

Vegetative not involved with sexual reproduction.

Vein vascular tissue within a leaf.

Venation the arrangement of veins in a leaf blade.

Vermiculite a soil substitute made from expanded mica.

Vesicle saclike food storage fungal cell produced within a root cell or in the space between cells; also a descriptive term for any saclike structure in a cell such as Golgi vesicle.

Vessel a group of vessel elements attached end-to-end forming a tube in the xylem tissue for conduction of water and minerals.

Vessel element one type of water-conducting cell in xylem, characterized by its relatively large diameter and lack of end walls.

Virulence the degree to which an organism will cause disease.

Wax a solid fat usually found on the outer surfaces of leaves and stems as a waterproofing layer.

Whisk fern *Psilotum* sp.; a member of the division Psilophyta.

Whorled an arrangement of leaves in which three or more leaves are present at a nodal region.

Woody containing large amounts of secondary xylem (wood).

Woody plant a plant that contains large amounts of secondary tissues, particularly xylem.

X-ray diffraction a technique that produces a characteristic diffraction pattern of an object based on its structure when a beam of focused X-rays is allowed to strike the object.

Xerophyte (xerophytic, adj.) a plant capable of growing under dry conditions.

Xylem ray row(s) of parenchyma cells in the xylem.

Xylem specialized water-conducting tissue.

Yeast a unicellular fungus belonging to the class Ascomycetes.

Zoologist one who studies animals.

Zooplankton free-floating or motile microscopic animals.

Zoospore an asexual, motile spore.

Zygomycetes a fungal class characterized by nonseptate hyphae and nonflagellate spores.

Zygospore a thick-walled, diploid spore developed from a zygote with the ability to survive under adverse conditions.

Zygote the cell that results from the fusion of sex cells.

Index